SCHRÖDINGER OPERATORS: EIGENVALUES AND LIEB–THIRRING INEQUALITIES

The analysis of eigenvalues of Laplace and Schrödinger operators is an important and classical topic in mathematical physics with many applications. This book presents a thorough introduction to the area, suitable for masters and graduate students, and includes an ample amount of background material on the spectral theory of linear operators in Hilbert spaces and on Sobolev space theory.

Of particular interest is a family of inequalities by Lieb and Thirring on eigenvalues of Schrödinger operators, which they used in their proof of stability of matter. The final part of this book is devoted to the active research on sharp constants in these inequalities and contains state-of-the-art results, serving as a reference for experts and as a starting point for further research.

Rupert L. Frank holds a chair in applied mathematics at LMU Munich and is doing research primarily in analysis and mathematical physics. He is an invited speaker at the 2022 International Congress of Mathematics.

Ari Laptev is Professor at Imperial College London. His research interests include different aspects of spectral theory and functional inequalities. He is a member of the Royal Swedish Academy of Science, a Fellow of EurASc and a member of Academia Europaea.

Timo Weidl is Professor at the University of Stuttgart. He works on spectral theory and mathematical physics.

Schrödinger Operators: Eigenvalues and Lieb–Thirring Inequalities

RUPERT L. FRANK
Ludwig-Maximilians-Universität München

ARI LAPTEV
Imperial College of Science, Technology and Medicine, London

TIMO WEIDL
Universität Stuttgart

CAMBRIDGE
UNIVERSITY PRESS

CAMBRIDGE
UNIVERSITY PRESS

University Printing House, Cambridge CB2 8BS, United Kingdom

One Liberty Plaza, 20th Floor, New York, NY 10006, USA

477 Williamstown Road, Port Melbourne, VIC 3207, Australia

314–321, 3rd Floor, Plot 3, Splendor Forum, Jasola District Centre,
New Delhi – 110025, India

103 Penang Road, #05–06/07, Visioncrest Commercial, Singapore 238467

Cambridge University Press is part of the University of Cambridge.

It furthers the University's mission by disseminating knowledge in the pursuit of
education, learning, and research at the highest international levels
of excellence.

www.cambridge.org
Information on this title: www.cambridge.org/9781009218467
DOI: 10.1017/9781009218436

First published 2023

A catalogue record for this publication is available from the British Library.

ISBN 978-1-009-21846-7 Hardback

To

Semra, Sami and Sima

To

Marilyn, Maria, Vanya, Katya and Eugenia

To

Galia, Adi and Alex

Contents

Preface

Ever since E. H. Lieb and W. Thirring published their celebrated work *Inequalities for the moments of the eigenvalues of the Schrödinger Hamiltonian and their relation to Sobolev inequalities* (Lieb and Thirring, 1976), the branch of spectral theory related to such inequalities has flourished and these bounds are now named for them.

In their 1976 paper Lieb and Thirring developed a family of inequalities, of which they had used a special case in an earlier (1975) paper, to prove stability of matter. Their approach simplified and improved the work by Lenard and Dyson (1968), and introduced fundamental ideas in the analysis of fermionic quantum many-body systems.

Lieb–Thirring inequalities come in two different versions, namely a spectral form and a kinetic (or dual) form. The kinetic form is most directly applicable to quantum many-body systems and provides a lower bound on the total kinetic energy in terms of certain simple effective characteristic of the state. It is a mathematical expression of the uncertainty and Pauli exclusion principles and a far-reaching generalization of Sobolev inequalities. The spectral form of the Lieb–Thirring inequalities concerns the negative eigenvalues $-E_j$ of the one-particle Schrödinger operator

$$H = -\Delta - V$$

and provides upper bounds on the Riesz means

$$\sum_j E_j^\gamma, \qquad \gamma > 0,$$

that only involve an L^p norm of V. In the special case $\gamma = 1$, these bounds are equivalent to the kinetic form of the Lieb–Thirring inequalities.

It turned out that this form is also very useful in the study of the dimension of attractors for the Navier–Stokes and other non-linear evolution equations.

Originally, it was calculated by exploiting the available Sobolev inequalities and then employing the relationship between that dimension and the prevailing Lyapunov exponents, as first conjectured by Farmer et al. (1983). This effective, but somewhat cumbersome approach was significantly simplified by the use of the Lieb–Thirring improvement of the Sobolev inequalities (Lieb, 1984; Constantin et al., 1985; Temam, 1997).

Lieb–Thirring inequalities are also important in the study of properties of the essential spectrum of Schrödinger operators. Deift and Killip (1999) were able to obtain a sharp result on the absolute continuity of the positive spectrum for one-dimensional Schrödinger operators using a trace formula by Zaharov and Faddeev (1971); see also Killip and Simon (2009). Such trace formulas provide identities between characteristics of the spectrum (eigenvalues and scattering data) and some functionals involving electric potentials of Schrödinger operators. In addition, they describe integrals of motions of the Korteweg–De Vries (KdV) equation. It is worth noting that one such trace formula was used by Gardner et al. (1974) to prove a special case of what came to be called the Lieb–Thirring inequalities. Later a version of this trace formula for matrix-valued potentials played an important role in obtaining sharp constants in Lieb–Thirring inequalities for multi-dimensional Schrödinger operators (Laptev and Weidl, 2000b). It is remarkable that some sharp constants in Lieb–Thirring inequalities are related to soliton-type potentials appearing in the theory of the KdV equation.

Spectral inequalities in the special case $\gamma = 0$, which correspond to bounds on the number of negative eigenvalues of Schrödinger operators, have an even longer history. After initial results by Bargmann, Birman, Schwinger, Calogero and others, a systematic investigation was started in the Russian school of M. Birman and M. Solomyak and in the US by B. Simon and E. Lieb. The respective estimates are known as Cwikel–Lieb–Rozenblum (CLR) inequalities (Cwikel, 1977; Lieb, 1976, 1980; Rozenbljum, 1972a, 1976). One of the main motivations for proving such inequalities was to obtain necessary and sufficient conditions on the potentials for admitting Weyl asymptotics. By now, there are at least seven proofs of the CLR bound using rather different tools from mathematical analysis.

It has been more than four decades since the paper of Lieb and Thirring was published, and the theory of Lieb–Thirring inequalities is still a very dynamically developing area of functional analysis and mathematical physics. The main conjecture on the sharp Lieb–Thirring constant for $\gamma = 1$ in dimension three remains open. This area has become a well-established part of analysis that generates beautiful new ideas and that finds applications beyond the spectral theory of Schrödinger operators.

This book is aimed at presenting the current state of the art of some parts of spectral theory of partial differential equations. Our intention was to write a book that covers some new results connected to spectral properties of Schrödinger and Laplace operators that have been obtained during the last few decades. Most of them are focused around our own interests related to Lieb–Thirring inequalities. While writing the text and when giving courses based on preliminary versions of the book, we faced the problem that we needed a lot of material from the general spectral theory of self-adjoint operators in Hilbert spaces and a number of standard functional inequalities. Including all this has substantially increased the size of the book and forced us to postpone some core material for the future. We now hope that the book might be useful not only to our colleagues who are interested in spectral inequalities, but also to students specializing in analysis.

Acknowledgements

We would like to express our appreciation to M. Birman, E. Lieb, B. Simon and M. Solomyak who profoundly influenced our mathematical taste. We have been fortunate to have had the opportunity to collaborate with so many individuals on the topic of Lieb–Thirring and spectral inequalities. We would like to thank our teachers and students, colleagues and friends for all the inspiring discussions.

We owe a great debt to J. Peteranderl who read a preliminary version of this book line by line and found countless minor and major inaccuracies. For valuable remarks, suggestions and corrections we are very grateful to C. Dietze, F. Gesztesy, A. Ilyin, V. Kußmaul, S. Larson, M. Lewin, E. Lieb, G. Rozenblum, L. Schimmer and H. Siedentop. Our special thanks go to E. Lieb and B. Simon for their interest in this book and their constant encouragement.

We are grateful to David Tranah and the team at CUP for their assistance when publishing the book.

Parts of this book were written at Caltech, Imperial College London, KTH Stockholm, LMU Munich, Mittag–Leffler Institute, Oberwolfach Research Institute, Stuttgart University, and Tsinghua University, and we would like to thank all these institutions for their hospitality.

Partial support through US National Science Foundation grants PHY-1068285, PHY-1347399, DMS-1363432 and DMS-1954995 and through the Deutsche Forschungsgemeinschaft EXC-2111-390814868 (R.L.F.) is acknowledged.

Munich, London, Stuttgart,
August 2022

Rupert L. Frank
Ari Laptev
Timo Weidl

Overview

The aim of this book is to present qualitative and quantitative results on the discrete spectrum of Schrödinger operators and on Laplacians in Euclidean domains that have appeared in research papers during the last five decades.

If V is a real-valued function on $\mathbb{R}^d$, $d \geq 1$, that is sufficiently regular and tends, at least in some averaged sense, to zero at infinity (we will be more precise about these assumptions later in this book), then the spectrum of the Schrödinger operator $-\Delta - V$ in $L^2(\mathbb{R}^d)$ can be divided into the essential spectrum, which is equal to $[0, \infty)$, and the discrete spectrum, which consists of an at most countable number of negative eigenvalues with finite multiplicities. For general V, these eigenvalues cannot be computed explicitly and, motivated by applications, one is interested in qualitative and quantitative information on their distribution.

A basic result on the negative eigenvalues is Weyl's formula about their asymptotic distribution in the strong coupling limit. More precisely, we consider the family of Schrödinger operators $-\Delta - \alpha V$ with a parameter $\alpha > 0$ and denote their negative eigenvalues, in non-decreasing order and repeated according to multiplicities, by $-E_j(\alpha)$. Then Weyl's law states that, under some assumptions on V and for any $\gamma \geq 0$,

$$\lim_{\alpha \to \infty} \alpha^{-\gamma - d/2} \sum_j E_j(\alpha)^\gamma = L_{\gamma,d}^{\mathrm{cl}} \int_{\mathbb{R}^d} V(x)_+^{\gamma + d/2} \, dx,$$

where, for $\gamma = 0$, the sum is interpreted as the number of negative eigenvalues of $-\Delta - \alpha V$. Moreover, we wrote $t_+ := \max\{t, 0\}$. The constant $L_{\gamma,d}^{\mathrm{cl}}$ appearing in this formula is explicit and has an interpretation in terms of a classical phase space integral.

In many applications one also needs, in addition to the asymptotic statement of Weyl's law, non-asymptotic bounds that capture the correct order of magnitude in the asymptotic regime, even if they do not reproduce the correct asymptotic constant. Thus we are interested in bounds of the form

$$\alpha^{-\gamma-d/2} \sum_j E_j(\alpha)^\gamma < L_{\gamma,d} \int_{\mathbb{R}^d} V(x)_+^{\gamma+d/2} \, dx,$$

with a constant $L_{\gamma,d}$ that may depend on γ and d, but is independent of V. Note that it presents no loss of generality to take $\alpha = 1$ in this inequality. The validity of this bound for $\gamma > 1/2$ in $d = 1$ and for $\gamma > 0$ in $d \geq 2$ is a celebrated result of Lieb and Thirring. The case $\gamma = 0$ in $d \geq 3$ is due to Cwikel, Lieb and Rozenblum and the case $\gamma = 1/2$ in $d = 1$ is due to Weidl. This completely settles the validity of this inequality.

Ideally, one would like to know the optimal values of the constants $L_{\gamma,d}$. Some of them are known, but the sharp value of the most important constant $L_{1,3}$, namely for the sum of the negative eigenvalues of the Schrödinger operator in three dimensions, is not. The (still open) Lieb–Thirring conjecture states that the constant $L_{1,3}$ coincides with the constant $L_{1,3}^{\mathrm{cl}}$ that appears in Weyl's asymptotic formula.

In this book, we describe different techniques that establish spectral inequalities of semiclassical type and Weyl's asymptotics for Schrödinger and Laplace operators.

Structure of this book

This book is divided into three parts. The first one contains background material on the spectral theory of linear operators in Hilbert spaces and on Sobolev space theory. Our selection of material here is far from comprehensive and it is guided by what is needed in the later parts of the book. Our presentation in this first part is at times fast-paced and with hardly any illustrative examples, since we expect that most of the readers will have seen at least the basics of the material treated here. More advanced readers may wish to skip Part One and only return to it when some specific technical results in the later parts are needed.

The second part of the book contains the basics of the theory of Laplace and Schrödinger operators, with a special emphasis on their discrete spectra. We do not require any previous acquaintance with these operators and develop the theory starting from what is recalled in Part One. In particular, we will see a close connection between spectral inequalities and Sobolev-type inequalities. In Part Two we prove, among other things, Weyl's asymptotic formula mentioned above and versions of the Cwikel–Lieb–Rozenblum and Lieb–Thirring inequalities, without paying too much attention to the constants.

The third part of the book is devoted to the quest for the sharp constants in the Cwikel–Lieb–Rozenblum and Lieb–Thirring inequalities. This part is more specific and sometimes more technical than the previous two, but, in our opinion, contains some beautiful ideas. In Part Three we prove all the sharp Lieb–Thirring inequalities that are presently known and we also give the proofs of the currently best known results. We hope that this part, besides giving a snapshot of the state of the art, serves the purpose of inviting new researchers to this exciting area of research.

Since every chapter has an introduction describing its content, here we only briefly present the structure of the book.

Chapter 1 contains standard material regarding self-adjoint operators in Hilbert spaces, including the spectral theorem, the variational characterization of eigenvalues and the important relation between lower semibounded quadratic forms and lower semibounded self-adjoint operators. We also discuss the Birman–Schwinger principle, which allows us to reduce problems of counting eigenvalues in the discrete spectrum of unbounded operators to the study of the spectrum of compact operators.

In Chapter 2 we review some material from Sobolev space theory. In particular, we prove some important functional inequalities such as the Sobolev, Gagliardo–Nirenberg, Friedrichs, Poincaré, and Hardy inequalities. As we mentioned before, these inequalities are major tools in the study of spectral inequalities.

In Chapter 3 we introduce the Laplacian with Dirichlet or Neumann boundary conditions on an open set in Euclidean space as a self-adjoint operator. We prove, among other things, Weyl's formula for the asymptotic distribution of its eigenvalues and we present some (sharp) spectral inequalities due to Pólya, Berezin, Li, and Yau.

Chapter 4 is devoted to semibounded Schrödinger operators. After covering explicitly solvable models, we prove Weyl's asymptotic formula, Cwikel–Lieb–Rozenblum and Lieb–Thirring inequalities.

In Chapter 5 we begin our discussion of sharp constants in the latter. We include some general facts about these constants and mention the original conjecture due to Lieb and Thirring. Then we proceed to the proof of two cases of sharp inequalities in one dimension.

In Chapter 6 we derive Lieb–Thirring inequalities with the sharp, semiclassical constant in higher dimension for $\gamma \geq 3/2$. This is achieved by the so-called lifting argument with respect to dimension. The input for this argument is a one-dimensional Lieb–Thirring inequality with matrix-valued potentials, for which we give two different proofs.

In Chapter 7 we collect some additional results concerning sharp Lieb–

Thirring inequalities, including a sharp version of the CLR inequality for radial potentials in dimension 4, as well as counterexamples to parts of the original Lieb–Thirring conjecture.

In Chapter 8 we use again the lifting argument in order to prove the so-far best known constants in Lieb–Thirring inequalities in cases where the sharp constants are not known. We end with a section summarizing the most important results discussed in this book.

Finally, we would like to mention that, while some of the proofs in this book are new and have not been published before, most of them have previously appeared in the literature. We provide references in the comments section at the end of each chapter. We use these sections also to give references to additional, related results and, sometimes, to discuss particularly important results that we did not include in the main text.

Notation and conventions
We use the convention $\mathbb{N} = \{1, 2, 3, \ldots\}$. By $\mathbb{R}$ and $\mathbb{C}$ we denote the real and complex numbers and by $\mathbb{R}^d$ and $\mathbb{C}^d$ their d-fold Cartesian products. Typically, we write $x \in \mathbb{R}^d$ as $(x_1, \ldots, x_d)$ with $x_1, \ldots, x_d \in \mathbb{R}$. The symbol $|\cdot|$ denotes the Euclidean norm and, for $a \in \mathbb{R}^d$ and $r > 0$, we write

$$B_r(a) := \{x \in \mathbb{R}^d : |x - a| < r\}$$

for the open ball of radius r centered at a. Moreover, $\mathbb{R}_+ := \{x \in \mathbb{R} : x > 0\}$.

We use the symbols $\subset$ and $\supset$ in the non-strict sense; that is, allowing for the case of equality. Strict inclusions are denoted by $\subsetneq$ and $\supsetneq$.

We use the terms 'positive' and 'negative' in the strict sense; for instance, a real number is positive if it is non-negative and not zero. Similarly, for a function the terms 'increasing' and 'decreasing' are used in the strong sense; for instance, a function is increasing if it is non-decreasing and not constant on any non-trivial subinterval. Sometimes, for special emphasis, we write 'strictly positive' or 'strictly increasing' instead of 'positive' and 'increasing'.

Other notations will be introduced as we proceed; see also the index.

PART ONE

BACKGROUND MATERIAL

1

Elements of Operator Theory

In this chapter, we collect some material from the theory of self-adjoint operators. While the main focus of this book is on the specific cases of Laplace and Schrödinger operators, already the definition of these operators and the formulation of the questions we ask requires language from operator theory. At the same time, this theory provides the foundations of the spectral analysis of the operators of interest.

One of our main goals in this chapter is to prove the variational principle, which will be the essential tool in our proofs of spectral inequalities. It transforms the problem of counting eigenvalues into an optimization problem for quadratic forms. In this principle, and therefore in all our presentation, quadratic forms will play a prominent role, for the most part even more than the underlying operator.

Operator theory also provides a framework for perturbation theory, which considers operators given, for instance, as a sum of an 'unperturbed' operator, which is in some sense well understood, and a perturbation, which is in some sense small. This point of view will be particularly relevant when dealing with Schrödinger operators.

Let us now give a brief overview of the content of this chapter. We begin by recalling some basic facts from Hilbert space theory and, in particular, the notion of the spectrum of an operator and that of self-adjointness of an operator. Next, in §1.1.6, we state without proof the spectral theorem for self-adjoint operators and discuss the resulting functional calculus. As a consequence of the spectral theorem, we prove Weyl's theorem (Theorem 1.14) on the stability of the essential spectrum.

The second section is devoted to quadratic forms. In §1.2.1 we present the fundamental connection between lower semibounded self-adjoint operators and lower semibounded closed quadratic forms. In our applications to Laplace and Schrödinger operators, we use this connection to define the relevant operators.

In §§1.2.3 and 1.2.5 we formulate the variational principle for eigenvalues and their sums, which are naturally formulated in the language of quadratic forms and, as we already mentioned, are fundamental for the developments in the following chapters.

In §1.2.6 we discuss Riesz means, which will play an important role in the study of Lieb–Thirring inequalities. As an application of the variational principle, we prove various continuity results that are frequently used in applications.

Finally, in §§1.2.7 and 1.2.8 we discuss perturbations of operators in terms of quadratic forms and the Birman–Schwinger principle. The latter translates the problem of counting negative eigenvalues of unbounded operators to a related problem for compact operators.

While we try to be rather self-contained concerning the material on the theory of unbounded operators, it is probably advantageous if the reader has had some previous exposure to a standard course on functional analysis and measure theory; see, for instance, Brezis (2011), Friedman (1970), Folland (1999), Lax (2002), and Rudin (1991). For detailed expositions of operator theory and spectral analysis, we refer to the references in §1.3.

1.1 Hilbert spaces, self-adjoint operators and the spectral theorem

In this section we briefly recall the theory of self-adjoint operators in Hilbert spaces and we use this opportunity to fix our notation.

1.1.1 Bounded operators

Let $\mathcal{V}$ be a complex vector space together with an inner product $(\cdot,\cdot)$; that is, a complex-valued function on $\mathcal{V} \times \mathcal{V}$ satisfying

$$(f,g) = \overline{(g,f)} \qquad\qquad \text{for all } f,g \in \mathcal{V},$$
$$(\alpha_1 f_1 + \alpha_2 f_2, g) = \alpha_1(f_1,g) + \alpha_2(f_2,g) \qquad \text{for all } \alpha_1,\alpha_2 \in \mathbb{C},\ f_1,f_2,g \in \mathcal{V},$$
$$(f,f) > 0 \qquad\qquad \text{for all } 0 \neq f \in \mathcal{V}.$$

Here and in all the following, our sesquilinear forms are linear in the first and anti-linear in the second argument.

If $(\cdot,\cdot)$ is an inner product on $\mathcal{V}$, then

$$\|f\| := \sqrt{(f,f)}$$

defines a norm on $\mathcal{V}$. The vector space $\mathcal{V}$, together with its inner product, is called a *Hilbert space* if it is complete with respect to the corresponding norm. It is called *separable* if it has a countable orthonormal basis.

From now on, let $\mathcal{H}$ be a separable Hilbert space.

We say that a sequence $(f_n) \subset \mathcal{H}$ *converges* to $f \in \mathcal{H}$ if $\|f_n - f\| \to 0$ as $n \to \infty$. We say that $(f_n) \subset \mathcal{H}$ *converges weakly* to $f \in \mathcal{H}$ if, for every $g \in \mathcal{H}$, $(f_n, g) \to (f, g)$ as $n \to \infty$. If we want to emphasize the difference between convergence (in norm) and weak convergence, we also call the former *strong convergence*.

The following facts about weak convergence are standard results from functional analysis. First, if $(f_n) \subset \mathcal{H}$ is weakly convergent, then $(\|f_n\|)$ is bounded. Second, the closed unit ball in $\mathcal{H}$ is weakly sequentially compact; that is, if $(\|f_n\|)$ is bounded, then there is a subsequence (f_{n_m}) that converges weakly to some $f \in \mathcal{H}$.

Another standard result from functional analysis is the Riesz representation theorem. It states that, if ℓ is a bounded, linear functional on $\mathcal{H}$, then there is a unique $g \in \mathcal{H}$ such that $\ell(f) = (f, g)$ for all $f \in \mathcal{H}$. Moreover, $\|g\| = \|\ell\|$.

Next, we discuss operators on $\mathcal{H}$. A continuous, linear map $T : \mathcal{H} \to \mathcal{H}$ is called a *bounded* (linear) operator. This name comes from the fact that a linear map T is continuous if and only if

$$\|T\| := \sup_{\|f\|=1} \|Tf\| < \infty .$$

The set of bounded operators is complete with respect to the above norm.

By the Riesz representation theorem, for any bounded operator T on $\mathcal{H}$ one can define a unique bounded operator T^* on $\mathcal{H}$, called the *adjoint of T*, such that

$$(T^*f, g) = (f, Tg) \qquad \text{for all } f, g \in \mathcal{H} .$$

This implies, in particular, that

$$\|T^*\| = \|T\|$$

and that, if $f_n \to f$ weakly in $\mathcal{H}$, then $Tf_n \to Tf$ weakly in $\mathcal{H}$.

An operator K on a Hilbert space $\mathcal{H}$ is called *compact* if the image of the closed unit ball in $\mathcal{H}$ is relatively compact; that is, if any sequence (f_n) with $\|f_n\| \le 1$ has a subsequence (f_{n_m}) such that (Kf_{n_m}) converges. Clearly, compact operators are bounded and the product of a compact operator with a bounded operator is compact.

The following lemma characterizes compactness in terms of weak convergence.

Lemma 1.1 *A bounded operator K is compact if and only if it transforms every weakly convergent sequence into a strongly convergent sequence.*

Proof First, assume that K is compact and that $f_n \to 0$ weakly. (We may assume, without loss of generality, that the weak limit is zero.) Then, as recalled above, $K f_n \to 0$ weakly. Moreover, as also mentioned before, $\sup \|f_n\| < \infty$. We choose a subsequence (f_{n_m}) such that $\lim_{m \to \infty} \|K f_{n_m}\| = \lim \sup_{n \to \infty} \|K f_n\| =: a$. Since (f_n) is bounded, the compactness of K implies that $(K f_{n_m})$ is relatively compact and, therefore, there is a $g \in \mathcal{H}$ and a further subsequence $(f_{n_{m_l}})$ such that $K f_{n_{m_l}} \to g$ strongly in $\mathcal{H}$ as $l \to \infty$. Since $K f_n \to 0$ weakly, we have $g = 0$ and therefore $\|K f_{n_{m_l}}\| \to 0$ as $l \to \infty$. Thus $a = 0$ which means that $K f_n \to 0$ strongly as $n \to \infty$.

To prove the converse implication, let (f_n) be a sequence with $\|f_n\| \le 1$. Then, by the weak sequential compactness of the unit ball, there is a subsequence (f_{n_m}) that converges weakly. By assumption, $(K f_{n_m})$ converges strongly. This shows that the image of the closed unit ball is relatively compact, as claimed. □

Lemma 1.2 *Let K be a bounded operator. Then K is compact if and only if K^* is compact if and only if $K^* K$ is compact.*

Proof The product of a bounded and a compact operator is compact, so if K is compact, then so is $K^* K$. Conversely, assume that $K^* K$ is compact. To show that K is compact, let (f_n) be a sequence converging weakly to zero. By Lemma 1.1, $(K^* K f_n)$ converges strongly to zero, so, since (f_n) is bounded, $\|K f_n\|^2 = (K^* K f_n, f_n) \le \|K^* K f_n\| \|f_n\| \to 0$. By Lemma 1.1 again, this means that K is compact.

Now assume that K^* is compact. Then $K^* K$ is compact as a product of a compact and a bounded operator, and therefore, by what we have just shown, K is compact. Applying this to $K = (K^*)^*$, we see that compactness of K implies that of K^*. □

Let us conclude this subsection by discussing modes of convergence for operators. Let (T_n) be a sequence of bounded operators on $\mathcal{H}$ and let T be a bounded operator on $\mathcal{H}$. We say that T is the *weak limit* of (T_n) and write

$$T = \text{w-}\lim_{n \to \infty} T_n \quad \text{if} \quad \lim_{n \to \infty} (T_n f, g) = (T f, g) \qquad \text{for all } f, g \in \mathcal{H}.$$

We say that T is the *strong limit* of (T_n) and write

$$T = \text{s-}\lim_{n \to \infty} T_n \quad \text{if} \quad \lim_{n \to \infty} \|(T_n - T) f\| = 0 \qquad \text{for all } f \in \mathcal{H}.$$

Finally, we say that T is the *norm limit* of (T_n) and write $T = \lim_{n \to \infty} T_n$ if

$$\lim_{n \to \infty} \|T_n - T\| = 0.$$

Lemma 1.3   *Let (K_n) be a sequence of compact operators and let K be a bounded operator such that $\lim_{n\to\infty} K_n = K$. Then K is compact.*

Proof   Let (f_n) be a sequence that converges weakly to zero. According to Lemma 1.1, we need to prove that (Kf_n) tends strongly to zero. For any n, m we have the bound

$$\|Kf_n\| \le \|K_m f_n\| + \|K_m - K\|\|f_n\|.$$

Since K_m is compact, we have $K_m f_n \to 0$ strongly as $n \to \infty$. Thus,

$$\limsup_{n\to\infty} \|Kf_n\| \le \|K_m - K\| \sup_n \|f_n\|.$$

The supremum on the right side is finite, as recalled before. Letting $m \to \infty$, we obtain $\limsup_{n\to\infty} \|Kf_n\| = 0$, which proves the claimed convergence.   □

Lemma 1.4   *Let K be a compact operator, let T and S be bounded operators and let $(T_n), (S_n)$ be sequences of bounded operators with $T = $ s-$\lim_{n\to\infty} T_n$ and $S = $ s-$\lim_{n\to\infty} S_n$. Then $\lim_{n\to\infty} T_n K S_n^* = TKS^*$.*

Proof   *Step 1.* We first prove the assertion in the case where $S_n = S$ for all n and we abbreviate $L := KS^*$. Assume the stated convergence would not hold. Then there are $\varepsilon > 0$ and, passing to a subsequence if necessary, $f_n \in \mathcal{H}$ such that $\|f_n\| = 1$ and $\|(T_n - T)Lf_n\| \ge \varepsilon$. By weak compactness, passing to another subsequence if necessary, we may assume that $f_n \to f$ weakly for some f. Since L is compact as the product of a bounded and a compact operator, we infer from Lemma 1.1 that $Lf_n \to Lf$ strongly. Writing

$$\|(T_n - T)Lf_n\| \le \|(T_n - T)Lf\| + \|(T_n - T)L(f_n - f)\|$$
$$\le \|(T_n - T)Lf\| + (\|T_n\| + \|T\|)\|Lf_n - Lf\|,$$

we see that the first term on the right side tends to zero by strong convergence of T_n and the second term tends to zero since $\|T_n\|$ remains bounded by the uniform boundedness principle (see, for instance, Friedman, 1970, Theorem 4.5.1). Thus, $\|(T_n - T)Lf_n\| \to 0$, a contradiction.

Step 2. We now prove the assertion in the case where $T_n = T$ for all n and we abbreviate $M := TK$. Since M is compact as the product of a bounded and a compact operator, we infer from Lemma 1.2 that M^* is compact. Therefore, Step 1 implies that $S_n M^* \to SM^*$ in norm. Since the norm of the adjoint equals the norm of the operator itself, this implies that $MS_n^* \to MS^*$ in operator norm, which proves the assertion in this case.

Step 3. To prove the assertion in the general case, we write

$$T_n K S_n^* - TKS^* = (T_n - T)KS_n^* + TK(S_n^* - S^*).$$

The first term on the right side tends to zero in operator norm by Step 1 (applied with S replaced by the identity) and the fact that $\|S_n^*\| = \|S_n\|$ is uniformly bounded, and the second term tends to zero in operator norm by Step 2. This concludes the proof of the proposition.   □

1.1.2 Unbounded operators

We now extend the notion of a bounded linear operator to that of a not necessarily bounded operator, often called an *unbounded* operator. In the following, an operator in $\mathcal{H}$ is, for us, a linear map T from its domain $\operatorname{dom} T$, a subspace of $\mathcal{H}$, to $\mathcal{H}$. (We emphasize that in this book subspaces are *not* necessarily assumed to be closed.) In particular, two operators T and S coincide, by definition, if $\operatorname{dom} S = \operatorname{dom} T$ and $Tu = Su$ for all $u \in \operatorname{dom} S = \operatorname{dom} T$.

The operator T is called *closed* if $\operatorname{dom} T$ is complete with respect to the norm $(\|Tu\|^2 + \|u\|^2)^{1/2}$. Clearly, every bounded operator defined on a closed subspace of $\mathcal{H}$ is closed. If the kernel, $\ker T$, of a closed operator T is trivial, then its inverse T^{-1}, defined on its range, $\operatorname{ran} T$, is also closed.

The operator T is called *densely defined* if $\operatorname{dom} T$ is dense in $\mathcal{H}$. For such a T we now define the *adjoint* T^* as follows. Its domain is

$$\operatorname{dom} T^* := \{v \in \mathcal{H} : \text{there is a } g \in \mathcal{H} \text{ such that}$$
$$\text{for all } u \in \operatorname{dom} T \text{ one has } (g, u) = (v, Tu)\} \,.$$

Since T is densely defined, for $v \in \operatorname{dom} T^*$ the corresponding g is unique and we define $T^*v := g$. For bounded T, this coincides with the definition given above.

One can show that T^* is always closed. If T is closed, then T^* is densely defined and $T^{**} := (T^*)^* = T$. Moreover, one has

$$\ker T^* = (\operatorname{ran} T)^\perp \,, \tag{1.1}$$

which, in turn, implies that

$$(\ker T^*)^\perp = \overline{\operatorname{ran} T} \,. \tag{1.2}$$

If T is densely defined, has trivial kernel and dense range, then, by (1.1), T^* has trivial kernel and its inverse is given by

$$(T^*)^{-1} = \left(T^{-1}\right)^* \,. \tag{1.3}$$

For a closed operator T, the *resolvent set* $\rho(T)$ is defined by

$$\rho(T) := \{z \in \mathbb{C} : T - z \text{ is a bijection from } \operatorname{dom} T \text{ onto } \mathcal{H}$$
$$\text{with a bounded inverse}\}$$

and the operator $(T - z)^{-1}$ is called the *resolvent* of T at $z \in \rho(T)$. Using the closed graph theorem, we could deduce the boundedness of the inverse from the fact that the range of $T - z$ is equal to $\mathcal{H}$, although we will not use this fact explicitly in this book. We note too that here and in what follows, in the notation $T - z$ we identify the number z with z times the identity operator on $\mathcal{H}$.

The *spectrum* $\sigma(T)$ is defined by

$$\sigma(T) := \mathbb{C} \setminus \rho(T)$$

and the *point spectrum* is defined by

$$\sigma_p(T) := \{z \in \sigma(T) \colon \ker(T - z) \neq \{0\}\} .$$

A number $z \in \sigma_p(T)$ is called an *eigenvalue* of T and any $0 \neq u \in \ker(T - z)$ is called a corresponding *eigenvector*. Moreover, $\dim \ker(T - z)$ is called the *(geometric) multiplicity* of the eigenvalue z of T.

We now discuss the orthogonal sum of operators. Let N be a countable (possibly finite) index set and assume that, for each $n \in N$, $\mathcal{H}_n$ is a separable Hilbert space with a norm $\| \cdot \|_{\mathcal{H}_n}$ and an inner product $(\cdot, \cdot)_{\mathcal{H}_n}$. We recall that the Hilbert space

$$\mathcal{H} := \bigoplus_{n \in N} \mathcal{H}_n$$

is the space of all elements $f = (f_n)_{n \in N}$ in the Cartesian product $\prod_{n \in N} \mathcal{H}_n$ (that is, $f_n \in \mathcal{H}_n$ for all $n \in N$) such that

$$\|f\|_{\mathcal{H}} := \left(\sum_{n \in N} \|f_n\|_{\mathcal{H}_n}^2 \right)^{1/2} < \infty .$$

This is a separable Hilbert space with inner product

$$(f, g)_{\mathcal{H}} := \sum_{n \in N} (f_n, g_n)_{\mathcal{H}_n} .$$

For each $n \in N$, let T_n be an operator in $\mathcal{H}_n$. We define an operator, denoted by $\bigoplus_{n \in N} T_n$, in $\mathcal{H}$ with domain

$$\mathrm{dom}\left(\bigoplus_{n \in N} T_n \right) := \left\{ u \in \mathcal{H} \colon u_n \in \mathrm{dom}\, T_n \text{ for all } n \in N, \ \sum_{n \in N} \|T_n u_n\|_{\mathcal{H}_n}^2 < \infty \right\}$$

by

$$\left(\bigoplus_{n \in N} T_n u \right)_m := T_m u_m \quad \text{for all } m \in N, \ u \in \mathrm{dom}\left(\bigoplus_{n \in N} T_n \right).$$

If all the T_n are bounded, then

$$\left\| \bigoplus_{n \in N} T_n \right\| = \sup_{n \in N} \|T_n\|, \tag{1.4}$$

where the right side may or may not be finite. In particular, $\bigoplus_{n\in N} T_n$ is bounded if and only if the T_n are uniformly bounded. If all the T_n are closed, then $\bigoplus_{n\in N} T_n$ is closed. If all the T_n are densely defined, then $\bigoplus_{n\in N} T_n$ is densely defined, and in this case one finds for the adjoint

$$\left(\bigoplus_{n\in N} T_n\right)^* = \bigoplus_{n\in N} T_n^* .$$

The relation between the spectrum of $\bigoplus_{n\in N} T_n$ and the spectra of the T_n is as follows.

Lemma 1.5 *Assume that the T_n are closed. Then*

$$\rho\left(\bigoplus_{n\in N} T_n\right) = \left\{ z \in \bigcap_{n\in N} \rho(T_n) : \sup_{n\in N} \left\| (T_n - z)^{-1} \right\| < \infty \right\}. \qquad (1.5)$$

Moreover,

$$\sigma_p\left(\bigoplus_{n\in N} T_n\right) = \bigcup_{n\in N} \sigma_p(T_n)$$

and, for each $z \in \mathbb{C}$,

$$\dim \ker\left(\bigoplus_{n\in N} T_n - z\right) = \sum_{n\in N} \dim \ker(T_n - z) .$$

Proof Writing $T := \bigoplus_{n\in N} T_n$, we first note that

$$\ker(T - z) = \bigoplus_{n\in N} \ker(T_n - z) .$$

This immediately implies the assertion about the point spectrum of T and the multiplicity of eigenvalues. To prove the assertion about the resolvent set, let ρ denote the set on the right side of (1.5). If $z \in \rho$, then the operator

$$R(z) := \bigoplus_{n\in N} (T_n - z)^{-1}$$

is well defined and bounded by (1.4). Moreover, one easily checks that

$$(T - z)R(z) = 1 \quad \text{and} \quad R(z)(T - z) \text{ is the identity on dom } T.$$

Thus, $z \in \rho(T)$.

Conversely, if $z \in \rho(T)$, then $(T - z)^{-1}$ is well defined and bounded. This means that $z \in \rho(T_n)$ for each n and that $(T - z)^{-1} = R(z)$. Again by (1.4), boundedness of $R(z)$ implies $\sup_{n\in N} \left\| (T_n - z)^{-1} \right\| < \infty$. Thus $z \in \rho$. $\qquad \square$

1.1.3 Self-adjoint operators and their spectra

A densely defined operator A is called *symmetric* if $\operatorname{dom} A \subset \operatorname{dom} A^*$ and $A^* u = Au$ for every $u \in \operatorname{dom} A$. Equivalently, A is symmetric if and only if

$$(Au, v) - (u, Av) \quad \text{for all } u, v \in \operatorname{dom} A.$$

This, in turn, is equivalent to

$$(Au, u) \in \mathbb{R} \quad \text{for all } u \in \operatorname{dom} A.$$

Note that eigenvalues of a symmetric operator are necessarily real since $Au = zu$ implies

$$z\|u\|^2 = (Au, u) = (u, Au) = \overline{z}\|u\|^2.$$

Moreover, if u and v are eigenvectors of a symmetric operator A corresponding to distinct eigenvalues z and ζ, then $(u, v) = 0$. Indeed,

$$z(u, v) = (Au, v) = (u, Av) = \overline{\zeta}(u, v) = \zeta(u, v).$$

An operator A is called *self-adjoint* if it is densely defined, symmetric and $\operatorname{dom} A = \operatorname{dom} A^*$. Clearly, any self-adjoint operator is closed.

A symmetric operator A that is bounded (and defined on all of $\mathcal{H}$) is self-adjoint.

Lemma 1.6 *Let A be self-adjoint. Then $\sigma(A) \subset \mathbb{R}$. Moreover, $z \in \rho(A)$ if and only if there is an $\varepsilon > 0$ such that*

$$\|(A - z)u\| \geq \varepsilon\|u\| \quad \text{for all } u \in \operatorname{dom} A. \tag{1.6}$$

Note that the second part of this lemma says that $z \in \sigma(A)$ if and only if there is a sequence $(u_n) \subset \operatorname{dom} A$ with $\|u_n\| = 1$ for all n and $\|(A - z)u_n\| \to 0$ as $n \to \infty$.

Proof Indeed, for any symmetric operator A,

$$\|(A - z)u\|^2 = \|(A - \operatorname{Re} z)u\|^2 + (\operatorname{Im} z)^2\|u\|^2 \geq (\operatorname{Im} z)^2\|u\|^2. \tag{1.7}$$

Hence, if $\operatorname{Im} z \neq 0$ the operator $A - z$ is injective and its inverse, defined on $\operatorname{ran}(A - z)$, is bounded. If, in addition, A is closed, the inequality above implies that $\operatorname{ran}(A - z)$ is closed. Hence, if A is self-adjoint, by (1.2) for $\operatorname{Im} z \neq 0$ we get

$$\operatorname{ran}(A - z) = \overline{\operatorname{ran}(A - z)} = (\ker(A^* - \overline{z}))^\perp = (\ker(A - \overline{z}))^\perp = \mathcal{H},$$

and therefore $z \in \rho(A)$. Thus, we have shown $\sigma(A) \subset \mathbb{R}$, as claimed.

The same argument implies that, for all $z \in \mathbb{R}$ for which (1.6) holds, one has $z \in \rho(A)$.

Conversely, if $z \in \sigma_p(A)$, then (1.6) clearly fails for eigenvectors u. If $z \in \sigma(A) \backslash \sigma_p(A) \subset \mathbb{R}$, then $A - z$ is invertible. We show that its inverse is unbounded, which contradicts (1.6). Indeed, as above, we have $\overline{\mathrm{ran}(A - z)} = (\ker(A - z))^\perp = \mathcal{H}$. Since A is closed, the inverse of $A - z$ is closed as well. For a bounded inverse of $A - z$ this would mean that its domain $\mathrm{ran}(A - z)$ is closed in $\mathcal{H}$ and $\mathrm{ran}(A - z) = \mathcal{H}$. But then $z \in \rho(A)$, which contradicts $z \in \sigma(A)$. We conclude that the inverse of $A - z$ is unbounded. □

1.1.4 The spectrum of a multiplication operator

Let X be a set, $\mathcal{A}$ a sigma-algebra on X and μ a non-negative measure on $(X, \mathcal{A})$. The measure space $(X, \mathcal{A}, \mu)$ is called *separable* if there is a countable subset $\mathcal{B} \subset \mathcal{A}$ such that for any $E \in \mathcal{A}$ with $\mu(E) < \infty$ and any $\varepsilon > 0$ there is a $B \in \mathcal{B}$ with $\mu(B \Delta E) \le \varepsilon$. In that case, the Hilbert space $L^2(X, \mathcal{A}, \mu)$ is separable (in the sense of having a countable orthonormal basis).

In this subsection we assume that $(X, \mathcal{A}, \mu)$ is separable and sigma-finite.

Let φ be an $\mathcal{A}$-measurable, complex-valued and μ-a.e. finite function on X, and consider the multiplication operator T_φ in $L^2(X)$ defined by

$$T_\varphi u := \varphi u, \qquad \mathrm{dom}\, T_\varphi := \left\{ u \in L^2(X) : \varphi u \in L^2(X) \right\}.$$

The completeness of $L^2(X, (1 + |\varphi|^2)\mu)$ implies that T_φ is a closed operator. Moreover, T_φ is densely defined because, for any $u \in L^2(X)$, the functions $\chi_{\{|\varphi| \le n\}} u \in \mathrm{dom}\, T_\varphi$ converge to u as $n \to \infty$ by dominated convergence, using the μ-a.e. finiteness of φ. The adjoint of T_φ is given by

$$T_\varphi^* = T_{\overline{\varphi}}.$$

In particular, T_φ is self-adjoint if and only if φ is real-valued μ-a.e. The operator T_φ is bounded if and only if φ is μ-essentially bounded, and in this case

$$\|T_\varphi\| = \|\varphi\|_{L^\infty(X, \mathcal{A}, \mu)}.$$

Let us characterize the spectrum of the operator T_φ.

Theorem 1.7 *Let φ be an $\mathcal{A}$-measurable, complex-valued and μ-a.e. finite function on X. Then*

$$\sigma(T_\varphi) = \left\{ z \in \mathbb{C} : \mu(\varphi^{-1}(\{\zeta \in \mathbb{C} : |\zeta - z| \le \varepsilon\})) > 0 \text{ for all } \varepsilon > 0 \right\} \quad (1.8)$$

and

$$\sigma_p(T_\varphi) = \left\{ z \in \mathbb{C} : \mu(\varphi^{-1}(\{z\})) > 0 \right\}, \quad (1.9)$$

where $\varphi^{-1}(E) := \{ x \in X : \varphi(x) \in E \}$ denotes the pre-image of E under φ.

Proof We begin with proving (1.9). If $\varphi u = zu$ μ-a.e. on X for some $u \neq 0$, then $\varphi = z$ μ-a.e. on the set $\{x \in X : u(x) \neq 0\}$, which has positive measure. Conversely, if $Y \subset \varphi^{-1}(\{z\})$ is measurable with $0 < \mu(Y) < \infty$ (such a set exists by sigma-finiteness), then $0 \neq \chi_Y \in \operatorname{dom} T_\varphi$ and $T_\varphi \chi_Y = z\chi_Y$.

We turn to (1.8). By what we have just shown, if $z \in \sigma_p(T_\varphi)$, then z belongs to the set on the right side of (1.8). If $z \notin \sigma_p(T_\varphi)$, then the operator $T_\varphi - z$ is invertible on $\operatorname{ran} T_\varphi$ and the inverse is given by T_{ψ_z} with

$$\psi_z(x) := \frac{1}{\varphi(x) - z} \qquad \text{for all } x \in X .$$

As we have noticed before, this operator is bounded if and only if ψ_z is μ-essentially bounded; that is, if and only if there is an $\varepsilon > 0$ such that for μ-a.e. $x \in X$ one has $|\varphi(x) - z| \geq \varepsilon$. This means that z does not belong to the right side in (1.8). □

1.1.5 Functional calculus

A bounded operator Π on a Hilbert space $\mathcal{H}$ is called an *orthogonal projection* if $\Pi = \Pi^* = \Pi^2$. A *projection-valued measure* is a map $P : \omega \mapsto P_\omega$ on the Borel sigma-algebra on $\mathbb{R}$ taking values in the set of orthogonal projections such that

(a) If $(\omega_n)_{n \in N}$, $N \subset \mathbb{N}$, is a finite or countable family of disjoint Borel sets, then

$$P_{\bigcup_n \omega_n} = \underset{N \to \infty}{\text{s-lim}} \sum_{n \leq N} P_{\omega_n} .$$

(b) $P_{\mathbb{R}} = 1$.

One can deduce from these properties that $P_\emptyset = 0$ and that

$$P_{\omega_1} P_{\omega_2} = P_{\omega_1 \cap \omega_2} \qquad \text{for all Borel sets } \omega_1, \omega_2 . \tag{1.10}$$

Notions for scalar measures have natural analogues for projection-valued measures. The support of P is

$$\operatorname{supp} P := \left\{ \lambda \in \mathbb{R} : P_{(\lambda-\varepsilon, \lambda+\varepsilon)} \neq 0 \text{ for all } \varepsilon > 0 \right\} . \tag{1.11}$$

A property hold P-a.e. if it holds outside of a Borel set ω with $P_\omega = 0$. The P-essential supremum of a real-valued measurable function φ on $\mathbb{R}$ is

$$P\text{-sup}_\lambda \varphi(\lambda) := \inf \left\{ a \in \mathbb{R} : P_{\{\varphi > a\}} = 0 \right\}$$

and a measurable function φ is P-bounded if $P\text{-sup}_\lambda |\varphi(\lambda)| < \infty$.

Given a projection-valued measure P and $f, g \in \mathcal{H}$, then $\omega \mapsto (P_\omega f, g)$ is a complex Borel measure on $\mathbb{R}$ and we denote integration with respect to this measure by $d(P_\lambda f, g)$. If $f = g$, the measure $\omega \mapsto (P_\omega f, f)$ is non-negative and, since $P_\mathbb{R} = 1$, we have

$$\int_\mathbb{R} d(P_\lambda f, f) = \|f\|^2 .$$

The next result provides us with the existence and fundamental properties of a *functional calculus*.

Theorem 1.8 *Let P be a projection-valued measure.*

(a) *For every measurable, P-a.e. finite function φ on $\mathbb{R}$ there is a unique operator J_φ in $\mathcal{H}$ satisfying*

$$\operatorname{dom} J_\varphi = \left\{ u \in \mathcal{H} : \int_\mathbb{R} |\varphi(\lambda)|^2 \, d(P_\lambda u, u) < \infty \right\}$$

and

$$(J_\varphi u, g) = \int_\mathbb{R} \varphi(\lambda) \, d(P_\lambda u, g) \qquad \text{for all } u \in \operatorname{dom} J_\varphi \, , \, g \in \mathcal{H} \, . \quad (1.12)$$

The operator J_φ is closed and densely defined.

(b) *If φ, ψ are measurable, P-a.e. finite functions on $\mathbb{R}$, then*

$$J_\varphi^* = J_{\overline{\varphi}} \, , \tag{1.13}$$

$$J_1 = 1 \, , \tag{1.14}$$

$$\left\| J_\varphi u \right\|^2 = \int_\mathbb{R} |\varphi(\lambda)|^2 \, d(P_\lambda u, u) \quad \text{for all } u \in \operatorname{dom} J_\varphi \, , \tag{1.15}$$

$$\left\| J_\varphi \right\| = P\text{-}\sup_\lambda |\varphi(\lambda)| \, . \tag{1.16}$$

If $\alpha, \beta \in \mathbb{C}$ and if $|\varphi| + |\psi| \leq C(1 + |\alpha\varphi + \beta\psi|)$ P-a.e. for some $C < \infty$, then

$$\alpha J_\varphi + \beta J_\psi = J_{\alpha\varphi + \beta\psi} \, , \tag{1.17}$$

and, if $|\varphi| + |\psi| \leq C(1 + |\varphi\psi|)$ P-a.e. for some $C < \infty$, then

$$J_\varphi J_\psi = J_{\psi\varphi} \, . \tag{1.18}$$

(c) *If φ_n are measurable, P-bounded functions that converge pointwise P-a.e. to a P-a.e. finite function φ and satisfy $|\varphi_n| \leq C(1 + |\varphi|)$ P-a.e. for all n, then $J_{\varphi_n} u \to J_\varphi u$ for all $u \in \operatorname{dom} J_\varphi$.*

In the following we will sometimes use the notation

$$J_\varphi =: \int_{\mathbb{R}} \varphi(\lambda)\, dP_\lambda\,.$$

We note that the conditions for (1.17) and (1.18) are satisfied, in particular, if φ and ψ are P-bounded. Hence, Theorem 1.8 shows that the map $\varphi \mapsto J_\varphi$ restricts to an isometric isomorphism from the C^*-Banach algebra of all measurable, P-bounded functions on $\mathbb{R}$ (with the norm P-sup$_\lambda$, the constant function 1 as unit, and the involution $\varphi \mapsto \overline{\varphi}$) onto a commutative subalgebra of the C^*-Banach algebra of bounded operators on $\mathcal{H}$ (with the operator norm, the identity as unit, and the involution $T \mapsto T^*$).

For general, measurable, P-a.e. finite functions φ and ψ the identities (1.17) and (1.18) hold pointwise on the domains of the operator expressions on the left side and, moreover, these domains are dense in the operator norm of the right sides.

Proof   We only sketch the main steps of the construction and refer for details to, for instance, Birman and Solomjak (1987, §§5.3 and 5.4) or Teschl (2014, §3.1). We begin by defining the operator J_φ for simple functions $\varphi = \sum_{n=1}^N c_n \chi_{\omega_n}$ with $c_n \in \mathbb{C}$ and disjoint Borel $\omega_n \subset \mathbb{R}$ by

$$J_\varphi := \sum_n c_n P_{\omega_n}\,.$$

Using the properties of a projection-valued measure one easily verifies the assertions in parts (a) and (b) of the theorem. By the density of simple functions in the set of measurable, P-bounded functions with respect to the P-essential supremum, the operator $\varphi \mapsto J_\varphi$ can be uniquely extended to the latter class of functions φ. The assertions in parts (a) and (b) carry over to this extension. Moreover, the assertion in (c) in the case where the φ_n are uniformly P-bounded follows easily by dominated convergence.

For general measurable, P-a.e. finite functions φ one first verifies that the set dom J_φ in part (a) is, indeed, a dense subspace. On the other hand, using dominated convergence, one sees that, for $u \in$ dom J_φ and for every sequence (φ_n) as in (c), the elements $J_{\varphi_n} u$ converge to a limit that is independent of the choice of the (φ_n). This defines a linear operator J_φ as in (a), except for the claimed closedness, which will be established momentarily. The assertion in (c) holds by construction and the properties in (b) follow by approximation from the corresponding properties in the bounded cases. Finally, (1.13) implies that $J_\varphi = (J_{\overline{\varphi}})^*$, so J_φ is closed as an adjoint operator. This completes our sketch of the proof of Theorem 1.8.   □

1.1.6 The spectral theorem

The content of the spectral theorem for self-adjoint operators is that any such operator can be obtained via the procedure, in the previous subsection, of integration against a spectral measure. This theorem generalizes the diagonalization of a Hermitian matrix and plays a fundamental role in the theory of self-adjoint operators.

Theorem 1.9 *Let A be a self-adjoint operator in $\mathcal{H}$. Then there is a unique projection-valued measure P on $\mathcal{H}$ such that*

$$A = \int_{\mathbb{R}} \lambda \, dP_\lambda .$$

Proof Various proofs can be found in the textbooks mentioned in §1.3. Here we briefly sketch the main lines of the argument in Teschl (2014, §3.1).

Let $f \in \mathcal{H}$. Since, by Lemma 1.6, $\mathbb{C}_+ \subset \rho(A)$, we can consider the function $\mathbb{C}_+ \ni z \mapsto ((A - z)^{-1} f, f)$, and, by (1.7), we obtain the bound

$$\left| ((A - z)^{-1} f, f) \right| \le \left\| (A - z)^{-1} \right\| \|f\|^2 \le |\operatorname{Im} z|^{-1} \|f\|^2 .$$

By a Neumann series expansion we see that the function is analytic with respect to z. Moreover, using

$$(A - z)^{-1} - (A - \zeta)^{-1} = (z - \zeta)(A - \zeta)^{-1}(A - z)^{-1}, \qquad z, \zeta \in \rho(A), \quad (1.19)$$

with $\zeta = \bar{z}$ and, by (1.3), $((A - z)^{-1})^* = (A - \bar{z})^{-1}$, we find

$$\operatorname{Im} \left((A - z)^{-1} f, f \right) = (2i)^{-1} \left(((A - z)^{-1} f, f) - \overline{((A - z)^{-1} f, f)} \right)$$

$$= (2i)^{-1} \left(((A - z)^{-1} - (A - \bar{z})^{-1}) f, f \right)$$

$$= (\operatorname{Im} z) \left((A - \bar{z})^{-1}(A - z)^{-1} f, f \right)$$

$$= (\operatorname{Im} z) \left\| (A - z)^{-1} f \right\|^2 . \qquad (1.20)$$

Thus, by the Herglotz representation theorem (see, e.g., Teschl, 2014, Theorem 3.20), there is a non-negative Borel measure μ_f on $\mathbb{R}$ with $\mu_f(\mathbb{R}) \le \|f\|^2$ such that

$$((A - z)^{-1} f, f) = \int_{\mathbb{R}} \frac{d\mu_f(\lambda)}{\lambda - z} \quad \text{for all } z \in \mathbb{C}_+ .$$

For $f, g \in \mathcal{H}$ we define a complex Borel measure

$$\mu_{f,g} := \frac{1}{4} \left(\mu_{f+g} - \mu_{f-g} + i\mu_{f+ig} - i\mu_{f-ig} \right)$$

and obtain

$$((A - z)^{-1} f, g) = \int_{\mathbb{R}} \frac{d\mu_{f,g}(\lambda)}{\lambda - z}. \tag{1.21}$$

The right side is the Stieltjes transform of $\mu_{f,g}$. By a uniqueness theorem for this transform (see, e.g., Teschl, 2014, Theorem 3.21), the measure $\mu_{f,g}$ depends linearly on f and anti-linearly on g. For any Borel $\omega \subset \mathbb{R}$, $f \mapsto \mu_f(\omega)$ is a non-negative quadratic form that is bounded by $\mu_f(\omega) \leq \mu_f(\mathbb{R}) \leq \|f\|^2$. Thus, by the Riesz representation theorem, there is a self-adjoint operator P_ω in $\mathcal{H}$ with $\|P_\omega\| \leq 1$ such that

$$(P_\omega f, g) = \int_\omega d\mu_{f,g}(\lambda) \qquad \text{for all } f, g \in \mathcal{H}.$$

We will show that P is a projection-valued measure. The defining property (a) of such a measure follows, with a weak limit instead of a strong limit, from the sigma-additivity of the measures $\mu_{f,g}$. Once we have shown that P is projection-valued, the weak limit can be replaced by a strong limit. To prove the defining property (b) of such a measure, let $f \in \ker P_{\mathbb{R}}$. Then $0 = (P_{\mathbb{R}} f, f) = \mu_f(\mathbb{R})$ and therefore $((A - z)^{-1} f, f) = 0$ for all $z \in \mathbb{C}_+$. By (1.20), this implies $f = 0$. Thus, $\ker P_{\mathbb{R}} = \{0\}$ and, once we have shown that $P_{\mathbb{R}}$ is an orthogonal projection, we deduce that $P_{\mathbb{R}} = 1$, as claimed.

Let us show that $P_\omega^2 = P_\omega$. For all $z, \zeta \in \mathbb{C}_+$ with $z \neq \zeta$, by (1.19),

$$\int_{\mathbb{R}} \frac{d\mu_{f,g}(\lambda)}{(\lambda - z)(\lambda - \zeta)} = \frac{1}{z - \zeta} \left(\int_{\mathbb{R}} \frac{d\mu_{f,g}(\lambda)}{\lambda - z} - \int_{\mathbb{R}} \frac{d\mu_{f,g}(\lambda)}{\lambda - \zeta} \right)$$

$$= \frac{1}{z - \zeta} \left(((A - z)^{-1} - (A - \zeta)^{-1}) f, g \right) = \left((A - \zeta)^{-1} f, (A - \bar{z})^{-1} g \right)$$

$$= ((A - z)^{-1}(A - \zeta)^{-1} f, g) = \int_{\mathbb{R}} \frac{d\mu_{f,(A-\bar{z})^{-1}g}(\lambda)}{\lambda - \zeta}.$$

By a uniqueness theorem for the Stieltjes transform, this implies that

$$\frac{d\mu_{f,g}(\lambda)}{\lambda - z} = d\mu_{f,(A-\bar{z})^{-1}g}(\lambda).$$

Using this formula, we find for any Borel $\omega \subset \mathbb{R}$,

$$\int_\omega \frac{d\mu_{f,g}(\lambda)}{\lambda - z} = \int_\omega d\mu_{f,(A-\bar{z})^{-1}g}(\lambda) = \left(P_\omega f, (A - \bar{z})^{-1} g \right)$$

$$= ((A - z)^{-1} P_\omega f, g) = \int_{\mathbb{R}} \frac{d\mu_{P_\omega f,g}(\lambda)}{\lambda - z}.$$

Again by the uniqueness theorem for the Stieltjes transform, this implies that

$$\chi_\omega(\lambda) d\mu_{f,g}(\lambda) = d\mu_{P_\omega f,g}(\lambda).$$

Multiplying this identity by $\chi_\omega(\lambda)$ and integrating, we find

$$(P_\omega f, g) = \int_\omega d\mu_{f,g}(\lambda) = \int_\omega d\mu_{P_\omega f,g}(\lambda) = (P_\omega^2 f, g).$$

Thus, $P_\omega = P_\omega^2$. We have shown that P is a projection-valued measure.

In terms of this measure, (1.21) means that $(A - z)^{-1} = \int_\mathbb{R} (\lambda - z)^{-1} dP_\lambda$ for $z \in \mathbb{C}_+$. By the functional calculus (Theorem 1.8), this implies $A = \int_\mathbb{R} \lambda \, dP_\lambda$, as claimed. Uniqueness of the projection-valued measure follows from (1.21) and the uniqueness of the Stieltjes transform. □

In the situation of Theorem 1.9 the projection-valued measure is also called the *spectral measure* of A. Its relation to the spectrum is clarified in Corollary 1.10 below.

Given the spectral measure of a self-adjoint operator A one can apply the construction in the previous subsection and define functions of a self-adjoint operator. More precisely, if A is a self-adjoint operator, P its spectral measure, and if φ is a measurable, P-a.e. finite function on $\mathbb{R}$, then, according to Theorem 1.8, there is an operator

$$\varphi(A) := \int_\mathbb{R} \varphi(\lambda) \, dP_\lambda$$

with domain

$$\operatorname{dom} \varphi(A) = \left\{ u \in \mathcal{H} : \int_\mathbb{R} |\varphi(\lambda)|^2 d(P_\lambda u, u) < \infty \right\} \tag{1.22}$$

satisfying

$$(\varphi(A)u, g) = \int_\mathbb{R} \varphi(\lambda) \, d(P_\lambda u, g) \qquad \text{for all } u \in \operatorname{dom} \varphi(A), \, g \in \mathcal{H}. \tag{1.23}$$

Note, in particular, that by (1.18), the operator $\varphi(A)$ for $\varphi(\lambda) = \lambda^n$, $n \in \mathbb{N}$, coincides with the n-fold product of A defined in the sense of operator products. Moreover, if $z \in \mathbb{C} \setminus \sigma_p(A)$, then $(A - z)^{-1}$, defined by the functional calculus, coincides with the resolvent defined as a possibly unbounded operator. The spectral projections P_ω of Borel sets ω reappear through the characteristic functions χ_ω, namely, $P_\omega = \chi_\omega(A)$.

Recall that the support of a projection-valued measure was defined in (1.11).

Corollary 1.10 *We have* $\sigma(A) = \operatorname{supp} P$ *and, for all* $z \in \rho(A)$,

$$\left\| (A - z)^{-1} \right\| = \operatorname{dist}(z, \sigma(A))^{-1}. \tag{1.24}$$

Moreover $\sigma_p(A) = \{\lambda \in \mathbb{R} : P_{\{\lambda\}} \neq 0\}$ *and* $\ker(A - \lambda) = \operatorname{ran} P_{\{\lambda\}}$, *for any* $\lambda \in \sigma_p(A)$.

Proof The proof is based on the fact that, for $u \in \mathrm{dom}\, A$, $z \in \mathbb{C}$, and Borel $\omega \subset \mathbb{R}$, we have $P_\omega u \in \mathrm{dom}\, A$ and

$$\|(A - z)P_\omega u\|^2 = \int_\omega |\mu - z|^2 d(P_\mu u, u). \tag{1.25}$$

Indeed, by (1.15) and (1.18),

$$\|(A - z)P_\omega u\|^2 = \|(A - z)\chi_\omega(A)u\|^2 = \|((\cdot - z)\chi_\omega)(A)u\|^2$$
$$= \int_\omega |\mu - z|^2 \, d(P_\mu u, u).$$

First, assume that $z \in \mathbb{C} \setminus \mathrm{supp}\, P$. Then, by (1.25) and the fact that the support of the measure $\omega \mapsto (P_\omega u, u)$ is contained in the support of the projection-valued measure P, for all $u \in \mathrm{dom}\, A$,

$$\|(A - z)u\|^2 = \int_\mathbb{R} |\mu - z|^2 d(P_\mu u, u) = \int_{\mathrm{supp}\, P} |\mu - z|^2 d(P_\mu u, u)$$
$$\geq \mathrm{dist}(z, \mathrm{supp}\, P)^2 \int_{\mathrm{supp}\, P} d(P_\mu u, u) = \mathrm{dist}(z, \mathrm{supp}\, P)^2 \|u\|^2.$$

Thus, by Lemma 1.6, $z \in \rho(A)$ and

$$\|(A - z)^{-1}\| \leq \mathrm{dist}(z, \mathrm{supp}\, P)^{-1}.$$

This proves, in particular, that $\sigma(A) \subset \mathrm{supp}\, P$.

Conversely, if $\lambda \in \mathrm{supp}\, P$, there is a sequence (u_n) with $0 \neq u_n \in \mathrm{ran}\, P_{[\lambda - 1/n, \lambda + 1/n]}$ for all n. Then, by (1.25), for any $z \in \mathbb{C}$,

$$\|(A - z)u_n\|^2 = \int_{[\lambda - 1/n, \lambda + 1/n]} |\mu - z|^2 d(P_\mu u_n, u_n)$$
$$\leq \left((|\mathrm{Re}\, z - \lambda| + n^{-1})^2 + (\mathrm{Im}\, z)^2 \right) \int_{[\lambda - 1/n, \lambda + 1/n]} d(P_\mu u_n, u_n)$$
$$= \left((|\mathrm{Re}\, z - \lambda| + n^{-1})^2 + (\mathrm{Im}\, z)^2 \right) \|\chi_{[\lambda - 1/n, \lambda + 1/n]}(A)u_n\|^2$$
$$= \left((|\mathrm{Re}\, z - \lambda| + n^{-1})^2 + (\mathrm{Im}\, z)^2 \right) \|u_n\|^2.$$

In particular, with the choice $z = \lambda$ we see that the bound (1.6) is violated and therefore $\lambda \in \sigma(A)$. Thus, $\mathrm{supp}\, P \subset \sigma(A)$, which completes the proof of the first assertion.

Applying the above bound for $z \in \rho(A)$, we deduce that

$$\|(A - z)^{-1}\|^2 \geq \left((|\mathrm{Re}\, z - \lambda| + n^{-1})^2 + (\mathrm{Im}\, z)^2 \right)^{-1}.$$

Since n is arbitrary, we find

$$\|(A - z)^{-1}\|^2 \geq \left((\mathrm{Re}\, z - \lambda)^2 + (\mathrm{Im}\, z)^2 \right)^{-1}$$

and, taking the supremum over all $\lambda \in \text{supp } P$, we obtain the bound

$$\|(A-z)^{-1}\|^2 \geq \text{dist}(z, \text{supp } P)^{-2}.$$

This proves the formula for the norm of the resolvent.

Finally, by (1.25) with $\omega = \mathbb{R}$, $u \in \ker(A-z)$ if and only $d(P_\mu u, u)$ is a point measure of mass $\|u\|^2$ at $\mu = z$. The latter is equivalent to $u = P_{\{z\}}u$. This completes the proof. $\quad\square$

For a symmetric operator A we set

$$m_A := \inf_{0 \neq u \in \text{dom } A} \frac{(Au, u)}{\|u\|^2}.$$

An operator A is called *lower semibounded* if it is symmetric and $m_A > -\infty$. An operator A is called *non-negative* if it is symmetric and $m_A \geq 0$.

Corollary 1.11 *Let A be a self-adjoint operator. Then*

$$m_A = \inf \sigma(A).$$

Assuming that $m_A > -\infty$, we have that m_A is an eigenvalue of A if and only if the infimum $\inf_{0 \neq u \in \text{dom } A}(Au, u)/\|u\|^2$ is a minimum, and the eigenvectors are precisely those vectors for which the infimum is attained.

Proof We assume that $m_A > -\infty$ and omit the minor modifications needed for $m_A = -\infty$. First note that, for $\lambda < m_A$ and $u \in \text{dom } A$,

$$(m_A - \lambda)\|u\|^2 \leq (Au, u) - \lambda\|u\|^2 = ((A-\lambda)u, u) \leq \|(A-\lambda)u\|\|u\|,$$

and therefore, by Lemma 1.6, $\lambda \in \rho(A)$. Hence, $\sigma(A) \subset [m_A, +\infty)$ and, in particular, $m_A \leq \inf \sigma(A)$. Conversely, by the spectral theorem (Theorem 1.9) and Corollary 1.10,

$$(Au, u) = \int_{\mathbb{R}} \lambda d(P_\lambda u, u) = \int_{\sigma(A)} \lambda d(P_\lambda u, u) \geq \inf \sigma(A)\|u\|^2 \text{ for all } u \in \text{dom } A.$$

Thus, $m_A \geq \inf \sigma(A)$ which proves the first assertion. Equality in the previous bound is attained if and only if $d(P_\lambda u, u)$ is a point measure of mass $\|u\|^2$ at $\lambda = m_A$. As in the proof of Corollary 1.10, this is equivalent to $u \in \ker(A - m_A)$. $\quad\square$

According to the *spectral mapping theorem*, the original spectral measure $P =: P(A)$ of A and the spectral measure $P(\varphi(A))$ of $\varphi(A)$, where φ is a measurable, real-valued function, are related by

$$P_\omega(\varphi(A)) = P_{\varphi^{-1}(\omega)}(A) \qquad \text{for all Borel sets } \omega \subset \mathbb{R}. \tag{1.26}$$

Indeed, the right side defines a projection-valued measure $\tilde{P}$ and for all $u \in \mathcal{H}$, by a change of variables,

$$\int_{\mathbb{R}} \mu^2 \, d(\tilde{P}_\mu u, u) = \int_{\mathbb{R}} \varphi(\lambda)^2 \, d(P_\lambda u, u).$$

Moreover, for all u for which this is finite one has, by the same change of variables,

$$\int_{\mathbb{R}} \mu \, d(\tilde{P}_\mu u, g) = \int_{\mathbb{R}} \varphi(\lambda) \, d(P_\lambda u, g) = (\varphi(A)u, g) \qquad \text{for all } g \in \mathcal{H}.$$

By the uniqueness assertion in Theorem 1.9, $\tilde{P}$ is the spectral measure of $\varphi(A)$, as claimed in (1.26). In particular, (1.26) implies that for all functions φ that are also continuous on supp P one has

$$\sigma(\varphi(A)) = \overline{\varphi(\sigma(A))}.$$

We now return to the study of orthogonal sums of operators. Let $\mathcal{N}$ be a countable (possibly finite) index set and, for each $n \in \mathcal{N}$, assume that $\mathcal{H}_n$ is a separable Hilbert space. Moreover, for each $n \in \mathcal{N}$, let A_n be a self-adjoint operator in $\mathcal{H}_n$. According to our discussion in §1.1.2, we know that $\bigoplus_{n \in \mathcal{N}} A_n$ is self-adjoint in $\bigoplus_{n \in \mathcal{N}} \mathcal{H}_n$. We denote by $P(A_n)$ the spectral measure of A_n and define a map $\bigoplus_{n \in \mathcal{N}} P(A_n)$ from Borel sets in $\mathbb{R}$ to operators on $\bigoplus_{n \in \mathcal{N}} \mathcal{H}_n$ by

$$\left(\bigoplus_{n \in \mathcal{N}} P(A_n) \right)_\omega := \bigoplus_{n \in \mathcal{N}} P_\omega(A_n) \qquad \text{for any Borel set } \omega \subset \mathbb{R}.$$

This is clearly a projection-valued measure.

Lemma 1.12 *One has*

$$\sigma\left(\bigoplus_{n \in \mathcal{N}} A_n \right) = \overline{\bigcup_{n \in \mathcal{N}} \sigma(A_n)}$$

and $\bigoplus_{n \in \mathcal{N}} P(A_n)$ *is the spectral measure of* $\bigoplus_{n \in \mathcal{N}} A_n$.

Proof According to (1.24), we have $\|(A_n - z)^{-1}\| = (\text{dist}(z, \sigma(A_n)))^{-1}$, so by Lemma 1.5 the first assertion follows from the fact that

$$\sup_{n \in \mathcal{N}} \left(\text{dist}(z, \sigma(A_n)) \right)^{-1} = \left(\inf_{n \in \mathcal{N}} \text{dist}(z, \sigma(A_n)) \right)^{-1} = \left(\text{dist}\left(z, \overline{\bigcup_{n \in \mathcal{N}} \sigma(A_n)} \right) \right)^{-1},$$

which is valid for all $z \in \bigcap_{n \in \mathcal{N}} \rho(A_n)$. For the second assertion we first check that $\bigoplus_{n \in \mathcal{N}} P(A_n)$ is a projection-valued measure. We then verify the defining relations in part (a) of Theorem 1.8. $\qquad \square$

1.1.7 The essential spectrum and Weyl's theorem

Let A be a self-adjoint operator in a Hilbert space $\mathcal{H}$ with spectral measure P. Let us define the *essential spectrum* $\sigma_{\mathrm{ess}}(A)$ and the *discrete spectrum* $\sigma_{\mathrm{disc}}(A)$ of A by

$$\sigma_{\mathrm{ess}}(A) := \left\{ \lambda \in \mathbb{R} : \dim \operatorname{ran} P_{(\lambda - \varepsilon, \lambda + \varepsilon)} = \infty \text{ for all } \varepsilon > 0 \right\},$$

$$\sigma_{\mathrm{disc}}(A) := \sigma(A) \setminus \sigma_{\mathrm{ess}}(A)$$

$$= \left\{ \lambda \in \sigma(A) : \dim \operatorname{ran} P_{(\lambda - \varepsilon, \lambda + \varepsilon)} < \infty \text{ for some } \varepsilon > 0 \right\}.$$

It is easy to prove that $\lambda \in \sigma_{\mathrm{disc}}(A)$ if and only if λ is an isolated point of $\sigma(A)$ (i.e., $\sigma(A) \cap (\lambda - \varepsilon, \lambda + \varepsilon) = \{\lambda\}$ for some $\varepsilon > 0$) and λ is an eigenvalue of finite multiplicity. Moreover, one can prove that $\lambda \in \sigma_{\mathrm{ess}}(A)$ if and only if one or more of the following statements hold: $\operatorname{ran}(A - \lambda)$ is not closed; or λ is an accumulation point of eigenvalues; or λ is an eigenvalue of infinite multiplicity.

In practice, it is useful to have a criterion of whether a point belongs to the essential spectrum of A that is not expressed through the spectral measure of A, but rather through A itself. To state such a criterion, we shall say that a sequence $(u_n) \subset \mathcal{H}$ is a *singular sequence* for A at a point $\lambda \in \mathbb{R}$ if the following conditions are satisfied:

$$\inf_n \|u_n\| > 0, \tag{1.27}$$

$$u_n \to 0 \text{ weakly in } \mathcal{H}, \tag{1.28}$$

$$u_n \in \operatorname{dom} A, \tag{1.29}$$

$$(A - \lambda)u_n \to 0 \text{ strongly in } \mathcal{H}. \tag{1.30}$$

Lemma 1.13 *A point $\lambda \in \mathbb{R}$ belongs to $\sigma_{\mathrm{ess}}(A)$ if and only if there is a singular sequence for A at λ.*

Proof First, let $\lambda \in \sigma_{\mathrm{ess}}(A)$ and let (ε_n) be a decreasing sequence of positive numbers tending to zero. Then, by the definition of the essential spectrum, there is an orthonormal system (u_n) with $u_n \in \operatorname{ran} P_{(\lambda - \varepsilon_n, \lambda + \varepsilon_n)}$ for all n. Then (1.27), (1.28) and (1.29) are clearly satisfied and (1.30) follows from

$$\|(A - \lambda)u_n\|^2 = \int_{(\lambda - \varepsilon_n, \lambda + \varepsilon_n)} (\mu - \lambda)^2 d(P_\mu u_n, u_n)$$

$$\leq \varepsilon_n^2 \int_{(\lambda - \varepsilon_n, \lambda + \varepsilon_n)} d(P_\mu u_n, u_n) = \varepsilon_n^2 \to 0.$$

Conversely, assume that there is a singular sequence (u_n) for A at λ. We argue by contradiction assuming that $\lambda \notin \sigma_{\mathrm{ess}}(A)$. Then $\dim \operatorname{ran} P_{(\lambda - \varepsilon, \lambda + \varepsilon)} < \infty$ for some $\varepsilon > 0$.

Let $v_n := u_n - P_{(\lambda-\varepsilon,\lambda+\varepsilon)}u_n$. Then, by (1.25),

$$\varepsilon\|v_n\| \leq \|(A - \lambda)v_n\| = \|P_{\mathbb{R}\setminus(\lambda-\varepsilon,\lambda+\varepsilon)}(A - \lambda)u_n\| \leq \|(A - \lambda)u_n\| .$$

Using (1.30), we deduce that $v_n \to 0$ in $\mathcal{H}$. Since $P_{(\lambda-\varepsilon,\lambda+\varepsilon)}$ has finite rank, (1.28) implies that $P_{(\lambda-\varepsilon,\lambda+\varepsilon)}u_n \to 0$ in $\mathcal{H}$. Thus $u_n = v_n + P_{(\lambda-\varepsilon,\lambda+\varepsilon)}u_n \to 0$ in $\mathcal{H}$. This contradicts (1.27) and completes the proof. $\qquad\square$

The following theorem is due to Weyl and states the stability of the essential spectrum under certain perturbations. This is very useful in practice since it reduces the computation of the essential spectrum for general operators to that for certain model operators. It is an example of a perturbation-theoretic result.

Theorem 1.14 *Let A_1 and A_2 be self-adjoint operators and assume that, for some $z \in \rho(A_1) \cap \rho(A_2)$,*

$$(A_1 - z)^{-1} - (A_2 - z)^{-1} \qquad \text{is compact}. \tag{1.31}$$

Then $\sigma_{\mathrm{ess}}(A_1) = \sigma_{\mathrm{ess}}(A_2)$.

Note that assumption (1.31) holds for bounded A_1, A_2 with $A_1 - A_2$ compact, since

$$(A_1 - z)^{-1} - (A_2 - z)^{-1} = -(A_2 - z)^{-1}(A_1 - A_2)(A_1 - z)^{-1}, \qquad z \in \rho(A_1) \cap \rho(A_2). \tag{1.32}$$

In our applications it is crucial, however, that it suffices that the compactness holds only for the resolvent difference, rather than the operator difference. A convenient way to verify this condition in terms of quadratic forms will be given in Theorem 1.51.

We note that if (1.31) holds for *some* $z \in \rho(A_1) \cap \rho(A_2)$, then it holds for *any* such z. Indeed, $D(\zeta) := (A_1 - \zeta)^{-1} - (A_2 - \zeta)^{-1}$ satisfies

$$D(z') = \big((A_1 - z)(A_1 - z')^{-1}\big)D(z)\big((A_2 - z)(A_2 - z')^{-1}\big).$$

Since the factors $(A_j - z)(A_j - z')^{-1}$ are bounded, compactness of $D(z)$ implies compactness of $D(z')$.

Proof Since the assertion is symmetric in the operators A_1 and A_2, it suffices to prove that $\sigma_{\mathrm{ess}}(A_1) \subset \sigma_{\mathrm{ess}}(A_2)$. Let $\lambda \in \sigma_{\mathrm{ess}}(A_1)$. Then, by Lemma 1.13, there is a singular sequence (u_n) for A_1 at λ. With z from (1.31), let

$$v_n := (A_2 - z)^{-1}(A_1 - z)u_n,$$

which is well defined because of (1.29). We would like to show that (v_n) is a singular sequence for A_2 at λ. Once this is done, the theorem follows again from Lemma 1.13.

Obviously, (1.29) is satisfied for v_n and A_2.

Let us verify (1.27) and (1.28). With $K := (A_2 - z)^{-1} - (A_1 - z)^{-1}$, we have

$$v_n = K(A_1 - z)u_n + u_n = K(A_1 - \lambda)u_n + (\lambda - z)Ku_n + u_n .$$

Because of (1.28) and (1.30) for u_n and A_1, and Lemma 1.1, we have $v_n - u_n \to 0$ strongly in $\mathcal{H}$. Therefore, (1.27) and (1.28) for u_n imply (1.27) and (1.28) for v_n.

Finally, in order to verify (1.30), we compute

$$(A_2 - \lambda)v_n = (A_1 - z)u_n + (z - \lambda)v_n = (A_1 - \lambda)u_n + (z - \lambda)(v_n - u_n) .$$

By (1.30) for u_n and A_1, and by the fact that $v_n - u_n \to 0$ strongly in $\mathcal{H}$, we obtain (1.30) for v_n and A_2. This completes the proof. $\square$

Here is a consequence of the spectral theorem for compact operators.

Lemma 1.15 *Let K be a bounded, self-adjoint operator in $\mathcal{H}$ with $\dim \mathcal{H} = \infty$. Then K is compact if and only if $\sigma_{\mathrm{ess}}(K) = \{0\}$. In this case, $\mathcal{H}$ has an orthonormal basis consisting of eigenvectors of K.*

Proof First, assume that K is compact. By (1.32), the assumption (1.31) in Weyl's theorem holds with $A_1 = K$ and $A_2 = 0$. Thus, the theorem implies that $\sigma_{\mathrm{ess}}(K) = \sigma_{\mathrm{ess}}(0) = \{0\}$. This means that the spectrum of K away from zero consists of isolated eigenvalues of finite multiplicities. Thus if $(\lambda_n^{\pm})$ denotes the positive and negative eigenvalues of K, *not* repeated according to multiplicities, then by Corollary 1.10, the spectral measure P of K has support

$$\{\lambda_n^+ : n\} \cup \{\lambda_n^- : n\} \cup \{0\}$$

and $\operatorname{ran} P_{\lambda_n^{\pm}}$ are the eigenspaces corresponding to non-zero eigenvalues. The defining properties of a projection-valued measure give that, in the sense of strong convergence,

$$P_{\{0\}} + \sum_n P_{\{\lambda_n^+\}} + \sum_n P_{\{\lambda_n^-\}} = 1 .$$

This identity implies the completeness of an orthonormal basis obtained by combining orthonormal bases in the range of each of these projections.

Conversely, assume that $\sigma_{\mathrm{ess}}(K) = \{0\}$. To prove compactness of K, assume that $u_n \to 0$ weakly in $\mathcal{H}$. Let $\tau > 0$ and $P := \chi_{(-\tau,\tau)}(K)$. Since $\sigma_{\mathrm{ess}}(K) = \{0\}$, the operator $P^{\perp} := 1 - P$ has finite rank. Therefore, the weak convergence $u_n \to 0$ implies the strong convergence $P^{\perp}Ku_n \to 0$. On the other hand, by the spectral theorem, $\|PKu_n\| \leq \tau \|u_n\|$. Since, by orthogonality,

$$\|Ku_n\|^2 = \|P^{\perp}Ku_n\|^2 + \|PKu_n\|^2,$$

we find that $\limsup_{n \to \infty} \|Ku_n\|^2 \leq \tau^2 C$ with $C := \sup_n \|u_n\|^2$. Note that C is finite by uniform boundedness. Since $\tau > 0$ can be chosen arbitrarily small, we conclude that $Ku_n \to 0$ strongly in $\mathcal{H}$. Thus, by Lemma 1.1, K is compact. □

1.2 Semibounded operators and forms, and the variational principle

1.2.1 Semibounded operators and forms

A *sesquilinear form* $a[\cdot, \cdot]$ in $\mathcal{H}$ is a map from its domain $d[a] \times d[a]$ that is linear in its first and anti-linear in its second argument, where $d[a]$ is a subspace of $\mathcal{H}$. It is called *symmetric* if

$$a[u, v] = \overline{a[v, u]} \text{ for all } u, v \in d[a],$$

and it is called *densely defined* if $d[a]$ is dense in $\mathcal{H}$.

The *quadratic form* $a[\cdot]$ associated to a sesquilinear form $a[\cdot, \cdot]$ is defined by

$$a[u] := a[u, u] \text{ for all } u \in d[a].$$

The sesquilinear form $a[\cdot, \cdot]$ can be recovered from its quadratic form in view of the polarization identity

$$a[u, v] = \frac{1}{4} \left(a[u + v] - a[u - v] + ia[u + iv] - ia[u - iv] \right) \text{ for all } u, v \in d[a].$$

The quadratic form $a[\cdot]$ is real-valued if and only if $a[\cdot, \cdot]$ is symmetric.

A quadratic form $a = a[\cdot]$ with domain $d[a]$ is called *lower semibounded* in a Hilbert space $\mathcal{H}$ if it is real-valued and

$$m_a := \inf_{0 \neq u \in d[a]} \frac{a[u]}{\|u\|^2} > -\infty.$$

In this case, for each $m > -m_a$, the expression

$$a[u, v] + m(u, v)$$

defines an inner product on $d[a]$ and for different $m > -m_a$ the corresponding norms are equivalent.

A lower semibounded quadratic form a with domain $d[a]$ is called *closed* in a Hilbert space $\mathcal{H}$ if, for some (and hence any) $m > -m_a$, the set $d[a]$ is complete with respect to the norm

$$\left(a[u] + m\|u\|^2 \right)^{1/2}.$$

A *form core* is a subspace $F \subset d[a]$ that is dense with respect to the norm $\left(a[u] + m\|u\|^2 \right)^{1/2}$ in $d[a]$.

Let us discuss the relation between lower semibounded quadratic forms and lower semibounded operators. The latter notion was introduced before Corollary 1.11 and we recall the notation

$$m_A := \inf_{0 \neq u \in \text{dom } A} \frac{(Au, u)}{\|u\|^2} > -\infty.$$

Theorem 1.16 *Let A be a self-adjoint, lower semibounded operator. Then there is a unique lower semibounded, closed quadratic form a satisfying*

$$\text{dom } A \subset d[a], \tag{1.33}$$

$$a[u, v] = (Au, v) \quad \text{for all } u \in \text{dom } A, \, v \in d[a]. \tag{1.34}$$

The form a is densely defined, $\text{dom } A$ is a form core of a, and $m_a = m_A$. Moreover, for all $m \geq -m_A$, one has $d[a] = \text{dom}(A + m)^{1/2}$ and

$$a[u, v] = ((A + m)^{1/2} u, (A + m)^{1/2} v) - m(u, v) \quad \text{for all } u, v \in d[a], \tag{1.35}$$

and, if P is the spectral measure of A, then

$$a[u, v] = \int_{\mathbb{R}} \lambda \, d(P_\lambda u, v) \quad \text{for all } u, v \in d[a]. \tag{1.36}$$

In the following, for brevity, we refer to a described in Theorem 1.16 as the quadratic form *corresponding* to A.

We begin with a technical lemma.

Lemma 1.17 *Let A be a self-adjoint, lower semibounded operator and let a be a lower semibounded, closed quadratic form satisfying (1.33) and (1.34). Then a is densely defined, $\text{dom } A$ is a form core of a, and $m_a = m_A$.*

Proof Since A is densely defined, (1.33) implies that a is densely defined. By (1.33) and (1.34) we have $a[u] = (Au, u)$ for all $u \in \text{dom } A$. Thus, by enlarging the set over which the infimum is taken, we see that $m_A \geq m_a$.

Let us show that $\text{dom } A$ is a form core of a. Indeed, if $v \in d[a]$ satisfies $a[u, v] + m(u, v) = 0$ for all $u \in \text{dom } A$ and some fixed $m > -m_a$, then, by (1.34), $v \in (\text{ran}(A + m))^\perp = \ker(A + m) = \{0\}$. Here we used (1.1), the assumption that A is self-adjoint, and the fact that, by Corollary 1.11, $m > -m_a \geq -m_A = -\inf \sigma(A)$.

Since the infimum defining m_a remains the same when restricted to a form core, we conclude from the above facts that $m_A = m_a$. $\square$

Proof of Theorem 1.16 We fix $m \geq -m_A$ and use (1.35) to define a symmetric sesquilinear form a with domain $d[a] := \text{dom}(A + m)^{1/2}$. Note that by Corollary 1.11 we have $\inf \sigma(A + m) \geq 0$, and therefore the square root $(A + m)^{1/2}$ is well defined as a self-adjoint, non-negative operator by the functional calculus

(Theorem 1.8). The quadratic form a is closed since the operator $(A + m)^{1/2}$ is closed.

Let us show (1.33) and (1.34). The former follows from (1.22) applied to $\varphi(\lambda) = \sqrt{\lambda + m}$ and $\varphi(\lambda) = \lambda$. Before proving (1.34), we note that (1.36) follows from (1.15) for $u = v$ and then by polarization for general u, v. The identity in (1.34) follows from (1.36) by the spectral theorem (Theorem 1.9).

The facts that a is densely defined, that dom A is a form core, and that $m_a = m_A$ all follow from Lemma 1.17.

To show uniqueness, and, in particular, independence of the parameter $m \geq -m_A$, let $\tilde{a}$ be a lower semibounded, closed quadratic form satisfying (1.33) and (1.34). Then, by these properties for a and $\tilde{a}$,

$$\tilde{a}[u,v] = (Au,v) = a[u,v] \qquad \text{for all } u, v \in \text{dom } A. \tag{1.37}$$

By Lemma 1.17, dom A is a form core of both a and $\tilde{a}$. Therefore, (1.37) and the closedness of a and $\tilde{a}$ imply that $d[a] = d[\tilde{a}]$ and $\tilde{a}[u,v] = a[u,v]$ for all u, v from this set. Thus $a = \tilde{a}$. □

The usefulness of lower semibounded quadratic forms comes from the fact that the above construction can be reversed. That is, each densely defined, lower semibounded, and closed quadratic form gives rise to a unique self-adjoint operator. The precise statement is the following.

Theorem 1.18 *Let a be a densely defined, lower semibounded, and closed quadratic form. Then there is a unique self-adjoint operator A satisfying (1.33) and (1.34). Moreover, A is lower semibounded with $m_A = m_a$ and the domain of A is given by*

$$\text{dom } A = \{u \in d[a]: \text{ there exists } f \in \mathcal{H} \text{ such that for all } v \in d[a],$$
$$a[u,v] = (f,v)\}.$$

In the following, we briefly refer to A described in Theorem 1.18 as the operator *corresponding* to a.

Proof We fix $m > -m_a$. We have for any $g \in \mathcal{H}$ and any $v \in d[a]$,

$$|(g,v)| \leq \|g\|\|v\| \leq (m_a + m)^{-1/2}\|g\| \left(a[v] + m\|v\|^2\right)^{1/2}. \tag{1.38}$$

This means that for fixed $g \in \mathcal{H}$, $v \mapsto (g,v)$ is a bounded, anti-linear functional on $d[a]$ endowed with the norm $(a[\cdot] + m\|\cdot\|^2)^{1/2}$. Therefore, by the Riesz representation theorem, there is a unique $u_g \in d[a]$ such that

$$(g,v) = a[u_g,v] + m(u_g,v) \quad \text{for all } v \in d[a]. \tag{1.39}$$

The uniqueness of u_g implies that the map $g \mapsto u_g$ is linear. Moreover, taking $v = u_g$ in both (1.38) and (1.39), we obtain

$$a[u_g] + m\|u_g\|^2 \le (m_a + m)^{-1/2}\|g\| \left(a[u_g] + m\|u_g\|^2\right)^{1/2};$$

that is, $a[u_g] + m\|u_g\|^2 \le (m_a + m)^{-1}\|g\|^2$. This means that the map $g \mapsto u_g$ is bounded from $\mathcal{H}$ to $d[a]$. Since the embedding $d[a] \to \mathcal{H}$ is continuous, we can consider the mapping $g \mapsto Bg := u_g$ as a bounded linear operator on $\mathcal{H}$. Taking $v = u_g$ in (1.39) gives

$$(g, Bg) = a[u_g] + m\|u_g\|^2 \ge (m_a + m)\|u_g\|^2 \ge 0, \qquad (1.40)$$

and therefore B is a self-adjoint operator on $\mathcal{H}$. Its kernel is trivial since, by (1.39), $u_g = 0$ implies $g = 0$. Moreover, since, by (1.2), $\overline{\operatorname{ran} B} = (\ker B)^{\perp} = \mathcal{H}$, its range is dense in $\mathcal{H}$.

These facts imply that $A := B^{-1} - m$ can be defined as a possibly unbounded operator with domain $\operatorname{dom} A := \operatorname{ran} B$. By (1.3), A is self-adjoint. By construction, we have $\operatorname{dom} A \subset d[a]$ and, for $u \in \operatorname{dom} A$ and $v \in d[a]$, we can apply (1.39) to $g := (A + m)u$ and get, using $Bg = u$,

$$(Au, v) = (g, v) - m(u, v) = a[u_g, v] + m(u_g, v) - m(u, v)$$
$$= a[Bg, v] + m(Bg, v) - m(u, v) = a[u, v].$$

This means that a is associated to A in the sense of (1.33) and (1.34).

It follows from (1.33), (1.34), and the lower semiboundedness of a that A is semibounded, and then from Lemma 1.17 that $m_A = m_a$. The formula for $\operatorname{dom} A$ follows easily from this construction. Indeed, the element g in $\operatorname{dom} A = \{u_g : g \in \mathcal{H}\}$ and the element f in the formula in the theorem are related by $f = g - m u_g$.

Finally, to show uniqueness, let A_1 and A_2 be self-adjoint, lower semibounded operators satisfying (1.33) and (1.34), and let $v \in \operatorname{dom} A_1$. Then on the one hand, by (1.34) and the symmetry of a, we have for all $u \in d[a]$, $(u, A_1 v) = a[u, v]$ and, on the other hand, by Theorem 1.16, we have $d[a] = \operatorname{dom}(A_2 + m)^{1/2}$ and for all u from this set, $\left((A_2 + m)^{1/2}u, (A_2 + m)^{1/2}v\right) - m(u, v) = a[u, v]$. Thus,

$$\left(u, (A_1 + m)v\right) = \left((A_2 + m)^{1/2}u, (A_2 + m)^{1/2}v\right) \quad \text{for all } u \in \operatorname{dom}(A_2 + m)^{1/2}.$$

By the definition of the adjoint operator, this means that $(A_2 + m)^{1/2}v \in \operatorname{dom}\left((A_2 + m)^{1/2}\right)^*$ and $\left((A_2 + m)^{1/2}\right)^*(A_2 + m)^{1/2}v = (A_1 + m)v$. Since A_2 is self-adjoint, using (1.18) we find $v \in \operatorname{dom} A_2$ and $(A_2 + m)v = (A_1 + m)v$; that is, $A_2 v = A_1 v$. Interchanging the roles of A_1 and A_2, we also infer that, if $v \in \operatorname{dom} A_2$, then $v \in \operatorname{dom} A_1$ and $A_1 v = A_2 v$. Thus, $A_1 = A_2$, as claimed. $\qquad\square$

It is important in applications to describe spectral properties of A in terms of the quadratic form a. The following lemma concerns the bottom of the spectrum of A.

Lemma 1.19 *Let A be a self-adjoint, lower semibounded operator with corresponding quadratic form a. Then*

$$\inf \sigma(A) = m_a .$$

Moreover, m_a is an eigenvalue of A if and only if the infimum

$$\inf_{0 \neq u \in d[a]} a[u]/\|u\|^2$$

is a minimum, and eigenvectors are precisely those vectors for which the infimum is attained.

Proof The equality follows immediately from $m_a = m_A$ and Corollary 1.11. Moreover, given the characterization of minimizers there, it remains to show that, if $u_0 \in d[a]$ is a minimizer for $\inf_{0 \neq u \in d[a]} a[u]/\|u\|^2$, then $u_0 \in \operatorname{dom} A$. This follows from (1.36), which implies $u_0 = P_{\{m_a\}} u_0$ and, therefore, $u_0 \in \operatorname{dom} A$. $\square$

Next, we characterize the bottom of the essential spectrum.

Lemma 1.20 *Let A be a self-adjoint, lower semibounded operator with corresponding quadratic form a. Then*

$$\inf \sigma_{\mathrm{ess}}(A)$$

$$= \inf \left\{ \liminf_{n \to \infty} a[u_n] : (u_n) \subset d[a], \|u_n\| = 1, u_n \to 0 \text{ weakly in } \mathcal{H} \right\}$$

with the convention that $\inf \emptyset = +\infty$.

Proof Let us denote the left and right sides by Λ and $\tilde{\Lambda}$, respectively. For the proof of $\Lambda \geq \tilde{\Lambda}$ we may assume that $\Lambda < \infty$. Then, since $\sigma_{\mathrm{ess}}(A)$ is closed, $\Lambda \in \sigma_{\mathrm{ess}}(A)$ and, by Lemma 1.13, there is a sequence $(u_n) \subset \operatorname{dom} A \subset d[a]$ such that $\|u_n\| = 1$, $u_n \to 0$ weakly in $\mathcal{H}$, and $(A - \Lambda)u_n \to 0$ strongly in $\mathcal{H}$. Thus $a[u_n] - \Lambda = ((A - \Lambda)u_n, u_n) \to 0$, which implies $\tilde{\Lambda} \leq \Lambda$.

For the proof of $\Lambda \leq \tilde{\Lambda}$, let $(u_n) \subset d[a]$ with $\|u_n\| = 1$ and $u_n \to 0$ weakly in $\mathcal{H}$. We will use the fact that for any $u \in d[a]$ and any Borel $\omega \subset \mathbb{R}$, one has $P_\omega u \in d[a]$ and

$$a[P_\omega u] = \int_\omega \mu \, d(P_\mu u, u) . \tag{1.41}$$

Indeed, this follows from (1.15), (1.18) and Theorem 1.16. Let $\lambda \in \mathbb{R}$ with $\lambda < \Lambda$ and set $P := P_{(-\infty,\lambda)}$ and $P^\perp := 1 - P$. Then, by the above fact,

$$a[u_n] = a[Pu_n] + \int_{[\lambda,\infty)} \mu \, d(\Gamma_\mu u_n, u_n) \geq a[Pu_n] + \lambda \|P^\perp u_n\|^2$$
$$= ((PAP)u_n, u_n) + \lambda(1 - (Pu_n, u_n)).$$

Since $\lambda < \Lambda$, the operators PAP and P have finite rank and, thus, by the assumed weak convergence, $((PAP)u_n, u_n) \to 0$ and $(Pu_n, u_n) \to 0$. This proves that $\liminf_{n\to\infty} a[u_n] \geq \lambda$. Since $\lambda < \Lambda$ is arbitrary, this proves $\tilde{\Lambda} \geq \Lambda$. $\quad\square$

We say that the operator A has *discrete spectrum* if the essential spectrum of A is empty. This is a slight abuse of terminology and means not only that the set $\sigma(A)$ is a discrete subset of $\mathbb{R}$, but also that the corresponding eigenvalue multiplicities are finite. In other words, the operator A has discrete spectrum if and only if $\dim \operatorname{ran} P_\omega < \infty$ for any compact interval $\omega \subset \mathbb{R}$. We say that A has a discrete spectrum in an interval $\omega \subset \mathbb{R}$ if $\sigma_{\text{ess}}(A) \cap \omega = \emptyset$.

According to Lemma 1.20, A has discrete spectrum if and only if any sequence $(u_n) \subset d[a]$, with $\|u_n\| = 1$ and $u_n \to 0$ weakly in $\mathcal{H}$, satisfies $a[u_n] \to \infty$. The following corollary gives a necessary and sufficient condition in terms of the *embedding operator* $\mathcal{J} : d[a] \to \mathcal{H}$, which maps every $u \in d[a]$ to itself as an element in $\mathcal{H}$. This is a bounded operator when $d[a]$ is equipped with its norm $\sqrt{a[u] + m\|u\|^2}$ for some $m > -m_a$. By definition, this operator is compact if the closed unit ball in $d[a]$ is relatively compact in $\mathcal{H}$. As in Lemma 1.1, this compactness is equivalent to the assertion that every weakly convergent sequence in $d[a]$ converges strongly in $\mathcal{H}$.

Corollary 1.21 *Let A be a self-adjoint, lower semibounded operator with corresponding quadratic form a. Then A has discrete spectrum if and only if the embedding from $d[a]$ to $\mathcal{H}$ is compact.*

Proof Assuming that A has discrete spectrum and that $(v_n) \subset d[a]$ with $v_n \to 0$ weakly in $d[a]$, we need to show that $v_n \to 0$ strongly in $\mathcal{H}$. If this was not the case, then, after passing to a subsequence, we may assume that $(\|v_n\|)$ converges to a positive constant. Then $u_n := v_n/\|v_n\| \to 0$ weakly in $d[a]$ and, since the embedding from $d[a]$ to $\mathcal{H}$ is continuous, also weakly in $\mathcal{H}$. By Lemma 1.20, we infer that $a[u_n] \to \infty$, which contradicts the boundedness of a weakly convergent sequence in $d[a]$.

Conversely, assuming that the embedding from $d[a]$ to $\mathcal{H}$ is compact and that $(u_n) \subset d[a]$ with $\|u_n\| = 1$ and $u_n \to 0$ weakly in $\mathcal{H}$, by Lemma 1.20, we need to show that $a[u_n] \to \infty$. If this was not the case, then, after passing to a subsequence, we may assume that $(a[u_n])$ converges. In particular, (u_n)

is bounded in $d[a]$ and, by the compactness of the embedding, after passing to a subsequence, we may assume that (u_n) converges strongly in $\mathcal{H}$. Its limit coincides necessarily with its weak limit, which is zero, but this contradicts $\|u_n\| = 1$ for all n. □

As a final topic in this subsection, we discuss orthogonal sums of operators via quadratic forms. Let N be a countable (possibly finite) index set and, for each $n \in N$, let $\mathcal{H}_n$ be a separable Hilbert space. For each $n \in N$, let a_n be a densely defined, lower semibounded and closed quadratic form in $\mathcal{H}_n$ with lower bound m_{a_n} satisfying

$$M := \inf_{n \in N} m_{a_n} > -\infty.$$

Then we define a quadratic form a in $\bigoplus_{n \in N} \mathcal{H}_n$ with domain

$$d[a] := \left\{ u \in \bigoplus_{n \in N} \mathcal{H}_n : u_n \in d[a_n] \text{ for all } n \in N, \ \sum_{n \in N} a[u_n] < \infty \right\}$$

by

$$a[u] := \sum_{n \in N} a[u_n] \qquad \text{for all } u \in d[a].$$

Note that the sum converges absolutely since $M > -\infty$. It is easy to see that a is densely defined, lower semibounded with lower bound $m_a = M$, and closed. Therefore, by Theorem 1.18, it generates a unique self-adjoint and lower semibounded operator A in $\bigoplus_{n \in N} \mathcal{H}_n$. Similarly, for each $n \in N$, a_n generates a unique self-adjoint and lower semibounded operator A_n in $\mathcal{H}_n$. By verifying (1.33) and (1.34), we see that

$$A = \bigoplus_{n \in N} A_n. \tag{1.42}$$

1.2.2 The operators T^*T and TT^*, and the polar decomposition

Let T be a densely defined, closed operator in a Hilbert space $\mathcal{H}$. Consider the quadratic form $a[u] := \|Tu\|^2$ with domain $d[a] := \operatorname{dom} T$. This form is densely defined, non-negative and closed. Hence, by Theorem 1.18, it induces a self-adjoint non-negative operator A with

$$\operatorname{dom} A = \big\{ u \in \operatorname{dom} T : \text{there exists } f \in \mathcal{H} \text{ such that for all } v \in \operatorname{dom} T,$$
$$(Tu, Tv) = (f, v) \big\}.$$

The f here is unique and one has $f = Au$. This means $Tu \in \operatorname{dom} T^*$ and $Au = T^*Tu$ for all $u \in \operatorname{dom} A$, which means that $A = T^*T$ in the sense of composition of unbounded operators. We write $A = T^*T$ in the following.

Since T^*T is self-adjoint and non-negative, its square root is defined by the spectral theorem. The operator

$$|T| = (T^*T)^{1/2}$$

is called the *absolute value* of T. It is the unique self-adjoint, non-negative operator on $\mathcal{H}$ with

$$\||T|f\| = \|Tf\| \quad \text{for any} \quad f \in \text{dom}\, T = \text{dom}\, |T|. \tag{1.43}$$

Indeed, since by its definition $|T|^2 = T^*T$, the corresponding quadratic forms $\||T|f\|^2$ and $\|Tf\|^2$ coincide, including equality of their respective domains. The uniqueness follows from the uniqueness of the positive square root of a non-negative operator and the uniqueness in Theorem 1.18.

Any complex number z has a polar representation $z = e^{i\varphi}r$ with $r = |z| \geq 0$ and $\varphi = \arg z \in [0, 2\pi)$. In this subsection we present an analogous representation of an operator in a Hilbert space.

The operator $|T|$ will correspond to the 'radial part' in the polar decomposition of T. The following theorem describes the 'angular part' of this decomposition.

Proposition 1.22 *Let T be a densely defined, closed operator. Then there is a unique bounded operator U such that $T = U|T|$ and*

$$\|Uf\| = \|f\| \quad \text{for all} \quad f \in (\ker T)^\perp, \tag{1.44}$$

$$\ker U = \ker T, \tag{1.45}$$

$$\text{ran}\, U = \overline{\text{ran}\, T}. \tag{1.46}$$

Proof We define $U : \text{ran}\, |T| \to \text{ran}\, T$ by $U|T|f := Tf$. According to (1.43), the operator U is well defined, norm-preserving and maps onto $\text{ran}\, T$. Thus, U extends to a unitary operator from $\overline{\text{ran}\, |T|}$ to $\overline{\text{ran}\, T}$ and, extending U by zero to $(\text{ran}\, |T|)^\perp$, we obtain a bounded operator on $\mathcal{H}$ satisfying (1.46).

To prove (1.44) and (1.45), it remains to show that $\overline{\text{ran}\, |T|} = (\ker T)^\perp$. Indeed, using the self-adjointness of $|T|$, we have $\overline{\text{ran}\, |T|} = (\ker |T|)^\perp$. Again by (1.43), we have $\ker |T| = \ker T$. This proves the existence of U with the claimed properties.

Uniqueness of U comes from the fact that the extension of U from $\text{ran}\, |T|$ to $\overline{\text{ran}\, |T|}$ is unique. $\square$

Because T is densely defined and closed, we have $T^{**} = T$, and the same construction as at the beginning of this subsection leads to the self-adjoint operator TT^* defined on $\{u \in \text{dom}\, T^* : T^*u \in \text{dom}\, T\}$ corresponding to the quadratic form $\|T^*u\|^2$ defined on $\text{dom}\, T^*$. Our next result compares the operators T^*T and TT^* away from their respective kernels.

Proposition 1.23 *Let T be a densely defined, closed operator. The operator T^*T restricted to $(\ker T^*T)^\perp$ is unitarily equivalent to the operator TT^* restricted to $(\ker TT^*)^\perp$.*

Proof Let $T = U|T|$ be the polar decomposition of T. First, we observe that $T^* = |T|U^*$ with $\operatorname{dom} T^* = \{f \in \mathcal{H} : U^*f \in \operatorname{dom} |T|\}$. Indeed, by the definition of the adjoint, we have

$$\operatorname{dom} T^* = \{f \in \mathcal{H} : \text{there is a } g \in \mathcal{H} \text{ such that}$$
$$\text{for all } h \in \operatorname{dom} |T| \text{ one has } (g, h) = (U^*f, |T|h)\}$$
$$= \{f \in \mathcal{H} : U^*f \in \operatorname{dom} |T|^*\}\,.$$

Since $|T|^* = |T|$, this means $\operatorname{dom} |T|U^* = \operatorname{dom} T^*$ and $T^*f = |T|U^*f$ for f in this set.

Note that, because of $\operatorname{dom} |T| = \operatorname{dom} T$, we also have $\operatorname{dom} T^* = \operatorname{dom} TU^*$. By the formula for T^* and by (1.43), we have

$$\|T^*u\| = \|\,|T|U^*u\| = \|TU^*u\| \quad \text{for all } u \in \operatorname{dom} T^* = \operatorname{dom} TU^*\,.$$

This means $TT^* = (TU^*)^*TU^*$. The same argument as before implies $(UT^*)^* = TU^*$. Since UT^* is closed, this gives $(TU^*)^* = UT^*$, and therefore

$$TT^* = UT^*TU^*\,. \tag{1.47}$$

According to Proposition 1.22, the operator U can be restricted to a unitary operator $V : (\ker T)^\perp \to \overline{\operatorname{ran} T}$. Consequently, we have a unitary operator $V^* : \overline{\operatorname{ran} T} \to (\ker T)^\perp$. Since $\ker TT^* = \ker T^* = (\operatorname{ran} T)^\perp$ and $\ker T^*T = \ker T$, we can restrict (1.47) to

$$TT^*|_{(\ker TT^*)^\perp} = V\left(T^*T|_{(\ker T^*T)^\perp}\right)V^*\,.$$

This provides the claimed unitary equivalence. $\qquad\qquad\square$

Corollary 1.24 *Let T be a densely defined, closed operator and $\lambda \neq 0$. Then*

$$\dim \ker(T^*T - \lambda) = \dim \ker(TT^* - \lambda)\,.$$

The above proof shows that, if u is an eigenvector for T^*T corresponding to an eigenvalue $\lambda \neq 0$, then Tu is non-zero and an eigenvector for TT^* corresponding to the eigenvalue λ.

1.2.3 The variational principle

The goal of this subsection is to prove the variational principle. It translates the problem of counting eigenvalues below the bottom of the essential spectrum of a self-adjoint, lower semibounded operator into a problem for the corresponding quadratic form.

Let A be a self-adjoint, lower semibounded operator. Then the spectrum of A below $\inf \sigma_{\text{ess}}(A) \in (-\infty, \omega]$ is discrete. If it is not empty, this portion of the spectrum consists of eigenvalues of finite multiplicities that may accumulate only at the value $\inf \sigma_{\text{ess}}(A)$. Therefore, these eigenvalues can be enumerated in non-decreasing order

$$\lambda_1(A) \le \lambda_2(A) \le \cdots \le \lambda_n(A) \le \cdots,$$

where each eigenvalue is repeated according to its multiplicity.

Instead of the eigenvalues it is sometimes more convenient to consider their counting function. For a self-adjoint operator A with spectral measure P and for $\mu \in \mathbb{R}$, let

$$N(\mu, A) := \dim \operatorname{ran} P_{(-\infty, \mu)},$$

which can be a natural number or infinite. The function $\mu \mapsto N(\mu, A)$ is clearly non-decreasing. It is referred to as the *spectral counting function*. For a given $\mu \in \mathbb{R}$, one has $N(\mu, A) < \infty$ if and only if the spectrum of A in $(-\infty, \mu)$ consists of finitely many eigenvalues with finite multiplicities, and in this case, $N(\mu, A)$ is equal to the total multiplicity of these eigenvalues. Consequently, one has $\sigma_{\text{ess}}(A) \ne \emptyset$ if and only if there is a $\mu \in \mathbb{R}$ with $N(\mu, A) = \infty$, and in this case one has $\inf \sigma_{\text{ess}}(A) = \inf\{\mu : N(\mu, A) = \infty\}$. We write

$$N_A := \begin{cases} N(\inf \sigma_{\text{ess}}(A), A) & \text{if } \sigma_{\text{ess}}(A) \ne \emptyset, \\ \dim \mathcal{H} & \text{if } \sigma_{\text{ess}}(A) = \emptyset. \end{cases}$$

We begin with a version of the variational principle that is sometimes called *Glazman's lemma*.

Theorem 1.25 *Let A be a self-adjoint, lower semibounded operator with corresponding quadratic form a. Then for any $\mu \in \mathbb{R}$,*

$$N(\mu, A) = \sup \big\{ \dim F : F \subset d[a] \text{ is a subspace such that}$$
$$\text{for all } 0 \ne u \in F \text{ one has } a[u] < \mu\|u\|^2 \big\} \quad (1.48)$$

and

$$N(\mu, A) + \dim \ker(A - \mu)$$
$$= \sup \big\{ \dim F : F \subset d[a] \text{ is a subspace such that for all } u \in F$$
$$\text{one has } a[u] \le \mu\|u\|^2 \big\}. \quad (1.49)$$

If $\mathcal{F}$ is a form core of a, then in (1.48) it suffices to take the supremum only over $F \subset \mathcal{F}$.

In particular, if the right side in (1.48) is finite, then it coincides with the number of eigenvalues of A that are strictly less than μ (counting multiplicities).

Proof Let $F := \operatorname{ran} P_{(-\infty,\mu)}$ and note that this is contained in $d[a]$. Moreover, by Theorem 1.16 (see also (1.41)), for any $0 \neq u \in F$,

$$a[u] = \int_{(-\infty,\mu)} \lambda \, d(P_\lambda u, u) < \mu \int_{(-\infty,\mu)} d(P_\lambda u, u) = \mu\|u\|^2,$$

which proves $\leq$ in (1.48).

To prove the reverse inequality, we consider an arbitrary subspace $F \subset d[a]$ with $\dim F > N(\mu, A)$. Since $\dim \operatorname{ran} P_{(-\infty,\mu)} = N(\mu, A)$, there is a $0 \neq u_0 \in F \cap \left(\operatorname{ran} P_{(-\infty,\mu)}\right)^\perp$. Then, again by Theorem 1.16 (see also (1.41)), we have

$$a[u_0] = \int_{[\mu,\infty)} \lambda \, d(P_\lambda u_0, u_0) \geq \mu \int_{[\mu,\infty)} d(P_\lambda u_0, u_0) = \mu\|u_0\|^2.$$

Therefore, this subspace F is not admissible in the right side of (1.48).

An analogous argument proves the second identity (1.49).

The proof when F is restricted to lie in a dense subspace $\mathcal{F}$ of $d[a]$ follows from approximating, in the form norm, a finite linear system from $\operatorname{ran} P_{(-\infty,\mu)}$ by elements from $\mathcal{F}$. This preserves the strict inequality $a[u] < \mu\|u\|^2$, $u \neq 0$, on the span of these elements and completes the proof. $\square$

Here is another useful form of the variational principle, also called Glazman's lemma.

Theorem 1.26 *Let A be a self-adjoint, lower semibounded operator with corresponding quadratic form a. Then for any $\mu \in \mathbb{R}$,*

$$N(\mu, A) = \inf \big\{ \dim F : F \subset \mathcal{H} \text{ is a subspace such that}$$
$$\text{for all } u \in F^\perp \cap d[a], \ a[u] \geq \mu\|u\|^2 \big\} \qquad (1.50)$$

and

$$N(\mu, A) + \dim \ker(A - \mu)$$
$$= \inf \big\{ \dim F : F \subset \mathcal{H} \text{ is a subspace such that for all } 0 \neq u \in F^\perp \cap d[a],$$
$$a[u] > \mu\|u\|^2 \big\}. \qquad (1.51)$$

Proof We first show that $N \geq \inf$, and then $N \leq \inf$ in (1.50). For $F := \operatorname{ran} P_{(-\infty,\mu)}$, we have $N(\mu, A) = \dim F$ and, by Theorem 1.16 (see also (1.41)), $a[u] \geq \mu\|u\|^2$ for all $u \in F^\perp \cap d[a]$. This proves $\geq$ in (1.50).

We now consider some arbitrary subspace $F \subset \mathcal{H}$ with $N(\mu, A) > \dim F$. Then there exists some $u_0 \neq 0$ with $u_0 \in \operatorname{ran} P_{(-\infty,\mu)} \subset d[a]$ and $u_0 \in F^\perp$. But for this u_0, the opposite inequality $a[u_0] < \mu\|u_0\|^2$ holds. Hence, this subspace is not admissible in the right side of (1.50), and we are done.

An analogous argument proves the second identity (1.51).   □

As well as the variational principles for the counting function in Theorems 1.25 and 1.26, there are closely related variational principles for individual eigenvalues that play a very important role in applications. While we will not use them in this book, it is worth stating them and sketching their proof.

The following theorem is sometimes referred to as the *Courant–Fischer–Weyl min–max principle*. We recall the definition of the $\lambda_n(A)$ and of N_A at the beginning of this subsection.

Theorem 1.27   *Let A be a self-adjoint, lower semibounded operator with corresponding quadratic form a. Then, for all $n \in \mathbb{N}$,*

$$\sup_{u_1,\ldots,u_{n-1}\in\mathcal{H}} \quad \inf_{0\neq u\in d[a]\cap\{u_1,\ldots,u_{n-1}\}^\perp} \frac{a[u]}{\|u\|^2} = \begin{cases} \lambda_n(A) & \text{if } n \leq N_A, \\ \inf \sigma_{\text{ess}}(A) & \text{if } n > N_A. \end{cases}$$

If $n \leq N_A$, the supremum is attained if $u_1,\ldots,u_{n-1}$ are orthonormal eigenvectors corresponding to the eigenvalues $\lambda_1(A),\ldots,\lambda_{n-1}(A)$ and, in this case, the infimum is attained if u is an eigenvector corresponding to $\lambda_n(A)$.

If $\dim \mathcal{H} < \infty$, then both sides of the equality in Theorem 1.27 are interpreted as $+\infty$ for $n > N_A = \dim \mathcal{H}$.

Proof   For each $n \in \mathbb{N}$ we denote by μ_n the left side of the assertion and, fixing orthonormal eigenvectors e_m corresponding to the $\lambda_m(A)$, we define, for $n \leq N_A$,

$$\tilde{\mu}_n := \inf_{0\neq u\in d[a]\cap\{e_1,\ldots,e_{n-1}\}^\perp} \frac{a[u]}{\|u\|^2}.$$

We prove that

$$\mu_n = \tilde{\mu}_n = \lambda_n(A) \qquad \text{for all } n \leq N_A. \tag{1.52}$$

Clearly, $\mu_n \geq \tilde{\mu}_n$ for $n \leq N_A$. Let P be the spectral measure of A. Since $\{e_1,\ldots,e_{n-1}\}^\perp \subset \operatorname{ran} P_{[\lambda_n(A),\infty)}$, we have, as in the proof of Theorem 1.25, that $a[u] \geq \lambda_n(A)\|u\|^2$ for every $u \in d[a] \cap \{e_1,\ldots,e_{n-1}\}^\perp$. Note that equality holds if $u = e_n$. Thus, we have $\tilde{\mu}_n \geq \lambda_n(A)$. On the other hand, for any $u_1,\ldots,u_{n-1} \in \mathcal{H}$, there is a $0 \neq v \in \operatorname{span}\{e_1,\ldots,e_n\} \cap \{u_1,\ldots,u_{n-1}\}^\perp$. Since $v \in \operatorname{span}\{e_1,\ldots,e_n\}$, we have $v \in d[a]$ and, again as in the proof of Theorem 1.25, $a[v] \leq \lambda_n(A)\|v\|^2$. Hence,

$$\inf_{0\neq u\in d[a]\cap\{u_1,\ldots,u_{n-1}\}^\perp} \frac{a[u]}{\|u\|^2} \leq \frac{a[v]}{\|v\|^2} \leq \lambda_n(A).$$

Thus, since $u_1, \ldots, u_{n-1}$ are arbitrary, we have $\mu_n \leq \lambda_n(A)$. This completes the proof of (1.52), and thereby of all conclusions in the theorem for $n \leq N_A$.

To prove the remaining conclusions, we may assume that $\dim \mathcal{H} = \infty$, $\kappa := \inf \sigma_{\mathrm{ess}}(A) < \infty$, and $N_A < \infty$. Then $\dim \operatorname{ran} P_{(-\infty, \kappa+\varepsilon)} = \infty$ for any $\varepsilon > 0$ and therefore, for any given $\varepsilon > 0$, we can extend the finite orthonormal system $(e_n)_{n=1}^{N_A}$ to an infinite orthonormal system $(e_n)_{n=1}^{\infty} \subset \operatorname{ran} P_{(-\infty, \kappa+\varepsilon)}$. Defining $\tilde{\mu}_n$ for $n > n_A$ by the same formula as before, we still have $\mu_n \geq \tilde{\mu}_n$. Moreover, by repeating the above reasoning, we get $\tilde{\mu}_n \geq \kappa$ and $\mu_n \leq \kappa + \varepsilon$ for $n > n_A$. Since $\varepsilon > 0$ is arbitrary, this proves that $\mu_n = \kappa$, as claimed. This completes the proof. $\qquad\square$

There is another version where the inf and the sup are interchanged.

Theorem 1.28 *Let A be a self-adjoint, lower semibounded operator with corresponding quadratic form a. Then, for all $n \in \mathbb{N}$,*

$$
\inf_{\substack{u_1, \ldots, u_n \in d[a] \\ \text{lin. independent}}} \quad \sup_{0 \neq u \in \mathrm{span}\{u_1, \ldots, u_n\}} \frac{a[u]}{\|u\|^2} = \begin{cases} \lambda_n(A) & \text{if } n \leq N_A, \\ \inf \sigma_{\mathrm{ess}}(A) & \text{if } n > N_A. \end{cases}
$$

If $n \leq N_A$, the infimum is attained if $u_1, \ldots, u_n$ are orthonormal eigenvectors corresponding to the eigenvalues $\lambda_1(A), \ldots, \lambda_n(A)$ and, in this case, the supremum is attained if u is an eigenvector corresponding to $\lambda_n(A)$.

The proof of this theorem is similar to those of the others in this subsection and is omitted; see, e.g., Davies (1995, Theorems 4.5.1 and 4.5.2).

1.2.4 Applications of the variational principle

We begin by introducing the important notion of comparison of two operators. Let A and B be self-adjoint, lower semibounded operators and let a and b be the corresponding quadratic forms with form domains $d[a]$ and $d[b]$, respectively. We say that A is *greater or equal than* B, in symbols

$$
A \geq B, \quad \text{or} \quad B \leq A, \tag{1.53}
$$

if the following two conditions are satisfied

$$
d[a] \subset d[b], \tag{1.54}
$$

$$
a[u] \geq b[u] \quad \text{for all } u \in d[a]. \tag{1.55}
$$

Note that the case $d[a] = d[b]$ in (1.54) may occur, for instance, if both operators A and B are bounded. Furthermore, the definition is meaningful if we have the special case $a[u] = b[u]$ for all $u \in d[a]$ in (1.55). In applications this occurs, for instance, for differential operators where a strict inclusion in (1.54) corresponds to different choices of boundary conditions for A and B.

Trivially, it holds that $A \geq A$. Note further that $A \geq B$ and $B \geq C$ implies $A \geq C$. Finally, if $A \geq B$ and $B \geq A$, then $A = B$. Therefore, this comparison defines a partial order relation.

The following is a consequence of the variational principle (Theorem 1.25).

Proposition 1.29 *Let A and B be self-adjoint, lower semibounded operators satisfying $A \geq B$. Then*

$$N(\mu, A) \leq N(\mu, B) \quad \text{for all } \mu \in \mathbb{R}.$$

In particular, $\inf \sigma_{\text{ess}}(A) \geq \inf \sigma_{\text{ess}}(B)$ *and*

$$\lambda_n(A) \geq \lambda_n(B) \quad \text{for all } n \leq \min\{N_A, N_B\}.$$

The second application of the variational principle concerns eigenvalues of sums of operators defined in the form sense.

Consider two self-adjoint, lower semibounded operators A and B with corresponding quadratic forms a and b. Assume that $d[a] \cap d[b]$ is dense in $\mathcal{H}$, and let c be the quadratic form $c[u] := a[u] + b[u]$ with domain $d[c] := d[a] \cap d[b]$. This form is lower semibounded and closed. It induces a corresponding self-adjoint operator C, which we formally write as $A + B$. The identification of C with $A + B$ has to be understood in the form sense, not in the sense of a sum of operators.

Corollary 1.30 *Let A and B be self-adjoint, lower semibounded operators and assume that $d[a] \cap d[b]$ is dense in $\mathcal{H}$. Then*

$$N(0, A + B) \leq N(0, A) + N(0, B).$$

Proof Let $P(A)$ and $P(B)$ denote the spectral measures for A and B, respectively, and let $L := \operatorname{ran} P_{(-\infty, 0)}(A)$ and $M := \operatorname{ran} P_{(-\infty, 0)}(B)$. By the spectral theorem,

$$a[u] \geq 0 \text{ for all } u \in d[a] \cap L^{\perp} \quad \text{and} \quad b[v] \geq 0 \text{ for all } v \in d[b] \cap M^{\perp}.$$

Thus,

$$c[w] = a[w] + b[w] \geq 0 \text{ for all } w \in d[a] \cap d[b] \cap (L + M)^{\perp}.$$

By Theorem 1.26, we deduce

$$N(0, C) \leq \dim(L + M) \leq \dim L + \dim M = N(0, A) + N(0, B),$$

as claimed. □

We conclude this subsection with two somewhat technical applications of the variational principle that will be needed later on.

Let A be a self-adjoint, lower semibounded operator and let a be the corresponding quadratic form. Consider a bounded operator T such that $\{u \subset \mathcal{H} : Tu \in d[u]\}$ is dense in $\mathcal{H}$, and define

$$a_T[u] := a[Tu] \qquad \text{for } u \in d[a_T] := \{v \in \mathcal{H} : Tv \in d[a]\}.$$

This form is lower semibounded and closed. It induces a self-adjoint operator, which we write formally as T^*AT. We emphasize that this notation is not understood in the sense of a product of operators.

Corollary 1.31 *If A is self-adjoint and lower semibounded and if T is bounded with $\{u \in \mathcal{H} : Tu \in d[a]\}$ dense in $\mathcal{H}$, then $N(0, T^*AT) \le N(0, A)$.*

Proof Let P be the spectral measure of A and set $L := \operatorname{ran} P_{(-\infty,0)}$. Then, for all $u \in (T^*L)^{\perp} \cap d[a_T]$, one has $Tu \in L^{\perp} \cap d[a]$. Thus, by the spectral theorem, we have $a_T[u] = a[Tu] \ge 0$. By the variational principle (Theorem 1.26), this implies

$$N(0, T^*AT) \le \dim T^*L \le \dim L = N(0, A),$$

as claimed. $\square$

We shall also need another version of this corollary. Let P be an orthogonal projection and, as before, let A be a self-adjoint, lower semibounded operator with quadratic form a. Assuming that $d[a] \cap \operatorname{ran} P$ is dense in $\operatorname{ran} P$, the set $\{u \in \mathcal{H} : Pu \in d[a]\}$ is dense in $\mathcal{H}$, and then PAP is defined as above. Let $\tilde{A}_P$ be the restriction of PAP to the Hilbert space $\operatorname{ran} P$. The following corollary compares the spectrum of A on $\mathcal{H}$ with that of $\tilde{A}_P$ on $\operatorname{ran} P$.

Corollary 1.32 *If A is self-adjoint and lower semibounded and if P is an orthogonal projection with $d[a] \cap \operatorname{ran} P$ dense in $\operatorname{ran} P$, then $N(\lambda, \tilde{A}_P) \le N(\lambda, A)$ for all $\lambda \in \mathbb{R}$.*

Proof Let $B := A - \lambda$ in $\mathcal{H}$. Then, by Corollary 1.31, we have $N(\lambda, A) = N(0, B) \ge N(0, PBP)$. Since $N(0, PBP)$ counts the strictly negative eigenvalues and since $PBP = \tilde{B}_P \oplus 0$ on $\mathcal{H} = \operatorname{ran} P \oplus (\operatorname{ran} P)^{\perp}$, we get $N(0, PBP) = N(0, \tilde{B}_P)$. On $\operatorname{ran} P$ we have $\tilde{B}_P = \tilde{A}_P - \lambda$, and thus $N(0, \tilde{B}_P) = N(\lambda, \tilde{A}_P)$. $\square$

1.2.5 Variational principle for sums of eigenvalues and the trace

In the previous two subsections, we have discussed a variational principle for the eigenvalues of a self-adjoint, lower semibounded operator. In this subsection, we prove a corresponding principle for sums of eigenvalues and, along the way, introduce the trace of a self-adjoint, non-negative operator.

We recall that, for a self-adjoint, lower semibounded operator A, we denote by $\lambda_j(A)$ the eigenvalues below the bottom of its essential spectrum, in non-decreasing order and repeated according to multiplicities. This list contains N_A elements and may be empty, finite or infinite.

Here is a variational characterization for partial sums of these eigenvalues.

Proposition 1.33 *Let A be a self-adjoint, lower semibounded operator with corresponding quadratic form a. Then, for any finite number $N \leq N_A$,*

$$\sum_{j=1}^{N} \lambda_j(A)$$

$$= \inf \left\{ \sum_{n=1}^{N} a[u_n] \colon (u_n)_{n=1}^{N} \subset d[a] \text{ and } (u_n, u_m) = \delta_{n,m} \text{ for all } 1 \leq n, m \leq N \right\}.$$

If $\mathcal{F}$ is a form core of a, then in the infimum on the right side it suffices to consider $(u_n)_{n=1}^{N} \subset \mathcal{F}$.

Proof We first prove that the left side $\geq$ the right side. In fact this follows immediately by choosing u_n to be orthonormal eigenvectors corresponding to $\lambda_n(A)$.

Let us prove the opposite bound. For a given finite number $N \leq N_A$ and given $u_1, \ldots, u_N \in d[a]$ with $(u_n, u_m) = \delta_{n,m}$, we define the orthogonal projection $P := \sum_{n=1}^{N} (\cdot, u_n) u_n$. Then, according to Corollary 1.32,

$$\lambda_j(A) \leq \lambda_j(\tilde{A}_P) \quad \text{for all } 1 \leq j \leq N.$$

Thus,

$$\sum_{j=1}^{N} \lambda_j(A) \leq \sum_{j=1}^{N} \lambda_j(\tilde{A}_P) = \sum_{n=1}^{N} a[u_n],$$

where we used the fact from linear algebra that the sum of all eigenvalues of a matrix in a finite-dimensional space can be evaluated by the sum of the diagonal entries.

The statement concerning a form core follows by a simple approximation argument in the form norm. This proves the proposition. □

Proposition 1.33 implies the following somewhat technical result, which will be of importance in §§5.3 and 8.2. Let Ξ be a probability space. We denote integration with respect to the underlying probability measure by $d\xi$. Let $\Xi \ni \xi \mapsto U(\xi)$ be a measurable function taking values in the unitary operators on $\mathcal{H}$, where measurability is understood in the form sense. Let A be a self-adjoint, lower semibounded operator with corresponding quadratic form a. We assume that ran $U(\xi) \subset d[a]$ for a.e. $\xi \in \Xi$ and that $\{u \in \mathcal{H} : \int_\Xi a[U(\xi)u] \, d\xi < \infty\}$ is dense in $\mathcal{H}$. Then the quadratic form

$$a^U[u] := \int_\Xi a[U(\xi)u] \, d\xi \quad \text{for } u \in d[a^U] := \left\{v \in \mathcal{H} : \int_\Xi a[U(\xi)v] \, d\xi < \infty\right\}$$

is lower semibounded and closed. It defines a self-adjoint, lower semibounded operator, which we denote by A^U and sometimes also by $\int_\Xi U(\xi)^* A U(\xi) \, d\xi$.

Corollary 1.34 *Let A be a self-adjoint, lower semibounded operator and let U be as above. Then for any finite number $N \leq \min\{N_A, N_{A^U}\}$,*

$$\sum_{j=1}^N \lambda_j(A^U) \geq \sum_{j=1}^N \lambda_j(A).$$

Proof Let $(u_n)_{n=1}^N$ be a orthonormal system of eigenvectors of A^U corresponding to the eigenvalues $\lambda_1(A^U), \ldots, \lambda_N(A^U)$. Then

$$\sum_{j=1}^N \lambda_j(A^U) = \sum_{n=1}^N a^U[u_n] = \sum_{n=1}^N \int_\Xi a[U(\xi)u_n] \, d\xi = \int_\Xi \sum_{n=1}^N a[U(\xi)u_j] \, d\xi.$$

Since for a.e. $\xi \in \Xi$ the system $(U(\xi)u_n)_{n=1}^N$ belongs to $d[a]$ and is orthonormal, the variational principle from Proposition 1.33 implies

$$\sum_{n=1}^N a[U(\xi)u_n] \geq \sum_{j=1}^N \lambda_j(A).$$

Since $d\xi$ is a probability measure, it remains to integrate both sides in this inequality to obtain the desired conclusion. $\qquad\square$

The next result is a version of Proposition 1.33 where N is not fixed. For this purpose, the following notion of the trace of a non-negative bounded operator T will be useful. If the essential spectrum of T is empty or consists only of the point 0, the negative spectrum of $-T$ is discrete and consists of the eigenvalues $\lambda_j(-T)$, which are enumerated counting multiplicities as explained above. We set

$$\operatorname{Tr} T := \begin{cases} +\infty & \text{if } \sup \sigma_{\text{ess}}(T) > 0, \\ -\sum_j \lambda_j(-T) & \text{otherwise.} \end{cases}$$

Even in the second case, this value can be infinite. On the other hand, if $\operatorname{Tr} T$ is finite, then Lemma 1.15 implies that T is compact.

In the statement of the following corollary, the operator A_- is defined by the functional calculus as $\varphi(A)$ with $\varphi(\lambda) = \lambda_- = \max\{-\lambda, 0\}$. Clearly, $A_- \geq 0$.

Corollary 1.35 *Let A be a self-adjoint, lower semibounded operator. Then*

$$- \operatorname{Tr} A_- = \inf \left\{ \sum_{n=1}^{N} a[u_n] : N \in \mathbb{N}, (u_n)_{n=1}^{N} \subset d[a] \right.$$

$$\left. \text{and } (u_n, u_m) = \delta_{n,m} \text{ for all } 1 \leq n, m \leq N \right\}.$$

If $\mathcal{F}$ is a form core of a, then in the infimum on the right side it suffices to consider $(u_n)_{n=1}^{N} \subset \mathcal{F}$.

Proof If $\sigma_{\mathrm{ess}}(A) = \emptyset$ or if $\inf \sigma_{\mathrm{ess}}(A) \geq 0$, then

$$- \operatorname{Tr} A_- = \sum_j \lambda_j(-A_-) = - \sum_j (\lambda_j(A))_- = \inf_{N \in \mathbb{N}} \sum_{j=1}^{N} \lambda_j(A),$$

and therefore the corollary follows from Proposition 1.33. By contrast, if $\kappa := \inf \sigma_{\mathrm{ess}}(A) < 0$, then $\operatorname{Tr} A_- = \infty$ and, with the spectral measure P of A, $\dim \operatorname{ran} P_{(-\infty, \kappa + \varepsilon)} = \infty$ for any $\varepsilon > 0$. In particular, if $\kappa + \varepsilon < 0$, there is an infinite sequence of orthonormal (u_n) with $a[u_n] \leq \kappa + \varepsilon$, which can also be assumed to lie in a form core. Thus, the right side in the corollary is also equal to $-\infty$. □

As a consequence of Corollary 1.35, we now prove a fundamental property of the trace. Note that we do not assume that $\operatorname{Tr} T$ is finite or even that T is compact.

Lemma 1.36 *Let T be a bounded and non-negative operator. Then for any complete orthonormal system (u_n),*

$$\operatorname{Tr} T = \sum_n (T u_n, u_n).$$

Proof We first assume that $\operatorname{Tr} T < \infty$ and we choose a complete orthonormal system (v_j) such that $T v_j = -\lambda_j(-T) v_j$. Then

$$\operatorname{Tr} T = - \sum_j \lambda_j(-T) = \sum_j \|T^{1/2} v_j\|^2 = \sum_j \sum_n |(T^{1/2} v_j, u_n)|^2$$

$$= \sum_n \sum_j |(v_j, T^{1/2} u_n)|^2 = \sum_n \|T^{1/2} u_n\|^2 = \sum_n (T u_n, u_n),$$

as claimed. (The interchange of summations here is justified since all terms are non-negative.) On the other hand, when $\operatorname{Tr} T = \infty$, by Corollary 1.35 applied to $A = -T$, for any $M > 0$ there is an $N \in \mathbb{N}$ and an orthonormal system $(v_j)_{j=1}^{N}$

such that $\sum_{j=1}^{N} \|T^{1/2}v_j\|^2 \geq M$. Similar to before, we have

$$\sum_{i=1}^{N} \|T^{1/2}v_j\|^2 = \sum_{j=1}^{N} \sum_n |(T^{1/2}v_j, u_n)|^2 = \sum_n \sum_{j=1}^{N} |(v_j, T^{1/2}u_n)|^2$$
$$\leq \sum_n \|T^{1/2}u_n\|^2 = \sum_n (Tu_n, u_n).$$

Since M is arbitrary this means that $\sum_n (Tu_n, u_n) = \infty$. $\qquad\square$

The next corollary is the analogue of Corollary 1.30 for sums of eigenvalues.

Corollary 1.37 *Let A and B be self-adjoint, lower semibounded operators with corresponding quadratic forms a and b. Assume that $d[a] \cap d[b]$ is dense in $\mathcal{H}$ and let A + B be defined in the form sense. Then*

$$\mathrm{Tr}(A + B)_- \leq \mathrm{Tr}\,A_- + \mathrm{Tr}\,B_- .$$

Proof Let $N \in \mathbb{N}$ and let $(u_n)_{n=1}^{N} \subset d[a] \cap d[b]$ be orthonormal functions. Then, by Corollary 1.35 for the operators A and B,

$$\sum_{n=1}^{N} (a+b)[u_n] = \sum_{n=1}^{N} a[u_n] + \sum_{n=1}^{N} b[u_n] \geq -\mathrm{Tr}\,A_- - \mathrm{Tr}\,B_- .$$

Taking the infimum over all N and $(u_n)_{n=1}^{N}$ as above and, using again Corollary 1.35, we obtain the claimed inequality. $\qquad\square$

As a brief digression, we will use Corollary 1.35 to compute the trace of a certain class of operators that appear frequently in applications. We recall the notion of a separable measure space from §1.1.4 needed in the following result.

Lemma 1.38 *Let X,Y be separable, sigma-finite measure spaces and let $N, M \in \mathbb{N}$. Let K be a measurable function on $X \times Y$ taking values in the complex $N \times M$ matrices such that*

$$\iint_{X \times Y} \mathrm{Tr}_{\mathbb{C}^M}\left((K(x,y))^* K(x,y)\right) dx\, dy < \infty .$$

Then the operator K from $L^2(Y, \mathbb{C}^M)$ to $L^2(X, \mathbb{C}^N)$, defined by

$$(Kv)(x) := \int_Y K(x,y)v(y)\, dy \qquad \text{for a.e. } x \in X,\ v \in L^2(Y, \mathbb{C}^M),$$

is compact and satisfies

$$\mathrm{Tr}\,K^*K = \iint_{X \times Y} \mathrm{Tr}_{\mathbb{C}^M}\left((K(x,y))^* K(x,y)\right) dx\, dy .$$

Proof For the sake of simplicity, we begin with the case $N = M = 1$. For any $u \in L^2(X)$, $v \in L^2(Y)$, we have, by the Cauchy–Schwarz inequality,

$$
\begin{aligned}
|(Kv, u)| &= \left| \iint_{X \times Y} K(x, y) v(y) \overline{u(x)} \, dx \, dy \right| \\
&\leq \left(\iint_{X \times Y} |K(x, y)|^2 \, dx \, dy \right)^{1/2} \left(\iint_{X \times Y} |v(y) \overline{u(x)}|^2 \, dx \, dy \right)^{1/2} \\
&= \|K\|_{L^2(X \times Y)} \|v\| \|u\| .
\end{aligned}
$$

This implies that K is a bounded operator from $L^2(Y)$ to $L^2(X)$.

Let $(v_n)_{n=1}^N$ be an orthonormal system in $L^2(Y)$ and let (u_j) be a complete orthonormal system in $L^2(X)$. Then

$$
\sum_{n=1}^N \|K v_n\|^2 = \sum_{n=1}^N \sum_j |(K v_n, u_j)|^2 = \sum_{n=1}^N \sum_j |(K, u_j \otimes \overline{v_n})|^2 ,
$$

where the inner product on the right side is in $L^2(X \times Y)$ and where the function $u_j \otimes \overline{v_n}$ is defined by

$$
\left(u_j \otimes \overline{v_n} \right)(x, y) = u_j(x) \overline{v_n(y)} \qquad \text{for all } (x, y) \in X \times Y .
$$

Since the $u_j \otimes \overline{v_n}$ are orthonormal in $L^2(X \times Y)$, we find, by Bessel's inequality, that

$$
\sum_{n=1}^N \sum_j |(K, u_j \otimes \overline{v_n})|^2 \leq \|K\|_{L^2(X \times Y)}^2 .
$$

By the variational principle in Corollary 1.35 (with $A = -K^*K$) we conclude that

$$
\operatorname{Tr} K^*K \leq \|K\|_{L^2(X \times Y)}^2 .
$$

In particular, finiteness of the trace implies that K^*K is compact, so, by Lemma 1.2, K is compact. Repeating the argument with a *complete* orthonormal system (v_n) and using Parseval's identity instead of Bessel's inequality we find that, in fact, $\operatorname{Tr} K^*K = \|K\|_{L^2(X \times Y)}^2$. The proves the lemma for $N = M = 1$.

The case of arbitrary N and M can be reduced to the previous case by identifying $L^2(X, \mathbb{C}^N)$ with $L^2(X \times \{1, \ldots, N\})$ and $L^2(Y, \mathbb{C}^M)$ with $L^2(Y \times \{1, \ldots, M\})$. Indeed, fixing a basis $(e_n)_{n=1}^N$ in $\mathbb{C}^N$, the operator U, defined by $(Uu)(x, n) := (u(x), e_n)_{\mathbb{C}^N}$ from $L^2(X, \mathbb{C}^N)$ to $L^2(X \times \{1, \ldots, N\}, \mathbb{C})$, is unitary. Using a similar unitary operator V from $L^2(Y, \mathbb{C}^M)$ to $L^2(Y \times \{1, \ldots, M\}, \mathbb{C})$ and applying the scalar result to the operator $V K U^*$, we obtain the lemma for arbitrary N and M. $\qquad\square$

1.2.6 Riesz means

In this subsection, we study *Riesz means* of order $\gamma > 0$ of a self-adjoint, lower semibounded operator A; that is, the quantities

$$\text{Tr } A_-^\gamma .$$

If this quantity is finite, then the negative spectrum of A consists of eigenvalues of finite multiplicities and, if $\lambda_1(A) \leq \lambda_2(A) \leq \cdots$ is an enumeration of those, we have

$$\text{Tr } A_-^\gamma = \sum_j \left(\lambda_j(A)\right)_-^\gamma .$$

A very useful identity connects the Riesz means to the spectral counting function.

Lemma 1.39 *Let A be a self-adjoint, lower semibounded operator and let $\gamma > 0$. Then*

$$\text{Tr } A_-^\gamma = \gamma \int_0^\infty N(-\tau, A)\, \tau^{\gamma-1}\, d\tau .$$

Proof Since

$$a_-^\gamma = \gamma \int_0^\infty \chi_{\{a < -\tau\}} \tau^{\gamma-1}\, d\tau \qquad \text{for } a \in \mathbb{R},$$

the identity follows from Lemma 1.36. □

One of the consequences of this formula is a generalization of Corollary 1.30 to the case of Riesz means.

Proposition 1.40 *Let A and B be self-adjoint, lower semibounded operators with corresponding quadratic forms a and b, assume that $d[a] \cap d[b]$ is dense in $\mathcal{H}$, and let $A + B$ be defined in the form sense. If $\gamma > 0$ then*

$$\left(\text{Tr}(A + B)_-^\gamma\right)^{\frac{1}{\gamma+1}} \leq \left(\text{Tr } A_-^\gamma\right)^{\frac{1}{\gamma+1}} + \left(\text{Tr } B_-^\gamma\right)^{\frac{1}{\gamma+1}}$$

and, for all $0 < \theta < 1$,

$$\text{Tr}(A + B)_-^\gamma \leq \theta^{-\gamma}\, \text{Tr } A_-^\gamma + (1 - \theta)^{-\gamma}\, \text{Tr } B_-^\gamma .$$

Proof According to Corollary 1.30, for any $\tau > 0$ and for any $0 < \theta < 1$ we have

$$\begin{aligned}
N(-\tau, A + B) &= N(0, (A + \theta\tau) + (B + (1 - \theta)\tau)) \\
&\leq N(0, A + \theta\tau) + N(0, B + (1 - \theta)\tau) \\
&= N(-\theta\tau, A) + N(-(1 - \theta)\tau, B) .
\end{aligned}$$

Therefore, by Lemma 1.39,

$$\begin{aligned}
\mathrm{Tr}(A+B)^\gamma_- &= \gamma \int_0^\infty N(-\tau, A+B)\tau^{\gamma-1}\,d\tau \\
&\le \gamma \int_0^\infty N(-\theta\tau, A)\tau^{\gamma-1}\,d\tau + \gamma \int_0^\infty N(-(1-\theta)\tau, B)\tau^{\gamma-1}\,d\tau \\
&= \theta^{-\gamma}\,\mathrm{Tr}\,A^\gamma_- + (1-\theta)^{-\gamma}\,\mathrm{Tr}\,B^\gamma_-,
\end{aligned}$$

as claimed. The other assertion follows by optimizing over θ.   □

The remainder of this subsection contains improvements of the constants appearing in the bound in Proposition 1.40. These are not necessary for the applications in this book and can be omitted in a first reading.

The fact that for $\gamma = 1$ an improvement is possible can be seen from Corollary 1.37, which contains a bound like that in Proposition 1.40, but without the prefactors $\theta^{-\gamma}$ and $(1-\theta)^{-\gamma}$.

The first improvements concern the case $\gamma > 1$. In this case, we start from the following variant of Lemma 1.39.

Lemma 1.41 *Let A be a self-adjoint, lower semibounded operator and let $\gamma > 1$. Then*

$$\mathrm{Tr}\,A^\gamma_- = \gamma(\gamma-1) \int_0^\infty \mathrm{Tr}(A+\tau)_-\,\tau^{\gamma-2}\,d\tau.$$

Proof Since

$$a^\gamma_- = \gamma(\gamma-1) \int_0^\infty (a+\tau)_-\tau^{\gamma-2}\,d\tau \qquad \text{for } a \in \mathbb{R},$$

the identity follows from Lemma 1.36.   □

Using this lemma, we obtain the following bound.

Proposition 1.42 *Let A and B be self-adjoint, lower semibounded operators with corresponding quadratic forms a and b. Assume that $d[a] \cap d[b]$ is dense in $\mathcal{H}$ and let $A+B$ be defined in the form sense. If $\gamma > 1$ then*

$$\left(\mathrm{Tr}(A+B)^\gamma_-\right)^{\frac{1}{\gamma}} \le \left(\mathrm{Tr}\,A^\gamma_-\right)^{\frac{1}{\gamma}} + \left(\mathrm{Tr}\,B^\gamma_-\right)^{\frac{1}{\gamma}}$$

and, for all $0 < \theta < 1$,

$$\mathrm{Tr}(A+B)^\gamma_- \le \theta^{-\gamma+1}\,\mathrm{Tr}\,A^\gamma_- + (1-\theta)^{-\gamma+1}\,\mathrm{Tr}\,B^\gamma_-.$$

Proof According to Corollary 1.37, for any $\tau > 0$ and for any $0 < \theta < 1$ we have

$$\text{Tr}(A + B + \tau)_- = \text{Tr}\left((A + \theta\tau) + (B + (1 - \theta)\tau)\right)_-$$
$$\leq \text{Tr}(A + \theta\tau)_- + \text{Tr}(B + (1 - \theta)\tau)_- .$$

Therefore, by Lemma 1.41,

$$\text{Tr}(A + B)_-^{\gamma} = \gamma(\gamma - 1) \int_0^{\infty} \text{Tr}(A + B + \tau)_- \tau^{\gamma-2} \, d\tau$$
$$\leq \gamma(\gamma - 1) \int_0^{\infty} \text{Tr}(A + \theta\tau)_- \tau^{\gamma-2} \, d\tau$$
$$+ \gamma(\gamma - 1) \int_0^{\infty} \text{Tr}(B + (1 - \theta)\tau)_- \tau^{\gamma-2} \, d\tau$$
$$= \theta^{-\gamma+1} \text{Tr} \, A_-^{\gamma} + (1 - \theta)^{-\gamma+1} \text{Tr} \, B_-^{\gamma},$$

as claimed. The other assertion follows by optimizing over θ. $\qquad\qquad\square$

Even for numbers $a, b \in \mathbb{R}$ the constants $\theta^{-\gamma+1}$ and $(1-\theta)^{-\gamma+1}$ in the inequality

$$(a + b)_-^{\gamma} \leq \theta^{-\gamma+1} a_-^{\gamma} + (1 - \theta)^{-\gamma+1} b_-^{\gamma}$$

cannot be improved for $\gamma > 1$. On the other hand, for $0 < \gamma < 1$, the inequality for numbers holds without the factors of $\theta^{-\gamma+1}$ and $(1 - \theta)^{-\gamma+1}$. This motivates the next result, which is an improvement of Proposition 1.40 in the case $0 < \gamma < 1$. It is a special case of a theorem by Rotfel'd (1967, 1968).

Proposition 1.43 *Let A and B be self-adjoint, lower semibounded operators with corresponding quadratic forms a and b. Assume that $d[a] \cap d[b]$ is dense in $\mathcal{H}$ and let $A + B$ be defined in the form sense. If $0 < \gamma < 1$, then*

$$\text{Tr}(A + B)_-^{\gamma} \leq \text{Tr} \, A_-^{\gamma} + \text{Tr} \, B_-^{\gamma} .$$

For the proof of this proposition, we need several preliminary results. The first one is an extension of Corollary 1.31 to Riesz means. We refer to the discussion before that corollary for the precise definition of the operator T^*AT.

Lemma 1.44 *If A is self-adjoint and lower semibounded and if T is bounded with $\{u \in \mathcal{H}: Tu \in d[a]\}$ dense in $\mathcal{H}$, then for any $\gamma > 0$,*

$$\text{Tr} \, (T^*AT)_-^{\gamma} \leq \|T\|^{2\gamma} \, \text{Tr} \, A_-^{\gamma} .$$

Proof According to Lemma 1.39, we have

$$\text{Tr} \, (T^*AT)_-^{\gamma} = \gamma \int_0^{\infty} N(-\tau, T^*AT)\tau^{\gamma-1} \, d\tau .$$

For any $\tau > 0$, we have $T^*AT + \tau \geq T^*(A + \|T\|^{-2}\tau)T$, and therefore

$$N(-\tau, T^*AT) = N(0, T^*AT + \tau) \leq N(0, T^*(A + \|T\|^{-2}\tau)T).$$

We now apply Corollary 1.31, which implies that

$$N(0, T^*(A + \|T\|^{-2}\tau)T) \leq N(0, A + \|T\|^{-2}\tau) = N(-\|T\|^{-2}\tau, A),$$

and therefore

$$\text{Tr}\,(T^*AT)_-^\gamma \leq \gamma \int_0^\infty N(-\|T\|^{-2}\tau, A)\tau^{\gamma-1}\,d\tau = \|T\|^{2\gamma}\,\text{Tr}\,A_-^\gamma,$$

as claimed.   □

We now bound the change in eigenvalues under a rank-one perturbation.

Lemma 1.45 *Let A be a self-adjoint, lower semibounded operator and let B be a self-adjoint, non-positive rank-one operator. Then*

$$\lambda_1(A + B) \leq \lambda_1(A) \leq \lambda_2(A + B) \leq \lambda_2(A) \leq \cdots,$$

where $\lambda_n(A)$ and $\lambda_n(A + B)$ denote the eigenvalues of A and A + B, respectively, below $\inf \sigma_{\text{ess}}(A) = \inf \sigma_{\text{ess}}(A + B)$, in non-decreasing order and counting multiplicities.

Proof The equality $\inf \sigma_{\text{ess}}(A) = \inf \sigma_{\text{ess}}(A + B)$ follows from Weyl's theorem (Theorem 1.14). From the variational principle (Proposition 1.29) we obtain $\lambda_n(A + B) \leq \lambda_n(A)$ for all n. We will show the remaining inequality $\lambda_n(A) \leq \lambda_{n+1}(A + B)$ by proving that, for any $\mu < \inf \sigma_{\text{ess}}(A)$,

$$N(\mu, A + B) \leq N(\mu, A) + 1. \tag{1.56}$$

Choosing here $\mu = \lambda_n(A)$, so that $N(\mu, A) \leq n - 1$, we deduce the claimed inequality.

We prove (1.56) by contradiction, assuming $N(\mu, A + B) > N(\mu, A) + 1$. Let $P(A)$ and $P(A + B)$ denote the spectral measures of A and $A + B$. Since the space $\text{ran}\,P_{(-\infty,\mu)}(A + B) \cap \ker B$ has dimension at least $N(\mu, A + B) - 1$, which, by assumption, is larger than the dimension of the space $\text{ran}\,P_{(-\infty,\mu)}(A)$, there is a $0 \neq u \in \text{ran}\,P_{(-\infty,\mu)}(A + B) \cap \ker B \cap (\text{ran}\,P_{(-\infty,\mu)}(A))^\perp$. The conditions $0 \neq u \in \text{ran}\,P_{(-\infty,\mu)}(A + B)$ and $u \in (\text{ran}\,P_{(-\infty,\mu)}(A))^\perp$ imply, respectively,

$$a[u] + b[u] < \mu\|u\|^2 \qquad \text{and} \qquad a[u] \geq \mu\|u\|^2.$$

Since $u \in \ker B$ implies $b[u] = 0$, this is a contradiction.   □

The final ingredient in the proof of Proposition 1.43 is the following rearrangement inequality.

Lemma 1.46 *Let $0 < \gamma < 1$. Then for any set $E \subset [0, \infty)$ of finite measure*

$$\int_E t^{\gamma-1} dt \le \int_0^{|E|} t^{\gamma-1} dt \,.$$

Proof Since

$$t^{\gamma-1} = (1-\gamma) \int_t^\infty s^{\gamma-2} ds = (1-\gamma) \int_0^\infty s^{\gamma-2} \chi_{(0,s)}(t) ds \,,$$

we have

$$\int_E t^{\gamma-1} dt = (1-\gamma) \int_0^\infty \left(\int_0^\infty \chi_E(t) \chi_{(0,s)}(t) dt \right) s^{\gamma-2} ds$$

$$= (1-\gamma) \int_0^\infty |E \cap (0,s)| s^{\gamma-2} ds \,.$$

Obviously, $|E \cap (0,s)| \le \min\{|E|, s\}$. Note that equality holds here if $E = (0, |E|)$. Inserting this into the above identity, we obtain

$$\int_E t^{\gamma-1} dt \le (1-\gamma) \left(\int_0^{|E|} s^{\gamma-1} ds + |E| \int_{|E|}^\infty s^{\gamma-2} ds \right) = \frac{|E|^\gamma}{\gamma} = \int_0^{|E|} t^{\gamma-1} dt \,,$$

as claimed. $\square$

Proof of Proposition 1.43 We may assume that $\operatorname{Tr} A_-^\gamma + \operatorname{Tr} B_-^\gamma < \infty$. We begin with the case where B_- is rank-one and denote its positive eigenvalue by β. Moreover, we write the negative eigenvalues of A and $A+B$ as $-\alpha_1 \le -\alpha_2 \le \cdots$ and $-\lambda_1 \le -\lambda_2 \le \cdots$, respectively. According to Lemma 1.45, we have

$$-\lambda_n \le -\alpha_n \le -\lambda_{n+1} \le -\alpha_{n+1} \le \cdots \qquad \text{for all } n. \tag{1.57}$$

Then

$$\operatorname{Tr}(A+B)_-^\gamma = \sum_n \lambda_n^\gamma = \gamma \sum_n \int_0^{\lambda_n} t^{\gamma-1} dt$$

$$= \gamma \sum_n \int_0^{\alpha_n} t^{\gamma-1} dt + \gamma \sum_n \int_{\alpha_n}^{\lambda_n} t^{\gamma-1} dt$$

$$= \operatorname{Tr} A_-^\gamma + \gamma \int_E t^{\gamma-1} dt \,, \tag{1.58}$$

where $E := \bigcup_n (\alpha_n, \lambda_n)$ and where we used the fact that, by (1.57), these intervals are disjoint. Furthermore, again by the disjointness,

$$|E| = \sum_n \lambda_n - \sum_n \alpha_n = \operatorname{Tr}(A+B)_- - \operatorname{Tr} A_- \le \operatorname{Tr} B_- = \beta \,,$$

where the inequality comes from Corollary 1.37. Note that the assumed finiteness of $\operatorname{Tr} A_-^\gamma$ implies the finiteness of $\operatorname{Tr} A_-$, and therefore there is no cancellation of infinities in the above computation. Now Lemma 1.46 implies that

$$\gamma \int_E t^{\gamma-1}\, dt \leq \gamma \int_0^{|E|} t^{\gamma-1}\, dt \leq \gamma \int_0^\beta t^{\gamma-1}\, dt = \beta^\gamma = \operatorname{Tr} B_-^\gamma\,.$$

Inserting this into (1.58) yields the claimed inequality in the case where B_- is rank-one.

The case where B_- is of finite rank follows by iterating the inequality in the rank-one case.

We finally deal with the case of an arbitrary operator B with $\operatorname{Tr} B_-^\gamma < \infty$. For $\varepsilon > 0$ we write $P_\varepsilon := \chi_{(-\infty,-\varepsilon)}(A + B)$. According to Lemma 1.44, we have

$$\operatorname{Tr}(P_\varepsilon A P_\varepsilon)_-^\gamma \leq \operatorname{Tr} A_-^\gamma < \infty \qquad \text{and} \qquad \operatorname{Tr}(P_\varepsilon B P_\varepsilon)_-^\gamma \leq \operatorname{Tr} B_-^\gamma < \infty\,.$$

We know from the non-sharp bound of Proposition 1.40 that under the assumption $\operatorname{Tr} A_-^\gamma + \operatorname{Tr} B_-^\gamma < \infty$ we have $\operatorname{Tr}(A + B)_-^\gamma < \infty$, and therefore, in particular, P_ε has finite rank. Thus, $P_\varepsilon B P_\varepsilon$ has finite rank as well and, by what we have shown so far, we know

$$\operatorname{Tr}(P_\varepsilon A P_\varepsilon + P_\varepsilon B P_\varepsilon)_-^\gamma \leq \operatorname{Tr}(P_\varepsilon A P_\varepsilon)_-^\gamma + \operatorname{Tr}(P_\varepsilon B P_\varepsilon)_-^\gamma \leq \operatorname{Tr} A_-^\gamma + \operatorname{Tr} B_-^\gamma\,.$$

On the other hand,

$$\operatorname{Tr}(P_\varepsilon A P_\varepsilon + P_\varepsilon B P_\varepsilon)_-^\gamma = \operatorname{Tr}(P_\varepsilon(A + B)P_\varepsilon)_-^\gamma = \sum_{|\lambda_n(A+B)|>\varepsilon} |\lambda_n(A + B)|^\gamma$$

and, by monotone convergence, this converges to $\sum_n |\lambda_n(A+B)|^\gamma = \operatorname{Tr}(A+B)_-^\gamma$ as $\varepsilon \to 0$. This proves the claimed inequality. $\qquad\qquad\square$

1.2.7 Perturbations of quadratic forms

In this subsection, we take on a perturbation-theoretic point of view. That is, there will be a quadratic form a that is densely defined, lower semibounded and closed, and corresponds to an operator A, and we will study self-adjointness of $A + B$, where B is, in a sense to be made precise, small with respect to A. We formulate this smallness in the sense of quadratic forms.

The following simple lemma is sometimes useful for verifying that a perturbation of a lower semibounded, closed quadratic form is also lower semibounded and closed.

Lemma 1.47 *Assume that a is a lower semibounded, closed quadratic form* *with domain $d[a]$ and assume that b is a real-valued quadratic form on $d[a]$ such that for some $\theta \in [0,1)$ and some $C \in \mathbb{R}$,*

$$|b[u]| \leq \theta\, a[u] + C\,\|u\|^2 \qquad \text{for all } u \in d[a]. \tag{1.59}$$

Then the quadratic form $a + b$ with domain $d[a]$ is lower semibounded and closed. Moreover, any form core of a is also a form core of $a + b$.

Proof By assumption, we have the inequalities

$$(1 - \theta)a[u] - C\|u\|^2 \leq a[u] + b[u] \leq (1 + \theta)a[u] + C\|u\|^2.$$

The inequality on the left shows the lower semiboundedness of $a + b$, and the proof of closedness, as well as the form core property, follows easily from this two-sided bound. □

As a consequence of Lemma 1.47 and Theorem 1.18, when $d[a]$ is dense, the quadratic form $a + b$ with domain $d[a]$ generates a self-adjoint operator in $\mathcal{H}$. Often we shall denote this operator by $A + B$, but we emphasize that this is an abuse of notation since, in general, there need not be a well-defined, self-adjoint operator B given by the difference of $A + B$ and A.

Next, we will discuss a version of the resolvent identity for operators $A + B$ defined via quadratic forms. To motivate the formula we are seeking, we recall that, if the real quadratic form b corresponds to a bounded self adjoint operator B, then, according to (1.32),

$$(A + B - z)^{-1} - (A - z)^{-1} = -(A - z)^{-1} B (A + B - z)^{-1}.$$

We can write the right side as

$$
-\left[(A + m)^{1/2}(A - z)^{-1}\right]\left[(A + m)^{-1/2} B (A + m)^{-1/2}\right]
$$
$$
\times \left[(A + m)^{1/2}(A + B - z)^{-1}\right]
$$

for some $m > -m_a$. As we will show, the analogue of each of the three factors in square brackets is well defined under assumption (1.59), and therefore will lead to a version of the resolvent formula when $A + B$ is defined via quadratic forms.

It follows from (1.59) that for any $m > -m_a$ there is a constant $C' < \infty$ such that

$$|b[u]| \leq C'\left(a[u] + m\|u\|^2\right) \qquad \text{for all } u \in d[a].$$

Let us fix an $m > -m_a$ and a corresponding norm $\left(a[u] + m\|u\|^2\right)^{1/2}$ on $d[a]$.

By the Riesz representation theorem, there is a bounded operator $\mathcal{B}_a$ on $d[a]$ such that

$$b[u, v] = a[\mathcal{B}_a u, v] + m(\mathcal{B}_a u, v) \qquad \text{for all } u, v \in d[a]. \tag{1.60}$$

Clearly, the operator $\mathcal{U} : \mathcal{H} \to d[a]$, $f \mapsto (A + m)^{-1/2} f$ is unitary. Thus,

$$\hat{\mathcal{B}}_a := \mathcal{U}^* \mathcal{B}_a \mathcal{U}$$

is a bounded operator on $\mathcal{H}$. Note that in the case where b comes from a bounded operator B in $\mathcal{H}$, we have

$$\hat{\mathcal{B}}_a = (A + m)^{-1/2} B (A + m)^{-1/2}.$$

This operator appears in the formula for the resolvent difference.

Proposition 1.48 *Let a be a densely defined, lower semibounded and closed quadratic form and let b be a real quadratic form satisfying (1.59) for some $\theta \in [0, 1)$ and $C \in \mathbb{R}$. Then, for all $z \in \rho(A) \cap \rho(A + B)$,*

$$(A + B - z)^{-1} - (A - z)^{-1}$$
$$= -\left[(A + m)^{1/2}(A - z)^{-1}\right] \hat{\mathcal{B}}_a \left[(A + m)^{1/2}(A + B - z)^{-1}\right].$$

Each one of the three factors on the right side is a bounded operator.

Proof Step 1. We begin by showing the last assertion, namely, that each one of the three factors on the right side is a bounded operator. For the first factor this is clear from the spectral theorem, and for the second factor this was discussed before the proposition. For any $M > -m_a$ we write the third factor as

$$(A + m)^{1/2}(A + B - z)^{-1}$$
$$= \left[(A + m)^{1/2}(A + M)^{-1/2}\right]\left[(A + M)^{1/2}(A + B + M)^{-1/2}\right]$$
$$\times \left[(A + B + M)^{1/2}(A + B - z)^{-1}\right].$$

Here the first and the third factors are bounded by the same arguments as before. In the remainder of this step we will show that for any $M > C/\theta$ the operator $(A + M)^{1/2}(A + B + M)^{-1/2}$ is bounded with

$$\left\|(A + M)^{1/2}(A + B + M)^{-1/2}\right\| \le (1 - \theta)^{-1/2}. \tag{1.61}$$

This proves the boundedness of the third factor in the proposition.

To prove (1.61), we first note that, since the left side of (1.59) is non-negative, we have $a[u] + (C/\theta)\|u\|^2 \geq 0$ for all $u \in d[a]$, and therefore $C/\theta \geq -m_a$. Again by (1.59), for any $M > C/\theta$ and any $u \in d[a]$ we have

$$a[u]+b[u]+M\|u\|^2 \geq (1-\theta)\left(a[u] + \frac{M-C}{1-\theta}\|u\|^2\right) \geq (1-\theta)\left(a[u] + M\|u\|^2\right).$$

Since $M > C/\theta \geq -m_a$, the right side is not smaller than a positive constant times $\|u\|^2$. This proves that the operator $(A+B+M)^{-1/2}$ exists and is bounded. Setting $u = (A + B + M)^{-1/2}f$ with $f \in \mathcal{H}$, the previous inequality becomes

$$\|f\|^2 \geq (1 - \theta)\|(A + M)^{1/2}(A + B + M)^{-1/2}f\|^2.$$

This implies (1.61).

Step 2. We show that for any $f, g \in \mathcal{H}$,

$$\left(\left((A + B - z)^{-1} - (A - z)^{-1}\right)f, g\right)$$
$$= -\left(\left[(A + m)^{1/2}(A - z)^{-1}\right]\hat{\mathcal{B}}_a\left[(A + m)^{1/2}(A + B - z)^{-1}\right]f, g\right).$$

This proves the formula in the proposition. Since $z \in \rho(A) \cap \rho(A + B)$, we can define

$$u := (A + B - z)^{-1}f, \qquad v := (A - \bar{z})^{-1}g.$$

Then $u \in \mathrm{dom}(A + B)$ and $v \in \mathrm{dom}\,A$ and, in particular, $u, v \in d[a]$. Thus, by the definition of an operator corresponding to a quadratic form, the left side above is

$$\left(\left((A + B - z)^{-1} - (A - z)^{-1}\right)f, g\right) = (u, (A - \bar{z})v) - ((A + B - z)u, v)$$
$$= a[u, v] - (a[u, v] + b[u, v]) = -b[u, v].$$

For the right side above, using

$$\left((A + m)^{1/2}(A - z)^{-1}\right)^* = (A + m)^{1/2}(A - \bar{z})^{-1},$$

we have

$$\left(\left[(A + m)^{1/2}(A - z)^{-1}\right]\hat{\mathcal{B}}_a\left[(A + m)^{1/2}(A + B - z)^{-1}\right]f, g\right)$$
$$= (\hat{\mathcal{B}}_a(A + m)^{1/2}u, (A + m)^{1/2}v) = (\mathcal{U}^*\mathcal{B}_a\mathcal{U}(A + m)^{1/2}u, (A + m)^{1/2}v)$$
$$= (\mathcal{U}^*\mathcal{B}_au, \mathcal{U}^*v) = a[\mathcal{B}_au, v] + m(\mathcal{B}_au, v) = b[u, v].$$

Thus, both sides of the identity coincide, which completes the proof. $\qquad\square$

We now introduce a relative compactness condition that is important in applications. Let $\mathcal{G}$ be a Hilbert space with norm $\|\cdot\|_*$ and let b be a real-valued quadratic form with $d[b] = \mathcal{G}$. Then b is said to be *compact* in $\mathcal{G}$ if there is a constant C such that $|b[u]| \leq C\|u\|_*^2$ for all $u \in \mathcal{G}$ and if the bounded operator in $\mathcal{G}$ induced by the form b is compact. Here we apply the Riesz representation theorem.

Lemma 1.49 *Let $\mathcal{G}$ be a Hilbert space with norm $\| \cdot \|_*$ and let b be a real-valued quadratic form with $d[b] = \mathcal{G}$ such that, for a constant C, one has $|b[u]| \le C\|u\|_*^2$ for all $u \in \mathcal{G}$. If b is compact in $\mathcal{G}$, then, for any sequence $(u_n) \subset \mathcal{G}$ that converges weakly in $\mathcal{G}$ to zero, we have $b[u_n] \to 0$. Conversely, if $b[u] \ge 0$ for all $u \in \mathcal{G}$ and if, for any sequence $(u_n) \subset \mathcal{G}$ that converges weakly in $\mathcal{G}$ to zero, we have $b[u_n] \to 0$, then b is compact in $\mathcal{G}$.*

Proof Let $(u_n) \subset \mathcal{G}$ be a sequence converging weakly to zero in $\mathcal{G}$. Denote by $\mathcal{B}$ the compact operator in $\mathcal{G}$ induced by the form b. Then, by Lemma 1.1, $(\mathcal{B}u_n)$ converges strongly to zero in $\mathcal{G}$, and therefore $b[u] = (\mathcal{B}u_n, u_n)_* \to 0$, as claimed.

To prove the converse, assume that $b[u] \ge 0$ for all $u \in \mathcal{G}$. By assumption, if $(u_n) \subset \mathcal{G}$ converges weakly to zero in $\mathcal{G}$, then $\|\mathcal{B}^{1/2}u_n\|_*^2 = b[u_n] \to 0$; that is, $\mathcal{B}^{1/2}u_n$ tends strongly to zero in $\mathcal{G}$. By Lemma 1.1, this implies that $\mathcal{B}^{1/2}$ is compact in $\mathcal{G}$. Thus $\mathcal{B} = (\mathcal{B}^{1/2})^2$ is compact, as claimed. $\square$

The most frequent use of this relative compactness notion is when $\mathcal{G} = d[a]$, endowed with the norm $(a[u] + m\|u\|^2)^{1/2}$ for some $m > -m_a$.

Lemma 1.50 *Let a be a lower semibounded, closed quadratic form and assume that b is a real-valued quadratic form that is compact in $d[a]$ with the norm $(a[u] + m\|u\|^2)^{1/2}$ for some $m > -m_a$. Then for any $\theta > 0$ there is a C such that (1.59) holds.*

Proof We argue by contradiction. If the assertion of the lemma was false, there would be a sequence $(u_n) \subset d[a]$ with $a[u_n] + m\|u_n\|^2 = 1$ (for some $m > -m_a$) and

$$|b[u_n]| \ge \theta + n\|u_n\|^2. \tag{1.62}$$

By weak compactness of the unit ball in $d[a]$, there is a subsequence (u_{n_m}) that converges weakly in $d[a]$ to some u. Since b is compact in $d[a]$, we have, by Lemma 1.1, $\mathcal{B}_a u_{n_m} \to \mathcal{B}_a u$ strongly in $d[a]$ and therefore $b[u_{n_m}] \to b[u]$. Thus, (1.62) implies that $n_m\|u_{n_m}\|^2$ is bounded, and hence $u_{n_m} \to 0$ in $\mathcal{H}$. Since $d[a]$ is continuously embedded into $\mathcal{H}$, we deduce that $u = 0$, and then (1.62) leads to a contradiction since $\theta > 0$. $\square$

We end this subsection with a variant of Weyl's theorem (Theorem 1.14) for operators defined via quadratic forms.

Theorem 1.51 *Let a be a densely defined, lower semibounded and closed*

quadratic form and let b be a real-valued quadratic form that is compact in
$d[a]$. *Then the operators A and $A + B$ corresponding to a and $a + b$ satisfy*
$\sigma_{\mathrm{ess}}(A) = \sigma_{\mathrm{ess}}(A + B)$.

Proof We want to apply Theorem 1.14 and therefore have to show that

$$(A + B - z)^{-1} - (A - z)^{-1}$$

is compact. This follows immediately from the resolvent identity in Proposition
1.48 since the middle factor $\hat{\mathcal{B}}_a$ is compact by assumption and the outer factors
are bounded. This completes the proof. □

1.2.8 The Birman–Schwinger principle

In this subsection, we work under the following assumptions, depending on a
number $\alpha > 0$.

(H_α) Let a be a densely defined, non-negative, closed quadratic form in a
Hilbert space $\mathcal{H}$ with domain $d[a]$ satisfying

$$a[u] > 0 \quad \text{for all } 0 \neq u \in d[a] . \tag{1.63}$$

Let b be a real-valued quadratic form satisfying, for some $M < \infty$,

$$d[b] \supset d[a] \quad \text{and} \quad |b[u]| \leq M\, a[u] \quad \text{for all } u \in d[a] . \tag{1.64}$$

Assume that the quadratic form $a - \alpha b$ is lower semibounded and closed
in $\mathcal{H}$, and denote by $A - \alpha B$ the corresponding self-adjoint operator.

Assumption (1.63) implies that $\sqrt{a[\cdot]}$ is a norm in $d[a]$. Let $\mathcal{H}_a$ be the
completion of $d[a]$ with respect to this norm and let $\hat{a}$ be the extension of a by
continuity to $\mathcal{H}_a$.

If (1.63) is replaced by the stronger assumption that there is an $\varepsilon > 0$ with

$$a[u] \geq \varepsilon \|u\|^2 \quad \text{for all } u \in d[a], \tag{1.65}$$

then $\mathcal{H}_a$ coincides with $d[a]$ (with equivalent norms). In general, this need not
be the case and $\mathcal{H}_a$ may not be a subset of $\mathcal{H}$.

For practical purposes it is useful to note that, if $\mathcal{F} \subset d[a]$ is dense in
$d[a]$ (with respect to the norm $\sqrt{a[\cdot] + m\|\cdot\|^2}$ for some $m > -m_a$), then $\mathcal{H}_a$
coincides with the completion of $\mathcal{F}$ with respect to $\sqrt{a[\cdot]}$.

Assumption (1.64) implies that b can be extended by continuity to a quadratic
form $\hat{b}$ on $\mathcal{H}_a$. This extended quadratic form defines a bounded, self-adjoint
operator $\mathcal{B}_a$ on $\mathcal{H}_a$. We emphasize that the operator $\mathcal{B}_a$ is related to, but
different from, the operator $\mathcal{B}_a$ appearing in (1.60). Indeed, the operator in

(1.60) is defined on $d[a]$ with norm $\sqrt{a[\cdot] + m\|\cdot\|^2}$, where $m > -m_a$. Now we allow for $m = -m_a$, which is why we have to introduce the possibly larger space $\mathcal{H}_a$.

The parameter $\alpha > 0$ in assumption (II_α) is not really necessary, since (H_α) holds if and only if (H_1) holds with b replaced by αb. Nevertheless, in applications it is sometimes convenient to have this parameter present. Conditions that guarantee that $a - \alpha b$ is lower semibounded and closed in $\mathcal{H}$ are given in Lemmas 1.47 and 1.50. The notation $A - \alpha B$ is, in general, an abuse of notation since there need not be a well-defined, self-adjoint operator αB given by the difference of A and $A - \alpha B$.

We denote the spectral measures of $A - \alpha B$ and $\mathcal{B}_a$ by $P(A - \alpha B)$ and $P(\mathcal{B}_a)$, respectively.

The following result is called the *Birman–Schwinger principle*.

Theorem 1.52　*Assume* (H_α) *for some* $\alpha > 0$. *Then*

$$\dim P_{(-\infty,0)}(A - \alpha B)\mathcal{H} = \dim P_{(\alpha^{-1},\infty)}(\mathcal{B}_a)\mathcal{H}_a . \tag{1.66}$$

If, in addition, (1.65) *holds, then*

$$\dim P_{(-\infty,0]}(A - \alpha B)\mathcal{H} = \dim P_{[\alpha^{-1},\infty)}(\mathcal{B}_a)\mathcal{H}_a \tag{1.67}$$

and

$$\dim \ker(A - \alpha B) = \dim \ker(\mathcal{B}_a - \alpha^{-1}) . \tag{1.68}$$

Proof　It follows from Glazman's lemma (Theorem 1.25) applied to the operator $-\mathcal{B}_a$ in the Hilbert space $\mathcal{H}_a$ that, for any $\lambda \in \mathbb{R}$,

$$\dim P_{(\lambda,\infty)}(\mathcal{B}_a)\mathcal{H}_a$$
$$= \sup\left\{\dim F : F \subset d[a] \text{ and } b[u] > \lambda a[u] \text{ for all } 0 \neq u \in F\right\} .$$

Here we used the fact that $d[a]$ is dense in $\mathcal{H}_a$. On the other hand, applying Theorem 1.25 to the operator $A - \alpha B$ in the Hilbert space $\mathcal{H}$, we find that

$$\dim P_{(-\infty,0)}(A - \alpha B)\mathcal{H}$$
$$= \sup\left\{\dim F : F \subset d[a] \text{ and } b[u] > \alpha^{-1} a[u] \text{ for all } 0 \neq u \in F\right\} .$$

Choosing $\lambda = \alpha^{-1}$, we obtain (1.66).

Now, assuming (1.65), let us prove (1.68). Note that, once this is proved, (1.67) follows by adding (1.66) and (1.68).

To show (1.68), we show that $\ker(A - \alpha B) = \ker(\mathcal{B}_a - \alpha^{-1})$. Note that, under (1.65), $\mathcal{H}_a$ can be considered as a subset of $\mathcal{H}$. We have $u \in \ker(A - \alpha B)$ if and only if

$$b[u, v] = \alpha^{-1} a[u, v] \quad \text{for all } v \in d[a] .$$

By the definition of $\mathcal{B}_a$, this means

$$(\mathcal{B}_a u, v)_{\mathcal{H}_a} = \alpha^{-1}(u, v)_{\mathcal{H}_a} \quad \text{for all } v \in \mathcal{H}_a,$$

which is equivalent to $u \in \ker(\mathcal{B}_a - \alpha^{-1})$, as claimed.

As an aside, we mention that (1.67) can be proved using the second identity in Theorem 1.25. Here it is important that under assumption (1.65), $d[a]$ is not only dense in $\mathcal{H}_a$, but actually equal to $\mathcal{H}_a$. If $\dim P_{(-\infty,0)}(A - \alpha B)\mathcal{H} < \infty$, then (1.68) follows by subtracting (1.66) from (1.67). $\qquad\square$

We emphasize that (1.67) and (1.68) may fail if assumption (1.65) does not hold. Elements of $\ker(\mathcal{B}_a - \alpha^{-1})$ may belong to $\mathcal{H}_a$ without belonging to $\mathcal{H}$; see the reference in §1.3.

The reason why Theorem 1.52 is useful is that it relates the number of negative eigenvalues of a lower semibounded operator to the number of eigenvalues of a bounded (and typically compact) operator. For example, by Lemma 1.15, an immediate consequence of Theorem 1.52 is the following.

Corollary 1.53 *Assume* (H_α) *for every* $\alpha > 0$. *Then*

$$\dim P_{(-\infty,0)}(A - \alpha B)\mathcal{H} < \infty \qquad \text{for all } \alpha > 0$$

if and only if $(\mathcal{B}_a)_+$ *is compact in* $\mathcal{H}_a$. *For this, it is sufficient that* $\hat{b}$ *is compact in* $\mathcal{H}_a$. *If* $b \geq 0$, *the compactness of* $\hat{b}$ *in* $\mathcal{H}_a$ *is also necessary.*

For applications it is useful to reformulate Theorem 1.52 in terms of operators acting on the original Hilbert space $\mathcal{H}$ rather than on $\mathcal{H}_a$. We note that assumption (1.63) guarantees that $A^{-1/2}$ is a densely defined operator in $\mathcal{H}$.

Lemma 1.54 *Assume* (1.63) *and* (1.64). *Then the quadratic form*

$$b[A^{-1/2}f], \qquad f \in \mathrm{dom}\, A^{-1/2},$$

is real-valued, densely defined and bounded in $\mathcal{H}$. *The corresponding bounded self-adjoint operator in* $\mathcal{H}$ *is unitarily equivalent to the operator* $\mathcal{B}_a$ *in* $\mathcal{H}_a$.

In the formulation and proof of the lemma we use the fact that a real-valued, densely defined and bounded quadratic form generates a bounded, self-adjoint operator. This is a consequence of the Riesz representation theorem.

Proof We consider the operator $A^{-1/2}$, defined on $\mathrm{dom}\, A^{-1/2}$, as a mapping from a dense subset of $\mathcal{H}$ into $d[a] = \mathrm{dom}\, A^{1/2}$ equipped with the norm $\sqrt{a[\cdot]}$. This mapping is isometric since $a[A^{-1/2}f] = \|f\|^2$ for all $f \in \mathrm{dom}\, A^{-1/2}$. Moreover, $\mathrm{ran}\, A^{-1/2} = d[a]$. Since $\mathcal{H}_a$ is the completion of $d[a]$ with respect to $\sqrt{a[\cdot]}$, the above mapping extends to a unitary operator $U : \mathcal{H} \to \mathcal{H}_a$.

By definition of $\mathcal{B}_a$, we have (recall $\hat{a}$, $\hat{b}$ from the discussion of (H_α))

$$\hat{a}[\mathcal{B}_a\varphi, \varphi] = \hat{b}[\varphi] \qquad \text{for all } \varphi \in \mathcal{H}_a .$$

We apply this identity to $\varphi = Uf$ with $f \subset \text{dom } A^{-1/2}$. Then, since $Uf = A^{-1/2}f \in d[a] \subset d[b]$ for $f \in \text{dom } A^{-1/2}$, we have $\hat{b}[\varphi] = b[A^{-1/2}f]$, and therefore

$$\hat{a}[\mathcal{B}_a Uf, Uf] - b[A^{-1/2}f] \quad \text{for all } f \in \text{dom } A^{-1/2} .$$

Since $U : \mathcal{H} \to \mathcal{H}_a$ is unitary, the previous identity implies that

$$(U^*\mathcal{B}_a Uf, f) = b[A^{-1/2}f] \qquad \text{for all } f \in \text{dom } A^{-1/2} .$$

By (1.64), the operator $\mathcal{B}_a$ is bounded on $\mathcal{H}_a$, and therefore $U^*\mathcal{B}_a U$ is bounded on $\mathcal{H}$. This implies the assertions of the lemma. □

We shall denote the operator in Lemma 1.54 by $A^{-1/2}BA^{-1/2}$ and its spectral measure by $P(A^{-1/2}BA^{-1/2})$.

Writing $A^{-1/2}BA^{-1/2}$ is an abuse of notation, which, however, is motivated by the following. Under assumption (1.65), the operator $A^{-1/2}$ is bounded and, under the additional assumption that the quadratic form b is bounded in $\mathcal{H}$, it generates a bounded operator B in $\mathcal{H}$. In this case, the quadratic form $b[A^{-1/2}f]$, $f \in \text{dom } A^{-1/2} = \mathcal{H}$, is closed and the corresponding operator is simply the product of the three bounded operators $A^{-1/2}BA^{-1/2}$.

In view of Lemma 1.54, Theorem 1.52 can be reformulated as follows.

Theorem 1.55 *Assume* (H_α) *for some* $\alpha > 0$. *Then*

$$\dim P_{(-\infty, 0)}(A - \alpha B)\mathcal{H} = \dim P_{(\alpha^{-1}, \infty)}(A^{-1/2}BA^{-1/2})\mathcal{H} .$$

If, in addition, (1.65) *holds, then*

$$\dim P_{(-\infty, 0]}(A - \alpha B)\mathcal{H} = \dim P_{[\alpha^{-1}, \infty)}(A^{-1/2}BA^{-1/2})\mathcal{H}$$

and

$$\dim \ker(A - \alpha B) = \dim \ker(A^{-1/2}BA^{-1/2} - \alpha^{-1}) .$$

Finally, we discuss the case where b is non-negative and where there is an operator Q in $\mathcal{H}$ such that

$$\text{dom } Q \supset \text{dom } A^{-1/2} \quad \text{and} \quad b[A^{-1/2}f] = \|Qf\|^2 \text{ for all } f \in \text{dom } A^{-1/2} . \tag{1.69}$$

Assuming (1.63) and (1.64), it follows from Lemma 1.54 that any operator Q satisfying (1.69) is densely defined and bounded. In particular, it has a unique extension to a bounded operator defined on all of $\mathcal{H}$. Therefore, we may assume without loss of generality that Q is closed and defined on all of $\mathcal{H}$.

By (1.69), the operator $A^{-1/2}BA^{-1/2}$ is equal to Q^*Q. We denote the operator QQ^* by $B^{1/2}A^{-1}B^{1/2}$ and its spectral measure by $P(B^{1/2}AB^{1/2})$. Writing $B^{1/2}A^{-1}B^{1/2}$ is an abuse of notation that can be motivated as in the discussion preceding Theorem 1.55. The operator $B^{1/2}A^{-1}B^{1/2}$ is called the *Birman–Schwinger operator*.

By Proposition 1.23 and Corollary 1.24, Theorem 1.55 yields the following.

Theorem 1.56 *Assume* (H_α) *for some* $\alpha > 0$ *and* (1.69). *Then*

$$\dim P_{(-\infty,0)}(A - \alpha B)\mathcal{H} = \dim P_{(\alpha^{-1},\infty)}(B^{1/2}A^{-1}B^{1/2})\mathcal{H}. \tag{1.70}$$

If, in addition, (1.65) *holds, then*

$$\dim P_{(-\infty,0]}(A - \alpha B)\mathcal{H} = \dim P_{[\alpha^{-1},\infty)}(B^{1/2}A^{-1}B^{1/2})\mathcal{H} \tag{1.71}$$

and

$$\dim \ker(A - \alpha B) = \dim \ker(B^{1/2}A^{-1}B^{1/2} - \alpha^{-1}). \tag{1.72}$$

Corollary 1.57 *Assume* (H_α) *for some* $\alpha > 0$, (1.69) *and that the negative spectrum of* $A - \alpha B$ *is discrete. Let* $(-E_n)$ *be its negative eigenvalues, in nondecreasing order and repeated according to multiplicities. For fixed* n, *let* (μ_m) *be the eigenvalues of* $B^{1/2}(A + E_n)B^{1/2}$ *greater than or equal to* α^{-1} *in nonincreasing order and repeated according to multiplicities. Then* $\mu_n = \alpha^{-1}$.

Note that, by (1.71) with $A + E_n$ instead of A, the total spectral multiplicity of $B^{1/2}(A + E_n)B^{1/2}$ in $[\alpha^{-1},\infty)$ is finite. Therefore, the μ_m in the corollary are well defined. Here we use the fact that for $A + E_n$, assumption (1.65) holds (with $\varepsilon = E_n$).

Proof Let K be the multiplicity of the eigenvalue $-E_n$ of $A - \alpha B$ and let k be such that $E_n = E_k = \cdots = E_{k+K-1}$. That is, $k = n$ if $-E_n$ is simple, and otherwise k is the minimal index for which $E_k = E_n$. Note also that $k \leq n \leq k + K - 1$.

By (1.72) (applied with $A+E_n$ instead of A), the operator $B^{1/2}(A+E_n)^{-1}B^{1/2}$ has an eigenvalue α^{-1} and its multiplicity is K. Thus, there is an ℓ such that $\alpha^{-1} = \mu_\ell = \cdots = \mu_{\ell+K-1}$ and, if $\ell \geq 2$, $\mu_{\ell-1} > \alpha^{-1}$.

We observe that $\ell = k$ since by (1.70) (applied with $A + E_n$ instead of A),

$$k - 1 = \dim P_{(-\infty,-E_n)}(A-\alpha B)\mathcal{H} = \dim P_{(\alpha^{-1},\infty)}(B^{1/2}(A+E_n)B^{1/2})\mathcal{H} = \ell - 1.$$

Since $k \leq n \leq k + K - 1$, this shows $\ell \leq n \leq \ell + K - 1$, and thus $\mu_n = \alpha^{-1}$. $\qquad\square$

1.3 Comments

We have not given a full account of spectral theory in this chapter, but rather selected material that is needed for our applications to Laplace or Schrödinger operators. Further material can be found in such textbooks as Akhiezer and Glazman (1963), Birman and Solomjak (1987), Davies (1995), Helffer (2013), Reed and Simon (1972, 1975, 1978, 1979), and Teschl (2014), as well as in the monograph of Kato (1980).

Section 1.1: Hilbert spaces, self-adjoint operators, and the spectral theorem
For the history of the development of the notion of unbounded operators we refer to Simon (2015c, §7.1). A first version of the spectral theorem for bounded self-adjoint operators goes back to Hilbert (1906). For further references on early developments, see Simon (2015c, §5.1). The spectral theorem in the unbounded case is due to von Neumann (1930). The above-mentioned references give several different proofs of this theorem. We also refer to the above-mentioned textbooks for the notion of a normal operator and for the corresponding spectral theorem.

There is a proof of the formula (1.24) for the norm of the resolvent that does not use the spectral theorem, but instead the fact that for a bounded, normal operator, the spectral radius is equal to the norm (Simon, 2015c, Theorems 2.2.10 and 2.2.11); see also Kato (1980, (V.3.16)) and Edmunds and Evans (2018, Lemma 3.4.3).

Theorem 1.14 on the stability of the essential spectrum is from Weyl (1909). The quadratic form version in Theorem 1.51 appears in Birman (1959).

Section 1.2: Semibounded operators and forms and the variational principle
The proof of Theorem 1.16 that we presented uses the square root of a non-negative self-adjoint operator, which we defined using the spectral theorem and the functional calculus. For an alternative construction of the square root, see Kato (1980, §V.3.11).

Theorem 1.18, which associates to each lower semibounded, closed quadratic form a self-adjoint operator, is due to Friedrichs (1934a). Applications to second-order differential operators are given in Friedrichs (1934b). For a proof that does not require self-adjointness, see Kato (1980, §VI.2).

Lemma 1.20 is well known to specialists, but hard to locate in the literature; see Lewin (2022, Théorème 5.14). One half of Corollary 1.21, namely that compactness of an embedding implies discreteness of the spectrum, appears in

Friedrichs (1934a, Zusatz 18) abstracting the arguments in Rellich (1930). An operator-domain version of the lemma appears in Rellich (1942).

Different versions of the variational principle go back to Rayleigh (1877), Fischer (1905), Ritz (1908), Weyl (1912a,b) and Courant (1920). An operator version of Glazman's lemma (Theorem 1.25) appears in Glazman (1966, §3, Theorem 9bis). The version for quadratic forms is mentioned, for instance, in Birman (1961).

Proposition 1.33 (at least in finite dimension) is due to Fan (1949). Proposition 1.33 and Corollary 1.35 remain valid when the orthogonality condition $(u_n, u_m) = \delta_{nm}$ is replaced by the sub-orthogonality condition that the eigenvalues of the matrix (u_n, u_m) are between 0 and 1; see, for instance, Simon (2011, Proposition 15.13). This is sometimes useful in applications and appears implicitly in §7.4.

Proposition 1.43 is from Rotfel'd (1967, 1968).

The Birman–Schwinger principle, discussed in various forms in §1.2.8, appears in Birman (1961) and Schwinger (1961); for a detailed presentation, see also Birman and Solomyak (1992). This abstract principle will be spelled out for Schrödinger operators in §4.3.3. In §4.9 we give references for concrete applications of this principle.

Let us already here give a typical example of the setting of §1.2.8, using freely the notation in the next chapter. In the Hilbert space $\mathcal{H} = L^2(\mathbb{R}^d)$, $d \geq 1$, and with a parameter $\tau \geq 0$ we consider the quadratic form $a[u] := \int_{\mathbb{R}^d} \left(|\nabla u|^2 + \tau |u|^2 \right) dx$ with form domain $d[a] := H^1(\mathbb{R}^d)$. Then (1.63) is satisfied, and (1.65) is satisfied if and only if $\tau > 0$. For $\tau > 0$, one has $\mathcal{H}_a = H^1(\mathbb{R}^d)$ (with equivalent norms). For $\tau = 0$ and $d \geq 3$, one has $\mathcal{H}_a = \dot{H}^1(\mathbb{R}^d)$, the homogeneous Sobolev space, while for $\tau = 0$ and $d = 1, 2$, the space $\mathcal{H}_a$ is not a space of (a.e. equivalence classes of) functions and is typically avoided; see §2.7.2. In applications to Schrödinger operators with sufficiently regular $V \geq 0$, one chooses $b[u] := \int_{\mathbb{R}^d} V|u|^2 \, dx$. Then the Birman–Schwinger principle, for instance in its form in Theorem 1.56, relates the number of eigenvalues less than $-\tau < 0$ of $-\Delta - V$ to the number of eigenvalues greater than 1 of the Birman–Schwinger operator $V^{1/2}(-\Delta + \tau)^{-1}V^{1/2}$. As mentioned after Theorem 1.52, the equalities (1.68) (or, equivalently, (1.72)) need not be true if (1.65) fails. In the concrete case of Schrödinger operators in dimensions $d \geq 3$, this occurs if there is a solution $0 \neq u \in \dot{H}^1(\mathbb{R}^d)$ of $-\Delta u - Vu = 0$ that does not belong to $L^2(\mathbb{R}^d)$. Such solutions are counted in the right side of (1.72), but not in the left side. One speaks of a 'zero energy resonance' or a 'virtual level'; see, for instance, Jensen and Kato (1979), Jensen (1980, 1984) and Klaus and Simon (1980) for more on this phenomenon.

2

Elements of Sobolev Space Theory

In this chapter, we collect some material from the theory of Sobolev spaces. These spaces appear naturally when defining Laplace and Schrödinger operators as self-adjoint operators in terms of quadratic forms and they play a crucial role in the spectral properties that interest us.

For the definition of Sobolev spaces we require the notion of a weak derivative, which generalizes the notion of a classical derivative but has a crucial closedness property under taking limits. The definition of weak derivatives and their basic properties are discussed in §2.1 and the corresponding Sobolev spaces are introduced in §2.2. Throughout this book we will only need first-order, L^2-based Sobolev spaces and we limit our discussion to those.

In §2.3 we will prove Rellich's theorem about the compactness of the embedding of Sobolev spaces into L^2 spaces for suitable domains. As applications of this theorem we will deduce in the next chapter the discreteness of the spectrum of the Dirichlet Laplacian on a set of finite measure, as well as the discreteness of the negative spectrum of Schrödinger operators with decaying potentials.

Sections 2.4, 2.5, and 2.6 are devoted to various functional inequalities, including such major ones as the Sobolev, Poincaré, and Hardy inequalities. On the one hand, these inequalities play an important role when discussing embeddings between different function spaces. On the other hand, as we will see later, they are closely related to spectral inequalities and will be crucial in our proofs in Chapter 4. For instance, the Keller–Lieb–Thirring inequalities in Proposition 5.3, which give a lower bound on the lowest eigenvalue of a Schrödinger operator in terms of an L^p norm of its potential, are equivalent to the family of Gagliardo–Nirenberg inequalities. Another example is Hardy's inequality which, in particular, appears in the question of whether the negative spectrum of a Schrödinger operator is finite or infinite; see §4.3.2. In this connection, homogeneous Sobolev spaces also arise: we discuss them in §2.7.

In §2.8 we discuss the Sobolev extension property, which is a useful notion when reducing certain questions about Sobolev spaces on an open set to the case of the whole space.

Most of the facts that we collect in this chapter, and also their proofs, can be found in standard references on the topic, like, for instance, Adams and Fournier (2003), Brezis (2011), Evans (2010), Gilbarg and Trudinger (1998) and Lieb and Loss (2001). We refer to these textbooks also for a more detailed treatment of Sobolev spaces.

2.1 Weak derivatives

2.1.1 Definition and basic properties

Throughout this subsection, $\Omega \subset \mathbb{R}^d$ is an open set. All functions considered in this chapter are complex-valued.

Let $L^1_{\mathrm{loc}}(\Omega)$ be the space of (a.e. equivalence classes of) functions whose restriction to any compact set $K \subset \Omega$ belongs to $L^1(K)$.

Let $C^m(\Omega)$ be the space of functions in Ω that are continuous together with their partial derivatives up to order $m \in \mathbb{N}_0$. Let $C^\infty(\Omega) = \bigcap_{m=0}^\infty C^m(\Omega)$, and let $C_0^\infty(\Omega)$ be the subset of functions in $C^\infty(\Omega)$ whose support is a compact subset of Ω. For $\eta \in C^1(\Omega)$, we denote by $\partial_e \eta(x) = \lim_{t \to 0}(\eta(x + te) - \eta(x))/t$ its derivative in direction $e \in \mathbb{S}^{d-1}$.

Let $u \in L^1_{\mathrm{loc}}(\Omega)$. We say that $v \in L^1_{\mathrm{loc}}(\Omega)$ is a *weak derivative* of u in direction $e \in \mathbb{S}^{d-1}$ if

$$\int_\Omega u \partial_e \eta \, dx = -\int_\Omega v\eta \, dx \qquad \text{for all } \eta \in C_0^\infty(\Omega).$$

It is well known that, if $w \in L^1_{\mathrm{loc}}(\Omega)$ and $\int_\Omega w\eta \, dx = 0$ for all $\eta \in C_0^\infty(\Omega)$, then $w = 0$ almost everywhere in Ω. As a consequence, a weak directional derivative of u, if it exists, is unique almost everywhere. It is considered as an element in $L^1_{\mathrm{loc}}(\Omega)$ and only defined up to sets of measure zero. By integration by parts, the weak directional derivative for $u \in C^1(\Omega)$ coincides almost everywhere with its classical directional derivative. We denote the weak derivative of u in direction $e \in \mathbb{S}^{d-1}$ by $\partial_e u$.

Taking weak directional derivatives is linear; that is, if $u_1, u_2 \in L^1_{\mathrm{loc}}(\Omega)$ have weak derivatives in direction e and if $c_1, c_2 \in \mathbb{C}$, then $c_1 u_1 + c_2 u_2$ has a weak derivative in direction e given by $\partial_e (c_1 u_1 + c_2 u_2) = c_1 \partial_e u_1 + c_2 \partial_e u_2$. Moreover, weak derivatives are local in the sense that, if v is the weak derivative of u in direction e in Ω, then for any open $\Omega' \subset \Omega$ the restriction of v to Ω' is the weak derivative in direction e of the restriction of u to Ω'.

The next lemma concerns approximation of weak directional derivatives by classical derivatives.

Lemma 2.1   *If* $u \in I^1_{loc}(\Omega)$ *has a weak derivative* $\partial_e u \in L^1_{loc}(\Omega)$ *in direction* e, *then there is a sequence* $(u_n) \subset C^\infty(\Omega)$ *such that* $u_n \to u$ *in* $L^1_{loc}(\Omega)$ *and* $\partial_e u_n \to \partial_e u$ *in* $L^1_{loc}(\Omega)$.

For the proof, we recall the definition of the convolution of two functions f and g on $\mathbb{R}^d$:

$$(f * g)(x) := \int_{\mathbb{R}^d} f(x - y)g(y)\, dy, \quad x \in \mathbb{R}^d.$$

This is well defined, for instance, if $f \in L^1(\mathbb{R}^d)$ and $g \in L^\infty(\mathbb{R}^d)$.

Proof   Let (Ω_n) be a sequence of bounded, open sets in $\mathbb{R}^d$ such that

$$\overline{\Omega_n} \subset \Omega_{n+1} \quad \text{for all } n \quad \text{and} \quad \bigcup_n \Omega_n = \Omega.$$

Let $\tilde{u}_n(x) := u(x)$ if $x \in \Omega_n$ and $\tilde{u}_n(x) := 0$ if $x \in \mathbb{R}^d \setminus \Omega_n$. Then $\tilde{u}_n \in L^1(\mathbb{R}^d)$. Let $0 \le \omega \in C_0^\infty(\mathbb{R}^d)$ have support in the unit ball and $\int_{\mathbb{R}^d} \omega\, dx = 1$, and let $\omega_\rho(x) := \rho^{-d}\omega(x/\rho)$. For a sequence (ρ_n) of positive numbers tending to zero, we note that

$$u_n(x) := (\omega_{\rho_n} * \tilde{u}_n)(x) = \rho_n^{-d} \int_{\mathbb{R}^d} \omega((x - y)/\rho_n)\tilde{u}_n(y)\, dy, \quad x \in \mathbb{R}^d,$$

belongs to $C^\infty(\mathbb{R}^d)$ and, using standard inequalities for convolutions, one sees that $u_n \to u$ in $L^1_{loc}(\Omega)$ and $\partial_e u_n \to \partial_e u$ in $L^1_{loc}(\Omega)$, as claimed.   □

Remark 2.2   The proof shows that, if $u \in L^1_{loc}(\Omega)$ has weak derivatives $\partial_e u \in L^1_{loc}(\Omega)$ in several directions $e \in E$, where $E \subset \mathbb{S}^{d-1}$, then the *same* sequence (u_n) satisfies $\partial_e u_n \to \partial_e u$ in $L^1_{loc}(\Omega)$ for all $e \in E$.

Lemma 2.3   *Let* $u \in L^1_{loc}(\Omega)$ *and* $(u_n) \subset L^1_{loc}(\Omega)$ *with* $u_n \to u$ *in* $L^1_{loc}(\Omega)$. *If each* u_n *has a weak derivative* $\partial_e u_n \in L^1_{loc}(\Omega)$ *in direction* e *and if* $\partial_e u_n \to v$ *in* $L^1_{loc}(\Omega)$ *for some* $v \in L^1_{loc}(\Omega)$, *then* v *is the weak derivative of* u *in direction* e.

Proof   For any $\eta \in C_0^\infty(\Omega)$, we have

$$\int_\Omega u_n \partial_e \eta\, dx = - \int_\Omega (\partial_e u_n)\eta\, dx.$$

From the compact support of η and the assumed L^1_{loc} convergence, the left side tends to $\int_\Omega u\partial_e \eta\, dx$ and the right side tends to $-\int_\Omega v\eta\, dx$. Therefore these two expressions are equal, which proves that v is the weak derivative of u in direction e.   □

Let $e_1, \ldots, e_d \in \mathbb{R}^d$ be the coordinate unit vectors. We call a function $u \in L^1_{\mathrm{loc}}(\Omega)$ *weakly differentiable* if it has weak derivatives $\partial_{e_j} u \in L^1_{\mathrm{loc}}(\Omega)$ for all $j = 1, \ldots, d$. We write $\partial_j u := \partial_{e_j} u$ for brevity and set

$$\nabla u :- (\partial_j u)_{j=1}^d \subset L^1_{\mathrm{loc}}(\Omega, \mathbb{C}^d).$$

A weakly differentiable u has weak derivatives in every direction e, given by $\partial_e u = e \cdot \nabla u$. In particular, the notion of weak differentiability is independent of the chosen basis in $\mathbb{R}^d$.

Lemma 2.4 *Assume that $u \in L^1_{\mathrm{loc}}(\Omega)$ satisfies*

$$|u(x) - u(x')| \le L|x - x'| \qquad \text{for all } x, x' \in \Omega.$$

Then u is weakly differentiable and $|\nabla u| \le L$ almost everywhere in Ω.

Proof Let $\Omega' \subset \mathbb{R}^d$ be a bounded, open set with $\overline{\Omega'} \subset \Omega$. Fix $1 \le j \le d$ and for $h \in \mathbb{R}$ with $|h| < \mathrm{dist}(\Omega', \Omega^c)$ consider

$$v_h(x) := h^{-1}(u(x + h e_j) - u(x)), \qquad x \in \Omega'.$$

By assumption, $|v_h| \le L$ pointwise in Ω'. By weak compactness in $L^2(\Omega')$, there is a sequence (h_n) tending to zero and a $v \in L^2(\Omega')$ such that $v_{h_n} \to v$ weakly in $L^2(\Omega')$. Thus, for any $\eta \in C_0^\infty(\Omega')$,

$$\int_{\Omega'} (\partial_j \eta) u \, dx = -\lim_{n \to \infty} \int_{\Omega'} h_n^{-1}(\eta(x - h_n e_j) \quad \eta(x)) u(x) \, dx$$

$$= -\lim_{n \to \infty} \int_{\Omega'} \eta(x) v_{h_n}(x) \, dx$$

$$= -\int_{\Omega'} \eta v \, dx.$$

This proves that u has a weak derivative $\partial_j u = v$ in Ω'. Since Ω' and j are arbitrary, we infer that u is weakly differentiable in Ω.

We now show $|\nabla u| \le L$. Let $0 \le \omega \in C_0^\infty(\mathbb{R}^d)$ have support in the unit ball and $\int_{\mathbb{R}^d} \omega \, dx = 1$, and set $\omega_\rho(x) := \rho^{-d} \omega(x/\rho)$. Let Ω' be as before and for $0 < \rho < \mathrm{dist}(\Omega', \Omega^c)$ consider

$$u_\rho(x) := (\omega_\rho * u)(x) = \rho^{-d} \int_\Omega \omega((x - y)/\rho) u(y) \, dy, \qquad x \in \Omega'.$$

Fix $x \in \Omega'$ and $\rho < \mathrm{dist}(\Omega', \Omega^c) =: \delta$. Then for any $x' \in B_{\delta - \rho}(x)$ and $y \in B_\rho(x)$ one has $y + x' - x \in B_\delta(x) \subset \Omega$, and therefore one can change variables and write

$$u_\rho(x) - u_\rho(x') = \rho^{-d} \int_{B_\rho(x)} \omega((x - y)/\rho)(u(y) - u(y + x' - x)) \, dy.$$

Using the assumption to bound the integral on the right side, one finds

$$|u_\rho(x) - u_\rho(x')| \le \rho^{-d} \int_{B_\rho(\lambda)} \omega((x-y)/\rho)L|x'-x|\,dy = L|x'-x|\,.$$

Recalling that u_ρ is (classically) differentiable, we obtain $|\nabla u_\rho(x)| \le L$. Meanwhile, since u is weakly differentiable, $\nabla u_\rho \to \nabla u$ in $L^1(\Omega')$ as $\rho \to 0$, and therefore $|\nabla u| \le L$ almost everywhere in Ω'. This completes the proof. □

For the next lemma we recall that an open set $\Omega \subset \mathbb{R}^d$ is called *connected* if it is not the disjoint union of two non-empty, open sets.

Lemma 2.5 *Let Ω be connected and let $u \in L^1_{\text{loc}}(\Omega)$ be weakly differentiable with $\nabla u = 0$ in Ω. Then there is a $c \in \mathbb{C}$ such that $u = c$ almost everywhere.*

Proof By a standard topological argument it suffices to show that for any bounded, open and connected $\Omega' \subset \mathbb{R}^d$ with $\overline{\Omega'} \subset \Omega$ there is a constant $c_{\Omega'}$ such that $u = c_{\Omega'}$ almost everywhere on Ω'.

To prove this statement we choose a bounded, open set $\Omega'' \subset \mathbb{R}^d$ such that

$$\overline{\Omega'} \subset \Omega'' \subset \overline{\Omega''} \subset \Omega\,.$$

Then $\rho_0 := \text{dist}(\Omega', (\Omega'')^c) > 0$. Let $\omega \subset C_0^\infty(\mathbb{R}^d)$ be non-negative with support in the unit ball and $\int_{\mathbb{R}^d} \omega\,dx = 1$. As above, for $\rho < \rho_0$ we set

$$u_\rho(x) := \rho^{-d} \int_{\Omega''} \omega((x-y)/\rho)u(y)\,dy\,.$$

Then, as in the proof of Lemma 2.1, $u_\rho \in C^\infty(\mathbb{R}^d)$ and

$$(\partial_j u_\rho)(x) = \rho^{-d-1} \int_{\Omega''} (\partial_j \omega)((x-y)/\rho)u(y)\,dy\,.$$

Moreover, for $x \in \Omega'$ the function $y \mapsto \rho^{-d}\omega((x-y)/\rho)$ belongs to $C_0^\infty(\Omega'')$, and therefore, by the definition of the weak derivative of u and the assumption in the lemma,

$$(\partial_j u_\rho)(x) = -\rho^{-d} \int_{\Omega''} \omega((x-y)/\rho)(\partial_j u)(y)\,dy = 0\,.$$

Since u_ρ is smooth and Ω' is connected, we infer that $u_\rho = c_{\rho,\Omega'}$ in Ω' for a $c_{\rho,\Omega'} \in \mathbb{C}$. Moreover, as in Lemma 2.1, $u_\rho \to u$ in $L^1(\Omega')$. This implies that there is a $c_{\Omega'} \in \mathbb{C}$ such that $u = c_{\Omega'}$ almost everywhere on Ω', as claimed. □

2.1.2 The one-dimensional case

The next lemma concerns the one-dimensional case, where we show that the fundamental theorem of calculus holds for weakly differentiable functions. We write $u' := \partial_1 u$ for the weak derivative of u, if it exists, in dimension $d = 1$.

Lemma 2.6 *Let $I \subset \mathbb{R}$ be an open interval. If $f \in L^1_{loc}(I)$ and $y \in I$, then*

$$v(x) := \int_y^x f(t)\, dt, \qquad x \in I,$$

defines a continuous and weakly differentiable function on I with $v' = f$. Moreover, if $u \in L^1_{loc}(I)$ is weakly differentiable, then there is a continuous function $\tilde{u}$ on I such that $u = \tilde{u}$ almost everywhere on I and

$$\tilde{u}(x) - \tilde{u}(x') = \int_{x'}^x u'(t)\, dt \qquad \text{for all } x, x' \in I. \tag{2.1}$$

The integral in the definition of v is the Lebesgue integral $\int_{(y,x)} f(t)\, dt$ if $x > y$ and $-\int_{(x,y)} f(t)\, dt$ if $x < y$, and similarly for the integral in (2.1).

In classical analysis, there is a notion of absolute continuity and one can show that the fundamental theorem of calculus holds if and only if a function is absolutely continuous; see Rudin (1987, Definition 7.17 and Theorem 7.18) or Royden (1963, §5.4). Therefore, as a consequence of Lemma 2.6, in one dimension a function is weakly differentiable if and only if it is locally absolutely continuous.

Proof The fact that v is continuous follows by dominated convergence. To show that it is weakly differentiable, we let $\eta \in C_0^\infty(I)$ and compute, using the notation $I = (a, b)$ with $-\infty \le a < b \le +\infty$,

$$\int_I v\eta'\, dx = -\int_a^y \int_x^y f(t)\, dt\, \eta'(x)\, dx + \int_y^b \int_y^x f(t)\, dt\, \eta'(x)\, dx$$

$$= -\int_a^y f(t) \int_a^t \eta'(x)\, dx\, dt + \int_y^b f(t) \int_t^b \eta'(x)\, dx\, dt$$

$$= -\int_a^y f(t)\eta(t)\, dt - \int_y^b f(t)\eta(t)\, dt = -\int_a^b f(t)\eta(t)\, dt.$$

This shows that v is weakly differentiable with $v' = f$.

To prove the second part, fix $y \in I$, put $f = u'$ and define v as before. Then, by the first part, $v' = u'$ on I, and therefore, by Lemma 2.5, $u = v + c$ almost everywhere on I for some $c \in \mathbb{C}$. Then $\tilde{u} := v + c$ is easily seen to satisfy the claimed properties. $\qquad\square$

Remark 2.7   For a general weakly differentiable function u, Lemma 2.6 makes
no assertion about the boundary behavior of u. If, however, u satisfies the global
integrability condition $u' \in L^1(I)$, then it is an easy consequence of the lemma
that the function $\tilde{u}$ extends continuously to finite endpoints of I and has limits
at infinite endpoints of I. In this case, formula (2.1) holds for $x, x' \in \bar{I}$ and, in
an obvious sense, for the infinite endpoints.

Remark 2.8   The following representation formula is sometimes useful. If
$I = (a, b)$ is a bounded, open interval and $u \in L^1(I)$ is weakly differentiable
with $u' \in L^1(I)$, then for all $x \in [a, b]$,

$$\tilde{u}(x) = \frac{1}{b-a} \int_a^b u(t)\, dt - \frac{1}{b-a} \int_x^b (b-t)u'(t)\, dt + \frac{1}{b-a} \int_a^x (t-a)u'(t)\, dt .$$

$$(2.2)$$

This follows by integrating (2.1) with respect to x' over I. This formula can
also be used to define the continuous representative $\tilde{u}$ of u.

Remark 2.9   The definition of the boundary values in Remark 2.7 leads to
the following integration by parts formula. If $I = (a, b)$ is a bounded, open
interval and $u \in L^1_{\text{loc}}(I)$ is weakly differentiable with $u' \in L^1(I)$, then for any
$\eta \in C^\infty(I) \cap C(\bar{I})$ with $\eta' \in L^1(I)$,

$$\int_I u\eta'\, dx = - \int_I u'\eta\, dx + \tilde{u}(b)\eta(b) - \tilde{u}(a)\eta(a) .$$

This follows by applying Lemma 2.6 to the function ηu. The latter function is
weakly differentiable on I with weak derivative $(\eta u)' = \eta'u + \eta u'$, as follows
from the definition of weak differentiability together with the product rule for
C^∞ functions. We will discuss more precise versions of the product rule in
Lemma 2.14.

2.1.3 The chain rule

We return to the case of arbitrary dimension d and prove the chain rule for
functions with a weak directional derivative.

Lemma 2.10   *Let $\varphi : \mathbb{R}^N \to \mathbb{R}$ be a continuously differentiable function
with bounded derivative, and assume that $u^{(1)}, \ldots, u^{(N)} \in L^1_{\text{loc}}(\Omega)$ are real
and have weak derivatives $\partial_e u^{(1)}, \ldots, \partial_e u^{(N)} \in L^1_{\text{loc}}(\Omega)$ in direction e. Then
$\varphi(u^{(1)}, \ldots, u^{(N)})$ has a weak derivative in direction e and it is given by*

$$\partial_e \varphi(u^{(1)}, \ldots, u^{(N)}) = \sum_{n=1}^N (\partial_n \varphi)(u^{(1)}, \ldots, u^{(N)}) \partial_e u^{(n)} .$$

Proof By Lemma 2.1, for every $n = 1, \ldots, N$ there is a sequence $(v_k^{(n)}) \subset C^\infty(\Omega)$ with $v_k^{(n)} \to u^{(n)}$ in $L^1_{\text{loc}}(\Omega)$ and $\partial_e v_k^{(n)} \to \partial_e u^{(n)}$ in $L^1_{\text{loc}}(\Omega)$. We shall use the short hand $u := (u^{(1)}, \ldots, u^{(N)})$ and $v_k := (v_k^{(1)}, \ldots, v_k^{(N)})$. By the ordinary chain rule applied to the C^1 function $\varphi(v_k)$, we have, for any $\eta \in C_0^\infty(\Omega)$,

$$\int_\Omega (\partial_e \eta) \varphi(v_k)\, dx = -\sum_{n=1}^N \int_\Omega \eta\, (\partial_n \varphi)\, (v_k)\, \partial_e v_k^{(n)}\, dx. \tag{2.3}$$

Since the derivatives of φ are bounded, we have, for some $C < \infty$,

$$|\varphi(s) - \varphi(s')| \le C|s - s'| \quad \text{for all } s, s' \in \mathbb{R}^N.$$

Thus,

$$\left| \int_\Omega (\partial_e \eta) \varphi(v_k)\, dx - \int_\Omega (\partial_e \eta) \varphi(u)\, dx \right| \le C \int_\Omega |\partial_e \eta| |v_k - u|\, dx$$

and, by the L^1_{loc} convergence of v_k to u, the right side tends to zero as $k \to \infty$. Similarly, by the boundedness of $\partial_n \varphi$,

$$\left| \int_\Omega \eta\, (\partial_n \varphi)\, (v_k)\, \partial_e v_k^{(n)}\, dx - \int_\Omega \eta\, (\partial_n \varphi)\, (v_k)\, \partial_e u^{(n)}\, dx \right|$$

$$\le C \int_\Omega |\eta| \left| \partial_e v_k^{(n)} - \partial_e u^{(n)} \right| dx$$

and, by the L^1_{loc} convergence of $\partial_e v_k^{(n)}$ to $\partial_e u^{(n)}$, the right side tends to zero as $k \to \infty$. Finally, after passing to a subsequence, we may assume that $v_k \to u$ almost everywhere. Therefore, by continuity of $\partial_n \varphi$, we obtain $\partial_n \varphi(v_k) \to \partial_n \varphi(u)$ almost everywhere and, by boundedness of $\partial_n \varphi$ and dominated convergence, we conclude

$$\int_\Omega \eta\, (\partial_n \varphi)\, (v_k)\, \partial_e u^{(n)}\, dx \to \int_\Omega \eta\, (\partial_n \varphi)\, (u)\, \partial_e u^{(n)}\, dx.$$

Thus, passing to the limit in (2.3), we find

$$\int_\Omega (\partial_e \eta) \varphi(u)\, dx = -\sum_{n=1}^N \int_\Omega \eta\, (\partial_n \varphi)\, (u)\, \partial_e u^{(n)}\, dx.$$

This proves the lemma. $\qquad \square$

Lemma 2.11 *Assume that $u \in L^1_{\text{loc}}(\Omega)$ has a weak derivative $\partial_e u \in L^1_{\text{loc}}(\Omega)$ in direction e. Then $|u|$ has a weak derivative in direction e and it is given by*

$$\partial_e |u| = \chi_{\{u \ne 0\}} \frac{\text{Re}\, (\overline{u} \partial_e u)}{|u|}.$$

In particular,

$$|\partial_e |u|| \le |\partial_e u| \qquad a.e. \ in \ \Omega .$$

Proof We apply Lemma 2.10 with $\varphi_\varepsilon(s_1, s_2) := \sqrt{s_1^2 + s_2^2 + \varepsilon^2} - \varepsilon$ to the pair of functions $(\operatorname{Re} u, \operatorname{Im} u)$. We obtain for any $\eta \in C_0^\infty(\Omega)$,

$$\int_\Omega (\partial_e \eta) \, \varphi_\varepsilon(\operatorname{Re} u, \operatorname{Im} u) \, dx = \int_\Omega \eta \frac{(\operatorname{Re} u)\partial_e \operatorname{Re} u + (\operatorname{Im} u)\partial_e \operatorname{Im} u}{\sqrt{|u|^2 + \varepsilon^2}} \, dx .$$

Since $\varphi_\varepsilon(\operatorname{Re} u, \operatorname{Im} u) \to |u|$ pointwise as $\varepsilon \to 0$ and $0 \le \varphi_\varepsilon(\operatorname{Re} u, \operatorname{Im} u) \le |u|$, dominated convergence gives

$$\int_\Omega (\partial_e \eta) \, \varphi_\varepsilon(\operatorname{Re} u, \operatorname{Im} u) \, dx \to \int_\Omega (\partial_e \eta) \, |u| \, dx .$$

Meanwhile, $\sqrt{|u|^2 + \varepsilon^2} \to |u|$ pointwise as $\varepsilon \to 0$ and

$$\left| \frac{(\operatorname{Re} u)\partial_e \operatorname{Re} u + (\operatorname{Im} u)\partial_e \operatorname{Im} u}{\sqrt{|u|^2 + \varepsilon^2}} \right| \le \sqrt{|\partial_e \operatorname{Re} u|^2 + |\partial_e \operatorname{Im} u|^2} ,$$

so again by dominated convergence,

$$\int_\Omega \eta \frac{(\operatorname{Re} u)\partial_e \operatorname{Re} u + (\operatorname{Im} u)\partial_e \operatorname{Im} u}{\sqrt{|u|^2 + \varepsilon^2}} \, dx$$
$$\to \int_\Omega \eta \chi_{\{u \ne 0\}} \frac{(\operatorname{Re} u)\partial_e \operatorname{Re} u + (\operatorname{Im} u)\partial_e \operatorname{Im} u}{|u|} \, dx .$$

Thus we have shown that

$$\int_\Omega (\partial_e \eta) \, |u| \, dx = - \int_\Omega \eta \chi_{\{u \ne 0\}} \frac{(\operatorname{Re} u)\partial_e \operatorname{Re} u + (\operatorname{Im} u)\partial_e \operatorname{Im} u}{|u|} \, dx .$$

This is the assertion since $(\operatorname{Re} u)\partial_e \operatorname{Re} u + (\operatorname{Im} u)\partial_e \operatorname{Im} u = \operatorname{Re}(\bar{u}\partial_e u)$. $\qquad\square$

Corollary 2.12 *Assume that $u, v \in L_{\text{loc}}^1(\Omega)$ are real and have weak derivatives $\partial_e u, \partial_e v \in L_{\text{loc}}^1(\Omega)$ in direction e. Then $\min\{u, v\}$ and $\max\{u, v\}$ have weak derivatives in direction e and they are given by*

$$\partial_e \min\{u, v\} = \chi_{\{u \ge v\}}\partial_e v + \chi_{\{u < v\}}\partial_e u = \chi_{\{u > v\}}\partial_e v + \chi_{\{u \le v\}}\partial_e u ,$$
$$\partial_e \max\{u, v\} = \chi_{\{u \ge v\}}\partial_e u + \chi_{\{u < v\}}\partial_e v = \chi_{\{u > v\}}\partial_e u + \chi_{\{u \le v\}}\partial_e v .$$

Proof We observe that for $a, b \in \mathbb{R}$,

$$\min\{a, b\} = \frac{1}{2}(a + b - |a - b|), \qquad \max\{a, b\} = \frac{1}{2}(a + b + |a - b|), \quad (2.4)$$

and therefore, by Lemma 2.11, $\min\{u,v\}$ and $\max\{u,v\}$ have weak derivatives:

$$\partial_e \min\{u,v\} = \chi_{\{u>v\}}\partial_e v + \chi_{\{u=v\}}\tfrac{1}{2}(\partial_e u + \partial_e v) + \chi_{\{u<v\}}\partial_e u,$$
$$\partial_e \max\{u,v\} = \chi_{\{u>v\}}\partial_e u + \chi_{\{u=v\}}\tfrac{1}{2}(\partial_e u + \partial_e v) + \chi_{\{u<v\}}\partial_e v.$$

Thus, it remains to show that $\partial_e u = \partial_e v$ almost everywhere on the set $\{u = v\}$. To prove this, we consider $w := \max\{u - v, 0\}$. Using again the second identity in (2.4), we infer from Lemma 2.11 that w has a weak derivative

$$\partial_e w = \chi_{\{u>v\}}(\partial_e u - \partial_e v) + \chi_{\{u=v\}}\tfrac{1}{2}(\partial_e u - \partial_e v). \tag{2.5}$$

Meanwhile, since $w \geq 0$, we have $w = |w|$ and so, by Lemma 2.11, we have

$$\partial_e w = \partial_e |w| = \chi_{\{w \neq 0\}}\frac{w\partial_e w}{|w|} = \chi_{\{w>0\}}\partial_e w = \chi_{\{u>v\}}\partial_e w. \tag{2.6}$$

Comparing (2.5) and (2.6) yields $\partial_e u = \partial_e v$ almost everywhere on the set $\{u = v\}$, as claimed. $\qquad\square$

Corollary 2.13 *Assume that $u \in L^1_{\mathrm{loc}}(\Omega)$ has a weak derivative $\partial_e u \in L^1_{\mathrm{loc}}(\Omega)$ in direction e. Then $\partial_e u = 0$ almost everywhere on $\{x \in \Omega \colon u(x) = 0\}$.*

Proof By writing $u = \operatorname{Re} u + i\operatorname{Im} u$, we see that it suffices to consider real u. In this case, the assertion follows from Corollary 2.12 with $v = 0$. $\qquad\square$

2.1.4 The product rule

In this subsection we prove the product rule for weakly differentiable functions.

Lemma 2.14 *Assume that $u, v \in L^1_{\mathrm{loc}}(\Omega)$ have weak derivatives $\partial_e u, \partial_e v \in L^1_{\mathrm{loc}}(\Omega)$ in direction e and that $uv, u\partial_e v + v\partial_e u \in L^1_{\mathrm{loc}}(\Omega)$. Then uv has a weak derivative in direction e and it is given by*

$$\partial_e(uv) = u\partial_e v + v\partial_e u.$$

Note that we do *not* require that $u\partial_e v$ and $v\partial_e u$ are individually in $L^1_{\mathrm{loc}}(\Omega)$; we only require this for their sum. Due to this rather weak assumption the proof is somewhat lengthy. For most applications, simpler versions proved in Steps 1 and 2 suffice.

Proof *Step 1.* We prove the lemma under the additional assumption $u, v \in L^\infty_{\mathrm{loc}}(\Omega)$.

To do so, we recall from Lemma 2.1 that there is a sequence $(u_n) \subset C^\infty(\Omega)$ such that $u_n \to u$ in $L^1_{\mathrm{loc}}(\Omega)$ and $\partial_e u_n \to \partial_e u$ in $L^1_{\mathrm{loc}}(\Omega)$. Passing to a subsequence, if necessary, we may assume that $u_n \to u$ almost everywhere. From the proof of Lemma 2.1 we see that the assumption $u \in L^\infty_{\mathrm{loc}}(\Omega)$ guarantees that the functions u_n satisfy $\sup_n \|u_n\|_{L^\infty(\Omega')} < \infty$ for any bounded, open set Ω' with $\overline{\Omega'} \subset \Omega$.

Let $\eta \in C_0^\infty(\Omega)$. Then, since $(\partial_e \eta)v$ is bounded and compactly supported in Ω, the L^1_{loc} convergence of u_n yields

$$\lim_{n\to\infty} \int_\Omega (\partial_e \eta) u_n v \, dx = \int_\Omega (\partial_e \eta) uv \, dx \, .$$

Moreover, since $u_n \eta$ is smooth, by the classical product rule, we have

$$\int_\Omega (\partial_e \eta) u_n v \, dx = \int_\Omega (\partial_e (\eta u_n)) v \, dx - \int_\Omega \eta (\partial_e u_n) v \, dx \, .$$

Concerning the first term on the right side, we use the fact that $\eta u_n \in C_0^\infty(\Omega)$ to obtain

$$\int_\Omega (\partial_e (\eta u_n)) v \, dx = - \int_\Omega \eta u_n (\partial_e v) \, dx \, .$$

Since $\eta(\partial_e v)$ is in $L^1(\Omega)$, and u_n converges almost everywhere and is uniformly bounded on the support of $\eta(\partial_e v)$, dominated convergence yields

$$\lim_{n\to\infty} \int_\Omega \eta u_n (\partial_e v) \, dx = \int_\Omega \eta u (\partial_e v) \, dx \, .$$

Finally, since $\eta v \in L^\infty(\Omega)$ with compact support, the $L^1_{\text{loc}}(\Omega)$ convergence of $\partial_e u_n$ implies

$$\lim_{n\to\infty} \int_\Omega \eta (\partial_e u_n) v \, dx = \int_\Omega \eta (\partial_e u) v \, dx \, .$$

Thus, we have shown that

$$\int_\Omega (\partial_e \eta) uv \, dx = - \int_\Omega \eta (u \partial_e v + v \partial_e u) \, dx \, ,$$

which proves weak differentiability and the claimed formula for $u, v \in L^\infty_{\text{loc}}(\Omega)$.

Step 2. We now drop the assumption that $v \in L^\infty_{\text{loc}}(\Omega)$ but still assume $u \in L^\infty_{\text{loc}}(\Omega)$.

Note that, in this case, the assumptions of the lemma simplify to $v \partial_e u \in L^1_{\text{loc}}(\Omega)$. Indeed, the assumption $uv \in L^1_{\text{loc}}(\Omega)$ is automatically satisfied and, since $u \partial_e v \in L^1_{\text{loc}}(\Omega)$, the assumption $u \partial_e v + v \partial_e u \in L^1_{\text{loc}}(\Omega)$ is equivalent to $v \partial_e u \in L^1_{\text{loc}}(\Omega)$.

First, assume, in addition, that v is real-valued. For $M > 0$, set $v_M := v$ on $\{|v| < M\}$ and $v_M := M$ on $\{|v| \geq M\}$. Since $v_M = \max\{\min\{v, M\}, -M\}$, it follows from Corollary 2.12 that the function v_M has a weak derivative $\partial_e v_M = \chi_{\{|v| < M\}} \partial_e v$.

Given $\eta \in C_0^\infty(\Omega)$, by Step 1 we have

$$\int_\Omega (\partial_e \eta) u v_M \, dx = -\int_\Omega \eta(u \partial_e v_M + v_M \partial_e u) \, dx \,.$$

By assumption and the preceding discussion, all three terms $(\partial_e \eta) uv$, $\eta u \partial_e v$, and $\eta v \partial_e u$ belong to $L^1(\Omega)$. The terms $(\partial_e \eta) u v_M$, $\eta u \partial_e v_M$, and $\eta v_M \partial_e u$ converge pointwise to these functions and are bounded, in absolute value, by their limits. Therefore, dominated convergence implies that

$$\int_\Omega (\partial_e \eta) uv \, dx = -\int_\Omega \eta(u \partial_e v + v \partial_e u) \, dx \,.$$

This proves the claim for real v and locally bounded u.

To obtain the result for complex v, we add the formulas that we just proved for its real and imaginary parts. Note that the assumption of the lemma, namely $v \partial_e u \in L_{\text{loc}}^1(\Omega)$, implies that $(\text{Re } v) \partial_e u, (\text{Im } v) \partial_e u \in L_{\text{loc}}^1(\Omega)$, so, indeed, the assumptions are satisfied for the two pairs $(u, \text{Re } v)$ and $(u, \text{Im } v)$.

Step 3. Finally, we will prove the lemma as stated without any additional assumptions. For $M, \varepsilon > 0$ we set

$$u_\varepsilon := \sqrt{|u|^2 + \varepsilon^2} \,, \qquad v_\varepsilon := \sqrt{|v|^2 + \varepsilon^2} \,, \qquad w_{M,\varepsilon} := \min\left\{ u_\varnothing, \frac{M}{v_\varepsilon} \right\} \frac{u}{u_\varepsilon} \,,$$

From Lemma 2.10 we know that u_ε and M/v_ε have weak derivatives

$$\partial_e u_\varepsilon = \frac{|u|}{u_\varepsilon} \partial_e |u| \qquad \text{and} \qquad \partial_e \frac{M}{v_\varepsilon} = -\frac{M|v|}{v_\varepsilon^3} \partial_e |v| \,.$$

Thus, by Corollary 2.12, $\min\{u_\varepsilon, M/v_\varepsilon\}$ has a weak derivative

$$\partial_e \min\left\{ u_\varepsilon, \frac{M}{v_\varepsilon} \right\} = \chi_{\{u_\varepsilon v_\varepsilon < M\}} \frac{|u|}{u_\varepsilon} \partial_e |u| - \chi_{\{u_\varepsilon v_\varepsilon \geq M\}} \frac{M|v|}{v_\varepsilon^3} \partial_e |v| \,.$$

The function $1/u_\varepsilon$ is bounded and, again by Lemma 2.10, it has a weak derivative given by $-|u|(\partial_e |u|)/u_\varepsilon^3$. In particular, $|u \partial_e (1/u_\varepsilon)| \leq u_\varepsilon^{-1} |\partial_e |u|| \in L_{\text{loc}}^1(\Omega)$. Therefore, by Step 2, u/u_ε has a weak derivative

$$\partial_e \frac{u}{u_\varepsilon} = u_\varepsilon^{-1} \partial_e u + u \partial_e u_\varepsilon^{-1} = u_\varepsilon^{-1} \partial_e u - \frac{u|u|}{u_\varepsilon^3} \partial_e |u| \,.$$

Note $M/v_\varepsilon \leq M\varepsilon^{-1}$ and $|u|/u_\varepsilon \leq 1$. Therefore, by Step 1, $w_{M,\varepsilon}$ has a weak derivative

$$\partial_e w_{M,\varepsilon}$$

$$= \frac{u}{u_\varepsilon} \partial_e \min\left\{u_\varepsilon, \frac{M}{v_\varepsilon}\right\} + \min\left\{u_\varepsilon, \frac{M}{v_\varepsilon}\right\} \partial_e \frac{u}{u_\varepsilon}$$

$$= \chi_{\{u_\varepsilon v_\varepsilon < M\}} \partial_e u + \chi_{\{u_\varepsilon v_\varepsilon \geq M\}} \left(-\frac{Mu|v|\partial_e|v|}{v_\varepsilon^3 u_\varepsilon} + \frac{M\partial_e u}{u_\varepsilon v_\varepsilon} - \frac{Mu|u|\partial_e|u|}{u_\varepsilon^3 v_\varepsilon}\right).$$

The inequalities $M/v_\varepsilon \leq M\varepsilon^{-1}$ and $|u|/u_\varepsilon \leq 1$ imply that $w_{M,\varepsilon}$ is bounded. Therefore, by Step 2, the product $w_{M,\varepsilon} v$ has a weak derivative given by

$$\partial_e(w_{M,\varepsilon} v) = \chi_{\{u_\varepsilon v_\varepsilon < M\}} (u\partial_e v + v\partial_e u) + \chi_{\{u_\varepsilon v_\varepsilon \geq M\}} \left(A + \varepsilon^2 B + \varepsilon^4 C\right),$$

where

$$A := \frac{M|u|^2|v|^2}{2u_\varepsilon^3 v_\varepsilon^3} (u\partial_e v + v\partial_e u) - \frac{Mu^2 v^2}{2u_\varepsilon^3 v_\varepsilon^3}\overline{(u\partial_e v + v\partial_e u)},$$

$$B := \frac{M}{u_\varepsilon^3 v_\varepsilon^3}\left(\left(|u|^2 + |v|^2\right)(u\partial_e v + v\partial_e u) - uv \operatorname{Re}\left(\overline{u}\partial_e u + \overline{v}\partial_e v\right)\right),$$

$$C := \frac{M}{u_\varepsilon^3 v_\varepsilon^3} (u\partial_e v + v\partial_e u).$$

Here we also used Lemma 2.11 to rewrite $|u|\partial_e|u|$ and $|v|\partial_e|v|$. On the set $\{u_\varepsilon v_\varepsilon \geq M\}$ we can bound

$$|A| \leq |u\partial_e v + v\partial_e u|,$$

$$\varepsilon^2 |B| \leq 2|u\partial_e v + v\partial_e u| + \varepsilon\left(|\partial_e u| + |\partial_e v|\right),$$

$$\varepsilon^4 |C| \leq |u\partial_e v + v\partial_e u|.$$

Thus, if $\eta \in C_0^\infty(\Omega)$, we have

$$\int_\Omega (\partial_e \eta) w_{M,\varepsilon} v \, dx = -\int_\Omega \chi_{\{u_\varepsilon v_\varepsilon < M\}} \eta (u\partial_e v + v\partial_e u) \, dx$$

$$- \int_\Omega \chi_{\{u_\varepsilon v_\varepsilon \geq M\}} \eta\left(A + \varepsilon^2 B + \varepsilon^4 C\right) dx.$$

Since

$$w_{M,\varepsilon} v = \chi_{\{u_\varepsilon v_\varepsilon < M\}} uv + \chi_{\{u_\varepsilon v_\varepsilon \geq M\}} \frac{M}{u_\varepsilon v_\varepsilon} uv$$

converges pointwise to uv as $M \to \infty$ and is bounded, in absolute value, by $|uv| \in L^1_{\text{loc}}(\Omega)$, dominated convergence shows that the left side converges to $\int_\Omega (\partial_e \eta) uv \, dx$ as $M \to \infty$. Moreover, according to the above bounds, $\eta\left(A + \varepsilon^2 B + \varepsilon^4 C\right)$ is bounded by an L^1 function independent of M. Therefore, by dominated convergence, the right side converges to $-\int_\Omega \eta(u\partial_e v + v\partial_e u)\,dx$ as $M \to \infty$. This completes the proof. $\qquad \square$

Corollary 2.15 *Assume that $u \in L^1_{\text{loc}}(\Omega)$ has a weak derivative $\partial_e u \in L^1_{\text{loc}}(\Omega)$ in direction e and, for $M > 0$, set*

$$u^{(M)} := \chi_{\{|u|<M\}} u + \chi_{\{|u|\ge M\}} \frac{Mu}{|u|}.$$

Then $u^{(M)}$ has a weak derivative $\partial_e u^{(M)} \in L^1_{\text{loc}}(\Omega)$ in direction e and it is given by

$$\partial_e u^{(M)} = \chi_{\{|u|<M\}} \partial_e u + \chi_{\{|u|\ge M\}} \left(\frac{M\partial_e u}{|u|} - \frac{Mu\partial_e|u|}{|u|^2} \right).$$

Proof For $\varepsilon > 0$ let

$$u_\varepsilon := \sqrt{|u|^2 + \varepsilon^2}, \qquad w_\varepsilon := \min\{u_\varepsilon, M\}\frac{u}{u_\varepsilon}.$$

Note that w_ε coincides with the function $w_{M\varepsilon,\varepsilon}$ from the third step of the proof of Lemma 2.14 with $v = 0$. In particular, $M\varepsilon$ plays the role of the parameter M in the previous proof. As shown in that proof, w_ε has a weak derivative

$$\partial_e w_\varepsilon = \chi_{\{u_\varepsilon<M\}} \partial_e u + \chi_{\{u_\varepsilon\ge M\}} \left(\frac{M\partial_e u}{u_\varepsilon} - \frac{Mu|u|\partial_e|u|}{u_\varepsilon^3} \right);$$

that is, for any $\eta \in C_0^\infty(\Omega)$,

$$\int_\Omega (\partial_e\eta)w_\varepsilon\,dx = -\int_{\{u_\varepsilon<M\}} \eta\partial_e u\,dx - \int_{\{u_\varepsilon\ge M\}} \eta\left(\frac{M\partial_e u}{u_\varepsilon} - \frac{Mu|u|\partial_e|u|}{u_\varepsilon^3} \right)dx.$$

We now pass to the limit $\varepsilon \to 0$. We have, in the sense of pointwise convergence, $u_\varepsilon \to |u|$, $\chi_{\{u_\varepsilon<M\}} \to \chi_{\{|u|<M\}}$, $\chi_{\{u_\varepsilon\ge M\}} \to \chi_{\{|u|\ge M\}}$ and $w_\varepsilon \to u^{(M)}$. Moreover, $1/u_\varepsilon \le 1/M$ and $|u|^2/u_\varepsilon^3 \le 1/M$ on $\{u_\varepsilon \ge M\}$, and, further, $|w_\varepsilon| \le |u^{(M)}|$. Therefore, dominated convergence implies that

$$\int_\Omega (\partial_e\eta)u^{(M)}\,dx = -\int_{\{|u|<M\}} \eta\partial_e u\,dx - \int_{\{|u|\ge M\}} \eta\left(\frac{M\partial_e u}{|u|} - \frac{Mu\partial_e|u|}{|u|^2} \right)dx,$$

which completes the proof of the corollary. $\qquad \square$

2.1.5 Separation of variables

Throughout this subsection, we assume that $d \ge 2$. We will discuss separation of variables for weakly differentiable functions and begin with the case of Cartesian coordinates.

Lemma 2.16 *Let $I \subset \mathbb{R}$ and $\omega \subset \mathbb{R}^{d-1}$ be open. Let $\Omega := I \times \omega$ and $u \in L^1_{loc}(\Omega)$. Then u has a weak partial derivative $\partial_1 u \in L^1_{loc}(\Omega)$ if and only if, for almost every $x' \in \omega$, the function $x_1 \mapsto u(x_1, x')$ is weakly differentiable on I and its derivative $(x_1, x') \mapsto (u(\cdot, x'))'(x_1)$ belongs to $L^1_{loc}(\Omega)$. In this case, $(\partial_1 u)(x) = (u(\cdot, x'))'(x_1)$ for a.e. $x \in \Omega$.*

Proof First assume that u has a weak partial derivative $\partial_1 u \in L^1_{loc}(\Omega)$. Then, by Lemma 2.1, there is a sequence $(u_n) \subset C^\infty(\Omega)$ such that $u_n \to u$ in $L^1_{loc}(\Omega)$ and $\partial_1 u_n \to \partial_1 u$ in $L^1_{loc}(\Omega)$. Fix an open set $I' \subset \mathbb{R}$ such that $\overline{I'} \subset I$ and let

$$f_n(x') := \int_{I'} (|u_n(x_1, x') - u(x_1, x')| + |\partial_1 u_n(x_1, x') - \partial_1 u(x_1, x')|)\, dx_1 \,.$$

Then, by Fubini's theorem, $f_n \in L^1_{loc}(\omega)$ and $f_n \to 0$ in $L^1_{loc}(\omega)$. Thus, there is a subsequence (f_{n_k}) such that, for all x' from a subset of ω of full measure, one has $f_{n_k}(x') \to 0$. That is, for all x' from this subset, one has $u_{n_k}(\cdot, x') \to u(\cdot, x')$ in $L^1(I')$ and $\partial_1 u_{n_k}(\cdot, x') \to \partial_1 u(\cdot, x')$ in $L^1(I')$. By Lemma 2.3, this implies that $u(\cdot, x')$ is weakly differentiable on I' and $u(\cdot, x')' = \partial_1 u(\cdot, x')$. Since I' is arbitrary, this proves the claim.

Conversely, assume that for almost every $x' \in \omega$ the function $x_1 \mapsto u(x_1, x')$ is weakly differentiable on I and its derivative $(x_1, x') \mapsto (u(\cdot, x'))'(x_1)$ belongs to $L^1_{loc}(\Omega)$. Let $\eta \in C^\infty_0(\Omega)$. Then, by Fubini's theorem,

$$\int_\Omega u \partial_1 \eta \, dx = \int_\omega \int_I u(x_1, x')(\partial_1 \eta)(x_1, x') \, dx_1 \, dx' \,.$$

By assumption, for almost every $x' \in \omega$,

$$\int_I u(x_1, x')(\partial_1 \eta)(x_1, x') \, dx_1 = - \int_I (u(\cdot, x'))'(x_1)\eta(x_1, x') \, dx_1 \,.$$

By Fubini's theorem and the L^1_{loc}-assumption, we have

$$\int_\omega \int_I (u(\cdot, x'))'(x_1)\eta(x_1, x') \, dx_1 \, dx' = \int_\Omega (u(\cdot, x'))'(x_1)\eta(x) \, dx \,.$$

This proves the existence of $\partial_1 u$ and the formula for it. $\square$

Remark 2.17 Note that, while the set of full measure in the lemma depends on the choice of a pointwise representative of its a.e. equivalence class in $L^1_{loc}(\Omega)$, the condition itself does not. Indeed, if u_1 and u_2 are two pointwise representatives of the same class, then, by Fubini's theorem, there is a subset of ω of full measure such that, for x' from this set, the functions $x_1 \mapsto u_1(x_1, x')$ and $x_1 \mapsto u_2(x_1, x')$ coincide almost everywhere on I and therefore they are weakly differentiable simultaneously and their weak derivatives coincide almost everywhere on I.

This lemma implies the following characterization of weak differentiability.

Corollary 2.18 *Let $I \subset \mathbb{R}$ be an open interval and let $u \in L^1_{loc}(I^d)$. Then u is weakly differentiable on I^d if and only if, for all $j = 1, \ldots, d$ and for almost every $(x_1, \ldots, x_{j-1}, x_{j+1}, \ldots, x_d) \in I^{d-1}$, the function $x_j \mapsto u(x_1, \ldots, x_d)$ is weakly differentiable on I and its derivative belongs to $L^1_{loc}(I^d)$. In this case, for all $j = 1, \ldots, d$, $(\partial_j u)(x) = (u(x_1, \ldots, x_{j-1}, \cdot, x_{j+1}, \ldots, x_d))'(x_j)$ for a.e. $x \in I^d$.*

Next, we turn to separation of variables in spherical coordinates. Our goal is to extend the formula

$$\nabla u = \omega\, \partial_r u + r^{-1} \nabla_{\mathbb{S}^{d-1}} u, \qquad x = r\omega,\ r > 0,\ \omega \in \mathbb{S}^{d-1},$$

valid for classically differentiable functions to weakly differentiable ones. We will need to introduce weak angular derivatives.

Before doing so, we recall several equivalent ways to define (classical) differentiability of a function u on $\mathbb{S}^{d-1}$. One is to extend u smoothly in the radial direction to a function on an open neighborhood of $\mathbb{S}^{d-1}$ in $\mathbb{R}^d$ and call u differentiable if its extension is. This notion can be shown to be independent of the chosen extension. Moreover, $\nabla_{\mathbb{S}^{d-1}} u$ can be defined as the tangential projection of ∇u. Another way is to introduce spherical coordinates on $\mathbb{S}^{d-1}$ and to define differentiability and the gradient in terms of these coordinates. More generally, one can define differentiability using abstract local coordinates by viewing $\mathbb{S}^{d-1}$ as a manifold. The gradient $\nabla_{\mathbb{S}^{d-1}}$ coincides with the covariant derivative in the sense of differential geometry.

We denote by $C^m(\mathbb{S}^{d-1})$ the space of functions on $\mathbb{S}^{d-1}$ that are continuous together with their derivatives up to order $m \in \mathbb{N}_0$, and set $C^\infty(\mathbb{S}^{d-1}) = \bigcap_{m=0}^\infty C^m(\mathbb{S}^{d-1})$. Moreover, we denote by $d\omega$ the surface measure on $\mathbb{S}^{d-1}$ induced by its canonical embedding into $\mathbb{R}^d$. A simple computation shows that, for $v \in C^1(\mathbb{S}^{d-1})$,

$$\int_{\mathbb{S}^{d-1}} \nabla_{\mathbb{S}^{d-1}} v\, d\omega = (d-1) \int_{\mathbb{S}^{d-1}} \omega v\, d\omega.$$

Applying this to the product $v = u\eta$ for $u, \eta \in C^1(\mathbb{S}^{d-1})$ and using the product rule $\nabla_{\mathbb{S}^{d-1}} v = u\nabla_{\mathbb{S}^{d-1}}\eta + \eta\nabla_{\mathbb{S}^{d-1}} u$, we see that

$$\int_{\mathbb{S}^{d-1}} u\nabla_{\mathbb{S}^{d-1}}\eta\, d\omega = -\int_{\mathbb{S}^{d-1}} (\nabla_{\mathbb{S}^{d-1}} u)\eta\, d\omega + (d-1)\int_{\mathbb{S}^{d-1}} \omega u\eta\, d\omega.$$

This motivates the following definition.

A function $u \in L^1(\mathbb{S}^{d-1})$ is called *weakly differentiable* on $\mathbb{S}^{d-1}$ if there is an $F \in L^1(\mathbb{S}^{d-1}, \mathbb{C}^d)$ with $\omega \cdot F(\omega) = 0$ for a.e. $\omega \in \mathbb{S}^{d-1}$ such that, for all

$\eta \in C^\infty(\mathbb{S}^{d-1})$,

$$\int_{\mathbb{S}^{d-1}} u \nabla_{\mathbb{S}^{d-1}} \eta \, d\omega = - \int_{\mathbb{S}^{d-1}} F\eta \, d\omega + (d-1) \int_{\mathbb{S}^{d-1}} \omega u \eta \, d\omega.$$

In this case, F is uniquely determined almost everywhere and we write $F =: \nabla_{\mathbb{S}^{d-1}} u$. As discussed, this notation extends the classical one.

Lemma 2.19 *Let $u \in L^1(\mathbb{S}^{d-1})$ and let $(u_n) \subset L^1(\mathbb{S}^{d-1})$ with $u_n \to u$ in $L^1(\mathbb{S}^{d-1})$. If each u_n is weakly differentiable and if $\nabla_{\mathbb{S}^{d-1}} u_n \to F$ in $L^1(\mathbb{S}^{d-1}, \mathbb{C}^d)$ for some $F \in L^1(\mathbb{S}^{d-1}, \mathbb{C}^d)$, then u is weakly differentiable and $\nabla_{\mathbb{S}^{d-1}} u = F$.*

The proof of this lemma is the same as that of Lemma 2.3 and is omitted.

We use the following notation in this subsection. If $I \subset \mathbb{R}_+$ is open, we set

$$A_I := \left\{ x \in \mathbb{R}^d : |x| \in I \right\}.$$

This is an open set and we note that $0 \notin A_I$.

Lemma 2.20 *Let $I \subset \mathbb{R}_+$ be open and $u \in L^1_{\text{loc}}(A_I)$. Then u is weakly differentiable in A_I if and only if*

(a) *for almost every $\omega \in \mathbb{S}^{d-1}$, $r \mapsto u(r\omega)$ is weakly differentiable on I and $x \mapsto \left(u(\cdot \frac{x}{|x|}) \right)'(|x|)$ belongs to $L^1_{\text{loc}}(A_I)$;*

(b) *for almost every $r \in I$, $\omega \mapsto u(r\omega)$ is weakly differentiable on $\mathbb{S}^{d-1}$ and $x \mapsto \left(\nabla_{\mathbb{S}^{d-1}} u(|x| \cdot) \right)\left(\frac{x}{|x|} \right)$ belongs to $L^1_{\text{loc}}(A_I)$.*

In this case,

$$(\nabla u)(x) = \left(u(\cdot \tfrac{x}{|x|}) \right)'(|x|) \tfrac{x}{|x|} + |x|^{-1} \left(\nabla_{\mathbb{S}^{d-1}} u(|x| \cdot) \right)\left(\tfrac{x}{|x|} \right) \quad a.e. \ x \in A_I.$$

Proof First, assume that u is weakly differentiable in A_I. Then, by Remark 2.2, there is a sequence $(u_n) \subset C^\infty(A_I)$ such that $u_n \to u$ in $L^1_{\text{loc}}(A_I)$ and $\nabla u_n \to \nabla u$ in $L^1_{\text{loc}}(A_I, \mathbb{C}^d)$. Fix an open set $I' \subset \mathbb{R}_+$ such that $\overline{I'} \subset I$. Then, by an argument similar to that in the proof of Lemma 2.16, there is a subsequence (u_{n_k}) such that, for all r from a subset of $\mathbb{R}_+$ of full measure, one has $u_{n_k}(r \cdot) \to u(r \cdot)$ in $L^1(\mathbb{S}^{d-1})$ and $\nabla u_{n_k}(r \cdot) \to \nabla u(r \cdot)$ in $L^1(\mathbb{S}^{d-1})$ and such that, for all ω from a subset of $\mathbb{S}^{d-1}$ of full measure, one has $u_{n_k}(\cdot \ \omega) \to u(\cdot \ \omega)$ in $L^1(I')$ and $\nabla u_{n_k}(\cdot \ \omega) \to \nabla u(\cdot \ \omega)$ in $L^1(I')$. By the orthogonality of the radial and the angular derivatives, this implies, in particular, for r and ω in the respective full measure sets,

$$\nabla_{\mathbb{S}^{d-1}} u_{n_k}(r \cdot) \to r((\nabla u)(r \cdot) - (\cdot)(\cdot) \cdot (\nabla u)(r \cdot)) \text{ in } L^1(\mathbb{S}^{d-1})$$

and

$$(u_{n_k}(\cdot \ \omega))' \to \omega \cdot (\nabla u)(\cdot \ \omega) \text{ in } L^1(I').$$

By Lemma 2.3 for I and Lemma 2.19 for $\mathbb{S}^{d-1}$, we conclude that for r and ω from the respective full measure sets, $u(r \cdot)$ is weakly differentiable on $\mathbb{S}^{d-1}$ and $u(\cdot \omega)$ is weakly differentiable on I' and

$$\nabla_{\mathbb{S}^{d-1}} u(r \cdot) = r((\nabla u)(r \cdot) - (\cdot)(\cdot) \cdot (\nabla u)(r \cdot)) \quad \text{and} \quad (u(\cdot \omega))' = \omega \cdot (\nabla u)(\cdot \omega).$$

This implies the claimed formula for ∇u and, since this gradient is by assumption in $L^1(A_{I'})$, also the claimed bound.

Conversely, assume now that (a) and (b) are satisfied. Let $\eta \in C_0^\infty(A_I)$. Decomposing its gradient in radial and angular parts, we have, by Fubini's theorem,

$$\int_{A_I} u\nabla \eta \, dx = \int_{\mathbb{S}^{d-1}} \int_I u(r\omega)((\eta(\cdot \omega))')(r) \, r^{d-1} \, dr \, \omega \, d\omega$$
$$+ \int_I \int_{\mathbb{S}^{d-1}} u(r\omega)(\nabla_{\mathbb{S}^{d-1}} \eta(r \cdot))(\omega) \, d\omega \, r^{d-2} \, dr \,.$$

By the assumed weak differentiability in I and the product rule, we have, for almost every $\omega \in \mathbb{S}^{d-1}$,

$$\int_I u(r\omega)((\eta(\cdot \omega))')(r) \, r^{d-1} \, dr$$
$$= - \int_I ((u(\cdot \omega))'(r) + (d-1)r^{-1}u(r\omega))\eta(r\omega) \, r^{d-1} \, dr$$

and, by the assumed weak differentiability on $\mathbb{S}^{d-1}$, we have, for almost every $r \in \mathbb{R}_+$,

$$\int_{\mathbb{S}^{d-1}} u(r\omega)(\nabla_{\mathbb{S}^{d-1}} \eta(r \cdot))(\omega) \, d\omega$$
$$= - \int_{\mathbb{S}^{d-1}} ((\nabla_{\mathbb{S}^{d-1}} u(r \cdot))(\omega) - (d-1)\omega u(r\omega))\eta(r\omega) \, d\omega \,.$$

By Fubini's theorem and the integrability assumptions, we deduce that

$$\int_{\mathbb{S}^{d-1}} \int_I u(r\omega)(\eta(\cdot \omega)')(r) \, r^{d-1} \, dr \, \omega \, d\omega$$
$$+ \int_I \int_{\mathbb{S}^{d-1}} u(r\omega)(\nabla_{\mathbb{S}^{d-1}} \eta(r \cdot))(\omega) \, d\omega \, r^{d-2} \, dr$$
$$= - \int_{A_I} ((u(\cdot \tfrac{x}{|x|}))'(|x|) + |x|^{-1}(\nabla_{\mathbb{S}^{d-1}} u(|x| \cdot))(\tfrac{x}{|x|}))\eta(x) \, dx \,.$$

This proves that u is weakly differentiable in A_I. $\qquad\square$

As in Remark 2.17, while the sets of full measure in the lemma depend on the choice of a pointwise representative of its a.e. equivalence class in $L_{\text{loc}}^1(A_I)$, the conditions (a) and (b) themselves do not.

In the following, for simplicity, we write in the situation of Lemma 2.20

$$(\partial_r u)(x) := \left(u\left(\cdot \, \tfrac{x}{|x|}\right)\right)'(|x|) \quad \text{and} \quad (\nabla_{\mathbb{S}^{d-1}} u)(x) := \left(\nabla_{\mathbb{S}^{d-1}} u(|x| \cdot \,)\right)\left(\tfrac{x}{|x|}\right).$$

Corollary 2.21 *Let $u \in L^1(\mathbb{S}^{d-1})$ be weakly differentiable. Then there is a sequence $(u_n) \subset C^\infty(\mathbb{S}^{d-1})$ such that $u_n \to u$ in $L^1(\mathbb{S}^{d-1})$ and $\nabla_{\mathbb{S}^{d-1}} u_n \to \nabla_{\mathbb{S}^{d-1}} u$ in $L^1(\mathbb{S}^{d-1}, \mathbb{C}^d)$.*

Proof Let $\tilde{u}(x) := u(x/|x|)$ for $x \in \mathbb{R}^d \setminus \{0\}$. Then, by Lemma 2.20, $\tilde{u}$ is weakly differentiable in $\mathbb{R}^d \setminus \{0\}$ and $(\nabla \tilde{u})(x) = |x|^{-1}(\nabla_{\mathbb{S}^{d-1}} u)(x/|x|)$ for a.e. $x \in \mathbb{R}^d \setminus \{0\}$. By Remark 2.2, there is a sequence $(u_n) \subset C^\infty(\mathbb{R}^d \setminus \{0\})$ such that $u_n \to \tilde{u}$ in $L^1_{\mathrm{loc}}(\mathbb{R}^d \setminus \{0\})$ and $\nabla u_n \to |x|^{-1}(\nabla_{\mathbb{S}^{d-1}} u)(x/|x|)$ in $L^1_{\mathrm{loc}}(\mathbb{R}^d \setminus \{0\})$. The latter implies, in particular, that $\nabla_{\mathbb{S}^{d-1}} u_n \to (\nabla_{\mathbb{S}^{d-1}} u)(x/|x|)$ in $L^1_{\mathrm{loc}}(\mathbb{R}^d \setminus \{0\})$. Arguing as in the proof of Lemma 2.16, there is a subsequence (u_{n_k}) such that, for all r from a subset of $\mathbb{R}_+$ of full measure, one has $u_{n_k}(r \,\cdot\,) \to u$ in $L^1(\mathbb{S}^{d-1})$ and $\nabla_{\mathbb{S}^{d-1}} u_{n_k}(r \,\cdot\,) \to \nabla_{\mathbb{S}^{d-1}} u$ in $L^1(\mathbb{S}^{d-1})$. Fixing an r from this set, we have found a sequence $(u_{n_k}(r \,\cdot\,)) \subset C^\infty(\mathbb{S}^{d-1})$ with the desired properties. □

2.1.6 Weak Laplacian

So far, we have considered weak derivatives of first order. The notion of weak derivatives can be generalized to higher-order derivatives, but the only instance of this generalization that we need is the following. Let $u \in L^1_{\mathrm{loc}}(\Omega)$. We say that $v \in L^1_{\mathrm{loc}}(\Omega)$ is a *weak Laplacian* of u if

$$\int_\Omega u \Delta \eta \, dx = \int_\Omega v \eta \, dx \qquad \text{for all } \eta \in C_0^\infty(\Omega).$$

Just as for weak directional derivatives, the weak Laplacian of u, if it exists, is unique almost everywhere, and we denote it by Δu. By integration by parts, we see that this definition extends the classical definition of the Laplacian for $u \in C^2(\Omega)$.

Before introducing the weak Laplace–Beltrami operator on the sphere $\mathbb{S}^{d-1}$, where $d \geq 2$, we briefly recall its classical definition for C^2 functions. Indeed there are at least three different and equivalent ways. First, it can be defined by extending functions on $\mathbb{S}^{d-1}$ to functions on $\mathbb{R}^d$ and using the formula

$$\Delta_{\mathbb{R}^d} = \partial_r^2 + (d-1)r^{-1}\partial_r + r^{-2}\Delta_{\mathbb{S}^{d-1}}.$$

Second, it can be expressed explicitly in spherical coordinates. Third, it can be defined in the sense of Riemannian geometry using local coordinates on $\mathbb{S}^{d-1}$. Let $u \in L^1(\mathbb{S}^{d-1})$. We say that $v \in L^1(\mathbb{S}^{d-1})$ is a *weak (spherical) Laplacian* of u if

$$\int_{\mathbb{S}^{d-1}} u \Delta_{\mathbb{S}^{d-1}} \eta \, d\omega = \int_{\mathbb{S}^{d-1}} v \eta \, d\omega \qquad \text{for all } \eta \in C^\infty(\mathbb{S}^{d-1}).$$

The weak (spherical) Laplacian of u, if it exists, is unique almost everywhere and we denote it by $\Delta_{\mathbb{S}^{d-1}} u$. In view of the classical integration by parts formula

$$\int_{\mathbb{S}^{d-1}} u \, \Delta_{\mathbb{S}^{d-1}} v \, d\omega = \int_{\mathbb{S}^{d-1}} (\Delta_{\mathbb{S}^{d-1}} u) \, v \, d\omega \quad \text{for all } u, v \in C^2(\mathbb{S}^{d-1}),$$

this definition extends the classical definition for C^2 functions.

2.2 Sobolev spaces

2.2.1 The Sobolev space $H^1(\Omega)$

Let $\Omega \subset \mathbb{R}^d$ be an open set. We denote by

$$H^1(\Omega)$$

the space of all (a.e. equivalence classes of) functions $u \in L^2(\Omega)$ that are weakly differentiable with $\partial_j u \in L^2(\Omega)$ for any $j = 1, \ldots, d$. This space is called the *Sobolev space* (of order 1 and with integrability exponent 2) on Ω. By definition, for every $u \in H^1(\Omega)$, the quantity

$$\|u\|_{H^1(\Omega)} := \left(\|u\|_{L^2(\Omega)}^2 + \sum_{j=1}^d \|\partial_j u\|_{L^2(\Omega)}^2 \right)^{1/2}$$

is finite, and it is easy to see that it defines a norm on $H^1(\Omega)$. This norm comes from the inner product

$$\langle u, v \rangle_{H^1(\Omega)} := (u, v)_{L^2(\Omega)} + \sum_{j=1}^d (\partial_j u, \partial_j v)_{L^2(\Omega)}.$$

Let us show that the Sobolev space is a separable Hilbert space.

Lemma 2.22 *Let $\Omega \subset \mathbb{R}^d$ be open. Then $H^1(\Omega)$ is complete and separable.*

Proof To prove completeness, let $(u_n) \subset H^1(\Omega)$ be Cauchy with respect to the norm $\| \cdot \|_{H^1(\Omega)}$. Then (u_n) and each $(\partial_j u_n)$, $j = 1, \ldots, d$, are Cauchy with respect to the norm $\| \cdot \|_{L^2(\Omega)}$ and therefore converge to u and v_j, respectively, in $L^2(\Omega)$. By Lemma 2.3 and the fact that $L^2(\Omega)$ convergence implies $L^1_{\text{loc}}(\Omega)$ convergence, we have $v_j = \partial_j u$ for all j. Thus, $u_n \to u$ in $H^1(\Omega)$, proving completeness.

The separability of $H^1(\Omega)$ follows similarly from that of $L^2(\Omega) \oplus \cdots \oplus L^2(\Omega)$ ($d + 1$ terms), which we can consider to contain $H^1(\Omega)$ as a subset. Then we use the fact that any subset of a separable metric space is separable. $\square$

The next proposition, due to Meyers and Serrin (1964), says that sufficiently nice functions are dense in $H^1(\Omega)$. We recall that the space $C^\infty(\Omega)$ consists of all functions that are infinitely often differentiable in Ω. Note that in this definition no restriction is imposed on the boundary behavior of these functions and their derivatives.

Proposition 2.23 *Let $\Omega \subset \mathbb{R}^d$ be open. Then the set*

$$\{u \in H^1(\Omega) : u \in C^\infty(\Omega) \cap L^\infty(\Omega), \text{ supp } u \text{ is bounded}\}$$

is dense in $H^1(\Omega)$.

Proof *Step 1.* We show that $H^1(\Omega) \cap L^\infty(\Omega)$ is dense in $H^1(\Omega)$.

Given $u \in H^1(\Omega)$ and $M > 0$, we define

$$u^{(M)} := \begin{cases} u & \text{if } |u| < M, \\ \dfrac{Mu}{|u|} & \text{if } |u| \geq M. \end{cases}$$

We recall from Corollary 2.15 that $u^{(M)}$ is weakly differentiable with gradient

$$\nabla u^{(M)} = \begin{cases} \nabla u & \text{if } |u| < M, \\ \dfrac{M\nabla u}{|u|} - \dfrac{Mu\nabla|u|}{|u|^2} & \text{if } |u| \geq M. \end{cases}$$

Since, by Lemma 2.11, $|u|$ is weakly differentiable with gradient satisfying $|\nabla|u|| \leq |\nabla u|$, we infer that $|u| \in H^1(\Omega)$ and also that $u^{(M)} \in H^1(\Omega)$. The above formula for $\nabla u^{(M)}$ implies $\nabla u^{(M)} \to \nabla u$ almost everywhere as $M \to \infty$ and

$$\left|\nabla\left(u^{(M)} - u\right)\right| = \chi_{\{|u| \geq M\}} \left|\left(\frac{M}{|u|} - 1\right)\nabla u - \frac{Mu}{|u|^2}\nabla|u|\right|$$

$$\leq \chi_{\{|u| \geq M\}} (|\nabla u| + |\nabla|u||),$$

so dominated convergence implies $\nabla u^{(M)} \to \nabla u$ in $L^2(\Omega)$ as $M \to \infty$. Moreover, $u^{(M)}$ is bounded and $u^{(M)} \to u$ in $L^2(\Omega)$ as $M \to \infty$.

Step 2. We show that

$$\{u \in H^1(\Omega) : u \in L^\infty(\Omega), \text{ supp } u \text{ is bounded}\}$$

is dense in $H^1(\Omega)$. Thanks to Step 1, only $u \in H^1(\Omega) \cap L^\infty(\Omega)$ needs to be approximated. Let $0 \leq \chi \leq 1$ be a compactly supported smooth function that is equal to 1 in a neighborhood of the origin. We put $u_R := \chi(\cdot/R)u$. Then $u_R \in H^1(\Omega)$ and

$$\nabla u_R = \chi(\cdot/R)\nabla u + R^{-1}u(\nabla\chi)(\cdot/R).$$

Therefore $\nabla u_R \to \nabla u$ pointwise as $R \to \infty$ and

$$|\nabla(u_R - u)| \le (1 - \chi(\cdot/R))|\nabla u| + R^{-1}\|\nabla\chi\|_\infty|u|,$$

so dominated convergence implies $\nabla u_R \to \nabla u$ in $L^2(\Omega)$ as $R \to \infty$. Moreover, u_R is bounded with bounded support and $u_R \to u$ in $L^2(\Omega)$ as $R \to \infty$.

Step 3. We complete the proof of the proposition. According to Step 2, we only need to approximate $u \in H^1(\Omega) \cap L^\infty(\Omega)$ with bounded support. Let $(\Omega_k)_{k \in \mathbb{N}}$ be an increasing sequence of bounded open sets such that $\overline{\Omega_k} \subset \Omega_{k+1}$ and $\bigcup_k \Omega_k = \Omega$. We set $\Omega_0 := \Omega_{-1} := \Omega_{-2} := \emptyset$. It is well known (Adams and Fournier, 2003, Theorem 3.15) that there are non-negative smooth functions χ_k such that

$$\sum_{k \in \mathbb{N}_0} \chi_k = 1 \quad \text{in } \Omega$$

and

$$\operatorname{supp} \chi_k \subset \Omega_{k+1} \setminus \Omega_{k-1} \quad \text{for all } k \in \mathbb{N}_0.$$

Let $0 \le \omega \in C_0^\infty(\mathbb{R}^d)$ with $\int_{\mathbb{R}^d} \omega\, dx = 1$, and set $\omega_\rho(x) := \rho^{-d}\omega(x/\rho)$. Let $\varepsilon > 0$ and fix $k \in \mathbb{N}_0$. For all sufficiently small $\rho > 0$, the function $\omega_\rho * (\chi_k u)$ is supported in $\Omega_{k+2} \setminus \Omega_{k-2}$ and, as in the proof of Lemma 2.1, converges to $\chi_k u$ in $H^1(\Omega_{k+2})$. Therefore there is a $\rho_k > 0$ such that

$$\left\|\omega_{\rho_k} * (\chi_k u) - \chi_k u\right\|_{H^1(\Omega)} = \left\|\omega_{\rho_k} * (\chi_k u) - \chi_k u\right\|_{H^1(\Omega_{k+2})} \le 2^{-k}\varepsilon.$$

We set

$$u_\varepsilon := \sum_{k \in \mathbb{N}_0} \omega_{\rho_k} * (\chi_k u).$$

For any bounded, open set Ω' with $\overline{\Omega'} \subset \Omega$, there is a $K \in \mathbb{N}_0$ such that $\Omega' \subset \Omega_{K-2}$, and therefore, by the support properties of $\omega_{\rho_k} * (\chi_k u)$, we have

$$u_\varepsilon = \sum_{k=0}^{K} \omega_{\rho_k} * (\chi_k u) \quad \text{in } \Omega'.$$

As each $\omega_{\rho_k} * (\chi_k u) \in C^\infty(\Omega)$, we have $u_\varepsilon \in C^\infty(\Omega')$. Since Ω' is arbitrary, we conclude that $u_\varepsilon \in C^\infty(\Omega)$. As u was assumed to have bounded support, the same is true for u_ε. Since $\|\omega_{\rho_k} * (\chi_k u)\|_\infty \le \|\chi_k u\|_\infty \le \|u\|_\infty$ and since for any

x there are at most five values of k such that $x \in \Omega_{k+2} \setminus \Omega_{k-2}$, we conclude that $\|u_\varepsilon\|_\infty \le 5\|u\|_\infty$.

Finally,

$$\|u_\varepsilon - u\|_{H^1(\Omega)} \le \sum_{k \in \mathbb{N}_0} \|\omega_{\rho_k} * (\chi_k u) - \chi_k u\|_{H^1(\Omega)} \le \sum_{k \in \mathbb{N}_0}' 2^{-k}\varepsilon = 2\varepsilon.$$

This shows that $u_\varepsilon \to u$ in $H^1(\Omega)$ as $\varepsilon \to 0$ and concludes the proof. $\qquad\square$

Remark 2.24   For any $0 \le w \in L^1_{\mathrm{loc}}(\Omega)$, the set in Proposition 2.23 is dense in

$$\left\{ u \in H^1(\Omega): \int_\Omega |u|^2 w \, dx < \infty \right\}$$

with respect to the norm

$$\left(\|u\|^2_{H^1(\Omega)} + \int_\Omega |u|^2 w \, dx \right)^{1/2}.$$

This follows from the fact that the three approximations used in the proof of Proposition 2.23 also converge with respect to $(\int_\Omega |u|^2 w \, dx)^{1/2}$. (For the third step we use $\|\omega_\rho * (\chi_k u)\|_\infty \le \|\chi_k u\|_\infty$, $\omega_\rho * (\chi_k u) \to \chi_k u$ a.e. as $\rho \to 0$ along a subsequence and $w \in L^1(\Omega_{k+2})$ to deduce by dominated convergence that $\int_\Omega |\omega_\rho * (\chi_k u) - \chi_k u|^2 w \, dx \to 0$ as $\rho \to 0$ along a subsequence.)

We now show that, in dimensions $d \ge 2$, functions vanishing near a given point are dense in $H^1(\Omega)$. We will see in the next subsection that this is not the case if $d = 1$.

Lemma 2.25   *Let $\Omega \subset \mathbb{R}^d$, $d \ge 2$, be open and $x_0 \in \Omega$. Then*

$$\left\{ u \in H^1(\Omega): u \in C^\infty(\Omega) \cap L^\infty(\Omega), \operatorname{supp} u \text{ bounded}, x_0 \notin \operatorname{supp} u \right\}$$

is dense in $H^1(\Omega)$.

Proof   By Proposition 2.23, it suffices to approximate

$$u \in H^1(\Omega) \cap C^\infty(\Omega) \cap L^\infty(\Omega) \quad \text{with bounded support in } H^1(\Omega)$$

by functions in $H^1(\Omega) \cap C^\infty(\Omega) \cap L^\infty(\Omega)$ with the additional property of not having x_0 in their support. After a translation, we may assume that $x_0 = 0$.

We begin with the case of dimension $d \ge 3$. Let $\eta \in C^\infty[0, \infty)$ with $0 \le \eta \le 1$ such that $\eta(t) = 0$ if $0 \le t \le 1/2$ and $\eta(t) = 1$ if $t \ge 1$. Let $u_\varepsilon(x) := \eta(|x|/\varepsilon)u(x)$. Then $u_\varepsilon \in H^1(\Omega) \cap C^\infty(\Omega) \cap L^\infty(\Omega)$, $\operatorname{supp} u_\varepsilon$ is bounded and $0 \notin \operatorname{supp} u_\varepsilon$. We have

$$\|u_\varepsilon - u\|_{L^2(\Omega)} \le \left(\int_{B_\varepsilon(0)} |u|^2 \, dx \right)^{1/2}$$

and, by dominated convergence, the right side tends to zero as $\varepsilon \to 0$. Moreover, $\partial_j u_\varepsilon = \eta(|x|/\varepsilon)\partial_j u + u\varepsilon^{-1}\eta'(|x|/\varepsilon)x_j/|x|$, and therefore,

$$\left\|\partial_j u_\varepsilon - \partial_j u\right\|_{L^2(\Omega)} < \left(\int_{B_\varepsilon(0)} |\partial_j u|^2 \, dx\right)^{1/2} + \left(\int_{B_\varepsilon(0)} |u|^2\varepsilon^{-2}|\eta'(|x|/c)|^2 \, dx\right)^{1/2}.$$

As $\varepsilon \to 0$, the first term on the right side tends to zero by dominated convergence and the second term does since we assume that $u \in L^\infty(\Omega)$ and $d \geq 3$.

We now turn to the case of dimension $d = 2$. We now define $u_\varepsilon(x) := \eta(|x|^\varepsilon)u(x)$ with η as before. Then, as before, $u_\varepsilon \in H^1(\Omega) \cap C^\infty(\Omega) \cap L^\infty(\Omega)$, supp u_ε is bounded, $0 \notin \operatorname{supp} u_\varepsilon$, and, by dominated convergence, we have $u_\varepsilon \to u$ in $L^2(\Omega)$. Moreover, $\partial_j u_\varepsilon = \eta(|x|^\varepsilon)\partial_j u + \varepsilon|x|^{\varepsilon-2}x_j\eta'(|x|^\varepsilon)u$ and therefore

$$\left\|\partial_j u_\varepsilon - \partial_j u\right\|_{L^2(\Omega)}$$
$$\leq \left(\int_\Omega (1 - \eta(|x|^\varepsilon))^2 |\partial_j u|^2 \, dx\right)^{1/2} + \left(\varepsilon^2 \int_\Omega \frac{\eta'(|x|^\varepsilon)^2|u|^2 \, dx}{|x|^{2-2\varepsilon}}\right)^{1/2}.$$

As $\varepsilon \to 0$, the first term on the right side tends to zero by dominated convergence. For the second term, we use the assumption that u is bounded and find

$$\varepsilon^2 \int_\Omega \eta'(|x|^\varepsilon)^2|u|^2 \frac{dx}{|x|^{2-2\varepsilon}} \leq \varepsilon^2 \|u\|_{L^\infty(\Omega)}^2 \|\eta'\|_{L^\infty}^2 \int_{B_1(0)} \frac{dx}{|x|^{2-2\varepsilon}}$$
$$= \varepsilon\pi\|u\|_{L^\infty(\Omega)}^2 \|\eta'\|_{L^\infty}^2.$$

This tends to zero. This completes the proof. □

2.2.2 The one-dimensional case

In the case where $\Omega = I$ is an interval, one has rather precise knowledge about functions in $H^1(I)$.

Lemma 2.26 *Let $I \subset \mathbb{R}$ be an open interval and $u \in H^1(I)$. Then there is a continuous function $\tilde{u}$ on $\bar{I}$ such that $u = \tilde{u}$ almost everywhere on I and, with the convention $|I|^{-1} = 0$ if I is unbounded,*

$$\sup_I |\tilde{u}|^2 \leq |I|^{-1} \int_I |u|^2 \, dx + 2 \left(\int_I |u'|^2 \, dx\right)^{1/2} \left(\int_I |u|^2 \, dx\right)^{1/2}. \tag{2.7}$$

Moreover, if I is unbounded, then

$$\lim_{|x|\to\infty, x\in I} \tilde{u}(x) = 0.$$

In the following, we always identify $\tilde{u}$ and u.

When $I = \mathbb{R}$, the factor of two in (2.7) can be omitted. This follows by a careful examination of the proof; see Theorem 2.48.

Proof The existence of $\tilde{u}$ and its continuity on $\bar{I}$ follow from Lemma 2.6 and Remark 2.7 (applied on any bounded, open interval $J \subset I$, if I is unbounded). Moreover, it follows from the product rule (Lemma 2.14) that $|u|^2$ is weakly differentiable on I with $(|u|^2)' = 2\,\mathrm{Re}(u'\bar{u})$. By the Cauchy–Schwarz inequality, this derivative belongs to $L^1(I)$. By Remark 2.8, for any bounded, open interval $J \subset I$ and $x \in J$, one has

$$|\tilde{u}(x)|^2 = |J|^{-1} \int_J |u(t)|^2 \, dt + \int_J k_J(x,t) 2\,\mathrm{Re}\left(u'(t)\overline{u(t)}\right) dt$$

with

$$k_{(a,b)}(x,t) := \frac{t-a}{b-a}\chi_{(a,x)}(t) - \frac{b-t}{b-a}\chi_{(x,b)}(t).$$

Using $|k_{(a,b)}| \le 1$ and the Cauchy–Schwarz inequality, we deduce

$$\|\tilde{u}\|_{L^\infty(J)}^2 \le |J|^{-1}\|u\|_{L^2(J)}^2 + 2\|u'\|_{L^2(J)}\|u\|_{L^2(J)}.$$

For bounded I we can choose $J = I$ and for unbounded I we choose an increasing sequence of Js that exhaust I to obtain the claimed bound on the supremum of $|\tilde{u}|$.

Finally, if I is unbounded, then the fact that $(|u|^2)' \in L^1(I)$ implies, by Remark 2.7, that $\lim_{|x|\to\infty,\, x\in I} |u(x)|^2$ exists. Since $|u|^2$ is integrable, this limit is necessarily equal to zero. $\square$

2.2.3 The Sobolev space $H_0^1(\Omega)$

For an open set $\Omega \subset \mathbb{R}^d$, we define

$$H_0^1(\Omega)$$

to be the closure of $C_0^\infty(\Omega)$ in $H^1(\Omega)$.

In dimension one, the space $H_0^1(I)$ with an open interval $I \subset \mathbb{R}$ can easily be characterized since, according to Lemma 2.26, functions in $H^1(I)$ extend continuously to ∂I.

Lemma 2.27 *Let $I \subset \mathbb{R}$ be an open interval and $u \in H^1(I)$. Then $u \in H_0^1(I)$ if and only if $u(x) = 0$ for $x \in \partial I$, where u is identified with its continuous extension.*

Proof First, let $u \in H_0^1(I)$ and let $(u_n) \subset C_0^\infty(I)$ such that $u_n \to u$ in $H^1(I)$. It follows from (2.7) that $u_n \to u$ uniformly in $\overline{I}$ and, in particular, $u_n(x) \to u(x)$ for all $x \in \partial I$. For such x, we have $u_n(x) = 0$, and therefore also $u(x) = 0$, as claimed.

Conversely, assume that $u(x) = 0$ for all $x \in \partial I$. By considering the real and imaginary part of u separately, we may assume that u is real. Let $\varphi \in C^1(\mathbb{R})$ such that $\varphi(w) = 0$ if $|w| \leq 1$, $\varphi(w) = w$ if $|w| \geq 2$ and $|\varphi(w)| \leq |w|$ for all $w \in \mathbb{R}$. Then, by Lemma 2.10, $u_n(x) := n^{-1}\varphi(nu(x))$ belongs to $H^1(I)$ and $u_n' = \varphi'(nu)u'$.

We claim that $\operatorname{supp} u_n$ is a compact subset of I for any n. Indeed, we have $\operatorname{supp} u_n \subset \{x \in I \colon |u(x)| \geq 1/n\}$. Since u is continuous and vanishes on ∂I, $\operatorname{supp} u_n$ is at a positive distance from ∂I. Moreover, if I is unbounded, then, by Lemma 2.26, $\lim_{|x| \to \infty, x \in I} u(x) = 0$, so $\operatorname{supp} u_n$ is bounded.

Using mollifiers as, for instance, in the proof of Lemma 2.1, it is easy to see that $u_n \in H_0^1(I)$ for any n. We claim that $u_n \to u$ in $H^1(I)$ as $n \to \infty$. Indeed, by dominated convergence, since $|n^{-1}\varphi(nu) - u| \leq 2|u|$,

$$\int_I |u_n - u|^2 \, dx = \int_{\{0 < |u| < 2/n\}} |n^{-1}\varphi(nu) - u|^2 \, dx \to 0 \qquad \text{as } n \to \infty$$

and, again by dominated convergence, since $|\varphi'(nu) - 1| \leq \text{const}$,

$$\int_I |u_n' - u'|^2 \, dx = \int_{\{0 < |u| < 2/n\}} |\varphi'(nu) - 1|^2 |u'|^2 \, dx \to 0 \qquad \text{as } n \to \infty,$$

Since $H_0^1(I)$ is closed in $H^1(I)$, we see that $u = \lim_{n \to \infty} u_n$ belongs to $H_0^1(I)$. $\square$

We now turn to the multi-dimensional context and discuss the extension of functions by zero.

Lemma 2.28 *Let* $\Omega \subset \tilde{\Omega} \subset \mathbb{R}^d$ *be open sets. Let* $u \in H_0^1(\Omega)$ *and define* $\tilde{u}(x) := u(x)$ *if* $x \in \Omega$ *and* $\tilde{u}(x) := 0$ *if* $x \in \tilde{\Omega} \setminus \Omega$. *Then* $\tilde{u} \in H_0^1(\tilde{\Omega})$ *and*

$$\nabla \tilde{u}(x) = \begin{cases} \nabla u(x) & \text{if } x \in \Omega, \\ 0 & \text{if } x \in \tilde{\Omega} \setminus \Omega. \end{cases}$$

Proof Let $(u_n) \subset C_0^\infty(\Omega)$ such that $u_n \to u$ in $H^1(\Omega)$ and denote by $\tilde{u}_n \in C_0^\infty(\tilde{\Omega})$ the extension of u_n to $\tilde{\Omega}$ by zero. Then $\tilde{u}_n \to \tilde{u}$ in $L^2(\tilde{\Omega})$ and $\partial_j \tilde{u}_n \to \tilde{v}_j$ in $L^2(\tilde{\Omega})$, where $\tilde{v}_j(x) := \partial_j u(x)$ if $x \in \Omega$ and $\tilde{v}_j(x) := 0$ if $x \in \tilde{\Omega} \setminus \Omega$. By Lemma 2.3, $\tilde{u}$ is weakly differentiable with $\partial_j \tilde{u} = \tilde{v}_j$. Since $\tilde{v}_j \in L^2(\tilde{\Omega})$, we infer that $\tilde{u} \in H^1(\tilde{\Omega})$. Since $\|\tilde{u}_n - \tilde{u}\|_{H^1(\tilde{\Omega})} = \|u_n - u\|_{H^1(\Omega)} \to 0$, one has even $\tilde{u} \in H_0^1(\tilde{\Omega})$. $\square$

The following lemma concerns integration by parts. It says that in the defining equation of a weak derivative we can take η not only in $C_0^\infty(\Omega)$, but even in $H_0^1(\Omega)$.

Lemma 2.29 *Let $u \subset H^1(\Omega)$ and $v \in H_0^1(\Omega)$. Then, for any $j = 1,\ldots,d$,*

$$\int_\Omega (\partial_j u) v \, dx = -\int_\Omega u \, \partial_j v \, dx .$$

Proof Let $(v_n) \subset C_0^\infty(\Omega)$ with $v_n \to v$ in $H^1(\Omega)$. Then, by the definition of a weak partial derivative,

$$\int_\Omega (\partial_j u) v_n \, dx = -\int_\Omega u \, \partial_j v_n \, dx .$$

The left side converges to $\int_\Omega (\partial_j u) v \, dx$ since

$$\left| \int_\Omega (\partial_j u)(v_n - v) \, dx \right| \leq \|\partial_j u\|_{L^2(\Omega)} \|v_n - v\|_{L^2(\Omega)} \to 0,$$

and the right side converges to $-\int_\Omega u \partial v \, dx$ since

$$\left| \int_\Omega u \, (\partial_j v_n - \partial_j v) \, dx \right| \leq \|u\|_{L^2(\Omega)} \|\partial_j v_n - \partial_j v\|_{L^2(\Omega)} \to 0 .$$

This proves the statement. $\square$

Remark 2.30 If $u \in H_0^1(\Omega)$ is real-valued and $0 \leq v \in H^1(\Omega)$, then $(u - v)_+ \in H_0^1(\Omega)$. For $v \equiv 0$, this can be shown by mollifying the positive parts of approximating $C_0^\infty(\Omega)$ functions. For $v \geq 0$, one uses the fact that the approximating smooth functions in the proof of Proposition 2.23 are non-negative.

2.2.4 Fourier characterization of $H^1(\mathbb{R}^d)$

We now characterize the Sobolev space $H^1(\mathbb{R}^d)$ through the Fourier transform,

$$\widehat{u}(\xi) = (2\pi)^{-d/2} \int_{\mathbb{R}^d} e^{-i\xi \cdot x} u(x) \, dx , \qquad \xi \in \mathbb{R}^d . \tag{2.8}$$

More precisely, the Fourier transform, originally defined for $u \in L^1(\mathbb{R}^d)$ by the above formula, has a unique continuous extension to all of $L^2(\mathbb{R}^d)$ from its dense subspace $L^1 \cap L^2(\mathbb{R}^d)$; see, e.g., Rudin (1987, Theorem 9.13).

Lemma 2.31 $H^1(\mathbb{R}^d) = \{u \in L^2(\mathbb{R}^d): (1 + |\xi|^2)^{1/2}\widehat{u} \in L^2(\mathbb{R}^d)\}$ *and*

$$\|u\|_{H^1(\mathbb{R}^d)} = \left\| (1 + |\xi|^2)^{1/2}\widehat{u} \right\|_{L^2(\mathbb{R}^d)} .$$

Moreover, for $u \in H^1(\mathbb{R}^d)$,

$$\widehat{\partial_j u} = i\xi_j \widehat{u} \qquad for\ j = 1, \ldots, d.$$

Proof First, let $u \in H^1(\mathbb{R}^d)$. Then for any $\eta \in C_0^\infty(\mathbb{R}^d)$, by applying Plancherel's theorem twice and recalling the definition of a weak derivative,

$$\int_{\mathbb{R}^d} \widehat{\partial_j u}\, \overline{\widehat{\eta}}\, d\xi = \int_{\mathbb{R}^d} (\partial_j u)\, \overline{\eta}\, dx = - \int_{\mathbb{R}^d} u\, \overline{\partial_j \eta}\, dx = - \int_{\mathbb{R}^d} \widehat{u}\, \overline{\widehat{\partial_j \eta}}\, d\xi.$$

A simple integration by parts shows that $\widehat{\partial_j \eta} = i\xi_j \widehat{\eta}$. Thus we obtain

$$\int_{\mathbb{R}^d} \widehat{\partial_j u}\, \overline{\widehat{\eta}}\, d\xi = i \int_{\mathbb{R}^d} \widehat{u}\, \xi_j\, \overline{\widehat{\eta}}\, d\xi.$$

Since, by Plancherel's theorem, the set $\{\widehat{\eta}: \eta \in C_0^\infty(\mathbb{R}^d)\}$ is dense in $L^2(\mathbb{R}^d)$, we conclude that $\widehat{\partial_j u} = i\xi_j \widehat{u}$, as claimed. Moreover, again by Plancherel's theorem,

$$\|u\|^2_{H^1(\mathbb{R}^d)} = \int_{\mathbb{R}^d} \left(|\widehat{u}|^2 + \sum_j |\widehat{\partial_j u}|^2 \right) d\xi = \int_{\mathbb{R}^d} \left(|\widehat{u}|^2 + \sum_j \xi_j^2 |\widehat{u}|^2 \right) d\xi,$$

as claimed.

Conversely, let $u \in L^2(\mathbb{R}^d)$ with $|\xi| \widehat{u} \in L^2(\mathbb{R}^d)$. Then, by assumption and by Plancherel's theorem, $v_j := (i\xi_j \widehat{u})^\vee \in L^2(\mathbb{R}^d)$ and, for any $\eta \in C_0^\infty(\mathbb{R}^d)$,

$$\int_{\mathbb{R}^d} u \overline{\partial_j \eta}\, dx = \int_{\mathbb{R}^d} \widehat{u}\, \overline{\widehat{\partial_j \eta}}\, d\xi = -i \int_{\mathbb{R}^d} \widehat{u} \xi_j \overline{\widehat{\eta}}\, d\xi = - \int_{\mathbb{R}^d} v_j \overline{\eta}\, dx.$$

This means that u has weak derivative $\partial_j u = v_j$. Thus $u \in H^1(\mathbb{R}^d)$. $\qquad\square$

2.2.5 Weighted one-dimensional Sobolev spaces

When working in spherical coordinates, one naturally arrives at one-dimensional Sobolev spaces with weights. Let $R \in (0, \infty]$ and $d \in [1, \infty)$. In applications, the parameter d will be an integer, namely the spatial dimension, but this is not important here. We denote by

$$L_d^2(R) := L^2((0, R), r^{d-1} dr)$$

the space of (a.e. equivalence classes of) measurable functions on $(0, R)$ that are square-integrable with respect to $r^{d-1} dr$. This is a Hilbert space with inner product

$$\int_0^R u\overline{v}\, r^{d-1}\, dr.$$

Moreover, we denote by

$$H_d^1(R)$$

the space of (a.e. equivalence classes of) functions $u \in L^2_d(R)$ that are weakly differentiable on $(0, R)$ with $u' \in L^2_d(R)$. This is a Hilbert space with inner product

$$\int_0^R \left(u'\overline{v'} + u\overline{v} \right) r^{d-1} dr .$$

We know from Lemma 2.26 that functions in $H^1_d(R)$ are (after modification on a set of measure zero) continuous in $(0, R)$ and, if $R < \infty$, extend continuously to $(0, R]$. The following pointwise bound is sometimes useful.

Lemma 2.32　*Let $d \in [1, \infty)$, $R \in (0, \infty]$ and $u \in H^1_d(R)$. Then, for a.e. $r \in (0, R)$,*

$$r^{d-1} |u(r)|^2 \leq \frac{d}{R} \int_0^R |u|^2 s^{d-1} \, ds + 2 \left(\int_0^R |u|^2 s^{d-1} \, ds \right)^{\frac{1}{2}} \left(\int_0^R |u'|^2 s^{d-1} \, ds \right)^{\frac{1}{2}} .$$

If $R = \infty$, we interpret $R^{-1} = 0$. If $R < \infty$, the bound extends to $r = R$ for the continuous representative of u.

Proof　We may assume that $R < \infty$. Let $u \in H^1_d(R)$. By Lemma 2.26, we may assume that u is continuous on $(0, R)$. As in the proof of this lemma, $r^{d-1} |u|^2$ is weakly differentiable on $(0, R)$ with derivative

$$(r^{d-1} |u|^2)' = 2 r^{d-1} \operatorname{Re}(\overline{u} u') + (d - 1) r^{d-2} |u|^2.$$

This derivative is integrable away from zero and, thus, by Remark 2.8, for $\varepsilon \leq r \leq R$,

$$
\begin{aligned}
r^{d-1} |u(r)|^2 &= \frac{1}{R - \varepsilon} \int_\varepsilon^R |u(s)|^2 s^{d-1} \, ds \\
&\quad - \frac{2}{R - \varepsilon} \int_r^R (R - s) \operatorname{Re} \left(\left(\overline{u'(s)} + \frac{d-1}{2s} \overline{u(s)} \right) u(s) \right) s^{d-1} \, ds \\
&\quad + \frac{2}{R - \varepsilon} \int_\varepsilon^r (s - \varepsilon) \operatorname{Re} \left(\left(\overline{u'(s)} + \frac{d-1}{2s} \overline{u(s)} \right) u(s) \right) s^{d-1} \, ds \\
&\leq \frac{1}{R - \varepsilon} \int_\varepsilon^R |u(s)|^2 s^{d-1} \, ds \\
&\quad + 2 \int_\varepsilon^R |u'(s)| |u(s)| \, s^{d-1} \, ds + \frac{d-1}{R - \varepsilon} \int_\varepsilon^r |u(s)|^2 s^{d-1} \, ds .
\end{aligned}
$$

Letting $\varepsilon \to 0$ we obtain the claimed inequality.　□

Next we prove an analogue of Lemma 2.25.

Lemma 2.33　*Let $d \in [2, \infty)$ and $R \in (0, \infty]$. The restrictions to $(0, R)$ of functions in $C_0^\infty(\mathbb{R}_+)$ are dense in $H^1_d(R)$ and in $H^1_d(R) \cap L^2((0, R), r^{d-3} dr)$.*

Proof We only sketch the major steps of the argument. As in the proofs of Proposition 2.23 and Remark 2.24, one shows density, both in $H_d^1(R)$ and $H_d^1(R) \cap L^2((0, R), r^{d-3} dr)$, of bounded functions whose support is bounded. This step works for any $d \geq 1$. To show that one can approximate by functions vanishing near 0, we argue as in the proof of Lemma 2.25.

Finally, let $u \in L^\infty(0, R)$ with bounded support not containing 0. If $R < \infty$, we set $u(r) := u(R)(r - 2R)_+/R$ for $r > R$. (Note that the value $u(R)$ is well defined for the continuous representative of $u \in H_d^1(R)$ by Remark 2.7.) By a standard convolution argument, we obtain a sequence of $C_0^\infty(\mathbb{R}_+)$ functions that converges to u in $H^1(\mathbb{R}_+)$. The restrictions to $(0, R)$ converge in $H^1(0, R)$. If we use a compactly supported convolution kernel, then the approximants have support contained in a neighborhood of the support of u and there the norms in $H^1(0, R)$, $H_d^1(R)$ and in $H_d^1(R) \cap L^2((0, R), r^{d-3} dr)$ are equivalent. This proves the lemma. □

The following lemma, which relates functions on the half-line and functions on the whole-line, is sometimes useful.

Lemma 2.34 *Let $u \in L_{loc}^1(\mathbb{R}_+)$ and $w \in L_{loc}^1(\mathbb{R})$ such that $u(r) = r^{-(d-2)/2} w(\ln r)$. Then u is weakly differentiable if and only if w is weakly differentiable. Assume that this is the case.*

1. *If $d = 2$, then $u' \in L^2(\mathbb{R}_+, r\, dr)$ if and only if $w' \in L^2(\mathbb{R}, dt)$.*
2. *If $d > 2$, then $u' \in L^2(\mathbb{R}_+, r^{d-1} dr)$ with $\liminf_{r \to \infty} |u(r)| = 0$ if and only if $w \in H^1(\mathbb{R})$.*

Moreover, in this case,

$$\int_0^\infty |u'(r)|^2 r^{d-1}\, dr = \int_\mathbb{R} \left(|w'(t)|^2 + \left(\frac{d-2}{2} \right)^2 |w(t)|^2 \right) dt. \qquad (2.9)$$

The lim inf in the lemma is taken with the absolutely continuous representative of u, which exists by Lemma 2.6 under the assumption $u' \in L^2(\mathbb{R}_+, r^{d-1}\, dr)$. We note that the lemma is one-dimensional and remains valid for any real parameter $d \geq 2$.

Proof The weak differentiability statement follows from the product and chain rules, which also give the identity

$$u'(r) = r^{-d/2} \left(w'(\ln r) - \frac{d-2}{2} w(\ln r) \right). \qquad (2.10)$$

If $d = 2$, this implies (2.9) by the change of variables $t = \ln r$.

We now turn to proving the assertion for $d > 2$. We first assume that

$w \in H^1(\mathbb{R})$. This implies, by the same change of variables, that $r^{-d/2}w'(\ln r)$ and $r^{-d/2}w(\ln r)$ are square-integrable with respect to $r^{d-1}dr$, and therefore, by (2.10), u' is square-integrable with respect to $r^{d-1}dr$. Moreover, the square-integrability of w implies that $\lim\inf_{t\to\infty}|w(t)| = 0$, which implies $\lim\inf_{r\to\infty}|u(r)| = 0$. Moreover, by (2.10), we obtain, with $t = \ln r$,

$$\int_0^\infty |u'(r)|^2 r^{d-1}\, dr = \int_0^\infty \left| w'(\ln r) - \frac{d-2}{2}w(\ln r) \right|^2 \frac{dr}{r}$$

$$= \int_{\mathbb{R}} \left| w'(t) - \frac{d-2}{2}w(t) \right|^2 dt \, .$$

We have, pointwise on $\mathbb{R}$,

$$\left| w'(t) - \frac{d-2}{2}w(t) \right|^2 = |w'(t)|^2 - \frac{d-2}{2}\left(|w|^2\right)'(t) + \left(\frac{d-2}{2}\right)^2 |w(t)|^2 \, .$$

Note that the term $2\,\mathrm{Re}(\overline{w}w') = \left(|w|^2\right)'$ is integrable by the Cauchy–Schwarz inequality and its integral vanishes by Lemma 2.26. Thus, we have

$$\int_{\mathbb{R}} \left| w'(t) - \frac{d-2}{2}w(t) \right|^2 dt = \int_{\mathbb{R}} \left(|w'(t)|^2 + \left(\frac{d-2}{2}\right)^2 |w(t)|^2 \right) dt \, .$$

This proves identity (2.9).

Now assume conversely that $u' \in L^2(\mathbb{R}_+, r^{d-1}dr)$ and $\lim\inf_{r\to\infty}|u(r)| = 0$. Then, by Hardy's inequality (which we will prove later as Theorem 2.65),

$$\int_0^\infty |u(r)|^2 r^{d-3}\, dr \le \left(\frac{2}{d-2}\right)^2 \int_0^\infty |u'(r)|^2 r^{d-1}\, dr \, .$$

By the change of variables, the square-integrability of $r^{-1}u = r^{-d/2}w(\ln r)$ with respect to $r^{d-1}dr$ implies that $w \in L^2(\mathbb{R})$. Moreover, by (2.10), it implies square-integrability of $r^{-d/2}w'(\ln r)$ with respect to $r^{d-1}dr$, which, again by the change of variables, is the same as $w' \in L^2(\mathbb{R})$. Thus, we have shown $w \in H^1(\mathbb{R})$. $\qquad\qquad\square$

2.3 Compact embeddings

By definition, the space $H^1(\Omega)$, and therefore also $H_0^1(\Omega)$, is contained in $L^2(\Omega)$ and $\|u\|_{L^2(\Omega)} \le \|u\|_{H^1(\Omega)}$ for any $u \in H^1(\Omega)$. This means that the inclusion operator $H^1(\Omega) \to L^2(\Omega)$, which maps a function to itself, is continuous. In this section we show that, under certain assumptions on Ω, this operator, or its restriction to $H_0^1(\Omega)$, is even compact.

2.3.1 Rellich's theorem for $H_0^1(\Omega)$

The following compactness result is known as *Rellich's theorem*. In order to formulate it, we recall the notion of weak convergence in $H_0^1(\Omega)$. We say that a sequence $(u_n) \subset H_0^1(\Omega)$ converges *weakly* in $H_0^1(\Omega)$ to $u \in H_0^1(\Omega)$ if, for all $v \in H_0^1(\Omega)$,

$$\int_\Omega (\nabla u_n \cdot \nabla v + u_n v)\, dx \to \int_\Omega (\nabla u \cdot \nabla v + uv)\, dx \qquad \text{as } n \to \infty.$$

Theorem 2.35 *Let $\Omega \subset \mathbb{R}^d$ be an open set of finite measure, let $u \in H_0^1(\Omega)$ and let $(u_n) \subset H_0^1(\Omega)$ be such that $u_n \to u$ weakly in $H_0^1(\Omega)$. Then $u_n \to u$ strongly in $L^2(\Omega)$.*

The proof of this theorem relies on a result that is interesting in its own right.

Proposition 2.36 *Let $u \in H^1(\mathbb{R}^d)$ and let $(u_n) \subset H^1(\mathbb{R}^d)$ be such that $u_n \to u$ weakly in $H^1(\mathbb{R}^d)$. Then, for any set $A \subset \mathbb{R}^d$ of finite measure, $\chi_A u_n \to \chi_A u$ in $L^2(\mathbb{R}^d)$.*

For the proof, we need the following lemma. We denote

$$G_t(x) := (4\pi t)^{-d/2} e^{-|x|^2/(4t)}, \qquad x \in \mathbb{R}^d,\, t > 0.$$

Lemma 2.37 *Let $u \in H^1(\mathbb{R}^d)$. Then, for all $t > 0$,*

$$\|u - G_t * u\|_{L^2(\mathbb{R}^d)} \leq \sqrt{t}\,\|\nabla u\|_{L^2(\mathbb{R}^d)}.$$

Proof We begin by noting that

$$G_t(x) = (2\pi)^{-d} \int_{\mathbb{R}^d} e^{i\xi \cdot x} e^{-t|\xi|^2}\, d\xi.$$

This can be verified by a straightforward integration. (Indeed, the one-dimensional integral can be computed using complex analysis, and the one-dimensional result implies the d-dimensional result.) It follows that

$$\widehat{G_t * u} = e^{-t|\xi|^2} \hat{u},$$

and therefore, by Plancherel's theorem,

$$\|u - G_t * u\|_{L^2(\mathbb{R}^d)}^2 = \int_{\mathbb{R}^d} \left(1 - e^{-t|\xi|^2}\right)^2 |\hat{u}(\xi)|^2\, d\xi.$$

Since $(1 - e^{-a})^2 \leq a$ for all $a \geq 0$, we have

$$\int_{\mathbb{R}^d} \left(1 - e^{-t|\xi|^2}\right)^2 |\hat{u}(\xi)|^2\, d\xi \leq t \int_{\mathbb{R}^d} |\xi|^2 |\hat{u}(\xi)|^2\, d\xi = t \int_{\mathbb{R}^d} |\nabla u(x)|^2\, dx.$$

In the last identity we used Lemma 2.31. This completes the proof. $\qquad \square$

Proof of Proposition 2.36 By the boundedness of weakly convergent sequences and the weak lower semicontinuity of the norm, we have

$$C := \sup \|u_n\|_{H^1(\mathbb{R}^d)} < \infty \quad \text{and} \quad \|u\|_{H^1(\mathbb{R}^d)} \leq C.$$

Let $A \subset \mathbb{R}^d$ be a set of finite measure and let $\varepsilon > 0$. We decompose the difference $\chi_A(u_n - u)$ and apply Lemma 2.37 to get

$$
\begin{aligned}
\|\chi_A(u_n &- u)\|_{L^2(\mathbb{R}^d)} \\
&\leq \|\chi_A(u_n - G_t * u_n)\|_{L^2(\mathbb{R}^d)} + \|\chi_A G_t * (u_n - u)\|_{L^2(\mathbb{R}^d)} \\
&\quad + \|\chi_A(u - G_t * u)\|_{L^2(\mathbb{R}^d)} \\
&\leq \|u_n - G_t * u_n\|_{L^2(\mathbb{R}^d)} + \|\chi_A G_t * (u_n - u)\|_{L^2(\mathbb{R}^d)} + \|u - G_t * u\|_{L^2(\mathbb{R}^d)} \\
&\leq \sqrt{t}\left(\|\nabla u_n\|_{L^2(\mathbb{R}^d)} + \|\nabla u\|_{L^2(\mathbb{R}^d)}\right) + \|\chi_A G_t * (u_n - u)\|_{L^2(\mathbb{R}^d)} \\
&\leq 2C\sqrt{t} + \|\chi_A G_t * (u_n - u)\|_{L^2(\mathbb{R}^d)}.
\end{aligned}
$$

In the remainder of the proof, we shall show that

$$\|\chi_A G_t * (u_n - u)\|_{L^2(\mathbb{R}^d)} \to 0 \qquad \text{for any fixed } t > 0. \tag{2.11}$$

This will then imply that

$$\limsup_{n\to\infty} \|\chi_A(u_n - u)\|_{L^2(\mathbb{R}^d)} \leq 2C\sqrt{t},$$

and, since $t > 0$ is arbitrary, we obtain the assertion of the proposition.

Let us prove (2.11). Since $G_t(\cdot - x) \in L^2(\mathbb{R}^d)$ for any x and since $u_n \to u$ weakly in $L^2(\mathbb{R}^d)$, we have $G_t * u_n \to G_t * u$ pointwise. This will prove (2.11) by dominated convergence, provided we can find an integrable majorant. To do this, we observe that by the Cauchy–Schwarz inequality,

$$|G_t * (u_n - u)(x)| \leq (4\pi t)^{-d/2}\left(\int_{\mathbb{R}^d} e^{-|x-y|^2/(2t)}\, dy\right)^{1/2} \|u_n - u\|_{L^2(\mathbb{R}^d)} \leq c_t 2C$$

with

$$c_t := (4\pi t)^{-d/2}\left(\int_{\mathbb{R}^d} e^{-|y|^2/2t}\, dy\right)^{1/2}.$$

Thus, $|\chi_A G_t * (u_n - u)|$ is bounded by the integrable function $c_t 2C\chi_A$. This concludes the proof. □

For the next lemma, we recall that any $u \in H_0^1(\Omega)$ can be extended by zero to a function $\widetilde{u} \in H^1(\mathbb{R}^d)$; see Lemma 2.28. We want to compare weak convergence of a sequence (u_n) in $H_0^1(\Omega)$ with weak convergence of the extended sequence $(\widetilde{u_n})$ in $H^1(\mathbb{R}^d)$. We show that in the definition of weak convergence in $H_0^1(\Omega)$

we may choose test functions that are restrictions to Ω of functions in $H^1(\mathbb{R}^d)$. Note that such functions do not necessarily belong to $H_0^1(\Omega)$.

Lemma 2.38 *Let $\Omega \subset \mathbb{R}^d$ be an open set, let $u \in H_0^1(\Omega)$ and let $(u_n) \subset H_0^1(\Omega)$ be such that $u_n \to u$ weakly in $H_0^1(\Omega)$. Then $\widetilde{u_n} \to \widetilde{u}$ weakly in $H^1(\mathbb{R}^d)$.*

Proof The result follows from the weak continuity of the extension operator $E : H_0^1(\Omega) \to H^1(\mathbb{R}^d)$ given by $w \mapsto \widetilde{w}$. Indeed, for all $v \in H^1(\mathbb{R}^d)$,

$$\langle \widetilde{u_n}, v \rangle_{H^1(\mathbb{R}^d)} = \langle Eu_n, v \rangle_{H^1(\mathbb{R}^d)} = \langle u_n, E^*v \rangle_{H_0^1(\Omega)} \to \langle u, E^*v \rangle_{H_0^1(\Omega)}$$
$$= \langle \widetilde{u}, v \rangle_{H^1(\mathbb{R}^d)} .$$

Here we used the fact that the bounded operator $E : H_0^1(\Omega) \to H^1(\mathbb{R}^d)$ has a bounded adjoint $E^* : H^1(\mathbb{R}^d) \to H_0^1(\Omega)$ and we wrote $\langle \cdot, \cdot \rangle_{H_0^1(\Omega)}$ for the restriction of the inner product in $H^1(\Omega)$ to its closed subspace $H_0^1(\Omega)$. $\qquad \square$

Finally, we can prove the main result of this subsection.

Proof of Theorem 2.35 We apply Proposition 2.36 with $A = \Omega$ to the sequence $(\widetilde{u_n})$, which by Lemma 2.38 converges weakly to $\widetilde{u}$ in $H^1(\mathbb{R}^d)$. $\qquad \square$

Remark 2.39 Combining Theorem 2.35 with the weak compactness of the unit ball in $H_0^1(\Omega)$ as in Lemma 1.1, we obtain that, for an open set $\Omega \subset \mathbb{R}^d$ of finite measure, any bounded sequence in $H_0^1(\Omega)$ contains a subsequence that converges weakly in $H_0^1(\Omega)$ and strongly in $L^2(\Omega)$.

2.3.2 Rellich's theorem for $H^1(\Omega)$

Our next goal is to extend Theorem 2.35 from $H_0^1(\Omega)$ to $H^1(\Omega)$. The assumption that Ω has finite measure is not sufficient for the compactness of the embedding, and we will prove compactness under the following, somewhat implicit assumption. We say that an open set $\Omega \subset \mathbb{R}^d$ has the *extension property* if there is a bounded, linear operator $E : H^1(\Omega) \to H^1(\mathbb{R}^d)$ such that for every $u \in H^1(\Omega)$ one has $(Eu)(x) = u(x)$ for almost every $x \in \Omega$. More precisely, this is the H^1-extension property, but this is the only extension property relevant for us.

We defer a detailed discussion of the extension property to §2.8. There, in Theorem 2.92, we show that sets with a uniformly Lipschitz continuous boundary have the extension property. In particular, balls and cubes have the extension property.

We say that a sequence $(u_n) \subset H^1(\Omega)$ *converges weakly* in $H^1(\Omega)$ to $u \in$

$H^1(\Omega)$ if, for all $v \in H^1(\Omega)$,

$$\int_\Omega (\nabla u_n \cdot \nabla v + u_n v)\, dx \longrightarrow \int_\Omega (\nabla u \cdot \nabla v + uv)\, dx \qquad \text{as } n \to \infty.$$

Theorem 2.40 *Let $\Omega \subset \mathbb{R}^d$ be an open set of finite measure with the extension property, let $u \in H^1(\Omega)$ and let $(u_n) \subset H^1(\Omega)$ be such that $u_n \to u$ weakly in $H^1(\Omega)$. Then $u_n \to u$ strongly in $L^2(\Omega)$.*

Proof Let E be as in the definition of the extension property. As in Lemma 2.38, using the weak continuity of $E : H^1(\Omega) \to H^1(\mathbb{R}^d)$, one sees that $Eu_n \to Eu$ weakly in $H^1(\mathbb{R}^d)$. Therefore, by Theorem 2.36, $\chi_\Omega Eu_n \to \chi_\Omega Eu$ strongly in $L^2(\mathbb{R}^d)$, that is, $u_n \to u$ strongly in $L^2(\Omega)$.   □

Remark 2.41   Combining Theorem 2.40 with the weak compactness of the unit ball in $H^1(\Omega)$ as in Lemma 1.1, we obtain that, for an open set $\Omega \subset \mathbb{R}^d$ of finite measure with the extension property, any bounded sequence in $H^1(\Omega)$ contains a subsequence that converges weakly in $H^1(\Omega)$ and strongly in $L^2(\Omega)$.

The following lemma shows that no analogue of Theorem 2.40 can hold for sets of infinite measure.

Lemma 2.42 *Let $\Omega \subset \mathbb{R}^d$ be an open set of infinite measure. Then there is a sequence $(u_n) \subset H^1(\Omega)$ such that $\|u_n\| = 1$ for all n and, as $n \to \infty$, $\|\nabla u_n\| \to 0$ and $u_n \to 0$ weakly in $L^2(\Omega)$.*

Proof   Step 1. Let $\Omega(R) := \Omega \cap \{|x| < R\}$ and $A(R) := \Omega \cap \{R \le |x| < R+1\}$. We claim that

$$\liminf_{R \to \infty} \frac{|A(R)|}{|\Omega(R)|} = 0.$$

Indeed, if this were not so, there would be an $R_0 < \infty$ and an $\varepsilon > 0$ such that

$$|A(R)| \ge \varepsilon |\Omega(R)| \quad \text{for all } R \ge R_0.$$

Increasing R_0 if necessary, we may assume that $|\Omega(R_0)| > 0$. Iterating

$$|\Omega(R+1)| \ge (1+\varepsilon)|\Omega(R)| \quad \text{for all } R \ge R_0,$$

we find, for all $k \in \mathbb{N}_0$,

$$|\Omega(R_0 + k)| \ge (1+\varepsilon)|\Omega(R_0 + k - 1)| \ge \cdots \ge (1+\varepsilon)^k |\Omega(R_0)|.$$

Note that the right side grows exponentially with k. By contrast,

$$|\Omega(R_0 + k)| \le |\{|x| < R_0 + k\}| = \text{const} (R_0 + k)^d.$$

This is a contradiction for large k and proves the claim.

Step 2. According to the previous step, there is an increasing sequence $R_n \to \infty$ such that $|A(R_n)|/|\Omega(R_n)| \to 0$ as $n \to \infty$. We may assume that $|\Omega(R_1)| > 0$. Let

$$v_n := \chi_{\Omega(R_n)} + (R_n + 1 - |x|)\chi_{A(R_n)} .$$

Then $v_n \in H^1(\Omega)$,

$$\|\nabla v_n\|^2 = |A(R_n)| \qquad \text{and} \qquad |\Omega(R_n)| \le \|v_n\|^2 \le |\Omega(R_n + 1)| .$$

Thus, $u_n := v_n/\|v_n\|$ satisfies $\|u_n\| = 1$ and, by choice of R_n, $\|\nabla u_n\| \to 0$ as $n \to \infty$. Moreover, if $f \in L^1(\Omega)$, then

$$\left| \int_\Omega f u_n \, dx \right| \le \|v_n\|^{-1} \int_\Omega |f| \, dx \to 0 \qquad \text{as } n \to \infty .$$

Since $L^1(\Omega) \cap L^2(\Omega)$ is dense in $L^2(\Omega)$ and the u_n are uniformly bounded in $L^2(\Omega)$, this implies that $u_n \to 0$ weakly in $L^2(\Omega)$, as claimed. $\qquad\square$

2.3.3 Rellich's theorem for one-dimensional weighted Sobolev spaces

In this subsection, we prove a compactness result in a one-dimensional setting with weights that we will later apply to radial functions. Recall that the spaces $L_d^2(R)$ and $H_d^1(R)$, for $R \in (0, \infty]$ and $d \in [1, \infty)$, were introduced in §2.2.5. We say that a sequence $(u_n) \subset H_d^1(R)$ *converges weakly* in $H_d^1(R)$ to $u \in H_d^1(R)$ if, for all $v \in H_d^1(R)$,

$$\int_0^R (u_n'\overline{v'} + u_n\overline{v})\, r^{d-1}\, dr \to \int_0^R (u'\overline{v'} + u\overline{v})\, r^{d-1}\, dr \qquad \text{as } n \to \infty .$$

Lemma 2.43 *Let $R \in (0, \infty)$, $d \in [1, \infty)$, let $u \in H_d^1(R)$ and let $(u_n) \subset H_d^1(R)$ be such that $u_n \to u$ weakly in $H_d^1(R)$. Then $u_n \to u$ strongly in $L_d^2(R)$.*

For integer d, this lemma is a consequence of Theorem 2.40 and the fact that a ball has the extension property (which is a consequence of Theorem 2.92, but can also be proved directly). It is nevertheless of value to have a direct and more elementary proof.

Proof We begin with several reductions. First, by considering $u_n - u$ instead of u_n, we may assume that $u = 0$. Second, it suffices to show that the claimed convergence in $L_d^2(R)$ holds along a subsequence. This implies the full assertion, for, if it failed, there would be a subsequence (u_{n_k}) and an $\varepsilon > 0$ such that $\|u_{n_k}\|_{L_d^2(R)} \ge \varepsilon$ for all k. Then, applying the assertion that we will prove to the subsequence (u_{n_k}), we obtain a further subsequence $(u_{n_{k_l}})$ such that $u_{n_{k_l}} \to 0$ in $L^2_d(R)$ as $l \to \infty$. This is a contradiction.

We now begin with the main part of the proof. It follows from Lemma 2.32 that $v \mapsto v(R)$ is a continuous linear functional on $H^1_d(R)$. Thus, by the Riesz representation theorem, there is a $w \in H^1_d(R)$ such that

$$v(R) = \int_0^R \left(v' \overline{w'} + v\overline{w} \right) r^{d-1}\, dr \qquad \text{for all } v \in H^1_d(R).$$

Therefore the weak convergence $u_n \to 0$ in $H^1_d(R)$ implies $u_n(R) \to 0$.

By Lemma 2.6, we know that, for any $0 < r' < r$,

$$|u_n(r) - u_n(r')| = \left| \int_{r'}^r u'_n(s)\, ds \right| \leq \left(\int_{r'}^r s^{-d+1}\, ds \right)^{1/2} \left(\int_0^R |u'_n(s)|^2 s^{d-1}\, ds \right)^{1/2}.$$

We fix a parameter $\varepsilon > 0$ and find that, if $r' \geq \varepsilon$, then

$$|u_n(r) - u_n(r')| \leq \varepsilon^{-(d-1)/2} |r - r'|^{1/2} \left(\int_0^R |u'_n(s)|^2 s^{d-1}\, ds \right)^{1/2}.$$

Since weakly convergent sequences are bounded, the right side is bounded uniformly in n. This shows that the sequence (u_n) is equicontinuous on $[\varepsilon, R]$. Moreover, taking $r = R$ and recalling $u_n(R) \to 0$, we see that (u_n) is uniformly bounded on $[\varepsilon, R]$. Therefore, by the Arzelà–Ascoli theorem, after passing to a subsequence if necessary, we may assume that (u_n) converges uniformly to a continuous function $\tilde{u}$ on $[\varepsilon, R]$. Since u_n converges weakly in $L^2_d(R)$ to zero, we must have $\tilde{u} = 0$. Thus,

$$\lim_{n \to \infty} \int_\varepsilon^R |u_n|^2 r^{d-1}\, dr = 0.$$

Combining this with the bound, obtained by integrating that in Lemma 2.32,

$$\int_0^\varepsilon |u_n(r)|^2 r^{d-1}\, dr \leq \varepsilon \left(dR^{-1} m_n + 2m_n^{1/2} t_n^{1/2} \right),$$

where $m_n := \int_0^R |u_n|^2 s^{d-1}\, ds$ and $t_n := \int_0^R |u'_n|^2 s^{d-1}\, ds$, we find

$$\limsup_{n \to \infty} \int_0^R |u_n|^2 r^{d-1}\, dr \leq \varepsilon \sup_n \left(dR^{-1} m_n + 2m_n^{1/2} t_n^{1/2} \right).$$

The supremum on the right side is finite. Since $\varepsilon > 0$ is arbitrary, this proves the claimed convergence. $\qquad\square$

2.4 Sobolev inequalities on the whole space

In this section we will discuss the fact that weak differentiability implies an improved local integrability. This is usually expressed in terms of Sobolev inequalities, which will be the topic of the next few sections. In this section we will deal with the case of the whole space $\mathbb{R}^d$. The optimal values of the constants in these inequalities will be of interest in our applications later.

2.4.1 Sobolev inequalities

The following crucial result is from Soboleff (1938)[1].

Theorem 2.44 (Sobolev inequality) *Let $d \geq 3$. Then there is a constant $S_d > 0$ such that for any $u \in H^1(\mathbb{R}^d)$,*

$$\int_{\mathbb{R}^d} |\nabla u|^2 \, dx \geq S_d \left(\int_{\mathbb{R}^d} |u|^{2d/(d-2)} \, dx \right)^{(d-2)/d}. \tag{2.12}$$

We will later show in Proposition 2.82 that inequality (2.12) extends to functions u from the larger, so-called homogeneous Sobolev space. The optimal value of the constant S_d will be determined in Theorem 2.49 below.

Proof The proof relies on the fact that for $u \in H^1(\mathbb{R}^d)$ one has $\xi\widehat{u} \in L^2(\mathbb{R}^d)$ and $\|\nabla u\|_{L^2(\mathbb{R}^d)} = \|\xi\widehat{u}\|_{L^2(\mathbb{R}^d)}$; see Lemma 2.31. We put $q := 2d/(d-2)$ for brevity and write

$$\int_{\mathbb{R}^d} |u|^q \, dx = q \int_0^\infty |\{|u| > \tau\}| \tau^{q-1} \, d\tau. \tag{2.13}$$

We then decompose in the integrand, for each $\tau > 0$, $u = (u - u_\tau) + u_\tau$, where

$$u_\tau := \left(\chi_{\{|\xi|<\Lambda_\tau\}} \widehat{u} \right)^\vee$$

for some parameter $\Lambda_\tau > 0$ to be specified later. (We emphasize that the above expression for $\int_{\mathbb{R}^d} |u|^q \, dx$ is well defined, although possibly equal to infinity. Our proof will rule this out.) Then

$$|\{|u| > \tau\}| \leq |\{|u - u_\tau| > \tau/2\}| + |\{|u_\tau| > \tau/2\}|. \tag{2.14}$$

---

[1] Here we use the transliteration Soboleff from the original paper, while elsewhere we use the current convention for the spelling of his name.

We have

$$
\begin{aligned}
|u_\tau(x)| &= (2\pi)^{-d/2} \left| \int_{\{|\xi| < \Lambda_\tau\}} e^{i\xi \cdot x} \widehat{u}(\xi) \, d\xi \right| \\
&\leq (2\pi)^{-d/2} \int_{\{|\xi| < \Lambda_\tau\}} |\widehat{u}(\xi)| \, d\xi \\
&\leq (2\pi)^{-d/2} \left(\int_{\{|\xi| < \Lambda_\tau\}} |\xi|^{-2} \, d\xi \right)^{1/2} \left(\int_{\mathbb{R}^d} |\xi|^2 \, |\widehat{u}(\xi)|^2 \, d\xi \right)^{1/2} \\
&= C_d \Lambda_\tau^{(d-2)/2} \|\nabla u\|_{L^2(\mathbb{R}^d)}
\end{aligned}
$$

with

$$
C_d := (2\pi)^{-d/2} \left(\int_{\{|\xi| < 1\}} |\xi|^{-2} \, d\xi \right)^{1/2} .
$$

For given $\tau > 0$ we choose Λ_τ such that

$$
C_d \Lambda_\tau^{(d-2)/2} \|\nabla u\|_{L^2(\mathbb{R}^d)} = \tau/2 . \tag{2.15}
$$

Thus, $|\{|u_\tau| > \tau/2\}| = 0$, and from (2.14) we obtain

$$
|\{|u| > \tau\}| \leq |\{|u - u_\tau| > \tau/2\}| \leq \frac{4}{\tau^2} \|u - u_\tau\|_{L^2(\mathbb{R}^d)}^2 .
$$

By (2.13), Fubini's theorem and Plancherel's identity,

$$
\begin{aligned}
\int_{\mathbb{R}^d} |u|^q \, dx &\leq 4q \int_0^\infty \|u - u_\tau\|_{L^2(\mathbb{R}^d)}^2 \tau^{q-3} \, d\tau \\
&= 4q \int_{\mathbb{R}^d} |\widehat{u}(\xi)|^2 \int_0^\infty \chi_{\{|\xi| \geq \Lambda_\tau\}} \tau^{q-3} \, d\tau \, d\xi .
\end{aligned}
$$

Recalling (2.15), we compute

$$
\begin{aligned}
4q \int_0^\infty \chi_{\{|\xi| \geq \Lambda_\tau\}} \tau^{q-3} \, d\tau &= 4q \int_0^\infty \chi_{\{\tau \leq 2C_d \|\nabla u\|_{L^2(\mathbb{R}^d)} |\xi|^{(d-2)/2}\}} \tau^{q-3} \, d\tau \\
&= C_d' \|\nabla u\|_{L^2(\mathbb{R}^d)}^{q-2} |\xi|^{(q-2)(d-2)/2}
\end{aligned}
$$

with

$$
C_d' := \frac{4q}{q-2} (2C_d)^{q-2} .
$$

By the definition of q we find $q - 2 = 4/(d-2)$ and $(q-2)(d-2)/2 = 2$. Thus, we have shown that

$$
\int_{\mathbb{R}^d} |u|^q \, dx \leq C_d' \|\nabla u\|_{L^2(\mathbb{R}^d)}^{q-2} \int_{\mathbb{R}^d} |\widehat{u}(\xi)|^2 |\xi|^2 \, d\xi = C_d' \|\nabla u\|_{L^2(\mathbb{R}^d)}^q ,
$$

as claimed. $\qquad \square$

Remark 2.45   Following the same steps as in the above proof of Theorem 2.44, one sees that for any $0 < s < d/2$ there is a constant $S_d^{(s)} > 0$ such that

$$\int_{\mathbb{R}^d} |\xi|^{2s} |\widehat{u}|^2 \, d\xi \geq S_d^{(s)} \left(\int_{\mathbb{R}^d} |u|^{2d/(d-2s)} \, dx \right)^{(d-2s)/d}$$

for all $u \in L^2(\mathbb{R}^d)$ for which the left side is finite.

2.4.2 Gagliardo–Nirenberg inequalities

The following inequality (Gagliardo, 1959; Nirenberg, 1959) extends Theorem 2.44 to values of q that are smaller than $2d/(d-2)$ at the expense of involving an additional L^2 norm on the left side. The inequalities are also valid in dimensions $d = 1$ and $d = 2$.

Theorem 2.46 (Gagliardo–Nirenberg inequalities)   *Let $2 < q \leq \infty$ if $d = 1$ and $2 < q < 2d/(d-2)$ if $d \geq 2$. Then there is a constant $S_{q,d} > 0$ such that, for all $u \in H^1(\mathbb{R}^d)$,*

$$\left(\int_{\mathbb{R}^d} |\nabla u|^2 \, dx \right)^{\theta} \left(\int_{\mathbb{R}^d} |u|^2 \, dx \right)^{1-\theta} \geq S_{q,d} \left(\int_{\mathbb{R}^d} |u|^q \, dx \right)^{2/q}, \qquad (2.16)$$

where $\theta \in (0,1)$ is defined by $\theta(d-2) + (1-\theta)d = 2d/q$.

Here and similarly in what follows, for $q = \infty$ and $d = 1$, the inequality (2.16) is understood as

$$\left(\int_{\mathbb{R}} |\nabla u|^2 \, dx \right)^{1/2} \left(\int_{\mathbb{R}} |u|^2 \, dx \right)^{1/2} \geq S_{\infty,1} \, \text{ess-sup}_{\mathbb{R}} |u|^2 \qquad \text{for all } u \in H^1(\mathbb{R}).$$

The sharp values of the constants $S_{q,d}$ play a role in bounds on the lowest eigenvalue of Schrödinger operators in terms of L^p norms of their potentials. We will discuss them in the next two subsections.

Proof   The inequality in the case $q = \infty$ for $d = 1$ follows from Lemma 2.26. For all other values of q in the theorem, we observe that θ satisfies $0 < \theta < \min\{1, d/2\}$ and $q = 2d/(d - 2\theta)$. Then, by Remark 2.45,

$$\int_{\mathbb{R}^d} |\xi|^{2\theta} |\widehat{u}|^2 \, d\xi \geq S_d^{(\theta)} \left(\int_{\mathbb{R}^d} |u|^q \, dx \right)^{2/q}.$$

The inequality in the theorem now follows from

$$\int_{\mathbb{R}^d} |\xi|^{2\theta} |\widehat{u}|^2 \, d\xi \leq \left(\int_{\mathbb{R}^d} |\xi|^2 |\widehat{u}|^2 \, d\xi \right)^{\theta} \left(\int_{\mathbb{R}^d} |\widehat{u}|^2 \, d\xi \right)^{1-\theta}$$

$$= \left(\int_{\mathbb{R}^d} |\nabla u|^2 \, dx \right)^{\theta} \left(\int_{\mathbb{R}^d} |u|^2 \, dx \right)^{1-\theta} .$$

This completes the proof.   □

We note that for $d \geq 3$ one can also deduce (2.16) directly from (2.12) via Hölder's inequality without going through Remark 2.45. Similarly, for $d = 1$ one can deduce (2.16) from Lemma 2.26 via Hölder's inequality.

Remark 2.47   Inequality (2.16) is equivalent to the inequality

$$\int_{\mathbb{R}^d} \left(|\nabla u|^2 + |u|^2 \right) dx \geq \tilde{S}_{q,d} \left(\int_{\mathbb{R}^d} |u|^q \, dx \right)^{2/q} \tag{2.17}$$

and the best constants $S_{q,d}$ and $\tilde{S}_{q,d}$ in these inequalities are related by

$$S_{q,d} = \theta^{\theta} (1 - \theta)^{1-\theta} \tilde{S}_{q,d}$$

with θ as in (2.16). Indeed, (2.16) implies (2.17) in view of the elementary inequality $ab \leq p^{-1} a^p + p'^{-1} b^{p'}$. Conversely, applying (2.17) to $u(\ell x)$ gives

$$\ell^{-d+2+2d/q} \int_{\mathbb{R}^d} \left(|\nabla u|^2 + \ell^{-2} |u|^2 \right) dx \geq \tilde{S}_{q,d} \left(\int_{\mathbb{R}^d} |u|^q \, dx \right)^{2/q} \tag{2.18}$$

for all $\ell > 0$, and (2.16) follows by optimizing the left side with respect to ℓ.

2.4.3 Sharp Gagliardo–Nirenberg inequalities in one dimension

The following theorem, due to Sz. Nagy (1941), gives the sharp constant in the family of one-dimensional Gagliardo–Nirenberg inequalities.

Theorem 2.48   *Let $d = 1$. Then inequality (2.16) holds with the constant*

$$S_{q,1} = \begin{cases} \left(\dfrac{(q+2)^{q+2}}{(q-2)^{q-2} \, 2^{2(q+2)}} \right)^{\frac{1}{2q}} \left(\sqrt{\pi} \, \dfrac{\Gamma(\frac{q}{q-2})}{\Gamma(\frac{q}{q-2} + \frac{1}{2})} \right)^{\frac{q-2}{q}} & \text{if } 2 < q < \infty, \\ 1 & \text{if } q = \infty. \end{cases}$$

These constants are optimal and equality in (2.16) is attained for $u(x) = \cosh^{-\frac{2}{q-2}} x$ for $q < \infty$ and $u(x) = e^{-|x|}$ for $q = \infty$.

Proof Let $u \in H^1(\mathbb{R})$. We know from Lemma 2.26 that, after modification on a set of measure zero, u is continuous on $\mathbb{R}$ and tends to zero at infinity.

We begin with the case $q = \infty$ of the theorem. As shown in the proof of Lemma 2.26, $|u|^2$ is weakly differentiable with derivative $(|u|^2)' = 2\operatorname{Re}(\bar{u}u')$. Applying (2.1) and sending $x' \to \pm\infty$ using $\lim_{x \to \pm\infty} u(x) = 0$, we obtain the two representations

$$|u(x)|^2 = -\int_x^\infty 2\operatorname{Re}\left(\overline{u(y)}u'(y)\right) dy \quad \text{and} \quad |u(x)|^2 = \int_{-\infty}^x 2\operatorname{Re}\left(\overline{u(y)}u'(y)\right) dy \,.$$

Thus

$$|u(x)|^2 = \int_{-\infty}^x \operatorname{Re}\left(\overline{u(y)}u'(y)\right) dy - \int_x^\infty \operatorname{Re}\left(\overline{u(y)}u'(y)\right) dy \,,$$

and therefore, by the Cauchy–Schwarz inequality,

$$|u(x)|^2 \leq \left(\int_{\mathbb{R}} |u|^2 \, dy\right)^{1/2} \left(\int_{\mathbb{R}} |u'|^2 \, dy\right)^{1/2} \,.$$

Since $x \in \mathbb{R}$ is arbitrary, this proves the claimed bound for $q = \infty$.

Now let $2 < q < \infty$. Multiplying u by a constant and recalling that u is bounded, we may assume, without loss of generality, that $\|u\|_\infty = 1$. Moreover, translating u, we may also assume that $|u(0)| = 1$. Set, for $0 \leq s \leq 1$,

$$\varphi(s) := (s^2 - s^q)^{1/2} \quad \text{and} \quad \Phi(s) := \int_0^s \varphi(\sigma) \, d\sigma \,.$$

Since the derivative of Φ is bounded on $[0,1]$ and since $|u| \in H^1(\mathbb{R})$ with derivative given in Lemma 2.11, we can apply Lemma 2.10 and infer that $\Phi(|u|)$ is weakly differentiable with derivative

$$\Phi(|u|)' = \chi_{\{u \neq 0\}} \varphi(|u|) |u|^{-1} \operatorname{Re}\left(\bar{u}u'\right) \,.$$

By the Cauchy–Schwarz inequality, we have

$$\int_{\mathbb{R}} \varphi(|u|) |u'| \, dx \leq \left(\int_{\mathbb{R}} \varphi(|u|)^2 \, dx\right)^{1/2} \left(\int_{\mathbb{R}} |u'|^2 \, dx\right)^{1/2} \,.$$

We bound the left side from below, using $\lim_{x\to\pm\infty}\Phi(|u(x)|)=\Phi(0)=0$,

$$\int_{\mathbb{R}}\varphi(|u|)|u'|\,dx$$

$$=\int_{\mathbb{R}}\varphi(|u|)|u'|\,\chi_{\{u\neq0\}}\,dx$$

$$\geq\int_{-\infty}^{0}\varphi(|u|)|u|^{-1}\operatorname{Re}(\bar{u}u')\,\chi_{\{u\neq0\}}\,dx-\int_{0}^{+\infty}\varphi(|u|)|u|^{-1}\operatorname{Re}(\bar{u}u')\,\chi_{\{u\neq0\}}\,dx$$

$$=\int_{-\infty}^{0}\Phi(|u|)'\,dx-\int_{0}^{+\infty}\Phi(|u|)'\,dx=2\,\Phi(|u(0)|)=2\,\Phi(1)\,.$$

Thus we have shown that

$$2\,\Phi(1)\leq\left(\int_{\mathbb{R}}(|u|^{2}-|u|^{q})\,dx\right)^{1/2}\left(\int_{\mathbb{R}}|u'|^{2}\,dx\right)^{1/2},$$

which we can rewrite as

$$2\,\Phi(1)\left(\int_{\mathbb{R}}|u|^{q}\,dx\right)^{\frac{2}{q-2}}\leq v^{\frac{2}{q-2}}(1-v)^{\frac{1}{2}}\left(\int_{\mathbb{R}}|u|^{2}\,dx\right)^{\frac{1}{2}+\frac{2}{q-2}}\left(\int_{\mathbb{R}}|u'|^{2}\,dx\right)^{\frac{1}{2}}$$

with

$$v:=\frac{\int_{\mathbb{R}}|u|^{q}\,dx}{\int_{\mathbb{R}}|u|^{2}\,dx}\,.$$

Since $|u|\leq1$, we have $0\leq v\leq1$, and therefore, by a simple computation,

$$v^{\frac{2}{q-2}}(1-v)^{\frac{1}{2}}\leq\sup_{0\leq t\leq1}t^{\frac{2}{q-2}}(1-t)^{\frac{1}{2}}=\frac{\left(\frac{2}{q-2}\right)^{\frac{2}{q-2}}\left(\frac{1}{2}\right)^{\frac{1}{2}}}{\left(\frac{2}{q-2}+\frac{1}{2}\right)^{\frac{2}{q-2}+\frac{1}{2}}}\,.$$

This proves the Gagliardo–Nirenberg inequality with the constant

$$S_{q,1}=\left(2\,\Phi(1)\frac{\left(\frac{2}{q-2}+\frac{1}{2}\right)^{\frac{2}{q-2}+\frac{1}{2}}}{\left(\frac{2}{q-2}\right)^{\frac{2}{q-2}}\left(\frac{1}{2}\right)^{\frac{1}{2}}}\right)^{2\theta}=\left(2\,\Phi(1)\frac{\left(\frac{2}{q-2}+\frac{1}{2}\right)^{\frac{2}{q-2}+\frac{1}{2}}}{\left(\frac{2}{q-2}\right)^{\frac{2}{q-2}}\left(\frac{1}{2}\right)^{\frac{1}{2}}}\right)^{\frac{q-2}{q}}\,.$$

It remains to compute $\Phi(1)$. Substituting $\tau=s^{q-2}$ and using (3.16), one obtains

$$\Phi(1)=\int_{0}^{1}(s^{2}-s^{q})^{\frac{1}{2}}\,ds=\frac{1}{q-2}\int_{0}^{1}(1-\tau)^{\frac{1}{2}}\tau^{\frac{2}{q-2}-1}\,d\tau=\frac{1}{q-2}\frac{\Gamma(\frac{3}{2})\,\Gamma(\frac{2}{q-2})}{\Gamma(\frac{2}{q-2}+\frac{3}{2})}$$

$$=\frac{\sqrt{\pi}}{4}\frac{\Gamma(\frac{q}{q-2})}{\Gamma(\frac{q}{q-2}+\frac{1}{2})}\,.$$

This leads to the constant in the statement of the theorem.

The fact that equality holds for $u(x) = \cosh^{-\frac{2}{q-2}} x$ for $q < \infty$ and $u(x) = e^{-|x|}$ for $q = \infty$ follows either by a direct computation (using, for instance, the formulas from the proof of Lemma 5.15) or by verifying that the above inequalities all become equalities for this function. This concludes the proof of the theorem. $\qquad\square$

An alternative and perhaps more intuitive approach to the proof of Theorem 2.48 is to prove the existence of a minimizer of the corresponding variational problem and to compute this minimizer using the Euler–Lagrange equation. This route is taken by Lieb and Thirring (1976), who were apparently unaware of Sz. Nagy's work; see also Lieb (1983d, Theorem 4.2). Due to the translation invariance of the minimization problem, the proof of the existence of an optimizer is not completely straightforward and would invoke either symmetrization techniques or ideas from concentration compactness. Solving the corresponding Euler–Lagrange equation shows that, in fact, $\cosh^{-\frac{2}{q-2}} x$, together with its translates, dilates and multiples, are the only minimizers.

2.4.4 Sharp Sobolev inequalities in higher dimensions

As an application of Theorem 2.48 we now give a proof of the Sobolev inequality (2.12) with optimal constant. The argument is based on the technique of symmetric decreasing rearrangement, which we do not fully cover and for which we refer, for instance, to Lieb and Loss (2001).

Theorem 2.49 *Let $d \geq 3$. Then (2.12) holds with*

$$S_d = \frac{d(d-2)}{4} 2^{2/d} \pi^{1+1/d} \Gamma((d+1)/2)^{-2/d}, \tag{2.19}$$

and this constant is best possible.

The optimal constant S_d was found independently by Rodemich (1966), Rosen (1971), Aubin (1976), and Talenti (1976). The cases of equality in (2.12) with constant (2.19) can be characterized; see the references in §2.9.

Proof Let $u \in H^1(\mathbb{R}^d)$ and let u^* be its symmetric decreasing rearrangement; that is, the function on $\mathbb{R}^d$ defined by

$$u^*(x) := \int_0^\infty \chi_{\{|x| < r_u(t)\}} \, dt \quad \text{with } r_u(t) := (|\{|u| > t\}|/\omega_d)^{1/d},$$

where ω_d is the measure of the unit ball in $\mathbb{R}^d$. This is a non-negative, spherically symmetric function that is non-increasing with respect to $|x|$. Moreover, it is not difficult to see that $|\{u^* > t\}| = |\{|u| > t\}|$ for all $t > 0$, and therefore, for any $p > 0$,

$$\int_{\mathbb{R}^d} (u^*)^p \, dx = \int_{\mathbb{R}^d} |u|^p \, dx \,. \tag{2.20}$$

For these facts we refer to Lieb and Loss (2001, §3.3). A property that is crucial for us is that u^* is weakly differentiable with

$$\int_{\mathbb{R}^d} |\nabla u^*|^2 \, dx \le \int_{\mathbb{R}^d} |\nabla u|^2 \, dx \,; \tag{2.21}$$

see Lieb and Loss (2001, Theorem 7.17). Because of (2.20) and (2.21), it suffices to prove the claimed Sobolev inequality for spherically symmetric functions.

For a spherically symmetric function u, written with a slight abuse of notation as a function of the radius r, we define a function w on $\mathbb{R}$ by

$$u(r) = r^{-(d-2)/2} w(\ln r) \qquad \text{for all } r > 0 \,.$$

Then one can show that

$$\int_{\mathbb{R}^d} |u|^{2d/(d-2)} \, dx = |\mathbb{S}^{d-1}| \int_{\mathbb{R}} |w|^{2d/(d-2)} \, dt$$

and

$$\int_{\mathbb{R}^d} |\nabla u|^2 \, dx = |\mathbb{S}^{d-1}| \int_{\mathbb{R}} \left(|w'|^2 + \left(\tfrac{d-2}{2} \right)^2 |w|^2 \right) dt \,.$$

The first formula follows simply by a change of variables, while the second one follows by the same change of variables and an integration by parts: see Lemma 2.34. To summarize, we see that the optimal constant S_d in (2.12) coincides with $|\mathbb{S}^{d-1}|^{2/d} \tilde{\sigma}_d$, where $\tilde{\sigma}_d$ is the optimal constant in the inequality

$$\int_{\mathbb{R}} \left(|w'|^2 + \left(\tfrac{d-2}{2} \right)^2 |w|^2 \right) dt \ge \tilde{\sigma}_d \left(\int_{\mathbb{R}} |w|^{2d/(d-2)} \, dt \right)^{(d-2)/d} \,.$$

Next, we use the same scaling argument as in Remark 2.47 to see that $\tilde{\sigma}_d$ is related to the optimal constant σ_d in the inequality

$$\left(\int_{\mathbb{R}} |w'|^2 \, dt \right)^{1/d} \left(\int_{\mathbb{R}} |w|^2 \, dt \right)^{(d-1)/d} \ge \sigma_d \left(\int_{\mathbb{R}} |w|^{2d/(d-2)} \, dt \right)^{(d-2)/d}$$

by

$$\tilde{\sigma}_d = \left(\tfrac{d-2}{2} \right)^{(2d-2)/d} d(d-1)^{-(d-1)/d} \sigma_d \,.$$

Finally, according to Theorem 2.48, σ_d is given by

$$\sigma_d = \left(\frac{(d-1)^{d-1}}{4(d-2)^{d-2}}\right)^{\frac{1}{d}} \left(\sqrt{\pi} \, \frac{\Gamma(\frac{d}{2})}{\Gamma(\frac{d+1}{2})}\right)^{\frac{2}{d}}.$$

Recalling that $|\mathbb{S}^{d-1}| = 2\pi^{d/2}/\Gamma(d/2)$, we obtain

$$S_d = \left(\frac{d-2}{2}\right)^{2-\frac{2}{d}} d(d-1)^{-1+\frac{1}{d}} |\mathbb{S}^{d-1}|^{\frac{2}{d}} \left(\frac{(d-1)^{d-1}}{4(d-2)^{d-2}}\right)^{\frac{1}{d}} \left(\sqrt{\pi} \, \frac{\Gamma(\frac{d}{2})}{\Gamma(\frac{d+1}{2})}\right)^{\frac{2}{d}}$$

$$= \frac{d(d-2)}{4} \left(\pi^{d/2} \sqrt{\pi} \, \frac{2}{\Gamma(\frac{d+1}{2})}\right)^{\frac{2}{d}},$$

which proves the inequality with the claimed constant S_d.

A computation shows that equality is reached for $U(x) = (1 + |x|^2)^{-(d-2)/2}$. (Note that this function corresponds under the above logarithmic change of variables to the function $\cosh^{-2/(q-2)} t$ appearing in Theorem 2.48.) If $d \geq 5$, then $U \in H^1(\mathbb{R}^d)$ and the claimed optimality of S_d is proved. If $d = 3, 4$, then $U \notin L^2(\mathbb{R}^d)$, but multiplying by a cut-off function one easily sees that the values of the integrals in (2.12) can get arbitrarily close to those for U. This proves Theorem 2.49. $\qquad\square$

We conclude this subsection with some results about the the optimal constants $S_{q,d}$ in (2.16). For $d \geq 2$ the sharp values of the constants are not known in closed analytic form. One can show, however, that there is a minimizer of the inequality which is unique up to translations, dilations and multiplication by a constant. This minimizer satisfies the Euler–Lagrange equation

$$-\Delta u - \lambda |u|^{q-2} u = -\mu u \qquad \text{in } \mathbb{R}^d, \tag{2.22}$$

where λ and μ arise as Lagrange multipliers. The minimizer can be shown to be radial and therefore (2.22) reduces to an ordinary differential equation. This can be solved numerically. Since the optimal constant $S_{q,d}$ is expressed in terms of integrals of this solution, this allows one to compute this constant numerically to a high accuracy.

Here, we prove a simple qualitative property of these constants. It is natural to set $S_{2,d} := 1$ for all $d \geq 1$ and $S_{\frac{2d}{d-2},d} := S_d$, the optimal constant in (2.12), for $d \geq 3$.

Lemma 2.50 *The optimal constants $S_{q,d}$ in (2.16) are continuous in q for $q \in [2, \infty]$ if $d = 1$, for $q \in [2, \infty)$ if $d = 2$, and for $q \in [2, \frac{2d}{d-2}]$ if $d \geq 3$.*

Proof For any $0 \not\equiv u \in H^1(\mathbb{R}^d)$ the expression

$$Q_q[u] := \ln \frac{\|\nabla u\|^{2\theta} \|u\|^{2(1-\theta)}}{\|u\|_q^2}, \qquad \theta = \frac{d}{2}\left(1 - \frac{2}{q}\right),$$

is continuous with respect to q on the indicated interval and concave with respect to $1/q$. Indeed, the logarithm of the numerator is an affine function in $1/q$ and for the denominator log-concavity follows from Hölder's inequality

$$\|u\|_q \leq \|u\|_{q_0}^{1-\alpha} \|u\|_{q_1}^{\alpha}, \qquad \frac{1}{q} = \frac{1-\alpha}{q_0} + \frac{\alpha}{q_1}.$$

Since the infimum over a family of continuous functions is upper semi-continuous, we conclude that $\ln S_{q,d} = \inf \{Q_q[u]: 0 \not\equiv u \in H^1(\mathbb{R}^d)\}$ is upper semicontinuous on the indicated interval. Moreover, since the infimum over a family of concave functions is concave, $\ln S_{q,d}$ is concave with respect to $1/q$. This concavity implies continuity in the interior of the indicated interval, and by the upper semicontinuity we then obtain continuity in the full interval. □

Remark 2.51 For later purposes, we record an upper bound for $d = 2$, $q = 4$,

$$S_{4,2} \leq \sqrt{\frac{\pi}{2} \frac{12 + 7\sqrt{3}}{3 + 2\sqrt{3}}} \simeq 2.4212.$$

To prove this, we consider the functions $u_\alpha(x) = (1 + |x|^2)^{-\alpha}$ with a parameter $\alpha > 1/2$. Elementary computations yield

$$\frac{\|\nabla u_\alpha\|^2 \|u_\alpha\|^2}{\|u_\alpha\|_4^4} = \frac{\left(\frac{2\pi\alpha}{2\alpha+1}\right)\left(\frac{\pi}{2\alpha-1}\right)}{\frac{\pi}{4\alpha-1}} = 2\pi \frac{\alpha(4\alpha - 1)}{4\alpha^2 - 1}.$$

The right side is minimized at $\alpha = 1 + \sqrt{3}/2$, which leads to the claimed bound.

2.5 Friedrichs and Poincaré inequalities

In this section we are interested in the question whether, for a given open set $\Omega \subset \mathbb{R}^d$ and a given class for functions $u \in H^1(\Omega)$, an inequality of the form

$$\int_\Omega |\nabla u|^2 \, dx \geq C_\Omega \int_\Omega |u|^2 \, dx \tag{2.23}$$

is valid. When Ω has finite measure, this inequality cannot hold for all $u \in H^1(\Omega)$; for instance, it is not true for the constant function. We will see, however, that the inequality is valid for u that vanish on the boundary of Ω in the sense of belonging to $H_0^1(\Omega)$, at least for open sets of finite measure. Inequalities of this type are called *Friedrichs inequalities*. We will also see that the inequality is valid for u with mean value zero, at least for open, connected sets of finite measure with the extension property. Inequalities of this type are called *Poincaré inequalities*.

2.5.1 The Friedrichs inequality on a cube

We begin with the case where the underlying set is a cube.

Lemma 2.52 *Let $Q \subset \mathbb{R}^d$ be a cube. Then*

$$\int_Q |\nabla u|^2 \, dx \geq \frac{d \pi^2}{|Q|^{2/d}} \int_Q |u|^2 \, dx \qquad \text{for all } u \in H_0^1(Q).$$

The constant in the lemma is best possible and attained if u is the product of the functions $\cos(\pi(x_j - a_j)/\ell)$, $j = 1, \ldots, d$, where $a = (a_1, \ldots, a_d)$ is the center of Q and $\ell = |Q|^{1/d}$ its side length. As we will see in the next chapter, the constant is equal to the first eigenvalue of the Dirichlet Laplacian on the cube Q, and the inequality in the lemma can be obtained by diagonalizing this operator explicitly. This is essentially the second proof of Lemma 2.52 that we give below. First, however, we present a more direct proof that does not involve a spectral decomposition.

First proof of Lemma 2.52 By translation and scaling, we may assume that $Q = (0, \pi)^d$, so that the constant in the claimed inequality becomes d. We prove the lemma by induction on d.

Step 1. First, let $d = 1$ and let $u \in H_0^1(0, \pi)$. We write

$$|u'|^2 - |u|^2 - |u' - u \cot x|^2 = -(1 + \cot^2 x)|u|^2 + 2 \operatorname{Re}(u' \overline{u}) \cot x = \frac{d}{dx} \left(|u|^2 \cot x \right).$$

We claim that

$$\lim_{x \to 0} |u(x)|^2 \cot x = \lim_{x \to \pi} |u(x)|^2 \cot x = 0. \tag{2.24}$$

Accepting this for the moment, we integrate the above identity over $(0, \pi)$ and notice that there is no contribution from the boundary. Thus we obtain

$$\int_0^\pi \left(|u'|^2 - |u|^2 - |u' - u \cot x|^2 \right) dx = 0.$$

Dropping the term $|u' - u \cot x|^2 \geq 0$ yields the claimed inequality.

To prove (2.24), we note that, by Remark 2.7 and Lemma 2.27,

$$|u(x)| = \left| \int_0^x u'(y) \, dy \right| \leq x^{1/2} \left(\int_0^x |u'(y)|^2 \, dy \right)^{1/2} \qquad \text{for all } x \in (0, \pi).$$

Since $\cot x \sim 1/x$ as $x \to 0$ and by dominated convergence, we have

$$\limsup_{x \searrow 0} |u(x)|^2 |\cot x| = \limsup_{x \to 0} |x|^{-1} |u(x)|^2 \le \limsup_{x \to 0} \int_0^x |u'(y)|^2 \, dy = 0,$$

which proves the first assertion in (2.24). The second one follows similarly.

Step 2. Now, let $d \ge 2$ and let $u \in H_0^1(Q)$. We denote coordinates by $x = (x', x_d)$ with $x' \in (0, \pi)^{d-1}$ and, correspondingly, $\nabla = (\nabla', \partial_d)$.

Using Corollary 2.18 we see that, for almost every $x' \in (0, \pi)^{d-1}$, we have $u(x', \cdot) \in H^1(0, \pi)$ with weak derivative $\partial_d(x', \cdot)$. Similarly, for almost every $x_d \in (0, \pi)$, we have $u(\cdot, x_d) \in H^1((0, \pi)^{d-1})$ with weak derivative $(\nabla' u)(\cdot, x_d)$. This holds, in fact, for every $u \in H^1(Q)$. In the present situation it is even true that $u \in H_0^1(Q)$ and therefore, for almost every $x' \in (0, \pi)^{d-1}$, it follows that $u(x', \cdot) \in H_0^1(0, \pi)$ and, for almost every $x_d \in (0, \pi)$, it follows that $u(\cdot, x_d) \in H_0^1((0, \pi)^{d-1})$. This follows by Fubini's theorem and the definition of the H_0^1 spaces.

Thus the induction hypothesis implies that

$$\int_0^\pi |\partial_d u(x', x_d)|^2 \, dx_d \ge \int_0^\pi |u(x', x_d)|^2 \, dx_d \qquad \text{for a.e. } x' \in (0, \pi)^{d-1}$$

and, for a.e. $x_d \in (0, \pi)$,

$$\int_{(0,\pi)^{d-1}} |\nabla' u(x', x_d)|^2 \, dx' \ge (d - 1) \int_{(0,\pi)^{d-1}} |u(x', x_d)|^2 \, dx'.$$

Integrating the previous two inequalities with respect to x' and x_d, respectively, and adding them, we obtain the claimed inequality in dimension d. □

Alternative proof of Lemma 2.52 By translation and scaling, we may assume that $Q = (0, \pi)^d$. It is well known that an orthonormal basis of $L^2(0, \pi)$ is given by the functions $a_n(x) := (2/\pi)^{1/2} \sin nx$ for $n \in \mathbb{N}$. Thus, a basis of $L^2(Q)$ is given by $a_n(x) := a_{n_1}(x_1) \cdots a_{n_d}(x_d)$ for $n = (n_1, \ldots, n_d) \in \mathbb{N}^d$ and, for any $u \in L^2(Q)$, we have

$$\int_Q |u|^2 \, dx = \sum_{n \in \mathbb{N}^d} |(u, a_n)|^2.$$

We claim that for $u \in H^1(Q)$ the integral $\int_Q |\nabla u|^2 \, dx$ can be expressed in terms of the coefficients (u, a_n) as

$$\int_Q |\nabla u|^2 \, dx = \sum_{n \in \mathbb{N}^d} |n|^2 |(u, a_n)|^2. \tag{2.25}$$

This will clearly imply the lemma, since

$$\sum_{n \in \mathbb{N}^d} |n|^2 |(u, a_n)|^2 \ge d \sum_{n \in \mathbb{N}^d} |(u, a_n)|^2.$$

We begin by proving (2.25) for $d = 1$. We recall that $b_0(x) := (1/\pi)^{1/2}$ and $b_n(x) := (2/\pi)^{1/2} \cos nx$ for $n \in \mathbb{N}$ constitutes another orthonormal basis of $L^2(0, \pi)$. Clearly, $b_n \in H^1(0, \pi)$ and $b'_n = -na_n$. Therefore, by Lemma 2.29, $(u', b_n) = -(u, b'_n) = n(u, a_n)$. Moreover, by Remark 2.7 and Lemma 2.27, $(u', b_0) = (u(\pi) - u(0))/\sqrt{\pi} = 0$. By Parseval's equality, we obtain

$$\int_0^\pi |u'|^2 \, dx = \sum_{n \in \mathbb{N}_0} |(u', b_n)|^2 = \sum_{n \in \mathbb{N}} n^2 |(u, a_n)|^2 .$$

This yields (2.25) for $d = 1$. In the case $d \geq 2$, we note that $u(x', \cdot) \in H_0^1(0, \pi)$ for almost every $x' \in (0, \pi)^{d-1}$, as discussed in the first proof of Lemma 2.52. Thus, by the one-dimensional equality, we conclude

$$\int_0^\pi |\partial_d u(x', x_d)|^2 \, dx_d = \sum_{n_d \in \mathbb{N}} n_d^2 \left| \int_0^\pi u(x', x_d) a_{n_d}(x_d) \, dx_d \right|^2 ,$$

for a.e. $x' \in (0, \pi)^{d-1}$. Integrating with respect to x' and using once again Parseval's equality, yields

$$\int_Q |\partial_d u|^2 \, dx = \sum_{n_d \in \mathbb{N}} n_d^2 \int_{(0,\pi)^{d-1}} \left| \int_0^\pi u(x', x_d) a_{n_d}(x_d) \, dx_d \right|^2 \, dx'$$

$$= \sum_{n \in \mathbb{N}^d} n_d^2 |(u, a_n)|^2 .$$

Repeating this argument with respect to the other partial derivatives, we finally obtain (2.25) for general d. This concludes the proof.   □

2.5.2 The Friedrichs inequality on a domain

As a consequence of the Gagliardo–Nirenberg inequalities (Theorem 2.46), we obtain the following.

Lemma 2.53 *Let $d \geq 1$. Then there is a constant $C_d > 0$ such that for any open set $\Omega \subset \mathbb{R}^d$ of finite measure and any $u \in H_0^1(\Omega)$,*

$$\int_\Omega |\nabla u|^2 \, dx \geq \frac{C_d}{|\Omega|^{2/d}} \int_\Omega |u|^2 \, dx . \tag{2.26}$$

The optimal value of the constant C_d will be determined in Theorem 2.54 below.

Proof By Lemma 2.28, the extension of u by zero to $\mathbb{R}^d$ belongs to $H^1(\mathbb{R}^d)$, and therefore we can apply the Gagliardo–Nirenberg inequality (2.16) to this extension (with some arbitrary $q > 2$ as in (2.16)). From this and Hölder's inequality we obtain

$$\left(\int_\Omega |\nabla u|^2 \, dx\right)^\theta \left(\int_\Omega |u|^2 \, dx\right)^{1-\theta} \geq S_{q,d} \left(\int_\Omega |u|^q \, dx\right)^{2/q}$$
$$\geq S_{q,d} |\Omega|^{-(q-2)/q} \int_\Omega |u|^2 \, dx \,.$$

This is the claimed inequality. $\square$

As another application of the technique of symmetric decreasing rearrangement, we now compute the sharp constant in Lemma 2.53. It is attained when Ω is a ball. The proof depends, in part, on some of the material in §3.8 below.

Theorem 2.54   *Let $d \geq 1$. Then (2.26) holds with*

$$C_d = j_{(d-2)/2,1}^2 \omega_d^{2/d} \,,$$

where $j_{(d-2)/2,1}$ is the first positive zero of the Bessel function $J_{(d-2)/2}$ and where ω_d is the measure of the unit ball in $\mathbb{R}^d$. This constant is best possible.

Note that $C_1 = \pi^2$ since $J_{-1/2}(r) = (2/(\pi r))^{1/2} \cos r$.

Proof   By Lemma 2.28, we can extend functions $u \in H_0^1(\Omega)$ by zero to functions in $H^1(\mathbb{R}^d)$, and to these functions we can then apply the rearrangement procedure described in the proof of Theorem 2.49. By definition, u^* is supported in the closure of Ω^*, defined as the centered, open ball of measure $|\Omega|$. A simple modification of the proof of Lemma 2.27 (see Lemma 3.51) implies then that $u^* \in H_0^1(\Omega^*)$. According to (2.20) and (2.21), it therefore suffices to compute the best constant C_d in the inequality

$$\int_{\Omega^*} |\nabla u|^2 \, dx \geq C_d |\Omega|^{-2/d} \int_{\Omega^*} |u|^2 \, dx$$

for radial functions $u \in H_0^1(\Omega^*)$. For $d \geq 2$, the latter problem is solved in §3.8.5; see, in particular, Lemma 3.60, where it is shown that the above inequality holds with $C_d |\Omega|^{-2/d} = R^{-2} j_{(d-2)/2,1}^2$, where $j_{(d-2)/2,1}$ is the first positive zero of the Bessel function $J_{(d-2)/2}$ and R is the radius of the ball Ω^*; that is, $R = (|\Omega|/\omega_d)^{1/d}$. This completes the proof for $d \geq 2$. For $d = 1$, see Lemma 2.52. $\square$

2.5.3 A Friedrichs inequality for convex domains

The material in this subsection will not be needed in what follows and can be skipped. In Lemma 2.53, we have shown that inequality (2.23) holds for any open set $\Omega \subset \mathbb{R}^d$ of finite measure with $C_\Omega \geq C_d\,|\Omega|^{-2/d}$ and, in Theorem 2.54, we have determined the sharp value of the constant $C_d > 0$. We now show that, in the special case where Ω is convex, the dependence of the constant on the shape of Ω can be improved to $C_\Omega \geq C_d'\,R_{\mathrm{in}}(\Omega)^{-2}$ with the *inradius*

$$R_{\mathrm{in}}(\Omega) := \sup\{\operatorname{dist}(x, \partial\Omega) : x \in \Omega\}. \tag{2.27}$$

Proposition 2.55 *Let $\Omega \subset \mathbb{R}^d$ be a convex, open set. Then*

$$\int_\Omega |\nabla u|^2\, dx \geq \frac{\pi^2}{4\,R_{\mathrm{in}}(\Omega)^2} \int_\Omega |u|^2\, dx \qquad \textit{for all } u \in H_0^1(\Omega).$$

Results of this type go back to Hersch (1960); see the references in §3.9.

For the proof we need the following lemma. We recall that a *convex polytope* is the convex hull of finitely many points.

Lemma 2.56 *Let $\Omega \subset \mathbb{R}^d$ be an open set such that $\overline{\Omega}$ is a convex polytope and let Γ_n, $n = 1,\ldots,N$, be the $(d-1)$-faces of $\overline{\Omega}$ with corresponding outer unit normals ν_n. We use the convention that the Γ_n are relatively open in $\partial\Omega$. Then there are functions $h_n : \Gamma_n \to [0, R_{\mathrm{in}}(\Omega)]$, $n = 1,\ldots,N$, such that the sets*

$$\{y - t\nu_n : y \in \Gamma_n,\ 0 < t < h_n(y)\} \quad \textit{are disjoint for all } n = 1,\ldots,N \tag{2.28}$$

and such that

$$\Omega \setminus \bigcup_{n=1}^N \{y - t\nu_n : y \in \Gamma_n,\ 0 < t < h_n(y)\} \quad \textit{has measure zero}. \tag{2.29}$$

Proof For $n = 1,\ldots,N$, let E_n be the affine hyperplane containing Γ_n and let

$$\Omega_n := \left\{x \in \Omega : \operatorname{dist}(x, E_n) < \min_{m \neq n} \operatorname{dist}(x, E_m)\right\}.$$

These sets are disjoint and, if $x \in \Omega \setminus \bigcup_{n=1}^N \Omega_n$, then x has equal distance to at least two hyperplanes and the set of such x has measure zero. Thus, to prove the lemma it suffices to show that there are functions $h_n : \Gamma_n \to [0, R_{\mathrm{in}}(\Omega)]$ such that $\Omega_n = \{y - t\nu_n : y \in \Gamma_n,\ 0 < t < h_n(y)\}$.

To find h_n, we first notice that Ω_n is convex. Indeed, since Ω is convex, for each $m \neq n$, the set $\{x \in \Omega : \operatorname{dist}(x, E_n) < \operatorname{dist}(x, E_m)\}$ is convex, and therefore Ω_n is convex as the intersection of these sets with $m \neq n$.

We claim that

$$\Omega_n \subset \{y - t\nu_n \in \mathbb{R}^d : y \in \Gamma_n,\ t > 0\} \quad \textit{for all } n. \tag{2.30}$$

Fix n and $x \in \Omega_n$, and write $x = y - t\nu_n$ with $y \in E_n$ and $t \in \mathbb{R}$. Since ν_n is the outer unit normal and Ω lies on one side of E_n, we have $t > 0$. We need to show $y \in \Gamma_n$.

Arguing by contradiction, we first assume $y \notin \overline{\Omega}$. Then there is a $\tau \in [0, t]$ such that $y - \tau \nu_n \in \partial \Omega$. Since $y \notin \overline{\Omega}$, we have $\tau > 0$. This and the fact that $y \in E_n$ implies that $y - \tau \nu_n \notin E_n$. Since $\partial \Omega \subset \bigcup_m E_m$, we have $y - \tau \nu_n \in E_m$ for some $m \neq n$. Therefore,

$$\text{dist}(x, E_m) \leq |x - (y - \tau \nu_n)| = t - \tau < t = \text{dist}(x, E_n),$$

which means that $x \notin \Omega_n$, contradicting our assumption.

Next, arguing again by contradiction, we assume $y \in \overline{\Omega} \setminus \Gamma_n$. Then there is an $m \neq n$ such that $y \in E_m$. Here we use the fact that Γ_n is the interior of $\overline{\Omega} \cap E_n$ in E_n. We find

$$\text{dist}(x, E_m) \leq |x - y| = t = \text{dist}(x, E_n),$$

which again means that $x \notin \Omega_n$, contradicting our assumption.

Thus, we infer that $y \in \Gamma_n$, which completes the proof of (2.30).

The fact that Ω_n is convex and open, together with (2.30), imply that Ω_n has the form given in (2.28) for a function $h_n : \Gamma_n \to [0, \infty)$.

Finally, note that $h_n(y) \leq R_{\text{in}}(\Omega)$ for all $y \in \Gamma_n$. Indeed, let $y \in \Gamma_n$ and consider $x = y - t\nu_n \in \Omega_n$. We have

$$R_{\text{in}}(\Omega) \geq \text{dist}(x, \partial \Omega) = \min_m \text{dist}(x, \Gamma_m) \geq \min_m \text{dist}(x, E_m) = \text{dist}(x, E_n) = t,$$

where we used $\Gamma_m \subset E_m$ and the definition of Ω_n. Taking the supremum over all t such that $x \in \Omega_n$, we conclude that $R_{\text{in}}(\Omega) \geq h_n(y)$, as claimed. $\qquad \square$

Proof of Proposition 2.55 **Step 1.** The inequality in the one-dimensional case is a special case of Lemma 2.52. Thus, from now on we assume $d \geq 2$.

Step 2. We now prove the bound when $\overline{\Omega}$ is a convex polytope.

Let h_n be as in Lemma 2.56 and denote the set in (2.28) by Ω_n. By (2.28) and (2.29), for any $0 \neq u \in H_0^1(\Omega)$,

$$\frac{\int_\Omega |\nabla u|^2 \, dx}{\int_\Omega |u|^2 \, dx} = \frac{\sum_{n=1}^N \int_{\Omega_n} |\nabla u|^2 \, dx}{\sum_{n=1}^N \int_{\Omega_n} |u|^2 \, dx} \geq \min_{n=1,\ldots,N} \frac{\int_{\Omega_n} |\nabla u|^2 \, dx}{\int_{\Omega_n} |u|^2 \, dx}.$$

This reduces the problem to a single set Ω_n. Let us introduce orthogonal coordinates (y, t) in Ω_n, where $y \in \Gamma_n$ and $0 < t < h_n(y)$. Then

$$\int_{\Omega_n} |\nabla u|^2 \, dx = \int_{\Gamma_n} \int_0^{h_n(y)} \left(|\nabla_y u|^2 + |\partial_t u|^2 \right) dt \, dy \geq \int_{\Gamma_n} \int_0^{h_n(y)} |\partial_t u|^2 \, dt \, dy.$$

For a.e. $y \in \Gamma_n$, we extend the function $u(y, \cdot)$ to an even function on the interval $(0, 2h_n(y))$. This function belongs to $H_0^1(0, 2h_n(y))$; see Lemmas 2.27 and 2.90.

Applying the one-dimensional inequality to this even function, we obtain

$$\int_0^{h_n(y)} |\partial_t u(y,t)|^2 \, dt \geq \frac{\pi^2}{4 \, h_n(y)^2} \int_0^{h_n(y)} |u(y,t)|^2 \, dt$$

$$\geq \frac{\pi^2}{4 \, R_{\text{in}}(\Omega)^2} \int_0^{h_n(y)} |u(y,t)|^2 \, dt \,.$$

Integrating the above inequality with respect to $y \in \Gamma_n$, we obtain

$$\frac{\int_{\Omega_n} |\nabla u|^2 \, dx}{\int_{\Omega_n} |u|^2 \, dx} \geq \frac{\pi^2}{4 \, R_{\text{in}}(\Omega)^2} \,,$$

which proves the bound for polytopes.

Step 3. We now complete the proof of the proposition.

It suffices to prove the inequality for $0 \neq u \in H_0^1(\Omega)$ with bounded support. Fix such a u and an $R > 0$ such that the closure of $\Omega_R := \Omega \cap B_R(0)$ contains the support of u. Let $\varepsilon > 0$. By compactness, there are finitely many cubes centered at points in Ω_R with side length ε that cover $\overline{\Omega_R}$. Let $\tilde{\Omega}$ be the interior of the convex hull of the vertices of these cubes. Then $\overline{\tilde{\Omega}}$ is a convex polytope with $\Omega_R \subset \tilde{\Omega}$ and, by Lemma 2.28, the extension $\tilde{u}$ of u by zero belongs to $H_0^1(\tilde{\Omega})$. Thus, by Step 2,

$$\frac{\int_\Omega |\nabla u|^2 \, dx}{\int_\Omega |u|^2 \, dx} = \frac{\int_{\Omega_R} |\nabla u|^2 \, dx}{\int_{\Omega_R} |u|^2 \, dx} = \frac{\int_{\tilde{\Omega}} |\nabla \tilde{u}|^2 \, dx}{\int_{\tilde{\Omega}} |\tilde{u}|^2 \, dx} \geq \frac{\pi^2}{4 \, R_{\text{in}}(\tilde{\Omega})^2} \,.$$

Since $\tilde{\Omega}$ is contained in an $\sqrt{d}\,\varepsilon/2$-neighborhood of Ω_R, we have

$$R_{\text{in}}(\tilde{\Omega}) \leq R_{\text{in}}(\Omega_R) + \sqrt{d}\,\varepsilon/2 \leq R_{\text{in}}(\Omega) + \sqrt{d}\,\varepsilon/2 \,.$$

Since $\varepsilon > 0$ is arbitrary, this proves the claimed bound. □

2.5.4 The Poincaré inequality on a cube

We begin our study of Poincaré inequalities with the case where the underlying set is a cube.

Lemma 2.57 *Let $Q \subset \mathbb{R}^d$ be a cube. Then*

$$\int_Q |\nabla u|^2 \, dx \geq \frac{\pi^2}{|Q|^{2/d}} \int_Q |u|^2 \, dx \qquad \text{for all } u \in H^1(Q) \text{ with } \int_Q u \, dx = 0 \,.$$

The constant in the lemma is best possible and attained if u is a linear combination of the functions $\sin(\pi(x_j - a_j)/\ell)$, $j = 1, \ldots, d$, where $a = (a_1, \ldots, a_d)$ is the center of Q and $\ell = |Q|^{1/d}$ its side length. As we will see in the next chapter, the constant is equal to the second eigenvalue of the Neumann Laplacian on the cube Q, and the inequality in the lemma can be obtained by diagonalizing this operator explicitly. This is essentially the second proof of Lemma 2.57 that we give below. First, however, we present a more direct proof that does not involve a spectral decomposition. We note its similarity to the first proof of Lemma 2.52 above.

First proof of Lemma 2.57   By translation and scaling, we may assume that $Q = (0, \pi)^d$, so that the constant in the claimed inequality becomes one. We prove the lemma by induction on d.

Step 1. First, let $d = 1$ and assume that $u \in H^1(0, \pi)$ has mean value zero. We extend u to an even function on $(-\pi, \pi)$ and this extended function belongs to $H^1(-\pi, \pi)$. (A similar argument appears in the proof of Lemma 2.90 and is omitted here.) By Lemma 2.26, u is continuous on $[0, \pi]$, and therefore its extension is continuous on $[-\pi, \pi]$. Since $u(x) - u(x - \pi)$ has opposite signs at $x = 0$ and $x = \pi$, it vanishes at some $x_0 \in [0, \pi)$. Set $c := u(x_0) = u(x_0 - \pi)$ and $v := u - c$. We write

$$|v'|^2 - |v|^2 - |v' - v \cot(x - x_0)|^2$$
$$= -\left(1 + \cot^2(x - x_0)\right)|v|^2 + 2 \cot(x - x_0) \operatorname{Re} v' \overline{v} = \frac{d}{dx}\left(|v|^2 \cot(x - x_0)\right).$$

The only poles of $\cot(x - x_0)$ in $[-\pi, \pi]$ are at x_0 and $x_0 - \pi$, which is where v vanishes. We claim that

$$\lim_{x \to x_0} |v(x)|^2 \cot(x - x_0) = \lim_{x \to x_0 - \pi} |v(x)|^2 \cot(x - x_0) = 0. \tag{2.31}$$

The proof of this is similar to that of (2.24) and is omitted. We integrate the above identity over $[-\pi, \pi]$. We claim that the integral of the right side vanishes. Indeed, (2.31) guarantees that there is no contribution from x_0 and $x_0 - \pi$, and we also notice that the boundary terms at $\pm \pi$ cancel each other. Thus, we obtain

$$\int_{-\pi}^{\pi} \left(|v'|^2 - |v|^2 - |v' - v \cot(x - x_0)|^2\right) dx = 0,$$

and therefore,

$$\int_{-\pi}^{\pi} \left(|v'|^2 - |v|^2\right) dx \geq 0. \tag{2.32}$$

Returning to $u = v + c$ and recalling that u has mean value zero, we have

$$
\begin{aligned}
\int_{-\pi}^{\pi} \left(|v'|^2 - |v|^2 \right) dx &= \int_{-\pi}^{\pi} \left(|u'|^2 - |u - c|^2 \right) dx \\
&= \int_{-\pi}^{\pi} \left(|u'|^2 - |u|^2 - |c|^2 \right) dx \\
&\leq \int_{-\pi}^{\pi} \left(|u'|^2 - |u|^2 \right) dx .
\end{aligned}
$$

This, together with (2.32) and the fact that u is even, implies the inequality for $d = 1$.

Step 2. Now, let $d \geq 2$. We denote coordinates by $x = (x', x_d)$ with $x' \in (0, \pi)^{d-1}$ and assume that $u \in H^1(Q)$ has mean value zero. We decompose $u = v + w$, where

$$
w(x_d) := \pi^{-d+1} \int_{(0,\pi)^{d-1}} u(x', x_d) \, dx' .
$$

Then, for almost every $x_d \in (0, \pi)$, $v(\cdot, x_d)$ is orthogonal in $L^2((0, \pi)^{d-1})$ to the constant $w(x_d)$ and therefore

$$
\int_{(0,\pi)^{d-1}} |u(x', x_d)|^2 \, dx' = \int_{(0,\pi)^{d-1}} |v(x', x_d)|^2 \, dx' + \pi^{d-1} |w(x_d)|^2 . \quad (2.33)
$$

Integrating with respect to x_d, we obtain

$$
\int_Q |u(x)|^2 \, dx = \int_Q |v(x)|^2 \, dx + \pi^{d-1} \int_0^\pi |w(x_d)|^2 \, dx_d . \quad (2.34)
$$

We recall from the proof of Lemma 2.51 that, for almost every $x' \in (0, \pi)^{d-1}$, one has $u(x', \cdot) \in H^1(0, \pi)$ with weak derivative $\partial_d u(x', \cdot)$ and, similarly, for almost every $x_d \in (0, \pi)$, one has $u(\cdot, x_d) \in H^1((0, \pi)^{d-1})$ with weak derivative $\nabla' u(\cdot, x_d)$, where $\nabla = (\nabla', \partial_d)$.

For almost every $x_d \in (0, \pi)$ one has $\nabla' u(\cdot, x_d) = \nabla' v(\cdot, x_d)$. Moreover, by the definition of w, $\int_{(0,\pi)^{d-1}} v(x', x_d) \, dx' = 0$. Thus, by the induction assumption, for almost every $x_d \in (0, \pi)$,

$$
\begin{aligned}
\int_{(0,\pi)^{d-1}} |\nabla' u(x', x_d)|^2 \, dx' &= \int_{(0,\pi)^{d-1}} |\nabla' v(x', x_d)|^2 \, dx' \\
&\geq \int_{(0,\pi)^{d-1}} |v(x', x_d)|^2 \, dx' .
\end{aligned}
$$

Integrating with respect to x_d, we obtain

$$
\int_Q |\nabla' u(x', x_d)|^2 \, dx' \geq \int_Q |v(x', x_d)|^2 \, dx' . \quad (2.35)
$$

Meanwhile, it is easy to see that $w \in H^1(0, \pi)$ with

$$w'(x_d) = \pi^{-d+1} \int_{(0,\pi)^{d-1}} \frac{\partial u}{\partial x_d}(x', x_d) \, dx' .$$

Consequently, $v(x', \cdot) \in H^1(0, \pi)$ for almost every $x' \in (0, \pi)^{d-1}$ and, arguing as in (2.33), one finds

$$\int_{(0,\pi)^{d-1}} |\partial_d u(x', x_d)|^2 \, dx' = \int_{(0,\pi)^{d-1}} |\partial_d v(x', x_d)|^2 \, dx' + \pi^{d-1} |w'(x_d)|^2 .$$

Integrating with respect to x_d and dropping a non-negative term, we obtain

$$\int_Q |\partial_d u(x', x_d)|^2 \, dx' \geq \pi^{d-1} \int_0^\pi |w'(x_d)|^2 \, dx_d .$$

Since u has mean value zero, we have $\int_0^\pi w(x_d) \, dx_d = 0$ and thus, again by the induction assumption,

$$\int_0^\pi |w'(x_d)|^2 \, dx_d \geq \int_0^\pi |w(x_d)|^2 \, dx_d .$$

Thus, we have shown that

$$\int_Q |\partial_d u(x', x_d)|^2 \, dx' \geq \pi^{d-1} \int_0^\pi |w(x_d)|^2 \, dx_d . \tag{2.36}$$

Adding inequalities (2.35) and (2.36) and taking (2.34) into account, we obtain the claimed inequality in dimension d. □

Alternative proof of Lemma 2.57 By translation and scaling, we may assume that $Q = (0, \pi)^d$. It is well known that an orthonormal basis of $L^2(0, \pi)$ is given by the functions $b_0(x) := (1/\pi)^{1/2}$ and $b_n(x) := (2/\pi)^{1/2} \cos nx$ for $n \in \mathbb{N}$. Thus a basis of $L^2(Q)$ is given by $b_n(x) := b_{n_1}(x_1) \cdots b_{n_d}(x_d)$ for $n = (n_1, \ldots, n_d) \in \mathbb{N}_0^d$ and, for any $u \in L^2(Q)$ we have

$$\int_Q |u|^2 \, dx = \sum_{n \in \mathbb{N}_0^d} |(u, b_n)|^2 .$$

We next claim that for $u \in H^1(Q)$ the integral $\int_Q |\nabla u|^2 \, dx$ can be expressed in terms of the coefficients (u, b_n) as

$$\int_Q |\nabla u|^2 \, dx = \sum_{n \in \mathbb{N}_0^d} |n|^2 |(u, b_n)|^2 . \tag{2.37}$$

This will clearly imply the lemma since

$$\sum_{n \in \mathbb{N}_0^d} |(u, b_n)|^2 \leq |(u, b_0)|^2 + \sum_{n \in \mathbb{N}_0^d} |n|^2 |(u, b_n)|^2$$

and $(u, b_0) = \pi^{-d/2} \int_Q u \, dx$.

We begin by proving (2.37) for $d = 1$. We recall that $a_n(x) := (2/\pi)^{1/2} \sin nx$ for $n \in \mathbb{N}$ constitutes another orthonormal basis of $L^2(0, \pi)$. By Lemma 2.27, $a_n \in H_0^1(0, \pi)$ and, clearly, $a_n' = nb_n$. Therefore, by Lemma 2.29, $(u', a_n) = -(u, a_n') = -n(u, b_n)$. By Parseval's equality, we obtain

$$\int_0^\pi |u'|^2 \, dx = \sum_{n \in \mathbb{N}} |(u', a_n)|^2 = \sum_{n \in \mathbb{N}_0} n^2 |(u, b_n)|^2 .$$

This yields (2.37) for $d = 1$. In the general case, we note that $u(x', \cdot) \in H^1(0, \pi)$ for almost every $x' \in (0, \pi)^{d-1}$, as discussed in the first proof of Lemma 2.57. Thus, by (2.37) in the case $d = 1$, we find

$$\int_0^\pi |\partial_d u(x', x_d)|^2 \, dx_d = \sum_{n_d \in \mathbb{N}_0} n_d^2 \left| \int_0^\pi u(x', x_d) b_{n_d}(x_d) \, dx_d \right|^2 .$$

Integrating with respect to the remaining variables and using once again Parseval gives

$$\int_Q |\partial_d u|^2 \, dx = \sum_{n_d \in \mathbb{N}_0} n_d^2 \int_{(0,\pi)^{d-1}} \left| \int_0^\pi u(x', x_d) b_{n_d}(x_d) \, dx_d \right|^2 dx'$$
$$= \sum_{n \in \mathbb{N}_0^d} n_d^2 |(u, b_n)|^2 .$$

Repeating this argument with respect to the other partial derivatives, we finally obtain (2.37) for general d. This concludes the proof. □

2.5.5 The Poincaré inequality on a domain

Next, we extend Lemma 2.57 to sets, $\Omega \subset \mathbb{R}^d$, more general than cubes. We recall that the notion of the extension property was introduced in §2.3.2; see also §2.8.

Proposition 2.58 (Poincaré inequality) *Let $\Omega \subset \mathbb{R}^d$ be an open connected set of finite measure with the extension property. Then there is a constant $P_\Omega > 0$ (the Poincaré constant) such that*

$$\int_\Omega |\nabla u|^2 \, dx \geq P_\Omega \int_\Omega |u|^2 \, dx \qquad \text{for all } u \in H^1(\Omega) \text{ with } \int_\Omega u \, dx = 0 . \quad (2.38)$$

Proof We argue by contradiction. If there was no such constant, there would be a sequence $(u_n) \in H^1(\Omega)$ with $\int_\Omega u_n \, dx = 0$ such that $\int_\Omega |u_n|^2 \, dx = 1$ and $\int_\Omega |\nabla u_n|^2 \, dx \to 0$. Since (u_n) is bounded in $H^1(\Omega)$, after passing to a subsequence, we may assume that $u_n \to u$ weakly in $H^1(\Omega)$ for some $u \in H^1(\Omega)$. Moreover, by weak convergence using $1 \in L^2(\Omega)$,

$$\int_\Omega u\, dx - \lim_{n\to\infty} \int_\Omega u_n \, dx - 0,$$

and by weak lower semicontinuity,

$$\int_\Omega |\nabla u|^2 \, dx \le \lim_{n\to\infty} \int_\Omega |\nabla u_n|^2 \, dx = 0.$$

By Lemma 2.5 and the connectedness of Ω, these two facts imply that $u \equiv 0$. On the other hand, by Theorem 2.40 we have $u_n \to u$ *strongly* in $L^2(\Omega)$, so

$$\int_\Omega |u|^2 \, dx = \lim_{n\to\infty} \int_\Omega |u_n|^2 \, dx = 1,$$

contradicting $u \equiv 0$. This completes the proof.　　　　　□

Remark 2.59　The statement of Proposition 2.58 can be slightly strengthened. If Ω is as in the proposition and $u \in L^1(\Omega)$ is weakly differentiable with $\nabla u \in L^2(\Omega)$, then $u \in L^2(\Omega)$ and, with the same constant as in (2.38),

$$\int_\Omega |\nabla u|^2 \, dx \ge P_\Omega \left(\int_\Omega |u|^2 \, dx - |\Omega|^{-1} \left| \int_\Omega u \, dx \right|^2 \right). \qquad (2.39)$$

To see this, let $u^{(M)}$ be as in the proof of Proposition 2.23. Then $u^{(M)} \in H^1(\Omega)$ and (2.38) applied to $u^{(M)} - |\Omega|^{-1} \int_\Omega u^{(M)} \, dx$ yields inequality (2.39) with u replaced by $u^{(M)}$. Since $|\nabla u^{(M)}| \le |\nabla u|$ and since $u^{(M)} \to u$ in $L^1(\Omega)$ as $M \to \infty$, we conclude by monotone convergence that $u \in L^2(\Omega)$ and that (2.39) holds for u.

　　The proof of Proposition 2.58 proceeds by compactness and does not yield an explicit value for the constant P_Ω. Besides cubes, treated in §2.5.4, and rectangles, which can be treated in the same way, one can also compute explicitly the Poincaré constant for balls, as described next.

Remark 2.60　Let $B_R \subset \mathbb{R}^d$ be a ball of radius $R > 0$. Then

$$\int_{B_R} |\nabla u|^2 \, dx \ge q_{1,d,1}^2 R^{-2} \int_{B_R} |u|^2 \, dx \quad \text{for all } u \in H^1(B_R) \text{ with } \int_{B_R} u \, dx = 0,$$

where $q_{1,d,1}$ is the first positive zero of the derivative of $t^{-(d-2)/2} J_{d/2}(t)$, and $J_{d/2}$ is a Bessel function (Watson, 1944; Abramowitz and Stegun, 1964). This inequality, as well as the optimality of the constant, follows via separation of variables in spherical coordinates (see §3.8.3) and the solution of the resulting one-dimensional problem (see §3.8.4 and, in particular, Lemma 3.55 and Remark 3.57).

We end this subsection with a somewhat technical lemma that can be skipped at first reading. It generalizes Lemma 2.53 to the case of functions that vanish only on some part of $\partial\Omega$. To state it, we introduce a new subspace of $H^1(\Omega)$, depending on a subset $\Gamma \subset \partial\Omega$, and which we denote by $H^1_{0,\Gamma}(\Omega)$. It is the closure in $H^1(\Omega)$ of all functions in $C^\infty(\Omega) \cap H^1(\Omega)$ that vanish in a neighborhood of Γ.

Lemma 2.61 *Let $Q \subset \mathbb{R}^d$ be an open cube and let $\Omega \subset Q$ be an open set. Then, for all $u \in H^1_{0,Q\cap\partial\Omega}(\Omega)$,*

$$\int_\Omega |\nabla u|^2 \, dx \geq \frac{\pi^2}{|Q|^{2/d}} \frac{|Q| - |\Omega|}{2|Q| - |\Omega|} \int_\Omega |u|^2 \, dx. \tag{2.40}$$

We note that the constant in (2.40) depends only on the measure and not on the shape of Ω. This will be important for our application of the lemma.

Proof We extend u by zero to all of Q and put $a := |Q|^{-1} \int_\Omega u \, dx$. Since the extended u belongs to $H^1(Q)$ (this can be shown in the same way as in the proof of Lemma 2.28) and since $\int_Q (u - a) \, dx = 0$, we can apply Lemma 2.57 to the function $u - a$ and obtain

$$\int_\Omega |\nabla u|^2 \, dx = \int_Q |\nabla(u - a)|^2 \, dx \geq \frac{\pi^2}{|Q|^{2/d}} \int_Q |u - a|^2 \, dx. \tag{2.41}$$

Thus, we need to find a lower bound on $\int_Q |u - a|^2 \, dx$.

On the one hand,

$$\int_\Omega |u|^2 \, dx = \int_\Omega |u - a|^2 \, dx + 2\operatorname{Re}\bar{a} \int_\Omega u \, dx - |a|^2 |\Omega|$$

$$= \int_\Omega |u - a|^2 \, dx + |a|^2 (2|Q| - |\Omega|). \tag{2.42}$$

On the other hand,

$$\int_Q |u - a|^2 \, dx = \int_\Omega |u - a|^2 \, dx + \int_{Q\setminus\Omega} |a|^2 \, dx$$

$$= \int_\Omega |u - a|^2 \, dx + |a|^2 (|Q| - |\Omega|). \tag{2.43}$$

If we eliminate the $|a|^2$ terms from (2.42) and (2.43), we find

$$\int_\Omega |u|^2 \, dx = \int_\Omega |u - a|^2 \, dx + \frac{2|Q| - |\Omega|}{|Q| - |\Omega|} \left(\int_Q |u - a|^2 \, dx - \int_\Omega |u - a|^2 \, dx \right)$$

$$= \frac{2|Q| - |\Omega|}{|Q| - |\Omega|} \int_Q |u - a|^2 \, dx - \frac{|Q|}{|Q| - |\Omega|} \int_\Omega |u - a|^2 \, dx.$$

Dropping the negative term on the right hand side, in view of (2.41) we arrive at (2.40). This completes the proof. $\qquad\square$

2.5.6 Poincaré–Sobolev inequalities

The following proposition gives a Sobolev inequality on domains. In general, such an inequality requires the L^2 norm of the function in addition to the L^2 norm of the gradient. By the Poincaré inequality, this term can be dropped when the set is connected and the function has mean value zero. The assumption on the boundary is encoded in the extension property; see §2.3.2.

Proposition 2.62 (Poincaré–Sobolev inequality) *Let $2 < q \leq \infty$ if $d = 1$, $2 < q < \infty$ if $d = 2$, and $2 < q \leq 2d/(d-2)$ if $d \geq 3$, and let $\Omega \subset \mathbb{R}^d$ be an open set with the extension property. Then there is a constant $\tilde{S}_{\Omega,q} > 0$ such that*

$$\int_\Omega |\nabla u|^2 \, dx + \int_\Omega |u|^2 \, dx \geq \tilde{S}_{\Omega,q} \left(\int_\Omega |u|^q \, dx \right)^{2/q} \qquad \text{for all } u \in H^1(\Omega).$$
(2.44)

Further, if Ω is connected and of finite measure, there is a constant $K_{\Omega,q} > 0$ such that

$$\int_\Omega |\nabla u|^2 \, dx \geq K_{\Omega,q} \left(\int_\Omega |u|^q \, dx \right)^{2/q} \qquad \text{for all } u \in H^1(\Omega) \text{ with } \int_\Omega u \, dx = 0.$$
(2.45)

Proof Let $E : H^1(\Omega) \to H^1(\mathbb{R}^d)$ be an extension operator. Then, by the Sobolev inequality (2.12) for $q = 2d/(d-2)$ and $d \geq 3$, or by the Gagliardo–Nirenberg inequality (2.16) (in the equivalent form (2.17)) for other values of q and for general $d \geq 1$,

$$\|u\|_{L^q(\Omega)} \leq \|Eu\|_{L^q(\mathbb{R}^d)} \leq (\tilde{S}_{q,d})^{-1/2} \|Eu\|_{H^1(\mathbb{R}^d)}$$
$$\leq (\tilde{S}_{q,d})^{-1/2} \|E\|_{H^1(\Omega) \to H^1(\mathbb{R}^d)} \|u\|_{H^1(\Omega)}.$$

This proves (2.44). If Ω is connected and of finite measure, and if $\int_\Omega u \, dx = 0$, then by the Poincaré inequality (2.38),

$$\|u\|_{H^1(\Omega)} \leq \left(1 + P_\Omega^{-1} \right)^{1/2} \|\nabla u\|_{L^2(\Omega)},$$

so (2.45) follows from (2.44). This proves the proposition.   □

Remark 2.63   In the situation of Proposition 2.62, but with $d \geq 3$, one has

$$\left(\int_\Omega |\nabla u|^2 \, dx + \int_\Omega |u|^2 \, dx \right)^\theta \left(\int_\Omega |u|^2 \, dx \right)^{1-\theta}$$
$$\geq S_{\Omega,q} \left(\int_\Omega |u|^q \, dx \right)^{2/q} \quad \text{for all } u \in H^1(\Omega)$$
(2.46)

with θ as in Theorem 2.46. This follows by Hölder's inequality from the proposition applied with $q = 2d/(d-2)$. A similar argument in $d = 1, 2$ leads to (2.46) with a larger value of θ than in Theorem 2.46. If the extension operator $E : H^1(\Omega) \to H^1(\mathbb{R}^d)$ extends to a bounded operator from $L^2(\Omega)$ to $L^2(\mathbb{R}^d)$, then (2.46) holds with θ as in Theorem 2.46. This follows by the same proof as in Proposition 2.62, but using the Gagliardo–Nirenberg inequality in the form (2.16). We note that this assumption on the extension operator is satisfied, for instance, for sets with uniformly Lipschitz continuous boundary; see Theorem 2.92.

Later, we will apply Proposition 2.62 with cubes of different sizes and, when doing this, we will need to track the dependence of the constants on the size of the cubes. By a simple scaling argument and the fact that cubes have the extension property (Theorem 2.92), we obtain from Proposition 2.62 the following consequence.

Corollary 2.64 (Poincaré–Sobolev inequality on cubes) *Let $2 < q \le \infty$ if $d = 1$, $2 < q < \infty$ if $d = 2$, and $2 < q \le 2d/(d-2)$ if $d \ge 3$. Then there are constants $S'_{q,d} > 0$ and $K_{q,d} > 0$ such that for any cube $Q \subset \mathbb{R}^d$ and any function $u \in H^1(Q)$,*

$$\int_Q |\nabla u|^2 \, dx + |Q|^{-2/d} \int_Q |u|^2 \, dx \ge S'_{q,d} |Q|^{-2\left(\frac{1}{q} - \frac{d-2}{2d}\right)} \left(\int_Q |u|^q \, dx \right)^{2/q}$$

and, if $\int_Q u \, dx = 0$,

$$\int_Q |\nabla u|^2 \, dx \ge K_{q,d} |Q|^{-2\left(\frac{1}{q} - \frac{d-2}{2d}\right)} \left(\int_Q |u|^q \, dx \right)^{2/q}.$$

2.6 Hardy inequalities

The Hardy inequalities that we consider in this section typically bound a weighted L^2 norm from above by a weighted L^2 norm of the gradient. We consider various versions of such inequalities, first in §2.6.1 on the half-line with power weights, then in §2.6.2 on the half-line with more general weights and finally in §2.6.3 on $\mathbb{R}^d$ with $d \ge 3$.

2.6.1 Hardy's inequality on the half-line

We remind readers that here and elsewhere we use the notation $\mathbb{R}_+ = (0, +\infty)$.

Theorem 2.65 *Let $\rho \in \mathbb{R} \setminus \{1\}$. Let u be weakly differentiable on $\mathbb{R}_+$ with $u' \in L^2(\mathbb{R}_+, r^\rho \, dr)$ and assume that*

$$\liminf_{r \to 0} |u(r)| = 0 \text{ if } \rho < 1, \qquad \liminf_{r \to \infty} |u(r)| = 0 \text{ if } \rho > 1.$$

Then

$$\int_0^\infty |u(r)|^2 r^{-2+\rho} \, dr \le \left(\frac{2}{\rho - 1} \right)^2 \int_0^\infty |u'(r)|^2 r^\rho \, dr. \tag{2.47}$$

The constant on the right side is optimal.

The lim inf in the theorem is taken with the continuous representative of u, which exists according to Lemma 2.6.

Proof We have, using twice the product rule for weakly differentiable functions,

$$\left| u' + \frac{\rho - 1}{2r} u \right|^2 r^\rho = |u'|^2 r^\rho + \frac{(\rho - 1)^2}{4} |u|^2 r^{-2+\rho} + \frac{\rho - 1}{2} (|u|^2)' r^{-1+\rho}$$

$$= |u'|^2 r^\rho - \frac{(\rho - 1)^2}{4} |u|^2 r^{-2+\rho} + \frac{\rho - 1}{2} \left(|u|^2 r^{-1+\rho} \right)'.$$

Thus, by Lemma 2.6 applied to the weakly differentiable function $|u|^2 r^{-1+\rho}$, we find for any $0 < \varepsilon < M < \infty$,

$$0 \le \int_\varepsilon^M \left| u' + \frac{\rho - 1}{2r} u \right|^2 r^\rho \, dr = \int_\varepsilon^M |u'|^2 r^\rho \, dr - \frac{(\rho - 1)^2}{4} \int_\varepsilon^M |u|^2 r^{-2+\rho} \, dr$$

$$+ \frac{\rho - 1}{2} |u(M)|^2 M^{-1+\rho} - \frac{\rho - 1}{2} |u(\varepsilon)|^2 \varepsilon^{-1+\rho}. \tag{2.48}$$

We first assume that $\rho < 1$ and note that we have the inequality

$$|u(r)|^2 \le (1 - \rho)^{-1} r^{1-\rho} \int_0^r |u'(s)|^2 s^\rho \, ds. \tag{2.49}$$

For, by Lemma 2.6, for any $0 < r' < r$,

$$|u(r) - u(r')| = \left| \int_{r'}^r u'(s) \, ds \right| \le \left(\int_{r'}^r |u'(s)|^2 s^\rho \, ds \right)^{1/2} \left(\int_{r'}^r s^{-\rho} \, ds \right)^{1/2}$$

$$\le \left(\int_0^r |u'(s)|^2 s^\rho \, ds \right)^{1/2} \left((1 - \rho)^{-1} r^{1-\rho} \right)^{1/2}.$$

Now using the assumption $\liminf_{r' \to 0} |u(r')| = 0$, we obtain (2.49).

We now return to (2.48) and want to let $M \to \infty$. Note that the term involving $u(M)$ has a favorable sign and can be dropped. Moreover, the first integral on the right side of (2.48) converges as $M \to \infty$ by assumption. We conclude from that

$$\int_\rho^\infty |u|^2 r^{-2+\rho}\, dr < \infty .$$

Thus we infer that

$$0 \le \int_\varepsilon^\infty \left| u' + \frac{\rho-1}{2r} u \right|^2 r^\rho\, dr = \int_\varepsilon^\infty |u'|^2 r^\rho\, dr - \frac{(\rho-1)^2}{4} \int_\varepsilon^\infty |u|^2 r^{-2+\rho}\, dr$$
$$- \frac{\rho-1}{2} |u(\varepsilon)|^2 \varepsilon^{-1+\rho}. \tag{2.50}$$

Finally, we let $\varepsilon \to 0$. The first term on the right side of (2.50) converges by assumption and, by dominated convergence, the same assumption together with (2.49) implies that $\lim_{\varepsilon\to 0} |u(\varepsilon)|^2 \varepsilon^{-1+\rho} = 0$. This implies that the second term on the right side of (2.50) converges as well and that we have

$$0 \le \int_0^\infty \left| u' + \frac{\rho-1}{2r} u \right|^2 r^\rho\, dr = \int_0^\infty |u'|^2 r^\rho\, dr - \frac{(\rho-1)^2}{4} \int_0^\infty |u|^2 r^{-2+\rho}\, dr .$$
$$\tag{2.51}$$

This proves, in particular, the claimed inequality in the case $\rho < 1$.

The case $\rho > 1$ is similar with the roles of 0 and ∞ swapped. Instead of (2.49) we use

$$|u(r)|^2 \le (\rho-1)^{-1} r^{1-\rho} \int_r^\infty |u'(s)|^2 s^\rho\, ds , \tag{2.52}$$

which is proved similarly. Using this inequality, we first let $\varepsilon \to 0$ and then $M \to \infty$. This shows that in this case (2.51) is valid as well with all three involved integrals being finite. Thus, we also obtain the claimed inequality in the case $\rho > 1$.

In order to prove sharpness of the constant in (2.47), we will construct a family of functions u_R satisfying the assumptions of the theorem and

$$\lim_{R\to\infty} \frac{\int_0^\infty \left| u_R' + \frac{\rho-1}{2r} u_R \right|^2 r^\rho\, dr}{\int_0^\infty |u_R|^2 r^{-2+\rho}\, dr} = 0 . \tag{2.53}$$

We shall choose $u_R := r^{(1-\rho)/2} \beta_R$ with a function β_R to be determined, satisfying $\lim_{r\to 0} \beta_R(r) = 0$ and $\lim_{r\to\infty} \beta_R(r) = 0$. Then

$$\frac{\int_0^\infty \left| u_R' + \frac{\rho-1}{2r} u_R \right|^2 r^\rho\, dr}{\int_0^\infty |u_R|^2 r^{-2+\rho}\, dr} = \frac{\int_0^\infty |\beta_R'|^2 r\, dr}{\int_0^\infty |\beta_R|^2 r^{-1}\, dr} .$$

Note that if, for $R \geq 1$,

$$\beta_R(r) := \left(1 - \frac{|\ln r|}{\ln R}\right),$$

then u_R is weakly differentiable with $u_R' \in L^2(\mathbb{R}_+, r^\rho \, dr)$. Moreover,

$$\int_0^\infty |\beta_R'|^2 r \, dr = (\ln R)^{-2} \int_{1/R}^R r^{-1} \, dr = \frac{2}{\ln R} \to 0 \qquad \text{as } R \to \infty$$

and

$$\int_0^\infty |\beta_R|^2 r^{-1} \, dr \geq \int_{1/\sqrt{R}}^{\sqrt{R}} (1/2)^2 r^{-1} \, dr = \frac{\ln R}{4} \to \infty \qquad \text{as } R \to \infty.$$

Hence, (2.53) holds, which implies the claimed sharpness using (2.51). $\qquad \square$

Remark 2.66 We claim that the assertions in Theorem 2.65 for different values of ρ are equivalent, in the sense that the theorem for any given $\rho_0 \in \mathbb{R} \setminus \{1\}$ implies the theorem for any other $\rho \in \mathbb{R} \setminus \{1\}$. Indeed, if u is weakly differentiable on $\mathbb{R}_+$ with $u' \in L^2(\mathbb{R}_+, r^\rho \, dr)$ and $\alpha := (\rho_0 - 1)/(\rho - 1)$, then $v(s) := u(s^\alpha)$ is weakly differentiable on $\mathbb{R}_+$ with $v'(s) = \alpha s^{\alpha-1} u'(s^\alpha)$ and

$$\int_0^\infty |v'(s)|^2 s^{\rho_0} \, ds = |\alpha| \int_0^\infty |u'(r)|^2 r^\rho \, dr,$$

$$\int_0^\infty |v(s)|^2 s^{-2+\rho_0} \, ds = |\alpha|^{-1} \int_0^\infty |u(r)|^2 r^{-2+\rho} \, dr.$$

Moreover, v satisfies the lim inf-condition corresponding to ρ_0 if u satisfies that corresponding to ρ. Therefore, the inequality in Theorem 2.65 for u follows from that for v. Note that this observation also 'explains' why in the previous proof in the cases $\rho < 1$ and $\rho > 1$ the roles of 0 and infinity are reversed.

Remark 2.67 Hardy's inequality can also be stated in integral form. Namely, if $f \in L^2(\mathbb{R}_+)$, then

$$\int_0^\infty r^{-2} \left| r^{\rho/2} \int_0^r s^{-\rho/2} f(s) \, ds \right|^2 dr \leq \left(\frac{2}{\rho - 1}\right)^2 \int_0^\infty |f(r)|^2 \, dr \qquad \text{if } \rho < 1,$$

$$\int_0^\infty r^{-2} \left| r^{\rho/2} \int_r^\infty s^{-\rho/2} f(s) \, ds \right|^2 dr \leq \left(\frac{2}{\rho - 1}\right)^2 \int_0^\infty |f(r)|^2 \, dr \qquad \text{if } \rho > 1.$$

This follows from Theorem 2.65 with $u(r) = \int_0^r s^{-\rho/2} f(s) \, ds$ and $u(r) = \int_r^\infty s^{-\rho/2} f(s) \, ds$, respectively, whose weak differentiability follows from Lemma 2.6.

Alternative proof of Theorem 2.65 We prove Hardy's inequality in the integral form of Remark 2.67. By Lemma 2.6, this integral inequality implies

Theorem 2.65 by setting $f(r) = r^{\rho/2} u'(r)$. Moreover, using the change of variables in Remark 2.66, it suffices to prove the inequality in Remark 2.67 for $\rho = 0$.

To prove that inequality, we note that for any $0 < \alpha < 1$,

$$\left| \int_0^r f(s)\,ds \right|^2 \le \int_0^r s^{-\alpha}\,ds \int_0^r s^\alpha |f(s)|^2\,ds = \frac{1}{1-\alpha} r^{1-\alpha} \int_0^r s^\alpha |f(s)|^2\,dr.$$

Multiplying by r^{-2}, integrating and interchanging the integrals we obtain

$$\int_0^\infty \left| \int_0^r f(s)\,ds \right|^2 \frac{dr}{r^2} \le \frac{1}{1-\alpha} \int_0^\infty t^\alpha |f(t)|^2 \int_t^\infty r^{-1-\alpha}\,dr\,dt$$

$$= \frac{1}{\alpha(1-\alpha)} \int_0^\infty |f(t)|^2\,dt.$$

Choosing $\alpha = 1/2$ yields the claim. $\qquad\square$

In Theorem 2.65 the value $\rho = 1$ is excluded and indeed, while proving (2.53), we showed that $\int_0^\infty |u|^2 r^{-1}\,dr$ cannot be bounded in terms of $\int_0^\infty |u'|^2 r\,dr$, even when u vanishes both near zero and infinity. The following proposition provides a replacement for the 'missing' case $\rho = 1$ in Theorem 2.65 with an additional logarithmic factor.

Proposition 2.68 *Let I be either* $(0,1)$ *or* $(1,\infty)$. *Let u be weakly differentiable on I with* $u' \in L^2(I, r\,dr)$ *and assume that*

$$\liminf_{r \to 1} |u(r)| = 0.$$

Then

$$\int_I |u(r)|^2 r^{-1} (\ln r)^{-2}\,dr \le 4 \int_I |u'(r)|^2 r\,dr.$$

The constant on the right side is optimal.

Proof For u as in the proposition and $t \in \mathbb{R}_+$, define $v(t) := u(e^{\pm t})$ with the $+$ sign if $I = (1, \infty)$ and the $-$ sign if $I = (0, 1)$. Then v is weakly differentiable on $\mathbb{R}_+$ with $v'(t) = \pm u'(e^{\pm t}) e^{\pm t}$ and $\liminf_{t \to 0} |v(t)| = 0$. Moreover, by a change of variables,

$$\int_0^\infty |v'(t)|^2\,dt = \int_0^\infty |u'(e^{\pm t})|^2 e^{\pm 2t}\,dt = \int_I |u'(r)|^2 r\,dr < \infty.$$

Thus, the Hardy inequality from Theorem 2.65 with $\rho = 0$ implies that

$$\int_I |u'(r)|^2 r\,dr$$

$$= \int_0^\infty |v'(t)|^2\,dt \ge \frac{1}{4} \int_0^\infty |v(t)|^2 t^{-2}\,dt = \frac{1}{4} \int_I |u(r)|^2 r^{-1} (\ln r)^{-2}\,dr,$$

as claimed. The optimality of the constant 4 follows from the optimality of this constant in Theorem 2.65.   □

2.6.2 Hardy's inequality on the half-line with weights

The material in this subsection is somewhat technical and can be skipped at first reading. We begin with a strengthening of Theorem 2.65 with a supremum under the integral on the left side.

Theorem 2.69 *Let $\rho \in \mathbb{R} \setminus \{1\}$. Let u be weakly differentiable on $\mathbb{R}_+$ with $u' \in L^2(\mathbb{R}_+, r^\rho \, dr)$ and assume that*

$$\liminf_{r \to 0} |u(r)| = 0 \text{ if } \rho < 1, \qquad \liminf_{r \to \infty} |u(r)| = 0 \text{ if } \rho > 1.$$

Then

$$\int_0^\infty \sup_{s \in \mathbb{R}_+} \min\{s^{-2+2\rho}, r^{-2+2\rho}\}|u(s)|^2 r^{-\rho} \, dr \leq \left(\frac{2}{\rho - 1}\right)^2 \int_0^\infty |u'(r)|^2 r^\rho \, dr .$$

$$(2.54)$$

This theorem implies Theorem 2.65 since, for each $r > 0$,

$$\sup_{s \in \mathbb{R}_+} \min\{s^{-2+2\rho}, r^{-2+2\rho}\}|u(s)|^2 r^{-\rho} \geq |u(r)|^2 r^{-2+\rho} .$$

This shows, in particular, that the constant $(2/(\rho - 1))^2$ in (2.54) is optimal.

Proof We prove the theorem only for $\rho = 0$. The case of general $\rho \neq 1$ can be deduced from this by the argument in Remark 2.66. Moreover, we shall prove the theorem in integral form as in Remark 2.67; that is, we show that for $f \in L^2(\mathbb{R}_+)$,

$$\int_0^\infty \sup_{s \in \mathbb{R}_+} \min\{s^{-2}, r^{-2}\} \left| \int_0^s f(t) \, dt \right|^2 dr \leq 4 \int_0^\infty |f(r)|^2 \, dr . \qquad (2.55)$$

Applying this to the function $f = u'$, we obtain the claimed theorem.

For the proof of (2.55) we denote by f^* the non-increasing rearrangement of f. This is a non-increasing, non-negative function on $\mathbb{R}_+$ such that

$$\{|f| > \tau\}| = |\{f^* > \tau\}| \quad \text{for all } \tau > 0.$$

We use the same notation as for the symmetric decreasing rearrangement in the proof of Theorem 2.49, although these two rearrangements are slightly different. Indeed, f^* here is the restriction to $\mathbb{R}_+$ of the symmetric decreasing rearrangement of the even extension of f to $\mathbb{R}$. For more on this rearrangement and, in particular, the following two simple properties that we will use, we refer, for instance, to Leoni (2017, §15.1). On the one hand, by the equimeasurability property,

$$\int_0^\infty |f(r)|^2 \, dr = \int_0^\infty (f^*(r))^2 \, dr \, .$$

On the other hand, by the simplest rearrangment inequality, for any $s > 0$,

$$\left| \int_0^s f(t) \, dt \right| \le \int_0^s |f(t)| \, dt \le \int_0^s f^*(t) \, dt \, .$$

Thus, for any $r > 0$,

$$\sup_{s \in \mathbb{R}_+} \min\{s^{-2}, r^{-2}\} \left| \int_0^s f(t) \, dt \right|^2 \le \sup_{s \in \mathbb{R}_+} \min\{s^{-2}, r^{-2}\} \left(\int_0^s f^*(t) \, dt \right)^2 .$$

As a consequence, if we can prove the inequality for f^*, it holds also for f.

The advantage of f^* is that the supremum can be computed. Indeed, since f^* is non-increasing, we have for all $r \le s$ and all $t > 0$, $f^*(t) \le f^*(rt/s)$, so

$$\frac{1}{s} \int_0^s f^*(t) \, dt \le \frac{1}{s} \int_0^s f^*(rt/s) \, dt = \frac{1}{r} \int_0^r f^*(u) \, du \, .$$

Thus,

$$\sup_{r \le s < \infty} \left(\frac{1}{s} \int_0^s f^*(t) \, dt \right)^2 = \left(\frac{1}{r} \int_0^r f^*(t) \, dt \right)^2$$

and, therefore,

$$\sup_{s \in \mathbb{R}_+} \min\{s^{-2}, r^{-2}\} \left(\int_0^s f^*(t) \, dt \right)^2 = r^{-2} \left(\int_0^r f^*(t) \, dt \right)^2 .$$

Thus (2.55) for f^* follows from Hardy's inequality in Remark 2.67. □

As a consequence, we obtain Hardy inequalities with more general weights.

Theorem 2.70 *Let $\rho \in \mathbb{R} \setminus \{1\}$. Let u be weakly differentiable on $\mathbb{R}_+$ with $u' \in L^2(\mathbb{R}_+, r^\rho \, dr)$ and assume that*

$$\liminf_{r \to 0} |u(r)| = 0 \text{ if } \rho < 1, \qquad \liminf_{r \to \infty} |u(r)| = 0 \text{ if } \rho > 1 .$$

Then for any $0 \le w \in L^1_{loc}(\mathbb{R}_+)$, the following two inequalities hold:

$$\int_0^\infty w(r) |u(r)|^2 r^\rho \, dr$$

$$\le \frac{4}{|\rho - 1|} \left(\sup_{s>0} s^{|\rho-1|} \int_s^\infty w(t) t^{-|\rho-1|+1} \, dt \right) \int_0^\infty |u'(r)|^2 r^\rho \, dr$$

and

$$\int_0^\infty w(r)|u(r)|^2 r^\rho \, dr$$

$$\leq \frac{4}{|\rho - 1|} \left(\sup_{s>0} s^{-|\rho-1|} \int_0^s w(t) t^{|\rho-1|+1} \, dt \right) \int_0^\infty |u'(r)|^2 r^\rho \, dr \, .$$

We note that, if $w(r) = r^{-2}$, then

$$s^{|\rho-1|} \int_s^\infty w(t) t^{-|\rho-1|+1} \, dt = |\rho - 1|^{-1} = s^{-|\rho-1|} \int_0^s w(t) t^{|\rho-1|+1} \, dt \, ,$$

and therefore Theorem 2.70 contains Theorem 2.65, including the (sharp) value of the constant, as a special case.

Proof   We prove the theorem only for $\rho = 0$. The case of general $\rho \neq 1$ can be deduced from this by the argument in Remark 2.66.

For the first inequality, we set $m(r) := \sup_{0 < s \leq r} |u(s)|^2$ and bound

$$\int_0^\infty w(r)|u(r)|^2 \, dr \leq \int_0^\infty w(r)m(r) \, dr \, .$$

Since m is non-decreasing, for each $\tau > 0$, the set $\{m > \tau\}$ is an interval of the form (a_τ, ∞). We bound

$$\int_0^\infty w(r)\chi_{\{m>\tau\}}(r) \, dr = \int_{a_\tau}^\infty w(r) \, dr \leq a_\tau^{-1} \sup_{s>0} s \int_s^\infty w(t) \, dt$$

$$= \int_0^\infty \chi_{\{m>\tau\}}(r) \frac{dr}{r^2} \sup_{s>0} s \int_s^\infty w(t) \, dt \, .$$

Integrating with respect to τ, we arrive at

$$\int_0^\infty w(r)m(r) \, dr \leq \int_0^\infty m(r) \frac{dr}{r^2} \sup_{s>0} s \int_s^\infty w(t) \, dt \, .$$

Since $m(r)r^{-2} \leq \sup_{s \in \mathbb{R}_+} \min\{s^{-2}, r^{-2}\}|u(s)|^2$, the claimed inequality follows from Theorem 2.69.

Similarly, for the proof of the second inequality we set

$$\tilde{m}(r) := \sup_{r \leq s < \infty} s^{-2}|u(s)|^2$$

and bound

$$\int_0^\infty w(r)|u(r)|^2 \, dr \leq \int_0^\infty w(r)r^2 \tilde{m}(r) \, dr \, .$$

Since $\tilde{m}$ is non-increasing, for each $\tau > 0$, the set $\{\tilde{m} > \tau\}$ is an interval of the form $(0, b_\tau)$. We bound

$$\int_0^\infty w(r)r^2 \chi_{\{\tilde{m}>\tau\}}(r)\,dr = \int_0^{b_\tau} w(r)r^2\,dr \le b_\tau \sup_{s>0} s^{-1} \int_0^s w(t)t^2\,dt$$

$$= \int_0^\infty \chi_{\{\tilde{m}>\tau\}}(r)\,dr \sup_{s>0} s^{-1} \int_0^s w(t)t^2\,dt\,.$$

Integrating with respect to τ, we arrive at

$$\int_0^\infty w(r)r^2\tilde{m}(r)\,dr \le \int_0^\infty \tilde{m}(r)\,dr \sup_{s>0} s^{-1} \int_0^s w(t)t^2\,dt\,.$$

Since $\tilde{m}(r) \le \sup_{s \in \mathbb{R}_+} \min\{s^{-2}, r^{-2}\}|u(s)|^2$, the claimed inequality now follows from Theorem 2.69. This completes the proof of the theorem. $\qquad\square$

Remark 2.71 We will use Theorem 2.70 in the following form. Let $\rho \in \mathbb{R} \setminus \{1\}$ and let $R > 0$. Let u be weakly differentiable on (R, ∞) with $u' \in L^2((R, \infty), r^\rho\,dr)$ and assume that

$$\liminf_{r \to \infty} |u(r)| = 0 \text{ if } \rho > 1\,.$$

Then for any $0 \le w \in L^1_{\mathrm{loc}}([R, \infty))$, if $\rho > 1$,

$$\int_R^\infty w(r)|u(r)|^2 r^\rho\,dr \le \frac{4}{\rho - 1}\left(\sup_{s>R} s^{\rho-1} \int_s^\infty w(t)t^{-\rho+2}\,dt\right) \int_R^\infty |u'(r)|^2 r^\rho\,dr \tag{2.56}$$

and, if $\rho < 1$,

$$\int_R^\infty w(r)|u(r)|^2 r^\rho\,dr \le \frac{4}{1 - \rho}\left(\sup_{s>R} s^{1-\rho} \int_s^\infty w(t)t^\rho\,dt\right)$$
$$\times \left(\int_R^\infty |u'(r)|^2 r^\rho\,dr + (1 - \rho)R^{-1+\rho}|u(R)|^2\right)\,. \tag{2.57}$$

As usual, the value $u(R)$ here refers to the continuous representative of u. Inequalities (2.56) and (2.57) follow by applying Theorem 2.70 to the weight w extended by zero to $(0, R)$ and the function $\tilde{u}$ that extends u to $(0, R)$ by $\tilde{u}(r) := u(R)$ if $\rho > 1$ and by $\tilde{u}(r) := u(R)(r/R)^{1-\rho}$ if $\rho < 1$. The fact that $\tilde{u}$ is weakly differentiable follows from Lemma 2.6.

The following proposition is a substitute for Theorem 2.70 when $\rho = 1$.

Proposition 2.72   *Let u be weakly differentiable on $(1, \infty)$ with $u' \in$*
$L^2((1, \infty), r\, dr)$ *and assume that*

$$\liminf_{r \to 1} |u(r)| = 0.$$

Then, for any $0 \le w \in L^1_{\text{loc}}((1, \infty))$,

$$\int_1^\infty w(r)|u(r)|^2 r\, dr \le 4 \left(\sup_{s>1} \ln s \int_s^\infty w(t)t\, dt \right) \int_1^\infty |u'(r)|^2 r\, dr$$

and

$$\int_1^\infty w(r)|u(r)|^2 r\, dr \le 4 \left(\sup_{s>1} (\ln s)^{-1} \int_1^s w(t)t(\ln t)^2\, dt \right) \int_1^\infty |u'(r)|^2 r\, dr .$$

We note that, if $w(r) = r^{-2}(\ln r)^{-2}$, then

$$\ln s \int_s^\infty w(t)t\, dt = 1 = (\ln s)^{-1} \int_1^s w(t)t(\ln t)^2\, dt ,$$

and therefore Proposition 2.72 contains Proposition 2.68, including the (sharp) value of the constant, as a special case.

Proof   For u as in the proposition, define $v(t) := u(e^t)$ for $t \in \mathbb{R}_+$ and $q(t) :=$ $e^{2t}w(e^t)$. Using the properties of v discussed in the proof of Proposition 2.68, together with the fact that $\int_0^\infty q(t)|v(t)|^2\, dt = \int_1^\infty w(r)|u(r)|^2 r\, dr$ and

$$T \int_T^\infty q(t)\, dt = \ln(e^T) \int_{e^T}^\infty w(r)r\, dr ,$$

$$T^{-1} \int_0^T q(t)t^2\, dt = (\ln(e^T))^{-1} \int_1^{e^T} w(r)r(\ln r)^2\, dr ,$$

the proposition follows from the case $\rho = 0$ of Theorem 2.70.   □

Remark 2.73   We will use Proposition 2.72 in the following form. Let $R > 1$ and let u be weakly differentiable on (R, ∞) with $u' \in L^2((R, \infty), r\, dr)$. Then for any $0 \le w \in L^1_{\text{loc}}([R, \infty))$,

$$\int_R^\infty w(r)|u(r)|^2 r\, dr$$

$$\le 4 \left(\sup_{s>R} \ln s \int_s^\infty w(t)t\, dt \right) \left(\int_R^\infty |u'(r)|^2 r\, dr + (\ln R)^{-1}|u(R)|^2 \right).$$

$$(2.58)$$

This follows by applying Proposition 2.72 to the weight w extended by zero to $(1, R)$ and the function $\tilde{u}$ that extends u to $(1, R)$ by $\tilde{u}(r) := (u(R)/\ln R)\ln r$. The fact that $\tilde{u}$ is weakly differentiable follows from Lemma 2.6.

2.6.3 Hardy's inequality on $\mathbb{R}^d$

We now use the one-dimensional Hardy inequalities to establish corresponding inequalities in higher dimensions.

Theorem 2.74 *Let $d \geq 3$. Assume that u is weakly differentiable on $\mathbb{R}^d$ with $\nabla u \in L^2(\mathbb{R}^d)$ and $|\{x \in \mathbb{R}^d : |u(x)| > \tau\}| < \infty$ for all $\tau > 0$. Then*

$$\int_{\mathbb{R}^d} \frac{|u(x)|^2}{|x|^2} \, dx \leq \left(\frac{2}{d-2}\right)^2 \int_{\mathbb{R}^d} |\nabla u(x)|^2 \, dx. \tag{2.59}$$

Proof The idea of the proof is to apply a one-dimensional Hardy inequality to the function $r \mapsto u(r\omega)$ with $\omega \in \mathbb{S}^{d-1}$ fixed. Let us show that for ω from a set of full measure, the assumptions of Theorem 2.65 are satisfied.

Given representatives of u and ∇u in their a.e. equivalence classes, we obtain from Lemma 2.20 a subset $\Sigma \subset \mathbb{S}^{d-1}$ of full measure such that for any $\omega \in \Sigma$, the function $r \mapsto u(r\omega)$ is weakly differentiable on $\mathbb{R}_+$ with weak derivative $\omega \cdot \nabla u(r\omega)$. Since $\nabla u \in L^2(\mathbb{R}^d)$ and $|\omega \cdot \nabla u(r\omega)| \leq |\nabla u(r\omega)|$, this weak derivative belongs to $L^2(\mathbb{R}_+, r^{d-1} dr)$. As

$$\int_{\mathbb{S}^{d-1}} \int_{\{r: |u(r\omega)|>\tau\}} r^{d-1} \, dr \, d\omega = |\{|u| > \tau\}| < \infty \quad \text{for all } \tau > 0,$$

for any $\tau > 0$ there is a set $\Sigma_\tau \subset \mathbb{S}^{d-1}$ of full measure such that

$$\int_{\{r: |u(r\omega)|>\tau\}} r^{d-1} \, dr < \infty \quad \text{for all } \omega \in \Sigma_\tau.$$

Taking the intersection $\widetilde{\Sigma}$ of the sets Σ_{τ_n} for a sequence $\tau_n \to 0$, we infer that $\widetilde{\Sigma}$ has full measure and so, for all $\omega \in \widetilde{\Sigma}$, we obtain

$$\int_{\{r: |u(r\omega)|>\tau\}} r^{d-1} \, dr < \infty \quad \text{for all } \tau > 0.$$

The latter property clearly implies that $\liminf_{r\to\infty} |u(r\omega)| = 0$ for all $\omega \in \widetilde{\Sigma}$. (Indeed, if for some $\omega \in \widetilde{\Sigma}$ there are $\rho < \infty$ and $\delta > 0$ such that $|u(r\omega)| > \delta$ for all $r \geq \rho$, then $\int_{\{r:|u(r\omega)|>\delta\}} r^{d-1} \, dr = \infty$.)

We apply Theorem 2.65 with $\rho = d - 1 > 1$ and obtain, for any $\omega \in \Sigma \cap \widetilde{\Sigma}$,

$$\int_0^\infty |u(r\omega)|^2 r^{-2} r^{d-1} \, dr \leq \left(\frac{2}{d-2}\right)^2 \int_0^\infty |\omega \cdot \nabla u(r\omega)|^2 r^{d-1} \, dr.$$

Integrating over the full measure set $\Sigma \cap \widetilde{\Sigma} \subset \mathbb{S}^{d-1}$, we obtain

$$\int_{\mathbb{R}^d} |u(x)|^2 |x|^{-2} \, dx \leq \left(\frac{2}{d-2}\right)^2 \int_{\mathbb{R}^d} \left| \frac{x}{|x|} \cdot \nabla u(x) \right|^2 dx.$$

The inequality $\left| \frac{x}{|x|} \cdot \nabla u(x) \right| \le |\nabla u(x)|$ yields the claimed bound. □

Note that the previous proof shows that the theorem continues to hold if the assumption of weak differentiability in $\mathbb{R}^d$ is replaced by that in $\mathbb{R}^d \setminus \{0\}$. We will see in Proposition 2.80 below, however, that if u is weakly differentiable in $\mathbb{R}^d \setminus \{0\}$ with $\nabla u \in L^2(\mathbb{R}^d)$, then u is weakly differentiable in $\mathbb{R}^d$.

Remark 2.75 Using inequality (2.56) instead of Theorem 2.65 in the previous proof, we obtain that, if $0 < R < \infty$ and u is weakly differentiable on $\mathbb{R}^d \setminus \overline{B_R}$ with $\nabla u \in L^2(\mathbb{R}^d \setminus \overline{B_R})$ and $|\{x \in \mathbb{R}^d \setminus \overline{B_R} : |u(x)| > \tau\}| < \infty$ for all $\tau > 0$, then

$$\int_{\mathbb{R}^d \setminus \overline{B_R}} \frac{|u(x)|^2}{|x|^2}\, dx \le \left(\frac{2}{d-2} \right)^2 \int_{\mathbb{R}^d \setminus \overline{B_R}} |\nabla u(x)|^2\, dx .$$

The following result concerns the two-dimensional case.

Proposition 2.76 *Let $d = 2$. Assume that u is weakly differentiable on $\mathbb{R}^2$ with $\nabla u \in L^2(\mathbb{R}^2)$ and assume that $u \equiv 0$ either in $\{x \in \mathbb{R}^2 : |x| < 1\}$ or in $\{x \in \mathbb{R}^2 : |x| > 1\}$. Then*

$$\int_{\mathbb{R}^2} \frac{|u(x)|^2}{|x|^2 (\ln |x|)^2}\, dx \le 4 \int_{\mathbb{R}^2} |\nabla u(x)|^2\, dx .$$

Proof The proposition follows by an argument similar to that in the proof of Theorem 2.74 from Proposition 2.68. □

We end this subsection with two somewhat technical inequalities in dimensions $d = 1$ and 2, which replace the missing Hardy inequalities in Theorem 2.74.

Proposition 2.77 *Let $d = 1, 2$. Then there is a constant C_d such that for all weakly differentiable $u \in L^1_{\mathrm{loc}}(\mathbb{R}^d)$ with $\nabla u \in L^2(\mathbb{R}^d)$ one has*

$$\int_{\mathbb{R}} \frac{|u|^2}{1 + x^2}\, dx \le C_1 \left(\int_{\mathbb{R}} |u'|^2\, dx + \int_B |u|^2\, dx \right) \quad \text{if } d = 1,$$

$$\int_{\mathbb{R}^2} \frac{|u|^2}{1 + |x|^2 (\ln |x|)_+^2}\, dx \le C_2 \left(\int_{\mathbb{R}^2} |\nabla u|^2\, dx + \int_B |u|^2\, dx \right) \quad \text{if } d = 2,$$

where B is the unit ball in $\mathbb{R}^d$ and $(\ln |x|)_+ = \max\{\ln |x|, 0\}$.

By the Poincaré inequality (see Remark 2.59), under the assumptions of the proposition, $u \in L^2(B)$, so the right sides in the inequalities in the proposition are finite. Note that, in contrast to Theorem 2.74, for the inequality to hold no assumption on the measure of the set $\{|u| > \tau\}$ is necessary.

Proof Arguing as in the proof of Theorem 2.74, it suffices to prove the following one-dimensional inequalities:

$$\int_0^\infty |u(r)|^2 w(r) r^{d-1}\, dr \le C_d \left(\int_0^\infty |u'(r)|^2 r^{d-1}\, dr + \int_0^1 |u(r)|^? r^{d-1}\, dr \right),$$

where

$$w(r) := \begin{cases} (1+r^2)^{-1} & \text{if } d = 1; \\ (1+r^2(\ln r)_+^2)^{-1} & \text{if } d = 2. \end{cases}$$

Let ζ be a smooth function on $\mathbb{R}_+$ with

$$\begin{aligned} 0 \le \zeta \le 1 &\quad \text{on } \mathbb{R}_+, \\ \zeta = 0 &\quad \text{on } (0, 1/2], \\ \zeta = 1 &\quad \text{on } [1, \infty). \end{aligned}$$

Then

$$\int_0^\infty |u|^2 w(r) r^{d-1}\, dr \le 2 \int_0^\infty (1-\zeta)^2 |u|^2 w(r) r^{d-1}\, dr + 2 \int_0^\infty \zeta^2 |u|^2 w(r) r^{d-1}\, dr.$$

The first term is controlled by

$$\begin{aligned} 2 \int_0^\infty (1-\zeta)^2 |u|^2 w(r) r^{d-1}\, dr &\le 2 \sup_{0 \le s \le 1} w(s) \int_0^1 |u|^2 r^{d-1}\, dr \\ &= 2 \int_0^1 |u|^2 r^{d-1}\, dr. \end{aligned}$$

To control the second term, we assume first that $d = 1$, and use $w(r) \le r^{-2}$ and Theorem 2.65 to get

$$\begin{aligned} 2 \int_0^\infty \zeta^2 |u|^2 w(r)\, dr &\le 2 \int_0^\infty |\zeta u|^2 r^{-2}\, dr \le 8 \int_0^\infty |(\zeta u)'|^2\, dr \\ &\le 16 \int_0^\infty (\zeta^2 |u'|^2 + (\zeta')^2 |u|^2)\, dr \\ &\le 16 \max\left\{1, \sup_{0 \le s \le 1} (\zeta'(s))^2\right\} \left(\int_0^\infty |u'|^2\, dr + \int_0^1 |u|^2\, dr \right). \end{aligned}$$

This proves the claimed inequality for $d = 1$.

For $d = 2$ we use the elementary fact that there is a $C < \infty$ such that $w(r) \le Cr^{-2}(\ln(2r))^{-2}$ if $r \ge 1/2$. This, together with a rescaled version of

Proposition 2.68, implies

$$2 \int_0^\infty \zeta^2 |u|^2 w(r) r^{d-1} \, dr$$

$$\leq 2C \int_{1/2}^\infty |\zeta u|^2 r^{-1} (\ln(2r))^{-2} \, dr \leq 8C \int_{1/2}^\infty |(\zeta u)'|^2 r \, dr$$

$$\leq 16C \int_{1/2}^\infty (\zeta^2 |u'|^2 + (\zeta')^2 |u|^2) r \, dr$$

$$\leq 16C \max \left\{ 1, \sup_{0 \leq s \leq 1} (\zeta'(s))^2 \right\} \left(\int_0^\infty |u'|^2 r \, dr + \int_0^1 |u|^2 r \, dr \right).$$

This proves the claimed inequality for $d = 2$ and completes the proof.   □

2.6.4 Weak differentiability and removable singularities

In some applications, for instance, when working in radial coordinates, one encounters the situation where a function defined on an open subset of $\mathbb{R}^d$ with $d \geq 2$ is weakly differentiable away from a point and the L^1 norm of its derivative is integrable in a neighborhood of that point. Under these assumptions we will show that the function is also weakly differentiable in a neighborhood of that point. To prove this result, we will use Hardy inequalities, although of a slightly different kind from those that we have discussed previously in this section, namely now for the integral of the absolute value of the derivatives rather than its square.

We begin with the following partial analogue of Theorem 2.65.

Proposition 2.78   *Let $\rho > 0$. Let u be weakly differentiable on $\mathbb{R}_+$ with $u' \in L^1(\mathbb{R}_+, r^\rho \, dr)$ and assume that $\liminf_{r \to \infty} |u(r)| = 0$. Then*

$$\int_0^\infty |u(r)| r^{-1+\rho} \, dr \leq \frac{1}{\rho} \int_0^\infty |u'(r)| r^\rho \, dr .$$

Proof   According to Lemma 2.6, we have for any $0 < r < R < \infty$,

$$u(R) - u(r) = \int_r^R u'(s) \, ds .$$

By assumption, we can let $R \to \infty$ along a subsequence along which u tends to zero. Moreover, the assumptions imply that $u' \in L^1((r, \infty), ds)$. In this way, we obtain

$$u(r) = - \int_r^\infty u'(s) \, ds$$

for all $r > 0$. Thus, by Fubini's theorem,

$$\int_0^\infty |u(r)|r^{-1+\rho}\,dr \leq \int_0^\infty \int_r^\infty |u'(s)|\,ds\,r^{-1+\rho}\,dr = \int_0^\infty |u'(s)| \int_0^s r^{-1+\rho}\,dr\,ds$$

$$= \frac{1}{\rho}\int_0^\infty |u'(s)|s^\rho\,ds\,,$$

which is the claimed inequality. □

Proposition 2.79 *Let $d \geq 2$. Assume that u is weakly differentiable on $\mathbb{R}^d \setminus \{0\}$ with $\nabla u \in L^1(\mathbb{R}^d)$ and $|\{x \in \mathbb{R}^d : |u(x)| > \tau\}| < \infty$ for all $\tau > 0$. Then*

$$\int_{\mathbb{R}^d} \frac{|u(x)|}{|x|}\,dx \leq \frac{1}{d-1}\int_{\mathbb{R}^d} |\nabla u(x)|\,dx\,.$$

Here we write $\nabla u \in L^1(\mathbb{R}^d)$ instead of the more precise $\nabla u \in L^1(\mathbb{R}^d \setminus \{0\})$.

Proof The proof proceeds analogously to that of Theorem 2.74 using Proposition 2.78 via Lemma 2.20. We omit the details. □

We now state and prove the result mentioned at the start of this subsection.

Proposition 2.80 *Let $d \geq 2$, $\Omega \subset \mathbb{R}^d$ be open and $x_0 \in \Omega$. If $u \in L^1_{\mathrm{loc}}(\Omega)$ is weakly differentiable in $\Omega \setminus \{x_0\}$ and if $\nabla u \in L^1_{\mathrm{loc}}(\Omega)$, then u is weakly differentiable in Ω.*

Proof We first note that we may assume that $\Omega = \mathbb{R}^d$. In fact, since $x_0 \in \Omega$, we can choose a function $\zeta \in C_0^\infty(\Omega)$ with $\zeta = 1$ near x_0. Decomposing $u = \zeta u + (1 - \zeta)u$, we see that it suffices to prove weak differentiability of the function ζu. This function, extended by zero to $\mathbb{R}^d$, is weakly differentiable in $\mathbb{R}^d \setminus \{0\}$ and has $\nabla(\zeta u) \in L^1(\mathbb{R}^d)$. Thus, from now on, we assume that $\Omega = \mathbb{R}^d$. Moreover, by a translation, we may also assume that $x_0 = 0$. To simplify notation, we write u instead of ζu.

Let $\eta \in C_0^\infty(\mathbb{R}^d)$ and let $\chi \in C^\infty(\mathbb{R}^d)$ with $\chi \equiv 0$ near the origin and $\chi \equiv 1$ outside a bounded set. Then, for any $\varepsilon > 0$, $\chi(\cdot/\varepsilon)\eta \in C_0^\infty(\mathbb{R}^d \setminus \{0\})$, and therefore, by the assumed weak differentiability of u in $\mathbb{R}^d \setminus \{0\}$,

$$\int_{\mathbb{R}^d} u\nabla(\chi(\cdot/\varepsilon)\eta)\,dx = -\int_{\mathbb{R}^d} \chi(x/\varepsilon)\eta\nabla u\,dx\,.$$

Concerning the right side, we have by dominated convergence, as $\varepsilon \to 0$,

$$\int_{\mathbb{R}^d} \chi(x/\varepsilon)\eta\nabla u\,dx \to \int_{\mathbb{R}^d} \eta\nabla u\,dx\,.$$

The left side is equal to

$$\int_{\mathbb{R}^d} u\nabla(\chi(\cdot/\varepsilon)\eta)\,dx = \int_{\mathbb{R}^d} \chi(x/\varepsilon)u\nabla\eta\,dx + \varepsilon^{-1}\int_{\mathbb{R}^d} \eta u(\nabla\chi)(x/\varepsilon)\,dx\,.$$

Similar to before, dominated convergence implies that, as $\varepsilon \to 0$,

$$\int_{\mathbb{R}^d} \chi(x/\varepsilon) u \nabla \eta \, dx \to \int_{\mathbb{R}^d} u \nabla \eta \, dx .$$

Since $\nabla \chi$ is bounded and compactly supported, we have $\varepsilon^{-1}(\nabla \chi)(x/\varepsilon) \to 0$ for any $x \in \mathbb{R}^d \setminus \{0\}$ and there is a $C < \infty$ such that, for any $x \in \mathbb{R}^d$ and $\varepsilon > 0$, we have $\varepsilon^{-1}|(\nabla \chi)(x/\varepsilon)| \le C|x|^{-1}$. By Theorem 2.79 and dominated convergence, as $\varepsilon \to 0$,

$$\varepsilon^{-1} \int_{\mathbb{R}^d} \eta u (\nabla \chi)(x/\varepsilon) \, dx \to 0 .$$

To summarize, we have shown that

$$\int_{\mathbb{R}^d} u \nabla \eta \, dx = - \int_{\mathbb{R}^d} \eta \nabla u \, dx .$$

Since $\eta \in C_0^\infty(\mathbb{R}^d)$ is arbitrary, this means that u is weakly differentiable. □

The assumption $d \ge 2$ in Proposition 2.80 is essential. Indeed, if $x \in \mathbb{R}$, the function $x \mapsto \operatorname{sgn} x$ is weakly differentiable on $\mathbb{R} \setminus \{0\}$, but not on $\mathbb{R}$.

2.7 Homogeneous Sobolev spaces

While in the definition of the usual Sobolev spaces $H^1(\mathbb{R}^d)$ the square integrability of the functions is required in addition to the square-integrability of their weak gradients, in some applications it is more natural to work only under the square integrability assumption on the gradient. This leads to the definition of homogeneous Sobolev spaces. In their discussion, one needs to distinguish between the cases of dimensions $d = 1$ and 2 on the one hand and dimensions $d \ge 3$ on the other.

2.7.1 Dimensions $d \ge 3$

By definition, for $d \ge 3$ the *homogeneous Sobolev space*

$$\dot{H}^1(\mathbb{R}^d)$$

consists of all (a.e. equivalence classes of) functions $u \in L^1_{\text{loc}}(\mathbb{R}^d)$ with $\nabla u \in L^2(\mathbb{R}^d)$ and $|\{x \in \mathbb{R}^d : |u(x)| > \tau\}| < \infty$ for all $\tau > 0$. This is an inner product space with respect to $\int_{\mathbb{R}^d} \nabla u \cdot \nabla v \, dx$.

Note that according to Theorem 2.74, Hardy's inequality (2.59) is valid for $u \in \dot{H}^1(\mathbb{R}^d)$. This is a crucial ingredient in the following proof.

Lemma 2.81 $C_0^\infty(\mathbb{R}^d)$ *is dense in* $\dot{H}^1(\mathbb{R}^d)$.

Proof   The proof has some similarities with the proof of Proposition 2.23. We begin by showing that functions with compact support are dense in $\dot{H}^1(\mathbb{R}^d)$.

Let $\zeta \in C^\infty[0, \infty)$ with $0 \le \zeta \le 1$ such that $\zeta = 1$ on $[0, 1]$ and $\zeta = 0$ on $[2, \infty)$. Let $u \in \dot{H}^1(\mathbb{R}^d)$ and put

$$u_R(x) := \zeta(|x|/R)u(x).$$

Then $u_R \in \dot{H}^1(\mathbb{R}^d)$ and

$$\partial_j u_R = \zeta(|x|/R)\partial_j u + uR^{-1}\zeta'(|x|/R)x_j/|x|.$$

Therefore

$$\|\partial_j u_R - \partial_j u\|_{L^2(\mathbb{R}^d)}$$

$$\le \left(\int_{\{|x|>R\}} |\partial_j u|^2 \, dx \right)^{1/2} + \left(\int_{\{|x|>R\}} |u|^2 R^{-2} |\zeta'(|x|/R)|^2 \, dx \right)^{1/2}.$$

The first term on the right side tends to zero as $R \to \infty$ by dominated convergence. For the second term we use the fact that $|\zeta'(r)| \le Cr^{-1}$ for some $C > 0$ and all $r > 0$. Thus, $R^{-2}|\zeta'(|x|/R)|^2 \le C^2|x|^{-2}$ for all $x \in \mathbb{R}^d$ and all $R > 0$. This bound, together with Hardy's inequality (Theorem 2.74), implies that the integrand in the second term is bounded by an integrable function independent of R. Since it tends to zero pointwise, we conclude by dominated convergence that the second term tends to zero as $R \to \infty$. This shows that we can approximate any $u \in \dot{H}^1(\mathbb{R}^d)$ by functions in $\dot{H}^1(\mathbb{R}^d)$ with compact support.

Convolution with a C_0^∞ function, as in the proof of Lemma 2.1, shows that the approximating functions can be chosen in $C_0^\infty(\mathbb{R}^d)$.   □

Proposition 2.82   *The Sobolev inequality* (2.12) *is valid for all* $u \in \dot{H}^1(\mathbb{R}^d)$.

Proof   According to Lemma 2.81, there is a sequence $(u_n) \subset C_0^\infty(\mathbb{R}^d)$ such that $u_n \to u$ in $\dot{H}^1(\mathbb{R}^d)$. In particular, (∇u_n) is a Cauchy sequence in $L^2(\mathbb{R}^d)$. It follows from the Sobolev inequality (2.12) applied to the functions $u_n - u_m \in C_0^\infty(\mathbb{R}^d)$ that (u_n) is a Cauchy sequence in $L^{2d/(d-2)}(\mathbb{R}^d)$. Therefore, there is a $v \in L^{2d/(d-2)}(\mathbb{R}^d)$ such that $u_n \to v$ in $L^{2d/(d-2)}(\mathbb{R}^d)$. It follows from Lemma 2.3 that $v = u$. Applying now the Sobolev inequality (2.12) to the functions $u_n \in C_0^\infty(\mathbb{R}^d)$ and passing to the limit $n \to \infty$, we obtain the Sobolev inequality for u.   □

Remark 2.83   The above arguments show that

$$\dot{H}^1(\mathbb{R}^d) = \{u \in L^2(\mathbb{R}^d, |x|^{-2} \, dx) \colon \nabla u \in L^2(\mathbb{R}^d)\}.$$

Indeed, if we denote the space on the right side temporarily by $\mathcal{H}$, then, by Hardy's inequality (Theorem 2.74), $\dot{H}^1(\mathbb{R}^d) \subset \mathcal{H}$. Conversely, by the same arguments as in Lemma 2.81 one shows that $C_0^\infty(\mathbb{R}^d)$ is dense in $\mathcal{H}$, endowed with its natural norm, and then by the same arguments as in Proposition 2.82 one shows that $\mathcal{H} \subset L^{2d/(d-2)}(\mathbb{R}^d)$, so, in particular, $|\{|u| > \tau\}| < \infty$ for all $\tau > 0$ if $u \in \mathcal{H}$.

Lemma 2.84 $\dot{H}^1(\mathbb{R}^d)$ *is complete.*

Proof If $(u_n) \subset \mathbb{R}^d$ is a Cauchy sequence in $\dot{H}^1(\mathbb{R}^d)$, then $(\partial_j u_n)$ is a Cauchy sequence in $L^2(\mathbb{R}^d)$ for every $j = 1, \ldots, d$ and converges to some $v_j \in L^2(\mathbb{R}^d)$. Moreover, by Sobolev's inequality (Proposition 2.82), (u_n) is a Cauchy sequence in $L^{2d/(d-2)}(\mathbb{R}^d)$ and converges to some $u \in L^{2d/(d-2)}(\mathbb{R}^d)$. Thus, from Lemma 2.3 we conclude that u is weakly differentiable and $\partial_j u = v_j$. In particular, $\nabla u \in L^2(\mathbb{R}^d)$. Moreover, since $u \in L^{2d/(d-2)}(\mathbb{R}^d)$, we have $|\{|u| > \tau\}| < \infty$ for every $\tau > 0$. Thus, $u \in \dot{H}^1(\mathbb{R}^d)$. $\qquad\square$

We next prove a version of Rellich's lemma for the homogeneous Sobolev space.

Proposition 2.85 *Let $d \geq 3$, $u \in \dot{H}^1(\mathbb{R}^d)$ and let $(u_n) \subset \dot{H}^1(\mathbb{R}^d)$ be such that $u_n \to u$ weakly in $\dot{H}^1(\mathbb{R}^d)$. Then, for any set $A \subset \mathbb{R}^d$ of finite measure, $\chi_A u_n \to \chi_A u$ in $L^2(\mathbb{R}^d)$.*

Proof We first note that Lemma 2.37 remains valid for $u \in \dot{H}^1(\mathbb{R}^d)$. (This follows from a simple approximation argument.) Then we can follow the proof of Proposition 2.36 (now $C := \sup \|\nabla u_n\|$) and we are again led to proving (2.11). For this, we use the fact that $u_n \to u$ weakly in $L^{2d/(d-2)}(\mathbb{R}^d)$ by Sobolev's inequality (Proposition 2.82) and that $G_t(\cdot - x) \in L^{2d/(d+2)}(\mathbb{R}^d)$. $\qquad\square$

2.7.2 Dimensions $d = 1$ and 2

In this subsection, we discuss a space which, in a certain sense, is a substitute for the homogeneous space $\dot{H}^1(\mathbb{R}^d)$ in dimensions $d = 1$ and 2. To illustrate the difference between these dimensions and when $d \geq 3$, we note that, as a consequence of the Hardy and Sobolev inequalities, if (u_n) is a sequence of weakly differentiable functions on $\mathbb{R}^d$, $d \geq 3$, with compact supports and if $\nabla u_n \to 0$ in $L^2(\mathbb{R}^d)$, then $u_n \to 0$ in $L^2(\mathbb{R}^d, |x|^{-2} dx)$ and $L^{2d/(d-2)}(\mathbb{R}^d)$ and, in particular, $u_n \to 0$ in $L^2_{\text{loc}}(\mathbb{R}^d)$.

The following example shows that this is not the case in dimensions $d = 1, 2$.

If $d = 1$, we define for $R > 0$,

$$\beta_R(x) := \begin{cases} 1 & \text{if } |x| \le R, \\ (2 - |x|/R)_+ & \text{if } |x| > R; \end{cases} \tag{2.60}$$

and if $d = 2$, we define, for $R \ge 1$,

$$\beta_R(x) := \begin{cases} 1 & \text{if } |x| \le R, \\ (1 - \ln(|x|/R)/\ln R)_+ & \text{if } |x| > R. \end{cases} \tag{2.61}$$

Then, as $R \to \infty$, $\nabla\beta_R \to 0$ in $L^2(\mathbb{R}^d)$ and $\beta_R \to 1$ in $L^2_{\text{loc}}(\mathbb{R}^d)$.

A consequence of this example is that in $d = 1, 2$, in the completion of compactly supported, weakly differentiable functions u on $\mathbb{R}^d$ with $\nabla u \in L^2(\mathbb{R}^d)$ with respect to $\int_{\mathbb{R}^d} |\nabla u|^2 \, dx$, functions that differ by constants are identified. In order to work with a space of functions, rather than of functions modulo constants, we will add a local L^2 norm to the gradient norm. By definition, for $d = 1, 2$, the space

$$\widetilde{H}^1(\mathbb{R}^d)$$

consists of all (a.e. equivalence classes of) functions $u \in L^1_{\text{loc}}(\mathbb{R}^d)$ with $\nabla u \in L^2(\mathbb{R}^d)$. Note that by the Poincaré inequality (Remark 2.59) we have $u \in L^2_{\text{loc}}(\mathbb{R}^d)$ for $u \in \widetilde{H}^1(\mathbb{R}^d)$. Consequently, $\widetilde{H}^1(\mathbb{R}^d)$ is an inner product with respect to $\int_{\mathbb{R}^d} \nabla u \cdot \nabla v \, dx + \int_B u\bar{v} \, dx$, where B denotes the unit ball in $\mathbb{R}^d$. Note also that in the definition of $\widetilde{H}^1(\mathbb{R}^d)$ there is no assumption on the measure of the sets $\{|u| > \tau\}$. In fact, $\widetilde{H}^1(\mathbb{R}^d)$ contains functions that tend to infinity at infinity.

According to Proposition 2.77, we have a modified Hardy inequality for the spaces $\widetilde{H}^1(\mathbb{R}^d)$. This is a crucial ingredient in the following proof.

Lemma 2.86 $C_0^\infty(\mathbb{R}^d)$ *is dense in* $\widetilde{H}^1(\mathbb{R}^d)$.

Proof The proof is similar to that of Lemma 2.81. One begins by proving that any function $u \in \widetilde{H}^1(\mathbb{R}^d)$ can be approximated by functions in $\widetilde{H}^1(\mathbb{R}^d)$ with compact support. We choose $u_R = \beta_R u$, where β_R is the family of functions defined before the lemma. We use the fact that, for some constant $C < \infty$ and all $R \ge 2$,

$$|\nabla\beta_R|^2 \le C\left(1 + x^2\right)^{-1} \chi_{\{R \le |x| \le 2R\}} \qquad \text{if } d = 1$$

and, with the notation $(\ln |x|)_+ = \max\{\ln |x|, 0\}$,

$$|\nabla\beta_R|^2 \le C\left(1 + |x|^2(\ln |x|)_+^2\right)^{-1} \chi_{\{R \le |x| \le R^2\}} \qquad \text{if } d = 2.$$

This, together with Hardy's inequality from Proposition 2.77 and dominated convergence, yields $\|\nabla(u_R - u)\|^2_{L^2(\mathbb{R}^d)} + \|u_R - u\|^2_{L^2(B)} \to 0$. By convolution with a C_0^∞ function, one can show that approximating functions can be chosen in $C_0^\infty(\mathbb{R}^d)$. □

Lemma 2.87 $\widetilde{H}^1(\mathbb{R}^d)$ *is complete.*

Proof The proof is similar to that of Lemma 2.84, using Hardy's inequality (Proposition 2.77) instead of Sobolev's inequality. □

In the proof of the next two results, we will make use of the fact that for any centered ball $B_R \subset \mathbb{R}^d$ there is a constant $c_R > 0$ such that, for all $u \in \widetilde{H}^1(\mathbb{R}^d)$,

$$\int_{\mathbb{R}^d} |\nabla u|^2 \, dx + \int_B |u|^2 \, dx \geq c_R \left(\int_{B_R} |\nabla u|^2 \, dx + \int_{B_R} |u|^2 \, dx \right). \quad (2.62)$$

This follows immediately from Proposition 2.77.

Proposition 2.88 *Let $\Omega \subset \mathbb{R}^d$ be a bounded, open set and let $2 < q \leq \infty$ if $d = 1$ and $2 < q < \infty$ if $d = 2$. Then there is a constant $C_{\Omega,q} > 0$ such that*

$$\int_{\mathbb{R}^d} |\nabla u|^2 \, dx + \int_B |u|^2 \, dx \geq C_{\Omega,q} \left(\int_\Omega |u|^q \, dx \right)^{2/q} \qquad \text{for all } u \in \widetilde{H}^1(\mathbb{R}^d).$$

Proof Let $B_R \subset \mathbb{R}^d$ be a centered ball containing Ω. Then the claimed inequality follows immediately by combining (2.62) with the Poincaré–Sobolev inequality (Proposition 2.62) for B_R. □

Proposition 2.89 *Let $u \in \widetilde{H}^1(\mathbb{R}^d)$ and let $(u_n) \subset \widetilde{H}^1(\mathbb{R}^d)$ be such that $u_n \to u$ weakly in $\widetilde{H}^1(\mathbb{R}^d)$. Then, for any bounded, measurable set $A \subset \mathbb{R}^d$, $\chi_A u_n \to \chi_A u$ in $L^2(\mathbb{R}^d)$.*

Proof Let $B_R \subset \mathbb{R}^d$ be a centered ball containing A. It follows from (2.62) that the restriction to B_R is continuous from $\widetilde{H}^1(\mathbb{R}^d)$ to $H^1(B_R)$. Therefore $u_n \to u$ weakly in $H^1(B_R)$ and, by Rellich's theorem (Theorem 2.40), $u_n \to u$ strongly in $L^2(B_R)$, and so in $L^2(A)$. □

2.8 The extension property

We recall that an open set $\Omega \subset \mathbb{R}^d$ has the extension property if there is a bounded, linear operator $E : H^1(\Omega) \to H^1(\mathbb{R}^d)$ such that $Eu(x) = u(x)$ for almost every $x \in \Omega$ and every $u \in H^1(\Omega)$. In earlier sections of this chapter we have used this notion in order to deduce properties of Sobolev functions on domains with the extension property from those of Sobolev functions defined on the whole space. Typical examples of this use are Rellich's embedding theorem in §2.3.2, the Poincaré inequality in §2.5.5 and the Poincaré–Sobolev inequality in §2.5.6.

In the following §2.8.1 we will have a closer look at the extension property and show, in particular, that sets with Lipschitz boundary have this property. In the subsequent §2.8.2 we prove the useful technical fact that the d-dimensional Lebesgue measure of the boundary of an extension domain is zero.

2.8.1 Lipschitz domains have the extension property

Our goal in this subsection is to introduce the notion of a uniformly Lipschitz continuous boundary and to show that sets with this boundary have the extension property.

The basic idea behind this proof is to extend a function to the complement by reflection. The following lemma about the halfspace $\mathbb{R}^d_+ = \{(x', x_d) \in \mathbb{R}^{d-1} \times \mathbb{R} : x_d > 0\}$ will not be needed in what follows, but illustrates this mechanism most clearly.

Lemma 2.90 *Let $u \in H^1(\mathbb{R}^d_+)$ and define*

$$(Eu)(x) := \begin{cases} u(x) & \text{if } x_d > 0, \\ u(x', -x_d) & \text{if } x_d < 0. \end{cases}$$

Then $Eu \in H^1(\mathbb{R}^d)$ and

$$\|Eu\|_{L^2(\mathbb{R}^d)} = \sqrt{2}\,\|u\|_{L^2(\mathbb{R}^d_+)} \qquad \text{and} \qquad \|\nabla Eu\|_{L^2(\mathbb{R}^d)} = \sqrt{2}\,\|\nabla u\|_{L^2(\mathbb{R}^d_+)}.$$

The norm equalities in the lemma imply, in particular, that E is a bounded operator from $H^1(\mathbb{R}^d_+)$ to $H^1(\mathbb{R}^d)$.

Proof Clearly, $Eu \in L^2(\mathbb{R}^d)$ with the claimed formula for its L^2 norm. Let us show that Eu is weakly differentiable in $\mathbb{R}^d$ with

$$(\partial_j Eu)(x) = \begin{cases} (\partial_j u)(x) & \text{if } x_d > 0, \\ (\partial_j u)(x', -x_d) & \text{if } x_d < 0, \end{cases}$$

for $j = 1, \ldots, d - 1$, and

$$(\partial_d Eu)(x) = \begin{cases} (\partial_d u)(x) & \text{if } x_d > 0, \\ -(\partial_d u)(x', -x_d) & \text{if } x_d < 0. \end{cases}$$

Once this formula is shown, we also obtain the claimed formula for the L^2 norm of ∇Eu.

Let $\eta \in C_0^\infty(\mathbb{R}^d)$ and let $j = 1, \ldots, d - 1$. Then, defining

$$\tilde{\eta}(x) := \eta(x) + \eta(x', -x_d),$$

we have

$$\int_{\mathbb{R}^d} (Eu)\partial_j \eta \, dx = \int_{\mathbb{R}^d_+} u \partial_j \tilde{\eta} \, dx \, .$$

Note that, in general, $\tilde{\eta}$ does not belong to $C^1_0(\mathbb{R}^d_+)$. Let $\zeta \in C^\infty(\mathbb{R})$ with $\zeta = 0$ near the origin and $\zeta = 1$ outside a neighborhood of the origin. Then $\zeta(x_d/\varepsilon)\tilde{\eta} \in C^\infty_0(\mathbb{R}^d_+)$, and therefore,

$$\int_{\mathbb{R}^d_+} u\zeta(x_d/\varepsilon)\partial_j \tilde{\eta} \, dx = \int_{\mathbb{R}^d_+} u\partial_j \left(\zeta(x_d/\varepsilon)\tilde{\eta}\right) dx = -\int_{\mathbb{R}^d_+} (\partial_j u)\zeta(x_d/\varepsilon)\tilde{\eta} \, dx \, .$$

By dominated convergence and the fact that $u, \partial_j u, \tilde{\eta}, \partial_j \tilde{\eta} \in L^2(\mathbb{R}^d_+)$ we obtain, in the limit $\varepsilon \to 0$,

$$\int_{\mathbb{R}^d_+} u\partial_j \tilde{\eta} \, dx = -\int_{\mathbb{R}^d_+} (\partial_j u)\tilde{\eta} \, dx = -\int_{\mathbb{R}^d_+} (\partial_j u)\eta \, dx - \int_{\mathbb{R}^d_-} (\partial_j u)(x', -x_d)\eta \, dx \, ,$$

which proves the claimed formula.

For $j = d$ we have, defining now $\tilde{\eta}(x) := \eta(x) - \eta(x', -x_d)$,

$$\int_{\mathbb{R}^d} (Eu)\partial_d \eta \, dx = \int_{\mathbb{R}^d_+} u\partial_d \tilde{\eta} \, dx \, .$$

With ζ as before, we get

$$\int_{\mathbb{R}^d_+} u\zeta(x_d/\varepsilon)\partial_d \tilde{\eta} \, dx + \varepsilon^{-1} \int_{\mathbb{R}^d_+} u\zeta'(x_d/\varepsilon)\tilde{\eta} \, dx = \int_{\mathbb{R}^d_+} u\partial_d \left(\zeta(x_d/\varepsilon)\tilde{\eta}\right) dx$$

$$= -\int_{\mathbb{R}^d_+} (\partial_d u)\zeta(x_d/\varepsilon)\tilde{\eta} \, dx \, .$$

Since $\eta \in C^\infty_0(\mathbb{R}^d)$ and $\tilde{\eta}(x', 0) = 0$, we have $|\tilde{\eta}(x)| \le C|x_d|$ for some $C < \infty$. Therefore

$$|\varepsilon^{-1}\zeta'(x_d/\varepsilon)\tilde{\eta}(x)| \le C\|t\zeta'\|_\infty \chi_{\mathrm{supp}\,\tilde{\eta}}(x).$$

This, together with the pointwise convergence $\varepsilon^{-1}\zeta'(x_d/\varepsilon) \to 0$ for every $x \in \mathbb{R}^d_+$, implies by dominated convergence that, as $\varepsilon \to 0$,

$$\varepsilon^{-1}\int_{\mathbb{R}^d_+} u\zeta'(x_d/\varepsilon)\tilde{\eta} \, dx \to 0 \, .$$

For the remaining terms we can argue as before and obtain, as $\varepsilon \to 0$,

$$\int_{\mathbb{R}^d_+} u\partial_d \tilde{\eta} \, dx = -\int_{\mathbb{R}^d_+} (\partial_d u)\tilde{\eta} \, dx = -\int_{\mathbb{R}^d_+} (\partial_d u)\eta \, dx + \int_{\mathbb{R}^d_-} (\partial u)(x', -x_d)\eta \, dx \, ,$$

which proves the claimed formula. $\qquad \square$

Next we construct an extension operator for a more general class of sets than halfspaces, namely sets that lie above the graph of a Lipschitz function.

Lemma 2.91 *Let $d \geq 2$, let $f : \mathbb{R}^{d-1} \to \mathbb{R}$ be Lipschitz with Lipschitz constant $L < \infty$ and let*

$$\Omega := \left\{ (x', x_d) \in \mathbb{R}^{d-1} \times \mathbb{R} : x_d > f(x') \right\}.$$

Let $u \in H^1(\Omega)$ and define

$$(Eu)(x) := \begin{cases} u(x) & \text{if } x_d > f(x'), \\ u(x', -x_d + 2f(x')) & \text{if } x_d < f(x'). \end{cases}$$

Then $Eu \in H^1(\mathbb{R}^d)$ and

$$\|Eu\|_{L^2(\mathbb{R}^d)} = \sqrt{2}\,\|u\|_{L^2(\Omega)} \quad \text{and} \quad \|\nabla Eu\|_{L^2(\mathbb{R}^d)} \leq \sqrt{1 + (1+2L)^2}\,\|\nabla u\|_{L^2(\Omega)}.$$

Proof Step 1. Let $u \in H^1(\Omega)$. We also assume that u has an extension to a continuously differentiable function on $\mathbb{R}^d$ with bounded derivative. Let Eu be as in the statement of the lemma. Clearly, Eu is weakly differentiable in Ω and its weak derivatives there coincide with those of u. Moreover, by the chain rule in Lemma 2.10, it is weakly differentiable in the interior of Ω^c and its weak derivatives there are

$$(\partial_j Eu)(x', x_d) = (\partial_j u)(x', -x_d + 2f(x')) + 2\,(\partial_j f)(x')\,(\partial_d u)(x', -x_d + 2f(x'))$$

if $j = 1, \ldots, d - 1$ and by

$$(\partial_d Eu)(x', x_d) = -(\partial_d u)(x', -x_d + 2f(x')).$$

For the application of Lemma 2.10 we used the fact that the real and imaginary parts of u are continuously differentiable on $\mathbb{R}^d$ with bounded derivative and that f is Lipschitz and therefore, by Lemma 2.4, weakly differentiable.

In the remainder of this step we will show that Eu is weakly differentiable in $\mathbb{R}^d$. This, together with the above formulas for its derivatives and the fact that $\partial\Omega$ has measure zero by Fubini's theorem, yields the claimed bounds on the L^2 norms of Eu and ∇Eu.

We will deduce the fact that Eu is weakly differentiable in $\mathbb{R}^d$ via Lemma 2.4 from the fact that Eu is Lipschitz continuous in $\mathbb{R}^d$. We note that Eu can be unambiguously defined on $\partial\Omega$ and, in this way, becomes a continuous function in $\mathbb{R}^d$. In the following we prove that, with $L' := (1 + 2L)\|\nabla u\|_\infty$,

$$|(Eu)(x) - (Eu)(y)| \leq L'\,|x - y| \qquad \text{for all } x, y \in \mathbb{R}^d. \tag{2.63}$$

To do this, for given $x, y \in \mathbb{R}^d$ we utilize the path $\gamma : [0, 1] \to \mathbb{R}^d$,

$$\gamma(t) := \begin{pmatrix} tx' + (1-t)y' \\ f(tx' + (1-t)y') + t(x_d - f(x')) + (1-t)(y_d - f(y')) \end{pmatrix}.$$

First, consider the case $x_d \geq f(x')$ and $y_d \geq f(y')$. Note that in this case $\gamma(t) \in \overline{\Omega}$ for all $t \in [0,1]$. The function $t \mapsto u(\gamma(t))$ is Lipschitz continuous as a composition of a Lipschitz function with a C^1 function with bounded derivative. Therefore, by Lemma 2.4, it is weakly differentiable and, by the chain rule in Lemma 2.10, its derivative is given by

$$\frac{d}{dt}u(\gamma(t)) = (x - y) \cdot \nabla u(\gamma(t))$$
$$+ \left((x' - y') \cdot (\nabla' f)(tx' + (1-t)y') - (f(x') - f(y'))\right) \partial_d u(\gamma(t)),$$

where $\nabla' f$ denotes the weak derivative of f with respect to x'. By Lemma 2.4, $\|\nabla f'\|_\infty \leq L$ and therefore

$$\left|\frac{d}{dt}u(\gamma(t))\right| \leq (\|\nabla u\|_\infty + 2\|\nabla' f\|_\infty \|\partial_d u\|_\infty) |x - y| \leq L' |x - y|.$$

By Lemma 2.6, we have

$$u(x) - u(y) = u(\gamma(1)) - u(\gamma(0)) = \int_0^1 \frac{d}{dt}u(\gamma(t)) \, dt$$

and inserting the above bound on the absolute value of the integrand yields the claimed bound (2.63) in the case $x_d \geq f(x')$ and $y_d \geq f(y')$.

The proof of (2.63) in the case $x_d < f(x')$ and $y_d < f(y')$ is similar, noting that $\gamma(t) \in \Omega^c$ for all $t \in [0,1]$.

Finally, in the case $x_d \geq f(x')$ and $y_d < f(y')$ (the case $x_d < f(x')$ and $y_d \geq f(y')$ is similar), we let $t = t_0 \in [0,1]$ be the unique solution of

$$t(x_d - f(x')) + (1-t)(y_d - f(y')) = 0$$

and $z := \gamma(t_0)$. Then $\gamma(t) \in \Omega^c$ for $t \in [0, t_0]$ and $\gamma(t) \in \overline{\Omega}$ if $t \in [t_0, 1]$. Applying the same argument as before to the two pieces of the path, we obtain again (2.63).

Step 2. As a preparation for the next step, we prove a simple geometric property of the set Ω. Let

$$\Gamma := \left\{(z', z_d) \in \mathbb{R}^{d-1} \times \mathbb{R} : z_d > L|z'|\right\}.$$

We claim that if $x \in \Omega$, then $x + \Gamma \subset \Omega$. Indeed, if $z \in \Gamma$, then $f(x' + z') \leq f(x') + L|z'| < x_d + z_d$, which means $x + z \in \Omega$.

Step 3. We claim that the set of $u \in H^1(\Omega)$ that have an extension to a continuously differentiable function on $\mathbb{R}^d$ with bounded derivative is dense in $H^1(\Omega)$. Indeed, let $\omega \in C_0^\infty(\mathbb{R}^d)$ be non-negative with $\int_{\mathbb{R}^d} \omega \, dx = 1$ and with support in the interior of $-\Gamma$. For $\rho > 0$ let $\omega_\rho(x) := \rho^{-d}\omega(x/\rho)$ and, for $u \in H^1(\Omega)$ and $x \in \mathbb{R}^d$, let

$$u_\rho(x) := \int_\Omega \omega_\rho(x - y)u(y)\,dy\,.$$

Then $u_\rho \in C^\infty(\mathbb{R}^d)$ and, by standard properties of convolutions, $u_\rho \to u$ in $L^2(\Omega)$. Moreover, $\|\nabla u_\rho\|_\infty \le \|\nabla \omega_\rho\|_2 \|u\|_2 < \infty$. Finally, for $x \in \Omega$, we have that

$$\partial_j u_\rho(x) = \int_\Omega (\partial_j \omega_\rho)(x - y)u(y)\,dy = \int_{\mathbb{R}^d} \omega_\rho(x - y)\partial_j u(y)\,dy\,.$$

Here we used the fact that for $x \in \Omega$ we have $\omega_\rho(x - \cdot) \in C_0^\infty(\Omega)$ by Step 2 and the support property of ω. The above formula implies, by standard properties of convolutions, that $\partial_j u_\rho \to \partial_j u$ in $L^2(\Omega)$ as $\rho \to 0$. Thus, $u_\rho \to u$ in $H^1(\Omega)$, as desired.

Step 4. Finally, it follows from the Steps 1 and 3 that the operator E, originally defined on functions $u \in H^1(\Omega) \cap C^1(\overline{\Omega})$ with $\nabla u \in L^\infty(\Omega)$, has a unique continuous extension to $H^1(\Omega)$ with the claimed bounds on its norms. To see that, for any $u \in H^1(\Omega)$, Eu is given by the formula in the lemma, we let $(u_n) \subset H^1(\Omega) \cap C^1(\overline{\Omega})$ be a sequence with $\nabla u_n \in L^\infty(\Omega)$ and $u_n \to u$ in $H^1(\Omega)$. Then, after passing to a subsequence, $u_n \to u$ almost everywhere in Ω, and therefore Eu_n converges almost everywhere in $\mathbb{R}^d$ to the expression for Eu given in the lemma. □

After these preliminaries we are now in the position to prove the existence of an extension operator for a rather larger class of sets, which is (in a more general form) due to Stein (1970, §VI.3) and which is the main result of this subsection. The idea of the proof is to localize and, in this way, to reduce the question to that solved in Lemma 2.91 for a special class of sets. While the localization is straightforward for bounded sets, in order to treat unbounded ones, we need to ensure a certain uniformity, which motivates the following definition.

Let $\Omega \subset \mathbb{R}^d$ be an open set. Its boundary $\partial\Omega$ is said to be *uniformly Lipschitz continuous* if there are $\varepsilon, L > 0, M \in \mathbb{N}$, and countably many open sets $\Omega_n \subset \mathbb{R}^d$ such that:

(a) If $x \in \partial\Omega$, then $B_\varepsilon(x) \subset \Omega_n$ for some n;
(b) No point in $\mathbb{R}^d$ is contained in more than M of the Ω_n;
(c) For each n there is an $O_n \in O(d)$ and a Lipschitz function $f_n : \mathbb{R}^{d-1} \to \mathbb{R}$ with Lipschitz constant bounded above by L such that

$$\Omega \cap \Omega_n = O_n\{(z', z_d) \in \mathbb{R}^{d-1} \times \mathbb{R} : z_d > f_n(z')\} \cap \Omega_n\,.$$

Theorem 2.92 *Let $\Omega \subset \mathbb{R}^d$ be an open set with uniformly Lipschitz continuous boundary. Then there is a bounded operator $E : H^1(\Omega) \to H^1(\mathbb{R}^d)$ such that $Eu = u$ a.e. in Ω for all $u \in H^1(\Omega)$. Moreover, in terms of the parameters in the definition of uniform Lipschitz continuity, Eu is supported in $\{\mathrm{dist}(\cdot, \Omega) \leq \varepsilon/2\}$ and satisfies*

$$\|Eu\|_{L^2(\Omega^c)} \leq M \|u\|_{L^2(\Omega)} \quad and \quad \|\nabla Eu\|_{L^2(\Omega^c)} \leq C_1 \|\nabla u\|_{L^2(\Omega)} + C_2 \|u\|_{L^2(\Omega)}$$

with $C_1 := M(1 + 2L)$ and $C_2 := 2\varepsilon^{-1} M \left(3 + 2L + 2\sqrt{M} \right)$.

Proof For each n, let

$$\lambda_n(x) := \begin{cases} 2\varepsilon^{-1}\mathrm{dist}(x, \Omega_n^c) & \text{if } \mathrm{dist}(x, \Omega_n^c) \leq \varepsilon/2, \\ 1 & \text{if } \mathrm{dist}(x, \Omega_n^c) > \varepsilon/2, \end{cases}$$

and let

$$\Lambda_+(x) := \begin{cases} 1 - 2\varepsilon^{-1}\mathrm{dist}(x, \partial\Omega) & \text{if } \mathrm{dist}(x, \partial\Omega) \leq \varepsilon/2, \\ 0 & \text{if } \mathrm{dist}(x, \partial\Omega) > \varepsilon/2, \end{cases}$$

and

$$\Lambda_-(x) := \begin{cases} 2\varepsilon^{-1}\mathrm{dist}(x, \Omega^c) & \text{if } \mathrm{dist}(x, \Omega^c) \leq \varepsilon/2, \\ 1 & \text{if } \mathrm{dist}(x, \Omega^c) > \varepsilon/2. \end{cases}$$

For each set $\{(z', z_d) \in \mathbb{R}^{d-1} \times \mathbb{R} : z_d > f_n(z')\}$ we have constructed an extension operator in Lemma 2.91. Composing this operator with an orthogonal matrix, we obtain, for each n, an extension operator E_n for the set

$$V_n := O_n\{(z', z_d) \in \mathbb{R}^{d-1} \times \mathbb{R} : z_d > f_n(z')\},$$

which satisfies, for each $u_n \in H^1(V_n)$,

$$\|E_n u_n\|_{L^2(\mathbb{R}^d)} = \sqrt{2}\|u_n\|_{L^2(V_n)} \quad and$$
$$\|\nabla E_n u_n\|_{L^2(\mathbb{R}^d)} \leq \sqrt{1 + (1 + 2L)^2}\|\nabla u_n\|_{L^2(V_n)}.$$

We now define for $u \in H^1(\Omega)$,

$$Eu := \Lambda_+ \frac{\sum_n \lambda_n E_n(\lambda_n u)}{\sum_m \lambda_m^2} + \Lambda_- u.$$

In this definition, we consider $\lambda_n u$ as an element of $H^1(V_n)$. Indeed, $\lambda_n u$ is originally defined in Ω and, by construction of λ_n, vanishes on $\Omega \cap \partial\Omega_n$. Since $\Omega \cap \Omega_n = V_n \cap \Omega_n$, we can extend $\lambda_n u$ by zero to $V_n \setminus \Omega$, and the resulting function belongs to $H^1(V_n)$.

Note also that $\sum_m \lambda_m^2 \geq 1$ on $\{\Lambda_+ > 0\}$. Indeed, if $\mathrm{dist}(x, \partial\Omega) \leq \varepsilon/2$, then there is a $y \in \partial\Omega$ with $|x - y| \leq \varepsilon/2$. By assumption, there is an n such that $B_\varepsilon(y) \subset \Omega_n$, and therefore

$$\operatorname{dist}(x, \Omega_n^c) \geq \operatorname{dist}(x, B_\varepsilon(y)^c) \geq \operatorname{dist}(y, B_\varepsilon(y)^c) - |x - y| \geq \varepsilon/2 \,.$$

Thus, $\lambda_n(x) = 1$ and, consequently, $\sum_m \lambda_m(x)^2 \geq 1$, as claimed.

Let us show that Eu is an extension operator; that is, $Eu = u$ a.e. in Ω. To prove this, we will show that for each n, $\lambda_n E_n(\lambda_n u) = \lambda_n^2 u$ a.e. in Ω. Indeed, by construction of E_n, this identity holds on V_n. On the other hand, the assumption $\Omega \cap \Omega_n = V_n \cap \Omega_n$ implies $V_n^c \cap \Omega \subset \Omega_n^c$, and therefore $\lambda_n = 0$ a.e. in $V_n^c \cap \Omega$. This proves that $\lambda_n E_n(\lambda_n u) = \lambda_n^2 u$ a.e. in Ω. The extension property then follows from the fact that $\Lambda_+ + \Lambda_- = 1$ in Ω.

Clearly, E is a bounded operator from $L^2(\Omega)$ to $L^2(\mathbb{R}^d)$. Indeed, as noted before, we have $0 \leq \Lambda_+ \lambda_n (\sum_m \lambda_m^2)^{-1} \leq \chi_{\Omega_n}$, and therefore, by the assumed bound on the multiplicity of the covering,

$$\|Eu\|_{L^2(\Omega^c)} \leq \sqrt{M} \left(\sum_n \int_{\Omega^c \cap \Omega_n} |E_n(\lambda_n u)|^2 \, dx \right)^{1/2}$$

$$\leq \sqrt{M} \left(\sum_n \int_{V_n} |\lambda_n u|^2 \, dx \right)^{1/2}$$

$$\leq M \, \|u\|_{L^2(\Omega)} \,.$$

For the bound on the extension operator E_n, we used $\Omega^c \cap \Omega_n \subset V_n^c$, and in the last inequality we used again the bound on the multiplicity of the covering.

Finally, we show that E is a bounded operator from $H^1(\Omega)$ to $L^2(\mathbb{R}^d)$. Note that the functions λ_n, Λ_+ and Λ_- are all Lipschitz, as is $(\sum_m \lambda_m^2)^{-1}$ on supp Λ_+. Therefore, by the product rule in Lemma 2.14, Eu is weakly differentiable for $u \in H^1(\Omega)$ and

$$\nabla Eu = (\nabla \Lambda_+) \frac{\sum_n \lambda_n E_n(\lambda_n u)}{\sum_m \lambda_m^2} + \Lambda_+ \frac{\sum_n ((\nabla \lambda_n) E_n(\lambda_n u) + \lambda_n \nabla E_n(\lambda_n u))}{\sum_m \lambda_m^2}$$
$$- 2\Lambda_+ \frac{\sum_n \lambda_n E_n(\lambda_n u)}{(\sum_m \lambda_m^2)^2} \sum_\ell \lambda_\ell \nabla \lambda_\ell + (\nabla \Lambda_-)u + \Lambda_- \nabla u \,.$$

For the term involving $\nabla E_n(\lambda_n u)$, we can argue exactly as in the proof of the L^2 boundedness and we obtain

$$\left\| \Lambda_+ \frac{\sum_n \lambda_n \nabla E_n(\lambda_n u)}{\sum_m \lambda_m^2} \right\|_{L^2(\Omega^c)} \leq \sqrt{M} \, (1 + 2L) \left(\sum_n \int_{V_n} |\nabla(\lambda_n u)|^2 \right)^{1/2} .$$

We now claim that

$$\left(\sum_n \int_{V_n} |\nabla(\lambda_n u)|^2\right)^{1/2} \le \sqrt{M}\left(\|\nabla u\|_{L^2(\Omega)} + 2\varepsilon^{-1}\|u\|_{L^2(\Omega)}\right).$$

To show this, we bound pointwise in Ω for any $\theta > 0$,

$$\sum_n |\nabla(\lambda_n u)|^2 \le \sum_n\left((1+\theta)\lambda_n^2|\nabla u|^2 + (1+\theta^{-1})|\nabla\lambda_n|^2|u|^2\right)$$

$$\le M\left((1+\theta)|\nabla u|^2 + (1+\theta^{-1})4\varepsilon^{-2}|u|^2\right).$$

Thus,

$$\left(\sum_n \int_{V_n} |\nabla(\lambda_n u)|^2\right)^{1/2} \le \sqrt{M}\left((1+\theta)\|\nabla u\|_{L^2(\Omega)}^2 + (1+\theta^{-1})4\varepsilon^{-2}\|u\|_{L^2(\Omega)}^2\right)^{1/2}$$

and the choice $\theta = 2\varepsilon^{-1}\|u\|_{L^2(\Omega)}\|\nabla u\|_{L^2(\Omega)}^{-1}$ gives the claimed bound.

The remaining terms in the formula for $\nabla E u$ in Ω^c are

$$(\nabla\Lambda_+)\frac{\sum_n \lambda_n E_n(\lambda_n u)}{\sum_m \lambda_m^2} + \Lambda_+\frac{\sum_n(\nabla\lambda_n)E_n(\lambda_n u)}{\sum_m \lambda_m^2} - 2\Lambda_+\frac{\sum_n \lambda_n E_n(\lambda_n u)}{(\sum_m \lambda_m^2)^2}\sum_\ell \lambda_\ell\nabla\lambda_\ell.$$

For the first two terms, we use the fact that, by Lemma 2.4, the length of the weak gradient of a distance function is bounded by 1, and obtain

$$|\nabla\Lambda_+| \le 2\varepsilon^{-1}\chi_{\text{supp}\,\Lambda_+} \qquad \text{and} \qquad |\nabla\lambda_n| \le 2\varepsilon^{-1}\chi_{\text{supp}\,\lambda_n}.$$

Proceeding as in the proof of the L^2 boundedness, we see that the $L^2(\Omega^c)$ norm of each of these terms is bounded by $2\varepsilon^{-1}M$ times the $L^2(\Omega)$ of u. For the last term, we use

$$\Lambda_+\frac{|\sum_\ell \lambda_\ell\nabla\lambda_\ell|}{\sum_m \lambda_m^2} \le \Lambda_+\frac{(\sum_\ell |\nabla\lambda_\ell|^2)^{1/2}}{(\sum_m \lambda_m^2)^{1/2}} \le 2\varepsilon^{-1}\sqrt{M}\,\Lambda_+$$

and, arguing again as before, we obtain the claimed bound.   □

Let us end by noting that the extension property does not hold for all domains.

Example 2.93   If $d \ge 2$ and $\alpha > 1$, then

$$\Omega := \{(x_1, x') \in \mathbb{R}^d : |x'|^{1/\alpha} < x_1 < 1\}$$

does *not* have the extension property. Indeed, the function $u(x) = x_1^{-\beta}$, $\beta > 0$, is weakly differentiable in Ω with $\partial_1 u(x) = -\beta x_1^{-\beta-1}$ and $\partial_j u(x) = 0$ for $j = 2, \ldots, d$. This function belongs to $H^1(\Omega)$ if and only if $\beta < (\alpha(d-1)-1)/2$. If Ω had the extension property, then it would follow from the Sobolev and Gagliardo–Nirenberg inequalities on $\mathbb{R}^d$ (Theorems 2.44 and 2.46) that $u \in$

$L^q(\Omega)$ for every $q < \infty$ if $d = 2$ and for $q = 2d/(d-2)$ if $d \geq 3$. A direct computation, however, shows that $u \in L^q(\Omega)$ if and only if $q < (\alpha(d-1)+1)/\beta$. This proves the claim for $d = 2$, and for $d \geq 3$ it remains to note that there is a β with $\beta < (\alpha(d-1)-1)/2$ and $(\alpha(d-1)+1)/\beta < 2d/(d-?)$

2.8.2 Extension domains have boundaries of measure zero

The fact that a domain has the extension property implies certain geometric properties. In particular, we shall show that such a domain has a boundary of zero measure and consists of at most finitely many connected components.

Lemma 2.94 *Let $\Omega \subset \mathbb{R}^d$ be an open set of finite measure with the extension property. Then Ω has at most finitely many connected components.*

Proof We argue by contradiction and assume that there are infinitely many distinct connected components Ω_n, $n \in \mathbb{N}$. The sequence of functions $u_n := |\Omega_n|^{-1/2}\chi_{\Omega_n}$ is bounded in $H^1(\Omega)$. Therefore, by Theorem 2.40 a subsequence converges in $L^2(\Omega)$ to a function $u \in L^2(\Omega)$. Since $\|u_n\| = 1$ it follows that $\|u\| = 1$, and therefore, in particular, $u \not\equiv 0$. Moreover, u_n converges pointwise to 0 because, for every fixed $x \in \Omega$, we have $u_n(x) = 0$ for all large enough n. This contradiction completes the proof. □

The following proposition is due to Hajłasz et al. (2008).

Proposition 2.95 *Let $\Omega \subset \mathbb{R}^d$ be an open set with the extension property. Then the d-dimensional Lebesgue measure of $\partial\Omega$ is zero.*

Proof The statement for $d = 1$ follows immediately from Lemma 2.94. In the following, we give the proof for $d \geq 3$. The idea for $d = 2$ is similar, but technically more involved and we refer to the above-mentioned paper for that case.

The strategy for the proof will be to show that there is a constant $c > 0$ (depending on Ω) such that for any $x \in \overline{\Omega}$ and any $0 < r \leq 1$ one has

$$|\Omega \cap B_r(x)| \geq c\,|B_r(x)|. \tag{2.64}$$

Before proving this, let us deduce from it the statement of the proposition. By Lebesgue's differentiation theorem (see, e.g., Rudin, 1987, Theorem 7.7; Stein and Shakarchi, 2005, §3.1),

$$\lim_{r \to 0} \frac{|\Omega \cap B_r(x)|}{|B_r(x)|} = \chi_\Omega(x) \quad \text{for a.e. } x \in \mathbb{R}^d.$$

In particular, since Ω is open, $\frac{|\Omega \cap B_r(x)|}{|B_r(x)|} \to 0$ for a.e. (with respect to d-dimensional Lebesgue measure) $x \in \partial\Omega$. On the other hand, by (2.64), the quotient is bounded away from zero for $x \in \partial\Omega$. Thus, $\partial\Omega$ has d-dimensional Lebesgue measure zero.

We now turn to the proof of (2.64). Fix $x \in \mathbb{R}^d$ and $0 < r \le 1$ with $|\Omega \cap B_r(x)| > 0$ and choose $\tilde{r} \in (0, r)$ such that

$$|\Omega \cap B_{\tilde{r}}(x)| = \frac{1}{2}|\Omega \cap B_r(x)|.$$

This is possible since, by dominated convergence, $\rho \mapsto |\Omega \cap B_\rho(x)|$ is a continuous non-decreasing function that vanishes at $\rho = 0$.

In a first step, we show that there is a constant $C < \infty$ such that, for any $x \in \mathbb{R}^d$ and $0 < r \le 1$ with $|\Omega \cap B_r(x)| > 0$, one has

$$r - \tilde{r} \le C |\Omega \cap B_r(x)|^{1/d}. \tag{2.65}$$

To prove this, consider the function

$$u(y) := \begin{cases} 1 & \text{if } |y - x| \le \tilde{r}, \\ \frac{r - |y-x|}{r - \tilde{r}} & \text{if } \tilde{r} < |y - x| < r, \\ 0 & \text{if } |y - x| \ge r. \end{cases}$$

The restriction of u to Ω belongs to $H^1(\Omega)$, and therefore the Sobolev inequality in Proposition 2.62 yields

$$\int_\Omega \left(|\nabla u|^2 + |u|^2 \right) dx \ge S \left(\int_\Omega |u|^{2d/(d-2)} \, dx \right)^{(d-2)/d}$$

with $S = S_{\Omega, 2d/(d-2)}$. We bound the right side from below by

$$\left(\int_\Omega |u|^{2d/(d-2)} \, dx \right)^{(d-2)/d} \ge |\Omega \cap B_{\tilde{r}}(x)|^{(d-2)/d} = 2^{-(d-2)/d} |\Omega \cap B_r(x)|^{(d-2)/d}$$

and the left side from above by

$$\int_\Omega \left(|\nabla u|^2 + |u|^2 \right) dx \le \frac{1}{(r - \tilde{r})^2} |\Omega \cap (B_r(x) \setminus B_{\tilde{r}}(x))| + |\Omega \cap B_r(x)|$$

$$= \left(\frac{1}{2(r - \tilde{r})^2} + 1 \right) |\Omega \cap B_r(x)|$$

$$\le \frac{3}{2(r - \tilde{r})^2} |\Omega \cap B_r(x)|.$$

In the last step we used $r \le 1$. Inserting these bounds into the Sobolev inequality gives

$$\frac{3}{2(r - \tilde{r})^2} |\Omega \cap B_r(x)| \ge S \, 2^{-(d-2)/d} |\Omega \cap B_r(x)|^{(d-2)/d},$$

which is the same as

$$r - \tilde{r} \le \left(\frac{3}{2^{2/d} S} \right)^{1/2} |\Omega \cap B_r(x)|^{1/d} .$$

This is the claimed inequality (2.65).

We now turn to the proof of (2.64). We assume that $x \in \overline{\Omega}$ and $0 < r \le 1$. Then $|\Omega \cap B_r(x)| > 0$ and we can recursively define $r_0 := r$ and $r_k > 0$ for $k \in \mathbb{N}$ by

$$|\Omega \cap B_{r_k}(x)| = \frac{1}{2} |\Omega \cap B_{r_{k-1}}(x)| .$$

Note that

$$|\Omega \cap B_{r_k}(x)| = 2^{-k} |\Omega \cap B_r(x)| . \tag{2.66}$$

We claim that $r_k \to 0$ as $k \to \infty$. Indeed, as a decreasing sequence of positive numbers, (r_k) has a limit $r_\infty \ge 0$ which satisfies $|\Omega \cap B_{r_\infty}(x)| = 0$ according to (2.66). Since $x \in \overline{\Omega}$ and Ω is open, this implies $r_\infty = 0$.

Thanks to (2.65), we have

$$r_k - r_{k+1} \le C |\Omega \cap B_{r_k}(x)|^{1/d} = C\, 2^{-k/d} |\Omega \cap B_r(x)|^{1/d} ,$$

and therefore, using $r_k \to 0$ as $k \to \infty$,

$$r = r_0 = \sum_{k=0}^{\infty} (r_k - r_{k+1}) \le C |\Omega \cap B_r(x)|^{1/d} \sum_{k=0}^{\infty} 2^{-k/d} = \frac{C}{1 - 2^{-1/d}} |\Omega \cap B_r(x)|^{1/d} .$$

This implies (2.64) and completes the proof of the proposition. $\square$

2.9 Comments

The selection of the material presented in this chapter is based on our needs in applications to Laplace and Schrödinger operators. For a more in-depth discussion of the theory of Sobolev spaces, we refer to the textbooks Brezis (2011), Evans (2010), Leoni (2017), Lieb and Loss (2001), Smirnov (1964), and the monographs Adams and Hedberg (1996), Adams and Fournier (2003), Gilbarg and Trudinger (1998), and Maz'ya (2011). Some of the proofs are taken from these books.

For the early history of Sobolev spaces and weak derivatives, including contributions by Levi, Tonelli, Fubini, Nikodym, Hilbert, Courant, Friedrichs, Rellich, Leray, Evans, Calkin, Morrey, and Sobolev, we refer to Naumann (2002).

Section 2.1: Weak derivatives

The definition of weak derivatives appears in the works of Sobolev (1935, 1936, 1938). It is closely related to the notion of distributional derivative. The theory of distributions goes back to Schwartz (1950, 1951) and is treated in many textbooks. It will not play a major role for us, however.

Relaxing the assumptions of continuous differentiability and boundedness of the derivative of φ in the chain rule in Lemma 2.10 is rather subtle, in particular, in the vectorial case $N \geq 2$; see, for instance, Leoni and Morini (2007) and references therein. The vectorial case with $N = 2$ is relevant, for instance, in the case of complex-valued functions appearing in the context of Schrödinger operators with magnetic fields. Note that the components of the map $\varphi : \mathbb{R}^2 \to \mathbb{R}^2$ implicit in the definition of the function $u^{(M)}$ in Corollary 2.15 (namely, $\varphi(u) = u$ if $|u| < M$ and $\varphi(u) = Mu/|u|$ for $|u| \geq M$) are not continuously differentiable, which explains in part its rather lengthy proof, based on Lemma 2.14.

The product rule in Lemma 2.14 often appears in the literature under stronger assumptions. In the case of real functions, it occurs in the present form in Gilbarg and Trudinger (1998, Exercise 7.4). Proving it under rather weak assumptions is again motivated by applications to Schrödinger operators with magnetic fields.

For us, the absolute continuity on lines (Lemma 2.16) plays a relatively minor role. It can be used to give alternative proofs of Lemma 2.11 and Corollaries 2.12 and 2.15; see Deny and Lions (1954, §I.3). The material on spherical coordinates in the setting of weak derivatives is probably standard, but, since it is not easy to locate a precise reference, we give a rather detailed account.

Section 2.2: Sobolev spaces

Sobolev spaces appear implicitly, for instance, in Friedrichs (1928, 1934b) through the closure of spaces of functions with classical derivatives with respect to the Dirichlet form, but were not studied systematically there. Their explicit definition and detailed study appears in the above mentioned papers of Sobolev and in his book (Sobolev, 1963). The equivalence between the definition via completion (originally denoted by $H^1(\Omega)$ or $H^{1,2}(\Omega)$) and that of Sobolev via weak derivatives (originally denoted by $W^{1,2}(\Omega)$) is proved in Proposition 2.23, which is from Meyers and Serrin (1964) with the title '$H = W$'; see also Deny and Lions (1954, Théorème I.2.3). This proposition is often stated in a weaker form, namely without the conclusion that approximating functions can be chosen in $L^\infty(\Omega)$. We note that the derivatives of the approximating functions in this proposition are not necessarily bounded up to the boundary. An approximation by restrictions to Ω of functions in $C_0^\infty(\mathbb{R}^d)$ holds if the boundary of Ω satisfies the so-called segment condition; see, for instance,

Adams and Fournier (2003, Theorem 3.22). For the proof of Lemma 2.27 we follow Brezis (2011, Theorem 8.12).

Section 2.3: Compact embeddings

Rellich's theorem is from Rellich (1930); see also Kondrachov (1945). For the proof of Proposition 2.36 we follow Lieb and Loss (2001, Theorem 8.6).

The condition of finite measure of Ω for the compactness of the embedding $H_0^1(\Omega) \subset L^2(\Omega)$ in Theorem 2.35 is convenient in applications, but not necessary. Necessary and sufficient conditions in terms of capacities were given by Molčanov (1953). This was further developed in Maz'ya (1974), Maz'ja and Otelbaev (1977), and Maz'ya and Shubin (2005b); see also Maz'ya (2007, §§16 and 18). Easier sufficient conditions can be found in Adams and Fournier (2003, Chapter 6).

Finiteness of the measure of Ω, or even boundedness of Ω, are not sufficient for compactness of the embedding $H^1(\Omega) \subset L^2(\Omega)$. Necessary and sufficient conditions are in Maz'ya (1969).

Section 2.4: Sobolev inequalities on the whole space

The Sobolev embedding theorems, including Theorem 2.44, go back to Soboleff (1938), who argued via mapping properties of fractional integral operators. The proof of Theorem 2.44 that we presented follows Chemin and Xu (1997). Other proofs of the inequality proceed via an isoperimetric-type inequality (Gagliardo, 1958; Nirenberg, 1959); see also Maz'ya (2011, 2003) for connections to isoperimetric and isocapacitary problems and the original papers (Maz'ya, 1960; Federer and Fleming, 1960).

Theorem 2.49, i.e., the version of Theorem 2.44 with the sharp constant, was obtained independently by Rodemich (1966), Rosen (1971), Aubin (1976), and Talenti (1976). Lieb (1983d) gives, essentially, two different proofs: the first goes through the Hardy–Littlewood–Sobolev inequality (Theorem 3.1 in Lieb, 1983d); the second, which is the one we present, goes through a one-dimensional inequality (Theorem 4.3 in Lieb, 1983d). This one-dimensional inequality (Theorem 4.2 in Lieb, 1983d) is essentially Theorem 2.48. Lieb, unaware of von Sz. Nagy's work, gives an alternative proof of the latter theorem. Instead of going via Theorem 2.48, Aubin and Talenti relied on a one-dimensional inequality due to Bliss (1930). Rodemich used logarithmic variables as in the proof that we presented. Further proofs that yield the sharp constant are in Cordero-Erausquin et al. (2004), Carlen et al. (2010); Frank and Lieb (2010), and Frank and Lieb (2012). For quantitative stability inequalities related to the Sobolev inequality in Theorem 2.49, we refer to Brezis and Lieb (1985) and Bianchi and Egnell (1991).

Most of the proofs of the Sobolev inequality with optimal constants rely on the technique of symmetric decreasing rearrangement. This technique was introduced by Faber and by Krahn in their proofs of the optimal lower bound on the eigenvalue of the Dirichlet Laplacian on sets of specified measure. We followed this strategy in the proof of Theorem 2.54. For references to works of Faber and of Krahn, we refer to the comments in §3.9.

Using the technique of symmetric decreasing rearrangement, one can deduce Sobolev's inequality (with non-sharp constant) from Hardy's inequality in Theorem 2.74 (Frank and Seiringer, 2008). In fact, this argument gives a strengthening in terms of Lorentz spaces of the Sobolev inequality with optimal constant (Alvino, 1977).

The Sobolev interpolation inequalities in Theorem 2.46 are named after Gagliardo (1959) and Nirenberg (1959). As we mentioned, except for the one-dimensional case (Theorem 2.48) due to Sz. Nagy (1941), the sharp constants in the Gagliardo–Nirenberg inequalities are not known in closed analytic form.

Let us summarize what is known concerning these constants. The existence of an optimizing function can be shown either using the technique of symmetric decreasing rearrangement (this proof is suggested in Lieb and Thirring (1976, §4, Remark A); see also Lieb (1976/77)) or using a collection of tools known collectively as concentration compactness (Lions, 1984a,b) (this proof is carried out in Weinstein, 1983). It is easy to see that any minimizer is (a constant multiple of) a positive function. Moreover, using symmetric decreasing rearrangement, one sees that at least one of the minimizers is a radially decreasing function. Using the method of moving planes, one can show that any positive solution of the corresponding Euler–Lagrange equation is radially decreasing (Gidas et al., 1981). Finally, one can show that there is a unique positive solution vanishing at infinity of the resulting ordinary differential equation. This was shown in Kwong (1989) after earlier work of Coffman (1972) and McLeod and Serrin (1987); see also McLeod (1993), and Serrin and Tang (2000). For a stability result for (an equivalent form of) the Gagliardo–Nirenberg inequality, see Carlen et al. (2014).

The bound on $S_{4,2}$ in Remark 2.51 is from Beckner (2004). For a numerical value, see Weinstein (1983).

Section 2.5: Friedrichs and Poincaré inequalities

The terminology of the Friedrichs and Poincaré inequalities is not completely standard. What we call the Friedrichs inequality is sometimes called the Friedrichs–Poincaré or Poincaré inequality. What we call the Poincaré inequality is sometimes called (in particular, in one dimension and for periodic functions) the Wirtinger inequality. The Friedrichs inequality goes back at least to Steklov (1896-97) and Friedrichs (1928), and the Poincaré inequality to Poincaré (1890, 1894, 1897); for references, see also Mitrinović (1970, §2.23.1).

The first proof of Lemma 2.52 in $d = 1$ that we presented is taken from Hardy et al. (1952, Theorem 25). The first proof of Lemma 2.57 in $d = 1$ is a variation of that of Hardy et al. (1952, Theorem 258), where it is attributed to H. Lewy. The second proofs of both lemmas are folklore.

Friedrichs inequalities are equivalent to lower bounds on the lowest eigenvalue of the Dirichlet Laplacian. For references on Theorem 2.54 (due to Faber and Krahn) and Proposition 2.55 (due to Hersch), see the comments in §3.9.

Poincaré inequalities are equivalent to lower bounds on the first positive eigenvalue of the Neumann Laplacian. For convex sets $\Omega \subset \mathbb{R}^d$ the Poincaré inequality in Proposition 2.58 holds with $P_\Omega \geq \pi^2(\text{diam}\,\Omega)^{-2}$. This bound is due to Payne and Weinberger (1960) (see also Bebendorf, 2003) and is optimal, as can be seen from cuboids $\Omega = (0, L_1) \times \cdots \times (0, L_d)$ with $L_1 \leq \cdots \leq L_{d-1} \ll L_d$. For a more general bound for convex sets with nonoptimal constant, see Gilbarg and Trudinger (1998, Equation (7.45)).

We shall use Lemma 2.61 in the proof of Weyl asymptotics for domains with rough boundaries. Our proof of this lemma is a modification of that in Netrusov and Safarov (2005). More precise inequalities of this type can be found, for instance, in Maz'ya (2011, Chapter 14).

Regarding the Poincaré–Sobolev inequality, we note that Proposition 2.62 holds also if the assumed extension property is replaced by the so-called weak cone-condition (Adams and Fournier, 2003, Theorem 4.12 and Remark 4.13).

Section 2.6: Hardy inequalities

More information on Hardy inequalities can be found, for instance, in the books of Opic and Kufner (1990), Kufner et al. (2007), and Balinsky et al. (2015). The inequality in Theorem 2.65 for $\rho = 0$ is stated in Hardy (1919), for functions vanishing in a neighborhood of the origin and in the equivalent integral formulation of Remark 2.67. A complete proof of the inequality for $\rho = 0$, even with general integrability exponent $1 < p < \infty$, appears in Hardy (1925). The inequality in Theorem 2.65 for general $\rho < 1$ is from Hardy (1928). A classical reference on this topic is Hardy et al. (1952, Theorems 253 and 327). We refer to Kufner et al. (2007) (see also Kufner et al., 2006) for a detailed account of the somewhat convoluted history of the inequality and the original contributions of Riesz, Schur, Landau and Pólya to the proof.

We have not seen the inequality in Theorem 2.69 in the literature. The first inequality in Theorem 2.70 with $\rho = 0$ appears in Kac and Kreĭn (1958); see also Birman (1961, Eq. (2.21)), where the inequality is generalized to the higher order case. The first inequality in Theorem 2.70 with $\rho < 1$ appears in Tomaselli (1969); see also the exposition in Talenti (1969). In fact, these two papers treat general weight functions in the gradient term and general integrability exponents $1 < p < \infty$. For an elegant proof, see also Muckenhoupt (1972). The second inequality in Theorem 2.70 with $\rho > 1$ follows from the first one with $\rho < 1$ via Remark 2.66. The second inequality in Theorem 2.70 with $\rho < 1$ appears in Tomaselli (1969); see also Talenti (1969, Eq. (27′)). Via Remark 2.66, this gives the first inequality for $\rho > 1$. The proof in Tomaselli (1969) is based on finding a positive solution to a second-order differential equation. We have not seen the proof we presented in the literature. The first inequality in Theorem 2.70 will later be crucial for obtaining the optimal version of Proposition 4.19 in all dimensions. For an extension of the inequality in Theorem 2.70 to exponents $p \neq 2$ see Frank et al. (2022).

One can prove Theorem 2.74 by first showing the inequality for smooth, compactly supported functions and then extending it by an approximation argument to the class given in the statement of the theorem. In contrast, we prove the inequality directly for all functions from this class, and therefore need Lemma 2.20 concerning separation of variables in spherical coordinates for Sobolev functions. This lemma is useful also at other places in this book.

For an alternative proof of (a strengthening of) Proposition 2.80, see Maz'ya (2011, Theorem 1.1.18) and the reference there.

Section 2.7: Homogeneous Sobolev spaces
The relevance of homogeneous Sobolev spaces to questions in spectral theory and also the role of the spaces $\widetilde{H}^1(\mathbb{R}^d)$ for $d = 1, 2$ is emphasized in Birman (1961).

One can define the spaces $\widetilde{H}^1(\mathbb{R}^d)$ in the same way for $d \geq 3$. These spaces are complete (this is essentially Deny and Lions, 1954, Corollaire 1.1), but $C_0^\infty(\mathbb{R}^d)$ is not dense in $\widetilde{H}^1(\mathbb{R}^d)$ if $d \geq 3$. In fact, one can show that if $u \in \widetilde{H}^1(\mathbb{R}^d)$ then there is a $c \in \mathbb{C}$ such that $u - c \in \dot{H}^1(\mathbb{R}^d)$. (This follows from Deny and Lions, 1954, Theorem 4.5, since, by the mean-value formula for the gradient, harmonic functions on $\mathbb{R}^d$ with square-integrable gradient are constant.)

Section 2.8: The extension property
Important extension theorems are due to Calderón (1961) (see also Adams and Fournier, 2003, Theorem 5.28) and Stein (1970, §VI.3). Calderón's assumption is a slight modification of the uniform cone condition. Stein's assumption is the same uniform Lipschitz continuity as in Theorem 2.92. Stein's proof, however, is different from the one we presented, in that his extension operator works also for higher-order Sobolev spaces. (Just like his, the extension operator constructed

in Theorem 2.92 also works for general integrability indices $1 \le p \le \infty$.) The proof of Lemma 2.90 that we presented is taken from Brezis (2011, Lemma 9.2). This lemma can be used to prove the extension property for sets with C^1 boundary; see, for instance, Brezis (2011, Theorem 9.7). Lemma 2.91 appears as Leoni (2017, Theorem 13.4), where it is deduced from what is our Lemma 2.90 by a Lipschitz change of variables, relying on Rademacher's theorem on almost everywhere differentiability of Lipschitz functions. Our proof of Lemma 2.91 is different and avoids the use of Rademacher's theorem. The approximation argument in Step 3 is from Stein (1970, §VI.3.2). For the proof of Theorem 2.92 we follow Stein (1970, §VI.3), but use more explicit localization functions.

The extension property is also discussed in Maz'ya (2011, §1.1.17 and §1.5). For extension operators for a larger class of sets than those with uniformly Lipschitz continuous boundary, see Jones (1981) and Rogers (2006) and references therein. Example 2.93 for $d = 2$ is Davies (1995, Example 7.1.9). It shows, in particular, that the Lipschitz assumption in Theorem 2.92 cannot be replaced by a Hölder condition with any exponent $1/\alpha < 1$.

Proposition 2.95, taken from Hajłasz et al. (2008), will simplify our proof of Weyl asymptotics in the Neumann case.

PART TWO

THE LAPLACE AND SCHRÖDINGER OPERATORS

3

The Laplacian on a Domain

After the preparations in the previous two chapters, we now come to the first main object of our study, the Laplace operator on an open set $\Omega \subset \mathbb{R}^d$. More precisely, we investigate self-adjoint realizations $-\Delta_\Omega^D$ and $-\Delta_\Omega^N$ of this operator in $L^2(\Omega)$ with Dirichlet and Neumann boundary conditions, respectively. We define these operators via quadratic forms. On the one hand, this allows us to define these operators for sets Ω without any assumptions on the regularity of their boundary and to prove results under rather minimal assumption. On the other hand, the approach via quadratic forms is the most natural one in view of the variational techniques that we will be using.

In the Dirichlet case we will be mostly concerned with open sets Ω of finite measure and in the Neumann case mostly with bounded, open sets with the extension property. In these cases, the spectra of $-\Delta_\Omega^D$ and $-\Delta_\Omega^N$ are discrete and consist of eigenvalues

$$0 < \lambda_1(-\Delta_\Omega^D) \le \lambda_2(-\Delta_\Omega^D) \le \lambda_3(-\Delta_\Omega^D) \le \cdots,$$

and

$$0 = \lambda_1(-\Delta_\Omega^N) \le \lambda_2(-\Delta_\Omega^N) \le \lambda_3(-\Delta_\Omega^N) \le \cdots,$$

which accumulate at infinity only. Eigenvalues are repeated according to their multiplicities.

In §§3.2 and 3.3 we study the asymptotic behavior of $\lambda_n(-\Delta_\Omega^D)$ and $\lambda_n(-\Delta_\Omega^N)$ as $n \to \infty$. In particular, we prove Weyl's theorem (Weyl, 1911, 1912a,b,c, 1913)

$$\lim_{n\to\infty} n^{-2/d} \lambda_n(-\Delta_\Omega^D) = \lim_{n\to\infty} n^{-2/d} \lambda_n(-\Delta_\Omega^N) = \left(L_{0,d}^{\text{cl}} |\Omega| \right)^{-2/d}$$

with

$$L_{0,d}^{\text{cl}} := \frac{1}{(4\pi)^{d/2}\Gamma(1 + d/2)}; \tag{3.1}$$

167

see Corollary 3.15 and Theorem 3.20. It is remarkable that the limiting behavior of the eigenvalues depends on the set Ω only through its measure. We prefer to state and prove this result in terms of the *counting functions*

$$N\left(\lambda, -\Delta_\Omega^D\right) := \# \left\{ n : \lambda_n(-\Delta_\Omega^D) < \lambda \right\}$$

and

$$N\left(\lambda, -\Delta_\Omega^N\right) := \# \left\{ n : \lambda_n(-\Lambda_\Omega^N) < \lambda \right\}.$$

Then Weyl's law takes the equivalent form

$$\lim_{\lambda \to \infty} \lambda^{-d/2} N\left(\lambda, -\Delta_\Omega^D\right) = \lim_{\lambda \to \infty} \lambda^{-d/2} N\left(\lambda, -\Delta_\Omega^N\right) = L_{0,d}^{cl} |\Omega|.$$

The proofs of Weyl's theorem that we present are based on the technique of *Dirichlet–Neumann bracketing* (Courant, 1920), which reduces the problem to an explicit computation of eigenvalues of the Laplacian on cubes. In the Dirichlet case we not only prove the asymptotics, but also obtain a remainder estimate which depends on the quality of the boundary of Ω.

Polya (1954) conjectured that the expression entering Weyl's law serves also as a universal, non-asymptotic spectral bound, namely,

$$n^{-2/d} \lambda_n(-\Delta_\Omega^D) \ge \left(L_{0,d}^{cl} |\Omega| \right)^{-2/d} \ge (n-1)^{-2/d} \lambda_n(-\Delta_\Omega^N)$$

for *all* $n \in \mathbb{N}$ and all Ω. Equivalently, in terms of the counting functions, this conjecture reads

$$\lambda^{-d/2} N\left(\lambda, -\Delta_\Omega^D\right) \le L_{0,d}^{cl} |\Omega| \le \lambda^{-d/2} N\left(\lambda, -\Delta_\Omega^N\right)$$

for all $\lambda > 0$ and all Ω. Note that the conjectured bound on the eigenvalues is a *lower bound* in the Dirichlet case and an *upper bound* in the Neumann case. Pólya (1961) proved these bounds for so-called tiling domains. We present this proof in §3.4. The conjecture in the general case remains open.

Related universal spectral inequalities for arbitrary domains have been obtained for so-called *Riesz means of eigenvalues*, namely for

$$\sum_n \left(\lambda_n(-\Delta_\Omega^D) - \lambda \right)_-^\gamma \quad \text{and} \quad \sum_n \left(\lambda_n(-\Delta_\Omega^N) - \lambda \right)_-^\gamma.$$

Here γ is a positive parameter and the case $\gamma = 0$ corresponds formally to the counting function considered before. Weyl's law implies that for any $\gamma > 0$ one has

$$\lim_{\lambda \to \infty} \lambda^{-\gamma-d/2} \sum_n \left(\lambda_n(-\Delta_\Omega^D) - \lambda \right)_-^\gamma = \lim_{\lambda \to \infty} \lambda^{-\gamma-d/2} \sum_n \left(\lambda_n(-\Delta_\Omega^N) - \lambda \right)_-^\gamma$$

$$= L_{\gamma,d}^{cl} |\Omega|$$

with

$$L_{\gamma,d}^{cl} := \frac{\Gamma(\gamma + 1)}{(4\pi)^{d/2}\Gamma(\gamma + 1 + d/2)}. \tag{3.2}$$

Berezin (1972b) proved that for $\gamma \geq 1$ these asymptotics are supplemented by the universal bound

$$\lambda^{-\gamma-d/2} \sum_n \left(\lambda_n(-\Delta_\Omega^D) - \lambda\right)_-^\gamma \leq L_{\gamma,d}^{cl}|\Omega|,$$

valid for all $\lambda > 0$ and for all open sets $\Omega \subset \mathbb{R}^d$ of finite measure. One can show that this bound with $\gamma = 1$ is equivalent to the lower bound

$$N^{-1-2/d} \sum_{n=1}^N \lambda_n(-\Delta_\Omega^D) \geq \frac{d}{d+2}\left(L_{0,d}^{cl}|\Omega|\right)^{-2/d},$$

valid for all $N \in \mathbb{N}$ and for all open sets $\Omega \subset \mathbb{R}^d$. The latter was obtained independently by Li and Yau (1983). The analogous reverse bound in the Neumann case is due to Kröger (1992) and states that

$$(M - 1)^{-1-2/d} \sum_{n=1}^M \lambda_n(-\Delta_\Omega^N) \leq \frac{d}{d+2}\left(L_{0,d}^{cl}|\Omega|\right)^{-2/d}$$

for all $M \in \mathbb{N}$ and all open sets $\Omega \subset \mathbb{R}^d$ for which $-\Delta_\Omega^N$ has discrete spectrum.

The bounds of Berezin–Li–Yau and of Kröger on sums of eigenvalues lead to the following bounds for individual eigenvalues:

$$n^{-2/d}\lambda_n(-\Delta_\Omega^D) \geq \frac{d}{d+2}\left(L_{0,d}^{cl}|\Omega|\right)^{-2/d}$$

and

$$(n - 1)^{-2/d}\lambda_n(-\Delta_\Omega^N) \leq \left(\frac{d+2}{2}\right)^{2/d}\left(L_{0,d}^{cl}|\Omega|\right)^{-2/d}.$$

These bounds are off by factors of $d/(d+2)$ and $((d+2)/2)^{2/d}$, respectively, from the conjectured bounds by Pólya. We present the bounds of Berezin–Li–Yau and of Kröger in §§3.5 and 3.6.

So far, the only class of domains other than the tiling ones for which one can prove Pólya's conjectured bound are domains of a Cartesian product structure. We present this bound in Theorem 3.36 and in its proof we apply for the first time the *lifting technique*, which plays an important role in later chapters.

It is often useful to look at the asymptotics and the bounds in this chapter from a phase space point of view. For instance, one has

$$L_{\gamma,d}^{cl}|\Omega|\,\lambda^{\gamma+d/2} = \iint_{\Omega\times\mathbb{R}^d}\left(|\xi|^2 - \lambda\right)_-^\gamma\frac{dx\,d\xi}{(2\pi)^d},$$

and therefore the Weyl asymptotics for Riesz means take the form

$$\lim_{\lambda \to \omega} \frac{\sum_n \left(\lambda_n(-\Delta_\Omega^D) - \lambda\right)_-^\gamma}{\iint_{\Omega \times \mathbb{R}^d} \left(|\xi|^2 - \lambda\right)_-^\gamma \frac{dx\,d\xi}{(2\pi)^d}} = 1 .$$

This will be discussed further in §3.7. In an appendix, §3.8, we will discuss the Laplacian is spherical coordinates and, in particular, discuss the spectrum of the Dirichlet and Neumann Laplacians on a ball.

3.1 The Dirichlet and Neumann Laplacians

In this section, we define the Dirichlet and Neumann Laplacians on a domain and discuss basic properties of their spectra.

3.1.1 Definition of the Dirichlet and Neumann Laplacians

Let $\Omega \subset \mathbb{R}^d$ be an open set and consider the quadratic forms

$$a^N[u] := \int_\Omega |\nabla u|^2 \, dx \qquad \text{and} \qquad a^D[u] := \int_\Omega |\nabla u|^2 \, dx$$

in $L^2(\Omega)$ with form domains

$$d[a^N] := H^1(\Omega) \qquad \text{and} \qquad d[a^D] := H_0^1(\Omega) .$$

Here, $H^1(\Omega)$ and $H_0^1(\Omega)$ denote the Sobolev spaces from §2.2.

Clearly, the forms a^N and a^D are non-negative. Lemma 2.22 implies that a^N is closed in $L^2(\Omega)$. The same lemma, together with the fact that $H_0^1(\Omega)$ is, by definition, a closed subspace of $H^1(\Omega)$, implies that a^D is closed in $L^2(\Omega)$. Therefore, by Theorem 1.18, there are unique self-adjoint operators A^N and A^D in $L^2(\Omega)$ corresponding to a^N and a^D, respectively, and these operators are non-negative. We denote them by $-\Delta_\Omega^N$ and $-\Delta_\Omega^D$ and call them, respectively, the *Neumann Laplacian* and the *Dirichlet Laplacian* on Ω.

In the case $\Omega = \mathbb{R}^d$, it follows from Proposition 2.23 that $-\Delta_{\mathbb{R}^d}^N = -\Delta_{\mathbb{R}^d}^D$, and usually we denote this operator simply by $-\Delta$ and call it the *Laplacian*.

By definition, $H_0^1(\Omega)$ is the closure of $C_0^\infty(\Omega)$ in $H^1(\Omega)$, which means that $C_0^\infty(\Omega)$ is a form core of the Dirichlet Laplacian. Moreover, by Proposition 2.23, the set

$$\left\{ u \in H^1(\Omega) : u \in C^\infty(\Omega) \cap L^\infty(\Omega), \text{ supp } u \text{ bounded} \right\}$$

is dense in $H^1(\Omega)$, which means that it is a form core of the Neumann Laplacian.

For the next result, we recall the notion of the weak Laplacian in §2.1.6.

Lemma 3.1 *We have*

$$\mathrm{dom}\left(-\Delta_\Omega^N\right) = \Big\{ u \in H^1(\Omega): u \text{ has a weak Laplacian } \Delta u \in L^2(\Omega)$$

$$\text{and for all } w \in H^1(\Omega),\ \int_\Omega \nabla u \cdot \nabla w\, dx = -\int_\Omega (\Delta u) w\, dx \Big\}$$

and

$$\mathrm{dom}\left(-\Delta_\Omega^D\right) = \{ v \in H_0^1(\Omega) : v \text{ has a weak Laplacian } \Delta v \in L^2(\Omega)\},$$

and, for $u \in \mathrm{dom}\left(-\Delta_\Omega^N\right)$ and $v \in \mathrm{dom}\left(-\Delta_\Omega^D\right)$, we also have $-\Delta_\Omega^N u = -\Delta u$ and $-\Delta_\Omega^D v = -\Delta v$.

The last part of the lemma motivates the notation $-\Delta_\Omega^N$ and $-\Delta_\Omega^D$ for the operators corresponding to the quadratic forms a^N and a^D, respectively.

Proof According to the representation theorem (Theorem 1.18), the operator domains of $-\Delta_\Omega^N$ and $-\Delta_\Omega^D$ are given by

$$\mathrm{dom}\left(-\Delta_\Omega^N\right) = \Big\{ u \in H^1(\Omega): \text{ there exists } f \in L^2(\Omega) \text{ such that for all}$$

$$w \in H^1(\Omega),\ \int_\Omega \nabla u \cdot \nabla w\, dx = \int_\Omega f w\, dx \Big\}$$

and

$$\mathrm{dom}\left(-\Delta_\Omega^D\right) = \Big\{ v \in H_0^1(\Omega): \text{ there exists } g \in L^2(\Omega) \text{ such that for all}$$

$$z \in H_0^1(\Omega),\ \int_\Omega \nabla v \cdot \nabla z\, dx = \int_\Omega g z\, dx \Big\}.$$

Moreover, for $u \in \mathrm{dom}\left(-\Delta_\Omega^N\right)$ and $v \in \mathrm{dom}\left(-\Delta_\Omega^D\right)$ the f and the g in these formulas are unique and given by $f = -\Delta_\Omega^N u$ and $g = -\Delta_\Omega^D u$, respectively. Taking $w, z \in C_0^\infty(\Omega)$, we see that all $u \in \mathrm{dom}\left(-\Delta_\Omega^N\right)$ and $v \in \mathrm{dom}\left(-\Delta_\Omega^D\right)$ have weak Laplacians, which satisfy $\Delta u \in L^2(\Omega)$ and $\Delta v \in L^2(\Omega)$, as well as $-\Delta u = f = -\Delta_\Omega^N u$ and $-\Delta v = g = -\Delta_\Omega^D v$. This proves the assertion in the Neumann case and the inclusion $\subset$ in the Dirichlet case. The reverse inclusion $\supset$ follows using the fact that $C_0^\infty(\Omega)$ is a form core of $-\Delta_\Omega^D$. □

We emphasize that so far we have made no assumptions about the boundary of Ω. The reason for the terms 'Neumann' and 'Dirichlet' comes from boundary conditions satisfied when the boundary is sufficiently smooth. The term 'Neumann' is used since, in the case where $\partial\Omega$ is sufficiently smooth (C^2, for instance), functions u in $\mathrm{dom}\left(-\Delta_\Omega^N\right)$ satisfy the boundary conditions

$\partial u/\partial v = 0$ on $\partial\Omega$, where v denotes the outer unit normal. The term 'Dirichlet' is used since, in the case where $\partial\Omega$ is sufficiently smooth (Lipschitz, for instance), functions v in $H_0^1(\Omega)$ satisfy the boundary condition $v = 0$ on $\partial\Omega$ in the sense of so-called Sobolev trace theorems. For more on operator domains, see §3.9. We stress, however, that an advantage of the approach via quadratic forms is that we do not need these facts and that we can work with general open sets Ω without any assumptions on their boundaries.

3.1.2 Examples

We begin with the case $\Omega = \mathbb{R}^d$ and recall that in this case $-\Delta_{\mathbb{R}^d}^N = -\Delta_{\mathbb{R}^d}^D = -\Delta$. By Lemma 2.31 and its proof, we know that

$$H^1(\mathbb{R}^d) = \{u \in L^2(\mathbb{R}^d) : (1 + |\xi|^2)^{1/2}\widehat{u} \in L^2(\mathbb{R}^d)\},$$

where $\widehat{u}$ denotes the Fourier transform of u, see (2.8), and that for all u from this set

$$\|\nabla u\|^2 = \int_{\mathbb{R}^d} |\xi|^2 |\widehat{u}(\xi)|^2 \, d\xi.$$

We conclude that $-\Delta$ is unitarily equivalent to multiplication by $|\xi|^2$ in $L^2(\mathbb{R}^d)$, with the unitary operator being the Fourier transform. As in §1.1.4 we deduce that $\sigma(-\Delta) = \sigma_{\mathrm{ess}}(-\Delta) = [0, \infty)$ and $\sigma_p(-\Delta) = \sigma_{\mathrm{disc}}(-\Delta) = \emptyset$.

Our second example is the case where Ω is a cube. Let $Q_L := (0, L)^d$ be the open cube in $\mathbb{R}^d$ of side length L with center $(L/2, \ldots, L/2)$. We claim that $\sigma_{\mathrm{ess}}(-\Delta_{Q_L}^N) = \sigma_{\mathrm{ess}}(-\Delta_{Q_L}^D) = \emptyset$ and that the eigenvalues are given by

$$\left(\frac{\pi}{L}\right)^2 \sum_{j=1}^d m_j^2 \quad \text{and} \quad \left(\frac{\pi}{L}\right)^2 \sum_{j=1}^d n_j^2,$$

where $m = (m_1, \ldots, m_d) \in \mathbb{N}_0^d$ in the Neumann case and $n = (n_1, \ldots, n_d) \in \mathbb{N}^d$ in the Dirichlet case. Moreover, the multiplicity of an eigenvalue λ in the two cases is

$$\#\left\{m \in \mathbb{N}_0^d : \left(\frac{\pi}{L}\right)^2 \sum_{j=1}^d m_j^2 = \lambda\right\} \quad \text{and} \quad \#\left\{n \in \mathbb{N}^d : \left(\frac{\pi}{L}\right)^2 \sum_{j=1}^d n_j^2 = \lambda\right\},$$

respectively. Here and elsewhere, we write $\mathbb{N} = \{1, 2, 3, \ldots\}$ and $\mathbb{N}_0 = \{0, 1, 2, \ldots\}$.

In order to prove this claim, for $n \in \mathbb{N}^d$ and $m \in \mathbb{N}_0^d$ let

$$\Psi_m(x) := \prod_{j=1}^d \psi_{m_j}(L^{-1}x_j) \quad \text{and} \quad \Phi_n(x) := \prod_{j=1}^d \varphi_{n_j}(L^{-1}x_j) \qquad \text{for } x \in Q_L,$$

where

$$\psi_{m_j}(y) := \cos(\pi m_j y) \quad \text{and} \quad \varphi_{n_j}(y) := \sin(\pi n_j y) \qquad \text{for } y \in (0,1).$$

The functions Ψ_m and Φ_n are smooth and, in particular, in $H^1(Q_L)$. Moreover, $\Phi_n \in H_0^1(Q_L)$, since $\varphi_{n_j} \in H_0^1(0,1)$ by Lemma 2.27 and since scaled products of functions in $C_0^\infty(0,1)$ approximating φ_{n_j} in $H^1(0,1)$ belong to $C_0^\infty(Q_L)$ and approximate Φ_n in $H^1(Q_L)$.

Next, we claim that

$$\int_{Q_L} \nabla\Psi_m \cdot \nabla w \, dx = \left(\frac{\pi}{L}\right)^2 \sum_{j=1}^d m_j^2 \int_{Q_L} \Psi_m w \, dx \qquad \text{for all } w \in H^1(Q_L) \quad (3.3)$$

and

$$\int_{Q_L} \nabla\Phi_n \cdot \nabla z \, dx = \left(\frac{\pi}{L}\right)^2 \sum_{j=1}^d n_j^2 \int_{Q_L} \Phi_n z \, dx \qquad \text{for all } z \in H_0^1(Q_L). \quad (3.4)$$

These equations can be justified using the same arguments as in the derivations of (2.25) and (2.37) and we omit the details.

It follows from (3.3) and (3.4) that $\Psi_m \in \mathrm{dom}(-\Delta_{Q_L}^N)$ and $\Phi_n \in \mathrm{dom}(-\Delta_{Q_L}^D)$, and that any number λ of the form $\left(\frac{\pi}{L}\right)^2 \sum_{j=1}^d m_j^2$ or $\left(\frac{\pi}{L}\right)^2 \sum_{j=1}^d n_j^2$ is a Neumann or Dirichlet eigenvalue, respectively, of at least the claimed multiplicity.

It is well known that both systems $(\Psi_m)_{m \in \mathbb{N}_0^d}$ and $(\Phi_n)_{n \in \mathbb{N}^d}$ are complete orthogonal (but not orthonormal) systems of functions in $L^2(Q_L)$. This is a standard result in Fourier analysis; see, for instance, Simon (2015a, Theorem 3.5.1) for the one-dimensional case. The higher-dimensional case follows from this by separation of variables. We conclude that we have determined the whole spectrum of $-\Delta_{Q_L}^D$ and $-\Delta_{Q_L}^N$, respectively. This proves the claim.

Remark 3.2 For a cuboid $\Omega = (0, L_1) \times \cdots \times (0, L_d)$ with $L_1, \dots, L_d > 0$ the eigenvalues of $-\Delta_\Omega^N$ and $-\Delta_\Omega^D$ are given by

$$\pi^2 \sum_{j=1}^d \left(\frac{m_j}{L_j}\right)^2, \ m \in \mathbb{N}_0^d, \quad \text{and} \quad \pi^2 \sum_{j=1}^d \left(\frac{n_j}{L_j}\right)^2, \ n \in \mathbb{N}^d,$$

respectively, with a similar statement about multiplicities as in the case of a cube.

As another example one can determine the spectrum of the Neumann and Dirichlet Laplacians on a ball in terms of zeros of special functions. The computations are somewhat lengthy and are deferred to the appendix in §3.8. Similar arguments apply in the case of spherical shells.

3.1.3 Basic spectral properties

After having discussed the cases of the whole space and of a cube, we now turn our attention to the spectral properties of $-\Delta_\Omega^N$ and $-\Delta_\Omega^D$ for general open sets Ω. In general, it is not possible to compute the eigenvalues explicitly and one of our principal aims is to find inequalities on these eigenvalues for general domains.

Since $H_0^1(\Omega) \subset H^1(\Omega)$, we have, in the sense of definition (1.53),

$$-\Delta_\Omega^N \leq -\Delta_\Omega^D .$$

Therefore, by Proposition 1.29,

$$N(\lambda, -\Delta_\Omega^N) \geq N(\lambda, -\Delta_\Omega^D) \qquad \text{for all } \lambda > 0 .$$

In particular, $\inf \sigma_{\text{ess}}(-\Delta_\Omega^N) \leq \inf \sigma_{\text{ess}}(-\Delta_\Omega^D)$ (with the convention that the infimum of the empty set is $+\infty$) and the eigenvalues, arranged in increasing order, counting multiplicities, satisfy

$$\lambda_n(-\Delta_\Omega^N) \leq \lambda_n(-\Delta_\Omega^D) \tag{3.5}$$

for all $n \leq \min\left\{ N(\inf \sigma_{\text{ess}}(-\Delta_\Omega^D), -\Delta_\Omega^D), N(\inf \sigma_{\text{ess}}(-\Delta_\Omega^N), -\Delta_\Omega^N) \right\}$. We will study more refined versions of inequality (3.5) in §3.1.4 below.

A particularly important case is when the spectrum is discrete. Corollary 1.21 allows us to state the following necessary and sufficient condition for this.

Lemma 3.3 *The spectrum of the Neumann Laplacian is discrete if and only if the embedding $H^1(\Omega) \subset L^2(\Omega)$ is compact. The spectrum of the Dirichlet Laplacian is discrete if and only if the embedding $H_0^1(\Omega) \subset L^2(\Omega)$ is compact.*

This lemma, together with Rellich's compactness theorem (Theorems 2.35 and 2.40), leads to convenient sufficient conditions for the spectrum to be discrete. We recall that the extension property was introduced in §2.3.2 and, by Theorem 2.92, domains with uniformly Lipschitz continuous boundary have this property.

Lemma 3.4 *If $\Omega \subset \mathbb{R}^d$ is an open set of finite measure, then the spectrum of the Dirichlet Laplacian is discrete. If $\Omega \subset \mathbb{R}^d$ is an open set of finite measure with the extension property, then the spectrum of the Neumann Laplacian is discrete.*

In general, the fact that Ω has finite measure is *not* sufficient for the discreteness of the spectrum of the Neumann Laplacian; see the references in §3.9.

Remark 3.5   If $\Omega \subset \mathbb{R}^d$ is an open set of infinite measure, then $0 \in \sigma_{\mathrm{ess}}(-\Delta_\Omega^N)$. To prove this, recall that by Lemma 2.42, there is a sequence $(u_n) \subset H^1(\Omega)$ with $\|u_n\| = 1$ for all n such that $\|\nabla u_n\| \to 0$ and $u_n \to 0$ weakly in $L^2(\Omega)$ as $n \to \infty$. Since $\|\nabla u_n\| = \left\| \sqrt{-\Delta_\Omega^N} u_n \right\|$, this is a Weyl sequence for $\sqrt{-\Delta_\Omega^N}$, and therefore, by Lemma 1.13, $0 \in \sigma_{\mathrm{ess}}(\sqrt{-\Delta_\Omega^N})$. By (1.26), this implies $0 \in \sigma_{\mathrm{ess}}(-\Delta_\Omega^N)$, as claimed. Alternatively, one can apply Lemma 1.20 to the above sequence (u_n) and arrive at the same conclusion.

The following property of the Dirichlet Laplacian is called 'domain monotonicity'.

Lemma 3.6   *Let $\Omega \subset \widetilde{\Omega} \subset \mathbb{R}^d$ be open sets. Then*

$$\inf \sigma_{\mathrm{ess}}\left(-\Delta_{\widetilde{\Omega}}^D \right) \le \inf \sigma_{\mathrm{ess}}\left(-\Delta_\Omega^D \right)$$

and, with $\tilde{N} := N\left(\inf \sigma_{\mathrm{ess}}\left(-\Delta_{\widetilde{\Omega}}^D \right), -\Delta_{\widetilde{\Omega}}^D \right)$ and $N := N\left(\inf \sigma_{\mathrm{ess}}\left(-\Delta_\Omega^D \right), -\Delta_\Omega^D \right)$,

$$\lambda_n\left(-\Delta_{\widetilde{\Omega}}^D \right) \le \lambda_n\left(-\Delta_\Omega^D \right) \qquad \text{for all } n \le \min\{\tilde{N}, N\}.$$

Moreover, if $\inf \sigma_{\mathrm{ess}}\left(-\Delta_{\widetilde{\Omega}}^D \right) < \infty$ and $N > \tilde{N}$, then

$$\inf \sigma_{\mathrm{ess}}\left(-\Delta_{\widetilde{\Omega}}^D \right) \le \lambda_n\left(-\Delta_\Omega^D \right) \qquad \text{for all } \tilde{N} < n \le N.$$

This lemma follows immediately from the variational principle (Proposition 1.29) and the fact that the extension by zero of a function in $H_0^1(\Omega)$ belongs to $H_0^1(\widetilde{\Omega})$ (Lemma 2.28).

Domain monotonicity can fail for the Neumann Laplacian, as illustrated by the following example.

Example 3.7   Let us consider the square $\widetilde{\Omega} := (-1/2, 1/2) \times (-1/2, 1/2) \subset \mathbb{R}^2$. Then, as we saw in §3.1.2, $\lambda_1\left(-\Delta_{\widetilde{\Omega}}^N \right) = 0$ and $\lambda_2\left(-\Delta_{\widetilde{\Omega}}^N \right) = \pi^2$. Now let $a \in (0, 1/2)$ and consider the 'tilted' open rectangle Ω, described by the corners $(a, 1/2)$, $(1/2, a)$, $(-a, -1/2)$ and $(-1/2, -a)$. Then clearly $\Omega \subset \widetilde{\Omega}$ and Ω has side lengths $\sqrt{2}(1/2 - a)$ and $\sqrt{2}(1/2 + a)$. By Remark 3.2, $\lambda_1\left(-\Delta_\Omega^N \right) = 0$ and $\lambda_2\left(-\Delta_\Omega^N \right) = \pi^2/(2(1/2 + a)^2)$. Thus, if $a > (1/2)(\sqrt{2} - 1)$, then $\lambda_2\left(-\Delta_\Omega^N \right) < \lambda_2\left(-\Delta_{\widetilde{\Omega}}^N \right)$.

If Ω has finite measure, then 0 is an eigenvalue of $-\Delta_\Omega^N$ and the multiplicity of this eigenvalue equals the number of connected components of Ω. Eigenfunctions are constant on these components. In the Dirichlet case, 0 is never an eigenvalue and, in fact, if Ω has finite measure, one can give a lower bound on the smallest eigenvalue, which only depends on the measure of Ω. Specifically, Theorem 2.54 together with the variational characterization of $\lambda_1\left(-\Delta_\Omega^D \right)$ (Lemma 1.19) imply the following bound.

Lemma 3.8 (Faber–Krahn inequality) *For any open set $\Omega \subset \mathbb{R}^d$ of finite measure the first eigenvalue $\lambda_1\big(-\Delta_\Omega^D \big)$ of the Dirichlet Laplacian satisfies*

$$\lambda_1\big(-\Delta_\Omega^D \big) \geq \frac{C_d}{|\Omega|^{2/d}} \tag{3.6}$$

with

$$C_d := j_{(d-2)/2,1}^2 \, \omega_d^{2/d},$$

where $j_{(d-2)/2,1}$ is the first positive zero of the Bessel function $J_{(d-2)/2}$ and where ω_d is the measure of the unit ball in $\mathbb{R}^d$. The constant C_d is optimal.

Equality in (3.6) with the optimal value of C_d is attained if and only if Ω is a ball. For references, see §2.9.

Lemma 3.8 gives a lower bound on $\lambda_1(-\Delta_\Omega^D)$ in terms of the measure of the set Ω. For some domains, for instance, whose extension in one direction is significantly smaller than in the remaining directions, these bounds are not particularly good. This motivates us to seek lower bounds on $\lambda_1(-\Delta_\Omega^D)$ in terms of other geometric characteristics of Ω than its measure. As an example of such a bound we will present an inequality in terms of the inradius $R_{in}(\Omega)$ (see (2.27)). Namely, Proposition 2.55 together with the variational characterization of $\lambda_1(-\Delta_\Omega^D)$ (Lemma 1.19) imply the following bound.

Lemma 3.9 (Hersch inequality) *Let $\Omega \subset \mathbb{R}^d$ be open and convex. Then*

$$\lambda_1(-\Delta_\Omega^D) \geq \frac{\pi^2}{4\,R_{in}(\Omega)^2} \tag{3.7}$$

with the convention $\lambda_1(-\Delta_\Omega^D) := \inf \sigma(-\Delta_\Omega^D)$.

Equality in (3.7) is attained for $\Omega = \mathbb{R}^{d-1} \times (0,l)$. For references, see §2.9.

3.1.4 Comparing Dirichlet and Neumann eigenvalues

Let $\Omega \subset \mathbb{R}^d$ be an open set such that $H^1(\Omega)$ is compactly embedded into $L^2(\Omega)$. Then, by Lemma 3.3, the spectra of the Dirichlet and Neumann Laplacians are discrete and, according to (3.5), the eigenvalues satisfy

$$\lambda_k(-\Delta_\Omega^N) \leq \lambda_k(-\Delta_\Omega^D) \qquad \text{for all} \quad k \in \mathbb{N}.$$

If $d = 1$ and Ω is an interval, then $\lambda_{k+1}(-\Delta_\Omega^N) = \lambda_k(-\Delta_\Omega^D)$ for all $k \in \mathbb{N}$. In higher dimensions $d \geq 2$ the following improved comparison holds.

Theorem 3.10 *Let $d \geq 2$ and let $\Omega \subset \mathbb{R}^d$ be an open set such that $H^1(\Omega)$ is compactly embedded into $L^2(\Omega)$. Then*

$$\lambda_{k+1}(-\Delta_\Omega^N) < \lambda_k(-\Delta_\Omega^D) \qquad \text{for all } k \in \mathbb{N}.$$

This theorem goes back to Friedlander (1991) and we refer to §3.9 for a discussion of earlier results and conjectures. Here, we present a proof from Filonov (2004), which works under the sole assumption that the embedding is compact. It slightly improves Friedlander's result.

As a first step in the proof, we note that an eigenfunction of the Neumann Laplacian cannot satisfy Dirichlet boundary conditions.

Lemma 3.11 *We have for any $\mu \geq 0$,*

$$\ker(-\Delta_\Omega^N - \mu) \cap H_0^1(\Omega) = \{0\}\,.$$

Proof Let $v \in \ker(-\Delta_\Omega^N - \mu) \cap H_0^1(\Omega)$ and $\eta \in C_0^\infty(\mathbb{R}^d)$. By Lemma 2.28, the trivial extension

$$\widetilde{v}(x) := \begin{cases} v(x) & \text{for } x \in \Omega, \\ 0 & \text{for } x \notin \Omega \end{cases}$$

belongs to $H^1(\mathbb{R}^d)$ and satisfies

$$\int_{\mathbb{R}^d} \nabla \widetilde{v} \cdot \overline{\nabla \eta}\, dx = \int_\Omega \nabla v \cdot \overline{\nabla \eta}\, dx\,.$$

On the other hand, since v is an eigenfunction of $-\Delta_\Omega^N$ corresponding to the eigenvalue μ and since $\eta|_\Omega \in H^1(\Omega)$, the form domain of $-\Delta_\Omega^N$, we have

$$\int_\Omega \nabla v \cdot \overline{\nabla \eta}\, dx = \int_\Omega (-\Delta_\Omega^N v)\overline{\eta}\, dx = \mu \int_\Omega v\overline{\eta}\, dx = \mu \int_{\mathbb{R}^d} \widetilde{v}\overline{\eta}\, dx\,.$$

Combining the last two identities, we infer that

$$\int_{\mathbb{R}^d} \nabla \widetilde{v} \cdot \overline{\nabla \eta}\, dx = \mu \int_{\mathbb{R}^d} \widetilde{v}\overline{\eta}\, dx\,.$$

This inequality holds for all $\eta \in C_0^\infty(\mathbb{R}^d)$ and, since $C_0^\infty(\mathbb{R}^d)$ is dense in $H^1(\mathbb{R}^d)$ by Proposition 2.23, it holds in fact for all $\eta \in H^1(\mathbb{R}^d)$. We conclude that $\widetilde{v} \in \mathrm{dom}(-\Delta_{\mathbb{R}^d})$ and $-\Delta_{\mathbb{R}^d}\widetilde{v} = \mu\widetilde{v}$. As we saw in §3.1.2, $-\Delta_{\mathbb{R}^d}$ does not have eigenvalues, so $\widetilde{v} = 0$, and therefore $v = 0$, as claimed. □

Proof of Theorem 3.10 Let $\mu > 0$ and $L := \mathrm{ran}\, P_{(-\infty,\mu]}$, where P is the spectral measure of the Dirichlet Laplacian. Then $L \subset H_0^1(\Omega)$,

$$\dim L = N(\mu, -\Delta_\Omega^D) + \dim \ker(-\Delta_\Omega^D - \mu) < \infty \tag{3.8}$$

and

$$\int_\Omega |\nabla u|^2\, dx \leq \mu \int_\Omega |u|^2\, dx \quad \text{for all} \quad u \in L\,. \tag{3.9}$$

By Lemma 3.11, $\ker(-\Delta_\Omega^N - \mu)$ has a trivial intersection with $H_0^1(\Omega)$ and therefore also with L. Hence, the sum

$$F := L + \ker(-\Delta_\Omega^N - \mu)$$

has dimension

$$\dim F = \dim L + \ker(-\Delta_\Omega^N - \mu).$$

Meanwhile, recall from Remark 3.5 that the assumption of the theorem implies that Ω has finite measure and, in particular, $e^{i\omega \cdot x} \in L^2(\Omega)$ for all $\omega \in \mathbb{R}^d$. The linear span of the functions $e^{i\omega \cdot x}|_\Omega$, where $\omega \in \mathbb{R}^d$ satisfies $|\omega| = \sqrt{\mu}$, is infinite-dimensional. Thus, there is an $\omega_0 \in \mathbb{R}^d$ with $|\omega_0| = \sqrt{\mu}$ such that $e^{i\omega_0 \cdot x} \notin F$. Hence, the sum

$$G := F + \operatorname{span}\{e^{i\omega_0 \cdot x}\}$$

is contained in $H^1(\Omega)$ and has dimension

$$\dim G = \dim F + 1 = \dim L + \dim \ker(-\Delta_\Omega^N - \mu) + 1. \tag{3.10}$$

Let us show that inequality (3.9) remains true for all $u \in G$. Indeed, for any $u = u_L + v + c e^{i\omega_0 \cdot x}$ with $u_L \in L$, $v \in \ker(-\Delta_\Omega^N - \mu)$ and $c \in \mathbb{C}$, we have

$$\int_\Omega |\nabla u|^2 \, dx = \int_\Omega |\nabla(u_L + v + c e^{i\omega_0 \cdot x})|^2 \, dx = I_1 + I_2$$

with

$$I_1 := \int_\Omega (|\nabla u_L|^2 + |\nabla v|^2 + |c\nabla e^{i\omega_0 \cdot x}|^2) \, dx \le \mu \int_\Omega (|u_L|^2 + |v|^2 + |c e^{i\omega_0 \cdot x}|^2) \, dx$$

and

$$I_2 := 2\operatorname{Re} \int_\Omega \left(\nabla v \cdot \overline{\nabla(u_L + c e^{i\omega_0 \cdot x})} + c\nabla e^{i\omega_0 \cdot x} \cdot \overline{\nabla u_L} \right) dx.$$

Since the eigenfunction v belongs to the operator domain of $-\Delta_\Omega^N$ and as $u_L + c e^{i\omega_0 \cdot x}$ belongs to the form domain $H^1(\Omega)$ of $-\Delta_\Omega^N$, and $u_L \in H_0^1(\Omega)$, we have

$$I_2 = 2\operatorname{Re} \int_\Omega \left((-\Delta_\Omega^N v)\,\overline{(u_L + c e^{i\omega_0 \cdot x})} + c(-\Delta e^{i\omega_0 \cdot x})\,\overline{u_L} \right) dx$$

$$= 2\mu \operatorname{Re} \int_\Omega \left(v\,\overline{(u_L + c e^{i\omega_0 \cdot x})} + c e^{i\omega_0 \cdot x}\,\overline{u_L} \right) dx.$$

This gives

$$\int_\Omega |\nabla u|^2 \, dx \le \mu \int_\Omega (|u_L|^2 + |v|^2 + |c e^{i\omega_0 \cdot x}|^2) \, dx + I_2$$

$$= \mu \int_\Omega |u_L + v + c e^{i\omega_0 \cdot x}|^2 \, dx = \mu \int_\Omega |u|^2 \, dx;$$

that is, inequality (3.9) remains true for all $u \in G$.

Therefore, the variational principle (Theorem 1.25) implies that

$$N\left(\mu, -\Delta_\Omega^N\right) + \dim \ker(-\Delta_\Omega^N - \mu) \geq \dim G.$$

Combining this inequality with identities (3.8) and (3.10), we arrive at

$$N\left(\mu, -\Delta_\Omega^N\right) \geq N\left(\mu, -\Delta_\Omega^D\right) + \dim \ker\left(-\Delta_\Omega^D - \mu\right) + 1. \tag{3.11}$$

This is essentially the result of the theorem in terms of spectral counting functions. We now translate it into an inequality for eigenvalues. We fix $k \in \mathbb{N}$ and choose $\mu = \lambda_k(-\Delta_\Omega^D)$. If $\ell \in \mathbb{N}_0$ is such that

$$\lambda_k(-\Delta_\Omega^D) = \cdots = \lambda_{k+\ell}(-\Delta_\Omega^D) < \lambda_{k+\ell+1}(-\Delta_\Omega^D),$$

then

$$N\left(\mu, -\Delta_\Omega^D\right) + \dim \ker\left(-\Delta_\Omega^D - \mu\right) = k + \ell.$$

Therefore, from (3.11) it follows that $N\left(\mu, -\Delta_\Omega^N\right) \geq k + \ell + 1$; that is,

$$\lambda_{k+\ell+1}(-\Delta_\Omega^N) < \mu = \lambda_k(-\Delta_\Omega^D).$$

Since $\lambda_{k+\ell+1}(-\Delta_\Omega^N) \geq \lambda_{k+1}(-\Delta_\Omega^N)$, this implies the claimed inequality. □

3.2 Weyl's asymptotic formula for the Dirichlet Laplacian on a domain

In this section we prove Weyl's formula (Weyl, 1911, 1912a,b,c, 1913)

$$\lim_{\lambda \to \infty} \lambda^{-d/2} N\left(\lambda, -\Delta_\Omega^D\right) = L_{0,d}^{cl} |\Omega|$$

for the eigenvalues of the Dirichlet Laplacian $-\Delta_\Omega^D$ on an open set $\Omega \subset \mathbb{R}^d$ of finite measure. The corresponding result

$$\lim_{\lambda \to \infty} \lambda^{-\gamma-d/2} \sum_n \left(\lambda_n(-\Delta_\Omega^D) - \lambda\right)_-^\gamma = L_{\gamma,d}^{cl} |\Omega|$$

for the Riesz means of order $\gamma > 0$ is obtained as a corollary.

The constants $L_{0,d}^{cl}$ and $L_{\gamma,d}^{cl}$ are defined in (3.1) and (3.2), respectively.

For the proof of Weyl's law we proceed in several steps. In §3.2.1 we start with the analysis of the Dirichlet and Neumann Laplacians on cubes, where the eigenvalues can be computed explicitly and where Weyl's law, even with a remainder bound, can be easily read off. In §3.2.2 we extend Weyl's law to polycubes (that is, finite unions of cubes with disjoint interiors) and then to Jordan measurable sets. This subsection serves as a first illustration of the technique of *Dirichlet–Neumann bracketing* in a relatively simple setting. The result, including an error bound, for general open sets of finite measure is stated and discussed in §3.2.3. The somewhat technical proof is deferred to §3.2.4.

3.2.1 Cubes

Let

$$Q_L := \{x \in \mathbb{R}^d : 0 < x_j < L, j = 1, \ldots, d\}$$

be the open cube in $\mathbb{R}^d$ of side length L with center $(L/2, \ldots, L/2)$. We consider the Dirichlet Laplacian $-\Delta_{Q_L}^D$ and the Neumann Laplacian $-\Delta_{Q_L}^N$ acting in $L^2(Q_L)$. As discussed in §3.1.2, their spectra are discrete and consist of the eigenvalues

$$\left(\frac{\pi}{L}\right)^2 \sum_{j=1}^d n_j^2, \quad n \in \mathbb{N}^d, \qquad \text{and} \qquad \left(\frac{\pi}{L}\right)^2 \sum_{j=1}^d m_j^2, \quad m \in \mathbb{N}_0^d,$$

respectively, and the multiplicity of an eigenvalue λ is equal to

$$\#\left\{n \in \mathbb{N}^d : \left(\frac{\pi}{L}\right)^2 \sum_{j=1}^d n_j^2 = \lambda\right\} \qquad \text{and} \qquad \#\left\{m \in \mathbb{N}_0^d : \left(\frac{\pi}{L}\right)^2 \sum_{j=1}^d m_j^2 = \lambda\right\},$$

respectively. Here, as always, we write $\mathbb{N} = \{1, 2, 3, \ldots\}$ and $\mathbb{N}_0 = \{0, 1, 2, \ldots\}$. Therefore, the counting functions $N(\lambda, -\Delta_{Q_L}^D)$ and $N(\lambda, -\Delta_{Q_L}^N)$ are given by the number of points in $\mathbb{N}^d$ (in the Dirichlet case) or $\mathbb{N}_0^d$ (in the Neumann case) inside the open ball of radius $\frac{L\sqrt{\lambda}}{\pi}$ centered at the origin. Since these points can all be associated with disjoint cubes of side length one, it is intuitively clear that for large λ the numbers $N(\lambda, -\Delta_{Q_L}^D)$ and $N(\lambda, -\Delta_{Q_L}^N)$ will approximately behave as the volume of the spherical sector

$$B_+ := \left\{x \in \mathbb{R}^d : |x| < \frac{L\sqrt{\lambda}}{\pi} \text{ and } x_j \geq 0 \text{ for } j = 1, \ldots, d\right\}.$$

We shall prove this now with a suitable remainder estimate.

Lemma 3.12 *There is a constant C_d such that for all $\lambda > 0$ and $L > 0$ one has*

$$0 \leq \lambda^{d/2} L_{0,d}^{cl} L^d - N(\lambda, -\Delta_{Q_L}^D) \leq C_d\left(1 + (L^2\lambda)^{(d-1)/2}\right), \qquad (3.12)$$

$$0 \leq N(\lambda, -\Delta_{Q_L}^N) - \lambda^{d/2} L_{0,d}^{cl} L^d \leq C_d\left(1 + (L^2\lambda)^{(d-1)/2}\right) \qquad (3.13)$$

with $L_{0,d}^{cl}$ as in (3.1).

Proof   To show the left inequality in (3.12), we note that any $n \in B_+ \cap \mathbb{N}^d$ comes with a cube $\{x \in \mathbb{R}^d : n_j - 1 < x_j < n_j\} \subset B_+$. Since all these cubes are disjoint, the number of such cubes does not exceed the volume of B_+, which equals 2^{-d} times the volume of the ball of radius $\frac{L\sqrt{\lambda}}{\pi}$, which, in turn, is equal to $\lambda^{d/2} L^{cl}_{0,d} L^d$. Similarly, the first inequality in (3.13) follows from the fact that, if we take for any $m \in B_+ \cap \mathbb{N}_0^d$ the cube $\{x \in \mathbb{R}^d : m_j < x_j < m_j + 1\}$, then any point from B_+ is contained in the closure of one of these cubes. Therefore, the number of these cubes exceeds the volume of B_+.

In order to prove the right inequalities in (3.12) and (3.13), it suffices to show that

$$N\left(\lambda, -\Delta_{Q_L}^N\right) - N\left(\lambda, -\Delta_{Q_L}^D\right) \leq C_d\left(1 + (L^2\lambda)^{(d-1)/2}\right).$$

To this end, we note that

$$N\left(\lambda, -\Delta_{Q_L}^N\right) - N\left(\lambda, -\Delta_{Q_L}^D\right) = \#\left\{n \in (\mathbb{N}_0^d \setminus \mathbb{N}^d) \cap B_+\right\} = \bigcup_{k=0}^{d-1} M_k \,,$$

where the M_k are the disjoint subsets of $\mathbb{N}_0^d \cap B_+$, containing those $n = (n_1, \ldots, n_d)$ for which exactly k components are different from zero. We have $\#M_0 = 1$ and, for $k \geq 1$, by the left inequality in (3.12) with k instead of d,

$$\#M_k = \binom{d}{k}\#\left\{n' \in \mathbb{N}^k : |n'| < \frac{L\sqrt{\lambda}}{\pi}\right\} \leq \binom{d}{k}L^{cl}_{0,k}L^k\lambda^{k/2}\,.$$

Summing over k we arrive at

$$N\left(\lambda, -\Delta_{Q_L}^N\right) - N\left(\lambda, -\Delta_{Q_L}^D\right) \leq \sum_{k=0}^{d-1}\binom{d}{k}L^{cl}_{0,k}L^k\lambda^{k/2}\,,$$

which completes the proof. $\square$

Since $|Q_L| = L^d$, Lemma 3.12 implies the asymptotics, as $\lambda \to \infty$,

$$N\left(\lambda, -\Delta_{Q_L}^D\right) = \lambda^{d/2}L^{cl}_{0,d}|Q_L| + O\left(\lambda^{(d-1)/2}\right) \tag{3.14}$$

and

$$N\left(\lambda, -\Delta_{Q_L}^N\right) = \lambda^{d/2}L^{cl}_{0,d}|Q_L| + O\left(\lambda^{(d-1)/2}\right). \tag{3.15}$$

This is the simplest form of Weyl's law of the asymptotic distribution of eigenvalues of the Laplace operator. The bounds from Lemma 3.12 can easily be generalized to Riesz means.

Lemma 3.13 *For any $\gamma > 0$ there is a constant $C_{\gamma,d}$ such that for all $\lambda > 0$ and $L > 0$ one has*

$$0 \le \lambda^{\gamma+d/2} L^{cl}_{\gamma,d} L^d - \mathrm{Tr}\left(-\Delta^D_{Q_L} - \lambda\right)^\gamma_- \le C_{\gamma,d}\lambda^\gamma\left(1 + (L^2\lambda)^{(d-1)/2}\right),$$

$$0 \le \mathrm{Tr}\left(-\Delta^N_{Q_L} - \lambda\right)^\gamma_- - \lambda^{\gamma+d/2} L^{cl}_{\gamma,d} L^d \le C_{\gamma,d}\lambda^\gamma\left(1 + (L^2\lambda)^{(d-1)/2}\right)$$

with $L^{cl}_{\gamma,d}$ as in (3.2).

A simple consequence of Lemma 3.13 are the asymptotics, as $\lambda \to \infty$,

$$\mathrm{Tr}\left(-\Delta^D_{Q_L} - \lambda\right)^\gamma_- = \lambda^{\gamma+d/2} L^{cl}_{\gamma,d}|Q_L| + O(\lambda^{\gamma+(d-1)/2})$$

and

$$\mathrm{Tr}\left(-\Delta^N_{Q_L} - \lambda\right)^\gamma_- = \lambda^{\gamma+d/2} L^{cl}_{\gamma,d}|Q_L| + O\left(\lambda^{\gamma+(d-1)/2}\right).$$

Proof We apply Lemma 1.39 with $A = -\Delta^D_{Q_L} - \lambda$ and use (3.12) for the integrand. To simplify the constants, we note that the beta function identity

$$\int_0^1 (1-s)^{a-1} s^{b-1}\, ds = \frac{\Gamma(a)\Gamma(b)}{\Gamma(a+b)}, \tag{3.16}$$

together with the explicit expressions in (3.1) and (3.2), implies that

$$\gamma \int_0^\lambda (\lambda - \kappa)^{d/2} \kappa^{\gamma-1}\, d\kappa\, L^{cl}_{0,d} = \lambda^{\gamma+d/2} L^{cl}_{\gamma,d}. \tag{3.17}$$

In this way, we obtain

$$0 \le \lambda^{\gamma+d/2} L^{cl}_{\gamma,d} L^d - \mathrm{Tr}\left(-\Delta^D_{Q_L} - \lambda\right)^\gamma_- \le C_d\lambda^\gamma\left(1 + c_{\gamma,d}(L^2\lambda)^{(d-1)/2}\right)$$

with

$$c_{\gamma,d} := \gamma \int_0^1 (1-s)^{(d-1)/2} s^{\gamma-1}\, ds = \gamma\, \frac{\Gamma((d+1)/2)\,\Gamma(\gamma)}{\Gamma(\gamma + (d+1)/2)}.$$

This proves the first pair of inequalities; the second is proved similarly. □

3.2.2 Polycubes and Jordan measurable sets

The expressions (3.14) and (3.15) are, in fact, special cases of a general result on the asymptotics of the spectral counting function of the Laplacian on arbitrary domains. To prove this, we first consider the case of *polycubes* Ω; that is, the interior of a finite union of the closures of disjoint open cubes $Q_{L_k}(a_k) = Q_{L_k} + a_k$ with certain $L_k > 0$ and $a_k \in \mathbb{R}^d$:

$$\Omega := \mathrm{int}\left(\bigcup_{k=1}^n \overline{Q_{L_k}(a_k)}\right).$$

Here, int A denotes the interior of a set A. We define

$$\Omega_0 := \bigcup_{k=1}^{n} Q_{L_k}(a_k).$$

Since the total measure of all boundaries $\partial Q_{L_k}(a_k)$ is zero, we have

$$L^2(\Omega) = L^2(\Omega_0) = \bigoplus_{k=1}^{n} L^2(Q_{L_k}(a_k)).$$

In view of the fact that

$$\bigoplus_{k=1}^{n} H_0^1(Q_{L_k}(a_k)) = H_0^1(\Omega_0) \subset H_0^1(\Omega) \subset H^1(\Omega) \subset H^1(\Omega_0)$$

$$= \bigoplus_{k=1}^{n} H^1(Q_{L_k}(a_k)),$$

we have the operator inequalities

$$\bigoplus_{k=1}^{n} \left(-\Delta_{Q_{L_k}(a_k)}^{\mathrm{N}}\right) = -\Delta_{\Omega_0}^{\mathrm{N}} \leq -\Delta_{\Omega}^{\mathrm{N}} \leq -\Delta_{\Omega}^{\mathrm{D}} \leq -\Delta_{\Omega_0}^{\mathrm{D}} = \bigoplus_{k=1}^{n} \left(-\Delta_{Q_{L_k}(a_k)}^{\mathrm{D}}\right).$$

By the variational principle (Proposition 1.29), this implies

$$\sum_{k=1}^{n} N\left(\lambda, -\Delta_{Q_{L_k}(a_k)}^{\mathrm{N}}\right) = N\left(\lambda, \bigoplus_{k=1}^{n}\left(-\Delta_{Q_{L_k}(a_k)}^{\mathrm{N}}\right)\right) = N(\lambda, -\Delta_{\Omega_0}^{\mathrm{N}})$$

$$\geq N\left(\lambda, -\Delta_{\Omega}^{\mathrm{N}}\right) \geq N\left(\lambda, -\Delta_{\Omega}^{\mathrm{D}}\right) \geq N(\lambda, -\Delta_{\Omega_0}^{\mathrm{D}})$$

$$\geq N\left(\lambda, \bigoplus_{k=1}^{n}\left(-\Delta_{Q_{L_k}(a_k)}^{\mathrm{D}}\right)\right) = \sum_{k=1}^{n} N\left(\lambda, -\Delta_{Q_{L_k}(a_k)}^{\mathrm{D}}\right).$$

Inserting the asymptotics (3.14) and (3.15) and noting that

$$|\Omega| = \sum_{k=1}^{n} |Q_{L_k}(a_k)|,$$

we obtain, as $\lambda \to \infty$,

$$N\left(\lambda, -\Delta_{\Omega}^{\mathrm{D}}\right) = \lambda^{d/2} L_{0,d}^{\mathrm{cl}} |\Omega| + O\left(\lambda^{(d-1)/2}\right) \tag{3.18}$$

and

$$N\left(\lambda, -\Delta_{\Omega}^{\mathrm{N}}\right) = \lambda^{d/2} L_{0,d}^{\mathrm{cl}} |\Omega| + O\left(\lambda^{(d-1)/2}\right).$$

This is Weyl's law for polycubes.

Next, let $\Omega \subset \mathbb{R}^d$ be a bounded, open set with *finite Jordan measure*. This means that, for any $\varepsilon > 0$, there are two polycubes $\Omega_\varepsilon^- \subset \Omega \subset \Omega_\varepsilon^+$ such that $|\Omega_\varepsilon^+ \setminus \Omega_\varepsilon^-| \le \varepsilon$. Then the boundary of Ω has (Lebesgue) measure zero and

$$\lim_{\varepsilon \to 0} |\Omega_\varepsilon^+| = \lim_{\varepsilon \to 0} |\Omega_\varepsilon^-| = |\Omega| \tag{3.19}$$

In view of the domain monotonicity of the Dirichlet Laplacian (Lemma 3.6), we have

$$N\left(\lambda, -\Delta_{\Omega_\varepsilon}^{\mathrm{D}}\right) \le N\left(\lambda, -\Delta_{\Omega}^{\mathrm{D}}\right) \le N\left(\lambda, -\Delta_{\Omega_\varepsilon^+}^{\mathrm{D}}\right).$$

From the asymptotics (3.18) for polycubes we conclude that

$$L_{0,d}^{\mathrm{cl}}|\Omega_\varepsilon^-| \le \liminf_{\lambda \to +\infty} \lambda^{-d/2} N\left(\lambda, -\Delta_\Omega^{\mathrm{D}}\right) \le \limsup_{\lambda \to +\infty} \lambda^{-d/2} N\left(\lambda, -\Delta_\Omega^{\mathrm{D}}\right) \le L_{0,d}^{\mathrm{cl}}|\Omega_\varepsilon^+|.$$

Passing to the limit $\varepsilon \to 0$ and recalling (3.19) we find that, asymptotically,

$$N\left(\lambda, -\Delta_\Omega^{\mathrm{D}}\right) = \lambda^{d/2} L_{0,d}^{\mathrm{cl}}|\Omega| + o(\lambda^{d/2}) \text{ as } \lambda \to \infty. \tag{3.20}$$

This is Weyl's law for bounded, open, Jordan measurable sets Ω.

3.2.3 Arbitrary open sets with finite measure

In the previous subsection, we used an elementary approximation argument to pass from polycubes to more general sets. But for this, we had to restrict ourselves to Jordan measurable sets and we obtained only a non-quantitative remainder bound in (3.20). A more subtle version of this approximation argument allows us to generalize (3.20) to arbitrary open sets of finite Lebesgue measure and, at the same time, to gain a quantitative remainder bound.

Let $\Omega \subset \mathbb{R}^d$ be an open set of finite Lebesgue measure. Note that, by Lemma 3.4, the Dirichlet Laplacian $-\Delta_\Omega^{\mathrm{D}}$ has discrete spectrum. In fact, in the following, we do not need to know this fact beforehand as it will be a consequence of our result.

The next theorem is the main result of this section. To state it, for $t > 0$, we write

$$\Omega(t) := \{x \in \Omega \colon \mathrm{dist}(x, \partial\Omega) < t\}.$$

Theorem 3.14 *Let $\Omega \subset \mathbb{R}^d$ be an open set of finite measure and denote by L_Ω the side length of the largest open cube contained in Ω. Then, for all $\lambda \ge \frac{\pi^2}{4} L_\Omega^{-2}$,*

$$\left| N\left(\lambda, -\Delta_\Omega^{\mathrm{D}}\right) - L_{0,d}^{\mathrm{cl}}|\Omega|\lambda^{\frac{d}{2}} \right| \le c_d \lambda^{\frac{d-1}{2}} \int_{2^{-1}\sqrt{d}\pi\lambda^{-1/2}}^{+\infty} |\Omega(t)| \frac{dt}{t^2}. \tag{3.21}$$

The constant c_d depends only on the dimension d.

Results of the type of Theorem 3.14 and some of the following corollaries go back to Courant (1920). The bound (3.21) appears in Netrusov and Safarov (2005), improving earlier results from van den Berg and Lianantonakis (2001).

We defer the proof of Theorem 3.14 to §3.2.4 and first discuss some consequences. Note that, for any open set $\Omega \subset \mathbb{R}^d$ of finite measure, the function $t \mapsto |\Omega(t)|$ is non-decreasing and, by dominated convergence, $\lim_{t \to 0} |\Omega(t)| = 0$. Hence

$$\int_{2^{-1}\sqrt{d}\pi\lambda^{-1/2}}^{+\infty} |\Omega(t)| \frac{dt}{t^2} = \sqrt{\lambda} \int_{2^{-1}\sqrt{d}\pi}^{+\infty} |\Omega(s/\sqrt{\lambda})| \frac{ds}{s^2} = o(\sqrt{\lambda}) \quad \text{as} \quad \lambda \to \infty.$$

We therefore deduce from Theorem 3.14 the following asymptotics.

Corollary 3.15 *Let $\Omega \subset \mathbb{R}^d$ be an open set of finite measure. Then*

$$N(\lambda, -\Delta_\Omega^{\mathrm{D}}) = \lambda^{\frac{d}{2}} L_{0,d}^{\mathrm{cl}} |\Omega| + o(\lambda^{\frac{d}{2}}) \qquad \text{as } \lambda \to \infty$$

and

$$\lambda_n(-\Delta_\Omega^{\mathrm{D}}) = (L_{0,d}^{\mathrm{cl}})^{-2/d} |\Omega|^{-2/d} n^{2/d} + o(n^{2/d}) \qquad \text{as } n \to \infty.$$

The remainder estimates in the corollary can be improved if the boundary of Ω has some regularity. Assume, for example, that for some $\alpha \in (0, 1]$,

$$|\Omega(t)| = O(t^\alpha) \quad \text{as} \quad t \to 0. \tag{3.22}$$

Then, as $\lambda \to +\infty$,

$$\int_{2^{-1}\sqrt{d}\pi\lambda^{-1/2}}^{+\infty} |\Omega(t)| \frac{dt}{t^2} = \begin{cases} O(\lambda^{\frac{1-\alpha}{2}}) & \text{if } \alpha \in (0,1), \\ O(\ln \lambda) & \text{if } \alpha = 1. \end{cases} \tag{3.23}$$

From Theorem 3.14 we conclude that, as $\lambda \to +\infty$,

$$N(\lambda, -\Delta_\Omega^{\mathrm{D}}) = \lambda^{\frac{d}{2}} L_{0,d}^{\mathrm{cl}} |\Omega| + \begin{cases} O(\lambda^{\frac{d-\alpha}{2}}) & \text{if } \alpha \in (0,1), \\ O(\lambda^{\frac{d-1}{2}} \ln \lambda) & \text{if } \alpha = 1. \end{cases} \tag{3.24}$$

In particular, if Ω is bounded with uniformly Lipschitz continuous boundary, as defined before Theorem 2.92, then (3.22) is satisfied with $\alpha = 1$. This gives the following.

Corollary 3.16 *Let $\Omega \subset \mathbb{R}^d$ be a bounded, open set with uniformly Lipschitz continuous boundary. Then*

$$N(\lambda, -\Delta_\Omega^{\mathrm{D}}) = \lambda^{\frac{d}{2}} L_{0,d}^{\mathrm{cl}} |\Omega| + O(\lambda^{\frac{d-1}{2}} \ln \lambda) \quad \text{as } \lambda \to +\infty.$$

Using different techniques, one can improve the remainder bound to $O(\lambda^{(d-1)/2})$ for domains with smoother boundary and, under an additional assumption on the geodesic flow, a second term can be derived; for references see §3.9, where we also give a reference showing that the error bound in (3.24), under assumption (3.22) with $0 < \alpha < 1$, is best possible. The latter is achieved for domains with fractal boundaries.

We now discuss similar questions for Riesz means.

Corollary 3.17 *Let $\Omega \subset \mathbb{R}^d$ be an open set of finite measure and let $\gamma > 0$. Then*

$$\operatorname{Tr}\left(-\Delta_\Omega^D - \lambda\right)_-^\gamma = L_{\gamma,d}^{cl}|\Omega|\lambda^{\gamma+d/2} + o\left(\lambda^{\gamma+d/2}\right) \qquad \text{as } \lambda \to \infty.$$

The idea behind the proof of this corollary is to integrate the asymptotics in Theorem 3.15.

Proof Let $\varepsilon > 0$. By Theorem 3.15 there is a $\lambda_\varepsilon < \infty$ such that

$$\left|N\left(\lambda, -\Delta_\Omega^D\right) - L_{0,d}^{cl}|\Omega|\lambda^{d/2}\right| \leq \varepsilon\lambda^{d/2} \qquad \text{for all } \lambda \geq \lambda_\varepsilon.$$

Therefore using the formula from Lemma 1.39 and (3.17) gives, for $\lambda \geq \lambda_\varepsilon$,

$$\left|\operatorname{Tr}\left(-\Delta_\Omega^D - \lambda\right)_-^\gamma - \lambda^{\gamma+\frac{d}{2}}L_{\gamma,d}^{cl}|\Omega|\right|$$

$$\leq \gamma \int_0^\lambda \left|N\left(\lambda - \kappa, -\Delta_\Omega^D\right) - L_{0,d}^{cl}|\Omega|(\lambda - \kappa)^{\frac{d}{2}}\right|\kappa^{\gamma-1}\, d\kappa$$

$$\leq \varepsilon\gamma \int_0^{\lambda-\lambda_\varepsilon} (\lambda - \kappa)^{\frac{d}{2}}\kappa^{\gamma-1}\, d\kappa$$

$$\qquad + \gamma \int_{\lambda-\lambda_\varepsilon}^\lambda \left(N\left(\lambda - \kappa, -\Delta_\Omega^D\right) + L_{0,d}^{cl}|\Omega|(\lambda - \kappa)^{\frac{d}{2}}\right)\kappa^{\gamma-1}\, d\kappa$$

$$\leq \varepsilon\gamma \int_0^\lambda (\lambda - \kappa)^{\frac{d}{2}}\kappa^{\gamma-1}\, d\kappa$$

$$\qquad + \gamma\left(N\left(\lambda_\varepsilon, -\Delta_\Omega^D\right) + L_{0,d}^{cl}|\Omega|\lambda_\varepsilon^{\frac{d}{2}}\right)\int_0^\lambda \kappa^{\gamma-1}\, d\kappa$$

$$\leq \varepsilon \frac{\Gamma(\gamma+1)\Gamma(\frac{d}{2}+1)}{\Gamma(\gamma+\frac{d}{2}+1)}\lambda^{\gamma+\frac{d}{2}} + \left(N\left(\lambda_\varepsilon, -\Delta_\Omega^D\right) + L_{0,d}^{cl}|\Omega|\lambda_\varepsilon^{\frac{d}{2}}\right)\lambda^\gamma.$$

Thus,

$$\limsup_{\lambda\to\infty} \lambda^{-\gamma-\frac{d}{2}}\left|\operatorname{Tr}\left(-\Delta_\Omega^D - \lambda\right)_-^\gamma - \lambda^{\gamma+\frac{d}{2}}L_{\gamma,d}^{cl}|\Omega|\right| \leq \varepsilon\frac{\Gamma(\gamma+1)\Gamma(\frac{d}{2}+1)}{\Gamma(\gamma+\frac{d}{2}+1)}.$$

Since $\varepsilon > 0$ is arbitrary, this proves the corollary. $\qquad\square$

The proof of this corollary demonstrates the general fact that from a given asymptotic behavior of the counting function one can deduce the corresponding behavior of the Riesz means for $\gamma > 0$.

One can also get a bound analogous to that of Theorem 3.14. A straightforward integration of (3.21) meets the problem that the latter is only valid for $\lambda \geq (\pi^2/4)L_\Omega^{-2}$. However, an obvious modification of the proof, which we leave to the reader, yields the following bound for Riesz means.

Corollary 3.18 *Let $\gamma > 0$ and let $\Omega \subset \mathbb{R}^d$ be an open set of finite measure and denote by L_Ω the side length of the largest open cube contained in Ω. Then, for all $\lambda \geq \frac{\pi^2}{4}L_\Omega^{-2}$,*

$$\left| \operatorname{Tr}\left(-\Delta_\Omega^D - \lambda \right)_-^\gamma - L_{\gamma,d}^{\mathrm{cl}}|\Omega|\lambda^{\gamma+\frac{d}{2}} \right| \leq c_{\gamma,d}\lambda^{\gamma+\frac{d-1}{2}} \int_{2^{-1}\sqrt{d}\pi\lambda^{-1/2}}^{+\infty} |\Omega(t)|\,\frac{dt}{t^2}\,.$$

The constant $c_{\gamma,d}$ depends only on γ and the dimension d.

Finally, we record the asymptotics of a somewhat different eigenvalue sum.

Corollary 3.19 *Let $\Omega \subset \mathbb{R}^d$ be an open set of finite measure and let $\sigma > 0$. Then, as $N \to \infty$,*

$$\sum_{n=1}^N \left(\lambda_n\left(-\Delta_\Omega^D \right) \right)^\sigma = \frac{d}{2\sigma+d}\left((2\pi)^2\omega_d^{-2/d} \right)^\sigma N^{\frac{2\sigma}{d}+1}|\Omega|^{-\frac{2\sigma}{d}}\,(1+o(1))\,,$$

where $\omega_d = d^{-1}|\mathbb{S}^{d-1}|$ denotes the volume of the unit ball in $\mathbb{R}^d$.

These asymptotics are a simple consequence of those in Corollary 3.15.

3.2.4 Proof of Theorem 3.14

We divide the proof into four steps.

In the *first step* we construct inductively a covering of the set Ω. We abbreviate by $L := L_\Omega$ the side length of the largest open cube contained in Ω. After a translation, we may assume that $(0,L)^d \subset \Omega$. For $j \in \mathbb{N}_0$, we denote by Q_j the set of all cubes $Q_{2^{-j}L}(2^{-j}Ln)$ with $n \in \mathbb{Z}^d$. Let

$$\mathcal{J}_0 := \{Q \in Q_0 : Q \subset \Omega\}, \qquad W_0 := \bigcup_{Q\in\mathcal{J}_0} \overline{Q}, \qquad \text{and} \qquad \Omega_0 := \Omega \setminus W_0\,.$$

By choice of L, the set $\mathcal{J}_0$ is non-empty and, since Ω has finite measure, it contains at most finitely many cubes. Iteratively, for $j \geq 1$, if $\Omega_{j-1} \neq \emptyset$ we let

$$\mathcal{J}_j := \{Q \in Q_j : Q \subset \Omega_{j-1}\}, \qquad W_j := \bigcup_{Q\in\mathcal{J}_j} \overline{Q}, \qquad \text{and} \qquad \Omega_j := \Omega_{j-1} \setminus W_j\,.$$

The set $\mathcal{J}_j$ is at most finite and possibly empty. If $\Omega_{j-1} = \emptyset$ the process stops. In this way, we cover Ω by the closures of disjoint, open, dyadic cubes. We have

$$\#\mathcal{J}_j \leq 2^{dj}L^{-d}|\Omega(\sqrt{d}2^{-j+1}L)| \qquad \text{for } j \geq 0\,. \tag{3.25}$$

For $j \geq 1$, this follows from

$$\operatorname{int} W_j \subset \Omega_{j-1} \subset \Omega(\sqrt{d}2^{-j+1}L) \qquad \text{if } j \geq 1. \tag{3.26}$$

On the other hand, since $\Omega(\sqrt{d}2L) = \Omega$, (3.25) holds for $j = 0$ as well.

Finally, in order to avoid too many case distinctions, we assume that there are infinitely many non-empty sets $\mathcal{J}_j$. Note that, in the case of only finitely many non-empty sets, $\mathcal{J}_i$ is a polycube and, combining the arguments in §3.2.2 with the bounds in §3.2.1 gives Weyl's law with a remainder estimate.

In the case of infinitely many non-empty sets $\mathcal{J}_j$, we have to stop this iterative procedure at some finite value $J \in \mathbb{N}$. Later on in the proof, we will choose J depending on λ, but for now it is an arbitrary parameter. The closure of all (finitely many) cubes $Q \in \bigcup_{j=0}^{J} \mathcal{J}_j$ forms a covering of the subset $\Omega \setminus \Omega_J = \bigcup_{j=0}^{J}(W_j \cap \Omega)$. Moreover, let

$$\tilde{\mathcal{J}}_j := \left\{ Q \in \mathcal{Q}_j : Q \not\subset \Omega \quad \text{and} \quad Q \cap \Omega \neq \emptyset \right\}.$$

This set is possibly infinite. The closure of all disjoint open cubes $Q \in \left(\bigcup_{j=0}^{J} \mathcal{J}_j \right) \cup \tilde{\mathcal{J}}_J$ is now a covering of the initial set Ω.

In the *second step* our strategy is as follows: we introduce additional Neumann boundary conditions at the boundary of all cubes, but at the same time preserve the already given Dirichlet boundary conditions at $\partial\Omega$. We then estimate the counting function $N(\lambda, -\Delta_\Omega^{\mathrm{D}})$ from above by the sum of the counting functions of the Laplacians on domains of the form $Q \cap \Omega$.

In particular, for the finitely many cubes $Q \in \bigcup_{j=0}^{J} \mathcal{J}_j$ we bound, in view of $Q \subset \Omega$, the counting functions $N(\lambda, -\Delta_Q^{\mathrm{N}})$ directly by Lemma 3.12, taking into account the different sizes and the number of such cubes in each generation. Moreover, for the (possibly infinitely many) cubes $Q \in \tilde{\mathcal{J}}_J$ the Laplacian on $Q \cap \Omega$ comes with mixed boundary conditions: namely, Neumann conditions on $\partial Q \setminus \partial\Omega$ and Dirichlet conditions on $Q \cap \partial\Omega$. A suitable coupling of the cutoff J with the parameter λ will then yield the final asymptotic formula.

We now turn to the details of the argument. Let $\widetilde{\Omega}_J$ be the interior of the set $\Omega \setminus \Omega_J = \bigcup_{j=0}^{J} W_j$. Then

$$N\big(\lambda, -\Delta_{\Omega_J}^{\mathrm{D}}\big) \leq N\big(\lambda, -\Delta_\Omega^{\mathrm{D}}\big) \leq N\big(\lambda, -\Delta_{\widetilde{\Omega}_J}^{\mathrm{N}}\big) + N\big(\lambda, -\Delta_{\Omega_J}^{\mathrm{N,D}}\big),$$

where $-\Delta_{\Omega_J}^{\mathrm{N,D}}$ is the Laplacian on Ω_J with Dirichlet boundary conditions at $\partial\Omega_J \cap \partial\Omega$ and Neumann boundary conditions at $\partial\Omega_J \setminus \partial\Omega$. Let us first estimate the counting functions on the set $\widetilde{\Omega}_J$. Introducing additional Dirichlet or Neumann conditions at the boundaries of the generating cubes, the variational principle, (3.12), and (3.13) yield

$$N\left(\lambda,-\Delta_{\widetilde{\Omega}_J}^{D}\right) \geq \sum_{j=0}^{J}\sum_{Q\in\mathcal{J}_j} N\left(\lambda,-\Delta_{Q}^{D}\right) \geq \left(\sum_{j=0}^{J}\sum_{Q\in\mathcal{J}_j}\lambda^{\frac{d}{2}}L_{0,d}^{\mathrm{cl}}|Q|\right) - R(J,\lambda),$$

$$N\left(\lambda,-\Delta_{\widetilde{\Omega}_J}^{N}\right) \leq \sum_{j=0}^{J}\sum_{Q\in\mathcal{J}_j} N\left(\lambda,-\Delta_{Q}^{N}\right) \leq \left(\sum_{j=0}^{J}\sum_{Q\in\mathcal{J}_j}\lambda^{\frac{d}{2}}L_{0,d}^{\mathrm{cl}}|Q|\right) + R(J,\lambda),$$

where

$$R(J,\lambda) := C_d \sum_{j=0}^{J}(\#\mathcal{J}_j)\left(1 + \lambda^{\frac{d-1}{2}}2^{-(d-1)j}L^{d-1}\right).$$

Note that $\sum_{j=0}^{J}\sum_{Q\in\mathcal{J}_j}|Q| = |\widetilde{\Omega}_J|$ and, therefore,

$$\lambda^{\frac{d}{2}}L_{0,d}^{\mathrm{cl}}|\widetilde{\Omega}_J| - R(J,\lambda) \leq N\left(\lambda,-\Delta_{\Omega}^{D}\right) \leq \lambda^{\frac{d}{2}}L_{0,d}^{\mathrm{cl}}|\widetilde{\Omega}_J| + R(J,\lambda) + N\left(\lambda,-\Delta_{\widetilde{\Omega}_J}^{N,D}\right).$$

Subtracting the expected Weyl term and using $|\widetilde{\Omega}_J| \leq |\Omega|$, we arrive at

$$-\lambda^{\frac{d}{2}}L_{0,d}^{\mathrm{cl}}\left(|\Omega| - |\widetilde{\Omega}_J|\right) - R(J,\lambda) \leq N\left(\lambda,-\Delta_{\Omega}^{D}\right) - \lambda^{\frac{d}{2}}L_{0,d}^{\mathrm{cl}}|\Omega|$$

$$\leq R(J,\lambda) + N\left(\lambda,-\Delta_{\widetilde{\Omega}_J}^{N,D}\right). \qquad (3.27)$$

In the *third step* we will estimate the three remainder terms in this bound; namely, $R(J,\lambda)$, $N\left(\lambda,-\Delta_{\widetilde{\Omega}_J}^{N,D}\right)$ and $\lambda^{\frac{d}{2}}L_{0,d}^{\mathrm{cl}}\left(|\Omega| - |\widetilde{\Omega}_J|\right)$, which (for an appropriate choice of J) will be $o(\lambda^{\frac{d}{2}})$ as $\lambda \to \infty$. Let us analyze these terms individually. First of all, in view of (3.25), the term $R(J,\lambda)$ satisfies the bound

$$R(J,\lambda) \leq C_d S(J,\lambda) \qquad (3.28)$$

with

$$S(J,\lambda) := L^{-d}\sum_{j=0}^{J}2^{dj}\left|\Omega(\sqrt{d}2^{-j+1}L)\right|\left(1 + \lambda^{\frac{d-1}{2}}2^{-(d-1)j}L^{d-1}\right).$$

We will see that the other remainder terms in (3.27) are small compared to $S(J,\lambda)$. In order to make this precise, we need the following lower bound, which simply follows by considering only the term $j = J$ in the definition of $S(J,\lambda)$,

$$S(J,\lambda) \geq L^{-d}2^{dJ}\left|\Omega(\sqrt{d}2^{1-J}L)\right| \geq L^{-d}2^{dJ}\left|\Omega(\sqrt{d}2^{-J}L)\right|. \qquad (3.29)$$

Now we turn our attention to the term $N\left(\lambda,-\Delta_{\widetilde{\Omega}_J}^{N,D}\right)$. For $Q \in \tilde{\mathcal{J}}_J$ let $-\Delta_{\Omega\cap Q}^{N,D}$

be the Laplacian on $\Omega \cap Q$ with Dirichlet boundary conditions at $\partial \Omega \cap \overline{Q}$ and Neumann boundary conditions at $\partial Q \cap \overline{\Omega}$. Then

$$N\left(\lambda, \ \Delta_{\Omega_J}^{N,D}\right) \le \sum_{Q \in \tilde{\mathcal{J}}_J} N\left(\lambda, -\Lambda_{\Omega \cap Q}^{N,D}\right).$$

Assume now that $\lambda > 0$ and $J \in \mathbb{N}$ satisfy the condition

$$\lambda L^2 \le 2^{2J} \pi^2. \tag{3.30}$$

Recall that the cubes $Q \in \tilde{\mathcal{J}}_J$ have sides of length $2^{-J} L$. Because of the explicit expressions for the Neumann eigenvalues on a cube of side length $2^{-J} L$, the second eigenvalue of $-\Delta_Q^N$ equals $\pi^2 2^{2J} L^{-2}$ (see §3.1.2), which, according to (3.30), is not smaller than λ. This, together with the variational principle, yields

$$N\left(\lambda, -\Delta_{\Omega \cap Q}^{N,D}\right) \le N\left(\lambda, -\Delta_Q^N\right) \le 1.$$

If λ and J satisfy the condition

$$\lambda L^2 \le 2^{2J-2} \pi^2, \tag{3.31}$$

which is stronger than (3.30), then for all $Q \in \tilde{\mathcal{J}}_J$ with

$$|\Omega \cap Q| \le |Q| \frac{2^{2J} \pi^2 - 2\lambda L^2}{2^{2J} \pi^2 - \lambda L^2}$$

we have

$$\lambda \le \frac{\pi^2}{|Q|^{2/d}} \frac{|Q| - |\Omega \cap Q|}{2|Q| - |\Omega \cap Q|}.$$

Therefore, by Lemma 2.61, we infer that $N\left(\lambda, -\Delta_{\Omega \cap Q}^{N,D}\right) = 0$. Put differently, the equality $N\left(\lambda, -\Delta_{\Omega \cap Q}^{N,D}\right) = 1$ is only possible if

$$|\Omega \cap Q| > |Q| \frac{2^{2J} \pi^2 - 2\lambda L^2}{2^{2J} \pi^2 - \lambda L^2}.$$

Since, by (3.26), all the disjoint sets $\Omega \cap Q$ with $Q \in \tilde{\mathcal{J}}_J$ are contained in $\Omega(\sqrt{d} 2^{-J} L)$, in view of (3.31) and (3.29) we have

$$\sum_{Q \in \tilde{\mathcal{J}}_J} N\left(\lambda, -\Delta_{\Omega \cap Q}^{N,D}\right) \le \#\left\{ Q \in \tilde{\mathcal{J}}_J : |\Omega \cap Q| > |Q| \frac{2^{2J} \pi^2 - 2\lambda L^2}{2^{2J} \pi^2 - \lambda L^2} \right\}$$

$$\le L^{-d} 2^{dJ} \frac{2^{2J} \pi^2 - \lambda L^2}{2^{2J} \pi^2 - 2\lambda L^2} \left|\Omega(\sqrt{d} 2^{-J} L)\right|$$

$$\le \tfrac{3}{2} L^{-d} 2^{dJ} \left|\Omega(\sqrt{d} 2^{-J} L)\right|$$

$$\le \tfrac{3}{2} S(J, \lambda). \tag{3.32}$$

Here we used the fact that $(1 - x)/(1 - 2x) \le 3/2$ if $0 \le x \le 1/4$.

Finally, note that in view of (3.26), (3.29) and (3.31),

$$\lambda^{\frac{d}{2}} L_{0,d}^{cl}(|\Omega| - |\widetilde{\Omega}_J|) \le \lambda^{\frac{d}{2}} L_{0,d}^{cl} |\Omega(\sqrt{d}2^{-J}L)| \le 2^{-d}\pi^d L_{0,d}^{cl} S(J,\lambda). \tag{3.33}$$

To summarize, the remainder bounds (3.28), (3.32) and (3.33) combined with (3.27) lead to

$$-(2^{-d}\pi^d L_{0,d}^{cl} + C_d)S(J,\lambda) \le N(\lambda, -\Delta_\Omega^D) - \lambda^{\frac{d}{2}} L_{0,d}^{cl} |\Omega|$$

$$\le (\tfrac{3}{2} + C_d)S(J,\lambda). \tag{3.34}$$

In the *fourth step* we now choose J depending on λ. We assume that $\lambda L^2 \ge 2^{-2}\pi^2$. Then there is a unique $J \ge 1$ such that

$$2^{2J-4}\pi^2 \le \lambda L^2 < 2^{2J-2}\pi^2.$$

Note that with this choice (3.31) is satisfied. We observe that the lower bound implies that

$$\lambda^{\frac{d-1}{2}} 2^{-(d-1)j} L^{d-1} \ge 2^{-2(d-1)}\pi^{d-1} \qquad \text{for any } 0 \le j \le J,$$

and therefore

$$S(J,\lambda) \le L^{-d}\left(2^{2(d-1)}\pi^{-d+1} + 1\right) \sum_{j-0}^{J} 2^{dj}|\Omega(\sqrt{d}2^{-j+1}L)|\lambda^{\frac{d-1}{2}} 2^{-(d-1)j} L^{d-1}$$

$$= L^{-1}\lambda^{\frac{d-1}{2}}\left(2^{2(d-1)}\pi^{-d+1} + 1\right) \sum_{j=0}^{J} 2^{j}|\Omega(\sqrt{d}2^{-j+1}L)|.$$

This is already a bound of the desired form, but we will transform it into a somewhat nicer form with an integral instead of a sum. Indeed, since $t \mapsto |\Omega(t)|$ is non-decreasing, we have, for any j,

$$2^{j}|\Omega(\sqrt{d}2^{-j+1}L)| \le 4\int_{2^{-j+1}}^{2^{-j+2}} |\Omega(\sqrt{d}sL)|\frac{ds}{s^2} = 4\sqrt{d}L \int_{2^{-j+1}\sqrt{d}L}^{2^{-j+2}\sqrt{d}L} |\Omega(t)|\frac{dt}{t^2}.$$

Thus

$$S(J,\lambda) \le 4\sqrt{d}\left(2^{2(d-1)}\pi^{-d+1} + 1\right)\lambda^{\frac{d-1}{2}} \int_{2^{-J+1}\sqrt{d}L}^{4\sqrt{d}L} |\Omega(t)|\frac{dt}{t^2}.$$

We extend the integral to infinity and use the definition of J to obtain

$$S(J,\lambda) \le 4\sqrt{d}\left(2^{2(d-1)}\pi^{-d+1} + 1\right)\lambda^{\frac{d-1}{2}} \int_{2^{-1}\sqrt{d}\pi\lambda^{-1/2}}^{\infty} |\Omega(t)|\frac{dt}{t^2}.$$

This, together with (3.34), completes the proof of Theorem 3.14. □

3.3 Weyl's asymptotic formula for the Neumann Laplacian on a domain

We recall that an open set $\Omega \subset \mathbb{R}^d$ is said to have the *extension property* if there is a bounded, linear operator $E : H^1(\Omega) \to H^1(\mathbb{R}^d)$ such that $Eu = u$ a.e. on Ω for every $u \in H^1(\Omega)$; see §2.3.2 and §2.8. As discussed there, the existence of such an operator depends on the regularity of the boundary of Ω and, in particular, if Ω has a uniformly Lipschitz continuous boundary, such an extension operator exists.

We know from Lemma 3.4 that if Ω is bounded and has the extension property then the Neumann Laplacian $-\Delta_\Omega^N$ has discrete spectrum. Our goal in this section is to study the asymptotics of the eigenvalues of this operator. We will see that the leading term in these asymptotics is the same as for the Dirichlet Laplacian.

Theorem 3.20 *Let $\Omega \subset \mathbb{R}^d$ be a bounded, open set with the extension property. Then*

$$N(\lambda, -\Delta_\Omega^N) = \lambda^{d/2} L_{0,d}^{cl} |\Omega| + o(\lambda^{d/2}) \qquad as \ \lambda \to \infty \qquad (3.35)$$

and

$$\lambda_n(-\Delta_\Omega^N) = (L_{0,d}^{cl})^{-2/d} |\Omega|^{-2/d} n^{2/d} + o(n^{2/d}) \qquad as \ n \to \infty.$$

Since $-\Delta_\Omega^N \leq -\Delta_\Omega^D$, we have $N(\lambda, -\Delta_\Omega^N) \geq N(\lambda, -\Delta_\Omega^D)$ and therefore, by Corollary 3.15, for *any* open set $\Omega \subset \mathbb{R}^d$ of finite measure,

$$\liminf_{\lambda \to \infty} \lambda^{-d/2} N(\lambda, -\Delta_\Omega^N) \geq L_{0,d}^{cl} |\Omega|.$$

Thus, the content of the theorem is that the extension property implies a similar asymptotic upper bound on $N(\lambda, -\Delta_\Omega^N)$.

In contrast to the Dirichlet case, Weyl's formula (3.35) does *not* hold for any bounded, open set $\Omega \subset \mathbb{R}^d$; see the references in §3.9.

The following proof uses the Weyl asymptotics for Schrödinger operators with continuous potential, which will be stated and proved later in §4.4.

Proof If $d = 1$, by Lemma 2.94, Ω is the disjoint union of at most finitely many bounded, open intervals. Therefore the counting function for $-\Delta_\Omega^N$ is the sum of the counting functions for the Neumann Laplacian on these intervals, which can be computed explicitly (§3.1.2). This settles the case $d = 1$.

For $d \geq 2$ we shall prove that there is a constant $c < \infty$, depending on Ω, such that, for any $\delta > 0$,

$$\limsup_{\lambda \to \infty} \lambda^{-d/2} N(\lambda, -\Delta_\Omega^N) \leq L_{0,d}^{cl} |\Omega| + c \left| \{x \in \mathbb{R}^d : \operatorname{dist}(x, \partial\Omega) \leq 4\delta\} \right|. \quad (3.36)$$

By dominated convergence, we have

$$|\{x \in \mathbb{R}^d : \operatorname{dist}(x, \partial\Omega) \leq 4\delta\}| \to |\partial\Omega| \text{ as } \delta \to 0,$$

where $|\partial\Omega|$ denotes the d-dimensional Lebesgue measure of the boundary. On the other hand, the fact that Ω has the extension property implies that $|\partial\Omega| = 0$; see Proposition 2.95. Thus, (3.36) implies the theorem.

Let us turn to the proof of (3.36). Let

$$\eta(t) := \begin{cases} 0 & \text{if } t \in [0,1], \\ 1 & \text{if } t \in [2,\infty], \\ (t-1)^2(3(t-1)^2 - 8(t-1) + 6) & \text{if } t \in (1,2). \end{cases}$$

We define

$$\chi_0(x) := \eta(\operatorname{dist}(x, \Omega^c)/\delta) \quad \text{and} \quad \chi_1(x) := \sqrt{1 - (\eta(\operatorname{dist}(x, \Omega^c)/\delta))^2}.$$

By construction, $\chi_0^2 + \chi_1^2 = 1$ and the supports of χ_0 and χ_1 are contained in $\{x \in \Omega : \operatorname{dist}(x, \partial\Omega) \geq \delta\}$ and $\Omega^c \cup \{x \in \Omega : \operatorname{dist}(x, \partial\Omega) \leq 2\delta\}$, respectively. Moreover, as we argued in the proof of Theorem 2.92, $\operatorname{dist}(x, \Omega^c)$ is weakly differentiable with $|\nabla \operatorname{dist}(x, \Omega^c)| \leq 1$ almost everywhere. By the chain rule (Lemma 2.10), we have

$$L(x) := |\nabla\chi_0(x)|^2 + |\nabla\chi_1(x)|^2 \leq \frac{\eta'(\operatorname{dist}(x, \Omega^c)/\delta)^2}{\delta^2 \left(1 - (\eta(\operatorname{dist}(x, \Omega^c)/\delta)^2)\right)} \leq \delta^{-2}L_0,$$

with $L_0 := \sup_{[0,\infty)}((\eta')^2/(1 - \eta^2)) < \infty$. On the other hand, a simple computation using the product rule (Lemma 2.14) shows that, for any $u \in H^1(\Omega)$,

$$\int_\Omega |\nabla u|^2 \, dx = \int_\Omega \left(|\nabla(\chi_0 u)|^2 - L|\chi_0 u|^2\right) dx + \int_\Omega \left(|\nabla(\chi_1 u)|^2 - L|\chi_1 u|^2\right) dx.$$

This formula is sometimes called the *IMS localization formula*. Therefore we have, in the sense of quadratic forms,

$$-\Delta_\Omega^N - \lambda = \chi_0 \left(-\Delta_\Omega^N - L - \lambda\right)\chi_0 + \chi_1 \left(-\Delta_\Omega^N - L - \lambda\right)\chi_1.$$

The operators on the right side are defined as explained before Corollary 1.31. By the variational principle (Corollary 1.30), we deduce

$$N\left(\lambda, -\Delta_\Omega^N\right) = N\left(0, -\Delta_\Omega^N - \lambda\right)$$
$$\leq N\left(0, \chi_0\left(-\Delta_\Omega^N - L - \lambda\right)\chi_0\right) + N\left(0, \chi_1\left(-\Delta_\Omega^N - L - \lambda\right)\chi_1\right).$$

In the remainder of the proof we treat the two terms on the right side separately.

The key observation is that, since the support of χ_0 is contained in the interior of Ω, we have $\chi_0\left(-\Delta_\Omega^N\right)\chi_0 = \chi_0\left(-\Delta_\Omega^D\right)\chi_0$. Moreover, by Corollary

1.31, we have $N(0, \chi_0 A \chi_0) \leq N(0, A)$ for any self-adjoint, lower semibounded operator A. Furthermore, $L \leq \delta^{-2} L_0$ so

$$N\left(0, \chi_0\left(\Delta_\Omega^N - L - \lambda\right)\chi_0\right) \leq N\left(0, \chi_0\left(-\Delta_\Omega^D - (\lambda + \delta^{-2}L_0)\right)\chi_0\right)$$
$$\leq N\left(0, -\Delta_\Omega^D - (\lambda + \delta^{-2}L_0)\right)$$
$$= N\left(\lambda + \delta^{-2}L_0, -\Delta_\Omega^D\right).$$

Thus, Corollary 3.15 implies that

$$\limsup_{\lambda \to \infty} \lambda^{-d/2} N\left(0, \chi_0\left(-\Delta_\Omega^N - L - \lambda\right)\chi_0\right) \leq \limsup_{\lambda \to \infty} \lambda^{-d/2} N\left(\lambda + \delta^{-2}L_0, -\Delta_\Omega^D\right)$$
$$\leq L_{0,d}^{cl} |\Omega|,$$

which yields the first term on the right side of (3.36).

To complete the proof of (3.36), we now prove that for some $c < \infty$ and all $\delta > 0$,

$$\limsup_{\lambda \to \infty} \lambda^{-d/2} N\left(0, \chi_1\left(-\Delta_\Omega^N - L - \lambda\right)\chi_1\right) \leq c|\{x \in \mathbb{R}^d : \text{dist}(x, \partial\Omega) \leq 4\delta\}|.$$
$$(3.37)$$

The first observation is that, pointwise on Ω,

$$(L + \lambda + 1)\chi_1^2 \leq (\delta^{-2}L_0 + \lambda + 1)V\chi_1^2 \qquad (3.38)$$

with

$$V(x) := 1 - \eta(\text{dist}(x, \partial\Omega)/(2\delta)), \qquad x \in \mathbb{R}^d.$$

Thus, again by Corollary 1.31,

$$N\left(0, \chi_1\left(-\Delta_\Omega^N - L - \lambda\right)\chi_1\right) \leq N\left(0, -\Delta_\Omega^N + 1 - (\lambda + \delta^{-2}L_0 + 1)V\right).$$

Since Ω has the extension property there is an operator $E: H^1(\Omega) \to H^1(\mathbb{R}^d)$ such that

$$\int_{\mathbb{R}^d} |\nabla(Eu)|^2 \, dx \leq C \int_\Omega \left(|\nabla u|^2 + |u|^2\right) dx \qquad \text{for all } u \in H^1(\Omega)$$

with $C := \|E\|_{H^1(\Omega) \to H^1(\mathbb{R}^d)}^2$. This implies, by the variational principle, that

$$N\left(0, -\Delta_\Omega^N + 1 - (\lambda + \delta^{-2}L_0 + 1)V\right) \leq N\left(0, C^{-1}(-\Delta_{\mathbb{R}^d}) - (\lambda + \delta^{-2}L_0 + 1)V\right)$$
$$= N\left(0, -\Delta_{\mathbb{R}^d} - C(\lambda + \delta^{-2}L_0 + 1)V\right).$$

The Weyl asymptotics for Schrödinger operators, which we will discuss in §4.4 below, imply that

$$\lim_{\lambda \to \infty} \lambda^{-d/2} N\left(0, -\Delta_{\mathbb{R}^d} - C(\lambda + \delta^{-2}L_0 + 1)V\right) = L_{0,d}^{cl} C^{d/2} \int_{\mathbb{R}^d} V^{d/2} \, dx.$$

Since

$$\int_{\mathbb{R}^d} V^{d/2} \, dx \le |\{x \in \mathbb{R}^d : \text{dist}(x, \partial\Omega) \le 4\delta\}|,$$

we obtain the desired bound (3.37). $\qquad\qquad\square$

From a conceptual point of view it is somewhat unsatisfactory that in the previous proof we had to consider a neighborhood of $\partial\Omega$ both in Ω and in $\mathbb{R}^d \setminus \Omega$. This should be compared with the proof of Weyl asymptotics in the Dirichlet case (Theorem 3.14), where only a neighborhood of $\partial\Omega$ that was *inside* Ω was considered. For instance, when proving Weyl-type asymptotics in the Neumann case for a more general class of domains than those with the extension property, it is desirable to work only with a one-sided neighborhood of $\partial\Omega$ in Ω; see, e.g., Netrusov and Safarov (2005).

We now present an alternative proof of Theorem 3.20, which only considers such a one-sided neighborhood. Instead of Weyl asymptotics for Schrödinger operators we now use non-asymptotic spectral inequalities for these operators, which are also the subject of the next chapter.

Alternative proof of Theorem 3.20 For $d \ge 3$ instead of (3.36) we show

$$\limsup_{\lambda \to \infty} \lambda^{-d/2} N(\lambda, -\Delta_\Omega^{\mathrm{N}}) \le L_{0,d}^{\mathrm{cl}} |\Omega| + c \, |\{x \in \Omega : \text{dist}(x, \partial\Omega) \le 2\delta\}|. \quad (3.39)$$

By dominated convergence, this implies Theorem 3.20. Note that now we do not need the fact that $|\partial\Omega| = 0$. The argument for $d = 2$ is similar, but with the second term on the right side of (3.39) replaced by

$$c \, |\{x \in \Omega : \text{dist}(x, \partial\Omega) \le 2\delta\}| \left(1 + \ln \frac{|\Omega|}{|\{x \in \Omega : \text{dist}(x, \partial\Omega) \le 2\delta\}|} \right),$$

which leads to the same conclusion.

For the proof of (3.39), we proceed as before. We need to prove the analogue of (3.37), but with the right side replaced by that in (3.39). Now instead of (3.38) we bound

$$(L + \lambda + 1)\chi_1^2 \le (\delta^{-2} L_0 + \lambda + 1)\chi \chi_1^2$$

with $\chi := \chi_{\{x \in \Omega : \text{dist}(x, \Omega^c) \le 2\delta\}}$ and obtain, again by Lemma 1.31 and the use of an extension operator,

$$N(0, \chi_1(-\Delta_\Omega^{\mathrm{N}} - L - \lambda)\chi_1) \le N(0, -\Delta_{\mathbb{R}^d} - C(\lambda + \delta^{-2} L_0 + 1)\chi).$$

Then a non-asymptotic spectral bound, called the CLR inequality, Theorem 4.31, which we state and prove in §4.5, implies that, if $d \geq 3$, then

$$N\left(0, -\Delta_{\mathbb{R}^d} - C(\lambda + \delta^{-2}L_0 + 1)\chi\right)$$
$$\leq c_d \, C^{d/2}(\lambda + \delta^{-2}L_0 + 1)^{d/2} \left|\{x \in \Omega: \text{ dist}(x, \partial\Omega) \leq 2\delta\}\right|$$

for a constant c_d depending only on d. Multiplying by $\lambda^{-d/2}$ and passing to the limit $\lambda \to \infty$, we obtain (3.39) with $c = c_d \, C^{d/2}$. For $d = 2$ we bound

$$N\left(0, \chi_1\left(-\Delta_\Omega^N - L - \lambda\right)\chi_1\right) \leq N\left(0, -\Delta_\Omega^N - (\lambda + \delta^{-2}L_0)\chi\right)$$

and use a non-asymptotic spectral inequality due to Solomyak (1994), which implies

$$N\left(0, -\Delta_\Omega^N - (\lambda + \delta^{-2}L_0)\chi\right) \leq 1 + C_\Omega(\lambda + \delta^{-2}L_0)\|\chi\|_\Omega$$

with

$$\|g\|_\Omega :=$$
$$\sup\left\{\left|\int_\Omega fg \, dx\right| : f : \Omega \to \mathbb{R} \text{ measurable}, \int_\Omega \left(e^{|f|} - 1 - |f|\right) dx \leq |\Omega|\right\}.$$

It is a simple exercise, using Jensen's inequality, to show that

$$\|\chi\|_\Omega \leq |\{x \in \Omega: \text{ dist}(x, \Omega^c) < 2\delta\}| \, \mathcal{A}^{-1}(|\Omega|/|\{x \in \Omega: \text{ dist}(x, \Omega^c) < 2\delta\}|),$$

where $\mathcal{A}^{-1}$ is the inverse function of $\mathcal{A} : \mathbb{R}_+ \to \mathbb{R}_+$, $t \mapsto e^t - 1 - t$. The bound $\mathcal{A}^{-1}(s) \leq c(1 + \ln s)$ for $s \geq 1$ leads to the claimed inequality. This completes the proof. $\qquad\qquad\square$

We can now derive asymptotics of Riesz means for the Neumann Laplacian.

Corollary 3.21 *Let* $\Omega \subset \mathbb{R}^d$ *be a bounded, open set with the extension property and let* $\gamma > 0$. *Then*

$$\text{Tr}\left(-\Delta_\Omega^N - \lambda\right)_-^\gamma = L_{\gamma,d}^{\text{cl}}|\Omega|\lambda^{\gamma+d/2} + o(\lambda^{\gamma+d/2}) \qquad \text{as } \lambda \to \infty.$$

This corollary is deduced from Theorem 3.20 in exactly the same way as Corollary 3.17 was deduced from Corollary 3.15.

Remark 3.22   In dimension $d \geq 3$, Theorem 3.20 and Corollary 3.21 remain valid if the boundedness of Ω is replaced by a finite measure assumption. This follows immediately from the alternative proof given above. The same extension is true in $d = 1$, but is trivial in view of Lemma 2.94. In contrast, Solomyak's inequality in $d = 2$ seems to require that Ω is bounded.

3.4 Pólya's inequality for tiling domains

In §3.2, we have proved the asymptotic formula

$$\lim_{\lambda \to \infty} \lambda^{-d/2} N\big(\lambda, -\Delta_\Omega^D\big) = L_{0,d}^{cl} |\Omega|,$$

where $-\Delta_\Omega^D$ is the Dirichlet Laplacian on an open set $\Omega \subset \mathbb{R}^d$ of finite measure. The famous conjecture of Polya (1954) states that $\lambda^{-d/2} N\big(\lambda, -\Delta_\Omega^D\big)$ is always less than its limit. In this section, we present the proof from Pólya (1961) of this universal bound for a special class of domains. The general case is still open.

An open set $\Omega \subset \mathbb{R}^d$ is called *tiling* if there are countable families of orthogonal $d \times d$ matrices $\{R_n\}$ and of vectors $\{a_n\}$ in $\mathbb{R}^d$ such that the sets

$$\Omega_n := \{R_n x + a_n : x \in \Omega\}$$

satisfy the following two properties

(1) $\Omega_n \cap \Omega_m = \emptyset$ if $n \neq m$,
(2) $\big|\mathbb{R}^d \setminus \bigcup_n \Omega_n\big| = 0$.

Of course, triangles, parallelograms and regular hexagons are examples of tiling sets, but there are more exotic examples; see, e.g., Grünbaum and Shephard (1987), Radin (1994), and references therein.

Here is Pólya's theorem (Pólya, 1961).

Theorem 3.23 *Let $\Omega \subset \mathbb{R}^d$ be a bounded, open, tiling set. Then, for any $\lambda > 0$,*

$$N\big(\lambda, -\Delta_\Omega^D\big) \leq L_{0,d}^{cl} |\Omega| \lambda^{d/2}.$$

Proof We use the sets Ω_n from the definition of the tiling property and recall that n ranges over a countable index set. For $L > 0$, let

$$Q_L := (-L/2, L/2)^d, \quad J_L := \{n : \Omega_n \subset Q_L\} \quad \text{and} \quad \Omega^L := \text{int}\bigg(\bigcup_{n \in J_L} \overline{\Omega_n}\bigg).$$

We note that $\Omega^L \subset Q_L$. Moreover, by second of the tiling properties, (2), above,

$$\lim_{L \to \infty} L^{-d} \# J_L = |\Omega|^{-1}. \tag{3.40}$$

By the inclusion of the form domains, one has the operator inequalities

$$-\Delta_{Q_L}^D \leq -\Delta_{\Omega^L}^D \leq \bigoplus_{n \in J_L}\big(-\Delta_{\Omega_n}^D\big).$$

The first inequality is understood in terms of the natural embedding $L^2(\Omega^L) \subset L^2(Q_L)$ by extension by zero and Lemma 2.28. Since the sets Ω_n are congruent to Ω, the Laplacians $-\Delta^D_{\Omega_n}$ are unitarily equivalent to $-\Delta^D_\Omega$. By the variational principle,

$$N(\lambda, -\Delta^D_{Q_L}) \geq N\left(\lambda, \bigoplus_{n \in J_L}(-\Delta^D_{\Omega_n})\right) = (\#J_L)\, N(\lambda, -\Delta^D_\Omega),$$

which can be rewritten as

$$N(\lambda, -\Delta^D_\Omega) \leq \frac{L^d}{\#J_L}\, \frac{N(\lambda, -\Delta^D_{Q_L})}{L^d}. \tag{3.41}$$

We would like to pass to the limit $L \to \infty$ using (3.40). In order to deal with the second factor on the right side, we note that, by scaling $x \mapsto x/L$, the operator $-\Delta^D_{Q_L}$ is unitarily equivalent to the operator $-L^{-2}\Delta^D_{Q_1}$ in $L^2(Q_1)$. Hence $N(\lambda, -\Delta^D_{Q_L}) = N(L^2\lambda, -\Delta^D_{Q_1})$. According to Weyl's law for cubes (Lemma 3.12), we have, as $L \to \infty$,

$$\lim_{L \to \infty} L^{-d} N(\lambda, -\Delta^D_{Q_L}) = \lambda^{d/2} \lim_{L \to \infty} (L^2\lambda)^{-d/2} N(L^2\lambda, -\Delta^D_{Q_1}) = \lambda^{d/2} L^{cl}_{0,d}.$$

This, together with (3.40) and (3.41), completes the proof of the theorem. $\quad\square$

Another class of domains for which Pólya's conjecture has been verified will be discussed in §3.5.5.

There is a version of Theorem 3.23 for the Neumann Laplacian with the *reversed* inequality. The following is due to Kellner (1966), improving an earlier argument by Pólya.

Theorem 3.24 *Let $\Omega \subset \mathbb{R}^d$ be a bounded, open, tiling set. Then, for any $\lambda > 0$,*

$$N(\lambda, -\Delta^N_\Omega) \geq L^{cl}_{0,d}\, |\Omega|\, \lambda^{d/2}.$$

Proof The argument is similar to before, but with J_L and Ω^L replaced by

$$\widehat{J}_L := \{n : \Omega_n \cap Q_L \neq \emptyset\} \quad \text{and} \quad \widehat{\Omega}^L := \text{int}\left(\bigcup_{n \in \widehat{J}_L} \overline{\Omega_n}\right),$$

so that now $\widehat{\Omega}^L \supset Q_L$. As before, we have

$$\lim_{L \to \infty} L^{-d} \#\widehat{J}_L = |\Omega|^{-1}.$$

Again by the inclusions of the form domains, we find the operator inequalities

$$-\Delta^D_{Q_L} \geq -\Delta^D_{\widehat{\Omega}^L} \geq \bigoplus_{n \in \widehat{J}_L}(-\Delta^N_{\Omega_n}),$$

where the first inequality is understood via extension by zero. We conclude that

$$N\left(\lambda, -\Delta_\Omega^N\right) \geq \frac{L^d}{\#\widehat{J}_L} \frac{N\left(\lambda, -\Delta_{Q_L}^D\right)}{L^d}$$

and argue as before to obtain the assertion. $\qquad\square$

3.5 Lower bounds for the eigenvalues of the Dirichlet Laplacian

In this section we discuss bounds on both the sums and Riesz means of the eigenvalues of the Laplacian with Dirichlet boundary conditions.

3.5.1 Berezin's inequality

Let us start with an argument by Berezin (1972b). Let $\Omega \subset \mathbb{R}^d$ be an open set of finite measure. Then, by Lemma 3.4, the spectrum of $-\Delta_\Omega^D$ is discrete. In this subsection we abbreviate by $\lambda_n := \lambda_n(-\Delta_\Omega^D)$ the non-decreasing sequence of its eigenvalues, repeated with multiplicities, and we denote by φ_n an associated $L^2(\Omega)$-orthonormal system of eigenfunctions.

According to Lemma 2.28, these eigenfunctions can be continued by zero outside of Ω to H^1 functions on $\mathbb{R}^d$. Further, we denote by $\widehat{\varphi_n}$ the Fourier transform of this extension. By Lemma 2.31 we have, for any $\Lambda > 0$,

$$\sum_n (\Lambda - \lambda_n)_+ = \sum_{\lambda_n < \Lambda} \int_\Omega (\Lambda |\varphi_n|^2 - |\nabla \varphi_n|^2)\, dx = \sum_{\lambda_n < \Lambda} \int_{\mathbb{R}^d} (\Lambda - |\xi|^2)|\widehat{\varphi_n}|^2\, d\xi$$

$$= \sum_n \int_{\mathbb{R}^d} (\Lambda - |\xi|^2)_+ |\widehat{\varphi_n}|^2\, d\xi - R_D$$

with

$$R_D := \sum_{\lambda_n \geq \Lambda} \int_{\mathbb{R}^d} (\Lambda - |\xi|^2)_+ |\widehat{\varphi_n}|^2\, d\xi + \sum_{\lambda_n < \Lambda} \int_{\mathbb{R}^d} (|\xi|^2 - \Lambda)_+ |\widehat{\varphi_n}|^2\, d\xi,$$

which is clearly non-negative. This implies the bound

$$\sum_n (\Lambda - \lambda_n)_+ \leq \int_{\mathbb{R}^d} (\Lambda - |\xi|^2)_+ \sum_n |\widehat{\varphi_n}|^2\, d\xi.$$

By Parseval's identity, we have pointwise on $\mathbb{R}^d$,

$$\sum_n |\widehat{\varphi_n}|^2 = \sum_n |(\varphi_n, (2\pi)^{-d/2} e^{ix\cdot\xi})_{L^2(\Omega)}|^2 = \left\|(2\pi)^{-d/2} e^{ix\cdot\xi}\right\|_{L^2(\Omega)}^2$$

$$= (2\pi)^{-d} |\Omega|.$$

Recall that $L_{1,d}^{cl}$ was introduced in (3.2) and equals $(4\pi)^{-\frac{d}{2}}\Gamma(2 + \frac{d}{2})^{-1}$. An explicit computation shows that

$$(2\pi)^{-d}\int_{\mathbb{R}^d}(\Lambda - |\xi|^2)_+ d\zeta = L_{1,d}^{cl}\Lambda^{1+\frac{d}{2}}. \qquad (3.42)$$

This proves the following theorem.

Theorem 3.25 *Let $\Omega \subset \mathbb{R}^d$ be an open set of finite measure. Then, for all $\Lambda \geq 0$,*

$$\sum_n (\Lambda - \lambda_n(-\Delta_\Omega^D))_+ \leq L_{1,d}^{cl}|\Omega|\Lambda^{1+\frac{d}{2}}. \qquad (3.43)$$

By Corollary 3.17, the constant on the right side of (3.43) is best possible.

3.5.2 The Li–Yau inequality

Let us now turn to a bound of Li and Yau (1983) on partial sum of the first eigenvalues. We assume again that $\Omega \subset \mathbb{R}^d$ is an open set of finite measure and let (φ_n) be an $L^2(\Omega)$-orthonormal basis of eigenfunctions corresponding to the non-decreasing sequence of eigenvalues (λ_n) of this operator. By Lemma 2.28, the φ_n can be extended by zero to H^1-functions on $\mathbb{R}^d$, where they remain an orthonormal system. Let $\widehat{\varphi_n}$ be the Fourier transforms of these extensions. By Plancherel's theorem, this system of functions is $L^2(\mathbb{R}^d)$-orthonormal. The non-negative function

$$F_N(\xi) := \sum_{n=1}^N |\widehat{\varphi_n}(\xi)|^2, \qquad \xi \in \mathbb{R}^d,$$

satisfies

$$\int_{\mathbb{R}^d} F_N(\xi)\, d\xi = \sum_{n=1}^N \int_{\mathbb{R}^d} |\widehat{\varphi_n}|^2\, d\xi = \sum_{n=1}^N \int_{\mathbb{R}^d} |\varphi_n|^2\, dx = N.$$

Moreover, by Parseval's identity, we have the pointwise bound

$$F_N(\xi) = \sum_{n=1}^N |\widehat{\varphi_n}(\xi)|^2 = \sum_{n=1}^N \left|\left\langle \varphi_n, \frac{e^{i\xi\cdot}}{(2\pi)^{d/2}}\right\rangle_{L^2(\Omega)}\right|^2 = \left\|\frac{e^{i\xi\cdot}}{(2\pi)^{d/2}}\right\|_{L^2(\Omega)}^2 - R_N(\xi),$$

where $R_N(\xi) := \sum_{n=N+1}^\infty |\widehat{\varphi_n}(\xi)|^2 \geq 0$. In particular, this implies

$$F_N(\xi) \leq (2\pi)^{-d}|\Omega|.$$

It follows from Lemma 2.31 that $F_N \in L^1(\mathbb{R}^d, |\xi|^2\, d\xi)$ and

$$\int_{\mathbb{R}^d} |\xi|^2 F_N(\xi)\, d\xi = \sum_{n=1}^N \int_{\mathbb{R}^d} |\xi|^2 |\widehat{\varphi_n}|^2\, d\xi = \sum_{n=1}^N \int_\Omega |\nabla\varphi_n|^2 dx = \sum_{n=1}^N \lambda_n.$$

Let

$$\Lambda := \left((2\pi)^{-d} d^{-1} |\mathbb{S}^{d-1}| \, |\Omega|\right)^{-2/d} N^{2/d}$$

and write

$$\int_{\mathbb{R}^d} |\xi|^2 F_N(\xi) \, d\xi = (2\pi)^{-d} |\Omega| \int_{|\xi|^2 < \Lambda} |\xi|^2 \, d\xi + R$$

with

$$R := \int_{|\xi|^2 < \Lambda} |\xi|^2 (F_N(\xi) - (2\pi)^{-d} |\Omega|) \, d\xi + \int_{|\xi|^2 \geq \Lambda} |\xi|^2 F_N(\xi) \, d\xi \, .$$

Using

$$\int_{\mathbb{R}^d} F_N(\xi) \, d\xi = N = (2\pi)^{-d} |\Omega| \int_{|\xi|^2 < \Lambda} d\xi \, ,$$

we can rewrite

$$R = \int_{|\xi|^2 < \Lambda} (|\xi|^2 - \Lambda)(F_N(\xi) - (2\pi)^{-d} |\Omega|) \, d\xi + \int_{|\xi|^2 \geq \Lambda} (|\xi|^2 - \Lambda) F_N(\xi) \, d\xi \, ,$$

and, as $0 \leq F_N(\xi) \leq (2\pi)^{-d} |\Omega|$, we conclude that $R \geq 0$. Therefore,

$$\sum_{n=1}^{N} \lambda_n = \int_{\mathbb{R}^d} |\xi|^2 F_N(\xi) \, d\xi \geq (2\pi)^{-d} |\Omega| \int_{|\xi|^2 < \Lambda} |\xi|^2 \, d\xi$$

$$= \frac{d}{d+2} (2\pi)^2 |\Omega|^{-2/d} (d^{-1} |\mathbb{S}^{d-1}|)^{-2/d} N^{1+2/d} \, .$$

We have therefore arrived at the following result (Li and Yau, 1983).

Theorem 3.26 *Let $\Omega \subset \mathbb{R}^d$ be an open set of finite measure. Then, for all $N \in \mathbb{N}$,*

$$\sum_{n=1}^{N} \lambda_n(-\Delta_\Omega^{\mathrm{D}}) \geq \frac{d}{d+2} (2\pi)^2 \omega_d^{-2/d} N^{1+\frac{2}{d}} |\Omega|^{-2/d} \, , \tag{3.44}$$

where $\omega_d = d^{-1} |\mathbb{S}^{d-1}|$ denotes the volume of the unit ball in $\mathbb{R}^d$.

By Corollary 3.19 with $\sigma = 1$, the constant on the right side of (3.44) is best possible.

3.5.3 Equivalence between the bounds of Berezin and Li–Yau

In this subsection we will show that Berezin's theorem and the Li–Yau theorem are equivalent in the sense that either can be formally derived from the other one. This follows from a general duality principle, given in the following proposition.

Proposition 3.27 *Let $(b_n)_{n \in \mathbb{N}}$ be a non-decreasing sequence of non-negative numbers that tends to infinity. Let $1 < p, q < \infty$ and let $K, L > 0$ be such that*

$$\frac{1}{p} + \frac{1}{q} = 1 \qquad and \qquad K^p L^q = 1.$$

Then one has

$$\sum_n (\lambda - b_n)_+ \leq \frac{1}{p} L \lambda^p \qquad for \ all \ \lambda \geq 0 \tag{3.45}$$

if and only if

$$\sum_{n=1}^{N} b_n \geq \frac{1}{q} K N^q \qquad for \ all \ N \in \mathbb{N}. \tag{3.46}$$

Let us apply this proposition with $b_n = \lambda_n(-\Delta_\Omega^D)$. For

$$p = 1 + \frac{d}{2} \qquad and \qquad L = L_{0,d}^{cl} |\Omega|,$$

inequality (3.45) is precisely Berezin's inequality from Theorem 3.25 and, for

$$q = 1 + \frac{2}{d} \qquad and \qquad K = (2\pi)^2 \omega_d^{-2/d} |\Omega|^{-2/d},$$

inequality (3.46) is precisely Li–Yau's inequality from Theorem 3.26. Therefore, Proposition 3.27 shows that these two inequalities are equivalent to each other.

The proof of Proposition 3.27 is based on a Legendre transform. The following lemma gives the formula for the Legendre transform of Riesz means of order 1.

Lemma 3.28 *Let $(b_n)_{n \in \mathbb{N}}$ be a non-decreasing sequence of non-negative numbers that tends to infinity. Then, for all $v \geq 0$,*

$$\sup_{\lambda \geq 0} \left(\lambda v - \sum_{n \in \mathbb{N}} (\lambda - b_n)_+ \right) = \sum_{n=1}^{[v]} b_n + \{v\} b_{[v]+1}, \tag{3.47}$$

where $v = [v] + \{v\}$ with $[v] \in \mathbb{N}_0$ and $0 \leq \{v\} < 1$. Moreover,

$$\sup_{v \geq 0} \left(\lambda v - \sum_{n=1}^{[v]} b_n - \{v\} b_{[v]+1} \right) = \sum_{n \in \mathbb{N}} (\lambda - b_n)_+. \tag{3.48}$$

Proof We set $b_0 := 0$ and define $\sum_{n=1}^{N} b_n = 0$ for $N = 0$. Then

$$\sup_{\lambda \geq 0} \left(\lambda v - \sum_{n \in \mathbb{N}} (\lambda - b_n)_+ \right) = \sup_{N \in \mathbb{N}_0} \sup_{b_N \leq \lambda < b_{N+1}} \left(\lambda v - \sum_{n \in \mathbb{N}} (\lambda - b_n)_+ \right).$$

For fixed $N \in \mathbb{N}_0$ and $\lambda \in [b_N, b_{N+1})$, we have

$$\lambda v - \sum_{n \in \mathbb{N}} (\lambda - b_n)_+ = \lambda(v - N) + \sum_{n=1}^{N} b_n .$$

Thus,

$$\sup_{b_N \leq \lambda < b_{N+1}} \left(\lambda v - \sum_{n \in \mathbb{N}} (\lambda - b_n)_+ \right) = \begin{cases} \sum_{n=1}^{N} b_n - (N - v)b_N & \text{if } N > v, \\ \sum_{n=1}^{N} b_n - (N - v)b_{N+1} & \text{if } N \leq v. \end{cases}$$

It follows that

$$\sup_{\lambda \geq 0} \left(\lambda v - \sum_{n \in \mathbb{N}} (\lambda - b_n)_+ \right)$$

$$= \max \left\{ \sup_{N > v} \left(\sum_{n=1}^{N} b_n - (N - v)b_N \right), \sup_{N \leq v} \left(\sum_{n=1}^{N} b_n - (N - v)b_{N+1} \right) \right\} .$$

Since b_n is non-decreasing, the sequences

$$\sum_{n=1}^{N} b_n - (N - v)b_N \qquad \text{and} \qquad \sum_{n=1}^{N} b_n - (N - v)b_{N+1}$$

are non-increasing, respectively non-decreasing, in N. Thus, the suprema are attained at $N = [v] + 1$, respectively $N = [v]$, and equal $\sum_{n=1}^{[v]} b_n + \{v\}b_{[v]+1}$. This proves the first formula in the lemma.

To prove the second formula, we write

$$\sup_{v \geq 0} \left(\lambda v - \sum_{n=1}^{[v]} b_n - \{v\}b_{[v]+1} \right) = \sup_{N \in \mathbb{N}_0} \sup_{N \leq v < N+1} \left(\lambda v - \sum_{n=1}^{[v]} b_n - \{v\}b_{[v]+1} \right) .$$

For fixed $N \in \mathbb{N}_0$ and $v \in [N, N + 1)$, we have

$$\lambda v - \sum_{n=1}^{[v]} b_n - \{v\}b_{[v]+1} = (\lambda - b_{N+1})v - \sum_{n=1}^{N} b_n + Nb_{N+1} .$$

Thus,

$$\sup_{N \leq v < N+1} \left(\lambda v - \sum_{n=1}^{[v]} b_n - \{v\}b_{[v]+1} \right)$$

$$= \begin{cases} -\sum_{n=1}^{N} b_n + Nb_{N+1} + (N + 1)(\lambda - b_{N+1}) & \text{if } \lambda \geq b_{N+1}, \\ -\sum_{n=1}^{N} b_n + Nb_{N+1} + N(\lambda - b_{N+1}) & \text{if } \lambda < b_{N+1}, \end{cases}$$

$$= \sum_{n=1}^{N+1} (\lambda - b_n)_+ .$$

Taking the supremum over N yields the second formula in the lemma. □

We also need the Legendre transforms of the right sides of (3.45) and (3.46). Indeed, it is elementary that, if p and q are related by $1/p + 1/q = 1$, then

$$\sup_{\lambda \geq 0}\left(\lambda v - \frac{1}{p}\lambda^p\right) = \frac{1}{q}v^q \quad \text{and} \quad \sup_{v \geq 0}\left(\lambda v - \frac{1}{q}v^q\right) = \frac{1}{p}\lambda^p. \quad (3.49)$$

Proof of Proposition 3.27 First, assume that (3.45) is valid. Then, for any $v \geq 0$, we have

$$\sup_{\lambda \geq 0}\left(\lambda v - \sum_n (\lambda - b_n)_+\right) \geq L \sup_{\lambda \geq 0}\left(\lambda(v/L) - \frac{1}{p}\lambda^p\right).$$

According to (3.47) and (3.49), this inequality is equivalent to

$$\sum_{n=1}^{[v]} b_n + \{v\}b_{[v]+1} \geq \frac{1}{q}L^{-1/(p-1)}v^q \quad \text{for all } v \geq 0.$$

In particular, for $v \in \mathbb{N}$ we obtain (3.46).

Conversely, assume that (3.46) holds. This implies that

$$\sum_{n=1}^{[v]} b_n + \{v\}b_{[v]+1} \geq \frac{1}{q}Kv^q \quad \text{for all } v \geq 0. \quad (3.50)$$

Indeed, this follows by convexity of $v \mapsto v^q$ as follows. We set $N = [v]$ and write $v = (1 - \{v\})N + \{v\}(N + 1)$. Then

$$\frac{1}{q}Kv^q \leq \frac{1}{q}K(1 - \{v\})N^q + \frac{1}{q}K\{v\}(N+1)^q$$

$$\leq (1 - \{v\})\sum_{n=1}^{N} b_n + \{v\}\sum_{n=1}^{N+1} b_n = \sum_{n=1}^{N} b_n + \{v\}b_{N+1},$$

as claimed.

Now (3.45) follows from (3.50) just as before, using (3.48) and (3.49). □

3.5.4 Consequences of the Berezin–Li–Yau theorem

Next, we discuss two consequences of Theorems 3.25 and 3.26. The first one is an extension to Riesz means of order $\gamma \geq 1$.

Corollary 3.29 *Let $\Omega \subset \mathbb{R}^d$ be an open set of finite measure and let $\gamma \geq 1$. Then, for all $\Lambda \geq 0$,*

$$\operatorname{Tr}(-\Delta_\Omega^D - \Lambda)_-^\gamma \leq L_{\gamma,d}^{\mathrm{cl}}|\Omega|\Lambda^{\gamma+\frac{d}{2}}. \quad (3.51)$$

By Corollary 3.17, the constant on the right side of (3.51) is best possible.

Proof By Theorem 3.25, we may assume $\gamma > 1$. Then, by Lemma 1.41,

$$\mathrm{Tr}(-\Delta_\Omega^D - \Lambda)_-^\gamma = \gamma(\gamma - 1) \int_0^\infty \mathrm{Tr}(-\Delta_\Omega^D - \Lambda + \kappa)_- \kappa^{\gamma-2}\, d\kappa .$$

Therefore, Berezin's inequality (Theorem 3.25) implies that

$$\mathrm{Tr}(-\Delta_\Omega^D - \Lambda)_-^\gamma \le L_{1,d}^{\mathrm{cl}} |\Omega| \gamma(\gamma - 1) \int_0^\infty (\Lambda - \kappa)_+^{1+\frac{d}{2}} \kappa^{\gamma-2}\, d\kappa .$$

Using the beta function identity (3.16), we find

$$\gamma(\gamma - 1) \int_0^\infty (\Lambda - \kappa)_+^{1+\frac{d}{2}} \kappa^{\gamma-2}\, d\kappa = \frac{\Gamma(\gamma + 1)\Gamma(\frac{d}{2} + 2)}{\Gamma(\gamma + 1 + \frac{d}{2})} \Lambda^{\gamma+\frac{d}{2}} .$$

The explicit expressions (3.2) for the semiclassical constants give

$$\frac{L_{1,d}^{\mathrm{cl}}}{L_{\gamma,d}^{\mathrm{cl}}} = \frac{\Gamma(\gamma + 1 + \frac{d}{2})}{\Gamma(\gamma + 1)\Gamma(2 + \frac{d}{2})} . \tag{3.52}$$

Combining the last three equations implies the corollary. □

The Berezin inequality can not only be used to obtain sharp inequalities for $\gamma > 1$, but also to prove universal (though probably non-sharp) inequalities for $\gamma \in [0, 1)$. The constants in these inequalities are the best currently known ones for general open sets.

Corollary 3.30 *Let $\Omega \subset \mathbb{R}^d$ be an open set of finite measure. Then, for all $\Lambda \ge 0$,*

$$N(\Lambda, -\Delta_\Omega^D) \le \left(1 + \frac{2}{d}\right)^{\frac{d}{2}} L_{0,d}^{\mathrm{cl}} |\Omega| \Lambda^{\frac{d}{2}} . \tag{3.53}$$

Moreover, if $0 < \gamma < 1$, then, for all $\Lambda \ge 0$,

$$\mathrm{Tr}(-\Delta_\Omega^D - \Lambda)_-^\gamma \le \frac{\gamma^\gamma (1 + \frac{d}{2})^{1+\frac{d}{2}}}{(\gamma + \frac{d}{2})^{\gamma+\frac{d}{2}}} \frac{\Gamma(\gamma + 1 + \frac{d}{2})}{\Gamma(\gamma + 1)\Gamma(2 + \frac{d}{2})} L_{\gamma,d}^{\mathrm{cl}} |\Omega| \Lambda^{\gamma+\frac{d}{2}} .$$

The proof of this corollary uses the following lemma.

Lemma 3.31 *Let $0 \le \gamma < \sigma$ and*

$$C(\gamma, \sigma) := \sigma^{-\sigma} \gamma^\gamma (\sigma - \gamma)^{\sigma-\gamma} ,$$

with the convention that $\gamma^\gamma = 1$ for $\gamma = 0$. Then, for all $\mu > \Lambda$ and all $E \in \mathbb{R}$,

$$(E - \Lambda)_-^\gamma \le C(\gamma, \sigma)(\mu - \Lambda)^{-\sigma+\gamma}(E - \mu)_-^\sigma .$$

Moreover, for all $\Lambda > 0$,

$$\inf_{\mu > \Lambda} (\mu - \Lambda)^{-\gamma} \mu^{\sigma} = C(\gamma, \sigma)^{-1} \Lambda^{\sigma - \gamma}.$$

Both parts of this lemma follow by elementary arguments from the fact that

$$C(\gamma, \sigma) = \max_{x > 0} \frac{x^{\gamma}}{(1 + x)^{\sigma}}.$$

Proof of Corollary 3.30 According to the first part of Lemma 3.31, we have, for all $\mu > \Lambda$,

$$\mathrm{Tr}(-\Delta_{\Omega}^{D} - \Lambda)_{-}^{\gamma} \leq C(\gamma, 1)(\mu - \Lambda)^{-1+\gamma} \mathrm{Tr}(-\Delta_{\Omega}^{D} - \mu)_{-},$$

where the left side is understood as $N(\Lambda, -\Delta_{\Omega}^{D})$ for $\gamma = 0$. Applying the Berezin inequality to the right side gives

$$\mathrm{Tr}(-\Delta_{\Omega}^{D} - \Lambda)_{-}^{\gamma} \leq C(\gamma, 1) L_{1,d}^{\mathrm{cl}} |\Omega| (\mu - \Lambda)^{-1+\gamma} \mu^{1+\frac{d}{2}}.$$

Applying the second part of Lemma 3.31, we find

$$\mathrm{Tr}(-\Delta_{\Omega}^{D} - \Lambda)_{-}^{\gamma} \leq \frac{C(\gamma, 1)}{C(1 - \gamma, 1 + \frac{d}{2})} L_{1,d}^{\mathrm{cl}} |\Omega| \Lambda^{\gamma + \frac{d}{2}}.$$

Combining the explicit expressions (3.52) for the semiclassical constants with the expression for the constant in Lemma 3.31, we obtain the corollary. $\quad\square$

So far we have discussed consequences of the Berezin inequality. Let us now turn to those of the Li–Yau inequality. Since $\lambda_N(-\Delta_{\Omega}^{D}) \geq N^{-1} \sum_{n=1}^{N} \lambda_n(-\Delta_{\Omega}^{D})$, Theorem 3.26 immediately implies the following bound.

Corollary 3.32 *Let $\Omega \subset \mathbb{R}^d$ be an open set of finite measure. Then, for any $N \in \mathbb{N}$,*

$$\lambda_N(-\Delta_{\Omega}^{D}) \geq \frac{d}{d+2} (2\pi)^2 \omega_d^{-\frac{2}{d}} N^{\frac{2}{d}} |\Omega|^{-\frac{2}{d}},$$

where $\omega_d = d^{-1}|\mathbb{S}^{d-1}|$ denotes the volume of the unit ball in $\mathbb{R}^d$.

Note that this bound is equivalent to (3.53). As already mentioned, this bound is probably non-sharp because of the excess factor of $d/(d + 2)$ compared with the asymptotic expression, but it is currently the best known bound for arbitrary domains.

In fact, Corollary 3.32 is a limiting case of the following inequality, which gives (again, probably non-sharp) bounds on sums of powers of the first N eigenvalues.

Corollary 3.33 *Let $\Omega \subset \mathbb{R}^d$ be an open set of finite measure and let $\sigma > 1$. Then, for any $N \in \mathbb{N}$,*

$$\sum_{n=1}^{N} \left(\lambda_n(-\Delta_\Omega^D)\right)^\sigma \geq \left(\frac{d}{d+2}(2\pi)^2 \omega_d^{-\frac{2}{d}}\right)^\sigma N^{\frac{2\sigma}{d}+1}|\Omega|^{-\frac{2\sigma}{d}}. \tag{3.54}$$

Compared with the asymptotics as $N \to \infty$ of the left side given in Corollary 3.19, the right side of (3.54) is too small by a factor of $(d/(d+2))^\sigma((d+2\sigma)/d)$.

Proof By Hölder's inequality,

$$\sum_{n=1}^{N} \lambda_n(-\Delta_\Omega^D) \leq N^{1-\frac{1}{\sigma}} \left(\sum_{n=1}^{N} \left(\lambda_n(-\Delta_\Omega^D)\right)^\sigma\right)^{\frac{1}{\sigma}}.$$

Bounding the left side by Theorem 3.26, we obtain the claimed inequality. □

Next, we turn our attention to $\sigma < 1$, where, in view of Corollary 3.19, we obtain an inequality with the optimal constant.

Corollary 3.34 *Let $\Omega \subset \mathbb{R}^d$ be an open set of finite measure and let $0 < \sigma < 1$. Then, for any $N \in \mathbb{N}$,*

$$\sum_{n=1}^{N} \left(\lambda_n(-\Delta_\Omega^D)\right)^\sigma \geq \frac{d}{d+2\sigma}\left((2\pi)^2 \omega_d^{-\frac{2}{d}}\right)^\sigma N^{\frac{2\sigma}{d}+1}|\Omega|^{-\frac{2\sigma}{d}}.$$

The proof relies on the following general lemma.

Lemma 3.35 *Let $(a_n)_{n\in\mathbb{N}}$ be a non-decreasing sequence of non-negative numbers that tends to infinity and assume that, for some $p > 1$,*

$$\sum_{n=1}^{N} a_n \geq \frac{1}{p}N^p \qquad \text{for all } N \in \mathbb{N}. \tag{3.55}$$

Then for all $0 < \sigma < 1$,

$$\sum_{n=1}^{N} a_n^\sigma \geq \frac{1}{\sigma(p-1)+1}N^{\sigma(p-1)+1} \qquad \text{for all } N \in \mathbb{N}. \tag{3.56}$$

Proof By Proposition 3.27 with $b_n = a_n$, respectively $b_n = a_n^\sigma$, and $K = L = 1$, we see that our assumption (3.55) is equivalent to

$$\sum_n (\lambda - a_n)_+ \leq \frac{1}{q}\lambda^q, \qquad \text{for all } \lambda \geq 0 \tag{3.57}$$

with $q = p/(p-1)$, while our conclusion (3.56) is equivalent to

$$\sum_n \left(\mu - a_n^\sigma\right)_+ \leq \frac{1}{r}\mu^r \qquad \text{for all } \mu \geq 0$$

with $r = (\sigma(p-1) + 1)/(\sigma(p-1))$. Setting $\mu = \lambda^\sigma$ and writing r in terms of q, we see that the above expression is equivalent to

$$\sum_n (\lambda^\sigma - u_n^\sigma)_+ \leq \frac{\sigma}{\sigma + q - 1} \lambda^{\sigma+q-1} \qquad \text{for all } \lambda \geq 0.$$

Using the fact that, for $a_n < \lambda$,

$$\lambda^\sigma - a_n^\sigma = \sigma \int_0^\infty \left(\min\{\lambda, s^{-1/(1-\sigma)}\} - a_n \right)_+ ds,$$

we obtain

$$\sum_n (\lambda^\sigma - a_n^\sigma)_+ \leq \sigma \int_0^\infty \sum_n \left(\min\{\lambda, s^{-1/(1-\sigma)}\} - a_n \right)_+ ds.$$

Applying (3.57) with λ replaced by $\min\{\lambda, s^{-1/(1-\sigma)}\}$, we see that the expression on the right side is bounded from above by

$$\frac{\sigma}{q} \int_0^\infty \left(\min\{\lambda, s^{-1/(1-\sigma)}\} \right)^q ds = \frac{\sigma}{q - 1 + \sigma} \lambda^{q-1+\sigma},$$

which is what we wanted to prove. $\square$

Proof of Corollary 3.34   According to the Li–Yau bound from Theorem 3.26, we have inequality (3.55) with $p = 1 + \frac{2}{d}$ and

$$a_n = (2\pi)^{-2} \omega_d^{2/d} |\Omega|^{2/d} \lambda_n(-\Delta_\Omega^D).$$

Lemma 3.35 now implies the conclusion for $\sigma < 1$. $\square$

3.5.5 Pólya's inequality for product domains

The following theorem of Laptev (1997) provides another class of domains for which Pólya's conjecture holds. The proof is based on separation of variables, which is a very special case of the lifting technique that we will see again in Chapter 6.

Theorem 3.36   *Let* $d = d_1 + d_2$ *with* $d_1 \geq 2$ *and* $d_2 \geq 1$. *Let* $\Omega_1 \subset \mathbb{R}^{d_1}$ *and* $\Omega_2 \subset \mathbb{R}^{d_2}$ *be open sets of finite measure and assume that*

$$N(\lambda, -\Delta_{\Omega_1}^D) \leq L_{0,d_1}^{cl} \lambda^{d_1/2} |\Omega_1| \qquad \text{for all } \lambda > 0.$$

Then, for $\Omega := \Omega_1 \times \Omega_2$,

$$N(\lambda, -\Delta_\Omega^D) \leq L_{0,d}^{cl} \lambda^{d/2} |\Omega| \qquad \text{for all } \lambda > 0.$$

Note that, according to Theorem 3.23, the assumption on Ω_1 is satisfied, for instance, if Ω_1 is a tiling domain.

Proof We denote by $\lambda_n^{(j)}$, $n \in \mathbb{N}$, the eigenvalues of $-\Delta_{\Omega_j}^D$, arranged in non-decreasing order and repeated according to multiplicities. Then the eigenvalues of $-\Delta_\Omega^D$ are given by $\lambda_n^{(1)} + \lambda_m^{(2)}$, for $(n,m) \in \mathbb{N} \times \mathbb{N}$, and

$$N(\lambda, -\Delta_\Omega^D) = \#\{(n,m) \in \mathbb{N} \times \mathbb{N} : \lambda_n^{(1)} + \lambda_m^{(2)} < \lambda\}$$
$$= \sum_{m \in \mathbb{N}} \#\{n \in \mathbb{N} : \lambda_n^{(1)} < \lambda - \lambda_m^{(2)}\} = \sum_{m \in \mathbb{N}} N(\lambda - \lambda_m^{(2)}, -\Delta_{\Omega_1}^D).$$

Thus, using the assumption, we find

$$N(\lambda, -\Delta_\Omega^D) \leq \sum_{m \in \mathbb{N}} L_{0,d_1}^{cl} (\lambda - \lambda_m^{(2)})_+^{d_1/2} |\Omega_1| = L_{0,d_1}^{cl} |\Omega_1| \operatorname{Tr}(-\Delta_{\Omega_2}^D - \lambda)_-^{d_1/2}.$$

Since $d_1 \geq 2$, Berezin's inequality (Corollary 3.29) yields

$$N(\lambda, -\Delta_\Omega^D) \leq L_{0,d_1}^{cl} L_{d_1/2,d_2}^{cl} |\Omega_1||\Omega_2| \lambda^{(d_1+d_2)/2} = L_{0,d}^{cl} |\Omega| \lambda^{d/2}.$$

The last identity follows from the explicit expression for $L_{\gamma,d}^{cl}$ in (3.2). $\qquad\square$

3.6 Upper bounds for the eigenvalues of the Neumann Laplacian

In the previous section we have studied lower bounds on eigenvalues of the Dirichlet Laplacian. In this section we show that similar, but *upper* bounds hold for the Neumann Laplacian.

The central result of this section is the following theorem of Laptev (1997).

Theorem 3.37 *Let $\Omega \subset \mathbb{R}^d$ be an open set of finite measure such that $-\Delta_\Omega^N$ has discrete spectrum. Then, for all $\Lambda \geq 0$,*

$$\sum_n \left(\Lambda - \lambda_n(-\Delta_\Omega^N)\right)_+ \geq L_{1,d}^{cl} |\Omega| \Lambda^{1+d/2}. \tag{3.58}$$

By the Weyl asymptotics from Corollary 3.21, the constant on the right side of (3.58) is best possible if Ω is bounded with the extension property. We also note that the right side of (3.58) coincides with the right side in Berezin's inequality (3.43), but that the inequality sign is reversed.

Proof We abbreviate by $\mu_n := \lambda_n(-\Delta_\Omega^N)$ the non-decreasing sequence of eigenvalues, repeated with multiplicities, and we denote by ψ_n an associated $L^2(\Omega)$-orthonormal system of eigenfunctions.

As in the proof of the Berezin and Li–Yau inequalities, we extend ψ_n by zero to $\mathbb{R}^d$. Now this function may no longer belong to $H^1(\mathbb{R}^d)$, but its Fourier transform $\widehat{\psi_n}$ is a well-defined function in $L^2(\mathbb{R}^d)$. We write, for any $\Lambda > 0$,

$$\sum_n (\Lambda - \mu_n)_+ = \sum_{\mu_n < \Lambda} \int_{|\xi|^2 < \Lambda} (\Lambda - \mu_n)|\widehat{\psi_n}|^2 d\xi + \sum_{\mu_n < \Lambda} \int_{|\xi|^2 \geq \Lambda} (\Lambda - \mu_n)|\widehat{\psi_n}|^2 d\xi$$

$$= \sum_n \int_{|\xi|^2 < \Lambda} (\Lambda - \mu_n)|\widehat{\psi_n}|^2 d\xi + R_N \qquad (3.59)$$

with

$$R_N := \sum_{\mu_n \geq \Lambda} \int_{|\xi|^2 < \Lambda} (\mu_n - \Lambda)|\widehat{\psi_n}|^2 d\xi + \sum_{\mu_n < \Lambda} \int_{|\xi|^2 \geq \Lambda} (\Lambda - \mu_n)|\widehat{\psi_n}|^2 d\xi,$$

which is clearly non-negative. To compute the first term in (3.59), we note that

$$\sum_n |\widehat{\psi_n}|^2 = \sum_n |(\psi_n, (2\pi)^{-d/2} e^{ix \cdot \xi})_{L^2(\Omega)}|^2 = (2\pi)^{-d}|\Omega| \qquad (3.60)$$

and, by the spectral mapping theorem,

$$\sum_n \mu_n |\widehat{\psi_n}|^2 = \sum_n |(\sqrt{\mu_n}\psi_n, (2\pi)^{-d/2} e^{ix \cdot \xi})_{L^2(\Omega)}|^2$$

$$= (2\pi)^{-d} \sum_n \left|(\sqrt{-\Delta_\Omega^N}\,\psi_n, e^{ix \cdot \xi})_{L^2(\Omega)}\right|^2 .$$

The function $e^{ix \cdot \xi}$ belongs to $H^1(\Omega)$, the form domain of $-\Delta_\Omega^N$, and therefore to the operator domain of the self-adjoint operator $\sqrt{-\Delta_\Omega^N}$. Consequently, we can continue:

$$\sum_n \mu_n |\widehat{\psi_n}|^2 = (2\pi)^{-d} \sum_n \left|(\psi_n, \sqrt{-\Delta_\Omega^N}\, e^{ix \cdot \xi})_{L^2(\Omega)}\right|^2$$

$$= (2\pi)^{-d} \left\|\sqrt{-\Delta_\Omega^N}\, e^{ix \cdot \xi}\right\|^2_{L^2(\Omega)}$$

$$= (2\pi)^{-d} \left\|\nabla e^{ix \cdot \xi}\right\|^2_{L^2(\Omega)}$$

$$= (2\pi)^{-d}|\xi|^2|\Omega| . \qquad (3.61)$$

If we insert (3.60) and (3.61) into (3.59), we find that

$$\sum_n (\Lambda - \mu_n)_+ = (2\pi)^{-d}|\Omega| \int_{|\xi|^2 < \Lambda} (\Lambda - |\xi|^2)\, d\xi + R_N .$$

In view of (3.42), this proves the theorem.   □

In the same way as for Corollary 3.29, we obtain the following result for Riesz means of order $\gamma \geq 1$.

Corollary 3.38   *Let $\Omega \subset \mathbb{R}^d$ be an open set of finite measure such that $-\Delta_\Omega^N$ has discrete spectrum and let $\gamma \geq 1$. Then, for all $\Lambda \geq 0$,*

$$\mathrm{Tr}(-\Delta_\Omega^N - \Lambda)_-^\gamma \geq L_{\gamma,d}^{\mathrm{cl}} |\Omega| \Lambda^{\gamma+d/2} . \tag{3.62}$$

By Corollary 3.21, the constant on the right side of (3.62) is best possible if Ω is bounded with the extension property.

Next, we deduce (probably non-sharp) inequalities for $\gamma \in [0,1)$.

Corollary 3.39   *Let $\Omega \subset \mathbb{R}^d$ be an open set of finite measure such that $-\Delta_\Omega^N$ has discrete spectrum. Then, for any $\Lambda \geq 0$,*

$$N(\Lambda, -\Delta_\Omega^N) \geq \frac{2}{2+d} L_{0,d}^{\mathrm{cl}} |\Omega| \Lambda^{\frac{d}{2}} . \tag{3.63}$$

Moreover, if $0 < \gamma < 1$, then, for all $\Lambda \geq 0$,

$$\mathrm{Tr}(-\Delta_\Omega^N - \Lambda)_-^\gamma \geq \frac{\Gamma(\gamma + 1 + \frac{d}{2})}{\Gamma(\gamma+1)\Gamma(2+\frac{d}{2})} L_{\gamma,d}^{\mathrm{cl}} |\Omega| \Lambda^{\gamma+\frac{d}{2}} .$$

Proof   Since $(E - \Lambda)_-^\gamma \geq \Lambda^{\gamma-1}(E-\Lambda)_-$ for any $E \geq 0$ and $0 \leq \gamma < 1$, we have

$$\mathrm{Tr}(-\Delta_\Omega^N - \Lambda)_-^\gamma \geq \Lambda^{\gamma-1}\, \mathrm{Tr}(-\Delta_\Omega^N - \Lambda)_- .$$

Inserting the bound of Theorem 3.37 on the right side and recalling (3.52), we obtain the claimed bounds, with $\gamma = 0$ corresponding to (3.63).   □

Inequality (3.63) implies directly the following bound on individual eigenvalues.

Corollary 3.40   *Let $\Omega \subset \mathbb{R}^d$ be an open set of finite measure such that $-\Delta_\Omega^N$ has discrete spectrum. Then, for any $n \in \mathbb{N}$,*

$$\lambda_n(-\Delta_\Omega^N) \leq \left(\frac{2+d}{2}\right)^{2/d} (2\pi)^2 \omega_d^{-2/d} (n-1)^{2/d} |\Omega|^{-2/d} .$$

This bound is due to Kröger (1992). It is probably non-sharp because of the excess factor of $((2+d)/2)^{2/d}$ compared with the asymptotic expression in Theorem 3.20, but currently is the best known bound for arbitrary domains.

We finally derive upper bounds for partial sums of Neumann eigenvalues.

Corollary 3.41   *Let $\Omega \subset \mathbb{R}^d$ be an open set of finite measure such that $-\Delta_\Omega^N$ has discrete spectrum. Then, for any $M \in \mathbb{N}$,*

$$\sum_{n=1}^M \lambda_n(-\Delta_\Omega^N) \leq \frac{d}{d+2}(2\pi)^2 \omega_d^{-\frac{2}{d}} (M-1)^{1+\frac{2}{d}} |\Omega|^{-\frac{2}{d}} .$$

This corollary follows from Theorem 3.37 by a duality argument as in Proposition 3.27. To be more precise, the same argument as in the proof of that proposition shows that the reverse inequality in (3.45) implies the reverse inequality in (3.46). Applying this to the sequence $b_n = \lambda_{n-1}(-\Delta_\Omega^N)$, we deduce the claimed bound. This corollary is originally due to Kröger (1992) with a different, albeit conceptually similar proof.

3.7 Phase space interpretation

It is often useful to associate to a differential operator in $L^2(\mathbb{R}^d)$ a function on the so-called *phase space*, consisting of points $(x, \xi) \in \mathbb{R}^d \times \mathbb{R}^d$. This association should be in such a way that the spectral asymptotics of the operator are related to properties of this function. We shall illustrate this general idea, which is closely related to the theory of pseudo-differential operators and microlocal analysis, in the simplest case of the Laplace operator on a domain.

In this toy case the function in question will be $|\xi|^2$. The choice of the function $|\xi|^2$ comes from the fact that, via the Fourier transform, the Laplacian $-\Delta_{\mathbb{R}^d}$ in $L^2(\mathbb{R}^d)$ is unitarily equivalent to multiplication by $|\xi|^2$ in $L^2(\mathbb{R}^d)$. Since we consider the Laplacian only in Ω, we will restrict the variable x in the phase space to the set Ω.

Note that, recalling the definition (3.1) of $L_{0,d}^{cl}$,

$$(2\pi)^{-d} |\{(x,\xi) \in \Omega \times \mathbb{R}^d : |\xi|^2 < \Lambda\}| = L_{0,d}^{cl} |\Omega| \Lambda^{d/2},$$

and therefore Weyl's law can be written as

$$N(\Lambda, -\Delta_\Omega^D) = (2\pi)^{-d} |\{(x,\xi) \in \Omega \times \mathbb{R}^d : |\xi|^2 < \Lambda\}| (1 + o(1))$$

and

$$N(\Lambda, -\Delta_\Omega^N) = (2\pi)^{-d} |\{(x,\xi) \in \Omega \times \mathbb{R}^d : |\xi|^2 < \Lambda\}| (1 + o(1))$$

as $\Lambda \to \infty$. Thus, the number of eigenvalues is given to leading order by the so-called phase space volume. In quantum mechanics this formula is often interpreted as saying that each particle occupies a volume of $(2\pi)^{-d}$ in phase space.

The correspondence between the Laplacian and the function $|\xi|^2$ becomes even clearer if we note that

$$N(\Lambda, -\Delta_\Omega^D) = \text{Tr } \chi_{[0,\Lambda)}(-\Delta_\Omega^D), \qquad N(\Lambda, -\Delta_\Omega^N) = \text{Tr } \chi_{[0,\Lambda)}(-\Delta_\Omega^N),$$

so that Weyl's law can be rewritten as

$$\text{Tr}\,\chi_{[0,\Lambda)}(-\Delta_\Omega^{\text{D}}) = (2\pi)^{-d} \iint_{\Omega\times\mathbb{R}^d} \chi_{[0,\Lambda)}(|\xi|^2)\,dx\,d\xi\,(1+o(1)) \tag{3.64}$$

and

$$\text{Tr}\,\chi_{[0,\Lambda)}(-\Delta_\Omega^{\text{N}}) = (2\pi)^{-d} \iint_{\Omega\times\mathbb{R}^d} \chi_{[0,\Lambda)}(|\xi|^2)\,dx\,d\xi\,(1+o(1)) \tag{3.65}$$

as $\Lambda \to \infty$. A computation based on the beta function identity (3.16) for the constants $L_{\gamma,d}^{\text{cl}}$ from (3.2) shows that

$$(2\pi)^{-d} \int_{\mathbb{R}^d} (|\xi|^2 - \Lambda)_-^\gamma\,d\xi = L_{\gamma,d}^{\text{cl}}\Lambda^{\gamma+d/2}\,. \tag{3.66}$$

Thus, we can write Weyl's law for Riesz means from Corollaries 3.17 and 3.21 in the form

$$\text{Tr}(-\Delta_\Omega^{\text{D}} - \Lambda)_-^\gamma = (2\pi)^{-d} \iint_{\Omega\times\mathbb{R}^d} (|\xi|^2 - \Lambda)_-^\gamma\,dx\,d\xi\,(1+o(1)) \tag{3.67}$$

and

$$\text{Tr}(-\Delta_\Omega^{\text{N}} - \Lambda)_-^\gamma = (2\pi)^{-d} \iint_{\Omega\times\mathbb{R}^d} (|\xi|^2 - \Lambda)_-^\gamma\,dx\,d\xi\,(1+o(1)) \tag{3.68}$$

as $\Lambda \to \infty$.

An observation that is useful in practice is that the phase space point of view allows for a computationally simpler and more conceptual proof of (3.67) and (3.68). In the proof of Corollaries 3.17 and 3.21 we used the beta function property (3.17) and above we used (3.66). However, if we use the identity

$$(E - \Lambda)_-^\gamma = \gamma \int_0^\infty \chi_{\{E<\Lambda-\kappa\}}\kappa^{\gamma-1}\,d\kappa$$

on *both* sides of (3.64) and (3.65), we obtain (3.67) and (3.68) directly, without ever using any beta function identities.

The same idea can be used to extend Weyl's law to a wide class of trace functionals.

Proposition 3.42 *Let* $\Omega \subset \mathbb{R}^d$ *be an open set of finite measure and let* $\varphi : (0,\infty) \to [0,\infty)$ *be a non-increasing function with* $\lim_{\lambda\to\infty}\varphi(\lambda) = 0$. *Then, as* $\Lambda \to \infty$,

$$\text{Tr}\,\varphi(-\Delta_\Omega^{\text{D}}/\Lambda) = (2\pi)^{-d}|\Omega| \int_{\mathbb{R}^d} \varphi(|\xi|^2/\Lambda)\,d\xi\,(1+o(1))\,.$$

Furthermore, if Ω *is bounded and has the extension property and if* $\varphi(0) :=$

$\lim_{\lambda \to 0} \varphi(\lambda)$ *is finite, then, as* $\Lambda \to \infty$,

$$\mathrm{Tr}\,\varphi(-\Delta_\Omega^N/\Lambda) = (2\pi)^{-d}|\Omega| \int_{\mathbb{R}^d} \varphi(|\xi|^2/\Lambda)\,d\xi\,(1 + o(1))\,.$$

We note that

$$(2\pi)^{-d}|\Omega| \int_{\mathbb{R}^d} \varphi(|\xi|^2/\Lambda)\,d\xi = (2\pi)^{-d}|\Omega| \int_{\mathbb{R}^d} \varphi(|\eta|^2)\,d\eta\,\Lambda^{d/2}\,. \qquad (3.69)$$

Therefore, the asymptotic formula reads

$$\lim_{\Lambda \to \infty} \Lambda^{-d/2}\,\mathrm{Tr}\,\varphi(-\Delta_\Omega^D/\Lambda) = (2\pi)^{-d}|\Omega| \int_{\mathbb{R}^d} \varphi(|\eta|^2)\,d\eta$$

and similarly for $-\Delta_\Omega^N$. We emphasize that it is part of the statement of Proposition 3.42 that, if $\int_{\mathbb{R}^d} \varphi(|\eta|^2)\,d\eta = \infty$, then

$$\Lambda^{-d/2}\,\mathrm{Tr}\,\varphi(-\Delta_\Omega^D/\Lambda) \to \infty \text{ as } \Lambda \to \infty$$

and similarly for $-\Delta_\Omega^N$.

Proof We begin with the Dirichlet case. Let φ_r and φ_l denote the right and left continuous versions of φ. Then $\varphi_l \geq \varphi \geq \varphi_r$ on $(0, \infty)$, and therefore

$$\mathrm{Tr}\,\varphi_l(-\Delta_\Omega^D/\Lambda) \geq \mathrm{Tr}\,\varphi(-\Delta_\Omega^D/\Lambda) \geq \mathrm{Tr}\,\varphi_r(-\Delta_\Omega^D/\Lambda)\,.$$

Since the three functions differ in at most countably many points, we have

$$\int_{\mathbb{R}^d} \varphi_l(|\xi|^2/\Lambda)\,d\xi = \int_{\mathbb{R}^d} \varphi(|\xi|^2/\Lambda)\,d\xi = \int_{\mathbb{R}^d} \varphi_r(|\xi|^2/\Lambda)\,d\xi$$

in the sense that, if one of the integrals is finite, then all of them are and they coincide.

The Lebesgue–Stieltjes construction (see, e.g., Folland, 1999, §1.5) yields a non-negative Borel measure ν on $[0, \infty)$ such that, for all $\lambda > 0$,

$$\varphi_l(\lambda) = \int_{[0,\infty)} \chi_{[\lambda,\infty)}(\mu)\,d\nu(\mu) = \int_{[0,\infty)} \chi_{[0,\mu]}(\lambda)\,d\nu(\mu)$$

and

$$\varphi_r(\lambda) = \int_{[0,\infty)} \chi_{(\lambda,\infty)}(\mu)\,d\nu(\mu) = \int_{[0,\infty)} \chi_{[0,\mu)}(\lambda)\,d\nu(\mu)\,.$$

Therefore, writing $N_\leq(\lambda, A) := N(\lambda, A) + \dim \ker(A - \lambda)$ for short, we have

$$\mathrm{Tr}\,\varphi_l(-\Delta_\Omega^D/\Lambda) = \int_{[0,\infty)} N_\leq(\mu, -\Delta_\Omega^D/\Lambda)\,d\nu(\mu)$$

and

$$\mathrm{Tr}\,\varphi_r(-\Delta_\Omega^D/\Lambda) = \int_{[0,\infty)} N(\mu, -\Delta_\Omega^D/\Lambda)\,d\nu(\mu)\,.$$

Let $0 < \varepsilon \leq 1$. By Weyl's theorem there is a $C_\varepsilon < \infty$ such that for all $\kappa \geq C_\varepsilon$,

$$N_\leq(\kappa, -\Delta_\Omega^D) \leq (1 + \varepsilon)(2\pi)^{-d}|\Omega| \int_{\mathbb{R}^d} \chi_{\{|\xi|^2 \leq \kappa\}} \, d\xi$$

and

$$N(\kappa, -\Delta_\Omega^D) \geq (1 - \varepsilon)(2\pi)^{-d}|\Omega| \int_{\mathbb{R}^d} \chi_{\{|\xi|^2 < \kappa\}} \, d\xi.$$

The second claim above follows directly from Theorem 3.15. To obtain the first claim, we bound $N_\leq(\kappa, -\Delta_\Omega^D) \leq N((1 + \delta)\kappa, -\Delta_\Omega^D)$ for an arbitrary $\delta > 0$ and again apply Theorem 3.15.

To prove the upper bound in the proposition, we recall that, by Corollary 3.30, there is a constant C such that for all $\kappa \geq 0$,

$$N_\leq(\kappa, -\Delta_\Omega^D) \leq C(2\pi)^{-d}|\Omega| \int_{\mathbb{R}^d} \chi_{\{|\xi|^2 \leq \kappa\}} \, d\xi.$$

Here, we use the fact that the bound in Corollary 3.30 for $N(\kappa, -\Delta_\Omega^D)$ yields, by continuity of the right side, the same bound for $N_\leq(\kappa, -\Delta_\Omega^D)$.

Moreover, to prove the upper bound, we may assume that $\int_{\mathbb{R}^d} \varphi(|\eta|^2) \, d\eta < \infty$. By combining the previous bounds, we conclude that

$$\begin{aligned}
\operatorname{Tr} \varphi_l(-\Delta_\Omega^D/\Lambda) &\leq (1 + \varepsilon)(2\pi)^{-d}|\Omega| \int_{[C_\varepsilon/\Lambda, \infty)} \int_{\mathbb{R}^d} \chi_{\{|\xi|^2/\Lambda \leq \mu\}} \, d\xi \, d\nu(\mu) \\
&\quad + C(2\pi)^{-d}|\Omega| \int_{[0, C_\varepsilon/\Lambda)} \int_{\mathbb{R}^d} \chi_{\{|\xi|^2/\Lambda \leq \mu\}} \, d\xi \, d\nu(\mu) \\
&= (1 + \varepsilon)(2\pi)^{-d}|\Omega| \int_{\mathbb{R}^d} \varphi_l(\max\{|\xi|^2/\Lambda, C_\varepsilon/\Lambda\}) \, d\xi \\
&\quad + C(2\pi)^{-d}|\Omega| \int_{\mathbb{R}^d} \left(\varphi_l(|\xi|^2/\Lambda) - \varphi_l(C_\varepsilon/\Lambda)\right)_+ \, d\xi \\
&\leq (1 + \varepsilon)(2\pi)^{-d}|\Omega| \int_{\mathbb{R}^d} \varphi_l(|\xi|^2/\Lambda) \, d\xi \\
&\quad + C(2\pi)^{-d}|\Omega| \int_{|\xi|^2 < C_\varepsilon} \varphi_l(|\xi|^2/\Lambda) \, d\xi.
\end{aligned}$$

By dominated convergence, as $\Lambda \to \infty$,

$$\int_{|\xi|^2 < C_\varepsilon} \varphi_l(|\xi|^2/\Lambda) \, d\xi = \Lambda^{d/2} \int_{|\eta|^2 < C_\varepsilon/\Lambda} \varphi_l(|\eta|^2) \, d\eta = o(\Lambda^{d/2}).$$

Since $\varepsilon > 0$ is arbitrary, we obtain, using (3.69),

$$\limsup_{\Lambda \to \infty} \Lambda^{-d/2} \operatorname{Tr} \varphi_l(-\Delta_\Omega^D/\Lambda) \leq \limsup_{\Lambda \to \infty} \Lambda^{-d/2}(2\pi)^{-d}|\Omega| \int_{\mathbb{R}^d} \varphi_l(|\xi|^2/\Lambda) \, d\xi,$$

which is the desired upper bound.

For the lower bound, we estimate $N(\mu, -\Delta_\Omega^D/\Lambda) \geq 0$ for $\mu \leq C_\varepsilon/\Lambda$. Therefore,

$$
\begin{aligned}
\operatorname{Tr} \varphi_r(-\Delta_\Omega^D/\Lambda) &\geq (1-\varepsilon)(2\pi)^{-d}|\Omega| \int_{(C_\varepsilon/\Lambda,\infty)} \int_{\mathbb{R}^d} \chi_{\{|\xi|^2/\Lambda \leq \mu\}} \, d\xi \, dv(\mu) \\
&= (1-\varepsilon)(2\pi)^{-d}|\Omega| \int_{\mathbb{R}^d} \varphi_r(\max\{|\xi|^2/\Lambda, C_\varepsilon/\Lambda\}) \, d\xi \\
&= (1-\varepsilon)(2\pi)^{-d}|\Omega| \int_{\mathbb{R}^d} \varphi_r(\max\{|\eta|^2, C_\varepsilon/\Lambda\}) \, d\eta \, \Lambda^{d/2}.
\end{aligned}
$$

By monotone convergence, we have

$$
\lim_{\Lambda \to \infty} \int_{\mathbb{R}^d} \varphi_r(\max\{|\eta|^2, C_\varepsilon/\Lambda\}) \, d\eta = \int_{\mathbb{R}^d} \varphi_r(|\eta|^2) \, d\eta.
$$

As $\varepsilon > 0$ is arbitrary, this proves the claimed lower bound.

We now turn to the Neumann case. Since, by the variational principle,

$$
\operatorname{Tr} \varphi\big(-\Delta_\Omega^N/\Lambda\big) \geq \operatorname{Tr} \varphi\big(-\Delta_\Omega^D/\Lambda\big),
$$

it suffices to prove an upper bound. Its proof is similar to that in the Dirichlet case except that, instead of relying on Corollary 3.30, we simply bound $N_\leq(\mu, -\Delta_\Omega^N/\Lambda) \leq N(C_\varepsilon, -\Delta_\Omega^N)$ for $\mu < C_\varepsilon/\Lambda$. Thus the contribution to the trace is bounded by

$$
\begin{aligned}
\int_{[0,C_\varepsilon/\Lambda)} N_\leq(\mu, -\Delta_\Omega^N/\Lambda) \, dv(\mu) &\leq N(C_\varepsilon, -\Delta_\Omega^N) \int_{[0,C_\varepsilon/\Lambda)} dv(\mu) \\
&= N(C_\varepsilon, -\Delta_\Omega^N)(\varphi(0) - \varphi_l(C_\varepsilon/\Lambda)).
\end{aligned}
$$

For fixed $\varepsilon > 0$, the right side tends to zero. This completes the proof. $\qquad\square$

Remark 3.43 Note that both sides of the asymptotics in Proposition 3.42 are linear in φ. This lets us extend the asymptotics to differences of functions of the form described there, thereby covering a large class of functions of bounded variation.

Remark 3.44 By a similar argument as in the previous proof, but using the bounds from Corollaries 3.30 and 3.39 instead of the Weyl asymptotics, we obtain, for any non-increasing function φ with $\lim_{\lambda\to\infty} \varphi(\lambda) = 0$, that

$$
\operatorname{Tr} \varphi(-\Delta_\Omega^D) \leq \left(1 + \frac{2}{d}\right)^{d/2} (2\pi)^{-d}|\Omega| \int_{\mathbb{R}^d} \varphi(|\xi|^2) \, d\xi
$$

and, if Ω is bounded with the extension property and $\varphi(0) := \lim_{\lambda\to 0} \varphi(\lambda)$ is finite,

$$
\operatorname{Tr} \varphi(-\Delta_\Omega^N) \geq \frac{2}{d+2}(2\pi)^{-d}|\Omega| \int_{\mathbb{R}^d} \varphi(|\xi|^2) \, d\xi.
$$

If, in addition, Ω is bounded and tiling, Pólya's bounds in Theorems 3.23 and 3.24 imply that the factors of $(1 + 2/d)^{d/2}$ and $2/(d + 2)$ can be removed.

The asymptotics in Proposition 3.42 and the universal bounds in the previous remark were obtained by integrating results for the 'elementary functions' $\chi_{[0,\Lambda)}$, $\Lambda \geq 0$. Let us illustrate this idea further by integrating with respect to a different set of 'elementary functions'. We recall that the inequalities by Berezin–Li–Yau and Kröger (Theorems 3.25 and 3.37) state that

$$\mathrm{Tr}(-\Delta_\Omega^{\mathrm{D}} - \Lambda)_- \leq (2\pi)^{-d}|\Omega| \int_{\mathbb{R}^d} (|\xi|^2 - \Lambda)_- \, d\xi \tag{3.70}$$

and

$$\mathrm{Tr}(-\Delta_\Omega^{\mathrm{N}} - \Lambda)_- \geq (2\pi)^{-d}|\Omega| \int_{\mathbb{R}^d} (|\xi|^2 - \Lambda)_- \, d\xi \tag{3.71}$$

for all $\Lambda \geq 0$. These bounds involve the family of functions $\lambda \mapsto (\lambda - \Lambda)_-$ parametrized by $\Lambda \geq 0$. We shall show that integration with respect to Λ yields the following bound.

Proposition 3.45 *Let $\Omega \subset \mathbb{R}^d$ be an open set of finite measure and let φ be a non-negative convex function on $(0, \infty)$ with $\lim_{\lambda \to \infty} \varphi(\lambda) = 0$. Then*

$$\mathrm{Tr}\, \varphi(-\Delta_\Omega^{\mathrm{D}}) \leq (2\pi)^{-d}|\Omega| \int_{\mathbb{R}^d} \varphi(|\xi|^2) \, d\xi \leq \mathrm{Tr}\, \varphi(-\Delta_\Omega^{\mathrm{N}}).$$

The Dirichlet part of this proposition appears in Berezin (1972b). In the Neumann case, we assume implicitly that $\varphi(0) := \lim_{\lambda \to 0} \varphi(\lambda)$ is finite. In this case, $\mathrm{Tr}\, \varphi(-\Delta_\Omega^{\mathrm{N}})$ is well defined whether or not $-\Delta_\Omega^{\mathrm{N}}$ has discrete spectrum. It is part of the statement that, if $\int_{\mathbb{R}^d} \varphi(|\xi|^2) \, d\xi = \infty$, then $\mathrm{Tr}\, \varphi(-\Delta_\Omega^{\mathrm{N}}) = \infty$.

The idea of the proof is again to write φ as a superposition with positive weight of functions for which the inequality is known; namely those functions $\varphi(\lambda) = (\lambda - \mu)_-$ for $\mu \geq 0$.

Lemma 3.46 *Let φ be a non-negative convex function on $(0, \infty)$ with $\lim_{\lambda \to \infty} \varphi(\lambda) = 0$. Then there is a non-negative Borel measure ν on $[0, \infty)$ such that*

$$\varphi(\lambda) = \int_{[0,\infty)} (\lambda - \mu)_- \, d\nu(\mu) \qquad \text{for all } \lambda > 0.$$

Proof By Simon (2011, Theorem 1.26),

$$D^+\varphi(\lambda) := \lim_{0 < \varepsilon \to 0} (\varphi(\lambda + \varepsilon) - \varphi(\lambda))/\varepsilon$$

exists for every $\lambda > 0$ and is non-decreasing and continuous from the right.

Moreover, by Simon (2011, Theorem 1.28), we have

$$\varphi(\lambda') - \varphi(\lambda) = \int_{\lambda}^{\lambda'} D^+\varphi(\kappa)\,d\kappa \qquad \text{for any } \lambda' > \lambda > 0.$$

Letting $\lambda' \to \infty$, we infer, in particular, that $\lim_{\kappa\to\infty} D^+\varphi(\kappa) = 0$ and that

$$\varphi(\lambda) = -\int_{\lambda}^{\infty} D^+\varphi(\kappa)\,d\kappa \qquad \text{for any } \lambda > 0.$$

As in the proof of Proposition 3.42, we find a non-negative Borel measure ν on $[0,\infty)$ with

$$D^+\varphi(\kappa) = -\int_{[0,\infty)} \chi_{(\kappa,\infty)}(\mu)\,d\nu(\mu) \qquad \text{for all } \kappa > 0.$$

We conclude that

$$\varphi(\lambda) = \int_{\lambda}^{\infty} \int_{[0,\infty)} \chi_{(\kappa,\infty)}(\mu)\,d\nu(\mu)\,d\kappa = \int_{[0,\infty)} (\lambda - \mu)_-\,d\nu(\mu),$$

as claimed. $\qquad\qquad\qquad\qquad\qquad\qquad\qquad\qquad\qquad\qquad\qquad\qquad\qquad\square$

Proof of Proposition 3.45 With the help of the measure ν from Lemma 3.46 and Berezin's inequality (3.70), we obtain

$$\begin{aligned}
\mathrm{Tr}\,\varphi(-\Delta_\Omega^D) &= \int_{[0,\infty)} \mathrm{Tr}(-\Delta_\Omega^D - \mu)_-\,d\nu(\mu) \\
&\le (2\pi)^{-d}|\Omega| \int_{[0,\infty)} \int_{\mathbb{R}^d} (|\xi|^2 - \mu)_-\,d\xi\,d\nu(\mu) \\
&= (2\pi)^{-d}|\Omega| \int_{\mathbb{R}^d} \varphi(|\xi|^2)\,d\xi,
\end{aligned}$$

as claimed. The bound in the Neumann case follows similarly using (3.71). $\quad\square$

As a final example of how the phase space point of view can be useful, we return to the proof of Theorem 3.36. In the last step of that proof we used the fact that $L_{0,d}^{\mathrm{cl}} = L_{0,d_1}^{\mathrm{cl}} L_{d_1/2,d_2}^{\mathrm{cl}}$. Using the integral representation

$$L_{\gamma,d}^{\mathrm{cl}} = \int_{\mathbb{R}^d} (|\xi|^2 - 1)_-^\gamma \frac{d\xi}{(2\pi)^d}$$

from (3.66), we obtain

$$
\begin{aligned}
L_{0,d}^{\mathrm{cl}} &= (2\pi)^{-d} |\{(\xi_1, \xi_2) \in \mathbb{R}^{d_1} \times \mathbb{R}^{d_2} : |\xi_1|^2 + |\xi_2|^2 < 1\}| \\
&= (2\pi)^{-d_2} \int_{\mathbb{R}^{d_2}} (2\pi)^{-d_1} |\{\xi_1 \in \mathbb{R}^{d_1} : |\xi_1|^2 < 1 - |\xi_2|^2\}| \, d\xi_2 \\
&= (2\pi)^{-d_2} \int_{\mathbb{R}^{d_2}} L_{0,d_1}^{\mathrm{cl}} (1 - |\xi_2|^2)_+^{d_1/2} \, d\xi_2 \\
&= L_{0,d_1}^{\mathrm{cl}} L_{d_1/2,d_2}^{\mathrm{cl}} ,
\end{aligned}
$$

as claimed. Note the similarity between the different steps in this computation with the treatment of the eigenvalue sums in the proof of Theorem 3.36.

Later on we will find useful a more general version of this identity; namely,

$$
L_{\gamma,d}^{\mathrm{cl}} = L_{\gamma,d_1}^{\mathrm{cl}} L_{\gamma+d_1/2,d_2}^{\mathrm{cl}} . \tag{3.72}
$$

Indeed,

$$
\begin{aligned}
L_{\gamma,d}^{\mathrm{cl}} &= (2\pi)^{-d} \int_{\mathbb{R}^d} (|\xi|^2 - 1)_-^{\gamma} \, d\xi \\
&= (2\pi)^{-d_2} \int_{\mathbb{R}^{d_2}} (2\pi)^{-d_1} \int_{\mathbb{R}^{d_1}} (|\xi_1|^2 - (1 - |\xi_2|^2))_-^{\gamma} \, d\xi_1 \, d\xi_2 \\
&= (2\pi)^{-d_2} \int_{\mathbb{R}^{d_2}} L_{\gamma,d_1}^{\mathrm{cl}} (1 - |\xi_2|^2)_+^{\gamma+d_1/2} \, d\xi_2 \\
&= L_{\gamma,d_1}^{\mathrm{cl}} L_{\gamma+d_1/2,d_2}^{\mathrm{cl}} ,
\end{aligned}
$$

as claimed. Of course, equality (3.72) can also be directly verified using identities for the Gamma function.

3.8 Appendix: The Laplacian in spherical coordinates

Throughout this section we assume that $d \geq 2$.

3.8.1 The Sobolev space on the sphere

Recall that in §2.1.5 we introduced the notion of weak differentiability of a function on $\mathbb{S}^{d-1}$ and the corresponding weak gradient $\nabla_{\mathbb{S}^{d-1}}$. We now define the *Sobolev space* (of order 1 with integrability exponent 2) on $\mathbb{S}^{d-1}$, denoted by

$$
H^1(\mathbb{S}^{d-1}),
$$

to be the space of all (a.e. equivalence classes of) functions $u \in L^2(\mathbb{S}^{d-1})$ that are weakly differentiable on $\mathbb{S}^{d-1}$ with $\nabla_{\mathbb{S}^{d-1}} u \in L^2(\mathbb{S}^{d-1}, \mathbb{C}^d)$. This is an inner product space with respect to

$$
\langle u, v \rangle_{H^1(\mathbb{S}^{d-1})} := (u, v)_{L^2(\mathbb{S}^{d-1})} + (\nabla_{\mathbb{S}^{d-1}} u, \nabla_{\mathbb{S}^{d-1}} v)_{L^2(\mathbb{S}^{d-1}, \mathbb{C}^d)} .
$$

We will need the following properties of this space.

Lemma 3.47 $H^1(\mathbb{S}^{d-1})$ *is complete and separable, and* $C^\infty(\mathbb{S}^{d-1})$ *is dense in* $H^1(\mathbb{S}^{d-1})$.

Proof Completeness and separability are proved in the same way as was Lemma 2.22 using Lemma 2.19. Density of smooth functions follows by the same argument as in the proof of Corollary 2.21, but using Proposition 2.23 instead of Remark 2.2.   □

3.8.2 The Laplace–Beltrami operator on the sphere

We recall that in §2.1.6 we introduced a weak notion of the Laplace–Beltrami operator on the sphere that extended the classical one. We now realize the Laplace–Beltrami operator on the sphere as a self-adjoint operator and discuss its spectrum.

In the Hilbert space $L^2(\mathbb{S}^{d-1})$, we consider the quadratic form

$$\int_{\mathbb{S}^{d-1}} |\nabla_{\mathbb{S}^{d-1}} u|^2 \, d\omega$$

with form domain $H^1(\mathbb{S}^{d-1})$. It is non-negative and, by Lemma 3.47, closed. It therefore generates a self-adjoint, non-negative operator in $L^2(\mathbb{S}^{d-1})$, which we denote by $-\Delta_{\mathbb{S}^{d-1}}$. By Lemma 3.47, $C^\infty(\mathbb{S}^{d-1})$ is a form core.

Lemma 3.48 *We have*

$$\mathrm{dom}(-\Delta_{\mathbb{S}^{d-1}}) = \left\{ u \in H^1(\mathbb{S}^{-1}) : u \text{ has a weak Laplacian } \Delta_{\mathbb{S}^{d-1}} u \in L^2(\mathbb{S}^{d-1}) \right\}$$

and, for $u \in \mathrm{dom}(-\Delta_{\mathbb{S}^{d-1}})$, $-\Delta_{\mathbb{S}^{d-1}} u$ *in the operator sense coincides with* $-\Delta_{\mathbb{S}^{d-1}} u$ *in the weak sense.*

The last part of this lemma motivates the notation $-\Delta_{\mathbb{S}^{d-1}}$.

Proof By definition and the form core property of $C^\infty(\mathbb{S}^{d-1})$, a function $u \in L^2(\mathbb{S}^{d-1})$ belongs to $\mathrm{dom}(-\Delta_{\mathbb{S}^{d-1}})$ if and only if there is an $f \in L^2(\mathbb{S}^{d-1})$ such that

$$\int_{\mathbb{S}^{d-1}} \nabla_{\mathbb{S}^{d-1}} u \cdot \nabla_{\mathbb{S}^{d-1}} v \, d\omega = \int_{\mathbb{S}^{d-1}} f v \, d\omega \qquad \text{for all } v \in C^\infty(\mathbb{S}^{d-1}).$$

In this case, f is unique and one has $f = -\Delta_{\mathbb{S}^{d-1}} u$. The lemma now follows since, for all $u \in H^1(\mathbb{S}^{d-1})$ and $v \in C^\infty(\mathbb{S}^{d-1})$,

$$\int_{\mathbb{S}^{d-1}} \nabla_{\mathbb{S}^{d-1}} u \cdot \nabla_{\mathbb{S}^{d-1}} v \, d\omega = \int_{\mathbb{S}^{d-1}} u(-\Delta_{\mathbb{S}^{d-1}} v) \, d\omega.$$

This formula when $u \in C^1(\mathbb{S}^{d-1})$ is well known and extends, by Lemma 3.47, to the general case. $\qquad\square$

The operator $-\Delta_{\mathbb{S}^{d-1}}$ can be explicitly diagonalized in terms of spherical harmonics and we summarize some facts about its spectrum in the following theorem.

Theorem 3.49 *The spectrum of the operator* $-\Delta_{\mathbb{S}^{d-1}}$ *in* $L^2(\mathbb{S}^{d-1})$ *consists precisely of the eigenvalues* $\ell(\ell + d - 2)$, $\ell \in \mathbb{N}_0$, *with multiplicity*

$$
\nu_\ell := \begin{cases} 1 & \text{if } d = 2,\, \ell = 0, \\ 2 & \text{if } d = 2,\, \ell \geq 1, \\ \frac{(2\ell+d-2)(\ell+d-3)!}{(d-2)!\,\ell!} & \text{if } d \geq 3. \end{cases}
$$

We omit the proof of this theorem and refer to, for instance, Stein and Weiss (1971, Chapter IV) and Simon (2015b, §3.5).

By Theorem 3.49 and the spectral theorem, $L^2(\mathbb{S}^{d-1})$ has an orthonormal basis $(Y_{\ell,m})$ consisting of eigenfunctions of this operator. Here, for each $\ell \in \mathbb{N}_0$, the index m runs through a set $\mathcal{M}_\ell$ of cardinality ν_ℓ. Note that, in particular, $Y_{\ell,m} \in H^1(\mathbb{S}^{d-1})$.

Any $f \in L^2(\mathbb{S}^{d-1})$ can be decomposed in this basis as follows. Defining the 'Fourier' coefficients

$$
f_{\ell,m} := \int_{\mathbb{S}^{d-1}} f(\omega)\overline{Y_{\ell,m}(\omega)}\,d\omega \qquad \text{for } \ell \in \mathbb{N}_0,\, m \in \mathcal{M}_\ell,
$$

we have, in the sense of L^2 convergence,

$$
f = \sum_{\ell,m} f_{\ell,m} Y_{\ell,m} \qquad \text{and} \qquad \int_{\mathbb{S}^{d-1}} |f|^2\,d\omega = \sum_{\ell,m} |f_{\ell,m}|^2. \tag{3.73}
$$

Since the $Y_{\ell,m}$ are eigenfunctions of the Laplacian corresponding to the eigenvalue $\ell(\ell + d - 2)$, we have, by the spectral theorem (see (1.36))

$$
\int_{\mathbb{S}^{d-1}} |\nabla_{\mathbb{S}^{d-1}} f|^2\,d\omega = \sum_{\ell,m} \ell(\ell + d - 2)|f_{\ell,m}|^2 \qquad \text{for any } f \in H^1(\mathbb{S}^{d-1}).
$$

$$\tag{3.74}$$

Conversely if, for a given $f \in L^2(\mathbb{S}^{d-1})$, the sum on the right side converges then $f \in H^1(\mathbb{S}^{d-1})$.

3.8.3 Spherical coordinates

For $R > 0$ we denote by B_R the ball in $\mathbb{R}^d$ centered at the origin. We allow the value $R = \infty$, where we interpret B_∞ as $\mathbb{R}^d$.

Let $u \in L^2(B_R)$ and introduce spherical coordinates $r = |x|$ and $\omega = x/|x|$. Then, by Fubini's theorem, $u(r \cdot) \in L^2(\mathbb{S}^{d-1})$ for almost every $r \in (0, R)$. Therefore, by the completeness of the $Y_{\ell,m}$ in $L^2(\mathbb{S}^{d-1})$, we have, for almost every $r \in (0, R)$,

$$u(x) = \sum_{\ell,m} u_{\ell,m}(r) Y_{\ell,m}(\omega), \qquad u_{\ell,m}(r) := \int_{\mathbb{S}^{d-1}} u(r\omega) \overline{Y_{\ell,m}(\omega)} \, d\omega, \quad (3.75)$$

with convergence in $L^2(\mathbb{S}^{d-1})$ and

$$\int_{\mathbb{S}^{d-1}} |u(r\omega)|^2 \, d\omega = \sum_{\ell,m} |u_{\ell,m}(r)|^2 \quad \text{for almost every } r \in (0, R).$$

It easy to see that the series in (3.75) also converges in $L^2(B_R)$.

In the next lemma, we characterize the property $u \in H^1(B_R)$ in terms of the functions $u_{\ell,m}$ on $(0, R)$.

Lemma 3.50 *If $u \in H^1(B_R)$, then the functions $u_{\ell,m}$, $\ell \in \mathbb{N}_0$, $m \in \mathcal{M}_\ell$, are weakly differentiable on $(0, R)$ and*

$$\sum_{\ell,m} \int_0^R \left(|u'_{\ell,m}|^2 + \frac{\ell(\ell + d - 2)}{r^2} |u_{\ell,m}|^2 \right) r^{d-1} \, dr = \|\nabla u\|^2_{L^2(B_R)}. \quad (3.76)$$

Conversely, if $u \in L^2(B_R)$ and if $u_{\ell,m}$ are weakly differentiable functions on $(0, R)$ such that the left side in (3.76) is finite, then $u \in H^1(B_R)$.

Proof Step 1. Let $u \in H^1(B_R)$. We know from Lemma 2.20 that, for almost every $\omega \in \mathbb{S}^{d-1}$, the function $r \mapsto u(r\omega)$ is weakly differentiable on $(0, R)$ with weak derivative $\partial_r u(r\omega) := (u(\cdot \, \omega))(r)' = \omega \cdot \nabla u(r\omega)$ belonging to $L^2(B_R)$.

We start by showing that, for any $\ell \in \mathbb{N}_0$, $m \in \mathcal{M}_\ell$, the function $u_{\ell,m}$ is weakly differentiable on $(0, R)$ with

$$u'_{\ell,m}(r) = \int_{\mathbb{S}^{d-1}} \partial_r u(r\omega) \overline{Y_{\ell,m}(\omega)} \, d\omega. \quad (3.77)$$

To prove this, let $\eta \in C_0^\infty(0, R)$ and write, using Fubini's theorem,

$$\int_0^R u_{\ell,m} \partial_r \eta \, dr = \int_{\mathbb{S}^{d-1}} \int_0^R u(r\omega) \partial_r \eta \, dr \, \overline{Y_{\ell,m}(\omega)} \, d\omega.$$

By the fact recalled above we have, for almost every $\omega \in \mathbb{S}^{d-1}$,

$$\int_0^R u(r\omega) \partial_r \eta \, dr = - \int_0^R \omega \cdot \nabla u(r\omega) \eta \, dr.$$

Another application of Fubini gives that

$$\int_0^R u_{\ell,m}\partial_r\eta \, dr = -\int_0^R \int_{\mathbb{S}^{d-1}} \omega \cdot \nabla u(r\omega)\overline{Y_{\ell,m}(\omega)} \, d\omega \, \eta \, dr$$

$$= -\int_0^R \int_{\mathbb{S}^{d-1}} \partial_r u(r\omega)\overline{Y_{\ell,m}(\omega)} \, d\omega \, \eta \, dr \, .$$

This proves the weak differentiability and (3.77).

It follows from (3.77) that $u'_{\ell,m} = (\partial_r u)_{\ell,m}$, and therefore, since the $Y_{\ell,m}$ form an orthonormal basis of $L^2(\mathbb{S}^{d-1})$, we have

$$\sum_{\ell,m} |u'_{\ell,m}(r)|^2 = \int_{\mathbb{S}^{d-1}} |\partial_r u(r\omega)|^2 \, d\omega \qquad \text{for almost every } r \in (0,R). \quad (3.78)$$

By Lemma 2.20, for almost every $r \in (0,R)$, the function $\omega \mapsto u(r\omega)$ is weakly differentiable on $\mathbb{S}^{d-1}$ with weak gradient

$$\nabla_{\mathbb{S}^{d-1}} u(r\omega) := (\nabla_{\mathbb{S}^{d-1}} u(r \cdot))(\omega) = r(\nabla u(r\omega) - \omega\,\omega \cdot \nabla u(r\omega)) \, .$$

Since this belongs to $H^1(\mathbb{S}^{d-1})$ for almost every $r \in (0,R)$, we have, by (3.74),

$$\int_{\mathbb{S}^{d-1}} |\nabla_{\mathbb{S}^{d-1}} u(r\omega)|^2 \, d\omega = \sum_{\ell,m} \ell(\ell+d-2)|u_{\ell,m}(r)|^2 \text{ for almost every } r \in (0,R) \, .$$

$$(3.79)$$

By (3.78), (3.79), and the $\mathbb{C}^d$-orthogonality of $\omega\partial_r u$ and $\nabla_{\mathbb{S}^{d-1}} u$ we have, for a.e. $r \in (0,R)$,

$$\int_{\mathbb{S}^{d-1}} |\nabla u(r\omega)|^2 \, d\omega = \sum_{\ell,m} \left(|u'_{\ell,m}(r)|^2 + \frac{\ell(\ell+d-2)}{r^2}|u_{\ell,m}(r)|^2 \right) \, .$$

Integrating with respect to r gives us the first part of the lemma.

Step 2. Now assume, conversely, that $u \in L^2(B_R)$, that $u_{\ell,m}$ is weakly differentiable on $(0,R)$ and that the left side of (3.76) is finite. It follows from Lemma 2.20 and the assumed finiteness that, for each $\ell \in \mathbb{N}_0$, $m \in \mathcal{M}_\ell$, the function $r\omega \mapsto u_{\ell,m}(r)Y_{\ell,m}(\omega)$ belongs to $H^1(B_R \setminus \{0\})$ and, by Proposition 2.80, to $H^1(B_R)$. Therefore, for each $L \in \mathbb{N}_0$, the function

$$u_L(x) := \sum_{\ell \leq L}\sum_{m \in \mathcal{M}_\ell} u_{\ell,m}(r)Y_{\ell,m}(\omega)$$

belongs to $H^1(B_R)$. Applying the first part of the lemma, which we have already proved, to the function $u_{L'} - u_L$ with $L' > L$ yields

$$\|\nabla u_{L'} - \nabla u_L\|^2 = \sum_{L<\ell\leq L'}\sum_{m\in\mathcal{M}_\ell} \int_0^R \left(|u'_{\ell,m}|^2 + \frac{\ell(\ell+d-2)}{r^2}|u_{\ell,m}|^2 \right) r^{d-1} \, dr \, .$$

By the assumed finiteness of (3.76), we see that (u_L) is Cauchy in $H^1(B_R)$ and thus convergent. As remarked before, u_L converges to u in $L^2(B_R)$. With Corollary 2.21 one can deduce that $u \in H^1(B_R)$. $\qquad \square$

If $R < \infty$ and $u \in H^1(B_R)$ we note that, by Lemma 2.26 applied on $I = (R/2, R)$, the boundary values $u_{\ell,m}(R)$ are well defined for all $\ell \in \mathbb{N}_0$, $m \in \mathcal{M}_\ell$.

Lemma 3.51 *Assume $R < \infty$ and $u \in H^1(B_R)$. Then $u \in H^1_0(B_R)$ if and only if $u_{\ell,m}(R) = 0$ for all $\ell \in \mathbb{N}_0$, $m \in \mathcal{M}_\ell$.*

Sketch of Proof The key ingredient is a modification of Lemma 2.27, which says that a weakly differentiable function $v \in L^2((0, R), r^{d-1} dr)$ for which

$$\int_0^R \left(|v'|^2 + \frac{\ell(\ell + d - 2)}{r^2} |v|^2 \right) r^{d-1} \, dr < \infty \tag{3.80}$$

satisfies $v(R) = 0$ if and only if it can be approximated in $L^2((0, R), r^{d-1} dr)$, and in the norm corresponding to (3.80), by smooth functions on $(0, R)$ vanishing near R. The proof of this fact proceeds along the same steps as that of Lemma 2.27 and is omitted.

Now, if $u \in H^1_0(B_R)$, let $(u^{(n)}) \subset C_0^\infty(B_R)$ be an H^1-approximating sequence. Then $(u^{(n)}_{\ell,m})_n$ is a sequence of smooth functions on $(0, R)$ vanishing in a neighborhood of R. Moreover, the $u^{(n)}_{\ell,m}$ converge to $u_{\ell,m}$ in $L^2((0, R), r^{d-1} dr)$ and, by (3.76), also in the norm corresponding to (3.80). Thus, by the above fact, $u_{\ell,m}(R) = 0$, as claimed.

Conversely, if $u_{\ell,m}(R) = 0$, the above fact shows that $u_{\ell,m}$ can be approximated, in $L^2((0, R), r^{d-1} dr)$ and in the norm corresponding to (3.80), by smooth functions on $(0, R)$ vanishing near R. Arguing as in Lemma 2.33, these functions can even be chosen in $C_0^\infty(0, R)$. Therefore, $u_{\ell,m}(r) Y_{\ell,m}(\omega)$ can be approximated in the norm of $H^1(B_R)$ by functions in $C_0^\infty(B_R)$, and hence belongs to $H^1_0(B_R)$. Therefore, in the notation of the previous proof, (u_L) is Cauchy in $H^1_0(B_R)$ and we conclude that $u \in H^1_0(B_R)$. $\qquad \square$

The remainder of this section will be needed only in the proof of Proposition 4.19 and can be skipped at first reading.

Next, still assuming $R < \infty$, we show that functions in $H^1(B_R)$ have a well defined boundary value on ∂B_R that belongs to $L^2(\partial B_R)$. To define this boundary value we recall that, as a consequence of Lemma 2.20, for any $u \in H^1(B_R)$ and any choice of pointwise representative of u, there is a subset $\Sigma \subset \mathbb{S}^{d-1}$ of full measure such that, for any $\omega \in \Sigma$, the function $r \mapsto u(r\omega)$ is weakly differentiable on $(0, R)$ with derivative in $L^2((0, R), r^{d-1} \, dr)$. Consequently, by Lemma 2.26 applied on $I = (R/2, R)$, the limit $\lim_{r \to R} u(r\omega) =: u(R\omega)$ exists for any $\omega \in \Sigma$. This defines a function on ∂B_R. Any two choices of pointwise representatives of the same $u \in H^1(B_R)$ lead to functions on ∂B_R that coincide almost everywhere with respect to the surface measure σ on ∂B_R. (Indeed, for pointwise representatives u_1 and u_2 of the same u, there is a set $\widetilde{\Sigma} \subset \mathbb{S}^{d-1}$ of full measure such that for every $\omega \in \widetilde{\Sigma}$, the functions $r \mapsto u_1(r\omega)$ and $r \mapsto u_2(r\omega)$ coincide almost everywhere on $(0, R)$.) Thus, the boundary value is well-defined almost everywhere on the boundary. We will now show that this function belongs to $L^2(\partial B_R)$.

Lemma 3.52 *Let $R < \infty$ and $u \in H^1(B_R)$. Then the restriction of u to ∂B_R, in the sense defined above, belongs to $L^2(\partial B_R)$ and satisfies*

$$\int_{\partial B_R} |u|^2 \, d\sigma \le dR^{-1} \int_{B_R} |u|^2 \, dx + 2 \left(\int_{B_R} |\nabla u|^2 \, dx \right)^{1/2} \left(\int_{B_R} |u|^2 \, dx \right)^{1/2}.$$

Proof For given pointwise representatives u, let $\Sigma \subset \mathbb{S}^{d-1}$ be as before the lemma and apply Lemma 2.32 (which extends to $r = R$) to each function $r \mapsto u(r\omega)$ with fixed $\omega \in \Sigma$. Integrating the resulting inequality with respect to $\omega \in \Sigma$, we obtain

$$R^{d-1} \int_{\mathbb{S}^{d-1}} |u(R\omega)|^2 \, d\omega$$

$$\le dR^{-1} \int_{B_R} |u|^2 \, dx + 2 \left(\int_{B_R} |\nabla u|^2 \, dx \right)^{1/2} \left(\int_{B_R} |u|^2 \, dx \right)^{1/2}.$$

Here, we used the Cauchy–Schwarz inequality as well as the bound $|\partial_r u| \le |\nabla u|$ from Lemma 2.20. This implies the claimed inequality. □

3.8.4 The spectrum of the Neumann Laplacian on a ball

We now discuss the Neumann Laplacian on the ball B_R. Throughout this subsection we assume that $R < \infty$.

We begin by rephrasing Lemma 3.50 in operator-theoretic terms. We set

$$\mathcal{H}_{\ell,m} := L^2((0, R), r^{d-1} dr) \qquad \text{for all } \ell \in \mathbb{N}_0, \, m \in \mathcal{M}_\ell$$

and

$$\mathcal{H} := \bigoplus_{\ell \in \mathbb{N}_0, \, m \in \mathcal{M}_\ell} \mathcal{H}_{\ell,m}.$$

For each $\ell \in \mathbb{N}_0$, $m \in \mathcal{M}_\ell$, we consider in $\mathcal{H}_{\ell,m}$ the quadratic form

$$b^{\mathrm{N}}_{\ell,m}[f] := \int_0^R (|f'|^2 + \ell(\ell + d - 2)r^{-2}|f|^2) r^{d-1} \, dr$$

with form domain

$$d[b^N_{\ell,m}] := \Big\{ f \in \mathcal{H}_{\ell,m} : f \text{ weakly differentiable on } (0,R),$$

$$\int_0^R (|f'|^2 + \ell(\ell + d - 2)r^{-2}|f|^2)r^{d-1}\, dr < \infty \Big\}.$$

Note that the forms $b^N_{\ell,m}$ are independent of m. These quadratic forms are non-negative and closed in $\mathcal{H}_{\ell,m}$ and, therefore, as in §1.2.1, lead to a non-negative and closed quadratic form b^N in $\mathcal{H}$ defined by

$$b^N[v] := \sum_{\ell \in \mathbb{N}_0,\, m \in M_\ell} b^N_{\ell,m}[v_{\ell,m}]$$

with

$$d[b^N] := \Big\{ v = (v_{\ell,m}) \in \mathcal{H} : v_{\ell,m} \in d[b^N_{\ell,m}], \sum_{\ell \in \mathbb{N}_0,\, m \in M_\ell} b^N_{\ell,m}[v_{\ell,m}] < \infty \Big\}.$$

According to (1.42), the operators B^N and $B^N_{\ell,m}$ corresponding to the quadratic forms b^N and $b^N_{\ell,m}$, respectively, satisfy

$$B^N = \bigoplus_{\ell \in \mathbb{N}_0,\, m \in M_\ell} B^N_{\ell,m}.$$

Let us define a unitary operator $U : L^2(B_R) \to \mathcal{H}$ by

$$(Uu)_{\ell,m} := u_{\ell,m}$$

with $u_{\ell,m}$ defined by (3.75). The unitarity of U follows from the fact that the $Y_{\ell,m}$ form an orthonormal basis in $L^2(\mathbb{S}^{d-1})$. As in §3.1.1, we denote by a^N the quadratic form of the Neumann Laplacian $-\Delta^N_{B_R}$; that is,

$$a^N[u] = \int_{B_R} |\nabla u|^2\, dx \quad \text{with} \quad d[a^N] = H^1(B_R).$$

Lemma 3.50 implies that

$$a^N[u] = b^N[Uu] \quad \text{for all } u \in d[a^N] = U^* d[b^N].$$

This means that the corresponding operator satisfies

$$-\Delta^N_{B_R} = U^* B^N U.$$

This identity reduces the spectral analysis of the operator $-\Delta^N_{B_R}$ to the spectral analysis of the individual operators $B^N_{\ell,m}$.

Proposition 3.53 *The spectrum of* $-\Delta_{B_R}^N$ *is discrete and coincides with the union of the spectra of the operators* $B_{\ell,m}^N$ *for* $\ell \in \mathbb{N}_0$, $m \in \mathcal{M}_\ell$. *Moreover, the multiplicity of an eigenvalue of* $-\Delta_{B_R}^N$ *coincides with the sum of the multiplicities of the corresponding eigenvalues of the operators* $B_{\ell,m}^N$.

The discreteness of the spectrum of $-\Delta_{B_R}^N$ was already observed in Lemma 3.4, but for the present case the proof that follows is somewhat more direct. It is based on the following lemma involving the operators $B_{\ell,m}^N$.

Lemma 3.54 *For each* $\ell \in \mathbb{N}_0$, $m \in \mathcal{M}_\ell$, *the spectrum of the operator* $B_{\ell,m}^N$ *is discrete. We have* $\inf \sigma(B_{\ell,m}^N) \geq 0$, *with equality if and only if* $\ell = 0$. *Moreover,* $\inf \sigma(B_{\ell,m}^N) \to \infty$ *as* $\ell \to \infty$.

Proof Discreteness of the spectrum follows from Lemma 2.43 by Corollary 1.21. For $\ell = 0$ we have $b_{0,m}^N[f] = 0$ if and only if f is constant, while for $\ell \geq 1$ we have

$$b_{\ell,m}^N[f] \geq \ell(\ell + d - 2)R^{-2} \int_0^R |f|^2 r^{d-1}\, dr,$$

which implies $\inf \sigma(B_{\ell,m}^N) \geq \ell(\ell + d - 2)R^{-2}$. This proves the lemma. $\square$

Proof of Proposition 3.53 According to Lemma 1.12, the spectrum of $-\Delta_{B_R}^N$ is the closure of the union of the spectra of the $B_{\ell,m}^N$. Since, by Lemma 3.54, each of these sets is discrete and has an infimum that tends to infinity with ℓ, the union is closed. Moreover, the fact that the infimum tends to infinity means that each eigenvalue of $-\Delta_{B_R}^N$ is an eigenvalue of at most finitely many $B_{\ell,m}^N$ and, since the latter have a discrete spectrum, $-\Delta_{B_R}^N$ has a discrete spectrum also. The statement about the multiplicity of the eigenvalues follows immediately from Lemma 1.5. $\square$

We will supplement the qualitative results in Proposition 3.53 by a more quantitative analysis. This is possible since the operators $B_{\ell,m}^N$ are differential operators on an interval and techniques from the theory of ordinary differential equations are available.

To that end, let us recall some facts about Bessel's differential equation

$$t^2 f'' + t f' + (t^2 - v^2)f = 0 \quad \text{in } (0, \infty),$$

where v is a non-negative parameter (Watson, 1944; Abramowitz and Stegun, 1964). This equation has two linearly independent solutions J_v and Y_v, which satisfy, as $r \to 0$,

$$J_v(t) = \frac{1}{\Gamma(v + 1)} \left(\frac{t}{2}\right)^v (1 + o(1))$$

and

$$Y_\nu(t) = \begin{cases} \frac{2}{\pi}(\ln t)(1 + o(1)) & \text{if } \nu = 0, \\ \frac{\Gamma(\nu)}{\pi}\left(\frac{2}{t}\right)^\nu (1 + o(1)) & \text{if } \nu > 0. \end{cases}$$

These asymptotics can be differentiated. We denote by $q_{\ell,d,n}$, $n \in \mathbb{N}$, the positive zeros of the derivative of $t^{-(d-2)/2} J_{\ell+(d-2)/2}(t)$ in increasing order.

Lemma 3.55 *For each $\ell \in \mathbb{N}_0$, $m \in M_\ell$, the spectrum of the operator $B^N_{\ell,m}$ consists precisely of the values 0 if $\ell = 0$ and $R^{-2}q^2_{\ell,d,n}$, $n \in \mathbb{N}$. Each such number is an eigenvalue of multiplicity 1.*

Proof We know from Lemma 3.54 that the spectrum of $B^N_{\ell,m}$ consists only of eigenvalues. Let λ be one of them and let f denote a corresponding eigenfunction. Then for all $\varphi \in d[b^N_{\ell,m}]$,

$$\int_0^R (f'\varphi' + \ell(\ell + d - 2)r^{-2}f\varphi)r^{d-1}\,dr = \lambda \int_0^R f\varphi r^{d-1}\,dr. \tag{3.81}$$

In particular, the fact that, for $\varphi \in C_0^1(0, R)$,

$$\int_0^R f'(r^{d-1}\varphi)'\,dr = \int_0^R (f'(d-1)r^{-1} - \ell(\ell + d - 2)r^{-2}f + \lambda f)\varphi r^{d-1}\,dr$$

implies that f' is weakly differentiable on $(0, R)$ and that

$$-f'' - (d-1)r^{-1}f' + \ell(\ell + d - 2)r^{-2}f = \lambda f \qquad \text{in } (0, R). \tag{3.82}$$

That f' is weakly differentiable on $(R/2, R)$ with $\int_{R/2}^R |f''|^2\,dr < \infty$ implies, by Lemma 2.26, that the boundary value $f'(R)$ exists. Using this fact, we can integrate by parts in (3.81) (see Remark 2.9) and use (3.82) to deduce that, for all $\varphi \in C^\infty(0, R) \cap C[0, R]$ that vanish in a neighborhood of the origin,

$$0 = \int_0^R (f'\varphi' + \ell(\ell + d - 2)r^{-2}f\varphi - \lambda f\varphi)r^{d-1}\,dr$$

$$= \int_0^R (-f''\varphi - (d-1)r^{-1}f + \ell(\ell + d - 2)r^{-2}f\varphi - \lambda f\varphi)r^{d-1}\,dr$$

$$\qquad + f'(R)\varphi(R)R^{d-1}$$

$$= f'(R)\varphi(R)R^{d-1}.$$

This means that

$$f'(R) = 0. \tag{3.83}$$

We have

$$\lambda \int_0^R |f|^2 r^{d-1}\,dr = \int_0^R (|f'|^2 + \ell(\ell + d - 2)r^{-2}|f|^2)r^{d-1}\,dr \geq 0,$$

and thus, $\lambda \geq 0$. Moreover, $\lambda = 0$ implies that f is constant and $\ell = 0$. Conversely, we see that $\lambda = 0$ is an eigenvalue of $B_{0,m}^{\mathrm{N}}$. In the following we assume that $\lambda > 0$.

We want to transform (3.82) into Bessel's equation. The properties of f imply that $g := r^{(d-2)/2} f$ is twice weakly differentiable and that

$$r^2 g'' + rg' + (\lambda r^2 - (\ell + (d-2)/2)^2)g$$
$$= r^{(d+2)/2}\left(f'' + (d-1)r^{-1}f' - \ell(\ell+d-2)r^{-2}f + \lambda f\right) = 0 \qquad \text{in } (0,R).$$

Let $k := R\sqrt{\lambda} > 0$. Then the function $h(t) := g(Rt/k)$ is twice weakly differentiable and solves Bessel's equation

$$t^2 h'' + t h' + \left(t^2 - (\ell + (d-2)/2)^2\right)h = 0 \qquad \text{in } (0,k).$$

By Lemma 2.6, h and h' are a.e. equal to continuous functions on $(0,k)$ and therefore, by the equation, so is h''. Thus, by Lemma 2.6 applied to h', after modification on a set of measure zero, h is twice continuously differentiable and the equation for h holds in the classical sense. The facts about Bessel's equation recalled above imply that there are constants $a, b \in \mathbb{C}$ such that

$$h(t) = a\, J_{\ell + \frac{d-2}{2}}(t) + b\, Y_{\ell + \frac{d-2}{2}}(t) \qquad \text{for all } t \in (0,k).$$

Thus

$$f(r) = a\, r^{-\frac{d-2}{2}} J_{\ell + \frac{d-2}{2}}(kr/R) + b\, r^{-\frac{d-2}{2}} Y_{\ell + \frac{d-2}{2}}(kr/R) \text{ for all } (0,R).$$

We argue that $b = 0$. Using the asymptotic behavior of Y_ν recalled above, we find

$$f'(r) = \begin{cases} b\frac{2}{\pi}r^{-1}(1 + o(1)) & \text{if } d = 2,\, \ell = 0, \\ b(\ell + d - 2)\frac{\Gamma(\ell + \frac{d-2}{2})}{\pi}\left(\frac{2R}{k}\right)^{\ell + \frac{d-2}{2}} r^{-\ell - d + 1}(1 + o(1)) & \text{otherwise}. \end{cases}$$

This belongs to $L^2((0,R), r^{d-1}dr)$ only if $b = 0$, which proves the claim.

Note that the fact that $b = 0$ implies, in particular, that the eigenvalue λ is simple. Moreover, inserting the expression for f into the boundary condition (3.83), we find that $\lambda = R^{-2}q_{\ell,d,n}^2$ for some n.

Conversely, if a number λ is of this form, then the same argument shows that $r^{-\frac{d-2}{2}} J_{\ell + \frac{d-2}{2}}(kr/R)$, with $k = R\sqrt{\lambda}$, is an eigenfunction of $B_{\ell,m}^{\mathrm{N}}$, as claimed. $\qquad \square$

Combining spectral-theoretic arguments with results about Bessel functions, we obtain the following description of the spectrum of the Neumann Laplacian on a ball.

Theorem 3.56 *The spectrum of the operator* $-\Delta_{B_R}^{\mathrm{N}}$ *consists precisely of* *the simple eigenvalue* 0 *and the eigenvalues* $R^{-2}q_{\ell,d,n}^2$, $\ell \in \mathbb{N}_0$, $n \in \mathbb{N}$, *of multiplicity* ν_ℓ.

Proof Proposition 3.53 and Lemma 3.55 imply that the spectrum consists precisely of 0 (which is clearly a simple eigenvalue) and the numbers $R^{-2}q_{\ell,d,n}^2$, $\ell \in \mathbb{N}_0$, $n \in \mathbb{N}$, and that the multiplicity of an eigenvalue $\lambda > 0$ of the latter form is equal to

$$\sum_{\ell \in \mathbb{N}_0} \nu_\ell \, \#\{n \in \mathbb{N}: \lambda = R^{-2}q_{\ell,d,n}^2\}.$$

Therefore, the assertion in the theorem follows from the fact that the derivatives of $t^{-(d-2)/2}J_{\ell+(d-2)/2}(t)$ and $t^{-(d-2)/2}J_{\ell'+(d-2)/2}(t)$ do not have any common positive zeros if $\ell \neq \ell'$; see Ashu (2013); Helffer and Persson Sundqvist (2016).   □

Remark 3.57   One can show that the lowest eigenvalues are $\lambda_1(-\Delta_{B_R}^N) = 0$ and

$$\lambda_2(-\Delta_{B_R}^N) = \cdots = \lambda_{d+1}(-\Delta_{B_R}^N) = R^{-2}q_{1,d,1}^2 \qquad \lambda_{d+2}(-\Delta_{B_R}^N) = R^{-2}q_{0,d,1}^2 \, ;$$

see Bandle (1980, Lemma 3.8 and the proof of Theorem 3.23 therein)

3.8.5 The spectrum of the Dirichlet Laplacian on a ball

We now discuss the Dirichlet Laplacian on the ball B_R, following closely the previous subsection.

We begin by rephrasing Lemma 3.51 in operator-theoretic terms. For each $\ell \in \mathbb{N}_0$, $m \in \mathcal{M}_\ell$ we define the quadratic form

$$b_{\ell,m}^D[f] := \int_0^R (|f'|^2 + \ell(\ell + d - 2)r^{-2}|f|^2)r^{d-1}\,dr$$

in $\mathcal{H}_{\ell,m}$ with form domain

$$d[b_{\ell,m}^D] := \Big\{ f \in \mathcal{H}_{\ell,m} : \ f \text{ weakly differentiable on } (0,R), \ f(R) = 0,$$

$$\int_0^R (|f'|^2 + \ell(\ell + d - 2)r^{-2}|f|^2)r^{d-1}\,dr < \infty \Big\}.$$

Note that the forms $b_{\ell,m}^D$ are independent of m. These quadratic forms are non-negative and closed in $\mathcal{H}_{\ell,m}$, and therefore, as in §1.2.1, lead to a non-negative and closed quadratic form b^D in $\mathcal{H}$ defined by

$$b^D[v] := \sum_{\ell \in \mathbb{N}_0, \, m \in \mathcal{M}_\ell} b_{\ell,m}^D[v_{\ell,m}]$$

with

$$d[b^D] := \left\{ v = (v_{\ell,m}) \in \mathcal{H} : v_{\ell,m} \in d[b^D_{\ell,m}], \sum_{\ell \in \mathbb{N}_0, \, m \in \mathcal{M}_\ell} b^D_{\ell,m}[v_{\ell,m}] < \infty \right\}.$$

According to (1.42), the operators B^D and $B^D_{\ell,m}$ corresponding to the quadratic forms b^D and $b^D_{\ell,m}$, respectively, satisfy

$$B^D = \bigoplus_{\ell \in \mathbb{N}_0, \, m \in \mathcal{M}_\ell} B^D_{\ell,m}.$$

As in §3.1.1, we denote by a^D the quadratic form of the Dirichlet Laplacian $-\Delta^D_{B_R}$; that is,

$$a^D[u] = \int_{B_R} |\nabla u|^2 \, dx \qquad \text{with} \qquad d[a^D] = H^1_0(B_R).$$

Lemmas 3.50 and 3.51 imply that

$$a^D[u] = b^D[Uu] \text{ for all } u \in d[a^D] = U^* d[b^D].$$

This means that the corresponding operators satisfy

$$-\Delta^D_{B_R} = U^* B^D U.$$

This identity reduces the spectral analysis of the operator $-\Delta^D_{B_R}$ to the spectral analysis of the individual operators $B^D_{\ell,m}$.

Proposition 3.58 *The spectrum of $-\Delta^D_{B_R}$ is discrete and coincides with the union of the spectra of the operators $B^D_{\ell,m}$ for $\ell \in \mathbb{N}_0$, $m \in \mathcal{M}_\ell$. Moreover, the multiplicity of an eigenvalue of $-\Delta^D_{B_R}$ coincides with the sum of the multiplicities of the corresponding eigenvalues of the operators $B^D_{\ell,m}$.*

The proof of this proposition is similar to that of Proposition 3.53 and is based on the following lemma.

Lemma 3.59 *For each $\ell \in \mathbb{N}_0$, $m \in \mathcal{M}_\ell$, the spectrum of the operator $B^D_{\ell,m}$ is discrete. We have $\inf \sigma(B^D_{\ell,m}) > 0$ for all ℓ, m and, additionally, $\inf \sigma(B^D_{\ell,m}) \to \infty$ as $\ell \to \infty$.*

Proof The first and the last assertion follow as in the Neumann case. Since $b^D_{\ell,m}[f] \geq 0$, we have $\inf \sigma(B^D_{\ell,m}) \geq 0$. If we had equality, then a corresponding eigenfunction f would have to be constant, which, in view of the boundary condition $f(R) = 0$, leads to a contradiction. This proves the lemma. □

Similar to the Neumann case, we now characterize the spectrum in terms of Bessel functions. We denote by $j_{\nu,n}$, $n \in \mathbb{N}$, the positive zeros of J_ν in increasing order.

Lemma 3.60   *For each $\ell \in \mathbb{N}_0$, $m \in \mathcal{M}_\ell$, the spectrum of the operator $B^D_{\ell,m}$ consists precisely of the values $R^{-2}j^2_{\ell+\frac{d-2}{2},n}$, $n \in \mathbb{N}$, and each such number is an eigenvalue of multiplicity 1.*

Proof   The proof is essentially the same as that of Lemma 3.55, except that the boundary condition (3.83) is replaced by the boundary condition $f(R) = 0$. We omit the details.   □

Combining spectral-theoretic arguments with results about Bessel functions, we obtain the following description of the spectrum of the Dirichlet Laplacian on a ball.

Theorem 3.61   *The spectrum of the operator $-\Delta^D_{B_R}$ consists precisely of the eigenvalues $R^{-2}j^2_{\ell+\frac{d-2}{2},n}$, $\ell \in \mathbb{N}_0$, $n \in \mathbb{N}$, of multiplicity ν_ℓ.*

Proof   Proposition 3.58 and Lemma 3.60 imply that the spectrum consists precisely of the numbers $R^{-2}j^2_{\ell+\frac{d-2}{2},n}$, $\ell \in \mathbb{N}_0$, $n \in \mathbb{N}$, and that the multiplicity of an eigenvalue $\lambda > 0$ of this form is equal to

$$\sum_{\ell \in \mathbb{N}_0} \nu_\ell \, \#\left\{ n \in \mathbb{N} : \lambda = R^{-2}j^2_{\ell+\frac{d-2}{2},n} \right\}.$$

Therefore the assertion follows from the fact that $J_{\ell+\frac{d-2}{2}}$ and $J_{\ell'+\frac{d-2}{2}}$ do not have any common positive zeros if $\ell \neq \ell'$; see Watson (1944, §15.28).   □

Numerical values of the zeros $j_{\ell+\frac{d-2}{2},n}$ can be found in Abramowitz and Stegun (1964, Tables 9.5 and 10.6).

3.9 Comments

Section 3.1: The Dirichlet and Neumann Laplacian

Grebenkov and Nguyen (2013) is a survey of eigenvalues and eigenfunctions of the Laplacian. Classical references that predate the notions of self-adjoint operators are Rayleigh (1877) and Pockels (1891). Quadratic form techniques for the solution of boundary value problems of elliptic equations have been used both before the origin of spectral theory and before the concept of weak derivatives was introduced; see, for instance, Courant (1920, 1922, 1925).

It is an advantage of the method of quadratic forms that one does not need any information about the operator domain. Let us now, nevertheless, discuss the operator domains of $-\Delta^N_\Omega$ or $-\Delta^D_\Omega$ in some more detail. It follows from (interior) elliptic regularity theory that any function in these operator domains is twice weakly differentiable in Ω and that all partial derivatives of order 2 belong to $L^2_{\text{loc}}(\Omega)$ (Gilbarg and Trudinger, 1998, Theorem 8.8). The question of whether they even belong to $L^2(\Omega)$ depends on the regularity of the boundary of Ω. In general, the answer is negative, as shown in the example (Grisvard, 1985, Preface) below. By contrast, the partial derivatives of order 2 do belong to $L^2(\Omega)$ if the boundary is $C^{1,1}$ (Grisvard, 1985, §2.2) or if Ω is convex (Grisvard, 1985, §3.2). (For proofs under the slightly stronger C^2 assumption, see Gilbarg and Trudinger, 1998, Theorem 8.12, and Mikhaĭlov, 1978, Theorem IV.2.4.) Thus, under either of these conditions,

$$\operatorname{dom}\left(-\Delta^N_\Omega\right) = \left\{u \in H^2(\Omega): \tfrac{\partial u}{\partial \nu} = 0 \text{ on } \partial\Omega\right\} ; \quad \operatorname{dom}\left(-\Delta^D_\Omega\right) = H^1_0(\Omega) \cap H^2(\Omega).$$

(Indeed, in the Neumann case, we use Green's formula in the form of Grisvard, 1985, Lemma 1.5.3.7.) Alternatively, when the boundary of Ω is $C^{1,1}$ or when Ω is convex, then one can *define* the Dirichlet and Neumann Laplacians on these operator domains and use the above elliptic regularity theorems to show that they are self-adjoint.

Example 3.62   For $0 < \omega < 2\pi$ let $\Omega := \{(r\cos\theta, r\sin\theta): r > 0, 0 < \theta < \omega\}$ and $u(r\cos\theta, r\sin\theta) = r^{\pi/\omega}\sin\frac{\pi\theta}{\omega}$. Then a simple computation shows that u is twice weakly differentiable in Ω and harmonic there. Moreover, if $\chi \in C^\infty_0(\mathbb{R}^d)$ with $\chi = 1$ near the origin, then $\chi u \in H^1_0(\Omega)$ and $\Delta(\chi u) \in L^2(\Omega)$, so $\chi u \in \operatorname{dom}(-\Delta^D_\Omega)$. However, if $\omega > \pi$, then $\partial^2_j(\chi u) \notin L^2(\Omega)$ for $j = 1, 2$, so $\chi u \notin H^2(\Omega)$.

We also mention that there are other self-adjoint realizations of the Laplacian: for instance, corresponding to Robin boundary conditions (also known as boundary conditions of the third type), or the Krein extension; see Alonso and Simon (1980), Ashbaugh et al. (2010a,b), and references therein.

The fact that the spectrum of the Dirichlet and Neumann Laplacians can be computed explicitly on cubes and balls is well known. It can also be explicitly computed for equilateral triangles (Lamé, 1833; Bérard, 1980; Pinsky, 1980, 1985) and for ellipses (Abramowitz and Stegun, 1964, Chapter 20).

The fact that the compactness of the embedding $H^1_0(\Omega) \subset L^2(\Omega)$ implies discreteness of the spectrum of the Dirichlet Laplacian appears in Rellich (1930). In Lemma 3.4 we stated that finiteness of the measure is sufficient for the discreteness of the Dirichlet spectrum but, while convenient, this is far from necessary. For instance, the spectrum is discrete for

$$\Omega = \{(x, y) \in (0, \infty) \times (0, \infty): y < f(x)\} \text{ as soon as } \lim_{x \to \infty} f(x) = 0,$$

no matter how slow this convergence is. Necessary and sufficient conditions for the discreteness of the spectrum of the Dirichlet Laplacian in terms of capacities were given in Molčanov (1953). For further developments, see the references in §2.9.

In Lemma 3.4 we have shown that the spectrum of the Neumann Laplacian is discrete under the assumption that Ω has the extension property. Necessary and sufficient conditions for the discreteness of the spectrum of the Neumann Laplacian are in Maz'ya (1969). The fact that there are bounded, open sets in $\mathbb{R}^2$ on which the Neumann Laplacian has non-empty essential spectrum was known much earlier; see Courant and Hilbert (1962, §VII.8.2). Remarkably, it was shown in Hempel et al. (1991) that for any d and any closed set $S \subset [0, \infty)$ there is a bounded, open and connected set $\Omega \subset \mathbb{R}^d$ such that the essential spectrum of $-\Delta_\Omega^N$ coincides with S. For examples of Neumann Laplacians on domains with absolutely continuous spectrum, see Davies and Simon (1992) or Simon (1992). There is also an interesting result (Colin de Verdière, 1987) about prescribing any discrete portion of the spectrum.

We also note that, if Ω is connected, then the first eigenvalue of the Dirichlet Laplacian and the first (trivial) eigenvalue of the Neumann Laplacian are simple. In the Dirichlet case this follows from the maximum principle. It is a generalization of the Perron–Frobenius theorem for matrices with positive entries.

There is a huge literature on the connection between the shape of domains (or manifolds) and the corresponding Dirichlet and Neumann eigenvalues. One type of problems of this kind is the optimization of certain spectral quantities given certain characteristics of the domain. We briefly comment on some representative problems and results in this area and refer for a more detailed introduction to the reviews of Ashbaugh and Benguria (2007), Benguria (2011), and Benguria et al. (2012), as well as the books of Bérard (1986) and Henrot (2006, 2017).

By looking at the eigenvalues of a rectangle (or its multi-dimensional generalizations), one sees that, under a volume constraint, the first Dirichlet eigenvalue is unbounded and the first non-trivial Neumann eigenvalue can become arbitrarily close to zero. For this reason, the only interesting problems under a volume constraint are to minimize Dirichlet eigenvalues and to maximize Neumann eigenvalues.

The prototypical problem in this area is that of minimizing the first eigenvalue of the Dirichlet Laplacian among all sets of a given measure. The solution of this problem is attained when the set is a ball (Lemma 3.8), as was conjectured by Lord Rayleigh (1877) and proved by Faber (1923) and Krahn (1925, 1926). One can show that equality in the Faber–Krahn inequality is attained only for the ball. Recently, there has been work on remainder terms in the Faber–Krahn inequality involving a certain distance between Ω and the closest ball of the same measure; namely,

$$\lambda_1(-\Delta_\Omega^D) \geq \frac{j_{(d-2)/2,1}^2 \, \omega_d^{2/d}}{|\Omega|^{2/d}} \left(1 + c_d \inf_{a \in \mathbb{R}^d} \frac{|\Omega \Delta B_r(a)|^2}{|\Omega|^2} \right),$$

where Δ denotes the symmetric difference and where r is such that $\omega_d r^d = |\Omega|$; see Brasco et al. (2015), Allen et al. (2021), and references therein.

Besides minimizing the first Dirichlet eigenvalue among sets of a given measure, another classical problem is to maximize the first non-trivial Neumann eigenvalue among sets of a given measure. Again the solution is a ball, as shown in Szegö (1954), and Weinberger (1956). Results about minimizing the second Dirichlet eigenvalue are in Krahn (1926), Hong (1954), and Pólya (1955, Final remark). Results about maximizing the second non-trivial Neumann eigenvalue are in Bucur and Henrot (2019).

Lemma 3.9 concerns the problem of minimizing the Dirichlet eigenvalue under a constraint on the inradius within the class of convex, open sets. Optimality is attained for a set enclosed by two parallel hyperplanes. This result is due to Hersch (1960) in two dimensions and it was indicated by Protter (1981) how this method of proof can be extended to higher dimensions. For details and an extension of this argument, we refer to Buttazzo et al. (2018). Moreover, using a different approach based on the maximum principle, the inequality was generalized in Payne and Stakgold (1973) (see also Li and Yau, 1980) to mean-convex domains in arbitrary dimension. Our proof via Lemma 2.55 is closer to the original proofs by Hersch and Protter, but uses an approximation argument to avoid finer properties of convex sets. Without the convexity assumption, the first Dirichlet eigenvalue cannot be bounded from below in terms of the inradius. A substitute was found in Lieb (1983c); see also Maz'ya and Shubin (2005a). For the case of simply connected domains in two dimensions, see Makai (1965).

Theorem 3.10, concerning the ordering of Dirichlet and Neumann eigenvalues, is due to Friedlander (1991) with an alternative proof and extension by Filonov (2004). This theorem settles a question by Payne (1955). The first result in this direction is due to Pólya (1952), who proved the inequality in Theorem 3.10 for $k = 1$. Before Friedlander, Aviles (1986) had established the bound in Theorem 3.10 under the additional assumption that the domain has $C^{2,\varepsilon}$-boundary and non-negative mean curvature. Stronger bounds than those in Theorem 3.10 are known for restricted classes of domains. Payne (1955) proved that

$$\lambda_{k+2}(-\Delta_\Omega^N) \leq \lambda_k(-\Delta_\Omega^D) \qquad \text{for all} \quad k \in \mathbb{N}$$

for two-dimensional convex domains Ω. Levine and Weinberger (1986) extended Payne's method to the higher-dimensional case and obtained

$$\lambda_{k+r}(-\Delta_\Omega^N) \le \lambda_k(-\Delta_\Omega^D) \qquad \text{for all} \quad k \in \mathbb{N}, \quad 1 \le r \le d, \qquad (3.84)$$

if certain conditions (depending on r) on the principal curvatures of $\partial\Omega$ are satisfied. The question of whether even the Levine–Weinberger result (3.84) holds without curvature assumptions remains open so far. For an extension of Theorem 3.10 to the setting of the Heisenberg group, see Frank and Laptev (2010).

Examples of other eigenvalue optimization problems are: maximizing the quotient between the second and the first Dirichlet eigenvalue (Ashbaugh and Benguria, 1992), and minimizing the difference between the second and the first Dirichlet eigenvalue for convex sets under a diameter constraint (Andrews and Clutterbuck, 2011).

All the previously discussed problems are 'direct' problems, in the sense that one tries to bound spectral quantities in terms of geometric characteristics of the domain. Another type of problem in spectral geometry is inverse in nature, where one investigates to which extent the spectrum determines a set. In a famous paper entitled 'Can one hear the shape of a drum?', M. Kac (1966) asked whether two bounded, open sets in $\mathbb{R}^2$ for which the Dirichlet Laplacians have the same eigenvalues are necessarily congruent to each other. This was answered in the negative in Gordon et al. (1992b,a); see also Sunada (1985). For an earlier counterexample to the corresponding question on manifolds, see Milnor (1964). For positive results under additional symmetry assumptions, see Zelditch (2004, 2009).

Section 3.2: Weyl's asymptotic formula for the Dirichlet Laplacian on a domain

An early reference for the study of asymptotics of eigenvalues of Dirichlet and Neumann Laplacians is the book of Lord Rayleigh (1877), where it is noted that the leading term in the case of parallelepipeds only depends on the volume. This fact is of relevance in connection with the problem of black-body radiation (Lorentz, 1910; Sommerfeld, 1910). The first mathematically rigorous results on the eigenvalue asymptotics for the Laplacian in higher dimensions are due to Weyl (1911, 1912a,b,c, 1913).

Weyl's original proof introduces the idea of bracketing (in the form of imposing additional Dirichlet boundary conditions) and reducing the problem to the explicitly solvable one on cubes, but his analysis is centered around the Green's function. Our presentation, like most other current ones, is closer to that of Courant (1920).

The results of Weyl and Courant make use of explicit and implicit assumptions on the regularity of the boundary. The fact that Weyl asymptotics are valid for arbitrary bounded sets without any assumption on their boundary appears in Ciesielski (1970) and Birman and Solomjak (1970). This was generalized to sets of finite measure by Rozenbljum (1971, 1972b). This result can also be obtained using the techniques in Berezin (1972b).

The problem of proving Weyl asymptotics under minimal assumptions on the domain and on the coefficients of the underlying differential operator was emphasized by Birman and Solomyak. In particular, they noted the importance of uniform, order-sharp a-priori bounds. One of the inspirations was an optimal result due to Kreĭn (1951) in the one-dimensional case.

Our proof of Weyl asymptotics under a finite measure assumption (Corollary 3.15) is based on the inequality in Theorem 3.14. A closely related inequality appears in Netrusov and Safarov (2005) and improves earlier results in van den Berg and Lianantonakis (2001). As we have shown in Corollary 3.16, the inequality in Theorem 3.14 leads to the error term $O(\lambda^{(d-1)/2} \ln \lambda)$, which was first derived by Courant (1920).

Another method of proving Weyl asymptotics appears in Berezin (1972b). Like the bracketing method, it is variational and applicable in situations of limited regularity. The method is closely related to the notion of coherent states. For a textbook presentation of a proof in this spirit, see Lieb and Loss (2001, §12.11).

A different, non-variational method for obtaining spectral asymptotics goes back to Carleman (1935, 1936) and is based on parametrix constructions for functions of the Laplacian and the use of Tauberian theorems. This method has the advantage of also providing results on the eigenfunctions, but the disadvantage of (typically) requiring more regularity assumptions on the boundary (and also on the coefficients of the differential operator); see also, for instance, Gårding (1951, 1953), Agmon (1965), Agmon and Kannai (1967), Agmon (1967/68), and Beals (1967). Using heat kernels, the Weyl law was generalized to Riemannian manifolds in Minakshisundaram and Pleijel (1949). Extensions of the heat kernel and, in particular, the wave kernel method were used in Levitan (1952), Avakumović (1956), and Hörmander (1968) to obtain the remainder estimate $O(\lambda^{(d-1)/2})$ in the case of closed manifolds, which is in general best possible; see Frank and Sabin (2022). Under an additional assumption on the underlying geodesic flow Duistermaat and Guillemin (1975) obtained a remainder estimate $o(\lambda^{(d-1)/2})$; see also Bérard (1977).

On domains in Euclidean space the presence of a boundary leads to additional complications. The optimal remainder $O(\lambda^{(d-1)/2})$ for domains with smooth boundary is due to Seeley (1978, 1980); see also Metivier (1982). The question of the existence of subleading corrections to the main term was mentioned in Weyl (1913), where he also quotes number-theoretic results which give a second term and an error bound in case of the Dirichlet Laplacian on a cube. For domains in Euclidean space with smooth boundary and satisfying a certain geometric condition, Ivriĭ (1980a,b)[1] obtained asymptotics with two terms.

Let us describe in detail Ivrii's geometric condition for a bounded, open set $\Omega \subset \mathbb{R}^d$, $d \geq 2$, with smooth boundary. Let

$$S_+\overline{\Omega} := \{(x,\xi) \in \overline{\Omega} \times \mathbb{S}^{d-1} : \text{either } x \in \Omega, \text{ or } x \in \partial\Omega, \, v_x \cdot \xi < 0\},$$

where v_x denotes the outer unit normal at $x \in \partial\Omega$. For $(x,\xi) \in S_+\overline{\Omega}$ we consider the trajectory $(x + t\xi, \xi)$, which belongs to $S_+\overline{\Omega}$ for all sufficiently small $t > 0$. If it hits the boundary at a time $t_0 > 0$, we reflect it according to the standard law of geometric optics and continue this procedure to obtain the *billiard flow* $\Phi_t : S_+\overline{\Omega} \to S_+\overline{\Omega}$. It is known that the maps Φ_t are well defined almost everywhere for all $t > 0$; see Cornfeld et al. (1982, Chapter 6) and Safarov and Vassiliev (1997). A point $(x,\xi) \in S_+\overline{\Omega}$ is called *periodic* if $\Phi_T(x,\xi) = (x,\xi)$ for some $T > 0$. The *non-periodicity condition* means that the measure of the subset of periodic points in $S_+\overline{\Omega}$ is zero. The measure here is the standard measure on $\overline{\Omega} \times \mathbb{S}^{d-1}$. Ivrii's result is the following.

Theorem 3.63 *Let $\Omega \subset \mathbb{R}^d$, $d \geq 2$, be a bounded, open set with smooth boundary satisfying the non-periodicity condition. Then, as $\lambda \to \infty$,*

$$N\big(\lambda, -\Delta_\Omega^D\big) = L_{0,d}^{cl}|\Omega|\lambda^{d/2} - \tfrac{1}{4}L_{0,d-1}^{cl}|\partial\Omega|\lambda^{(d-1)/2} + o\big(\lambda^{(d-1)/2}\big)$$

and

$$N\big(\lambda, -\Delta_\Omega^N\big) = L_{0,d}^{cl}|\Omega|\lambda^{d/2} + \tfrac{1}{4}L_{0,d-1}^{cl}|\partial\Omega|\lambda^{(d-1)/2} + o\big(\lambda^{(d-1)/2}\big).$$

Here, with an abuse of notation, we write $|\partial\Omega|$ for the surface measure of $\partial\Omega$ (while $|\Omega|$ denotes the d-dimensional Lebesgue measure of Ω).

Ivrii conjectured that the non-periodicity condition is fulfilled for all bounded Euclidean domains with smooth boundaries. This has only been proved in certain special cases, in particular, for convex sets with analytic boundaries; see, for instance, Vasil'ev (1984, 1986) and Petkov and Stojanov (1988). We also refer to Melrose (1980) and Ivriĭ (1980b). The proofs of Seeley, Melrose and Ivrii use techniques from microlocal analysis: we refer to the books of Safarov and Vassiliev (1997) and Ivrii (1998, 2019a,b,c,d) for more on these topics. For improved remainder estimates for certain classes of domains we

---

[1] Here we use the transliteration of the author's name from the original papers, while elsewhere we use the current convention for the spelling of his name.

refer to, for instance, Kuznetsov and Fedosov (1967) and Colin de Verdière (1977, 2010–2011).

Two-term asymptotics for Riesz means of order $\gamma = 1$ can be obtained using techniques in the spirit of coherent states under rather weak assumptions on the boundary of the domain (Frank and Geisinger, 2011, 2016; Frank and Larson, 2020).

Motivated, in part, by a conjecture of Berry (1980), there is a large literature on the order of the remainder term in the Weyl asymptotics for domains with fractal boundaries; see, for instance, Brossard and Carmona (1986), Fleckinger-Pellé and Vassiliev (1993), Lapidus and Pomerance (1993, 1996), Levitin and Vassiliev (1996), Molchanov and Vainberg (1997, 1998), and references therein. In particular, Molchanov and Vainberg (1998) has, for any $d \geq 1$ and any $0 < \alpha < 1$, an example of an open set $\Omega \subset \mathbb{R}^d$ such that $t^{-\alpha}|\Omega(t)|$ has a positive, finite limit as $t \to 0$ and $\lambda^{-(d-\alpha)/2}\big(N\big(\lambda, -\Delta_\Omega^{\mathrm{D}}\big) - \lambda^{d/2}L_{0,d}^{\mathrm{cl}}|\Omega|\big)$ has a negative, finite limit as $\lambda \to \infty$. In particular, the error bound in (3.24) under assumption (3.22) with $0 < \alpha < 1$ is best possible.

Eigenvalue asymptotics have also been investigated in the case where the Dirichlet Laplacian has discrete spectrum, but the measure of the domain is infinite. In this case, the counting function $N\big(\lambda, -\Delta_\Omega^{\mathrm{D}}\big)$ grows faster than $\lambda^{d/2}$ as $\lambda \to \infty$ and in some situations its leading-order asymptotic behavior can be determined; see, for instance, Rozenbljum (1973), Simon (1983), and van den Berg (1984, 1992).

For more information on spectral asymptotics and further references, we refer to the books of Birman and Solomjak (1980) and Rozenblyum et al. (1989) as well as the surveys by Clark (1967) and Birman and Solomjak (1977a).

Section 3.3: Weyl's asymptotic formula for the Neumann Laplacian on a domain

Some of the references just mentioned, starting with Weyl's papers, also address the Neumann case. In contrast to the Dirichlet case, the regularity of the boundary now plays an important role. Weyl asymptotics in the Neumann case, under the assumption of the extension property (Theorem 3.20), appears, for instance, in Birman and Solomjak (1980, Theorem 4.6).

For background and references on the IMS localization formula, see Cycon et al. (1987, §3).

Netrusov and Safarov (2005) proved Weyl asymptotics in the Neumann case under the assumption that the boundary of the domain is Hölder continuous with Hölder exponent $\alpha > 1 - \frac{1}{d}$ and they provided examples of domains with Hölder continuous boundary of Hölder exponent $\alpha < 1 - \frac{1}{d}$ for which Weyl asymptotics are violated.

Another class of examples of open sets of finite measure for which Weyl's asymptotic formula fails in the Neumann case is in Jakšić et al. (1992). More precisely, for any $\beta > 1$ the authors find an open set $\Omega \subset \mathbb{R}^2$ of finite measure such that $\lim_{\lambda \to \infty} \lambda^{-\beta} N(\lambda, -\Delta_\Omega^N)$ exists and is positive and finite. Moreover, they find a set $\Omega \subset \mathbb{R}^2$ of finite measure such that $\lim_{\lambda \to \infty} \lambda^{-1} N(\lambda, -\Delta_\Omega^N)$ exists and is finite, but strictly larger than $L_{0,2}^{cl} |\Omega|$. A similar phenomenon appears in the context of two-dimensional Schrödinger operators (Birman and Laptev, 1996); see also Solomyak (1998).

Section 3.4: Pólya's inequality for tiling domains
The theorems of Pólya and Kellner mentioned in §3.4 as well as the theorem in §3.5.5 seem to give the only existing classes of domains for which Pólya's conjecture has been verified. During the copy-editing phase of this book, two preprints (Levitin et al., 2022; Filonov, 2022) concerning the case of balls and of circular sectors have appeared on the arXiv.

A modification of Pólya's proof for non-tiling domains gives (probably non-sharp) bounds involving the packing density of a domain (Urakawa, 1984).

It was shown that the analogue of Pólya's conjecture fails in the presence of a constant magnetic field (Frank et al., 2009). This shows, in particular, that an attempt to prove Pólya's conjecture using heat kernels or Green's functions is probably futile since, by the diamagnetic inequality, the same proof would also apply in the presence of a magnetic field.

Pólya's proof immediately extends to the Dirichlet realization of integer powers of the Laplacian. Note that this operator is different from the corresponding power of the Dirichlet Laplacian.

Counterexamples to analogues of Pólya's conjecture for the Dirichlet realization of the fractional Laplacian appear in Kwaśnicki et al. (2019).

Section 3.5: Lower bounds for the eigenvalues of the Dirichlet Laplacian
An abstract trace inequality for convex functions of an operator, from which Theorem 3.25 can be deduced, was proved in Berezin (1972a); see also Laptev and Safarov (1996). Related abstract inequalities for coherent states were proved independently by Berezin (1972b) and by Lieb (1973), and are called Berezin–Lieb inequalities; see also Simon (1980).

The Berezin–Li–Yau inequalities have been generalized to different operators and different geometries. Among the long list of these generalizations, we only mention Strichartz (1996), Laptev (1997), Erdős et al. (2000), Hansson and Laptev (2008), and Il'in (2009).

There has been a significant amount of activity concerning remainder terms in the Berezin–Li–Yau inequalities, going back to Melas (2003). More recent results in this direction are, for instance, Weidl (2008), Kovařík et al. (2009), Geisinger et al. (2011), Kovařík and Weidl (2015), Larson (2017), and Frank and Larson (2020). Some of these results are based on operator-valued Lieb–Thirring inequalities, which we will discuss in Chapter 6 below.

Corollary 3.30 shows that $|\Omega|^{-1}\Lambda^{-d/2}N(\Lambda,-\Delta_\Omega^D)$ is bounded by a constant depending only on d, and the corollary gives the currently best-known value for this constant. The first bound of this type, but without mention of an explicit value of the constant, appears in Rozenbljum (1972b, Theorem 1).

The maximum principle for the heat equation shows that the integral kernel of $\exp(-t(-\Delta_\Omega^D))$, which can be shown to exist, is bounded by the integral kernel of $\exp(-t(\Delta_{\mathbb{R}^d}))$, which is $(4\pi t)^{-d/2}\exp(-|x-y|^2/(4t))$. When $-\Delta_\Omega^D$ has a discrete spectrum, restricting this bound to the diagonal, one obtains

$$\sum_n e^{-t\lambda_n(-\Delta_\Omega^D)}|\psi_n(x)|^2 \leq (4\pi t)^{-d/2}, \qquad x\in\Omega, \tag{3.85}$$

where the ψ_n form an orthonormal system of eigenfunctions corresponding to $\lambda_n(-\Delta_\Omega^D)$. This implies, in particular, that for any $\Lambda, t > 0$,

$$\sum_{\lambda_n(-\Delta_\Omega^D)<\Lambda} |\psi_n(x)|^2 \leq e^{t\Lambda}\sum_n e^{-t\lambda_n(-\Delta_\Omega^D)}|\psi_n(x)|^2 \leq e^{t\Lambda}(4\pi t)^{-d/2}, \qquad x\in\Omega,$$

and, by choosing $t = d/(2\Lambda)$, we finally obtain the pointwise bound

$$\sum_{\lambda_n(-\Delta_\Omega^D)<\Lambda} |\psi_n(x)|^2 \leq (e/(2\pi d))^{d/2}\Lambda^{d/2}, \quad x\in\Omega. \tag{3.86}$$

This bound has the correct order $\Lambda^{d/2}$, but probably not the optimal constant. In some applications, however, it is useful to have a pointwise bound of the correct order.

Note that integration of the bound (3.85) gives

$$\mathrm{Tr}\,\exp(-t(-\Delta_\Omega^D)) \leq (4\pi t)^{-d/2}|\Omega|. \tag{3.87}$$

The constant in this bound is optimal. The bound can also be obtained by integrating the Berezin–Li–Yau inequality (Theorem 3.25) with respect to Λ against $e^{-t\Lambda}$. Integration of (3.86) over Ω gives

$$N(\Lambda,-\Delta_\Omega^D) \leq (e/(2\pi d))^{d/2}|\Omega|\Lambda^{d/2},$$

which is a bound of the form (3.53), but with a worse constant.

Section 3.6: Upper bounds for the eigenvalues of the Neumann Laplacian
Theorem 3.37 and its proof are taken from Laptev (1997). Conceptually similar arguments appear in Berezin (1972b), Lieb (1973), and Kröger (1992). A

generalization to the case of a constant magnetic field is in Frank (2009a). Explicit remainder terms appear in Harrell and Stubbe (2018).

Section 3.7: Phase space interpretation

When discussing the phase space interpretation, we only scratch the surface of the relation between classical and quantum mechanics. This relation is at the basis of the theory of pseudodifferential operators, microlocal and semiclassical analysis. For more on this we refer to, for instance, the textbooks of Shubin (2001), Hörmander (1990, 1983, 1994a,b), Taylor (1981), Safarov and Vassiliev (1997), Dimassi and Sjöstrand (1999), Zworski (2012), and Ivrii (1998, 2019a,b,c,d).

The multiplication property (3.72) of the classical constants $L^{\mathrm{cl}}_{\gamma,d}$ is used in Laptev (1997) and Laptev and Weidl (2000b). Its elegant proof via phase space (in fact, here only momentum space) integrals is due to Hundertmark (2002).

Section 3.8: Appendix: The Laplacian in spherical coordinates

Our discussion of Sobolev spaces on the sphere takes advantage of the embedding of $\mathbb{S}^{d-1}$ into $\mathbb{R}^d$. One can also introduce Sobolev spaces on general manifolds, and we refer to Aubin (1998), Hebey (1996), and references therein for a detailed account of this theory.

The material concerning the Laplace–Beltrami operator on $\mathbb{S}^{d-1}$ and the Dirichlet and Neumann Laplacians on a ball is folklore, except possibly the fact that there are no accidental degeneracies in the Neumann case. The latter fact seems to have been first proved by Ashu (2013) for $d = 2$ and then by Helffer and Persson Sundqvist (2016) for $d \geq 3$. The proof of the fact that the derivatives of $t^{-\frac{d-2}{2}} J_{\ell+\frac{d-2}{2}}(t)$ and $t^{-\frac{d-2}{2}} J_{\ell'+\frac{d-2}{2}}(t)$ do not have any common positive zeros if $\ell \neq \ell'$ relies on deep results about transcendental numbers, as does the proof of the corresponding fact for $J_{\ell+\frac{d-2}{2}}$, used in the Dirichlet case; see Watson (1944), Ashu (2013), and Helffer and Persson Sundqvist (2016) for references.

Lemma 3.52 is a special case of a Sobolev trace theorem. Such theorems concern the unique and continuous definition of boundary values of functions in $H^1(\Omega)$ and they hold, for instance, if Ω has uniformly Lipschitz continuous boundary; see, for instance, Leoni (2017, Theorem 18.40).

Our definition of the trace makes use of the special geometry of the ball. Let us show that it coincides with the standard definition.

Let $u \in H^1(B_R)$. Applying Lemma 3.52 to a sequence $(u_n) \subset H^1(B_R) \cap C(\overline{B_R})$ that converges to u in $H^1(B_R)$ (such a sequence is easily seen to exist), we see that $(u_n|_{\partial B_R})$ is Cauchy in $L^2(\partial B_R)$ and, therefore, has a limit. It is easy to see that this limit is independent of the chosen sequence (u_n). This limit is usually defined as the trace of u on ∂B_R.

At the same time, by a variant of Lemma 2.20, we see that, for $\omega \in \Sigma$, the functions $u_{n_k}(\cdot\omega)$ and $\omega \cdot \nabla u_{n_k}(\cdot\omega)$ converge to $u(\cdot\omega)$ and $\omega \cdot \nabla u(\cdot\omega)$ in $L^2(R/2, R)$. By Lemma 2.26, this implies that the $u_{n_k}(\cdot\omega)$ converge to $u(\cdot\omega)$ uniformly on $[R/2, R]$ and, in particular, $u_{n_k}(R\omega) \to u(R\omega)$ for every $\omega \in \Sigma$. Thus, we have shown that, along a subsequence, $(u_n|_{\partial B_R})$ converges pointwise a.e. to the function that we defined to be the trace. Since the almost-everywhere limit coincides with the L^2 limit, both notions of a trace coincide.

4

The Schrödinger Operator

Having introduced the Laplace operators on open sets in the previous chapter, we now come to the second main object of our study, namely Schrödinger operators on $\mathbb{R}^d$ in dimension $d \geq 1$. These are operators of the form

$$-\Delta - V,$$

where V stands for the multiplication operator by the real-valued function denoted by the same symbol V. While in the physics literature the notation $-\Delta + V$ is predominant, it will be more convenient for us to change the sign of V.

We will see that under suitable conditions on V, the operator $-\Delta - V$ can be defined as a self-adjoint operator in $L^2(\mathbb{R}^d)$ via a lower semibounded, closed quadratic form. We will typically assume that the potential tends to zero at infinity, at least in some generalized sense, and in this case we will see that the essential spectrum of $-\Delta - V$ will be $[0, \infty)$. Therefore, the negative spectrum of $-\Delta - V$ consists of eigenvalues of finite multiplicities that can accumulate at zero only. We will enumerate them as

$$\lambda_1(-\Delta - V) \leq \lambda_2(-\Delta - V) \leq \lambda_3(-\Delta - V) \leq \cdots,$$

repeating each eigenvalue according to its multiplicity. These eigenvalues are the focus of our interest.

The eigenvalues $\lambda_j(-\Delta - V)$ depend non-trivially on the potential V and, except for a few special cases, they cannot be computed explicitly. One goal of this book is to describe bounds on the eigenvalues $\lambda_j(-\Delta - V)$ that depend on V only through simple quantitative characteristics of V such as its L^p norm. Of particular interest are bounds with optimal constants, which is the topic of the remaining chapters.

We start our discussion with some cases where the eigenvalues can be computed explicitly. Classical examples, which we recall in §4.2, are the Coulomb

Hamiltonian $-\Delta - \kappa|x|^{-1}$ in dimension $d \geq 2$ and the Pöschl–Teller Hamiltonian $-\frac{d^2}{dx^2} - \nu(\nu + 1)\cosh^{-2} x$ in dimension $d = 1$. Another classical example with eigenvalues tending to plus infinity is the harmonic oscillator $-\Delta + \omega^2|x|^2$ in arbitrary dimension $d \geq 1$. These model cases illustrate different possible scenarios of spectra of Schrödinger operators that are of interest to us and serve as benchmarks in the quest for optimal constants. The techniques that we use to compute the spectra in these specific cases, namely commutation methods, will play an important role in the next chapter in the proof of sharp spectral estimates.

Returning to general V, in §4.3 we study qualitative aspects of the spectrum of Schrödinger operators. Among other things, we discuss the question of whether the negative spectrum of a Schrödinger operator is discrete or finite and present the Birman–Schwinger principle.

Next, we study the negative eigenvalues of the family of Schrödinger operators $-\Delta - \alpha V$ in the limit of a large coupling constant $\alpha \to \infty$. We denote by $N(0, -\Delta - \alpha V)$ the number of negative eigenvalues of $-\Delta - \alpha V$, counting multiplicities. We then prove, for continuous, compactly supported V, the Weyl asymptotics

$$\lim_{\alpha \to \infty} \alpha^{-d/2} N(0, -\Delta - \alpha V) = L_{0,d}^{\mathrm{cl}} \int_{\mathbb{R}^d} V(x)_+^{d/2}\, dx \tag{4.1}$$

as well as, for $\gamma > 0$,

$$\lim_{\alpha \to \infty} \alpha^{-\gamma - d/2} \sum_j |\lambda_j(-\Delta - \alpha V)|^\gamma = L_{\gamma,d}^{\mathrm{cl}} \int_{\mathbb{R}^d} V(x)_+^{\gamma + d/2}\, dx\,; \tag{4.2}$$

see Theorems 4.28 and 4.29. Here $V_+ = \max\{V, 0\}$. In (4.1) and (4.2), the constant

$$L_{\gamma,d}^{\mathrm{cl}} = \frac{\Gamma(\gamma + 1)}{(4\pi)^{d/2}\Gamma(\gamma + 1 + d/2)}$$

is the same as that encountered in the Weyl asymptotics for the eigenvalues of the Laplacian in a domain. The proof of Weyl asymptotics for Schrödinger operators that we present proceeds similarly to the case of the Laplacian via Dirichlet–Neumann bracketing.

The main results of this section are the Cwikel–Lieb–Rozenblum (Cwikel, 1977; Lieb, 1976, 1980; Rozenbljum, 1972a, 1976) and Lieb–Thirring (Lieb and Thirring, 1976; Weidl, 1996) inequalities, which state that, if $d \geq 3$,

$$N(0, -\Delta - V) \leq L_{0,d} \int_{\mathbb{R}^d} V(x)_+^{d/2}\, dx \tag{4.3}$$

and that, if $\gamma \geq 1/2$ in $d = 1$ and $\gamma > 0$ in $d \geq 2$,

$$\sum_j |\lambda_j(-\Delta - V)|^\gamma \leq I_{\gamma,d} \int_{\mathbb{R}^d} V(x)_+^{\gamma+d/2} \, dx. \tag{4.4}$$

The constants in these inequalities are independent of V. These inequalities appear in Theorems 4.31 and 4.38, respectively.

To appreciate the Cwikel–Lieb–Rozenblum and Lieb–Thirring bounds, we note that, when V is replaced by αV, the right sides in these inequalities have the optimal order of growth as $\alpha \to \infty$ in view of the Weyl asymptotics (4.1) and (4.2). It is an important consequence of these inequalities that (4.1) and (4.2) are not only valid for continuous, compactly supported V, but for all $V \in L^{\gamma+d/2}(\mathbb{R}^d)$, provided γ satisfies the conditions of validity of (4.3) and (4.4). Moreover, we note that the Lieb–Thirring inequality implies

$$\inf \sigma(-\Delta - V) \geq -\left(L_{\gamma,d} \int_{\mathbb{R}^d} V(x)_+^{\gamma+d/2} \, dx \right)^{1/\gamma}.$$

This inequality, with a possibly different constant, will be deduced in Remark 4.5 from the Gagliardo–Nirenberg inequality (Theorem 2.46) and will be shown in §5.1.2 to be equivalent to it. Similarly, the CLR inequality implies a criterion for absence of negative eigenvalues that is equivalent to the Sobolev inequality.

For the proof of (4.3) and (4.4) we follow the original strategy of Rozenbljum (1972a, 1976), which is a refinement of the Dirichlet–Neumann bracketing used to prove the Weyl asymptotics (4.1) and (4.2). The refinement consists in allowing the cubes to have a controlled overlap.

We emphasize that our main goal in this chapter is to prove the validity of the inequalities with *some* constants. For this it suffices to rely on variational methods. The question of *optimal* constants in these inequalities will be discussed in the subsequent chapters and its resolution will require different methods.

Convention. In this and the next chapter we always assume that V is real-valued, even if this is not explicitly mentioned.

We will also use the notation

$$\mathrm{Tr}(-\Delta - V)^\gamma = \sum_j |\lambda_j(-\Delta - V)|^\gamma.$$

4.1 Definition of the Schrödinger operator

In this section, we define the Schrödinger operator $-\Delta - V$ with a real function V as a self-adjoint operator in $L^2(\mathbb{R}^d)$. When $V \in L^\infty(\mathbb{R}^d)$, the operator of multiplication by V is bounded and the operator $-\Delta - V$ can be defined directly on the domain of $-\Delta$. We would like to allow more singular V, and therefore we use the theory of quadratic forms to define the Schrödinger operator. The following proposition spells out the abstract approach described in §1.2.7. We recall that

$$V_\pm = \max\{\pm V, 0\}$$

denote the positive and negative parts of V. For $0 \le w \in L^1_{\text{loc}}(\mathbb{R}^d)$, we set

$$H^1(\mathbb{R}^d) \cap L^2(\mathbb{R}^d, w\, dx) := \left\{ u \in H^1(\mathbb{R}^d) \colon \int_{\mathbb{R}^d} |u|^2 \, w \, dx < \infty \right\}.$$

Proposition 4.1   *Let $V \in L^1_{\text{loc}}(\mathbb{R}^d)$ and assume that there are constants $\theta < 1$ and $C < \infty$ such that*

$$\int_{\mathbb{R}^d} V_+ |u|^2 \, dx \le \theta \int_{\mathbb{R}^d} \left(|\nabla u|^2 + V_- |u|^2 \right) dx + C \int_{\mathbb{R}^d} |u|^2 \, dx$$

$$\text{for all } u \in H^1(\mathbb{R}^d) \cap L^2(\mathbb{R}^d, V_- \, dx). \tag{4.5}$$

Then the quadratic form

$$\int_{\mathbb{R}^d} \left(|\nabla u|^2 - V |u|^2 \right) dx \tag{4.6}$$

with form domain $H^1(\mathbb{R}^d) \cap L^2(\mathbb{R}^d, V_- \, dx)$ is lower semibounded and closed in $L^2(\mathbb{R}^d)$, and $C_0^\infty(\mathbb{R}^d)$ is a form core.

Before proving this proposition, let us draw some important consequences.

Corollary 4.2   *Let V be as in Proposition 4.1. Then there is a unique, self-adjoint and lower semibounded operator H in $L^2(\mathbb{R}^d)$ that satisfies $\operatorname{dom} H \subset H^1(\mathbb{R}^d) \cap L^2(\mathbb{R}^d, V_- \, dx)$ and*

$$\int_{\mathbb{R}^d} (\nabla u \cdot \nabla \overline{v} - V u \overline{v}) \, dx = (Hu, v) \quad \text{for all } u \in \operatorname{dom} H,$$

$$v \in H^1(\mathbb{R}^d) \cap L^2(\mathbb{R}^d, V_- \, dx). \tag{4.7}$$

Its domain is given by

$$\operatorname{dom} H = \big\{ u \in H^1(\mathbb{R}^d) \cap L^2(\mathbb{R}^d, V_- \, dx) \colon u \text{ has a weak Laplacian}$$

$$\text{and } -\Delta u - Vu \in L^2(\mathbb{R}^d) \big\} \tag{4.8}$$

and for $u \in \operatorname{dom} H$ we have $Hu = -\Delta u - Vu$.

For the definition of the weak Laplacian, see §2.1.6. Note that, under the assumptions of the corollary, we have $|V|^{1/2} \in L^2_{loc}(\mathbb{R}^d)$ and if, additionally, we have $u \in H^1(\mathbb{R}^d) \cap L^2(\mathbb{R}^d, V_- dx)$, then $|V|^{1/2}u \in L^2(\mathbb{R}^d)$. It follows that the product Vu belongs to $L^1_{loc}(\mathbb{R}^d)$. Thus, if u has a weak Laplacian, the sum $-\Delta u - Vu$ belongs to $L^1_{loc}(\mathbb{R}^d)$ and the requirement in (4.8) is that this function belongs to $L^2(\mathbb{R}^d)$.

Because of the last part of the corollary, in what follows we abuse notation and denote the operator H by $-\Delta - V$, remembering that this does *not* mean the operators are summed.

Proof of Corollary 4.2 Proposition 4.1, together with Theorem 1.18, implies the existence and uniqueness of the operator H claimed in the corollary. This theorem also shows that

$$\text{dom } H =$$

$$\left\{ u \in H^1(\mathbb{R}^d) \cap L^2(\mathbb{R}^d, V_- dx) \colon \text{ there exists } f \in L^2(\mathbb{R}^d) \text{ such that} \right.$$

$$\left. \text{for all } v \in H^1(\mathbb{R}^d) \cap L^2(\mathbb{R}^d, V_- dx), \int_{\mathbb{R}^d} (\nabla u \cdot \nabla \overline{v} - Vu\overline{v}) \, dx = \int_{\mathbb{R}^d} f\overline{v} \, dx \right\}.$$

Clearly, if $u \in \text{dom } H$, we can take $v \in C_0^\infty(\mathbb{R}^d)$ in (4.7) and, since $Vu \in L^1_{loc}(\mathbb{R}^d)$ as discussed before the proof, see that u has a weak Laplacian and that $-\Delta u - Vu$ coincides with the L^2 function $f = Hu$.

Conversely, let $u \in H^1(\mathbb{R}^d) \cap L^2(\mathbb{R}^d, V_- dx)$ and assume that u has a weak Laplacian with $f := -\Delta u - Vu \in L^2(\mathbb{R}^d)$. Let $v \in H^1(\mathbb{R}^d) \cap L^2(\mathbb{R}^d, V_- dx)$. By the form core property in Proposition 4.1, there is a sequence $(v_n) \subset C_0^\infty(\mathbb{R}^d)$ such that $v_n \to v$ in $H^1(\mathbb{R}^d) \cap L^2(\mathbb{R}^d, V_- dx)$. For any n, we have

$$\int_{\mathbb{R}^d} (\nabla u \cdot \nabla \overline{v_n} - Vu\overline{v_n}) \, dx = \int_{\mathbb{R}^d} u(-\Delta - V)\overline{v_n} \, dx = \int_{\mathbb{R}^d} f\overline{v_n} \, dx.$$

Since $v_n \to v$ in $H^1(\mathbb{R}^d) \cap L^2(\mathbb{R}^d, V_- dx)$, we obtain, as $n \to \infty$,

$$\int_{\mathbb{R}^d} (\nabla u \cdot \nabla \overline{v} - Vu\overline{v}) \, dx = \int_{\mathbb{R}^d} f\overline{v} \, dx.$$

Thus, $u \in \text{dom } H$. This proves the corollary. $\qquad\square$

Proof of Proposition 4.1 The quadratic form

$$\int_{\mathbb{R}^d} \left(|\nabla u|^2 + V_-|u|^2 \right) dx$$

with domain $H^1(\mathbb{R}^d) \cap L^2(\mathbb{R}^d, V_- dx)$ is non-negative and closed. This follows from the completeness of $H^1(\mathbb{R}^d)$ (Lemma 2.22) and $L^2(\mathbb{R}^d, (V_- + 1) dx)$, together with the fact that both spaces are continuously embedded into $L^2(\mathbb{R}^d)$. Moreover, $C_0^\infty(\mathbb{R}^d)$ is a form core according to Remark 2.24.

Now Lemma 1.47 implies that the form stays lower semibounded and closed, and that $C_0^\infty(\mathbb{R}^d)$ stays a form core when we add the negative potential $-V_+$ satisfying (4.5). This completes the proof of the proposition. □

The definition of the Schrödinger operator in Corollary 4.2 hinges on inequality (4.5). We will now give some sufficient conditions for this inequality to hold. These conditions are natural in light of the applications in later chapters. For different and weaker conditions we refer to the comments in §4.9.

Proposition 4.3 *Let V be a function on $\mathbb{R}^d$ such that $V_- \in L^1_{\mathrm{loc}}(\mathbb{R}^d)$ and $V_+ \in L^\infty(\mathbb{R}^d) + L^p(\mathbb{R}^d)$, where*

$$
\begin{cases}
p = 1 & \text{if } d = 1, \\
p > 1 & \text{if } d = 2, \\
p = d/2 & \text{if } d \geq 3.
\end{cases}
\tag{4.9}
$$

Then for any $\theta > 0$ there is a $C > 0$ such that

$$
\int_{\mathbb{R}^d} V_+ |u|^2 \, dx \leq \theta \int_{\mathbb{R}^d} |\nabla u|^2 \, dx + C \int_{\mathbb{R}^d} |u|^2 \, dx \qquad \text{for all } u \in H^1(\mathbb{R}^d).
$$

Moreover, the quadratic form (4.6) defines a self-adjoint, lower semibounded operator $H = -\Delta - V$ in $L^2(\mathbb{R}^d)$ with form domain $H^1(\mathbb{R}^d) \cap L^2(\mathbb{R}^d, V_- \, dx)$ and form core $C_0^\infty(\mathbb{R}^d)$.

Since $L^\infty(\mathbb{R}^d) + L^q(\mathbb{R}^d) \subset L^\infty(\mathbb{R}^d) + L^p(\mathbb{R}^d)$ for $q > p$, the equality signs in (4.9) can be replaced by $\geq$ signs.

Note that, since the constant θ in this proposition can be chosen arbitrarily small, under its assumptions the operator $-\Delta - \alpha V$ is well defined, by Corollary 4.2, for any $\alpha > 0$. We will later study certain properties of these operators as a function of α: for instance, asymptotics of its negative eigenvalues as $\alpha \to \infty$.

Proof For every $\varepsilon > 0$ we can write $V_+ = W_1 + W_2$ with $W_2 \in L^\infty(\mathbb{R}^d)$ and $\|W_1\|_p \leq \varepsilon$. Then, by Hölder's inequality,

$$
\int_{\mathbb{R}^d} V_+ |u|^2 \, dx \leq \|W_1\|_p \left(\int_{\mathbb{R}^d} |u|^{2p'} \, dx \right)^{1/p'} + \|W_2\|_\infty \int_{\mathbb{R}^d} |u|^2 \, dx,
$$

where $p' = p/(p-1)$ if $d \geq 2$ and $p' = \infty$ if $d = 1$. By the Sobolev inequality (Theorem 2.44) for $d \geq 3$ and the Gagliardo–Nirenberg inequality (Theorem 2.46) for $d = 1, 2$, we learn that

$$
\left(\int_{\mathbb{R}^d} |u|^{2p'} \, dx \right)^{1/p'} \leq S_{2p',d}^{-1} \left(\int_{\mathbb{R}^d} |\nabla u|^2 \, dx \right)^\theta \left(\int_{\mathbb{R}^d} |u|^2 \, dx \right)^{1-\theta},
$$

where θ is given by $\theta(d-2)+(1-\theta)d = d/p'$. (Note that $\theta = 1$ and $S_{2p',d} = S_d$ if $d \geq 3$.) By the elementary inequality $a^\theta b^{1-\theta} \leq \theta a + (1-\theta)b$ we obtain

$$\left(\int_{\mathbb{R}^d} |u|^{2p'} dx\right)^{1/p'} \leq \theta S_{2p',d}^{-1} \int_{\mathbb{R}^d} |\nabla u|^2 dx + (1-\theta)S_{2p',d}^{-1} \int_{\mathbb{R}^d} |u|^2 dx.$$

To summarize, we have shown that

$$\int_{\mathbb{R}^d} V_+|u|^2 dx \leq \theta S_{2p',d}^{-1}\|W_1\|_p\|\nabla u\|^2 + \left((1-\theta)S_{2p',d}^{-1}\|W_1\|_p + \|W_2\|_\infty\right)\|u\|^2.$$
(4.10)

Since $\|W_1\|_p \leq \varepsilon$, we have shown the desired inequality.

The second part of the proposition follows from Corollary 4.2. □

Remark 4.4   Assume that, in addition to the assumptions of Proposition 4.3, one has $V_- \in L^\infty(\mathbb{R}^d) + L^p(\mathbb{R}^d)$ with p as in the proposition. Then the form domain of the operator in the proposition is $H^1(\mathbb{R}^d)$. Indeed, inequality (4.10) with V_- instead of V_+ implies that $H^1(\mathbb{R}^d) \cap L^2(\mathbb{R}^d, V_- dx) = H^1(\mathbb{R}^d)$.

Remark 4.5   Assume that $V_+ \in L^p(\mathbb{R}^d)$, where $p \geq 1$ if $d = 1$ and $p > d/2$ if $d \geq 2$. Then the same argument as in the proof of Proposition 4.3, but with $W_2 = 0$, implies that, in the notation of that proof,

$$\int_{\mathbb{R}^d} V_+|u|^2 dx \leq S_{2p',d}^{-1}\|V_+\|_p \left(\int_{\mathbb{R}^d} |\nabla u|^2 dx\right)^\theta \left(\int_{\mathbb{R}^d} |u|^2 dx\right)^{1-\theta}.$$

Using again the elementary inequality $a^\theta b^{1-\theta} \leq \theta a + (1-\theta)b$, we deduce that

$$\int_{\mathbb{R}^d} V_+|u|^2 dx \leq \varepsilon \int_{\mathbb{R}^d} |\nabla u|^2 dx + C_{p,d}\varepsilon^{-\theta/(1-\theta)}\|V_+\|_p^{1/(1-\theta)} \int_{\mathbb{R}^d} |u|^2 dx$$

for any $\varepsilon > 0$ and, in particular, for $\varepsilon = 1$,

$$\int_{\mathbb{R}^d} \left(|\nabla u|^2 - V|u|^2\right) dx \geq -C_{p,d}\|V_+\|_p^{1/(1-\theta)} \int_{\mathbb{R}^d} |u|^2 dx.$$

Thus, by variational characterization of the bottom of the spectrum (Lemma 1.19),

$$\inf \sigma(-\Delta - V) \geq -C_{p,d}\|V_+\|_p^{1/(1-\theta)} = -C_{p,d}\|V_+\|_p^{2p/(2p-d)}.$$
(4.11)

Similarly, assuming that $V_+ \in L^{d/2}(\mathbb{R}^d)$ and $d \geq 3$, then the proof of Proposition 4.3 implies that

$$\inf \sigma(-\Delta - V) \geq 0 \quad \text{if} \quad C_d\|V_+\|_{d/2} \leq 1.$$
(4.12)

We will study the optimal constants in (4.11) and (4.12) in detail in §5.1.2.

4.2 Explicitly solvable examples

In this section we discuss the three standard examples of Schrödinger operators whose spectrum can be computed explicitly: namely, the harmonic oscillator $-\Delta + \omega^2|x|^2$, the Pöschl–Teller Hamiltonian $-\frac{d^2}{dx^2} - \nu(\nu+1)\cosh^{-2} x$ and the Coulomb Hamiltonian $-\Delta - \kappa|x|^{-1}$.

These examples help to illustrate the general theory in the following sections and to develop some intuition. Moreover, we will return to these examples in the context of optimal constants in spectral inequalities and will use them to produce examples of optimality as well as counterexamples. The eigenvalues of all three operators treated here will be computed using a commutation method, which we will employ later for general Schrödinger operators to prove sharp spectral estimates.

We emphasize that our interest here is only in the discrete portion of the spectra of the operators. The techniques can also be modified to deduce properties of the continuous part of the spectra of the Pöschl–Teller and Coulomb Hamiltonians.

4.2.1 Example: The harmonic oscillator

In this subsection, we consider the (general, not necessarily isotropic) harmonic oscillator

$$H := -\Delta + \sum_{k=1}^{d} \omega_k^2 x_k^2 \qquad \text{in } L^2(\mathbb{R}^d),$$

where $\omega_k > 0$, $k = 1, \ldots, d$, are constants. According to Corollary 4.2 this defines a self-adjoint operator with form domain $H^1(\mathbb{R}^d) \cap L^2(\mathbb{R}^d, |x|^2\, dx)$.

Proposition 4.6 *The spectrum of H consists precisely of the eigenvalues*

$$\sum_{k=1}^{d} (2j_k - 1)\omega_k, \qquad j = (j_1, \ldots, j_d) \in \mathbb{N}^d,$$

and the multiplicity of an eigenvalue λ coincides with the number of $j \in \mathbb{N}^d$ such that $\lambda = \sum_{k=1}^{d} (2j_k - 1)\omega_k$.

Proof We will prove the proposition in dimension $d = 1$. The higher-dimensional case follows from the one-dimensional case by separation of variables similar to, for instance, the computation of the spectrum of the Laplacian on a cube in §3.1.2. Thus, from now on $d = 1$ and $\omega := \omega_1$.

Step 1. Let us define the operator

$$Q := \frac{d}{dx} + \omega x$$

with domain $\operatorname{dom} Q := H^1(\mathbb{R}) \cap L^2(\mathbb{R}, x^2\, dx)$. We claim that for all $u \in \operatorname{dom} Q$,

$$\int_{\mathbb{R}} |Qu|^2\, dx = \int_{\mathbb{R}} \left(|u'|^2 + \omega^2 x^2 |u|^2 - \omega |u|^2 \right) dx. \qquad (4.13)$$

Since functions with bounded support are dense in $H^1(\mathbb{R}) \cap L^2(\mathbb{R}, x^2\, dx)$ and both sides are continuous in $H^1(\mathbb{R}) \cap L^2(\mathbb{R}, x^2\, dx)$, it suffices to prove this identity for $u \in \operatorname{dom} Q$ with compact support. For such u we expand pointwise almost everywhere:

$$|Qu|^2 = |u'|^2 + \omega^2 x^2 |u|^2 + 2\omega x \operatorname{Re}(\bar{u} u') = |u'|^2 + \omega^2 x^2 |u|^2 + \omega x \, (|u|^2)'.$$

Since u has compact support, we can integrate by parts in the last term and obtain (4.13).

It follows from identity (4.13) that the operator Q is closed. Indeed, since the quadratic form on the right side of (4.13) is closed by Proposition 4.1, the identity (4.13) implies that the operator Q is closed.

The key consequence of identity (4.13) and the fact that the domain of Q coincides with the form domain of H is that

$$H = Q^*Q + \omega. \qquad (4.14)$$

Here, the product Q^*Q can be understood in the sense of the product of unbounded operators, as explained at the beginning of §1.2.2.

Moreover, since the left side of (4.13) is non-negative, we conclude that

$$H \geq \omega \qquad (4.15)$$

and that $\ker(H - \omega) = \ker Q$. Next, we claim that

$$\ker(H - \omega) = \operatorname{span}\{\varphi_1\} \qquad \text{with } \varphi_1(x) = e^{-\omega x^2/2}. \qquad (4.16)$$

Indeed, if $u \in \ker(H - \omega) = \ker Q \subset H^1(\mathbb{R})$ then, by Lemma 2.6, u is continuous and so $\omega x u$ is continuous as well. Thus, if $u \in \ker Q$, the weak derivative, u', is almost everywhere equal to a continuous function and hence, by Lemma 2.6, u is almost everywhere equal to a continuously differentiable function. Therefore the equation $Qu = 0$ holds in the classical sense and can be analyzed by standard ODE methods. This yields (4.16).

Step 2. Let us now determine Q^*, the adjoint of Q. We claim that

$$Q^* = -\frac{d}{dx} + \omega x$$

with domain dom $Q^* = H^1(\mathbb{R}) \cap L^2(\mathbb{R}, x^2\, dx)$ and that, for all $u \in \text{dom } Q^*$,

$$\int_{\mathbb{R}} |Q^*u|^2\, dx = \int_{\mathbb{R}} \left(|u'|^2 + \omega^2 x^2 |u|^2 + \omega |u|^2 \right) dx. \tag{4.17}$$

In order to prove this claim, we recall that, by definition of the adjoint, we have

$$\text{dom } Q^* =$$

$$\left\{ u \in L^2(\mathbb{R}): \text{ there exists } f \in L^2(\mathbb{R}) \text{ such that,} \right.$$

$$\left. \text{for all } v \in H^1(\mathbb{R}) \cap L^2(\mathbb{R}, x^2\, dx), \int_{\mathbb{R}} u(\overline{v'} + \omega x \overline{v})\, dx = \int_{\mathbb{R}} f \overline{v}\, dx \right\}$$

and that f here is unique and given by $f = Q^*u$.

We see that

$$H^1(\mathbb{R}) \cap L^2(\mathbb{R}, x^2\, dx) \subset \text{dom } Q^*.$$

Indeed, for $u \in H^1(\mathbb{R}) \cap L^2(\mathbb{R}, x^2\, dx)$ we can set $f = -u' + \omega x u$ and verify the identity in the above formula for dom Q^*.

It requires some work, however, to prove the opposite inclusion. For the proof, let $u \in \text{dom } Q^*$ and let f be as in the above expression for dom Q^*. We find, in particular, that, for any $\overline{v} = \eta \in C_0^\infty(\mathbb{R})$,

$$\int_{\mathbb{R}} \eta' u\, dx = \int_{\mathbb{R}} \eta(-\omega x u + f)\, dx.$$

This means that u is weakly differentiable with $u' = \omega x u - f$. Since $f = Q^*u$, this gives the claimed expression for the action of Q^*. We defer the proof of $u \in H^1(\mathbb{R}) \cap L^2(\mathbb{R}, x^2\, dx)$ and of (4.17) to Lemma 4.7 below and continue with the main argument.

The key consequence of identity (4.17) is that

$$H = QQ^* - \omega. \tag{4.18}$$

Here we also used the fact that the domain of Q coincides with the form domain of H and the fact that $Q^{**} = Q$ since Q is closed.

Step 3. To proceed, we use the fact (see Proposition 1.23 and Corollary 1.24) that the non-zero spectra of Q^*Q and QQ^* coincide. Therefore, it follows from (4.14) and (4.18) that

$$\sigma(H - \omega) \setminus \{0\} = \sigma(H + \omega) \setminus \{0\}$$

and

$$\dim \ker(H - \lambda) = \dim \ker(H + 2\omega - \lambda) \quad \text{for all } \lambda \neq \omega.$$

By a simple induction argument, we obtain that, for any $N \in \mathbb{N}$,

$$\sigma(H) \setminus \{(2n - 1)\omega : n = 1, \ldots, N\}$$
$$= (\sigma(H) + 2N\omega) \setminus \{(2n - 1)\omega : n = 1, \ldots, N\} \qquad (4.19)$$

and

$$\dim \ker(H - \lambda) = \dim \ker(H + 2N\omega - \lambda)$$
$$\text{for all } \lambda \in \mathbb{R} \setminus \{(2n - 1)\omega : n = 1, \ldots, N\}. \qquad (4.20)$$

It follows from (4.19), combined with (4.15), that

$$\sigma(H) \cap (-\infty, (2N + 1)\omega) \subset \{(2n - 1)\omega : n = 1, \ldots, N\}. \qquad (4.21)$$

Moreover, applying (4.20) with $\lambda = (2N + 1)\omega$ we obtain, in combination with (4.16),

$$\dim \ker(H - (2N + 1)\omega) = 1. \qquad (4.22)$$

Relations (4.21) and (4.22), for any $N \in \mathbb{N}$, and (4.16) complete the proof. $\quad\square$

In the above proof we used the following lemma to describe the adjoint Q^*.

Lemma 4.7 *Let $u \in L^2(\mathbb{R})$ be weakly differentiable with $u' - \omega x u \in L^2(\mathbb{R})$. Then $u \in H^1(\mathbb{R}) \cap L^2(\mathbb{R}, x^2 \, dx)$ and*

$$\int_{\mathbb{R}} \left(|u'|^2 + \omega^2 x^2 |u|^2 + \omega |u|^2 \right) dx = \int_{\mathbb{R}} |u' - \omega x u|^2 \, dx . \qquad (4.23)$$

Proof We set $f := -u' + \omega x u \in L^2(\mathbb{R})$ and note that

$$\frac{d}{dx} \left(e^{-\omega x^2/2} u \right) = -e^{-\omega x^2/2} f .$$

Thus, by Lemma 2.5, there is a $c \in \mathbb{C}$ such that, for all $x \geq 0$,

$$e^{-\omega x^2/2} u(x) = \int_x^\infty e^{-\omega y^2/2} f(y) \, dy + c .$$

Note that the integral here converges for any x, since $f \in L^2(\mathbb{R})$, and tends to zero as $x \to \infty$. Since $e^{-\omega x^2/2} u \in L^2(\mathbb{R})$, we conclude that $c = 0$. We bound

$$|u(x)| \leq e^{\omega x^2/2} \int_x^\infty e^{-\omega y^2/2} |f(y)| \, dy$$

$$\leq e^{\omega x^2/2} \left(\int_x^\infty e^{-\omega y^2} \, dy \right)^{1/2} \left(\int_x^\infty |f(y)|^2 \, dy \right)^{1/2} .$$

Since, for $x > 0$,

$$\int_x^\infty e^{-\omega y^2} \, dy \leq \frac{1}{x} \int_x^\infty e^{-\omega y^2} y \, dy = \frac{1}{2x} \int_{x^2}^\infty e^{-\omega t} \, dt = \frac{1}{2x\omega} e^{-\omega x^2} ,$$

we obtain

$$|u(x)| \leq (2x\omega)^{-1/2} \left(\int_x^\infty |f(y)|^2 \, dy \right)^{1/2}.$$

Similarly, we obtain for $x < 0$,

$$|u(x)| \leq (2|x|\omega)^{-1/2} \left(\int_{-\infty}^x |f(y)|^2 \, dy \right)^{1/2}.$$

From these bounds and the fact that $f \in L^2(\mathbb{R})$, we conclude that

$$\lim_{|x| \to \infty} |x| |u(x)|^2 = 0. \tag{4.24}$$

We now expand

$$|u' - \omega x u|^2 = |u'|^2 + \omega^2 x^2 |u|^2 - 2\omega x \, \mathrm{Re}(\bar{u}u') = |u'|^2 + \omega^2 x^2 |u|^2 - \omega x \, (|u|^2)'$$

and obtain, by integration by parts,

$$\int_{-R}^R |u' - \omega x u|^2 \, dx = \int_{-R}^R \left(|u'|^2 + \omega^2 x^2 |u|^2 + \omega |u|^2 \right) dx$$
$$- \omega R (|u(R)|^2 + |u(-R)|^2).$$

Because of (4.24) the boundary terms on the right side vanish as $R \to \infty$. Moreover, since $u' - \omega x u \in L^2(\mathbb{R})$, the left side converges, and therefore the integral on the right side needs to converge as well. Since it is a sum of three positive terms, we infer that $u \in H^1(\mathbb{R}) \cap L^2(\mathbb{R}, x^2 \, dx)$ and hence we have the identity (4.23). This concludes the proof of the lemma. $\qquad \square$

Remark 4.8   We were very careful in the proof of Proposition 4.6 when specifying the domains of the operators Q and Q^*. Failure to do so can produce wrong statements. For instance, following Langmann et al. (2006), let $Q = \frac{d}{dx} + x - \frac{1}{x}$, so $Q^* = -\frac{d}{dx} + x - \frac{1}{x}$. Does

$$0 \leq Q^*Q = \left(-\frac{d}{dx} + x - \frac{1}{x} \right) \left(\frac{d}{dx} + x - \frac{1}{x} \right) = -\frac{d^2}{dx^2} + x^2 - 3$$

imply the inequality $-\frac{d^2}{dx^2} + x^2 \geq 3$ for the harmonic oscillator? (Hint: What are the domains of the operators Q, Q^* and what is the form domain of $-\frac{d^2}{dx^2} + x^2$?)

In §§3.8.4, 3.8.5 and later, in §4.2.3, we use separation of variables in spherical coordinates to compute the spectra of the Laplacian on a ball and of the Coulomb Hamiltonian. It is an instructive exercise to apply these techniques to the operator H of this subsection with $\omega_1 = \cdots = \omega_d$.

4.2.2 Example: Pöschl–Teller potentials

In this subsection, we consider the Schrödinger operator

$$H_\nu := -\frac{d^2}{dx^2} - \frac{\nu(\nu+1)}{\cosh^2 x} \quad \text{in } L^2(\mathbb{R}),$$

where $\nu > 0$ is a constant. It is easy to infer from Corollary 4.2 that this defines a self adjoint operator with form domain $H^1(\mathbb{R})$. We now compute the negative spectrum of H_ν explicitly.

Proposition 4.9 *Let $\nu > 0$. The negative spectrum of H_ν consists precisely of the simple eigenvalues $-(\nu - n + 1)^2$, where $n = 1, \ldots, \lceil \nu \rceil$.*

Here, we use the notation $\lceil \nu \rceil = N$ if $\nu = N - \theta$ with $0 \le \theta < 1$ and $N \in \mathbb{Z}$.

Proof *Step 1.* Let us define the operator

$$Q_\nu := \frac{d}{dx} + \nu \tanh x$$

with domain $\text{dom } Q_\nu := H^1(\mathbb{R})$. Clearly, $\frac{d}{dx}$ is closed on $H^1(\mathbb{R})$ and, since tanh is bounded, the operator Q_ν is closed. We claim that for all $u \in \text{dom } Q_\nu$,

$$\int_{\mathbb{R}} |Q_\nu u|^2 \, dx = \int_{\mathbb{R}} \left(|u'|^2 - \frac{\nu(\nu+1)}{\cosh^2 x} |u|^2 + \nu^2 |u|^2 \right) dx . \tag{4.25}$$

Since functions with bounded support are dense in $H^1(\mathbb{R})$ and both sides are continuous in $H^1(\mathbb{R})$, it suffices to prove this identity for $u \in \text{dom } Q_\nu$ with compact support. We expand pointwise almost everywhere:

$$\begin{aligned} |Q_\nu u|^2 &= |u'|^2 + \nu^2 \tanh^2 x |u|^2 + 2\nu \tanh x \, \text{Re}(\overline{u} u') \\ &= |u'|^2 + \nu^2 \tanh^2 x |u|^2 + \nu \tanh x \, (|u|^2)' . \end{aligned}$$

Since u has compact support, we can integrate by parts in the last term and, using $\tanh' x = \cosh^{-2} x$ and $\tanh^2 x = 1 - \cosh^{-2} x$, we obtain (4.25).

A consequence of identity (4.25) and the fact that the domain of Q_ν coincides with the form domain of H_ν is that

$$H_\nu = Q_\nu^* Q_\nu - \nu^2 . \tag{4.26}$$

It follows from (4.25) that

$$H_\nu \ge -\nu^2$$

and that $\ker(H_\nu + \nu^2) = \ker Q_\nu$. Solving the first-order equation $Q_\nu u = 0$, we find that

$$\ker(H_\nu + \nu^2) = \text{span } \{\varphi_1\} \quad \text{with } \varphi_1(x) = \cosh^{-\nu} x . \tag{4.27}$$

Here, a similar remark as in the proof of Proposition 4.6 applies concerning the passage from weak to classical derivative.

Step 2. The adjoint of Q_ν is given by

$$Q_\nu^* = -\frac{d}{dx} + \nu \tanh x$$

with domain $\operatorname{dom} Q_\nu^* = H^1(\mathbb{R})$. Moreover, a computation similar to that in (4.25) yields

$$\int_{\mathbb{R}} |Q_\nu^* u|^2 \, dx = \int_{\mathbb{R}} \left(|u'|^2 - \frac{\nu(\nu-1)}{\cosh^2 x} |u|^2 + \nu^2 |u|^2 \right) dx \,.$$

The key consequence of this identity is that

$$H_{\nu-1} = Q_\nu Q_\nu^* - \nu^2 \,. \tag{4.28}$$

Here, for $\nu \in (0, 1]$, the operator $H_{\nu-1}$ is defined in the obvious way, and we also used the facts that the domain of Q_ν^* coincides with the form domain of $H_{\nu-1}$ and that $Q_\nu^{**} = Q_\nu$, since Q_ν is closed.

Step 3. To proceed, we use the fact (see Proposition 1.23 and Corollary 1.24) that the non-zero spectra of $Q_\nu^* Q_\nu$ and $Q_\nu Q_\nu^*$ coincide. Therefore, it follows from (4.26) and (4.28) that

$$\sigma(H_\nu) \setminus \{-\nu^2\} = \sigma(H_{\nu-1}) \setminus \{-\nu^2\} \tag{4.29}$$

and

$$\dim \ker(H_\nu - \lambda) = \dim \ker(H_{\nu-1} - \lambda) \quad \text{for all } \lambda \neq -\nu^2 \,. \tag{4.30}$$

By a simple induction argument we obtain from (4.29) and (4.30), respectively, that for any integer $N < \nu + 1$,

$$\sigma(H_\nu) \setminus \left\{ -(\nu - n + 1)^2 : n = 1, \dots, N \right\}$$
$$= \sigma(H_{\nu-N}) \setminus \left\{ -(\nu - n + 1)^2 : n = 1, \dots, N \right\} \tag{4.31}$$

and

$$\dim \ker(H_\nu - \lambda) = \dim \ker(H_{\nu-N} - \lambda)$$
$$\text{for all } \lambda \in \mathbb{R} \setminus \left\{ -(\nu - n + 1)^2 : n = 1, \dots, N \right\} \,. \tag{4.32}$$

We apply (4.31) with $N = \lceil \nu \rceil$ and note that, since $-1 < \nu - N \leq 0$, we have the obvious inequality $H_{\nu-N} \geq 0$. This gives

$$\sigma(H_\nu) \cap (-\infty, 0) \subset \left\{ -(\nu - n + 1)^2 : n = 1, \dots, \lceil \nu \rceil \right\} \,.$$

Now let $1 \leq N \leq \lceil \nu \rceil - 1$. Then (4.27) applied with ν replaced by $\nu - N > 0$ gives dim ker $\left(H_{v-N} + (v - N)^2\right) = 1$. Inserting this into (4.32) with $\lambda = -(v - N)^2$ gives

$$\dim \ker \left(H_v \mid (v - N)^2\right) - 1.$$

This completes the proof of Proposition 4.9.   □

Note that for $v \in \mathbb{N}$, after v steps in the above proof, one arrives at the operator $H_0 = -\frac{d^2}{dx^2}$.

4.2.3 Example: Coulomb potential

In this subsection, we consider the Schrödinger operator

$$-\Delta - \kappa |x|^{-1} \qquad \text{in } L^2(\mathbb{R}^d),$$

where $\kappa > 0$ is a constant and where the dimension is $d \geq 2$. Clearly $|x|^{-1} \in L^\infty(\mathbb{R}^d) + L^p(\mathbb{R}^d)$ for some $p > d/2$ and therefore, by Proposition 4.3, for any $\theta > 0$, there is a $C < \infty$ such that, for all $u \in H^1(\mathbb{R}^d)$,

$$\int_{\mathbb{R}^d} |x|^{-1} |u|^2 \, dx \leq \theta \int_{\mathbb{R}^d} |\nabla u|^2 \, dx + C \int_{\mathbb{R}^d} |u|^2 \, dx. \qquad (4.33)$$

Consequently, the operator $-\Delta - \kappa |x|^{-1}$ can be defined as a self-adjoint, lower semi-bounded operator in $L^2(\mathbb{R}^d)$ by means of the quadratic form

$$\int_{\mathbb{R}^d} \left(|\nabla u|^2 - \kappa |x|^{-1} |u|^2\right) dx$$

with form domain $H^1(\mathbb{R}^d)$. The following proposition describes its negative spectrum.

Proposition 4.10   *The negative spectrum of $-\Delta - \kappa |x|^{-1}$ consists precisely of the eigenvalues $-\kappa^2(2k + d - 1)^{-2}$, where $k \in \mathbb{N}_0$. The multiplicity of the eigenvalue $-\kappa^2(2k + d - 1)^{-2}$ is*

$$\frac{(d - 2 + k)! \, (d - 1 + 2k)}{(d - 1)! \, k!}.$$

We will prove this proposition by separating variables and studying the resulting half-line operators. The separation of variables is similar to the case of the Laplacian on a ball and we only sketch the differences. As in §3.8.2, for each $\ell \in \mathbb{N}_0$, we denote by $\mathcal{M}_\ell$ an index set of multiplicity v_ℓ, given in Theorem 3.49, and we let

$$\mathcal{H}_{\ell,m} := L^2(\mathbb{R}_+, r^{d-1} \, dr) \qquad \text{for all } \ell \in \mathbb{N}_0, \, m \in \mathcal{M}_\ell.$$

and

$$\mathcal{H} := \bigoplus_{\ell \in \mathbb{N}_0, \, m \in \mathcal{M}_\ell} \mathcal{H}_{\ell,m} .$$

For $\ell \in \mathbb{N}_0$, $m \in \mathcal{M}_\ell$, in the Hilbert space $\mathcal{H}_{\ell,m}$ we consider the quadratic form

$$b_{\ell,m}[f] := \int_0^\infty \left(|f'|^2 + \frac{\ell(\ell + d - 2)}{r^2} |f|^2 - \frac{\kappa}{r} |f|^2 \right) r^{d-1} \, dr$$

with form domain

$$d[b_{\ell,m}] := \Big\{ f \in \mathcal{H}_{\ell,m} : f \text{ weakly differentiable on } \mathbb{R}_+ ,$$
$$\int_0^\infty \left(|f'|^2 + \frac{\ell(\ell + d - 2)}{r^2} |f|^2 \right) r^{d-1} dr < \infty \Big\} .$$

The analogous quadratic form for $\kappa = 0$ is non-negative and closed. Using Lemma 1.47, we see that the same is true for $b_{\ell,m}$ as a consequence of the fact that for any $\theta > 0$ there is a $C < \infty$ such that for all $f \in d[b_{\ell,m}]$,

$$\int_0^\infty |f|^2 r^{d-2} \, dr$$
$$\leq \theta \int_0^\infty \left(|f'|^2 + \frac{\ell(\ell + d - 2)}{r^2} |f|^2 \right) r^{d-1} \, dr + C \int_0^\infty |f|^2 r^{d-1} \, dr .$$

The latter fact follows from (4.33), applied to the function $u(x) = f(|x|) Y_{\ell,m}(x/|x|)$ by means of Lemma 3.50 with $R = \infty$.

Note that $b_{0,0}$ is lower semibounded and that, for all $\ell \in \mathbb{N}_0$, $m \in \mathcal{M}_\ell$, one has, under the natural identification of $\mathcal{H}_{\ell,m}$ and $\mathcal{H}_{0,0}$, the inclusion $d[b_{\ell,m}] \subset d[b_{0,0}]$ and $b_{\ell,m}[f] \geq b_{0,0}[f]$ for all $f \in d[b_{\ell,m}]$. Consequently, the quadratic forms $b_{\ell,m}$ are uniformly bounded from below. Because of this uniform lower bound, as in §1.2.1, the quadratic forms $b_{\ell,m}$ lead to a non-negative and closed quadratic form b in $\mathcal{H}$ defined by

$$b[v] := \sum_{\ell \in \mathbb{N}_0, \, m \in \mathcal{M}_\ell} b_{\ell,m}[v_{\ell,m}]$$

with form domain

$$d[b] := \Big\{ v = (v_{\ell,m}) \in \mathcal{H} : v_{\ell,m} \in d[b_{\ell,m}], \sum_{\ell \in \mathbb{N}_0, \, m \in \mathcal{M}_\ell} b_{\ell,m}[v_{\ell,m}] < \infty \Big\} .$$

Thus, according to (1.42), the operators B and $B_{\ell,m}$ corresponding to the quadratic forms b and $b_{\ell,m}$ satisfy

$$B = \bigoplus_{\ell \in \mathbb{N}_0, \, m \in \mathcal{M}_\ell} B_{\ell,m} .$$

As we did in §3.8.4, but now with $R = \infty$, let us define a unitary operator $U : L^2(\mathbb{R}^d) \to \mathcal{H}$ by

$$(Uu)_{\ell,m} := u_{\ell,m}$$

with $u_{\ell,m}$ defined by (3.75). It follows by changing to spherical coordinates and using (3.73) on each sphere that, for any $u \in H^1(\mathbb{R}^d)$,

$$\int_{\mathbb{R}^d} |x|^{-1} |u|^2 \, dx = \sum_{\ell \in \mathbb{N}_0, \, m \in \mathcal{M}_\ell} \int_0^\infty r^{-1} |u_{\ell,m}|^2 r^{d-1} \, dr$$

Therefore, as a consequence of Lemma 3.50 with $R = \infty$, we have

$$d[b] = UH^1(\mathbb{R}^d)$$

and

$$b[Uu] = \int_{\mathbb{R}^d} \left(|\nabla u|^2 - \kappa |x|^{-1} |u|^2 \right) dx \qquad \text{for all } u \in H^1(\mathbb{R}^d).$$

This means that the corresponding operators satisfy

$$-\Delta - \kappa |x|^{-1} = U^* BU. \qquad (4.34)$$

This unitary equivalence reduces the proof of Proposition 4.10 to the proof of a corresponding property of the operators $B_{\ell,m}$.

Note that for fixed ℓ, the operators $B_{\ell,m}$ with $m \in \mathcal{M}_\ell$ all coincide. The role of the parameter m is only to take the multiplicity into account. Therefore we simplify the notation in the following and write H_ℓ instead of $B_{\ell,m}$. We also remark that a simple modification of the proof of Lemma 3.55 shows that for $f \in \text{dom } H_\ell$ we have

$$H_\ell f = -f'' - \frac{d-1}{r} f' + \frac{\ell(\ell+d-2)}{r^2} f - \frac{\kappa}{r} f.$$

Next we give the analogue of Proposition 4.10 for the individual operator H_ℓ.

Proposition 4.11 *The negative spectrum of H_ℓ consists precisely of the simple eigenvalues $-\kappa^2/(2(\ell+n)+d-1)^2$, $n \in \mathbb{N}_0$.*

Before proving this proposition, let us show that it implies Proposition 4.10.

Proof of Proposition 4.10 It follows from Proposition 4.11 that the negative spectrum of each operator $B_{\ell,m}$ is discrete and that $\inf \sigma(B_{\ell,m}) \to 0$ as $\ell \to \infty$. By Lemma 1.5 this implies that the negative spectrum of the operator B is discrete. The second part of that lemma implies that the point spectrum of B consists precisely of the points

$$\bigcup_{\ell \in \mathbb{N}_0} \left\{ -\frac{\kappa^2}{(2(\ell+n)+d-1)^2} : n \in \mathbb{N}_0 \right\} = \left\{ -\frac{\kappa^2}{(2k+d-1)^2} : k \in \mathbb{N}_0 \right\}.$$

Moreover, the multiplicity of an eigenvalue $-\kappa^2/(2k + d - 1)^2$ is equal to

$$\sum_{\ell \in \mathbb{N}_0,\, m \in \mathcal{M}_\ell\, n \in \mathbb{N}_0} \chi_{\{\ell+n=k\}} = \sum_{\ell=0}^{k} \nu_\ell = \frac{(d - 2 + k)!\,(d - 1 + 2k)}{(d - 1)!\,k!} \quad : \quad (4.35)$$

this last equality can be proved by induction in k. By the unitary equivalence (4.34), we obtain the assertions about the spectrum of $-\Delta - \kappa|x|^{-1}$. $\qquad\square$

Proof of Proposition 4.11 *Step 1.* In the proof we write $L_d^2 := \mathcal{H}_{\ell,m}$ and $H_{d,\ell}^1 := d[b_{\ell,m}]$. In the Hilbert space L_d^2, let us define the operator

$$Q_\ell := \frac{d}{dr} - \frac{\ell}{r} + \frac{\kappa}{2\ell + d - 1}$$

with domain $\mathrm{dom}\, Q_\ell := H_{d,\ell}^1$. We claim that, for all $u \in \mathrm{dom}\, Q_\ell$,

$$\int_0^\infty |Q_\ell u|^2 r^{d-1}\, dr$$
$$= \int_0^\infty \left(|u'|^2 + \frac{\ell(\ell + d - 2)}{r^2}|u|^2 - \frac{\kappa}{r}|u|^2 + \frac{\kappa^2}{(2\ell + d - 1)^2}|u|^2 \right) r^{d-1}\, dr \, .$$
$$(4.36)$$

Since functions whose support is a bounded subset of $(0, \infty)$ are dense in $H_{d,\ell}^1$ with its natural norm (Lemma 2.33) and since both sides are continuous with respect to this norm, it suffices to prove identity (4.36) for $u \in \mathrm{dom}\, Q_\ell$ whose support is a bounded subset of $(0, \infty)$. For such u we expand pointwise almost everywhere:

$$|Q_\ell u|^2 = |u'|^2 + \left(\frac{\ell}{r} - \frac{\kappa}{2\ell + d - 1} \right)^2 |u|^2 - 2 \left(\frac{\ell}{r} - \frac{\kappa}{2\ell + d - 1} \right) \mathrm{Re}\,\bar{u}u'$$
$$= |u'|^2 + \left(\frac{\ell}{r} - \frac{\kappa}{2\ell + d - 1} \right)^2 |u|^2 - \left(\frac{\ell}{r} - \frac{\kappa}{2\ell + d - 1} \right) \left(|u|^2 \right)' \, .$$

Since u has compact support, integrating by parts in the last term we obtain

$$\int_0^\infty |Q_\ell u|^2\, dr$$
$$= \int_0^\infty \left(|u'|^2 + \left(\left(\frac{\ell}{r} - \frac{\kappa}{2\ell + d - 1} \right)^2 + \left(\frac{\ell(d - 2)}{r^2} - \frac{\kappa(d - 1)}{(2\ell + d - 1)r} \right) \right) |u|^2 \right) r^{d-1}\, dr$$
$$= \int_0^\infty \left(|u'|^2 + \frac{\ell(\ell + d - 2)}{r^2}|u|^2 - \frac{\kappa}{r}|u|^2 + \frac{\kappa^2}{(2\ell + d - 1)^2}|u|^2 \right) r^{d-1}\, dr \, .$$

This proves the claimed identity (4.36).

It follows from identity (4.36) that the operator Q_ℓ is closed in L_d^2. Indeed, since, as discussed above, the quadratic form $b_{\ell,m}$ is closed, the form on the right side of (4.36) is closed and the identity (4.36) then implies that the operator Q_ℓ is closed.

Another consequence of identity (4.36), together with the fact that the domain of Q_ℓ coincides with the form domain of H_ℓ, is that

$$H_\ell = Q_\ell^* Q_\ell - \frac{\kappa^2}{(2\ell + d - 1)^2}. \tag{4.37}$$

Moreover, since the left side of (4.36) is non-negative, we conclude that

$$H_\ell \geq -\frac{\kappa^2}{(2\ell + d - 1)^2} \tag{4.38}$$

and that

$$\ker\left(H_\ell + \frac{\kappa^2}{(2\ell + d - 1)^2}\right) = \ker Q_\ell.$$

Solving the first-order equation $Q_\ell u = 0$, we find that

$$\ker\left(H_\ell + \frac{\kappa^2}{(2\ell + d - 1)^2}\right) = \operatorname{span}\{\varphi_1\} \qquad \text{with } \varphi_1(r) = r^\ell e^{-\kappa r/(2\ell + d - 1)}.$$
$$\tag{4.39}$$

A remark similar to one in the proof of Proposition 4.6 concerning the passage from weak to classical derivative applies here.

Step 2. Let us now consider Q_ℓ^*, the adjoint of Q_ℓ. We claim that

$$Q_\ell^* = -\frac{d}{dr} - \frac{\ell + d - 1}{r} + \frac{\kappa}{2\ell + d - 1}$$

with domain $\operatorname{dom} Q_\ell^* = H_{d,\ell+1}^1$ and that for all $u \in \operatorname{dom} Q_\ell^*$,

$$\int_0^\infty |Q_\ell^* u|^2 r^{d-1} \, dr$$
$$= \int_0^\infty \left(|u'|^2 + \frac{(\ell + 1)(\ell + d - 1)}{r^2}|u|^2 - \frac{\kappa}{r}|u|^2 + \frac{\kappa^2}{(2\ell + d - 1)^2}|u|^2\right) r^{d-1} \, dr.$$
$$\tag{4.40}$$

We recall that, by the definition of the adjoint,

$$\operatorname{dom} Q_\ell^* = \{u \in L_d^2 : \text{there exists } f \in L_d^2 \text{ such that}$$
$$\text{for all } w \in \operatorname{dom} Q_\ell : (u, Q_\ell w) = (f, w)\}$$

and that, for u from the right side, f is unique and given by $f = Q_\ell^* u$. We see that $H_{d,\ell+1}^1 \subset \operatorname{dom} Q_\ell^*$. Indeed, for $u \in H_{d,\ell+1}^1$ we can set

$$f := -u' - ((\ell + d - 1)/r)u + (\kappa/(2\ell + d - 1))u$$

and verify the identity in the above formula for $\operatorname{dom} Q_\ell^*$.

To prove the opposite inclusion, let $u \in \operatorname{dom} Q_\ell^*$ and let f be as in the above expression for $\operatorname{dom} Q_\ell^*$. In particular, for any $\overline{w} = \eta \in C_0^\infty(\mathbb{R}_+)$, we find that

$$\int_0^\infty \eta' u \, r^{d-1} \, dr = -\int_0^\infty \eta \left(-\frac{\ell}{r} u + \frac{\kappa}{2\ell + d - 1} u - f\right) r^{d-1} \, dr \, ;$$

that is,

$$\int_0^\infty (r^{d-1}\eta)' u \, dr = -\int_0^\infty (r^{d-1}\eta) \left(-\frac{\ell + d - 1}{r} u + \frac{\kappa}{2\ell + d - 1} u - f\right) dr \, .$$

Thus u is weakly differentiable with $u' = -((\ell+d-1)/r)u + (\kappa/(2\ell+d-1))u - f$. Since $f = Q_\ell^* u$, this gives the claimed expression for the action of Q_ℓ^*. We defer the proof of the fact that $u \in H_{d,\ell+1}^1$ and of identity (4.40) to Lemma 4.12 with $\mu = \ell + d - 1$ and $\alpha = \kappa/(2\ell + d - 1)$ and continue with the main argument.

The key consequence of identity (4.40) is that

$$H_{\ell+1} = Q_\ell Q_\ell^* - \frac{\kappa^2}{(2\ell + d - 1)^2} \, . \tag{4.41}$$

Here, we also used the fact that the domain of Q_ℓ^* coincides with the form domain of $H_{\ell+1}$ and that $Q_\ell^{**} = Q_\ell$ since Q_ℓ is closed.

Step 3. To proceed, we use the fact (see Proposition 1.23 and Corollary 1.24) that the non-zero spectra of $Q_\ell^* Q_\ell$ and $Q_\ell Q_\ell^*$ coincide. Therefore it follows from (4.37) and (4.41) that

$$\sigma(H_\ell) \setminus \left\{-\frac{\kappa^2}{(2\ell + d - 1)^2}\right\} = \sigma(H_{\ell+1}) \setminus \left\{-\frac{\kappa^2}{(2\ell + d - 1)^2}\right\}$$

and

$$\dim \ker(H_\ell - \lambda) = \dim \ker(H_{\ell+1} - \lambda) \quad \text{for all } \lambda \neq -\frac{\kappa^2}{(2\ell + d - 1)^2} \, .$$

By a simple induction argument we obtain from these two identities, respectively, that, for any $N \in \mathbb{N}$,

$$\sigma(H_\ell) \setminus \left\{-\frac{\kappa^2}{(2(\ell + n) + d - 1)^2} : n = 0, \ldots, N - 1\right\}$$

$$= \sigma(H_{\ell+N}) \setminus \left\{-\frac{\kappa^2}{(2(\ell + n) + d - 1)^2} : n = 0, \ldots, N - 1\right\} \tag{4.42}$$

and

$$\dim \ker(H_\ell - \lambda) = \dim \ker(H_{\ell+N} - \lambda)$$

$$\text{for all } \lambda \in \mathbb{R} \setminus \left\{-\frac{\kappa^2}{(2(\ell + n) + d - 1)^2} : n = 0, \ldots, N - 1\right\}. \tag{4.43}$$

It follows from (4.42), combined with (4.38) applied to $H_{\ell+N}$, that

$$\sigma(H_\ell) \cap \left(-\infty, -\frac{\kappa^2}{(2(\ell+N)+d-1)^2}\right)$$
$$\subset \left\{-\frac{\kappa^2}{(2(\ell+n)+d-1)^2} : n = 0, \ldots, N-1\right\}. \tag{4.44}$$

Moreover, applying (4.43) with $\lambda = -\kappa^2/(2(\ell+N)+d-1)^2$, we obtain, in combination with (4.39),

$$\dim \ker\left(H_\ell + \frac{\kappa^2}{(2(\ell+N)+d-1)^2}\right) = 1. \tag{4.45}$$

By (4.44) and (4.45), for any $N \in \mathbb{N}$, and (4.39) the claim follows. $\qquad\square$

In the above proof we used the next lemma to describe the adjoint of the operator Q_ℓ^*. Recall that the space L_d^2 was defined in the proof of Proposition 4.11.

Lemma 4.12 *Let $\mu \geq d/2$ and let $u \in L_d^2$ be weakly differentiable with $u' + \frac{\mu}{r}u \in L_d^2$. Then $u \in H_{d,\ell+1}^1$ and, for any $\alpha \in \mathbb{R}$,*

$$\int_0^\infty \left|u' + \frac{\mu}{r}u - \alpha u\right|^2 r^{d-1}\, dr$$
$$= \int_0^\infty \left(|u'|^2 + \frac{\mu(\mu-d+2)}{r^2}|u|^2 - \frac{\alpha(2\mu-d+1)}{r}|u|^2 + \alpha^2|u|^2\right) r^{d-1}\, dr. \tag{4.46}$$

Proof We set $f := -u' - \frac{\mu}{r}u + \alpha u \in L_d^2$ and note that

$$(r^\mu u)' = r^\mu\left(u' + \frac{\mu}{r}u\right) = r^\mu(\alpha u - f).$$

Thus, by Lemma 2.5, there is a $c \in \mathbb{C}$ such that, for all $r > 0$,

$$u(r) = r^{-\mu}\int_0^r s^\mu(\alpha u(s) - f(s))\, ds + cr^{-\mu}. \tag{4.47}$$

For any $r > 0$ the integral here converges since $u, f \in L_d^2$ and, by the Cauchy–Schwarz inequality,

$$\left|\int_0^r s^\mu(\alpha u(s) - f(s))\, ds\right| \leq \frac{r^{\mu-(d-2)/2}}{\sqrt{2\mu-d+2}}\|\alpha u - f\|_{L_d^2}.$$

Note that this term, multiplied by $r^{-\mu}$, is square-integrable on $(0,1)$ with respect to the weight r^{d-1}, while $cr^{-\mu}$ is not. (This is where we use the assumption $\mu \geq$

$d/2$.) Therefore (4.47) implies that $c = 0$. It follows from Hardy's inequality in Remark 2.67 that

$$\int_0^\infty \frac{|u|^2}{r^2} r^{d-1}\, dr = \int_0^\infty r^{-2} \left| r^{-\mu+(d-1)/2} \int_0^r s^\mu\, (\alpha u(s) - f(s))\, ds \right|^2 dr$$

$$\leq \frac{1}{(\mu - (d-2)/2)^2} \int_0^\infty |\alpha u(r) - f(r)|^2\, r^{d-1}\, dr < \infty.$$

Now, the formula $u' = -\frac{\mu}{r} u + \alpha u - f$ implies that $u' \in L_d^2$ and therefore $u \in H_{d,\ell+1}^1$.

Having proved $u \in H_{d,\ell+1}^1$, the proof of equality (4.46) proceeds in the same way as that of (4.36). $\qquad\square$

4.3 Basic spectral properties of Schrödinger operators

We now return to the case of arbitrary functions V and discuss the spectrum of the Schrödinger operator $-\Delta - V$. As in §4.1, we will approach this question from a perturbative point of view, treating $-\Delta - V$ as a perturbation of $-\Delta$. The spectrum of the latter operator is essential and coincides with $[0, \infty)$. One of the results in this section is that, if V is locally sufficiently regular and tends in some sense to zero at infinity, then the essential spectrum of $-\Delta - V$ will also be $[0, \infty)$, so the negative spectrum of this operator consists of eigenvalues of finite multiplicities with zero as the only possible accumulation point. Our main focus will be on this negative spectrum, which we think of as being generated by V.

In §§4.3.1 and 4.3.2 we discuss discreteness and finiteness of the negative spectrum of $-\Delta - V$. These results are closely related to compactness theorems in the theory of Sobolev spaces. Finally, in §4.3.3 we present the Birman–Schwinger principle, which is a powerful tool for translating questions about the negative spectrum of Schrödinger operators into the language of compact operators.

4.3.1 Discreteness of the negative spectrum

In this subsection, we discuss both necessary and sufficient conditions on V for the discreteness of the negative spectrum of the operator $-\Delta - V$. We use the theory in §§1.2.7 and 1.2.8 to do this. An important role will be played by the notion of compactness of a quadratic form, as introduced before Lemma 1.49. We verify this compactness property for a class of L^p potentials using Sobolev embedding theorems and Rellich's compactness theorem.

It follows from Lemma 1.50 that, if the quadratic form $\int_{\mathbb{R}^d} V_+ |u|^2 \, dx$ is compact in $H^1(\mathbb{R}^d) \cap L^2(\mathbb{R}^d, V_- \, dx)$ equipped with the norm

$$\left(\int_{\mathbb{R}^d} (|\nabla u|^2 + |u|^2 + V_- |u|^2) \, dx \right)^{1/2}, \tag{4.48}$$

then the inequality in Proposition 4.1 holds with arbitrarily small $\theta > 0$, and therefore the Schrödinger operator $-\Delta - V$ in $L^2(\mathbb{R}^d)$ is well defined as in Corollary 4.2 with form domain $H^1(\mathbb{R}^d) \cap L^2(\mathbb{R}^d, V_- \, dx)$. Similarly, if $\int_{\mathbb{R}^d} V |u|^2 \, dx$ is compact in $H^1(\mathbb{R}^d)$, then $-\Delta - V$ in $L^2(\mathbb{R}^d)$ is well defined with form domain $H^1(\mathbb{R}^d)$.

Proposition 4.13 *Let $V \in L^1_{\mathrm{loc}}(\mathbb{R}^d)$.*

(a) *If the form $\int_{\mathbb{R}^d} V_+ |u|^2 \, dx$ is compact in $H^1(\mathbb{R}^d) \cap L^2(\mathbb{R}^d, V_- \, dx)$, then the essential spectrum of $-\Delta - V$ coincides with the essential spectrum of $-\Delta + V_-$. In particular, the negative spectrum of $-\Delta - V$ is discrete.*

(b) *If the form $\int_{\mathbb{R}^d} V |u|^2 \, dx$ is compact in $H^1(\mathbb{R}^d)$, then the essential spectrum of $-\Delta - V$ coincides with $[0, \infty)$. In particular, the negative spectrum of $-\Delta - V$ is discrete.*

This proposition follows immediately from Weyl's theorem (Theorem 1.51) and the fact that the essential spectrum of $-\Delta$ in $L^2(\mathbb{R}^d)$ equals $[0, \infty)$ (§3.1.2).

We now give some sufficient conditions for the compactness assumption in the proposition. These conditions are natural in light of the applications we have in mind in the following chapters. For different and weaker conditions, see §4.9.

Proposition 4.14 *Let $V \in L^1_{\mathrm{loc}}(\mathbb{R}^d)$.*

(a) *Suppose that, for every $\varepsilon > 0$, there are non-negative functions V_1 and V_2 on $\mathbb{R}^d$ such that $V_+ = V_1 + V_2$, $\|V_1\|_\infty \leq \varepsilon$, and $V_2 \in L^p(\mathbb{R}^d)$ with p as in (4.9). Then the form $\int_{\mathbb{R}^d} V_+ |u|^2 \, dx$ is compact in $H^1(\mathbb{R}^d)$. In particular, the essential spectrum of $-\Delta - V$ coincides with the essential spectrum of $-\Delta + V_-$ and the negative spectrum of $-\Delta - V$ is discrete.*

(b) *Suppose that, for every $\varepsilon > 0$, there are functions V_1 and V_2 on $\mathbb{R}^d$ such that $V = V_1 + V_2$, $\|V_1\|_\infty \leq \varepsilon$, and $V_2 \in L^p(\mathbb{R}^d)$ with p as in (4.9). Then the form $\int_{\mathbb{R}^d} V |u|^2 \, dx$ is compact in $H^1(\mathbb{R}^d)$. In particular, the essential spectrum of $-\Delta - V$ coincides with $[0, \infty)$ and the negative spectrum of $-\Delta - V$ is discrete.*

Note that Proposition 4.14 is applicable, in particular, to the Pöschl–Teller and Coulomb potentials discussed in §§4.2.2 and 4.2.3. In addition to the discreteness of the negative spectrum, which we have shown directly, it also shows that in these cases the essential spectrum is $[0, \infty)$.

Proof We begin with part (a). By Proposition 4.3, the form $\int_{\mathbb{R}^d} V_+ |u|^2 \, dx$ is bounded in $H^1(\mathbb{R}^d)$. We use the criterion for form compactness in Lemma 1.49. We need to show that, if $(u_n) \subset H^1(\mathbb{R}^d)$ converges weakly to zero in $H^1(\mathbb{R}^d)$, then $\int_{\mathbb{R}^d} V_+ |u_n|^2 \, dx \to 0$ as $n \to \infty$.

Let $\varepsilon > 0$ and write $V_+ = V_1 + V_2$ as in the assumption. By monotone convergence, there are $R < \infty$ and $M < \infty$ such that $W := V_2 \chi_{\{|x|<R\}} \chi_{\{V_2<M\}}$ satisfies $\|W - V_2\|_p \leq \varepsilon$. Thus, by the Hölder and Sobolev inequalities (Theorem 2.12 and Remark 2.47),

$$
\begin{aligned}
0 \leq \int_{\mathbb{R}^d} (V_2 - W) |u_n|^2 \, dx &\leq \|V_2 - W\|_p \left(\int_{\mathbb{R}^d} |u_n|^{\frac{2p}{p-1}} \, dx \right)^{1 - \frac{1}{p}} \\
&\leq \varepsilon \tilde{S}^{-1}_{\frac{2p}{p-1}, d} \|u_n\|^2_{H^1(\mathbb{R}^d)},
\end{aligned}
$$

with $\tilde{S}_{\frac{2p}{p-1}, d} = S_d$ for $d \geq 3$. In contrast, by Rellich's lemma (Proposition 2.36),

$$
0 \leq \int_{\mathbb{R}^d} W |u_n|^2 \, dx \leq M \int_{\{|x|<R\}} |u_n|^2 \, dx \to 0 .
$$

Thus

$$
\limsup_{n \to \infty} \int_{\mathbb{R}^d} V_+ |u_n|^2 \, dx \leq \varepsilon \left(1 + \tilde{S}^{-1}_{\frac{2p}{p-1}, d} \right) \sup_n \|u_n\|^2_{H^1(\mathbb{R}^d)} .
$$

Since $\sup_n \|u_n\|_{H^1(\mathbb{R}^d)} < \infty$ by uniform boundedness and since $\varepsilon > 0$ is arbitrary, we conclude that $\int V_+ |u_n|^2 \, dx \to 0$. This shows that $\int_{\mathbb{R}^d} V_+ |u|^2 \, dx$ is compact in $H^1(\mathbb{R}^d)$.

Consequently, the form is also compact in $H^1(\mathbb{R}^d) \cap L^2(\mathbb{R}^d, V_- \, dx)$, and thus the remaining assertions in part (a) of the proposition follow from Proposition 4.13.

Part (b) of the proposition follows using the same argument but applied to V_- instead. $\qquad \square$

In the following corollary we illustrate how Proposition 4.14 can be used to prove discreteness of the spectrum for Schrödinger operators. This result is applicable, for instance, to the harmonic oscillator, for which we have proved discreteness of the spectrum directly in §4.2.1.

Corollary 4.15 *Assume that $V \in L^1_{\text{loc}}(\mathbb{R}^d)$ is bounded from above and that $|\{-V < \Lambda\}| < \infty$ for any $\Lambda \in \mathbb{R}$. Then the spectrum of $-\Delta - V$ in $L^2(\mathbb{R}^d)$ is discrete.*

Proof   For any $\Lambda > 0$, $(V + \Lambda)_+$ is bounded from above and zero off a set of finite measure. Thus, by part (a) in Proposition 4.14, the negative spectrum of $-\Delta - V - \Lambda$ is discrete; that is, the spectrum of $-\Delta - V$ below Λ is discrete. Since $\Lambda > 0$ is arbitrary, this implies the corollary.   □

We end this subsection by remarking that the relative compactness assumption on V in Proposition 4.13 is, in some sense, optimal.

Lemma 4.16   *Let $0 \le V \in L^1_{loc}(\mathbb{R}^d)$ and assume that for any $\theta > 0$ there is a constant C such that (4.5) holds. If the negative spectrum of $-\Delta - \alpha V$ is discrete for any $\alpha > 0$, then the quadratic form $\int_{\mathbb{R}^d} V|u|^2 \, dx$ is compact in $H^1(\mathbb{R}^d)$.*

Proof   Under the assumptions of the lemma, Corollary 4.2 allows one to define $-\Delta - \alpha V$ as a self-adjoint, lower semibounded operator for any $\alpha > 0$. The discreteness assumption implies, in particular, that the total spectral multiplicity of $-\Delta - \alpha V$ in $(-\infty, -1)$ is finite for any $\alpha > 0$. Therefore the lemma follows from Corollary 1.53 applied to the quadratic forms $a[u] = \int_{\mathbb{R}^d}(|\nabla u|^2 + |u|^2) \, dx$ and $b[u] = \int_{\mathbb{R}^d} V|u|^2 \, dx$.   □

4.3.2 Finiteness of the negative spectrum

Having studied the discreteness of the negative spectrum of Schrödinger operators in the previous subsection, we now turn to the question of the finiteness of the negative spectrum. We pursue two different aspects of this question. First, we discuss the connection between finiteness of the negative spectrum and the compactness of a certain quadratic form and use this to give sufficient conditions in terms of L^p norms. Second, we will see that there is a transition between finitely and infinitely many negative eigenvalues depending on whether V decays faster or slower than $|x|^{-2}$ at infinity.

We begin with the first aspect. We recall that in the previous subsection, when studying the question of discreteness of the negative spectrum, we discussed compactness properties of the form $\int_{\mathbb{R}^d} V_+|u|^2 \, dx$ on the space $H^1(\mathbb{R}^d) \cap L^2(\mathbb{R}^d, V_- \, dx)$ with norm (4.48). The following proposition will relate the question of finiteness of the negative spectrum again to the compactness properties of this quadratic form, but considered on a different space: namely, where $H^1(\mathbb{R}^d)$ is replaced by its homogeneous version. To facilitate the formulation of the result we set, for $0 \le W \in L^1_{loc}(\mathbb{R}^d)$,

$$\mathcal{G}_W := \begin{cases} \dot{H}^1(\mathbb{R}^d) \cap L^2(\mathbb{R}^d, W \, dx) & \text{if } d \ge 3, \\ \tilde{H}^1(\mathbb{R}^d) \cap L^2(\mathbb{R}^d, W \, dx) & \text{if } d = 1, 2. \end{cases}$$

The spaces $\dot{H}^1(\mathbb{R}^d)$ and $\tilde{H}^1(\mathbb{R}^d)$ were introduced in §2.7.

Note that if the form $\int_{\mathbb{R}^d} V_+|u|^2\,dx$ is compact in $\mathcal{G}_{V_-}$, then it is also compact in $H^1(\mathbb{R}^d) \cap L^2(\mathbb{R}^d, V_-\,dx)$, and therefore, as before, the operator $-\Delta - V$ in $L^2(\mathbb{R}^d)$ is well-defined with form domain $H^1(\mathbb{R}^d) \cap L^2(\mathbb{R}^d, V_-\,dx)$.

Proposition 4.17 *Let $V \in L^1_{\mathrm{loc}}(\mathbb{R}^d)$. If the form $\int_{\mathbb{R}^d} V_+|u|^2\,dx$ is compact in $\mathcal{G}_{V_-}$, then the negative spectrum of $-\Delta - V$ consists of at most finitely many eigenvalues.*

We emphasize that here and in what follows the assertion that the negative spectrum consists of at most finitely many eigenvalues is understood with multiplicities.

Proof We begin with the case $d \geq 3$, where $\mathcal{G}_{V_-} = \dot{H}^1(\mathbb{R}^d) \cap L^2(\mathbb{R}^d, V_-\,dx)$. We apply the Birman–Schwinger principle from §1.2.8 with the quadratic form $a[u] := \int_{\mathbb{R}^d}(|\nabla u|^2 + V_-|u|^2)\,dx$ and $d[a] := H^1(\mathbb{R}^d) \cap L^2(\mathbb{R}^d, V_-\,dx)$, which is closed and positive in the sense of (1.63). By Lemmas 2.81 and 2.84, the completion $\mathcal{H}_a$ of $d[a]$ with respect to the form a coincides with $\mathcal{G}_{V_-}$. By the assumed compactness, the quadratic form $b[u] := \int_{\mathbb{R}^d} V_+|u|^2\,dx$ with $d[b] := H^1(\mathbb{R}^d) \cap L^2(\mathbb{R}^d, V_-\,dx)$ satisfies (1.64) and so $\hat{b}[u] = \int_{\mathbb{R}^d} V_+|u|^2\,dx$ for $u \in \mathcal{G}_{V_-}$. Since, by assumption, this form is compact, the assertion for $d \geq 3$ follows from Corollary 1.53.

Let us turn to the case $d = 1, 2$. We argue similarly, but with $a[u] := \int_{\mathbb{R}^d}(|\nabla u|^2 + \chi_B|u|^2 + V_-|u|^2)\,dx$ and $d[a] := H^1(\mathbb{R}^d) \cap L^2(\mathbb{R}^d, V_-\,dx)$, which is closed and positive. By Lemmas 2.86 and 2.87, the completion $\mathcal{H}_a$ of $d[a]$ with respect to a coincides with $\mathcal{G}_{V_-}$. By the assumed compactness, the quadratic form $b[u] := \int_{\mathbb{R}^d}(V_+ + \chi_B)|u|^2\,dx$ with $d[b] := H^1(\mathbb{R}^d) \cap L^2(\mathbb{R}^d, V_-\,dx)$ satisfies (1.64) and we have $\hat{b}[u] = \int_{\mathbb{R}^d}(V_+ + \chi_B)|u|^2\,dx$ for $u \in \mathcal{G}_{V_-}$. Once we have shown that $\hat{b}$ is compact in $\mathcal{G}_{V_-}$, we will obtain the assertion for $d = 1, 2$ again by Corollary 1.53.

Let us show that the quadratic form $\hat{b}$ is compact in $\mathcal{G}_{V_-}$. By the assumed compactness, it suffices to verify that the form $\int_B |u|^2\,dx$ is compact in $\tilde{H}^1(\mathbb{R}^d)$. If $(u_n) \subset \tilde{H}^1(\mathbb{R}^d)$ converges weakly to zero in $\tilde{H}^1(\mathbb{R}^d)$ then, by Proposition 2.89, (u_n) converges strongly to zero in $L^2(B)$. By Lemma 1.49, this implies that $\int_B |u|^2\,dx$ is compact in $\tilde{H}^1(\mathbb{R}^d)$ thus proving the claimed compactness. $\square$

We now give some sufficient conditions for the compactness assumption in the proposition. At least in dimensions $d \geq 3$, these conditions are natural in light of the applications we have in mind in the following chapters. We will discuss different conditions in the remainder of this subsection as well as in the comments in §4.9.

Proposition 4.18 *Let $V \in L^1_{\mathrm{loc}}(\mathbb{R}^d)$ with $V_+ \in L^p(\mathbb{R}^d)$ for p as in (4.9).*

If $d = 1, 2$, *assume, in addition, that* V_+ *is compactly supported. Then the quadratic form* $\int_{\mathbb{R}^d} V_+ |u|^2 \, dx$ *is compact in* $\dot{H}^1(\mathbb{R}^d)$ *if* $d \geq 3$ *and in* $\tilde{H}^1(\mathbb{R}^d)$ *if* $d = 1, 2$. *In particular, the negative spectrum of* $-\Delta - V$ *consists of at most finitely many eigenvalues.*

Proof Note that $\mathcal{G}_0 = \dot{H}^1(\mathbb{R}^d)$ if $d \geq 3$ and $\mathcal{G}_0 = \tilde{H}^1(\mathbb{R}^d)$ if $d = 1, 2$. In order to prove the relative compactness, we use Lemma 1.49. Let $(u_n) \subset \mathcal{G}_0$ be a sequence such that $u_n \to 0$ weakly in $\mathcal{G}_0$. Let $\varepsilon > 0$ and write $V_+ = V_1 + V_2$, where $V_1 \geq 0$ is bounded and has compact support contained in some ball B_R and $V_2 \geq 0$ satisfies $\|V_2\|_p \leq \varepsilon$. Since $u_n \to 0$ weakly in $\mathcal{G}_0$, Propositions 2.85 and 2.89 imply that $u_n \to 0$ strongly in $L^2(B_R)$, and therefore

$$\int_{\mathbb{R}^d} V_1 |u_n|^2 \, dx \to 0 .$$

On the other hand, by the Sobolev inequalities in Propositions 2.82 and 2.88 (noting that V_2 is compactly supported if $d = 1, 2$), and with q defined by $2q^{-1} + p^{-1} = 1$, we have

$$\int_{\mathbb{R}^d} V_2 |u_n|^2 \, dx \leq \|V_2\|_p \|u_n\|_q^2 \leq C_d \varepsilon \sup_n \|u_n\|_{\mathcal{G}_0}^2 ,$$

where the supremum is finite by uniform boundedness. Thus

$$\limsup_{n \to \infty} \int_{\mathbb{R}^d} V_+ |u_n|^2 \, dx \leq C_d \varepsilon \sup_n \|u_n\|_{\mathcal{G}_0}^2$$

and, since $\varepsilon > 0$ is arbitrary, we conclude that $\int_{\mathbb{R}^d} V_+ |u_n|^2 \, dx \to 0$. $\square$

We now turn to the second aspect of this subsection, namely, we will relate the question of finitely or infinitely many eigenvalues to the behavior of V at infinity. The intuitive picture that emerges from this analysis is that the number of negative eigenvalues is finite if V decays faster than $|x|^{-2}$ and is infinite if it decays slower than $|x|^{-2}$. For the critical decay $|x|^{-2}$ there will be a critical constant which separates one regime from the other.

Proposition 4.17 concerns a subcritical case, in the sense that, if the assumptions of that proposition are satisfied for V, then they are also satisfied for αV for any $\alpha > 0$ and, consequently, the negative spectrum of $-\Delta - \alpha V$ is finite for any $\alpha > 0$. In contrast, the assumption in the following proposition depends on the value of the coupling constant. As we shall see in Proposition 4.21 below, the values of the constants in Proposition 4.19 are optimal.

Proposition 4.19 *Let V be as in Proposition 4.3 and assume, in addition, that* *there is an $R > 0$ such that, for all $r \geq R$,*

$$\sup_{\omega \in \mathbb{S}^{d-1}} \int_r^\infty V(s\omega)_+ \, s^{-|d-2|+1} \, ds \leq \frac{|d-2|}{4 \, r^{|d-2|}} \qquad \text{if } d \neq 2,$$

$$\sup_{\omega \in \mathbb{S}^1} \int_r^\infty V(s\omega)_+ \, s \, ds \leq \frac{1}{4 \ln r} \qquad \text{if } d = 2.$$

(4.49)

Then the negative spectrum of $-\Delta - V$ consists of at most finitely many eigenvalues.

The proof is based on the Hardy inequalities on the half-line from §2.6.2, together with the technique of Neumann bracketing.

Proof By the variational principle, it suffices to prove this proposition in the case $V \geq 0$, which we assume henceforth.

First, we assume that $d \geq 3$ and introduce the quadratic forms

$$h_i[u] := \int_{B_R} \left(|\nabla u|^2 - V|u|^2 \right) dx \quad \text{and} \quad h_o[u] := \int_{\overline{B_R}^c} \left(|\nabla u|^2 - V|u|^2 \right) dx$$

in $L^2(B_R)$ and $L^2(\overline{B_R}^c)$ with form domains $H^1(B_R)$ and $H^1(\overline{B_R}^c)$, respectively. Proceeding as in the proof of Proposition 4.3, but with the Sobolev inequality from Proposition 2.62, one sees that these two forms are lower semibounded and closed. For later purposes we record that this argument also shows that the form norm $(h_i[u] + m\|u\|_{L^2(B_R)}^2)^{1/2}$ for sufficiently large m is equivalent to the $H^1(B_R)$ norm.

Let H_i and H_o denote the corresponding operators in $L^2(B_R)$ and $L^2(\overline{B_R}^c)$, respectively. For any $u \in H^1(\mathbb{R}^d)$ we have $u|_{B_R} \in H^1(B_R)$ and $u|_{\overline{B_R}^c} \in H^1(\overline{B_R}^c)$, and

$$\int_{\mathbb{R}^d} \left(|\nabla u|^2 - V|u|^2 \right) dx = h_i[u|_{B_R}] + h_o[u|_{\overline{B_R}^c}].$$

By the variational principle (Corollary 1.30), this implies that

$$N(0, -\Delta - V) \leq N(0, H_i) + N(0, H_o).$$

Let us show that $N(0, H_o) = 0$. By Lemma 2.20, if $u \in H^1(\overline{B_R}^c)$, then for almost every $\omega \in \mathbb{S}^{d-1}$ the function $r \mapsto u(r\omega)$ is weakly differentiable on (R, ∞) and

$$h_o[u] \geq \int_{\mathbb{S}^{d-1}} \int_R^\infty \left(|\partial_r u(r\omega)|^2 - V(r\omega)|u(r\omega)|^2 \right) r^{d-1} \, dr \, d\omega.$$

Moreover, by the assumption on V and inequality (2.56) with $\rho = d - 1$, we infer that

$$\int_R^\infty \left(|\partial_r u(r\omega)|^2 - V(r\omega)|u(r\omega)|^2 \right) r^{d-1} \, dr \geq 0.$$

This shows that $h_o[u] \geq 0$ for every $u \in H^1(\overline{B_R}^c)$ and, consequently, that $N(0, H_o) = 0$.

Thus, it remains to be shown that $N(0, H_i)$ is finite. In fact, we claim that H_i has discrete spectrum. By Corollary 1.21, this follows from the compactness of the embedding $H^1(B_R) \subset L^2(B_R)$ (Theorem 2.40) and the fact that the form norm of h_i is equivalent to the norm in $H^1(B_R)$. This proves the proposition for $d \geq 3$.

Now let us discuss the necessary modifications in dimensions $d = 1, 2$, which are dictated by the fact that the analogues of inequality (2.56) involve a boundary term. We introduce the quadratic forms

$$h_i[u] := \int_{B_R} \left(|\nabla u|^2 - V|u|^2 \right) dx - c_R \int_{\partial B_R} |u|^2 \, d\sigma(x) \qquad \text{and}$$

$$h_o[u] := \int_{\overline{B_R}^c} \left(|\nabla u|^2 - V|u|^2 \right) dx + c_R \int_{\partial B_R} |u|^2 \, d\sigma(x)$$

in $L^2(B_R)$ and $L^2(\overline{B_R}^c)$, respectively, with form domains $H^1(B_R)$ and $H^1(\overline{B_R}^c)$, respectively. Here $c_R := R^{-1}$ if $d = 1$ and $c_R := (\ln R)^{-1}$ if $d = 2$. We use the fact that the boundary trace of a function in $H^1(B_R)$ and $H^1(\overline{B_R}^c)$ is well defined; see Lemma 3.52.

As before, we have for any $u \in H^1(\mathbb{R}^d)$,

$$\int_{\mathbb{R}^d} \left(|\nabla u|^2 - V|u|^2 \right) dx = h_i[u|_{B_R}] + h_o[u|_{\overline{B_R}^c}],$$

and therefore

$$N(0, -\Delta - V) \leq N(0, H_i) + N(0, H_o),$$

where again H_i and H_o are the operators generated by the forms h_i and h_o.

The fact that $N(0, H_o) = 0$ can be shown in the same way as for $d \geq 3$, using inequality (2.57) with $\rho = 0$ for $d = 1$ and (2.58) for $d = 2$. Here we use the fact that the restriction of u before Lemma 3.52 is defined in a ray-wise manner.

To show that $N(0, H_i)$ is finite, we again argue as before, but use, in addition, the fact that for any $\varepsilon > 0$ there is a $C_\varepsilon < \infty$ such that for all $u \in H^1(B_R)$,

$$\int_{\partial B_R} |u|^2 \, d\sigma(x) \leq \varepsilon \int_{B_R} |\nabla u|^2 \, dx + C_\varepsilon \int_{B_R} |u|^2 \, dx \, ;$$

see Lemma 3.52. This inequality shows that the form norm of h_i is equivalent to the norm in $H^1(B_R)$. This completes the proof of the proposition. □

As an immediate consequence of the previous proposition we get the following classical result under a pointwise assumption on V.

Corollary 4.20 *Let V be as in Proposition* 4.3 *and, in addition, assume that there is an R > 0 such that for all* $|x| \geq R$,

$$V(x) \leq \frac{(d-2)^2}{4|x|^2} \qquad \text{if } d \neq 2,$$

$$V(x) \leq \frac{1}{4|x|^2 (\ln|x|)^2} \qquad \text{if } d = 2.$$

Then the negative spectrum of $-\Delta - V$ *consists of at most finitely many eigenvalues.*

Next, we show that the constant in Corollary 4.20, and therefore in Proposition 4.19 as well, is best possible.

Proposition 4.21 *Let V be as in Proposition* 4.1 *and assume, in addition, that there are R > 0 and* $\varepsilon > 0$ *such that for all* $|x| \geq R$,

$$V(x) \geq (1+\varepsilon) \frac{(d-2)^2}{4|x|^2} \qquad \text{if } d \neq 2,$$

$$V(x) \geq (1+\varepsilon) \frac{1}{4|x|^2 (\ln|x|)^2} \qquad \text{if } d = 2.$$

Then $N(0, -\Delta - V) = \infty$; *that is, the subspace corresponding to the negative spectrum of* $-\Delta - V$ *is infinite-dimensional.*

If V is such that the negative spectrum of $-\Delta - V$ is discrete, we infer that it consists of infinitely many eigenvalues. This is the case, for instance, if V fulfills the assumptions of Proposition 4.14.

Proof First, let $d \neq 2$. By the proof of the optimality statement in the Hardy inequality in Theorem 2.65 with $\rho = d - 1$, for the given $\varepsilon > 0$ there is a weakly differentiable function u on $(0, \infty)$ satisfying supp $u \subset [\rho^{-1}, \rho]$ for some $\rho > 1$ and

$$\int_0^\infty \left(|u'|^2 - \frac{(d-2)^2}{4r^2} |u|^2 \right) r^{d-1} \, dr < \varepsilon \frac{(d-2)^2}{4} \int_0^\infty |u|^2 r^{d-3} \, dr \,.$$

(Indeed, $u = r^{-(d-2)/2} \beta_\rho$ in the notation of the proof of Theorem 2.65.) For $k \in \mathbb{N}$, let

$$\psi_k(x) := u(\rho^{-2k+1} R^{-1} |x|) \,.$$

Then $\psi_k \in H^1(\mathbb{R}^d)$, supp $\psi_k \subset \{\rho^{2(k-1)} R \leq |x| \leq \rho^{2k} R\}$, and

$$\int_{\mathbb{R}^d} \left(|\nabla \psi_k|^2 - \frac{(d-2)^2}{4|x|^2} |\psi_k|^2 \right) dx < \varepsilon \frac{(d-2)^2}{4} \int_{\mathbb{R}^d} \frac{1}{|x|^2} |\psi_k|^2 \, dx \,.$$

Because of our assumption on V, this implies that, for all $k \in \mathbb{N}$,

$$\int_{\mathbb{R}^d} \left(|\nabla \psi_k|^2 - V|\psi_k|^2 \right) dx < 0.$$

Since the ψ_k are orthogonal due to their support properties, we conclude, by the variational principle (Theorem 1.25), that $N(0, -\Delta - V) = \infty$.

The proof for $d = 2$ is similar. The corresponding functions ψ_k are most easily constructed from the above functions for $d = 1$ with the help of logarithmic coordinates, as in the proof of Proposition 2.68. We omit the details. $\qquad \square$

We end this subsection by remarking that the relative compactness assumption on V in Proposition 4.17 is, in some sense, optimal.

Lemma 4.22 *Let $0 \leq V \in L^1_{\mathrm{loc}}(\mathbb{R}^d)$. Assume that, for every $\theta > 0$, there is a $C < \infty$ such that (4.5) holds, and assume too that there is a constant $M > 0$ such that, for all $u \in G_0$,*

$$\int_{\mathbb{R}^d} V|u|^2 \, dx \leq M \times \begin{cases} \int_{\mathbb{R}^d} |\nabla u|^2 \, dx & \text{if } d \geq 3, \\ \int_{\mathbb{R}^d} |\nabla u|^2 \, dx + \int_B |u|^2 \, dx & \text{if } d = 1, 2. \end{cases}$$

If the negative spectrum of $-\Delta - \alpha V$ is finite for every $\alpha > 0$, then the quadratic form $\int_{\mathbb{R}^d} V|u|^2 \, dx$ is compact in G_0.

Proof If $d \geq 3$, the proposition follows immediately from Corollary 1.53 applied to the quadratic forms $a[u] = \int_{\mathbb{R}^d} |\nabla u|^2 \, dx$ and $b[u] = \int_{\mathbb{R}^d} V|u|^2 \, dx$. If $d = 1, 2$, then, by the variational principle, the assumption implies, in particular, that the total spectral multiplicity of $-\Delta + \chi_B - \alpha V$ in $(-\infty, 0)$ is finite for any $\alpha > 0$. Therefore the proposition follows again from Corollary 1.53 applied to the quadratic forms $a[u] = \int_{\mathbb{R}^d} \left(|\nabla u|^2 + \chi_B |u|^2 \right) dx$ and $b[u] = \int_{\mathbb{R}^d} V|u|^2 \, dx$. $\qquad \square$

Remark 4.23 If V is as in Proposition 4.19, but with (4.49) replaced by

$$\lim_{r \to \infty} \sup_{\omega \in \mathbb{S}^{d-1}} r^{|d-2|} \int_r^\infty V(s\omega)_+ s^{-|d-2|+1} \, ds = 0 \qquad \text{if } d \neq 2,$$

$$\lim_{r \to \infty} \sup_{\omega \in \mathbb{S}^1} \ln r \int_r^\infty V(s\omega)_+ s \, ds = 0 \qquad \text{if } d = 2,$$

then the quadratic form $\int_{\mathbb{R}^d} V_+|u|^2 \, dx$ is compact in G_{V_-}. Indeed, according to Proposition 4.19, the assumptions imply that $N(0, -\Delta - \alpha V_+) < \infty$ for any $\alpha > 0$. Therefore, Lemma 4.22 implies that $\int V_+|u|^2 \, dx$ is compact in G_0. Since the embedding $G_{V_-} \subset G_0$ is continuous, the form is also compact in G_{V_-}.

4.3.3 The Birman–Schwinger principle for Schrödinger operators

In previous subsections we have discussed sufficient conditions for $-\Delta - V$ to have discrete or finite negative spectrum. We now complement these qualitative results by quantitative ones. We show that, for every $\tau > 0$, the number of eigenvalues of $-\Delta - V$ less than $-\tau$ can be characterized in terms of eigenvalues of a certain bounded (and typically compact) operator. This is called the *Birman–Schwinger principle* for Schrödinger operators and, for $V \geq 0$, the operator $\sqrt{V}(-\Delta + \tau)^{-1}\sqrt{V}$ is called the *Birman–Schwinger operator*.

In the statement of the following theorem we recall that $P_\omega(A)$ denotes the spectral projection of a self-adjoint operator A corresponding to a Borel set $\omega \subset \mathbb{R}$.

Theorem 4.24 *Let $V \in L^1_{\mathrm{loc}}(\mathbb{R}^d)$ and $\tau \geq 0$ and assume that there are $\theta < 1$ and $M, C < \infty$ such that*

$$\int_{\mathbb{R}^d} V_+ |u|^2 \, dx \leq \theta \int_{\mathbb{R}^d} |\nabla u|^2 \, dx + C \int_{\mathbb{R}^d} |u|^2 \, dx \qquad \text{for all } u \in H^1(\mathbb{R}^d)$$

(4.50)

and

$$\left| \int_{\mathbb{R}^d} V |u|^2 \, dx \right| \leq M \int_{\mathbb{R}^d} \left(|\nabla u|^2 + \tau |u|^2 \right) dx \qquad \text{for all } u \in H^1(\mathbb{R}^d).$$ (4.51)

Then

$$\dim \operatorname{ran} P_{(-\infty, -\tau)}(-\Delta - V) = \dim \operatorname{ran} P_{(1,\infty)}((-\Delta + \tau)^{-1/2} V (-\Delta + \tau)^{-1/2}).$$

Moreover, if $\tau > 0$, then

$$\dim \operatorname{ran} P_{(-\infty, -\tau]}(-\Delta - V) = \dim \operatorname{ran} P_{[1,\infty)}((-\Delta + \tau)^{-1/2} V (-\Delta + \tau)^{-1/2})$$

and

$$\dim \ker(-\Delta - V + \tau) = \dim \ker((-\Delta + \tau)^{-1/2} V (-\Delta + \tau)^{-1/2} - 1).$$

Condition (4.51) implies that the quadratic form $\int_{\mathbb{R}^d} V |(-\Delta + \tau)^{-1/2} f|^2 \, dx$ initially defined on $\mathrm{dom}(-\Delta + \tau)^{-1/2}$ is real-valued, densely defined and bounded in $L^2(\mathbb{R}^d)$. Therefore the form extends uniquely to all of $L^2(\mathbb{R}^d)$. The operator $(-\Delta + \tau)^{-1/2} V (-\Delta + \tau)^{-1/2}$ should be understood, as in Lemma 1.54, as the bounded, self-adjoint operator in $L^2(\mathbb{R}^d)$ corresponding to this quadratic form.

Proof We apply Theorem 1.55 with $A = -\Delta + \tau$ and $b[u] = \int_{\mathbb{R}^d} V |u|^2 \, dx$. Clearly, (1.63) holds, and (1.64) holds by the second inequality assumed in the theorem. The operator $-\Delta - V$ is defined via the corresponding lower semibounded and closed quadratic form by Proposition 4.1. Moreover, if $\tau > 0$, then (1.65) is satisfied. Therefore Theorem 1.55 is applicable and yields the assertions in the theorem.   □

Theorem 4.25   *If, in addition to the conditions in Theorem 4.24, we have* $V \geq 0$, *then*

$$\dim \operatorname{ran} P_{(-\infty,-\tau)}(-\Delta - V) = \dim \operatorname{ran} P_{(1,\infty)}\left(\sqrt{V}(-\Delta + \tau)^{-1}\sqrt{V}\right).$$

Moreover, if $\tau > 0$, *then*

$$\dim \operatorname{ran} P_{(-\infty,-\tau]}(-\Delta - V) = \dim \operatorname{ran} P_{[1,\infty)}\left(\sqrt{V}(-\Delta + \tau)^{-1}\sqrt{V}\right)$$

and

$$\dim \ker(-\Delta - V + \tau) = \dim \ker\left(\sqrt{V}(-\Delta + \tau)^{-1}\sqrt{V} - 1\right).$$

Proof   This follows as in the previous proof, using Theorem 1.56 with $Q = V^{1/2}(-\Delta + \tau)^{-1/2}$.   □

Corollary 4.26   *Let* V *be a non-negative function on* $\mathbb{R}^d$ *and assume that for every* $\varepsilon > 0$ *there are non-negative functions* V_1 *and* V_2 *on* $\mathbb{R}^d$ *such that* $V = V_1 + V_2$, $\|V_1\|_\infty \leq \varepsilon$, *and* $V_2 \in L^p(\mathbb{R}^d)$ *for p as in* (4.9). *Then for any* $\tau > 0$,

$$\dim \operatorname{ran} P_{(-\infty,-\tau)}(-\Delta - V) = \dim \operatorname{ran} P_{(1,\infty)}(\sqrt{V}(-\Delta + \tau)^{-1}\sqrt{V})$$

and

$$\dim \ker(-\Delta - V + \tau) = \dim \ker(\sqrt{V}(-\Delta + \tau)^{-1}\sqrt{V} - 1).$$

Moreover, if $d \geq 3$ *and* $V \in L^{d/2}(\mathbb{R}^d)$, *then*

$$\dim \operatorname{ran} P_{(-\infty,0)}(-\Delta - V) = \dim \operatorname{ran} P_{(1,\infty)}(\sqrt{V}(-\Delta)^{-1}\sqrt{V}).$$

Note that, according to Proposition 4.14, under the conditions of the corollary $\dim \operatorname{ran} P_{(-\infty,-\tau)}(-\Delta - V)$ is finite for any $\tau > 0$. Moreover, by Proposition 4.18, the same is true for $\tau = 0$ in dimensions $d \geq 3$ with $V \in L^{d/2}(\mathbb{R}^d)$.

Proof   We deduce the corollary from Theorem 4.25. Inequality (4.50) was verified in Proposition 4.3 and, if $\tau > 0$, then (4.51) follows immediately from (4.50). If $d \geq 3$ and $V \in L^{d/2}(\mathbb{R}^d)$, inequality (4.51) follows as in the proof of Proposition 4.3 with $W_2 = 0$.   □

The idea of the Birman–Schwinger principle is to transfer questions about the spectrum of a Schrödinger operator to questions about the spectrum of a compact operator. On the one hand, this allows one to apply technical tools from the theory of compact operators to solve specific problems and, on the other hand, it presents a new point of view and unifies different problems. The Birman–Schwinger principle in the above form and in its generalizations has far-reaching consequences to several problems in the theory of Schrödinger operators, beyond those treated here. In this book we emphasize a more direct, variational approach and we invoke the Birman–Schwinger principle explicitly only in §§5.3 and 8.2.

We conclude this subsection with a simple and classical application of the Birman–Schwinger principle to an eigenvalue bound in dimension three, known as the *Birman–Schwinger bound*.

Proposition 4.27 *Let* $d = 3$. *Then, for all* $\tau \geq 0$ *and for all measurable* $0 \leq V \in L^1_{\mathrm{loc}}(\mathbb{R}^3)$ *for which the right side is finite,*

$$N(-\tau, -\Delta - V) \leq \frac{1}{(4\pi)^2} \iint_{\mathbb{R}^3 \times \mathbb{R}^3} \frac{V(x) \, e^{-2\sqrt{\tau}|x-y|} \, V(y)}{|x - y|^2} \, dx \, dy \, .$$

In the proof we will show that, if $0 \leq V \in L^1_{\mathrm{loc}}(\mathbb{R}^3)$ is such that the right side in the proposition is finite for some $\tau > 0$, then inequalities (4.50) and (4.51) hold and consequently, as discussed after Theorem 4.24, the operator $-\Delta - V$ can be defined in the sense of quadratic forms.

Proof We first note that if the right side is finite with $\tau = 0$, then it is also finite for every $\tau > 0$ and, by monotone convergence, we obtain the claimed inequality for $\tau = 0$ as a limit of the inequalities for $\tau > 0$. Thus, from now on we assume that $\tau > 0$ and that V is such that the corresponding double integral is finite.

We define the operator

$$\tilde{Q} := (-\Delta + \tau)^{-1/2} \sqrt{V}$$

with domain $\{f \in L^2(\mathbb{R}^3): \sqrt{V} f \in L^2(\mathbb{R}^3)\}$. Since $\sqrt{V} \in L^2_{\mathrm{loc}}(\mathbb{R}^3)$, the domain of $\tilde{Q}$ contains all bounded, compactly supported functions and, in particular, is dense in $L^2(\mathbb{R}^3)$. We now use the fact that

$$\int_{\mathbb{R}^3} \frac{e^{i\xi \cdot (x-y)}}{|\xi|^2 + \tau} \frac{d\xi}{(2\pi)^3} = \frac{e^{-\sqrt{\tau}|x-y|}}{4\pi \, |x - y|}, \qquad x, y \in \mathbb{R}^3 \, ,$$

where the Fourier integral on the left side is understood as the limit of the corresponding integrals over centered balls with diverging radii. To prove this formula, one first computes the angular integral and then the remaining radial integral using the residue theorem from complex analysis.

This formula implies that, for $f, g \in \mathrm{dom} \, \tilde{Q}$, we have

$$(\tilde{Q}f, \tilde{Q}g) = (Kf, g), \tag{4.52}$$

where K is the integral operator in $L^2(\mathbb{R}^3)$ with integral kernel

$$\frac{V(x)^{1/2} e^{-\sqrt{\tau}|x-y|} V(y)^{1/2}}{4\pi|x-y|}, \qquad x, y \in \mathbb{R}^3.$$

Then, according to Lemma 1.38, we have

$$\text{Tr}\, K^* K - \frac{1}{(4\pi)^2} \iint_{\mathbb{R}^3 \times \mathbb{R}^3} \frac{V(x) e^{-2\sqrt{\tau}|x-y|} V(y)}{|x-y|^2} \, dx\, dy < \infty, \tag{4.53}$$

In particular, K is compact and so, by Lemma 1.2, the operator $\tilde{Q}$ extends uniquely to a compact operator on $L^2(\mathbb{R})$.

Let us define $Q := (\tilde{Q})^*$. We claim that for all $f \in L^2(\mathbb{R}^3)$ we have

$$V^{1/2}(-\Delta + \tau)^{-1/2} f \in L^2(\mathbb{R}^3)$$

and

$$Qf = V^{1/2}(-\Delta + \tau)^{-1/2} f.$$

Indeed, we have for any $g \in \text{dom}\, \tilde{Q}$,

$$\int_{\mathbb{R}^3} ((-\Delta + \tau)^{-1/2} f)(x) V(x)^{1/2} \overline{g(x)} \, dx = ((-\Delta + \tau)^{-1/2} f, V^{1/2} g) = (f, \tilde{Q} g)$$

$$= (Qf, g),$$

and therefore

$$\left| \int_{\mathbb{R}^3} ((-\Delta + \tau)^{-1/2} f)(x) V(x)^{1/2} \overline{g(x)} \, dx \right| \le \|Qf\| \|g\|.$$

Since Q is bounded and because this bound holds for all g from a dense set in $L^2(\mathbb{R}^d)$, the Riesz representation theorem implies that $V^{1/2}(-\Delta + \tau)^{-1/2} f \in L^2(\mathbb{R}^3)$. Given this fact, the above computation also yields the claimed formula for Qf.

Since $\tilde{Q}$ is compact, Lemma 1.2 implies that Q is compact. Moreover, the boundedness of Q implies (4.51), and its compactness, in view of Lemma 1.50, implies (4.50) for every $\theta > 0$. We are, therefore, in the situation of Theorem 4.25 and deduce that, if μ_j denote the eigenvalues of QQ^* in non-increasing order and repeated according to multiplicities, then

$$N(-\tau, -\Delta - V) = \sum_{\mu_j > 1} 1 \le \sum_j \mu_j^2 = \text{Tr}(QQ^* QQ^*).$$

Recalling that $QQ^* = K$ by (4.52) and the explicit expression for $\text{Tr}\, K^* K$ from (4.53), we obtain the inequality in the proposition. $\square$

4.4 Weyl's asymptotic formula for Schrödinger operators

Since, in general, the number of negative eigenvalues of $-\Delta - V$ cannot be computed explicitly, it is of interest to study this number in asymptotic regimes. In this section, we are interested in the strong coupling regime; that is, where we replace V by αV and consider the limit $\alpha \to \infty$. Remarkably, for sufficiently regular and fast decaying V, the leading asymptotics of $N(0, -\Delta - \alpha V)$ can be determined, namely,

$$\lim_{\alpha \to \infty} \alpha^{-d/2} N(0, -\Delta - \alpha V) = L_{0,d}^{\mathrm{cl}} \int_{\mathbb{R}^d} V(x)_+^{d/2}\, dx. \tag{4.54}$$

Here, as always, we use the notation $v_+ := \max\{v, 0\}$. We see that the asymptotics is power-like and that the coefficient depends in a relatively simple manner on V. The negative part V_- does not contribute to the leading order of the asymptotics. Note also that two disjoint pieces of V contribute additively to the asymptotics. This locality property is of importance both in the proof of the bound and in its applications.

The idea of the following proof is to replace the potential by one which is piecewise constant on cubes and, in this way, to reduce the problem to that of large eigenvalue asymptotics of the Laplace operator on a cube. The latter was already solved in §3.2. To carry out the idea of making the potential locally constant, we assume here that V is continuous and compactly supported. Later, in Theorem 4.46, we will use an approximation argument and an a priori inequality to extend the asymptotics to a larger class of potentials in dimension $d \geq 3$.

Theorem 4.28 (Weyl asymptotics)   *Let V be a continuous function on $\mathbb{R}^d$ with compact support. Then (4.54) holds.*

Proof   For $L > 0$ and $a \in \mathbb{R}^d$ we write $Q_L(a) := (0, L)^d + a \subset \mathbb{R}^d$. We will show that for each $L > 0$,

$$\limsup_{\alpha \to \infty} \alpha^{-d/2} N(0, -\Delta - \alpha V) \leq L_{0,d}^{\mathrm{cl}} L^d \sum_{a \in L\mathbb{Z}^d} \sup_{Q_L(a)} V_+^{d/2} \tag{4.55}$$

and

$$\liminf_{\alpha \to \infty} \alpha^{-d/2} N(0, -\Delta - \alpha V) \geq L_{0,d}^{\mathrm{cl}} L^d \sum_{a \in L\mathbb{Z}^d} \inf_{Q_L(a)} V_+^{d/2}. \tag{4.56}$$

Since V_+ is uniformly continuous, we have

$$\lim_{L \to 0} L^d \sum_{a \in L\mathbb{Z}^d} \sup_{Q_L(a)} V_+^{d/2} = \lim_{L \to 0} L^d \sum_{a \in L\mathbb{Z}^d} \inf_{Q_L(a)} V_+^{d/2} = \int_{\mathbb{R}^d} V_+^{d/2}\, dx;$$

so proving (4.55) and (4.56) will prove the theorem.

Thus, let $L > 0$ be fixed and, for $a \in L\mathbb{Z}^d$, set $V^{(\max)}(a) := \sup_{Q_L(a)} V$ and $V^{(\min)}(a) := \inf_{Q_L(a)} V$. We identify the sequences $V^{(\max)}$ and $V^{(\min)}$ with the piecewise constant functions that take the values $V^{(\max)}(a)$ and $V^{(\min)}(a)$ on $Q_L(a)$. The proof proceeds by Dirichlet–Neumann bracketing, which leads to

$$\bigoplus_{a \in L\mathbb{Z}^d} \left(-\Delta^{\mathrm{N}}_{Q_L(a)} - \alpha V^{(\max)}(a)\right) \leq -\Delta - \alpha V \leq \bigoplus_{a \in L\mathbb{Z}^d} \left(-\Delta^{\mathrm{D}}_{Q_L(a)} - \alpha V^{(\min)}(a)\right).$$

$$(4.57)$$

Therefore, by the variational principle,

$$\sum_{a \in L\mathbb{Z}^d} N(0, -\Delta^{\mathrm{D}}_{Q_L(a)} - \alpha V^{(\min)}(a))$$

$$= N\left(0, \bigoplus_{a \in L\mathbb{Z}^d} \left(-\Delta^{\mathrm{D}}_{Q_L(a)} - \alpha V^{(\min)}(a)\right)\right) \leq N(0, -\Delta - \alpha V)$$

$$\leq N\left(0, \bigoplus_{a \in L\mathbb{Z}^d} \left(-\Delta^{\mathrm{N}}_{Q_L(a)} - \alpha V^{(\max)}(a)\right)\right)$$

$$= \sum_{a \in L\mathbb{Z}^d} N(0, -\Delta^{\mathrm{N}}_{Q_L(a)} - \alpha V^{(\max)}(a)).$$

$$(4.58)$$

We emphasize that both sums contain only finitely many terms since V has compact support. It follows from Lemma 3.12 that

$$\lim_{\alpha \to \infty} \alpha^{-d/2} N\left(0, -\Delta^{\mathrm{N}}_{Q_L(a)} - \alpha V^{(\max)}(a)\right) = L^{\mathrm{cl}}_{0,d} L^d \left(V^{(\max)}(a)\right)^{d/2}_+$$

and

$$\lim_{\alpha \to \infty} \alpha^{-d/2} N\left(0, -\Delta^{\mathrm{D}}_{Q_L(a)} - \alpha V^{(\min)}(a)\right) = L^{\mathrm{cl}}_{0,d} L^d \left(V^{(\min)}(a)\right)^{d/2}_+,$$

and consequently, by (4.58),

$$\limsup_{\alpha \to \infty} \alpha^{-d/2} N\left(0, -\Delta - \alpha V\right) \leq L^{\mathrm{cl}}_{0,d} L^d \sum_{a \in L\mathbb{Z}^d} \left(V^{(\max)}(a)\right)^{d/2}_+$$

and

$$\liminf_{\alpha \to \infty} \alpha^{-d/2} N\left(0, -\Delta - \alpha V\right) \geq L^{\mathrm{cl}}_{0,d} L^d \sum_{a \in L\mathbb{Z}^d} \left(V^{(\min)}(a)\right)^{d/2}_+.$$

Since

$$\left(V^{(\max)}(a)\right)_+ = \sup_{Q_L(a)} V_+ \quad \text{and} \quad \left(V^{(\min)}(a)\right)_+ = \inf_{Q_L(a)} V_+,$$

we obtain (4.55) and (4.56), as desired.   □

Next, we turn to the case of Riesz means of order $\gamma > 0$. Again, here we prove a result for continuous, compactly supported potentials, which later, in Theorem 4.46, will be extended to a larger class of potentials for γ within a certain range.

Theorem 4.29 *Let $\gamma > 0$ and let V be a continuous function on $\mathbb{R}^d$ with compact support. Then*

$$\lim_{\alpha \to \infty} \alpha^{-\gamma - d/2} \operatorname{Tr} (-\Delta - \alpha V)_-^\gamma = L_{\gamma,d}^{\mathrm{cl}} \int_{\mathbb{R}^d} V(x)_+^{\gamma + d/2} \, dx \, .$$

Proof We argue as in the previous proof. By the variational principle, the operator inequality (4.57) implies

$$\sum_{a \in L\mathbb{Z}^d} \operatorname{Tr} \left(-\Delta_{Q_L(a)}^{\mathrm{D}} - \alpha V^{(\min)}(a) \right)_-^\gamma$$

$$= \operatorname{Tr} \left(\bigoplus_{a \in L\mathbb{Z}^d} \left(-\Delta_{Q_L(a)}^{\mathrm{D}} - \alpha V^{(\min)}(a) \right) \right)_-^\gamma \leq \operatorname{Tr} (-\Delta - \alpha V)_-^\gamma$$

$$\leq \operatorname{Tr} \left(\bigoplus_{a \in L\mathbb{Z}^d} \left(-\Delta_{Q_L(a)}^{\mathrm{N}} - \alpha V^{(\max)}(a) \right) \right)_-^\gamma$$

$$= \sum_{a \in L\mathbb{Z}^d} \operatorname{Tr} \left(-\Delta_{Q_L(a)}^{\mathrm{N}} - \alpha V^{(\max)}(a) \right)_-^\gamma \, .$$

It follows from Lemma 3.13 that

$$\lim_{\alpha \to \infty} \alpha^{-\gamma - d/2} \operatorname{Tr} \left(-\Delta_{Q_L(a)}^{\mathrm{N}} - \alpha V^{(\max)}(a) \right)_-^\gamma = L_{\gamma,d}^{\mathrm{cl}} L^d \left(V^{(\max)}(a) \right)_+^{\gamma + d/2}$$

and

$$\lim_{\alpha \to \infty} \alpha^{-\gamma - d/2} \operatorname{Tr} \left(-\Delta_{Q_L(a)}^{\mathrm{D}} - \alpha V^{(\min)}(a) \right)_-^\gamma = L_{\gamma,d}^{\mathrm{cl}} L^d \left(V^{(\min)}(a) \right)_+^{\gamma + d/2} .$$

This allows us to conclude the proof as before. $\square$

We formulate a further generalization of these asymptotics in terms of the operator $-\hbar^2 \Delta - V$, where $\hbar$ plays the role of a normalized Planck constant. We are interested in the limit $\hbar \to 0$. For Riesz means, asymptotics of $-\Delta - \alpha V$ as $\alpha \to \infty$ are equivalent to those of $-\hbar^2 \Delta - V$ as $\hbar \to 0$. For more general trace functionals, however, it is more natural to consider the latter operator.

Proposition 4.30 *Let $\varphi : \mathbb{R} \to [0, \infty)$ be a non-increasing function that vanishes on $\mathbb{R}_+$ and let V is a continuous function on $\mathbb{R}^d$ with compact support. Then*

$$\lim_{\hbar \to 0} \hbar^d \operatorname{Tr} \varphi(-\hbar^2 \Delta - V) = \iint_{\mathbb{R}^d \times \mathbb{R}^d} \varphi(|\xi|^2 - V(x)) \frac{dx \, d\xi}{(2\pi)^d} \, .$$

Since both sides are linear in φ, the asymptotics are also valid for differences of functions of the described form, which covers a large class of functions of bounded variation. Note that for $\varphi(\lambda) = \chi_{(-\infty,0)}(\lambda)$ and $\varphi(\lambda) = \lambda_-^{\gamma}$ the proposition reduces to Theorems 4.28 and 4.29, respectively, when $h^2 = \alpha^{-1}$.

Proof We can argue as in the proof of Theorem 4.29, since, by Proposition 3.42,

$$\lim_{\hbar \to 0} \hbar^d \, \mathrm{Tr}\, \varphi \left(-\hbar^2 \Delta^{\mathrm{N}}_{Q_L(a)} - V^{(\mathrm{max})}(a) \right) = L^d \int_{\mathbb{R}^d} \varphi \left(|\xi|^2 - V^{(\mathrm{max})}(a) \right) \frac{d\xi}{(2\pi)^d}$$

and

$$\lim_{\hbar \to 0} \hbar^d \, \mathrm{Tr}\, \varphi \left(-\hbar^2 \Delta^{\mathrm{D}}_{Q_L(a)} - V^{(\mathrm{min})}(a) \right) = L^d \int_{\mathbb{R}^d} \varphi \left(|\xi|^2 - V^{(\mathrm{min})}(a) \right) \frac{d\xi}{(2\pi)^d} \,.$$

By dominated convergence,

$$\int_{\mathbb{R}^d} \varphi(|\xi|^2 - v)\, d\xi = \frac{1}{2} |\mathbb{S}^{d-1}| \int_0^{v_+} \varphi(-E)(v - E)^{\frac{d-2}{2}}\, dE$$

is continuous in v, which allows us to pass to the limit $L \to 0$. We omit the details. $\square$

4.5 The Cwikel–Lieb–Rozenblum inequality

In this section we state and prove the Cwikel–Lieb–Rozenblum (CLR) inequality. This bound says that in dimensions $d \geq 3$ the number $N(0, -\Delta - V)$ of negative eigenvalues of the Schrödinger operator $-\Delta - V$, counting multiplicities, satisfies

$$N(0, -\Delta - V) \leq L_{0,d} \int_{\mathbb{R}^d} V(x)_+^{d/2}\, dx \tag{4.59}$$

for all $V \in L^{d/2}(\mathbb{R}^d)$ with a constant $L_{0,d}$ depending only on the dimension d.

One remarkable feature of this inequality is that it is, in some sense, saturated both in the strong and in the weak coupling regime. Indeed, if V is replaced by αV, then the upper bound on $N(0, -\Delta - \alpha V)$ provided by (4.59) coincides with the leading order given by the Weyl asymptotic formula (4.54), except that the constant $L_{0,d}^{\mathrm{cl}}$ there is replaced by a possibly larger constant $L_{0,d}$. On the other hand, if $L_{0,d}\, \alpha^{d/2} \int_{\mathbb{R}^d} V(x)_+^{d/2}\, dx < 1$, then (4.59) implies that $-\Delta - \alpha V$ has no negative eigenvalues, so one recovers, up to the value of the constant, the result of (4.12), which is a consequence of the Sobolev inequality.

The main feature of the inequality, however, is that it not only correctly reflects the asymptotic regimes, but it holds uniformly for all V. This type of a priori bound is invaluable in many applications.

The CLR inequality was proved independently by Cwikel, Lieb and Rozenblum (Cwikel, 1977; Lieb, 1976, 1980; Rozenbljum, 1972a, 1976). They used rather different mathematical techniques and by now even more proofs have been found; see §4.9.

The proof we give here is essentially Rozenblum's and it is based on a bracketing argument, as was the proof of the Weyl asymptotics. The important difference, however, is that in the present case the cubes do not have the same size, but rather are adjusted to the local size of the potential. This leads to the technical difficulty of overlapping cubes.

The proof is based solely on the variational principle and, therefore, can be easily adjusted to other situations such as higher-order operators, operators on Riemannian manifolds, and also to non-local operators. Moreover, the same technique will allow us in the next section to derive the family of Lieb–Thirring estimates, including the critical case. A disadvantage of this method is that it typically does not give good control of the constants. For further information on the currently best values of the constants, which are due to Lieb (1976, 1980) in dimensions $d = 3, 4$, and to Hundertmark et al. (2018) in dimensions $d \geq 5$, we refer to Theorem 8.17. As of now, the sharp constants in the CLR inequality are unknown.

4.5.1 Statement and proof of the CLR inequality

Our goal in this section is to prove the following important result.

Theorem 4.31 (CLR inequality) *Let $d \geq 3$. Then there is a constant $L_{0,d} < \infty$ such that for all $V \in L^1_{\mathrm{loc}}(\mathbb{R}^d)$ with $V_+ \in L^{d/2}(\mathbb{R}^d)$,*

$$N(0, -\Delta - V) \leq L_{0,d} \int_{\mathbb{R}^d} V(x)_+^{d/2} \, dx . \tag{4.60}$$

The idea of the proof is to cover the support of V (which, by an approximation argument, may be assumed to be compact) by cubes, similar to the proof of Weyl asymptotics. The new idea is that the size of the cubes depends on the local size of the potential. In general, the cubes will overlap. To construct this covering we will use the following proposition, which is a variant of Besicovitch's covering theorem.

To state it, let us introduce some terminology. By a *cube* we always mean an open cube with edges parallel to the coordinate axes, and by a *covering* of a set $K \subset \mathbb{R}^d$ by cubes $Q_1, \ldots, Q_M$ we mean that $K \subset \bigcup_j \overline{Q_j}$. The *multiplicity* of such a covering is $\sup_{x \in \mathbb{R}^d} \#\{j : x \in Q_j\}$. We set $Q := (-1/2, 1/2)^d$ and note that $a + \ell Q$ is the cube centered at $a \in \mathbb{R}^d$ with side length $\ell > 0$. Finally, we recall that a function f on a set $K \subset \mathbb{R}^d$ is called *upper semicontinuous* if, for any $x \in K$ and any sequence $(x_n) \subset K$ with $x_n \to x$, we have $f(x) \geq \limsup_{n \to \infty} f(x_n)$.

Proposition 4.32 *Let $d \geq 1$, let $K \subset \mathbb{R}^d$ be a compact set and let ℓ be a positive, upper semicontinuous function on K. Then there is a (finite or infinite) sequence $(x_j) \subset K$ such that the cubes $Q_j := x_j + \ell(x_j)Q$, $j = 1, 2, 3, \ldots$, are a covering of K with multiplicity at most 2^d. Moreover, the sequence (Q_j) can be arranged into $4^d + 1$ sequences $\Xi^k := (Q_j^k)$ such that for any $1 \leq k \leq 4^d + 1$, the cubes in Ξ^k are disjoint; that is, $Q_{j_1}^k \cap Q_{j_2}^k = \emptyset$ if $j_1 \neq j_2$.*

Remark 4.33   The statement of this proposition remains valid if the upper semicontinuity assumption on ℓ and the compactness assumption on K are replaced by the boundedness of ℓ (de Guzmán, 1975). In this case, however, the constants 2^d and $4^d + 1$ may need to be replaced by larger constants (but again depending only on the dimension d). Our assumptions are satisfied in the applications that we have in mind and will allow us to present the idea of the proof more clearly.

For the proof of Theorem 4.31 we do not need the 'Moreover, . . .'-part of this proposition. It is included for the sake of completeness. We defer the proof of Proposition 4.32 to §4.5.4 and use it now to deduce the following important ingredient of Rozenblum's proof of Theorem 4.31.

Proposition 4.34 *Let $0 \leq W \in L^1(\mathbb{R}^d)$ with compact support. Then for any $0 < A \leq \int_{\mathbb{R}^d} W \, dx$ there is a covering of supp W by open cubes $Q_1, \ldots, Q_M$ of multiplicity at most 2^d such that*

$$\int_{Q_j} W \, dx = A \qquad \text{for all } j \,. \tag{4.61}$$

The number M of cubes is bounded by

$$M \leq A^{-1} 2^d \int_{\mathbb{R}^d} W \, dx \,. \tag{4.62}$$

Proof   We may assume that $0 < A < \int_{\mathbb{R}^d} W \, dx$. If we denote $K := \operatorname{supp} W$, then for any $x \in K$ the function $\ell \mapsto \int_{x+\ell Q} W \, dx$ is non-decreasing. Moreover, dominated convergence implies that it is continuous and vanishes as $\ell \to 0$. Since it is equal to $\int_{\mathbb{R}^d} W \, dx$ for sufficiently large ℓ, we conclude that for any $x \in K$ there is an $\ell > 0$ with

$$\int_{x+\ell Q} W \, dx = A \,.$$

To define $\ell = \ell(x)$ uniquely, we require that ℓ is maximal such that this equality holds.

Using dominated convergence, it is easy to see that $x \mapsto \ell(x)$ is upper semicontinuous. Indeed, if $x_n \to x \in K$ and $\ell_n \to \ell \in (0, \infty]$, then $\chi_{x_n + \ell_n Q} \to \chi_{x + \ell Q}$ pointwise (with the convention that $x + \ell Q = \mathbb{R}^d$ if $\ell = \infty$). Thus $\int_{x_n + \ell_n Q} W \, dx \to \int_{x + \ell Q} W \, dx$. If we apply this observation to $\ell_n = \ell(x_n)$, then we obtain that $A = \int_{x + \ell Q} W \, dx$, and therefore, by maximality, $\ell \leq \ell(x)$, which means that ℓ is upper semicontinuous.

We can now apply the Besicovitch theorem (Proposition 4.32) and obtain a countable covering of K by open cubes Q_j of multiplicity at most 2^d. Because of this bound on the multiplicity of the covering, we have, for any J,

$$ JA = \sum_{j=1}^{J} \int_{Q_j} W \, dx \leq 2^d \int_{\mathbb{R}^d} W \, dx . $$

This implies that the number of cubes is finite and obeys (4.62), as claimed. □

The second ingredient in Rozenblum's proof are Sobolev and Poincaré–Sobolev inequalities. Recall that the usual Sobolev inequality states that

$$ \int_{\mathbb{R}^d} |\nabla u|^2 \, dx \geq S_d \left(\int_{\mathbb{R}^d} |u|^{2d/(d-2)} \, dx \right)^{(d-2)/d} \qquad \text{for all } u \in H^1(\mathbb{R}^d); \quad (4.63) $$

see Theorem 2.44. Obviously, the inequality is not valid when the integrals on both sides are restricted to a bounded, positive measure subset of $\mathbb{R}^d$. (Consider a function that is constant on this subset.) However, it *does* remain valid if an additional orthogonality condition is imposed. More precisely, one has the Poincaré–Sobolev inequality

$$ \int_Q |\nabla u|^2 \, dx \geq \tilde{S}_d \left(\int_Q |u|^{2d/(d-2)} \, dx \right)^{(d-2)/d} $$

$$ \text{for all } u \in H^1(Q) \text{ with } \int_Q u \, dx = 0 ; \qquad (4.64) $$

see Corollary 2.64. By scaling, the constant $\tilde{S}_d$ is independent of the size of Q. Note that (4.63) and (4.64) require $d \geq 3$.

We are now in position to prove the main result in this section.

Proof of Theorem 4.31 By the variational principle, it suffices to consider $V \geq 0$. We first prove the theorem for $V \in L^{d/2}(\mathbb{R}^d)$ with compact support only. The generalization to arbitrary $V \in L^{d/2}(\mathbb{R}^d)$ will be provided in Theorem 4.45 below.

For $u \in H^1(\mathbb{R}^d)$ we have, by (4.63),

$$\int_{\mathbb{R}^d} \left(|\nabla u|^2 - V|u|^2 \right) dx$$

$$\geq \int_{\mathbb{R}^d} |\nabla u|^2 \, dx - \left(\int_{\mathbb{R}^d} V^{d/2} \, dx \right)^{2/d} \left(\int_{\mathbb{R}^d} |u|^{2d/(d-2)} \, dx \right)^{(d-2)/d}$$

$$\geq \int_{\mathbb{R}^d} |\nabla u|^2 \, dx \left(1 - S_d^{-1} \left(\int_{\mathbb{R}^d} V^{d/2} \, dx \right)^{2/d} \right).$$

Hence, if $\int_{\mathbb{R}^d} V^{d/2} \, dx \leq S_d^{d/2}$, then $N(0, -\Delta - V) = 0$. In the following, we assume

$$\int_{\mathbb{R}^d} V^{d/2} \, dx > S_d^{d/2}. \qquad (4.65)$$

We apply Proposition 4.34 with $W = V^{d/2}$ and

$$A := \min\{2^{-d}\tilde{S}_d, S_d\}^{d/2},$$

where $\tilde{S}_d$ denotes the constant in (4.64). By (4.65), we have $A < \int_{\mathbb{R}^d} V^{d/2} \, dx = \int_{\mathbb{R}^d} W \, dx$. Therefore we obtain a covering of supp V by cubes $Q_1, \ldots, Q_M$ with multiplicity $m \leq 2^d$, where

$$M \leq A^{-1} 2^d \int_{\mathbb{R}^d} V^{d/2} \, dx. \qquad (4.66)$$

For $u \in H^1(\mathbb{R}^d)$ we have

$$\int_{\mathbb{R}^d} \left(|\nabla u|^2 - V|u|^2 \right) dx \geq \sum_{j=1}^{M} \int_{Q_j} \left(m^{-1} |\nabla u|^2 - V|u|^2 \right) dx$$

$$= m^{-1} \sum_{j=1}^{M} \int_{Q_j} \left(|\nabla u|^2 - mV|u|^2 \right) dx.$$

If, for some $j = 1, \ldots, M$, $u \in H^1(\mathbb{R}^d)$ satisfies $\int_{Q_j} u \, dx = 0$, then, by (4.64),

$$\int_{Q_j} \left(|\nabla u|^2 - mV|u|^2 \right) dx$$

$$\geq \int_{Q_j} |\nabla u|^2 dx - m \left(\int_{Q_j} V^{d/2} \, dx \right)^{2/d} \left(\int_{Q_j} |u|^{2d/(d-2)} \, dx \right)^{(d-2)/d}$$

$$\geq \int_{Q_j} |\nabla u|^2 dx \left(1 - \tilde{S}_d^{-1} m \left(\int_{Q_j} V^{d/2} \, dx \right)^{2/d} \right).$$

Because of our choice of covering, we have $\tilde{S}_d^{-1} m \left(\int_{Q_j} V^{d/2} \, dx \right)^{2/d} \leq 1$ and the right side is non-negative. Thus,

$$\int_{\mathbb{R}^d} \left(|\nabla u|^2 - V|u|^2 \right) dx \geq 0 \quad \text{if} \quad u \perp \chi_{Q_j} \quad \text{for all} \quad j = 1, \ldots, M \,.$$

By the variational principle (Theorem 1.26), this implies $N(0, -\Delta - V) \leq M$. This, together with the bound (4.66), completes the proof. $\qquad\square$

4.5.2 A reformulation of the previous proof

As preparation for the proof of the Riesz means analogue of Theorem 4.31, in this subsection we reformulate the proof above in terms of Schrödinger operators on cubes.

Lemma 4.35 *Let $d \geq 1$ and let $0 \leq V \in L^1_{\mathrm{loc}}(\mathbb{R}^d)$, $d \geq 1$. Let (Q_j) be a covering of* supp V *by open sets with multiplicity at most $m \in \mathbb{N}$ and assume that there are $\theta < m^{-1}$ and $C < \infty$ such that, for all j,*

$$\int_{Q_j} V|u|^2 \, dx \leq \theta \int_{Q_j} |\nabla u|^2 \, dx + C \int_{Q_j} |u|^2 \, dx \quad \text{for all } u \in H^1(Q_j) \,.$$

Then, for all $\tau \geq 0$,

$$N(-\tau, -\Delta - V) \leq \sum_j N(-\tau, -\Delta^{\mathrm{N}}_{Q_j} - mV_j) \,,$$

where V_j is the restriction of V to Q_j.

The form boundedness condition on V in Lemma 4.35 ensures, by Lemma 1.47, that the operators $-\Delta^{\mathrm{N}}_{Q_j} - mV_j$ are well defined in the form sense on $L^2(Q_j)$. As we will see in the proof, this implies the existence of $-\Delta - V$ on $L^2(\mathbb{R}^d)$ in the form sense.

Proof For every $u \in H^1(\mathbb{R}^d)$ we have

$$\int_{\mathbb{R}^d} |\nabla u|^2 \, dx \geq m^{-1} \sum_j \int_{Q_j} |\nabla u|^2 \, dx \,, \tag{4.67}$$

$$\int_{\mathbb{R}^d} |u|^2 \, dx \geq m^{-1} \sum_j \int_{Q_j} |u|^2 \, dx \,, \tag{4.68}$$

$$\int_{\mathbb{R}^d} V|u|^2 \, dx \leq \sum_j \int_{Q_j} V_j|u|^2 \, dx \,. \tag{4.69}$$

In particular,

$$\int_{\mathbb{R}^d} V|u|^2 \, dx \leq m\theta \int_{\mathbb{R}^d} |\nabla u|^2 \, dx + mC \int_{\mathbb{R}^d} |u|^2 \, dx \quad \text{for all } u \in H^1(\mathbb{R}^d) \,.$$

By Lemma 1.47, the operator $-\Delta - V$ in $L^2(\mathbb{R}^d)$ is well defined in the form sense.

Now let us apply Glazman's lemma. To that end, let $\tau \geq 0$ and set $L_j :=$ $P^j_{(-\infty, -\tau)} L^2(Q_j)$, where P^j is the spectral measure of $-\Delta^N_{Q_j} - mV_j$ in $L^2(Q_j)$. Then

$$\int_{Q_j} \left(|\nabla u|^2 + \tau |u|^2 - mV_j |u|^2 \right) dx \geq 0 \quad \text{for all } u \in H^1(Q_j) \cap L_j^\perp .$$

Let $\tilde{L}_j$ be the subspace in $L^2(\mathbb{R}^d)$ that consists of all functions $u \in L_j \subset L^2(Q_j)$ extended by zero to $\mathbb{R}^d$. Let $L := \sum_j \tilde{L}_j$ be the linear span of all the spaces $\tilde{L}_j$. Then, by (4.67), (4.68) and (4.69), for all $u \in H^1(\mathbb{R}^d) \cap L^\perp$,

$$\int_{\mathbb{R}^d} \left(|\nabla u|^2 + \tau |u|^2 - V|u|^2 \right) dx \geq m^{-1} \sum_j \int_{Q_j} \left(|\nabla u|^2 + \tau |u|^2 - mV_j |u|^2 \right) dx$$

$$\geq 0 .$$

Therefore, by Glazman's lemma (Theorem 1.26), $N(-\tau, -\Delta - V) \leq \dim L$. Meanwhile, $\dim L \leq \sum_j \dim \tilde{L}_j$ and, for any j,

$$\dim \tilde{L}_j = \dim L_j = N(-\tau, -\Delta^N_{Q_j} - mV_j) .$$

This completes the proof of the lemma.   □

Next, we translate the Sobolev bounds (4.63) and (4.64) into bounds on the number of eigenvalues of Schrödinger operators. In the proof of Theorem 4.31 (see also (4.12)), we have shown that (4.63) implies that

$$N(0, -\Delta - V) = 0 \quad \text{if} \int_{\mathbb{R}^d} V(x)_+^{d/2} \, dx \leq S_d^{d/2} . \tag{4.70}$$

Similarly, using Glazman's lemma (Theorem 1.26), we see that (4.64) implies

$$N(0, -\Delta^N_Q - V) \leq 1 \quad \text{if} \int_Q V(x)_+^{d/2} \, dx \leq \tilde{S}_d^{d/2} . \tag{4.71}$$

Now we are ready to reformulate the proof of the CLR inequality.

Second proof of Theorem 4.31   Let $0 \leq V \in L^{d/2}(\mathbb{R}^d)$ with compact support. If $\|V\|_{d/2} \leq S_d$ then, by (4.70), $N(-\Delta - V) = 0$, and hence the assertion is true. Henceforth we assume that

$$\|V\|_{d/2} > S_d . \tag{4.72}$$

We apply Proposition 4.34 with $W = V^{d/2}$ and $A = \min\{2^{-d} \tilde{S}_d, S_d\}^{d/2}$. Then (4.72) guarantees that $A \leq \|V\|_{d/2}^{d/2}$, as required for the proof of Proposition 4.34.

Hence, we obtain a covering of supp V by finitely many cubes $Q_1, \ldots, Q_M$ of multiplicity at most 2^d such that

$$\int_{Q_j} V(x)^{d/2} \, dx \le \left(2^{-d} \tilde{S}_d \right)^{d/2} \qquad \text{for all } 1 \le j \le M \,.$$

By (4.71), this implies

$$N(0, -\Delta_{Q_j}^N - 2^d V_j) \le 1 \qquad \text{for all } 1 \le j \le M \,.$$

Thus Lemma 4.35 together with (4.62) implies the bound

$$N(0, -\Delta - V) \le \sum_{j=1}^M N(0, -\Delta_{Q_j}^N - 2^d V_j) \le M$$

$$\le \min\{2^{-d} \tilde{S}_d, S_d\}^{-d/2} \, 2^d \int_{\mathbb{R}^d} V^{d/2} \, dx \,.$$

(The assumption in Lemma 4.35 can be verified as in the proof of Proposition 4.3, but with (4.64) instead of Theorem 2.44.) This proves the theorem. □

4.5.3 The CLR inequality for Schrödinger operators on domains

For Schrödinger operators on domains with Dirichlet conditions, Theorem 4.31 together with the variational principle immediately implies the following bound.

Corollary 4.36 *Let $d \ge 3$. Then there is a constant $L_{0,d} < \infty$ such that, for all open sets $\Omega \subset \mathbb{R}^d$ and all $V \in L^1_{\mathrm{loc}}(\Omega)$ with $V_+ \in L^{d/2}(\Omega)$,*

$$N(0, -\Delta_\Omega^D - V) \le L_{0,d} \int_\Omega V(x)_+^{d/2} \, dx \,.$$

The case of Neumann boundary conditions requires a bit more effort. For the definition of the extension property, see §2.3.2 and §2.8.

Corollary 4.37 *Let $d \ge 3$ and let $\Omega \subset \mathbb{R}^d$ be an open and connected set of finite measure with the extension property. Then there is a constant $L_\Omega < \infty$ such that, for all $V \in L^1_{\mathrm{loc}}(\Omega)$ with $V_+ \in L^{d/2}(\Omega)$,*

$$N(0, -\Delta_\Omega^N - V) \le 1 + L_\Omega \int_\Omega V(x)_+^{d/2} \, dx \,.$$

Proof Let P denote the orthogonal projection in $L^2(\Omega)$ onto constant functions and let $P^\perp := 1 - P$. Then, by the variational principle,

$$N(0, -\Delta_\Omega^N - V) \le 1 + N(0, P^\perp(-\Delta_\Omega^N - V)P^\perp) \,.$$

By assumption, there is a bounded operator $E : H^1(\Omega) \to H^1(\mathbb{R}^d)$ with $Eu = u$ almost everywhere on Ω. Together with the Poincaré inequality (Proposition 2.58), this implies that for any $u \in H^1(\Omega)$ with $\int_\Omega u \, dx = 0$,

$$\int_{\mathbb{R}^d} |\nabla Eu|^2 \, dx \le \|E\|_{H^1(\Omega) \to H^1(\mathbb{R}^d)}^2 \|u\|_{H^1(\Omega)}^2 \le M_\Omega \int_\Omega |\nabla u|^2 \, dx$$

with $M_\Omega := \|E\|_{H^1(\Omega) \to H^1(\tilde\Omega)}^2 (1 + P_\Omega^{-1})$, where P_Ω is the Poincaré constant from Proposition 2.58. Thus, by the variational principle,

$$N(0, P^\perp(-\Delta_\Omega^N - V)P^\perp) \le N(0, M_\Omega^{-1}(-\Delta) - \widetilde{V}) = N(0, -\Delta - M_\Omega \widetilde{V}),$$

where $\widetilde{V}$ denotes the extension by zero of V to $\mathbb{R}^d$. Thus, from Theorem 4.31,

$$N(0, -\Delta_\Omega^N - V) \le 1 + N(0, -\Delta - M_\Omega \widetilde{V}) \le 1 + L_{0,d} M_\Omega^{d/2} \int_\Omega V(x)_+^{d/2} \, dx \, .$$

This completes the proof of the corollary. $\square$

4.5.4 Appendix: The Besicovitch covering theorem

In this subsection we prove the version of Besicovich's covering theorem in Proposition 4.32. In addition to the notations introduced before the statement of the theorem, we also write

$$|x|_\infty := \max\{|x_j| : 1 \le j \le d\} \quad \text{for } x = (x_1, \dots, x_d) \in \mathbb{R}^d.$$

Proof of Proposition 4.32 Since the function ℓ is upper semicontinuous on the compact set K, it attains its maximum at some point $x_1 \in K$. Now assume that, for some $m \in \mathbb{N}$, the points $x_1, \dots, x_m$ have already been chosen. If $K \setminus \bigcup_{j=1}^m Q_j = \emptyset$, then the selection process is finished. Otherwise, we take $x_{m+1} \in K \setminus \bigcup_{j=1}^m Q_j$ such that the maximum of ℓ over the compact $K \setminus \bigcup_{j=1}^m Q_j$ is attained at x_{m+1}. This procedure leads to a finite or infinite sequence of points x_j. Let us show that they have all the required properties.

We claim that

$$\left(x_i + \frac{\ell(x_i)}{2}Q\right) \cap \left(x_j + \frac{\ell(x_j)}{2}Q\right) = \emptyset \qquad \text{if } i \ne j. \qquad (4.73)$$

Indeed, to show this, we may assume that $i < j$. Then, by construction $x_j \notin Q_i$, and therefore, $|x_i - x_j|_\infty \ge \ell(x_i)/2$. By construction, we have $\ell(x_i) \ge \ell(x_j)$, and therefore, $|x_i - x_j|_\infty \ge (\ell(x_i) + \ell(x_j))/4$. This implies (4.73).

We also claim that

$$x_i \notin Q_j \qquad \text{if } i \ne j. \qquad (4.74)$$

Indeed, this is clear from the construction if $i > j$. On the other hand, if $i < j$, then, again from the construction, we have $\ell(x_i) \geq \ell(x_j)$ and $x_j \notin Q_i$, which implies that $|x_j - x_i|_\infty \geq \ell(x_i)/2 \geq \ell(x_j)/2$. Thus, (4.74) also holds in this case.

Using the compactness of K, one easily deduces from (4.73) that, if the sequence (Q_j) is infinite, then

$$\ell(x_j) \to 0 \qquad \text{as } j \to \infty. \tag{4.75}$$

We now prove that (Q_j) covers K. This is clear from the construction if the sequence (Q_j) is finite. If it is infinite we argue by contradiction and assume that there is an $x \in K \setminus \bigcup_j Q_j$. Then, because of (4.75), there is a j such that $\ell(x_j) < \ell(x)$. This, however, contradicts the construction of x_j.

Next, we show that the multiplicity of the covering is at most 2^d, that is, any point $x \in K$ belongs to at most 2^d of the cubes Q_j. To do this, we divide $\mathbb{R}^d$ into 2^d orthants with boundaries passing through x and parallel to the d coordinate hyperplanes. It suffices to show that in each closed orthant there is at most one point x_j such that $x \in Q_j$. We argue by contradiction and assume that there are two distinct points x_i and x_j in the same closed orthant with $x \in Q_i \cap Q_j$. We may assume that $|x_j - x|_\infty \geq |x_i - x|_\infty$. Since $x \in Q_j$, the set of all points y in the same orthant as x_j satisfying $|y - x|_\infty \leq |x_j - x|_\infty$ is contained in Q_j. In particular, we have $x_i \in Q_j$. This contradicts (4.74).

Finally, we have to rearrange the sequence into $4^d + 1$ disjoint sequences. We first claim that for any j there are at most 4^d cubes among the cubes $Q_1, \ldots, Q_{j-1}$ that have non-empty intersection with Q_j. To see this, note that, if $k < j$ and if $Q_k \cap Q_j \neq \emptyset$, then Q_k contains at least one of the 2^d vertices of Q_j. (This follows from the fact that $\ell(x_k) \geq \ell(x_j)$ for $k < j$.) However, by the bound on the multiplicity, any fixed vertex of Q_j is contained in at most 2^d cubes. Thus, there are at most $2^d \times 2^d$ cubes Q_k with $k < j$ that have non-empty intersection with Q_j.

We now use this fact to decompose our sequence. We are going to define inductively $r := 4^d + 1$ sequences $\Xi^1, \ldots, \Xi^r$ of cubes. To start, we set $Q_j \in \Xi^j$ for $j = 1, \ldots, r$. Now let $j \geq r + 1$ and assume that the families $\Xi^1, \ldots, \Xi^r$ contain all the cubes $Q_1, \ldots, Q_{j-1}$ and that each Ξ^k consists of disjoint cubes. By the above fact, Q_j can intersect at most $r - 1$ cubes among the cubes $Q_1, \ldots, Q_{j-1}$. Since there are r families of cubes in total, there must be a $k \in \{1, \ldots, r\}$ such that Q_j does not intersect any of the cubes in Ξ^k. We put $Q_j \in \Xi^k$. This defines inductively the claimed partitioning of the Q_j. The proof of the proposition is complete. $\qquad \square$

4.6 The Lieb–Thirring inequality

In this section, we state and prove the Lieb–Thirring (LT) inequalities. This family of inequalities states that for $\gamma \geq 1/2$ in dimension $d = 1$ and for $\gamma > 0$ in dimensions $d \geq 2$ the Riesz means $\mathrm{Tr}(-\Delta - V)^{\gamma}_{-}$ of the negative eigenvalues of the Schrödinger operator $-\Delta - V$ satisfy

$$\mathrm{Tr}(-\Delta - V)^{\gamma}_{-} \leq L_{\gamma,d} \int_{\mathbb{R}^d} V(x)^{\gamma+d/2}_{+} \, dx \, .$$

This is the natural extension of the CLR inequality in the previous section, which corresponds to $\gamma = 0$ in dimensions $d \geq 3$. Just like the CLR inequality, the Lieb–Thirring inequality is, in some sense, saturated in the strong coupling regime, as evidenced by the Weyl asymptotics in Theorem 4.29, and also implies, up to the value of the constant, the lower bound on the lowest eigenvalue in (4.11), which was derived using the Gagliardo–Nirenberg inequality.

4.6.1 Statement and proof of the Lieb–Thirring inequality

Our goal in this section is to prove

Theorem 4.38 (LT inequality) *Let*

$$\gamma \geq 1/2 \quad \text{if } d = 1,$$
$$\gamma > 0 \quad \text{if } d \geq 2.$$

Then there is a constant $L_{\gamma,d} < \infty$ such that, for any V with $V_+ \in L^{\gamma+d/2}(\mathbb{R}^d)$ and $V_- \in L^1_{\mathrm{loc}}(\mathbb{R}^d)$,

$$\mathrm{Tr}(-\Delta - V)^{\gamma}_{-} \leq L_{\gamma,d} \int_{\mathbb{R}^d} V(x)^{\gamma+d/2}_{+} \, dx \, . \tag{4.76}$$

This theorem, except for the case $\gamma = 1/2$ in dimension $d = 1$, is due to Lieb and Thirring (1976). The remaining endpoint case was settled by Weidl (1996).

We will present a proof using the bracketing technique similar to that in the works of Rozenbljum (1972a, 1976) and Weidl (1996). This technique covers the full range of parameters γ for which the inequality is valid. We do not here pay attention to the values of the constants $L_{\gamma,d}$ in (4.76): these will be the subject of the remaining chapters in this book.

Having already proved the CLR inequality (Theorem 4.31) in the previous section, we can deduce Theorem 4.38 in dimensions $d \geq 3$ as follows. By Lemma 1.39 and the variational principle, we have

$$\mathrm{Tr}(-\Delta-V)^{\gamma}_{-} = \gamma \int_0^{\infty} N(-\tau, -\Delta-V)\tau^{\gamma-1} \, d\tau \leq \gamma \int_0^{\infty} N(0, -\Delta-(V-\tau)_+)\tau^{\gamma-1} \, d\tau \, .$$

Therefore Theorem 4.31 yields

$$\mathrm{Tr}(-\Delta - V)_-^\gamma \le L_{0,d}\gamma \int_0^\infty \int_{\mathbb{R}^d} (V - \tau)_+^{d/2}\, dx\, \tau^{\gamma-1}\, d\tau$$

$$= L_{0,d}\frac{\Gamma(d/2 + 1)\,\Gamma(\gamma + 1)}{\Gamma(\gamma + d/2 + 1)} \int_{\mathbb{R}^d} V_+^{\gamma+d/2}\, dx\,.$$

We will revisit this argument, which is due to Aizenman and Lieb (1978), in §5.1.1, when discussing optimal constants in Lieb–Thirring inequalities. Thus, in what follows we may restrict ourselves to dimensions $d = 1$ and $d = 2$, but, since it presents no extra effort, we choose not to.

Just as for Theorem 4.31, the proof of Theorem 4.38 is based on the construction of an appropriate covering. We use the terminology introduced before Proposition 4.34.

Proposition 4.39 *Let $0 \le W \in L^1(\mathbb{R}^d)$ with compact support and $W \not\equiv 0$. Then, for any $A > 0$ and any $\alpha > 0$, there is a covering of* supp W *by open cubes $Q_1, Q_2, Q_3, \ldots$ of multiplicity at most 2^d such that*

$$|Q_j|^\alpha \int_{Q_j} W\, dx = A \qquad \text{for all } j\,.$$

Proof Let $K := $ supp W. Clearly, for any $x \in K$, the function $\ell \mapsto \ell^{\alpha d} \int_{x+\ell Q} W\, dx$, where $Q = (-1/2, 1/2)^d$ (so that $x + \ell Q$ is the cube centered at x with side length $\ell > 0$), is strictly increasing. Moreover, dominated convergence implies that it is continuous and vanishes as $\ell \to 0$. Since it tends to infinity as $\ell \to \infty$, we conclude that for any $x \in K$ there is a unique $\ell(x) > 0$ such that

$$\ell(x)^{\alpha d} \int_{x+\ell(x)Q} W\, dx = A\,.$$

It is easy to see that the function $x \mapsto \ell(x)$ is upper semicontinuous; see the proof of Proposition 4.34 for a similar argument.

We can now apply the Besicovitch-type Proposition 4.32 and obtain a countable covering of K by open cubes Q_j of multiplicity at most 2^d, as claimed. $\square$

The following is a consequence of Lemma 4.35.

Lemma 4.40 *Let $0 \le V \in L^1_{\mathrm{loc}}(\mathbb{R}^d)$. Let (Q_j) be a covering of* supp V *by open sets with multiplicity at most $m \in \mathbb{N}$ and assume that there are $\theta < m^{-1}$ and $C > 0$ such that, for all j,*

$$\int_{Q_j} V|u|^2\, dx \le \theta \int_{Q_j} |\nabla u|^2\, dx + C \int_{Q_j} |u|^2\, dx \quad \text{for all } u \in H^1(Q_j)\,.$$

Then, for all $\gamma > 0$,

$$\mathrm{Tr}(-\Delta - V)_-^\gamma \le \sum_j \mathrm{Tr}(-\Delta_{Q_j}^N - mV_j)_-^\gamma,$$

where V_j is the restriction of V to Q_j.

Proof By Lemma 1.39 we have

$$\mathrm{Tr}(-\Delta - V)_-^\gamma = \gamma \int_0^\infty N(-\tau, -\Delta - V)\tau^{\gamma-1}\, d\tau,$$

$$\mathrm{Tr}(-\Delta_{Q_j}^N - mV_j)_-^\gamma = \gamma \int_0^\infty N(-\tau, -\Delta_{Q_j}^N - mV_j)\, \tau^{\gamma-1}\, d\tau,$$

and it remains to apply Lemma 4.35. □

Apart from the covering technique, consisting of Proposition 4.39 and Lemma 4.40, the other ingredients in the proof of Theorem 4.38 are Gagliardo–Nirenberg–Sobolev inequalities for functions defined on cubes. We recall that we have stated and proved in Corollary 2.64 that, if $2 < q \le \infty$ for $d = 1$ and $2 < q < 2d/(d-2)$ for $d \ge 2$, then, for any cube $Q \subset \mathbb{R}^d$ and any function $u \in H^1(Q)$,

$$\int_Q |\nabla u|^2\, dx + |Q|^{-2/d} \int_Q |u|^2\, dx \ge S'_{q,d}|Q|^{-2\left(\frac{1}{q} - \frac{d-2}{2d}\right)} \left(\int_Q |u|^q\, dx\right)^{2/q}. \tag{4.77}$$

Moreover, if, in addition, $\int_Q u\, dx = 0$ then

$$\int_Q |\nabla u|^2\, dx \ge K_{q,d}|Q|^{-2\left(\frac{1}{q} - \frac{d-2}{2d}\right)} \left(\int_Q |u|^q\, dx\right)^{2/q}. \tag{4.78}$$

The constants $S'_{q,d} > 0$ and $K_{q,d} > 0$ here are independent of the cube. We now translate these bounds into information about the spectrum of Schrödinger operators on cubes.

Lemma 4.41 *Let*

$$\gamma \ge 1/2 \quad \text{if } d = 1,$$
$$\gamma > 0 \quad \text{if } d \ge 2.$$

Then there are constants $\sigma_{\gamma,d}, \tau_{\gamma,d} < \infty$ such that, for any cube $Q \subset \mathbb{R}^d$ and for any $0 \le V \in L^{\gamma + d/2}(Q)$,

$$N(-|Q|^{-2/d}, -\Delta_Q^N - V) = 0 \quad \text{if} \quad |Q|^{2\gamma/d} \int_Q V^{\gamma+d/2}\, dx \le \sigma_{\gamma,d} \tag{4.79}$$

and

$$N(0, -\Delta_Q^N - V) \le 1 \quad \text{if} \quad |Q|^{2\gamma/d} \int_Q V^{\gamma+d/2}\, dx \le \tau_{\gamma,d}. \tag{4.80}$$

Proof Given γ as in the proposition, we define q by $(\gamma + d/2)^{-1} + 2/q = 1$. Combining (4.77) with Hölder's inequality, we find, for all $u \in H^1(Q)$,

$$\int_Q V|u|^2 \, dx$$

$$\leq \left(\int_Q V^{\gamma+d/2} \, dx \right)^{\frac{1}{\gamma+d/2}} \left(\int_Q |u|^q \, dx \right)^{2/q}$$

$$\leq \left(\int_Q V^{\gamma+d/2} \, dx \right)^{\frac{1}{\gamma+d/2}} (S'_{q,d})^{-1} |Q|^{2\left(\frac{1}{q} - \frac{d-2}{2d}\right)} \int_Q \left(|\nabla u|^2 + |Q|^{-2/d} |u|^2 \right) dx$$

$$= \left(\int_Q V^{\gamma+d/2} \, dx \right)^{\frac{1}{\gamma+d/2}} (S'_{q,d})^{-1} |Q|^{\frac{2\gamma}{d(\gamma+d/2)}} \int_Q \left(|\nabla u|^2 + |Q|^{-2/d} |u|^2 \right) dx .$$

Thus, if $\int_Q V^{\gamma+d/2} \, dx \leq (S'_{q,d})^{\gamma+d/2} |Q|^{-2\gamma/d}$, then, for all $u \in H^1(Q)$,

$$\int_Q \left(|\nabla u|^2 - V|u|^2 \right) dx \geq -|Q|^{-2/d} \int_Q |u|^2 \, dx .$$

By the variational principle, this implies the first assertion. For the proof of the second one we argue similarly, but using (4.78). We find, for all $u \in H^1(Q)$ with $\int_Q u \, dx = 0$,

$$\int_Q V|u|^2 \, dx \leq \left(\int_Q V^{\gamma+d/2} \, dx \right)^{\frac{1}{\gamma+d/2}} (K_{q,d})^{-1} |Q|^{\frac{2\gamma}{d(\gamma+d/2)}} \int_Q |\nabla u|^2 \, dx .$$

Thus, if $\int_Q V^{\gamma+d/2} \, dx \leq (K_{q,d})^{\gamma+d/2} |Q|^{-2\gamma/d}$, then, for all $u \in H^1(Q)$ with $\int_Q u \, dx = 0$,

$$\int_Q \left(|\nabla u|^2 - V|u|^2 \right) dx \geq 0 .$$

Again by the variational principle, this implies the second assertion. $\qquad \square$

Now we are ready to prove the main result of this section.

Proof of Theorem 4.38 The variational principle implies we need only consider $V \geq 0$. By the approximation argument in Theorem 4.45 below, we may assume that V has compact support. We apply Proposition 4.39 with $W = V^{\gamma+d/2}$, $\alpha = 2\gamma/d$, and

$$A = 2^{-d(\gamma+d/2)} \min\{\sigma_{\gamma,d}, \tau_{\gamma,d}\} .$$

In view of Lemma 4.40 it suffices to bound $\mathrm{Tr} \left(-\Delta^N_{Q_j} - 2^d V_j \right)^\gamma_-$ for each j.

It follows from (4.79), (4.80) and the choice of A that for any j, $-\Delta_{Q_j}^{\mathrm{N}} - 2^d V_j$ has at most one negative eigenvalue and this eigenvalue is $\geq -|Q_j|^{-2/d}$. Thus,

$$\mathrm{Tr}\left(-\Delta_{Q_j}^{\mathrm{N}} - 2^d V_j\right)_-^\gamma \leq |Q_j|^{-2\gamma/d} = A^{-1} \int_{Q_j} V^{\gamma+d/2}\,dx\,.$$

Summing this inequality over j and using Lemma 4.40, we obtain

$$\mathrm{Tr}\,(-\Delta - V)_-^\gamma \leq \sum_j \mathrm{Tr}\left(-\Delta_{Q_j}^{\mathrm{N}} - 2^d V_j\right)_-^\gamma \leq A^{-1} \sum_j \int_{Q_j} V^{\gamma+d/2}\,dx$$

$$\leq A^{-1} 2^d \int_{\mathbb{R}^d} V^{\gamma+d/2}\,dx.$$

(The assumption in Lemma 4.40 can be verified as in the proof of Proposition 4.3, but with (4.77) instead of Theorem 2.46.) This completes the proof of the theorem. □

4.6.2 Optimality of the conditions on γ

The inequalities of Cwikel–Lieb–Rozenblum and Lieb–Thirring in Theorems 4.31 and 4.38 say that the negative eigenvalues of the Schrödinger operator $-\Delta - V$ in $L^2(\mathbb{R}^d)$ satisfy

$$\mathrm{Tr}\,(-\Delta - V)_-^\gamma \leq L_{\gamma,d} \int_{\mathbb{R}^d} V(x)_+^{\gamma+\frac{d}{2}}\,dx\,, \tag{4.81}$$

provided that

$$\gamma \geq \frac{1}{2} \quad \text{if} \quad d = 1, \qquad \gamma > 0 \quad \text{if} \quad d = 2, \qquad \gamma \geq 0 \quad \text{if} \quad d \geq 3. \tag{4.82}$$

Here the left side of (4.81) is understood as the number of negative eigenvalues, counting multiplicities, if $\gamma = 0$.

In this subsection, we show that assumptions (4.82) on γ are best possible, in the sense that for $0 \leq \gamma < 1/2$ in $d = 1$ and for $\gamma = 0$ in $d = 2$ there is no constant $L_{\gamma,d}$ such that (4.81) holds for all V.

In fact, we show that for $0 \leq \gamma < 1/2$ in $d = 1$ and for $\gamma = 0$ in $d = 2$ there is not even a constant $L_{\gamma,d}^{(1)}$ such that, for all $0 \leq V \in L^\infty(\mathbb{R}^d)$ with compact support,

$$(\inf \sigma(-\Delta - V))_-^\gamma \leq L_{\gamma,d}^{(1)} \int_{\mathbb{R}^d} V(x)_+^{\gamma+\frac{d}{2}}\,dx\,. \tag{4.83}$$

Clearly (4.83) is weaker than (4.81), so the latter cannot hold either. Here, when $\gamma = 0$, the left side of (4.83) is understood as 1 if $-\Delta - V$ has a negative eigenvalue, and as 0 otherwise. The validity of (4.83) under assumptions (4.82) was mentioned briefly in Remark 4.5 and will be discussed in more detail in §5.1.2.

A common feature of the proofs of non-validity of (4.83) is the appearance of the functions (2.60) and (2.61) from §2.7.2.

We begin with the non-validity of (4.83) for $0 < \gamma < 1/2$ in dimension $d = 1$.

Proposition 4.42 *Let $d = 1$ and $0 < \gamma < 1/2$. Then there is no constant $C_\gamma < \infty$ such that, for all $0 \leq V \in L^\infty(\mathbb{R})$ with compact support,*

$$\inf \sigma \left(-\frac{d^2}{dx^2} - V \right) \geq -\left(C_\gamma \int_{\mathbb{R}} V(x)_+^{\gamma+1/2} \, dx \right)^{1/\gamma}.$$

Proof Let us take $V = v\chi_{(-1,1)}$, where $v > 0$ is a parameter that we will eventually choose to be small. We claim that the lowest eigenvalue $-E_1$ of the corresponding Schrödinger operator satisfies

$$-E_1 \leq -\frac{3}{16}v^2. \tag{4.84}$$

Accepting this for the moment, we would conclude, if the inequality in the proposition were true, that

$$\frac{3}{16}v^2 \leq \left(2C_\gamma v^{\gamma+1/2} \right)^{1/\gamma}.$$

Since $0 < \gamma < 1/2$, we obtain a contradiction in the limit $v \to 0$.

To prove (4.84), we consider the functions β_R from (2.60) with $R > 0$. Then

$$\int_{\mathbb{R}} (\beta_R')^2 \, dx = \frac{2}{R} \quad \text{and} \quad \int_{\mathbb{R}} \beta_R^2 \, dx = \frac{8R}{3}, \tag{4.85}$$

and, for our choice of V if $R \geq 1$,

$$\int_{\mathbb{R}} V\beta_R^2 \, dx = 2v.$$

We will choose $R > 1/v$ and conclude from the variational principle that $-\frac{d^2}{dx^2} - V$ has a negative eigenvalue, $-E_1$, that satisfies

$$-E_1 \leq -\frac{3(vR - 1)}{4R^2}.$$

Taking $R = 2/v$, we obtain (4.84). This proves the proposition. $\qquad\square$

Next, we discuss the non-validity of (4.83) for $\gamma = 0$ in dimensions $d = 1, 2$.

Proposition 4.43 *Let $d = 1, 2$ and $V \in L^1(\mathbb{R}^d)$ with $\int_{\mathbb{R}^d} V \, dx > 0$. If $d = 2$, assume, in addition, that V is as in Proposition 4.1. Then*

$$N(0, -\Delta - V) \geq 1.$$

Proof Let $d = 1$ and consider β_R as in the proof of Proposition 4.42. By dominated convergence,

$$\int_{\mathbb{R}} V\beta_R^{'2} \, dx \to \int_{\mathbb{R}} V \, dx > 0 \qquad \text{as } R \to \infty.$$

This and the first identity in (4.85) imply that there is an $R_0 > 0$ such that

$$\int_{\mathbb{R}} \left((\beta'_{R_0})^2 - V\beta_{R_0}^2 \right) dx < 0.$$

Therefore, by the variational principle, $N(0, -\frac{d^2}{dx^2} - V) \geq 1$ as claimed.

Now let $d = 2$ and consider the functions β_R from (2.61) with $R > 0$. Then

$$\int_{\mathbb{R}^2} |\nabla\beta_R|^2 \, dx = 2\pi(\ln R)^{-1} \to 0 \qquad \text{as } R \to \infty.$$

Again, by dominated convergence,

$$\int_{\mathbb{R}^2} V\beta_R^2 \, dx \to \int_{\mathbb{R}^2} V \, dx > 0 \qquad \text{as } R \to \infty,$$

and one concludes the proof as before. $\qquad\qquad\qquad\qquad\qquad\qquad\square$

Remark 4.44 One may also wonder whether for some γ the *reverse* inequality to (4.81) holds. This will be the topic of §4.8. Here we note that, for $\gamma > 0$ in dimension $d = 1$, the reverse of inequality (4.83) does *not* hold. That is, there is no constant $C_\gamma < \infty$ such that, for all $V \in L^\infty(\mathbb{R})$ with compact support,

$$\inf \sigma\left(-\frac{d^2}{dx^2} - V\right) \leq -\left(C_\gamma \int_{\mathbb{R}} V_+^{\gamma+1/2} \, dx\right)^{1/\gamma}. \qquad (4.86)$$

To see this, we take again $V = v\chi_{(-1,1)}$ with a parameter $v > 0$. Since $\|V\|_\infty = v$, the variational principle immediately gives the bound $-E_1 \geq -v$ on the lowest eigenvalue $-E_1$. Thus, if (4.86) held, we would obtain $-v \leq -(2C_\gamma v^{\gamma+1/2})^{1/\gamma}$; that is, $v \leq (2C_\gamma)^{-2}$. This leads to a contradiction by letting $v \to \infty$.

4.7 Extending inequalities and asymptotics

In the previous sections in this chapter, we have proved spectral inequalities and asymptotics for potentials from a dense subset of certain L^p spaces. Now we show how one can "take the closure" of these spectral inequalities and asymptotics, by which we mean to extend them to all p-summable potentials.

Theorem 4.45 *Let*

$$\gamma \geq 1/2 \quad \text{if } d = 1,$$
$$\gamma > 0 \quad \text{if } d = 2,$$
$$\gamma \geq 0 \quad \text{if } d \geq 3.$$

Assume that there is a constant C and a dense subspace $\mathcal{D}$ of $L^{\gamma+\frac{d}{2}}(\mathbb{R}^d)$ such that, for all $V \in \mathcal{D}$,

$$\mathrm{Tr}\,(-\Delta - V)_-^{\gamma} \leq C \int_{\mathbb{R}^d} V_+^{\gamma+\frac{d}{2}}\, dx. \tag{4.87}$$

Then (4.87) holds for any V with $V_+ \in L^{\gamma+\frac{d}{2}}(\mathbb{R}^d)$ and $V_- \in L^1_{\mathrm{loc}}(\mathbb{R}^d)$.

Proof The proof relies on the following general fact. If $\gamma \geq 0$ and $\rho > 1$, then there is a constant $C_\rho < \infty$ such that for any self-adjoint, lower semibounded operators A and B one has

$$\mathrm{Tr}(A + B)_-^{\gamma} \leq \rho\,\mathrm{Tr}\,A_-^{\gamma} + C_\rho\,\mathrm{Tr}\,B_-^{\gamma}; \tag{4.88}$$

see Corollary 1.30 and Proposition 1.40. (In fact, if $0 \leq \gamma \leq 1$, one can take $\rho = 1$ and $C_\rho = 1$; see Corollaries 1.30 and 1.37, and Proposition 1.43.) Here, as usual, for $\gamma = 0$ we use the convention $\mathrm{Tr}\,H_-^0 = N(0, H)$.

We fix parameters $0 < \delta < 1$ and $\kappa > 0$ and apply inequality (4.88) with

$$A = -(1 - \delta)\Delta - W, \qquad B = -\delta\Delta - V + W + \kappa,$$

where $W \in \mathcal{D}$. Thus, using the assumed inequality (4.87) with V replaced by $(1 - \delta)^{-1}W \in \mathcal{D}$, we obtain

$$\mathrm{Tr}\,(-\Delta - V + \kappa)_-^{\gamma} \leq \rho(1 - \delta)^{-d/2}C \int_{\mathbb{R}^d} W_+^{\gamma+d/2}\, dx$$
$$+ C_\rho\,\mathrm{Tr}\,(-\delta\Delta - V + W + \kappa)_-^{\gamma}. \tag{4.89}$$

Let $(W_n) \subset \mathcal{D}$ be a sequence such that $W_n \to V_+$ in $L^{\gamma+d/2}(\mathbb{R}^d)$. We will show that

$$-\delta\Delta - V + W_n \geq -\kappa \qquad \text{for all sufficiently large } n, \text{ depending on } \delta \text{ and } \kappa. \tag{4.90}$$

From this, it will follow using (4.89) and the fact that $(W_n)_+ \to V_+$ in $L^{\gamma+d/2}(\mathbb{R}^d)$ that

$$\mathrm{Tr}\,(-\Delta - V + \kappa)_-^{\gamma} \leq \rho(1 - \delta)^{-d/2}C \int_{\mathbb{R}^d} V_+^{\gamma+d/2}\, dx.$$

Since $\rho > 1$ and $\delta \in (0, 1)$ are arbitrary, we find

$$\mathrm{Tr}\,(-\Delta - V + \kappa)_-^{\gamma} \leq C \int_{\mathbb{R}^d} V_+^{\gamma+d/2}\, dx.$$

Finally, we can let $\kappa \to 0$ and use monotone convergence to obtain the assertion of the theorem.

It remains to prove (4.90). First, we assume $\gamma \geq 1/2$ if $d = 1$ and $\gamma > 0$ if $d \geq 2$. Then the bound (4.11) implies that

$$-\delta\Delta - V + W_n \geq -\delta\Delta - V_+ + W_n$$

$$\geq -\delta^{\, d/(2\gamma)} C_{\gamma+d/2,d} \left(\int_{\mathbb{R}^d} (V_+ - W_n)_+^{\gamma+d/2} \, dx \right)^{1/\gamma}.$$

Thus the inequality in (4.90) holds provided that

$$\int_{\mathbb{R}^d} (V_+ - W_n)_+^{\gamma+d/2} \, dx \leq C_{\gamma+d/2,d}^{-\gamma} \delta^{d/2} \kappa^\gamma,$$

and, since $W_n \to V_+$ in $L^{\gamma+d/2}(\mathbb{R}^d)$, this is true if n is sufficiently large, depending on δ and κ.

In order to prove (4.90) for $\gamma = 0$ if $d \geq 3$, we apply (4.12) and infer that $-\delta\Delta - V + W_n \geq 0$ provided that

$$\int_{\mathbb{R}^d} (V_+ - W_n)_+^{d/2} \, dx \leq C_d^{-d/2} \delta^{d/2},$$

which again holds for all sufficiently large n, depending on δ. This completes the proof of the theorem. $\square$

Next, we show that the Cwikel–Lieb–Rozenblum and Lieb–Thirring inequalities allow us to extend Weyl-type asymptotics to a rather large class of potentials.

Theorem 4.46   *Let*

$$\gamma \geq 1/2 \quad \text{if } d = 1,$$
$$\gamma > 0 \quad \text{if } d = 2,$$
$$\gamma \geq 0 \quad \text{if } d \geq 3,$$

and let $V \in L^{\gamma+d/2}(\mathbb{R}^d)$. Then

$$\lim_{\alpha\to\infty} \alpha^{-\gamma-d/2} \operatorname{Tr}(-\Delta - \alpha V)_-^\gamma = L_{\gamma,d}^{\mathrm{cl}} \int_{\mathbb{R}^d} V_+^{\gamma+d/2} \, dx.$$

The assumptions on γ are precisely those under which the Cwikel–Lieb–Rozenblum and Lieb–Thirring inequalities hold. Remarkably, for $\gamma = 0$ in dimensions $d = 1, 2$, there are potentials in $L^{d/2}(\mathbb{R}^d)$ for which the asymptotics do *not* hold; see the references in §4.9. Finally, we remark that Theorem 4.46 remains valid under a weaker assumption on V, namely $V_+ \in L^{\gamma+d/2}(\mathbb{R}^d)$ and $V_- \in L^1_{\mathrm{loc}}(\mathbb{R}^d)$; see §4.9.

Proof We will use inequality (4.88), as before. Let $V \in L^{\gamma+d/2}(\mathbb{R}^d)$ and let W be a continuous function with compact support. With a parameter $\delta \in (0, 1)$ we decompose

$$-\Delta - \alpha V = (-(1 - \delta)\Delta - \alpha W) + (-\delta\Delta - \alpha V + \alpha W). \qquad (4.91)$$

Using (4.88), we find

$$\mathrm{Tr}(-\Delta - \alpha V)_-^\gamma \leq \rho\, \mathrm{Tr}(-(1 - \delta)\Delta - \alpha W)_-^\gamma + C_\rho\, \mathrm{Tr}(-\delta\Delta - \alpha V + \alpha W)_-^\gamma.$$

By the Cwikel–Lieb–Rozenblum and Lieb–Thirring inequalities from Theorems 4.31 and 4.38, we have, with $C := L_{\gamma,d}$,

$$\mathrm{Tr}(-\delta\Delta - \alpha V + \alpha W)_-^\gamma = \delta^\gamma\, \mathrm{Tr}(-\Delta - \alpha\delta^{-1}(V - W))_-^\gamma$$
$$\leq C\delta^{-d/2}\alpha^{\gamma+d/2} \int_{\mathbb{R}^d} (V - W)_+^{\gamma+d/2}\, dx$$

and, by the Weyl asymptotics from Theorems 4.28 and 4.29, we have

$$\lim_{\alpha\to\infty} \alpha^{-\gamma-d/2}\, \mathrm{Tr}(-(1 - \delta)\Delta - \alpha W)_-^\gamma$$
$$= \lim_{\alpha\to\infty} \alpha^{-\gamma-d/2}(1 - \delta)^\gamma\, \mathrm{Tr}(-\Delta - (1 - \delta)^{-1}\alpha W)_-^\gamma$$
$$= (1 - \delta)^{-d/2} L_{\gamma,d}^{\mathrm{cl}} \int_{\mathbb{R}^d} W_+^{\gamma+d/2}\, dx.$$

Thus,

$$\limsup_{\alpha\to\infty} \alpha^{-\gamma-d/2}\, \mathrm{Tr}(-\Delta - \alpha V)_-^\gamma \leq \rho(1 - \delta)^{-d/2} L_{\gamma,d}^{\mathrm{cl}} \int_{\mathbb{R}^d} W_+^{\gamma+d/2}\, dx$$
$$+ C_\rho C \delta^{-d/2} \int_{\mathbb{R}^d} (V - W)_+^{\gamma+d/2}\, dx.$$

Choosing a sequence (W_n) of continuous, compactly supported functions such that $W_n \to V$ in $L^{\gamma+d/2}(\mathbb{R}^d)$, we obtain

$$\limsup_{\alpha\to\infty} \alpha^{-\gamma-d/2}\, \mathrm{Tr}(-\Delta - \alpha V)_-^\gamma \leq \rho(1 - \delta)^{-d/2} L_{\gamma,d}^{\mathrm{cl}} \int_{\mathbb{R}^d} V_+^{\gamma+d/2}\, dx.$$

Finally, letting $\delta \to 0$ and $\rho \to 1$, gives

$$\limsup_{\alpha\to\infty} \alpha^{-\gamma-d/2}\, \mathrm{Tr}(-\Delta - \alpha V)_-^\gamma \leq L_{\gamma,d}^{\mathrm{cl}} \int_{\mathbb{R}^d} V_+^{\gamma+d/2}\, dx,$$

as claimed.

In order to prove an asymptotic lower bound, we decompose

$$-(1 + \delta)\Delta - \alpha W = (-\Delta - \alpha V) + (-\delta\Delta + \alpha V - \alpha W)$$

and find, again using (4.88),

$$\mathrm{Tr}(-\Delta - \alpha V)^\gamma_- \geq \rho^{-1} \mathrm{Tr}(-(1+\delta)\Delta - \alpha W)^\gamma_- - \rho^{-1} C_\rho \mathrm{Tr}(-\delta\Delta + \alpha V - \alpha W)^\gamma_- .$$

By the same argument as before, this implies

$$\liminf_{\alpha\to\infty} \alpha^{-\gamma-d/2} \mathrm{Tr}(-\Delta - \alpha V)^\gamma_- \geq \rho^{-1}(1+\delta)^{-d/2} L^{\mathrm{cl}}_{\gamma,d} \int_{\mathbb{R}^d} W^{\gamma+d/2}_+ \, dx$$
$$- \rho^{-1} C_\rho C \delta^{-d/2} \int_{\mathbb{R}^d} (W - V)^{\gamma+d/2}_+ \, dx ,$$

and, letting first $W_n \to V$ in $L^{\gamma+d/2}(\mathbb{R}^d)$ and then $\delta \to 0$ and $\rho \to 1$, we find

$$\liminf_{\alpha\to\infty} \alpha^{-\gamma-d/2} \mathrm{Tr}(-\Delta - \alpha V)^\gamma_- \geq L^{\mathrm{cl}}_{\gamma,d} \int_{\mathbb{R}^d} V^{\gamma+d/2}_+ \, dx .$$

This completes the proof. $\qquad\qquad\square$

Finally, we turn to a converse result. So far, we have shown that for potentials in $L^{\gamma+d/2}(\mathbb{R}^d)$ (for appropriate γ and d) one has Weyl asymptotics. Now we show that, at least for positive potentials, Weyl asymptotics imply inclusion in $L^{\gamma+d/2}(\mathbb{R}^d)$.

Proposition 4.47 *Let $0 \leq V \in L^1_{\mathrm{loc}}(\mathbb{R}^d)$ and assume that for any $\theta > 0$ there is a $C < \infty$ such that (4.5) holds. Let*

$$\begin{aligned} \gamma &\geq 1/2 & \text{if } d = 1, \\ \gamma &> 0 & \text{if } d = 2, \\ \gamma &\geq 0 & \text{if } d \geq 3, \end{aligned}$$

and assume that

$$\liminf_{\alpha\to\infty} \alpha^{-\gamma-d/2} \mathrm{Tr}(-\Delta - \alpha V)^\gamma_- < \infty .$$

Then $V \in L^{\gamma+d/2}(\mathbb{R}^d)$ and

$$\lim_{\alpha\to\infty} \alpha^{-\gamma-d/2} \mathrm{Tr}(-\Delta - \alpha V)^\gamma_- = L^{\mathrm{cl}}_{\gamma,d} \int_{\mathbb{R}^d} V^{\gamma+d/2}_+ \, dx .$$

Proof For $M, R > 0$, let $V_{M,R} := \min\{V, M\}\chi_{\{|\cdot| < R\}}$. Then $-\Delta - \alpha V_{M,R} \geq -\Delta - \alpha V$ for all $\alpha > 0$, and therefore, by the variational principle,

$$\mathrm{Tr}(-\Delta - \alpha V_{M,R})^\gamma_- \leq \mathrm{Tr}(-\Delta - \alpha V)^\gamma_- .$$

Thus, by assumption,

$$\liminf_{\alpha\to\infty} \alpha^{-\gamma-d/2} \mathrm{Tr}(-\Delta - \alpha V_{M,R})^\gamma_- \leq \liminf_{\alpha\to\infty} \alpha^{-\gamma-d/2} \mathrm{Tr}(-\Delta - \alpha V)^\gamma_- =: C < \infty .$$

But since $V_{M,R} \in L^{\gamma+d/2}(\mathbb{R}^d)$, Theorem 4.46 implies that

$$\lim_{\alpha \to \infty} \alpha^{-\gamma-d/2} \operatorname{Tr}(-\Delta - \alpha V_{M,R})_-^{\gamma} = L_{\gamma,d}^{\mathrm{cl}} \int_{\mathbb{R}^d} V_{M,R}^{\gamma+d/2} \, dx \,.$$

Thus, $\int_{\mathbb{R}^d} V_{M,R}^{\gamma+d/2} \, dx \leq C \left(L_{\gamma,d}^{\mathrm{cl}}\right)^{-1}$, uniformly in M and R. Monotone convergence as $M \to \infty$ and $R \to \infty$ now implies that $V \in L^{\gamma+d/2}(\mathbb{R}^d)$, and then the last assertion follows again from Theorem 4.46. □

4.8 Reversed Lieb–Thirring inequality

We can use the Cwikel–Lieb–Rozenblum and Lieb–Thirring inequalities to bound $\operatorname{Tr}(-\Delta - V)_-^{\gamma}$ from above in terms of an appropriate integral of V. Now we turn to the complementary problem of bounding this trace from below by the corresponding integral of V. We will focus on the one-dimensional case and briefly comment on the two-dimensional case at the end of this section.

Theorem 4.48 *Let $d = 1$ and $0 < \gamma \leq \frac{1}{2}$. Then there is a constant $L_{\gamma,1}^{(\mathrm{lower})} > 0$ such that, for all $0 \leq V \in L^{\gamma+1/2}(\mathbb{R})$ satisfying (4.5) for some $\theta < 1$ and $C < \infty$, we have*

$$\operatorname{Tr}\left(-\frac{d^2}{dx^2} - V\right)_-^{\gamma} > L_{\gamma,1}^{(\mathrm{lower})} \int_{\mathbb{R}} V(x)^{\gamma+1/2} \, dx \,. \tag{4.92}$$

In the case $\gamma = 1/2$ the lower bound (4.92) holds with constant $L_{\gamma,1}^{(\mathrm{lower})} = 1/4$, even for functions V that are not necessarily non-negative. This result is independently due to Glaser et al. (1978) and Schmincke (1978). We present their proofs in §§5.4 and 5.5.

The inequalities (4.92) for $0 < \gamma < 1/2$ were first proved by Damanik and Remling (2007). Below we give a unified proof which works both for $0 < \gamma < 1/2$ and for $\gamma = 1/2$ and which follows a strategy of Netrusov and Weidl (1996). It requires two preliminary lemmas.

The first states that from a covering as in Proposition 4.39 one can extract a subset of cubes with the properties that the doubles of these cubes are disjoint and such that the "mass" covered by the extracted cubes is still a positive fraction of the total "mass". Since it comes with no additional effort, we state the corresponding result in arbitrary dimension $d \geq 1$.

For an open cube $Q = a + (-\ell/2, \ell/2)^d$, we denote by

$$Q^* := a + (-\ell, \ell)^d$$

the cube with the same center as Q and twice its side length.

Lemma 4.49   *Let $0 \le W \in L^1(\mathbb{R}^d)$, let $A > 0$ and $\alpha > 0$ and let $(Q_j)_{j=1}^M$ be a finite covering of* supp W *by open cubes with multiplicity at most 2^d such that*

$$|Q_j|^\alpha \int_{Q_j} W \, dx = A \qquad \text{for all } 1 \le j \le M .$$

Then there is a subset $J \subset \{1, \ldots, M\}$ such that

$$Q_j^* \cap Q_k^* = \emptyset \qquad \text{for all } j, k \in J \text{ with } j \ne k \tag{4.93}$$

and

$$\sum_{j \in J} \int_{Q_j} W \, dx \ge \frac{1 - 2^{-\alpha}}{2^{4d}} \int_{\mathbb{R}^d} W \, dx . \tag{4.94}$$

Proof   Following Netrusov and Weidl (1996), the strategy of the proof is to always choose the smallest cube and to discard all those whose doubles intersect the double of the chosen cube.

More precisely, we construct the set J inductively. Put $I_0 := \{1, \ldots, M\}$ and $J_0 := \emptyset$. Let $0 \le m < M$ and assume that for $0 \le k \le m$ the sets I_k and J_k have been constructed. If $I_m = \emptyset$, we set $J := J_m$ and we finish the process. Otherwise, we choose $j_{m+1} \in I_m$ such that $\ell_{j_{m+1}} = \min_{j \in I_m} \ell_j$, where ℓ_j denotes the side length of Q_j, and set

$$I_{m+1} := \left\{ j \in I_m : Q_j^* \cap Q_{j_{m+1}}^* = \emptyset \right\} \qquad \text{and} \qquad J_{m+1} := J_m \cup \{j_{m+1}\} .$$

Since $j_{m+1} \in I_m \setminus I_{m+1}$, we have $\#I_{m+1} \le \#I_m - 1$ and so the process terminates after at most M steps.

Let us show that the set J has the claimed properties. Property (4.93) is satisfied by construction. In order to prove (4.94), we introduce the sets

$$K_m := \left\{ j \in I_{m-1} : Q_j^* \cap Q_{j_m}^* \ne \emptyset \right\} \qquad \text{for } 1 \le m \le \#J =: N$$

and note that we have the decomposition

$$\{1, \ldots, M\} = \bigcup_{m=1}^N K_m . \tag{4.95}$$

We fix $1 \le m \le N$. Since $\ell_j \ge \ell_{j_m}$ for every $j \in K_m$, we can decompose K_m as

$$K_m = \bigcup_{n=0}^\infty K_m^n, \qquad K_m^n := \left\{ j \in K_m : 2^n \ell_{j_m} \le \ell_j < 2^{n+1} \ell_{j_m} \right\} .$$

Our next goal is to derive an upper bound on the cardinality $\#K_m^n$ of K_m^n. Since $|Q_j| \ge 2^{nd} |Q_{j_m}|$ for $j \in K_m^n$, we find

$$\sum_{j \in K_m^n} |Q_j| \ge 2^{nd} |Q_{j_m}| \, \#K_m^n . \tag{4.96}$$

Meanwhile, by the definition of the sets K_m and K_m^n, for $j \in K_m^n$, we know that Q_j is contained in the cube with the same center as Q_{j_m} and side length $(2 + 3 \cdot 2^{n+1})\ell_{j_m}$. This, together with the fact that the multiplicity of the covering is at most 2^d, yields

$$\sum_{j \in K_m^n} |Q_j| \le 2^d (2 + 3 \cdot 2^{n+1})^d |Q_{j_m}|. \tag{4.97}$$

Combining (4.96) and (4.97), we obtain

$$\#K_m^n \le \frac{2^d(2 + 3 \cdot 2^{n+1})^d}{2^{nd}} \le 2^{4d}.$$

This is the desired upper bound on the cardinality of K_m^n.

Now, by the properties of the covering (Q_j) and the cardinality bound, we see that

$$\sum_{j \in K_m} \int_{Q_j} W \, dx = A \sum_{n=0}^{\infty} \sum_{j \in K_m^n} |Q_j|^{-\alpha} \le 2^{4d} A \sum_{n=0}^{\infty} 2^{-\alpha n} |Q_{j_m}|^{-\alpha}$$

$$= \frac{2^{4d}}{1 - 2^{-\alpha}} \int_{Q_{j_m}} W \, dx.$$

Summing these inequalities over m and using (4.95), we conclude that

$$\int_{\mathbb{R}^d} W \, dx \le \sum_{j=1}^{M} \int_{Q_j} W \, dx = \sum_{m=1}^{N} \sum_{j \in K_m} \int_{Q_j} W \, dx \le \frac{2^{4d}}{1 - 2^{-\alpha}} \sum_{m=1}^{N} \int_{Q_{j_m}} W \, dx,$$

as claimed. $\qquad \square$

Lemma 4.50 *Let $d = 1$ and $0 < \gamma \le \frac{1}{2}$. Then there are constants $A_\gamma < \infty$ and $C_\gamma > 0$ such that, for any bounded interval $Q \subset \mathbb{R}$ and any $0 \le V \in L^{\gamma+1/2}(Q^*)$ with $|Q|^{2\gamma} \int_Q V^{\gamma+1/2} \, dx = A_\gamma$, we have*

$$\mathrm{Tr}\left(-\Delta_{Q^*}^D - V\right)_-^\gamma \ge C_\gamma \int_Q V^{\gamma+1/2} \, dx.$$

Proof By scaling and translating we may assume that $Q = (-1/2, 1/2)$. We fix a function $0 \le u \in H_0^1(Q^*)$ with $u \equiv 1$ on Q and set

$$A_\gamma := \left(2 \int_{Q^*} |u'|^2 \, dx\right)^{\gamma+1/2}.$$

Using $\int_Q V^{\gamma+1/2} \, dx \ge A_\gamma$, followed by $\int_Q V^{\gamma+1/2} \, dx \le \left(\int_Q V \, dx\right)^{\gamma+1/2}$, we have

$$\frac{\int_{Q^*} (|u'|^2 - V|u|^2) \, dx}{\int_{Q^*} |u|^2 \, dx} \le \frac{\int_{Q^*} |u'|^2 \, dx - \int_Q V \, dx}{\int_{Q^*} |u|^2 \, dx} \le -\frac{\int_{Q^*} |u'|^2 \, dx}{\int_{Q^*} |u|^2 \, dx} =: -c_\gamma.$$

By the variational principle, this means that the operator $-\frac{d^2}{dx^2}\big|_{Q^*}^{\mathrm{D}} - V$ with Dirichlet boundary conditions has a negative eigenvalue $\leq -c_\gamma$. Thus, in view of the inequality $\int_Q V^{\gamma+1/2}\,dx \leq A_\gamma$,

$$\mathrm{Tr}\left(-\frac{d^2}{dx^2}\Big|_{Q^*}^{\mathrm{D}} - V\right)_-^\gamma \geq c_\gamma^\gamma \geq \frac{c_\gamma^\gamma}{A_\gamma}\int_Q V^{\gamma+1/2}\,dx\,.$$

This proves the assertion with the constant $C_\gamma := c_\gamma^\gamma/A_\gamma$.   □

Proof of Theorem 4.48   We first assume that $0 \leq V \in L^{\gamma+1/2}(\mathbb{R})$ has compact support. Let $A > 0$ be a parameter to be determined later. Then, by Proposition 4.39 with $W = V^{\gamma+1/2}$ and $\alpha = 2\gamma > 0$, there is a covering of supp V by open intervals $Q_1,\ldots,Q_M$ of multiplicity at most 2 such that

$$|Q_j|^{2\gamma}\int_{Q_j} V^{\gamma+1/2}\,dx = A \qquad \text{for all } 1 \leq j \leq M\,.$$

Then, by Lemma 4.49, there is a subset $J \subset \{1,\ldots,M\}$ such that (4.93) and (4.94) hold. The first property implies that, if we denote by V_j the restriction of V to Q_j^*,

$$-\frac{d^2}{dx^2} - V \leq \left(\bigoplus_{j\in J}\left(-\frac{d^2}{dx^2}\Big|_{Q_j^*}^{\mathrm{D}} - V_j\right)\right) \oplus \left(-\frac{d^2}{dx^2}\Big|_{\mathbb{R}\setminus\bigcup_{j\in J}\overline{Q_j^*}}^{\mathrm{D}}\right).$$

Thus, for any $\gamma \geq 0$,

$$\mathrm{Tr}\left(-\frac{d^2}{dx^2} - V\right)_-^\gamma \geq \sum_{j\in J}\mathrm{Tr}\left(-\frac{d^2}{dx^2}\Big|_{Q_j^*}^{\mathrm{D}} - V_j\right)_-^\gamma. \qquad (4.98)$$

Lemma 4.50 implies that, if $0 < \gamma \leq 1/2$ and if we choose $A = A_\gamma$, then for any $j \in J$,

$$\mathrm{Tr}\left(-\frac{d^2}{dx^2}\Big|_{Q_j^*}^{\mathrm{D}} - V_j\right)_-^\gamma \geq C_\gamma \int_{Q_j} V^{\gamma+1/2}\,dx\,.$$

Summing these inequalities over $j \in J$ and recalling (4.98) and (4.94), we obtain the claimed inequality.

For general $0 \leq V \in L^{\gamma+1/2}(\mathbb{R})$, not necessarily of compact support, and $R > 0$ we define $V_R := V\chi_{\{|\cdot|<R\}}$. Then $-\frac{d^2}{dx^2} - V \leq -\frac{d^2}{dx^2} - V_R$ and therefore, by the variational principle and the inequality in the compactly supported case,

$$\mathrm{Tr}\left(-\frac{d^2}{dx^2} - V\right)_-^\gamma \geq \mathrm{Tr}\left(-\frac{d^2}{dx^2} - V_R\right)_-^\gamma \geq L_{\gamma,1}^{(\mathrm{lower})}\int_{\mathbb{R}} V_R^{\gamma+1/2}\,dx\,.$$

By monotone convergence, we obtain the claimed assertion as $R \to \infty$. This completes the proof of Theorem 4.48.   □

Remark 4.51 The bound in Theorem 4.48 is invalid for $\gamma = 0$. To see this, let $V(x) := c \min\{\varepsilon^{-2}, |x|^{-2}\}\chi_{\{|x| \leq 1\}}$ with $0 < c \leq 1/4$. Then $\int_{\mathbb{R}} V(x)^{1/2} \, dx \to \infty$ as $\varepsilon \to 0$ but, by the variational principle, it is easy to see that

$$N\left(0, -\frac{d^2}{dx^2} - V\right) \leq 2N\left(0, -\frac{d^2}{dx^2}\Big|_{(0,\infty)}^{\mathrm{D}} - V\right) + 1 = 1,$$

where $-\frac{d^2}{dx^2}\Big|_{(0,\infty)}^{\mathrm{D}}$ denotes the Dirichlet Laplacian on the half-line and the equality follows from Hardy's inequality (Theorem 2.70) and the assumption $c \leq 1/4$.

Remark 4.52 The bound in Theorem 4.48 does not hold for $\gamma > 1/2$. To see this, let $0 \leq V \in L^\infty(\mathbb{R})$ with compact support be non-trivial. Then, for all sufficiently small $\alpha > 0$, the operator $-\frac{d^2}{dx^2} - \alpha V$ has exactly one negative eigenvalue $-E_1(\alpha)$ (see Simon, 1976b) and, by (4.11) with $p = 1$, it follows that $-E_1(\alpha) \geq -c_V \alpha^2$. Thus, if (4.92) is valid, we have

$$c_V^\gamma \alpha^{2\gamma} \geq L_{\gamma,1}^{\mathrm{lower}} \int_{\mathbb{R}} V(x)^{\gamma+1/2} \, dx \, \alpha^{\gamma+1/2}.$$

Letting $\alpha \to 0$, we deduce that $\gamma \leq 1/2$, as claimed.

So far in this section, we have been concerned with lower bounds on eigenvalue sums in one dimension. We now mention, without proof, a lower bound on the number of negative eigenvalues in two dimensions.

Theorem 4.53 *Let $d = 2$. Then, for all $V \in L^1(\mathbb{R}^2)$ satisfying (4.5) for some $\theta < 1$ and $C < \infty$,*

$$N(0, -\Delta - V) \geq \frac{1}{8\pi} \int_{\mathbb{R}^2} V(x) \, dx. \tag{4.99}$$

The constant $1/(8\pi)$ in inequality (4.99) is best possible, as discussed after the statement of Theorem 7.5 below.

In the special case of radial V, Theorem 4.53 is due to Glaser et al. (1978), and we present their proof in §7.2. The fact that a bound of the form (4.99) holds with some constant for not necessarily radial, but non-negative, V is due to Korevaar (1993) and Grigor'yan et al. (2004). Grigor'yan et al. (2016) showed that, if a bound as in the theorem holds with some constant for all non-negative V, then it holds with the same constant, for all V. The bound with the optimal constant $1/(8\pi)$ and, in fact, even a slightly stronger result, was proved by Karpukhin et al. (2021).

4.9 Comments

Section 4.1: Definition of the Schrödinger operator

The mathematical foundation of the theory of Schrödinger operators defined via the method of quadratic forms can be found in the paper of Friedrichs (1934b).

The assertion in Proposition 4.1 that $C_0^\infty(\mathbb{R}^d)$ is a form core appears in Kato (1978), and in Simon (1979b) in the case $V \le 0$, even in the presence of a magnetic field. Our proof via Remark 2.24 is essentially that in Leinfelder and Simader (1981).

The L^p conditions in Proposition 4.3 are convenient for our purposes, but Schrödinger operators can be defined in the sense of quadratic forms for different and larger classes of potentials; see, for instance, Aizenman and Simon (1982), Simon (1982), Birman and Solomyak (1992), Maz'ya and Verbitsky (2002, 2005), and Maz'ya (2007). We do not discuss the definition of Schrödinger operators in the sense of an operator sum, nor the question whether the operator, defined on $C_0^\infty(\mathbb{R}^d)$, has a unique self-adjoint extension. For these topics, we refer to Kato (1951, 1972, 1978), Stummel (1956), Kalf et al. (1975), Simon (1979b) and references therein.

In some cases, the Dirichlet Laplacian on a domain discussed in the previous chapter can be thought of as a Schrödinger operator with a potential that is zero inside, and infinite outside, the domain; see, for instance, Herbst and Zhao (1988).

Section 4.2: Explicitly solvable examples

The explicit spectral analysis in this section relies on the commutation method, which in a more general setting will appear again in §§5.4 and 6.3. The computations in the three examples are classical and can be found in most physics textbooks; see, for instance, Landau and Lifshitz (1958, §§23, 25, 36) or Flügge (1999, §§30–34, 39, 65, 67). The study of the special functions appearing as eigenfunctions for the harmonic oscillator and the Coulomb Hamiltonian goes back at least to the nineteenth century. In the context of the old quantum theory the expression for the eigenvalues of the Coulomb potential appear in Bohr (1913). Expressions for the eigenvalues of the harmonic oscillator are in Heisenberg (1925), Born and Jordan (1925), and Schrödinger (1926b); of the Pöschl–Teller potential in Pöschl and Teller (1933), Epstein (1930), and Eckart (1930); and of the Coulomb Hamiltonian in Pauli (1926) and Schrödinger (1926a). For the benefit of the reader we give a rather detailed exposition of the commutation method, as we have not been able to find one in the literature.

Section 4.3: Basic spectral properties of Schrödinger operators

Our discussion in this section is rooted in Birman (1961). An early reference, where discreteness of the negative spectrum of a certain class of Schrödinger operators is proved, is Friedrichs (1934b). Necessary and sufficient conditions for the discreteness of the negative spectrum for all coupling constants in terms of capacities are given in Maz'ja (1962, 1964); see also Maz'ya (2007, §9).

Corollary 4.15 gives a simple condition for discreteness of (all) the spectrum. Necessary and sufficient conditions for this, which are particularly simple in the one-dimensional case, appear in Molčanov (1953). For optimal conditions in terms of capacities, we refer to Maz'ya and Shubin (2005b); see also Maz'ya (2007, §§16 and 18).

Proposition 4.17 appears in Birman (1961). Necessary and sufficient conditions for both the finiteness and the infiniteness of the negative spectrum for all coupling constants in terms of capacities are given in Maz'ja (1962, 1964); see also Maz'ya (2007, §9).

The fact that r^{-2} behavior of the potential at infinity separates between a finite and an infinite number of negative eigenvalues is classical. In particular, Corollary 4.20 appears in a slightly weaker form in Courant and Hilbert (1953, §VI.5) and Glazman (1966, Chapter IV, Theorem 6), and Proposition 4.21 appears in a slightly weaker form in Courant and Hilbert (1953, §VI.5) and, in the present form, in Glazman (1966, Chapter IV, Theorem 6). Proposition 4.19 for $d = 1$ appears in Birman (1961, Theorem 4.3). This paper also implicitly contains variants for $d \geq 2$, which, however, are weaker than those stated in our Proposition 4.19.

Remarkably, as shown by Pavlov (1966, 1967), for Schrödinger operators with complex-valued potentials the threshold between finitely and infinitely many eigenvalues outside the essential spectrum is given by a decay rate of $\exp(-c\sqrt{r})$; see also Borichev et al. (2022).

The Birman–Schwinger principle is independently due to Birman (1961) and Schwinger (1961). In both papers this principle is used to derive the bound in Proposition 4.27, as well as to give an alternative proof of the following bound due to Bargmann (1952); see also Reed and Simon (1978, Theorem XIII.9) and Birman and Solomyak (1992, Eq. (3.16)).

Theorem 4.54 *Let $V \in L^1_{\mathrm{loc}}(\mathbb{R}_+)$ with $xV_+ \in L^1(\mathbb{R}_+)$. Then the number of negative eigenvalues of the Schrödinger operator $-\frac{d^2}{dx^2}\Big|_{\mathbb{R}_+}^{\mathrm{D}} - V$ in $L^2(\mathbb{R}_+)$ with a Dirichlet boundary condition at the origin satisfies*

$$N\left(0, -\frac{d^2}{dx^2}\Big|_{\mathbb{R}_+}^{\mathrm{D}} - V\right) \leq \int_{\mathbb{R}_+} x\,V(x)_+ \, dx \,.$$

Note that, if $xV_+ \in L^1(\mathbb{R}_+)$, then $x \int_x^\infty V(y)_+ dy \to 0$ as $x \to \infty$. (Indeed, the integrand $x\chi_{(x,\infty)}(y)V(y)_+$ is bounded by the integrable function $yV(y)_+$ and tends to zero pointwise as $x \to \infty$.) Therefore the fact that under this assumption the negative spectrum of $-\frac{d^2}{dx^2}\Big|_{\mathbb{R}_+}^D - V$ consists of at most finitely many eigenvalues follows from (the half-line version of) Proposition 4.19. The novel information of Theorem 4.54 is that it gives a quantitative bound on the number of negative eigenvalues.

As a simple consequence of Theorem 4.54, using the fact that imposing an additional Dirichlet boundary condition creates at most one additional negative eigenvalue, one obtains the bound

$$N\left(0, -\frac{d^2}{dx^2} - V\right) \leq 1 + \int_\mathbb{R} |x|\, V(x)_+\, dx \qquad (4.100)$$

provided that $(1 + |x|)V_+ \in L^1(\mathbb{R})$ and $V_- \in L^1_{\text{loc}}(\mathbb{R})$; see, for instance, Birman and Solomyak (1992, Eq. (3.18)).

The original proof of the Lieb–Thirring inequalities (Lieb and Thirring, 1976) used the Birman–Schwinger principle. As we mentioned, this principle is of fundamental importance in many different areas not covered in this book: see, for instance, Klaus (1982/83), Sigal (1983), Exner and Šeba (1989), Alama et al. (1989), Sobolev (1993, 1996), Safronov (1996), Abramov et al. (2001), Stepin (2001), Aizenman et al. (2006), Hainzl et al. (2008), Pushnitski (2009), Pushnitski (2011), Förster (2008), Latushkin and Sukhtayev (2010), Förster and Weidl (2011), Moser and Seiringer (2017), and Hänel and Weidl (2017). For other textbook presentations of the Birman–Schwinger principle, see Reed and Simon (1978), Simon (1979a), and Birman and Solomyak (1992).

Section 4.4: Weyl's asymptotic formula for Schrödinger operators

The asymptotics in Theorem 4.28 for large coupling constants are equivalent to the semiclassical asymptotics for $-\hbar^2\Delta - V$ as $\hbar \to 0$. An asymptotic formula was derived in the physics literature based on heuristic arguments; see, for instance, Landau and Lifshitz (1958, Chapter 7, §48).

A rigorous proof in the case of a bounded domain $\Omega \subset \mathbb{R}^3$ and of $V \geq 0$ appears in Birman (1961), where he deduces it using the Birman–Schwinger principle from earlier results of Courant (1920). The results were extended to the case of the whole space $\mathbb{R}^d$ and to a larger class of V by Birman and Borzov (1971), based on results by Birman and Solomjak (1970). Results in the one-dimensional case also appeared in Chadan (1968). Other early references are Robinson (1971), Martin (1972), and Tamura (1974b).

While in Theorem 4.28 the Weyl asymptotics were proved only for continuous compactly supported V, as shown in Theorem 4.46, the asymptotics hold for all $V \in L^{d/2}(\mathbb{R}^d)$ in dimension $d \geq 3$. The problem of extending these asymptotics to a class of V as large as possible was emphasized by Birman and Solomyak. So too was the method, discussed in §4.7, of doing this by proving order-sharp a priori bounds. This line of investigation culminated in Rozenblum's proof of Theorem 4.31. Remarkably, Weyl asymptotics can fail in dimensions $d = 1$ and $d = 2$, where such a priori bounds fail; see Birman and Laptev (1996); Naimark and Solomyak (1997).

Similar semiclassical formulas appear in different, but related problems of counting the number of eigenvalues below Λ of $-\Delta - V$ in asymptotic regimes, for instance, in the regime $\Lambda \to \infty$ for V such that $V(x) \to -\infty$ as $|x| \to \infty$ (see, e.g., de Wet and Mandl, 1950; Titchmarsh, 1958, Sec. 17.8) or in the regime $\Lambda \to 0$ for V such that $V(x) \to 0$ slowly as $|x| \to \infty$ (see, e.g., Brownell and Clark, 1961; McLeod, 1961; Tamura, 1974a). More on semiclassical techniques, including higher-order correction terms, can be found in Ivrii (1998, 2019a,b,c,d). Asymptotic formulas in the case where the semiclassical expression is infinite were investigated in, for instance, Laptev (1993), Birman and Solomyak (1991), and the references therein.

Section 4.5: The Cwikel–Lieb–Rozenblum inequality

The CLR inequality is named after the works by Cwikel (1977), Lieb (1976, 1980) (see also Reed and Simon, 1978 for a slightly different proof) and Rozenbljum (1972a, 1976). The work of Rozenblum was motivated by papers of Birman and Solomjak (1966, 1967, 1970, 1972, 1971). Cwikel's and Lieb's work was motivated by Simon (1976a). The proof we presented is Rozenblum's, but we work directly with Schrödinger operators instead of the Birman–Schwinger operator; see also Edmunds and Evans (2018, Theorem 11.5.4).

The proofs of Cwikel, Lieb and Rozenblum are all rather different from each other. Alternative proofs were found by Fefferman (1983), Li and Yau (1983), Conlon (1985) and Frank (2014). Some of these proofs have been extended to more general settings. Generalizations of Cwikel's proof are in, for instance, Birman et al. (1991), Weidl (1999a), and Hundertmark et al. (2018); of Lieb's proof in Rozenblyum and Solomyak (1997) and Molchanov and Vainberg (2010); of Li–Yau's proof in Levin and Solomyak (1997) and Blanchard et al. (1987); and of Fefferman's proof in Chang et al. (1985). In particular, in Levin and Solomyak (1997) it is proved that, under certain assumptions, the CLR inequality is equivalent to the Sobolev inequality; see also Frank et al. (2010).

We state Fefferman's strengthening of the CLR inequality (Fefferman, 1983).

Theorem 4.55 *Let $d \geq 3$ and $p > 1$. Then there are constants $C < \infty$ and*

$c > 0$ *such that, for all* $V \in L_{\text{loc}}^p(\mathbb{R}^d)$ *as in Proposition* 4.1 *and all* $N \in \mathbb{N}$ *with* $N(0, -\Delta - V) \geq CN$, *there are disjoint cubes* $Q_1, \ldots, Q_N$ *such that*

$$|Q_n|^{-1+2p/d} \int_{Q_n} V_+^p \, dx \geq c \qquad \text{for all } n - 1, \ldots, N.$$

Let us show that this theorem with $p = d/2$ (essentially) implies the CLR inequality. Indeed, if $N(0, -\Delta - V) \geq C$ we obtain disjoint cubes $Q_1, \ldots, Q_N$ with $N := [C^{-1}N(0, -\Delta - V)]$ such that $\int_{Q_n} V_+^{d/2} \, dx \geq c$ for $n = 1, \ldots, N$. Thus,

$$\int_{\mathbb{R}^d} V_+^{d/2} \, dx \geq \sum_{n=1}^N \int_{Q_n} V_+^{d/2} \, dx \geq cN \geq c2^{-1}C^{-1}N(0, -\Delta - V).$$

This is the CLR inequality under the assumption $N(0, -\Delta - V) \geq C$. On the other hand, since, by the Sobolev inequality, $\int_{\mathbb{R}^d} V_+^{d/2} \, dx \geq c'$ if $N(0, -\Delta - V) \geq 1$ (see (4.12)), we find that, if $N(0, -\Delta - V) < C$, then $N(0, -\Delta - V) \leq C(c')^{-1} \int_{\mathbb{R}^d} V_+^{d/2} \, dx$. In this way, the CLR inequality follows from Theorem 4.55.

The best-known value of the constant in the CLR inequality in three dimensions is due to Lieb (1976, 1980). For a more thorough discussion of what is known about CLR constants, we refer to §8.4.

Earlier bounds on the number of eigenvalues were given by, among others, Bargmann (1952), Birman (1961), Schwinger (1961) and Calogero (1965). There are different proofs of Bargmann's inequality either by bracketing with respect to zeros of the zero-energy solution (Bargmann, 1952) or via the Birman–Schwinger principle (Birman, 1961; Schwinger, 1961). Calogero's proof of the bound named after him also proceeds via bracketing with respect to zeros of eigenfunctions.

A related bound on the number of negative eigenvalues with a weighted L^p norm on the right side was established by Egorov and Kondratiev (1992, 1996).

Theorem 4.56 *Let* $d \geq 3$ *and* $\alpha > 0$. *Then, if* $V \in L_{\text{loc}}^1(\mathbb{R}^d)$ *with* $V_+ \in L^{(d+\alpha)/2}(\mathbb{R}^d, |x|^\alpha \, dx)$,

$$N(0, -\Delta - V) \leq C_{d,\alpha} \int_{\mathbb{R}^d} V(x)_+^{(d+\alpha)/2} |x|^\alpha \, dx.$$

There are similar bounds in dimensions $d = 1, 2$ and, in fact, the one-dimensional case is closely related to Bargmann's bound (Theorem 4.54). The proof of Theorem 4.56 by Egorov and Kondratiev proceeds by bracketing in the spirit of Rozenblum's proof of the CLR inequality. Birman and Solomyak (1992) strengthened this bound to have a certain weighted weak norm on the right side. The resulting inequality can be saturated for a class of long-range potentials.

There is a large literature on bounds on the number of negative eigenvalues in the two-dimensional case. An early important paper (Solomyak, 1994) contains bounds in Orlicz spaces in the case of Schrödinger operators on open sets of finite measure. Further results are, for instance, in Birman and Laptev (1996), Khuri et al. (2002), Chadan et al. (2003), Molchanov and Vainberg (2012), Shargorodsky (2014), Grigor'yan and Nadirashvili (2015), Laptev and Solomyak (2012), and Laptev and Solomyak (2013).

CLR-type bounds for potentials that are singular with respect to Lebesgue measure appear, for instance, in Frank and Laptev (2008), Rozenblum and Shargorodsky (2021), and Rozenblum and Tashchiyan (2021).

The CLR inequality extends to the case of higher-order and fractional Schrödinger operators, where $-\Delta$ is replaced by $(-\Delta)^s$ for some real (not necessarily integer) number $s > 0$.

Theorem 4.57 *Let $d > 2s > 0$. Then, if $V \in L^1_{\mathrm{loc}}(\mathbb{R}^d)$ with $V_+ \in L^{d/(2s)}(\mathbb{R}^d)$,*

$$N(0, (-\Delta)^s - V) \le C_{d,s} \int_{\mathbb{R}^d} V(x)_+^{d/(2s)}\, dx\,.$$

For integer s, this is due to Rozenbljum (1972a, 1976), whose proof can be modified for arbitrary s; see Frank (2018b) for a related bracketing argument. The result can also be obtained from Cwikel's bound (Cwikel, 1977) or, for $0 < s < 1$, from Lieb's method, as shown by Daubechies (1983).

The CLR inequality extends to the case of magnetic Schrödinger operators, where $-\Delta$ is replaced by $(-i\nabla + A)^2$ with $A \in L^2_{\mathrm{loc}}(\mathbb{R}^d, \mathbb{R}^d)$.

Theorem 4.58 *Let $d \ge 3$. Then, if $A \in L^2_{\mathrm{loc}}(\mathbb{R}^d, \mathbb{R}^d)$ and $V \in L^1_{\mathrm{loc}}(\mathbb{R}^d)$ with $V_+ \in L^{d/2}(\mathbb{R}^d)$,*

$$N(0, (-i\nabla + A)^2 - V) \le C_d \int_{\mathbb{R}^d} V(x)_+^{d/2}\, dx\,.$$

We emphasize that the constant in this theorem can be chosen independently of A. In fact the proof of the CLT inequality by Lieb (1976, 1980), which gives the best constant in dimension $d = 3$, applies in the magnetic case and gives the same constant. This observation is probably due to Avron et al. (1978); see also Rozenblyum and Solomyak (1997) and Frank (2009a) for an abstract comparison between "magnetic" and "non-magnetic" inequalities. It is unknown whether the best constant in the magnetic case (that is, allowing for an arbitrary $A \in L^2_{\mathrm{loc}}(\mathbb{R}^d, \mathbb{R}^d)$) agrees with the best constant in the non-magnetic case (that is, in Theorem 4.31).

We note that the CLR inequality, as well as the LT inequality, involve only V_+. In some applications it would be of interest to have spectral inequalities that take cancellations between V_+ and V_- into account. Such cancellations were discussed by Maz'ya and Verbitsky (2002, 2005) in connection with semiboundedness and infinitesimal form-boundedness, but we are not aware of corresponding results for the number or moments of negative eigenvalues.

Section 4.6: The Lieb–Thirring inequality

Our proof of the Lieb–Thirring inequality in Theorem 4.38 seems to be unpublished, but it is similar in spirit to those in Rozenbljum (1972a, 1976), Egorov and Kondratiev (1995), and Weidl (1996). The original proof by Lieb and Thirring (1976) (see also the exposition in Lieb and Seiringer, 2010) is based on the Birman–Schwinger principle and a trace inequality, which we state now.

Theorem 4.59 *Let* $1 \leq p < \infty$ *and let* A, B *be self-adjoint, non-negative operators such that* $\{x \in \mathrm{dom}\, A^p : A^p x \in \mathrm{dom}\, B^p\}$ *is dense. Then*

$$\mathrm{Tr}(AB^2A)^p \leq \mathrm{Tr}\, A^p B^{2p} A^p \,.$$

In addition to the original proof by Lieb and Thirring (1976) of this theorem there are alternatives in Simon (2005, Corollary 8.2 and its remark), Araki (1990), and Carlen (2010). In the special case where the underlying Hilbert space is $L^2(\mathbb{R}^d)$, and A and B are a multiplication operators in position and momentum space by non-negative functions a and b, respectively, the inequality in Theorem 4.59 turns into

$$\mathrm{Tr}(a(X)b(-i\nabla)^2 a(X))^p \leq (2\pi)^{-d} \int_{\mathbb{R}^d} |a(x)|^{2p}\, dx \int_{\mathbb{R}^d} |b(\xi)|^{2p}\, d\xi \,.$$

This is known as the Kato–Seiler–Simon inequality; see Seiler and Simon (1975) and Reed and Simon (1979, Notes on Theorem XI.20). For further spectral inequalities for operators of the form $a(X)b(-i\nabla)^2 a(X)$, see Birman and Solomjak (1977b), Simon (2005, Chapter 4), and references therein.

The original motivation for the Lieb–Thirring inequality came from the problem of the stability of matter (Lieb and Thirring, 1975), for which Lieb and Thirring used their inequality with $d = 3$ and $\gamma = 1$. More precisely, Lieb and Thirring proved the inequality in the form of Theorem 4.38 and then applied it to the stability of matter problem in an equivalent, dual form, which we will discuss in §7.4. Some alternative proofs for the inequality with $\gamma = 1$ that work directly in the dual setting have appeared since then by Eden and Foias (1991) (in $d = 1$), Kashin (2006) and Barsegyan (2007) (in $d = 2$), Rumin (2011), Sabin (2016), and Nam (2022). A proof by Lundholm and Solovej (2013) is written in two dimensions and in the case of a certain magnetic field, but it extends immediately to the Laplacian in higher dimensions.

While the Lieb–Thirring inequality in its usual form is not valid in $d =$

1 for $0 < \gamma < 1/2$, there are bounds containing two different terms that correspond, intuitively, to weak and strong coupling (Netrusov and Weidl, 1996). In dimension $d = 2$ there is no Lieb–Thirring inequality in the limiting case $\gamma = 0$, but there is a logarithmic replacement due to Kovařík et al. (2007).

Egorov and Kondratiev (1995) proved weighted versions of Lieb–Thirring inequalities, similar to their weighted version of the CLR inequality mentioned in the comments on the previous section.

Lieb–Thirring-type bounds for potentials that are singular with respect to Lebesgue measure appear, for instance, in Frank and Laptev (2008), Rozenblum (2021).

Lieb–Thirring inequalities remain valid for higher-order and fractional order versions of Schrödinger operators where $-\Delta$ is replaced by $(-\Delta)^s$.

Theorem 4.60 *Assume that*

$$\gamma \geq 1 - d/(2s) \quad \text{if } d < 2s,$$
$$\gamma > 0 \quad \text{if } d \geq 2s.$$

Then, if $V \in L^1_{\text{loc}}(\mathbb{R}^d)$ with $V_+ \in L^{\gamma + d/(2s)}(\mathbb{R}^d)$,

$$\operatorname{Tr}\left((-\Delta)^s - V\right)^{\gamma}_- \leq C_{d,\gamma,s} \int_{\mathbb{R}^d} V(x)^{\gamma + d/(2s)}_+ \, dx.$$

This theorem can be proved by the method used for Theorem 4.38. In fact, this is similar to how the endpoint case $\gamma = 1 - d/(2s)$ for $d < 2s$ and integer s is handled in Netrusov and Weidl (1996); see also Frank (2018b) for $0 < s < 1$. The non-endpoint cases can also be handled by the original method of Lieb and Thirring; see also Daubechies (1983) and Egorov and Kondratiev (1995). For another proof in the fractional case, based on an abstract inequality from Birman et al. (1975), see Lieb et al. (1997).

Lieb–Thirring inequalities extend to the case of magnetic Schrödinger operators where $-\Delta$ is replaced by $(-i\nabla + A)^2$.

Theorem 4.61 *Let*

$$\gamma \geq 1/2 \quad \text{if } d = 1,$$
$$\gamma > 0 \quad \text{if } d \geq 2.$$

Then, if $A \in L^2_{\text{loc}}(\mathbb{R}^d, \mathbb{R}^d)$ and $V \in L^1_{\text{loc}}(\mathbb{R}^d)$ with $V_+ \in L^{\gamma + d/2}(\mathbb{R}^d)$,

$$\operatorname{Tr}\left((-i\nabla + A)^2 - V\right)^{\gamma}_- \leq C_{\gamma,d} \int_{\mathbb{R}^d} V(x)^{\gamma + d/2}_+ \, dx.$$

This can be shown using the original proof of the Lieb–Thirring inequality together with the diamagnetic inequality as in Avron et al. (1978).

Variants of magnetic Schrödinger operators are Pauli operators. The question of the validity of analogues of Lieb–Thirring inequalities and the form of them is much more difficult; see Erdős (1995), Sobolev (1996), Erdős and Solovej (1999), Bugliaro et al. (1997), Erdős and Solovej (1997), Laptev and Weidl (2000b), and references therein.

The Hardy–Lieb–Thirring inequality (Ekholm and Frank, 2006) states that in dimension $d \geq 3$ for $\gamma > 0$,

$$\mathrm{Tr}\left(-\Delta - \frac{(d-2)^2}{4}\frac{1}{|x|^2} - V\right)_-^\gamma \leq C_{d,\gamma} \int_{\mathbb{R}^d} V(x)_+^{\gamma+d/2}\, dx\,;$$

see also Frank et al. (2008b), Ekholm and Frank (2008), Frank (2009b), and Laptev (2021).

As mentioned before, under certain assumptions one can show that not only does the CLR inequality imply the Sobolev inequality, but also, conversely, the Sobolev inequality implies the CLR inequality (Levin and Solomyak, 1997; Frank et al., 2010). A similar equivalence holds between the Lieb–Thirring and Gagliardo–Nirenberg inequalities (Frank et al., 2010). Apart from its abstract interest, this observation is also useful in applications (Frank et al., 2008b).

There is a version of the Lieb–Thirring inequality at positive density; that is, for potentials that tend to a constant at infinity (Frank et al., 2013; Frank and Pushnitski, 2015). Nothing is known about sharp constants in this case. For Lieb–Thirring inequalities in the presence of a periodic background potential, see Frank et al. (2008a) and Frank and Simon (2011).

Lieb–Thirring inequalities were generalized to the case of Schrödinger operators with complex-valued potentials in Frank et al. (2006). Since then, many results have been obtained, both for continuous and for discrete Schrödinger operators, and we refer to Borichev et al. (2009), Laptev and Safronov (2009), Hansmann (2011), Demuth et al. (2013), Frank and Sabin (2017a), Frank (2018a), Frank et al. (2016), Bögli (2017), Bögli and Štampach (2020), and the references therein. It is still not clear, however, what the most natural form of a Lieb–Thirring inequality for complex potentials would be.

As the arguments in §4.6.2 show, the non-validity of CLR and Lieb–Thirring inequalities for $0 \leq \gamma < 1/2$ in $d = 1$ and $\gamma = 0$ in $d = 2$ is closely related to properties of Schrödinger operators $-\Delta - \alpha V$ in the weak coupling limit $\alpha \to 0$. The fact that Schrödinger operators $-\Delta - \alpha V$ in one and two dimensions with $\int_{\mathbb{R}^d} V\, dx > 0$ have a negative eigenvalue for all $\alpha > 0$ is stated in Landau and Lifshitz (1958, §45), as is the asymptotics of the lowest eigenvalue as $\alpha \to 0$. A mathematical proof appeared in Simon (1976b); see also Blankenbecler et al. (1977) and Klaus (1977). The question whether there are weakly coupled bound states for a more general class of operators, including higher-order operators, two-dimensional Pauli operators and elliptic systems, has been investigated in Weidl (1999b); see also Arazy and Zelenko (2006) and Hoang et al. (2017).

Birman (1961) studied the relation between spectral properties, Sobolev embedding theorems and the validity of Hardy inequalities. In this relation, he noticed a fundamental difference between the case of dimensions 1 and 2 on one hand and 3 and higher on the other; for more on this we refer to Birman and Solomyak (1992).

Section 4.7: Extending inequalities and asymptotics
Extending Weyl asymptotics to situations of minimal regularity was one of the main goals of the group around Birman and Solomyak, and they emphasized the importance of uniform a priori bounds for achieving this goal.

It is shown in Frank (2022) that Theorem 4.46 remains valid if the assumptions on V are relaxed to $V_+ \in L^{\gamma + d/2}(\mathbb{R}^d)$ and $V_- \in L^1_{\mathrm{loc}}(\mathbb{R}^d)$. The proof uses the method of coherent states.

Remarkably, the conclusion of Theorem 4.46 fails for $\gamma = 0$ in dimensions $d = 1, 2$. More precisely, in these cases there are $0 \le V \in L^{d/2}(\mathbb{R}^d)$ for which

$$\lim_{\alpha \to \infty} \alpha^{-d/2} N(0, -\Delta - \alpha V) = C \int_{\mathbb{R}^d} V^{d/2}\, dx$$

with a finite constant $C > L^{\mathrm{cl}}_{0,d}$; see Birman and Laptev (1996) and Naimark and Solomyak (1997). For higher-order versions of these examples, see Birman et al. (1997). Whether such examples also exist in $d = 1$ for $0 < \gamma < 1/2$ is unclear.

Necessary and sufficient conditions both for the finiteness of the quantity $\limsup_{\alpha \to \infty} \alpha^{-d/2} N(0, -\Delta - \alpha V)$ and for Weyl asymptotics in the case of radial and non-negative V on $\mathbb{R}^2$ were derived in Laptev and Solomyak (2012).

Section 4.8: Reversed Lieb–Thirring inequality
The proof of Theorem 4.48 that we presented is an adaptation of the arguments in Netrusov and Weidl (1996), which concern $\mathrm{Tr}((-\Delta)^s - V)^{1-d/(2s)}_-$ for integer $s > d/2$.

Theorem 4.53 for $V \ge 0$ is due to Karpukhin et al. (2021). (In fact, they prove a slightly stronger inequality.) Arguing as in Grigor'yan et al. (2016) one sees that the bound also holds for V not necessarily non-negative. We note that the results in Karpukhin et al. (2021), and Grigor'yan et al. (2016) are formulated in terms of conformally equivalent Riemannian metrics on $\mathbb{S}^2$. The equivalence between this formulation on $\mathbb{S}^2$ and that for Schrödinger operators on $\mathbb{R}^2$ can be seen using the Birman–Schwinger principle and the stereographic projection as, for instance, in Morpurgo (2002, 1999); see also the proof in Frank (2021,

§4.6). For more on the problem of minimizing eigenvalues in a conformal class and the relation to harmonic maps, see, for instance, Montiel and Ros (1986) and El Soufi and Ilias (1986). For a one-dimensional analogue of Theorem 4.53 for $(-\Delta)^{1/2} - V$, see Shargorodsky (2013).

PART THREE

SHARP CONSTANTS IN LIEB–THIRRING INEQUALITIES

5

Sharp Lieb–Thirring Inequalities

In the previous chapter, we saw that the Cwikel–Lieb–Rozenblum and Lieb–Thirring inequalities

$$\text{Tr}(-\Delta - V)^{\gamma}_{-} \le L_{\gamma,d} \int_{\mathbb{R}^d} V(x)^{\gamma+d/2}_{+} \, dx \tag{5.1}$$

hold for $\gamma \ge 1/2$ if $d = 1$, $\gamma > 0$ if $d = 2$, and $\gamma \ge 0$ if $d \ge 3$. For $\gamma = 0$ the left side of (5.1) is to be understood as the number of negative eigenvalues of $-\Delta - V$. The constants $L_{\gamma,d}$ are independent of V.

In this and the following chapters we are interested in the optimal values of the constants $L_{\gamma,d}$. After discussing some basic properties of these constants, our focus in this chapter will be mainly on the one-dimensional case and we will compute the optimal constants $L_{\gamma,1}$ for $\gamma = 1/2$ and for $\gamma \ge 3/2$. The main result of the next chapter will be the identification of the sharp value of $L_{\gamma,d}$ for $\gamma \ge 3/2$ in arbitrary dimension $d \ge 1$. In all other cases and, in particular, in the physically most relevant case, $\gamma = 1$, $d = 3$, the sharp constant is still not known.

In more detail, the outline of this chapter is as follows. The Weyl-type asymptotic formulas (Theorems 4.28 and 4.29) imply that the constant $L_{\gamma,d}$ in (5.1) needs to satisfy

$$L_{\gamma,d} \ge L^{\text{cl}}_{\gamma,d},$$

where

$$L^{\text{cl}}_{\gamma,d} = \frac{\Gamma(\gamma + 1)}{(4\pi)^{\frac{d}{2}} \Gamma(\gamma + 1 + \frac{d}{2})}$$

is the constant appearing in the Weyl asymptotics. Moreover, an argument by Aizenman and Lieb (1978) implies that $L_{\gamma,d}/L^{\text{cl}}_{\gamma,d}$ is non-increasing in γ; see Lemma 5.2. This monotonicity has the consequence that if $L_{\gamma,d} = L^{\text{cl}}_{\gamma,d}$ for some $\gamma = \gamma_0$, then this identity will also hold for all $\gamma > \gamma_0$.

One of the main results in this chapter, due to Gardner et al. (1974, Equation 3.27) (see also Lieb and Thirring, 1976), is that

$$L_{\frac{3}{2},1} = L_{\frac{3}{2},1}^{\text{cl}} . \tag{5.2}$$

In Theorem 5.31 we will present a proof of this equality due to Benguria and Loss (2000), based on a commutator method. Equality in the corresponding Lieb–Thirring inequality is achieved for

$$V(x) = N(N+1)\cosh^{-2} x$$

for any $N \in \mathbb{N}$; see Lemma 5.16. In §5.5 we present the original proof of (5.2) based on trace formulas that appeared in Gardner et al. (1974); see also Lieb and Thirring (1976). It is interesting to note that the trace formulas play the role of conservation laws for the Korteweg–de Vries equation.

By the Aizenman–Lieb result mentioned before, we conclude from (5.2) that

$$L_{\gamma,1} = L_{\gamma,1}^{\text{cl}} \qquad \text{for all } \gamma \geq 3/2 .$$

Despite these results, it is *not* true that $L_{\gamma,d}$ always coincides with $L_{\gamma,d}^{\text{cl}}$. Indeed, following Lieb and Thirring (1976), in Lemma 5.15 we show that $L_{\gamma,1}$ is strictly larger than $L_{\gamma,1}^{\text{cl}}$ for $1/2 \leq \gamma < 3/2$ in dimension $d = 1$ and, following Helffer and Robert (1990b), in Corollary 5.23 we show that $L_{\gamma,d} > L_{\gamma,d}^{\text{cl}}$ for $0 \leq \gamma < 1$ in any dimension $d \geq 2$. The proof of Lemma 5.15 is based on the family $V(x) = \nu(\nu+1)\cosh^{-2} x$, whereas that of Corollary 5.23 is based on $V(x) = -\omega^2|x|^2 + \Lambda$. The spectra of both families can be explicitly computed; see §§4.2.1 and 4.2.2.

It was conjectured in Lieb and Thirring (1976) that when $d = 1$ the optimal constant for $1/2 < \gamma < 3/2$ is attained within the class of potentials

$$V(x) = \nu(\nu+1)\cosh^{-2} x, \quad \nu > 0,$$

and for $\gamma = 1/2$ in the limiting case of a delta potential. For $1/2 < \gamma < 3/2$ this conjecture is still open.

The Lieb–Thirring conjecture in the case $\gamma = 1/2$ was proved by Hundertmark, Lieb and Thomas (1998). So far, this is the only case where the optimal value of $L_{\gamma,d}$ is known in a situation where it is different from $L_{\gamma,d}^{\text{cl}}$. The Hundertmark–Lieb–Thomas result states that

$$L_{1/2,1} = 2 L_{1/2,1}^{\text{cl}} . \tag{5.3}$$

In Theorem 5.27 we present a proof of (5.3) due to Hundertmark, Laptev and Weidl (2000). It is based on a majorization result for a modified Birman–Schwinger operator.

5.1 Basic facts about Lieb–Thirring constants

In this chapter we consider the Lieb–Thirring inequalities for the negative eigenvalues $-E_j$ of the Schrödinger operator $-\Delta - V(x)$ in $L^2(\mathbb{R}^d)$. These inequalities have the form

$$\sum_j E_j^\gamma \le L_{\gamma,d} \int_{\mathbb{R}^d} V(x)_+^{\gamma + \frac{d}{2}} \, dx \tag{5.4}$$

for some constant $L_{\gamma,d}$, depending on γ and d, but not on V. For $\gamma = 0$, the left side of (5.4) is understood as the number of negative eigenvalues of $-\Delta - V$. We know from Theorems 4.31 and 4.38 that this inequality holds for

$$\gamma \ge \frac{1}{2} \quad \text{if} \quad d = 1; \qquad \gamma > 0 \quad \text{if} \quad d = 2; \qquad \gamma \ge 0 \quad \text{if} \quad d \ge 3. \tag{5.5}$$

Moreover, we have seen in Propositions 4.42 and 4.43 that there is no constant $L_{\gamma,d}$ such that (5.4) holds for $0 \le \gamma < 1/2$ if $d = 1$ and $\gamma = 0$ if $d = 2$.

In the following we will always denote by $L_{\gamma,d}$ the optimal constant in (5.4), which is the object of our interest. Also, we will often write

$$\sum_j E_j^\gamma = \operatorname{Tr}(-\Delta - V)_-^\gamma \ .$$

5.1.1 The semiclassical constant and the Aizenman–Lieb argument

We recall that by Theorem 4.29 for continuous and compactly supported V one has

$$\lim_{\alpha \to \infty} \alpha^{-\gamma - \frac{d}{2}} \operatorname{Tr}(-\Delta - \alpha V)_-^\gamma = L_{\gamma,d}^{\mathrm{cl}} \int_{\mathbb{R}^d} V(x)_+^{\gamma + \frac{d}{2}} \, dx \tag{5.6}$$

with the *semiclassical constant*

$$L_{\gamma,d}^{\mathrm{cl}} = \frac{\Gamma(\gamma + 1)}{(4\pi)^{\frac{d}{2}} \Gamma(\gamma + 1 + \frac{d}{2})}. \tag{5.7}$$

The asymptotic formula (5.6) immediately implies the following inequality.

Lemma 5.1 *For all γ satisfying (5.5) we have $L_{\gamma,d} \ge L_{\gamma,d}^{\mathrm{cl}}$.*

An important question in connection with Lieb–Thirring inequalities is whether for some γ and d one has equality in the lemma, and this question is the subject of most of the remaining chapters.

The following important observation of Aizenman and Lieb (1978) implies, in particular, that if $L_{\gamma,d} = L_{\gamma,d}^{\mathrm{cl}}$ for some γ, then $L_{\sigma,d} = L_{\sigma,d}^{\mathrm{cl}}$ for all $\sigma \ge \gamma$.

Lemma 5.2 *With $L_{\gamma,d}$ denoting the optimal constant in (5.4), for any $d \geq 1$, the quotient $L_{\gamma,d}/L^{cl}_{\gamma,d}$ is non-increasing in γ.*

To simplify some computations in the proof, the following formula will be useful,

$$L^{cl}_{\gamma,d} \int_{\mathbb{R}^d} V_+^{\gamma+\frac{d}{2}} \, dx = \iint_{\mathbb{R}^d \times \mathbb{R}^d} \left(|\xi|^2 - V(x) \right)_-^{\gamma} \frac{dx \, d\xi}{(2\pi)^d}. \tag{5.8}$$

This follows by performing the ξ-integration on the right side and using the beta function identity, (3.16).

Proof Let $\sigma > \gamma$. We shall show that for any $V \in L^{\sigma+\frac{d}{2}}(\mathbb{R}^d)$ one has

$$\text{Tr}\,(-\Delta - V)^\sigma_- \leq \frac{L_{\gamma,d}}{L^{cl}_{\gamma,d}} L^{cl}_{\sigma,d} \int_{\mathbb{R}^d} V(x)_+^{\sigma+\frac{d}{2}} \, dx. \tag{5.9}$$

Clearly, this implies that the sharp constant $L_{\sigma,d}$ satisfies $L_{\sigma,d} \leq \frac{L_{\gamma,d}}{L^{cl}_{\gamma,d}} L^{cl}_{\sigma,d}$, which is the claimed monotonicity statement.

For the proof of (5.9) we may assume that $-\Delta - V$ has at least one negative eigenvalue, since otherwise the bound holds trivially. Note that for $\sigma > \gamma$ and any real E we have

$$E_+^\sigma = \frac{1}{B(\gamma+1, \sigma-\gamma)} \int_0^\infty (E-t)_+^\gamma t^{\sigma-\gamma-1} dt. \tag{5.10}$$

If we use this identity for $E = E_j > 0$, sum over j and note that $-E_j + t$ are eigenvalues of the operator $-\Delta - V + t$, we arrive at

$$\text{Tr}\,(-\Delta - V)^\sigma_- = \frac{1}{B(\gamma+1, \sigma-\gamma)} \int_0^\infty \text{Tr}\,(-\Delta - V + t)^\gamma_- \, t^{\sigma-\gamma-1} dt. \tag{5.11}$$

Using (5.4) and (5.8), we obtain

$$\text{Tr}\,(-\Delta - V)^\sigma_-$$
$$\leq \frac{L_{\gamma,d}/L^{cl}_{\gamma,d}}{B(\gamma+1, \sigma-\gamma)} \int_0^\infty \iint_{\mathbb{R}^d \times \mathbb{R}^d} \left(|\xi|^2 - V(x) + t \right)_-^{\gamma} \frac{dx \, d\xi}{(2\pi)^d} t^{\sigma-\gamma-1} dt.$$

On the other hand, applying the identity (5.10) with $E = -|\xi|^2 + V(x)$ gives

$$\frac{1}{B(\gamma+1, \sigma-\gamma)} \int_0^\infty \left(|\xi|^2 - V(x) + t \right)_-^{\gamma} t^{\sigma-\gamma-1} dt = \left(|\xi|^2 - V(x) \right)_-^{\sigma}. \tag{5.12}$$

Inserting this into the previous equation and using again (5.8), we see that

$$\operatorname{Tr}(-\Delta - V)^{\sigma}_{-} \leq \frac{L_{\gamma,d}}{L^{cl}_{\gamma,d}} \iint_{\mathbb{R}^d \times \mathbb{R}^d} \left(|\xi|^2 - V(x)\right)^{\sigma}_{-} \frac{dx\, d\xi}{(2\pi)^d}$$

$$= \frac{L_{\gamma,d}}{L^{cl}_{\gamma,d}} L^{cl}_{\sigma,d} \int_{\mathbb{R}^d} V(x)^{\sigma+d/2}_{+}\, dx\,.$$

This proves the claimed bound (5.9). □

5.1.2 The one-particle constant

Keller (1961) raised the question of minimizing the lowest eigenvalue among all Schrödinger operators with potentials having a given L^p norm. Independently, Lieb and Thirring (1976) arrived at the same optimization problem and showed that it is intimately related to the problem of finding the sharp constant in Sobolev and Gagliardo–Nirenberg inequalities. The following two propositions, which are implicit in their work, answer Keller's question.

Proposition 5.3 *Let*

$$\gamma \geq 1/2 \quad \text{if } d = 1,$$
$$\gamma > 0 \quad \text{if } d \geq 2.$$

Let

$$L^{(1)}_{\gamma,d} := \frac{\gamma^\gamma (d/2)^{d/2}}{(\gamma + d/2)^{\gamma+d/2}} S^{-\gamma-d/2}_{q,d}, \qquad q := \frac{2(\gamma + d/2)}{\gamma + d/2 - 1}, \qquad (5.13)$$

where $S_{q,d}$ is the optimal constant in the Gagliardo–Nirenberg inequality (2.16). Then, if $\|V_+\|_{\gamma+d/2} < \infty$,

$$\inf \sigma(-\Delta - V) \geq - \left(L^{(1)}_{\gamma,d} \int_{\mathbb{R}^d} V(x)^{\gamma+d/2}_{+} dx \right)^{1/\gamma}. \qquad (5.14)$$

Moreover, $L^{(1)}_{\gamma,d}$ defined in (5.13) is the optimal constant in (5.14).

The proof that follows is essentially the argument used to arrive at (4.11), but following explicitly the dependence of constants.

Proof We note that q defined in (5.13) satisfies

$$\frac{1}{\gamma + d/2} + \frac{2}{q} = 1$$

and that the assumptions on γ imply that

$$2 < q \leq \infty \quad \text{if } d = 1,$$
$$2 < q < 2d/(d-2) \quad \text{if } d \geq 2.$$

Hence the Hölder and Gagliardo–Nirenberg inequalities (see Theorem 2.46) yield, for any $u \in H^1(\mathbb{R}^d)$,

$$\int_{\mathbb{R}^d} \left(|\nabla u|^2 - V|u|^2 \right) dx \geq \|\nabla u\|^2 - \|V_+\|_{\gamma + \frac{d}{2}} \|u\|_q^2$$

$$\geq T - S_{q,d}^{-1} \|V_+\|_{\gamma + \frac{d}{2}} T^\theta \|u\|^{2-2\theta} \tag{5.15}$$

with $T := \|\nabla u\|^2$ and $\theta(d-2) + (1-\theta)d = 2d/q$; that is, $\theta = (d/2)/(\gamma + d/2)$. An elementary calculation gives that for $a \geq 0$,

$$\inf_{t>0} \left(t - at^\theta \right) = -(1-\theta)\theta^{\theta/(1-\theta)} a^{1/(1-\theta)}. \tag{5.16}$$

Inserting this with $a = S_{q,d}^{-1} \|V_+\|_{\gamma+d/2} \|u\|^{2-2\theta}$ into (5.15), we obtain

$$\int_{\mathbb{R}^d} \left(|\nabla u|^2 - V|u|^2 \right) dx \geq -(1-\theta)\theta^{\theta/(1-\theta)} S_{q,d}^{-1/(1-\theta)} \|V_+\|_{\gamma+d/2}^{1/(1-\theta)} \|u\|^2. \tag{5.17}$$

By the variational characterization of the bottom of the spectrum, this implies the first assertion.

To prove the optimality of the constant in (5.13), let us assume that (5.14) holds with some constant L. Then we have, by the variational principle, for any $u \in H^1(\mathbb{R}^d)$ and any $0 \leq V \in L^{\gamma+d/2}(\mathbb{R}^d)$,

$$\int_{\mathbb{R}^d} \left(|\nabla u|^2 - V|u|^2 \right) dx \geq - \left(L \int_{\mathbb{R}^d} V^{\gamma+d/2} dx \right)^{1/\gamma} \|u\|^2.$$

Thus, since V is arbitrary,

$$\int_{\mathbb{R}^d} |\nabla u|^2 dx \geq \sup_{V \geq 0} \left(\int_{\mathbb{R}^d} V|u|^2 dx - \left(L \int_{\mathbb{R}^d} V^{\gamma+d/2} dx \right)^{1/\gamma} \|u\|^2 \right).$$

Computing the above supremum (V will be a constant times $|u|^{q-2}$), we obtain

$$\int_{\mathbb{R}^d} |\nabla u|^2 dx \geq S^{\frac{1}{\theta}} \|u\|^{-\frac{2-2\theta}{\theta}} \|u\|_q^{\frac{2}{\theta}}$$

with S satisfying

$$L S^{\gamma+d/2} = \frac{\gamma^\gamma (d/2)^{d/2}}{(\gamma + d/2)^{\gamma+d/2}}.$$

Since $S \leq S_{q,d}$, we obtain the claimed lower bound $L \geq L_{\gamma,d}^{(1)}$. $\quad\square$

Combining Proposition 5.3 with the optimal value of $S_{q,1}$ in dimension $d = 1$ from Theorem 2.48, we obtain the following result.

Corollary 5.4 *Let $d = 1$ and $\gamma \geq 1/2$. Then the sharp constant in (5.14) equals*

$$L^{(1)}_{\gamma,1} = \frac{1}{\sqrt{\pi}} \frac{\Gamma(\gamma + 1)}{\Gamma(\gamma + \frac{1}{2})} \frac{(\gamma - \frac{1}{2})^{\gamma - \frac{1}{2}}}{(\gamma + \frac{1}{2})^{\gamma + \frac{1}{2}}}, \tag{5.18}$$

with the convention that $(\gamma - 1/2)^{\gamma - 1/2} = 1$ for $\gamma = 1/2$.

We now discuss the cases of equality in inequality (5.14) in Proposition 5.3. As before, $S_{q,d}$ denotes the optimal constant in (2.16), so that $L^{(1)}_{\gamma,d}$, defined in (5.13), is the optimal constant in (5.14).

Proposition 5.5 *Let*

$$\gamma > 1/2 \quad \text{if } d = 1,$$
$$\gamma > 0 \quad \text{if } d \geq 2,$$

and let q be as in (5.13).

(a) *If $0 \not\equiv V_+ \in L^{\gamma + d/2}(\mathbb{R}^d)$ achieves equality in (5.14), then $\inf \sigma(-\Delta - V)$ is a negative eigenvalue of $-\Delta - V$ and any corresponding eigenfunction u achieves equality in (2.16). Moreover,*

$$V = \frac{2q}{d(q - 2)} \frac{\|\nabla u\|^2}{\|u\|^q_q} |u|^{q-2}. \tag{5.19}$$

(b) *Conversely, if $0 \not\equiv u \in H^1(\mathbb{R}^d)$ achieves equality in (2.16), then V, defined by (5.19), achieves equality in (5.14).*

Proof For the proof of part (a) we assume that $0 \not\equiv V_+ \in L^{\gamma + d/2}(\mathbb{R}^d)$ satisfies $\inf \sigma(-\Delta - V) = -\left(L^{(1)}_{\gamma,d} \int_{\mathbb{R}^d} V(x)^{\gamma + d/2}_+ dx\right)^{1/\gamma}$. Then $\inf \sigma(-\Delta - V) < 0$ and it follows from Proposition 4.14 that $\inf \sigma(-\Delta - V)$ is an eigenvalue of $-\Delta - V$.

We now repeat the first part of the proof of Proposition 5.3 with u chosen to be an eigenfunction corresponding to $\inf \sigma(-\Delta-V)$. Consequently, equality needs to hold in each inequality in that proof. In particular, by the characterization of equality in Hölder's inequality (Lieb and Loss, 2001, Theorem 2.3), equality in the first inequality in (5.15) implies that there is a constant $c > 0$ such that

$$V_+ = c|u|^{q-2}. \tag{5.20}$$

Note that this implies that $V \geq 0$. In fact, it is well known (see, e.g., Reed and Simon, 1978, §XIII.12 or Simon, 1982) that any eigenfunction corresponding to $\inf \sigma(-\Delta - V)$ is a complex multiple of a function that is positive almost everywhere. By (5.20) this implies that V_+ is positive almost everywhere, and therefore that $V_- \equiv 0$.

Next, equality in the second inequality in (5.15) implies that u achieves equality in (2.16); that is, with $\theta = d(q - 2)/(2q)$,

$$\frac{\|\nabla u\|^{2\theta}\|u\|^{2(1-\theta)}}{\|u\|_q^2} = S_{q,d} . \tag{5.21}$$

To complete the proof, it remains to compute the value of the constant c in (5.20). Since the infimum in (5.16) is attained exactly at $t = \theta^{1/(1-\theta)}a^{1/(1-\theta)}$, inequality in (5.17) implies that

$$\|\nabla u\|^2 = \theta^{1/(1-\theta)}S_{q,d}^{-1/(1-\theta)}\|V\|_{\gamma+d/2}^{1/(1-\theta)}\|u\|^2 .$$

Inserting (5.20) and (5.21) into this identity, we obtain the claimed value of c. This concludes the proof of part (a).

Now assume that $u \in H^1(\mathbb{R}^d)$ satisfies (5.21) and let V be defined in (5.19). Then by the same steps as in the proof of part (a) we find that

$$\|\nabla u\|^2 - \int_{\mathbb{R}^d} V|u|^2 \, dx = - \left(L_{\gamma,d}^{(1)} \int_{\mathbb{R}^d} V^{\gamma+d/2} \, dx \right)^{1/\gamma} \|u\|^2 .$$

This implies

$$\inf \sigma(-\Delta - V) \leq - \left(L_{\gamma,d}^{(1)} \int_{\mathbb{R}^d} V^{\gamma+d/2} \, dx \right)^{1/\gamma} ,$$

which means that V achieves equality in (5.14), as claimed. $\qquad\square$

Remark 5.6　As we mentioned before Lemma 2.50, for any $2 < q < \infty$ in $d = 1, 2$ and for any $2 < q < 2d/(d - 2)$ in $d \geq 3$, there is a $u \in H^1(\mathbb{R}^d)$ that achieves equality in (2.16). According to Proposition 5.5, this implies that, for any γ as in that proposition, there is an optimal potential $V \not\equiv 0$ achieving equality in (5.14).

We turn now to the analogue of Proposition 5.3 for $\gamma = 0$. It concerns absence of negative eigenvalues for small potentials and, more precisely, the problem of finding the smallest L such that

$$\inf \sigma(-\Delta - V) \geq 0 \qquad \text{if} \quad L \int_{\mathbb{R}^d} V(x)_+^{\frac{d}{2}} \, dx < 1 .$$

The next result says that such a constant exists in dimensions $d \geq 3$.

Proposition 5.7　*Let $d \geq 3$ and let*

$$L_{0,d}^{(1)} := S_d^{-d/2}, \tag{5.22}$$

where S_d is the optimal constant in the Sobolev inequality (2.12). If

$$L_{0,d}^{(1)} \int_{\mathbb{R}^d} V(x)_+^{d/2} \, dx \leq 1, \tag{5.23}$$

then

$$\inf \sigma(-\Delta - V) \geq 0 \,. \tag{5.24}$$

Moreover, $L_{0,d}^{(1)}$, defined in (5.22), is the optimal constant in (5.23) to imply (5.24).

Proof By the Hölder and Sobolev inequalities (Theorem 2.44),

$$\int_{\mathbb{R}^d} \left(|\nabla u|^2 - V|u|^2 \right) dx \geq \|\nabla u\|^2 - \|V_+\|_{d/2} \|u\|_{2d/(d-2)}^2$$

$$\geq \|\nabla u\|^2 \left(1 - S_d^{-1} \|V_+\|_{d/2} \right) \geq 0 \,.$$

By the variational principle, this implies the first part of the proposition. Conversely, if there is a constant C such that $-\Delta - V$ is non-negative for $\|V_+\|_{d/2} \leq C$, then for such V and all $u \in H^1(\mathbb{R}^d)$,

$$\int_{\mathbb{R}^d} \left(|\nabla u|^2 - V|u|^2 \right) dx \geq 0 \,.$$

Therefore, for all $u \in H^1(\mathbb{R}^d)$,

$$\int_{\mathbb{R}^d} |\nabla u|^2 \, dx \geq \sup_{\|V_+\|_{d/2} \leq C} \int_{\mathbb{R}^d} V|u|^2 \, dx \,.$$

Computing the above supremum (V will be a constant times $|u|^{4/(d-2)}$), we obtain

$$\int_{\mathbb{R}^d} |\nabla u|^2 \, dx \geq C \left(\int_{\mathbb{R}^d} |u|^{2d/(d-2)} \, dx \right)^{(d-2)/d} \,.$$

Thus, the sharp constant S_d satisfies $S_d \geq C$, which proves the claimed optimality assertion. □

Combining Proposition 5.7 with the optimal constant in the Sobolev inequality given in (2.19), we obtain the following result.

Corollary 5.8 *Let $d \geq 3$. Then the sharp constant in (5.23) equals*

$$L_{0,d}^{(1)} = \left(\frac{4}{d(d-2)} \right)^{d/2} 2^{-1} \pi^{-(d+1)/2} \Gamma(\tfrac{d+1}{2}) \,. \tag{5.25}$$

Remark 5.9 There is a characterization of optimizers in Proposition 5.7 that is similar to how Proposition 5.5 characterizes optimizers in Proposition 5.3. We state it without proof. If $L_{0,d}^{(1)} \int_{\mathbb{R}^d} V_+^{d/2} \, dx = 1$ and if $\inf \sigma(-\Delta + V_- - \alpha V_+) < 0$

for all $\alpha > 1$, then there is a $0 \not\equiv u \in \dot{H}^1(\mathbb{R}^d)$ satisfying $-\Delta u - Vu = 0$ that achieves equality in (2.12). Moreover,

$$V = \frac{\|\nabla u\|^2}{\|u\|_{2d/(d-2)}^{2d/(d-2)}} |u|^{4/(d-2)}. \qquad (5.26)$$

Conversely, if $0 \not\equiv u \in \dot{H}^1(\mathbb{R}^d)$ achieves equality in (2.12), then V, defined by (5.26), satisfies $L_{0,d}^{(1)} \int_{\mathbb{R}^d} V^{d/2} \, dx = 1$ and $\inf \sigma(-\Delta \quad \alpha V) < 0$ for all $\alpha > 1$.

Remark 5.10   The function $\gamma \mapsto L_{\gamma,d}^{(1)}$ is continuous for

$$\begin{aligned} \gamma &\in [1/2, \infty) \quad \text{if } d = 1, \\ \gamma &\in (0, \infty) \quad \text{if } d = 2, \\ \gamma &\in [d/2, \infty) \quad \text{if } d \geq 3. \end{aligned}$$

This follows from the definitions (5.13) and (5.22) by continuity of $q \mapsto S_{q,d}$; see Lemma 2.50.

Since the Lieb–Thirring inequality implies that

$$E_1^{\gamma} \leq L_{\gamma,d} \int_{\mathbb{R}^d} V(x)_+^{\gamma+d/2} \, dx \qquad \text{if} \quad \gamma > 0,$$

we conclude that

$$L_{\gamma,d} \geq L_{\gamma,d}^{(1)}. \qquad (5.27)$$

Similarly, since the Cwikel–Lieb–Rozenblum inequality implies that

$$N(0, -\Delta - V) = 0 \qquad \text{if} \quad L_{0,d} \int_{\mathbb{R}^d} V(x)_+^{d/2} \, dx < 1,$$

we conclude that (5.27) holds also for $\gamma = 0$. We summarize this discussion.

Lemma 5.11   *For all γ satisfying (5.5) one has $L_{\gamma,d} \geq L_{\gamma,d}^{(1)}$.*

In the context of the Lieb–Thirring inequalities, the constant $L_{\gamma,d}^{(1)}$ is sometimes referred to as the *one-particle constant*.

5.1.3 The Lieb–Thirring conjecture

It follows from Lemmas 5.1 and 5.11 that

$$L_{\gamma,d} \geq \max\left\{ L_{\gamma,d}^{cl}, L_{\gamma,d}^{(1)} \right\}. \qquad (5.28)$$

In Lieb and Thirring (1976) the authors conjectured that equality holds in (5.28); that is, the optimal constant $L_{\gamma,d}$ in the Lieb–Thirring inequality is equal to the maximum of $L_{\gamma,d}^{cl}$ and $L_{\gamma,d}^{(1)}$. For a further discussion of this conjecture, including references for examples and counterexamples, we refer to §8.4. Concerning the maximum appearing in this conjecture, we now prove the following result (Frank et al., 2021b) that was conjectured by Lieb and Thirring.

Proposition 5.12 *In dimensions $1 \leq d \leq 7$ there is a $\gamma_{c,d} > 0$ such that*

$$L_{\gamma,d}^{(1)} > L_{\gamma,d}^{\mathrm{cl}} \text{ if } \gamma < \gamma_{c,d};$$

$$L_{\gamma,d}^{(1)} = L_{\gamma,d}^{\mathrm{cl}} \text{ if } \gamma = \gamma_{c,d};$$

$$L_{\gamma,d}^{(1)} < L_{\gamma,d}^{\mathrm{cl}} \text{ if } \gamma > \gamma_{c,d}.$$

We have $\gamma_{c,1} = 3/2$ and $\gamma_{c,2} < 1$. Moreover in dimensions $d \geq 8$,

$$L_{\gamma,d}^{(1)} < L_{\gamma,d}^{\mathrm{cl}} \quad \text{for all } \gamma \geq 0.$$

The following table gives numerical values of the intersection points $\gamma_{c,d}$ (Lieb and Thirring, 1976; Frank et al., 2021b). One can prove analytically that $\gamma_{c,d} < \gamma_{c,d-1}$ if $2 \leq d \leq 7$; see §5.6.

d	1	2	3	4	5	6	7
$\gamma_{c,d}$	$\frac{3}{2}$	1.1654	0.8627	0.5973	0.3740	0.1970	0.0683

The proof of Proposition 5.12 is somewhat lengthy and will take up the remainder of this subsection. We begin with the case of one dimension, where the explicit expressions for $L_{\gamma,1}^{(1)}$ and $L_{\gamma,1}^{\mathrm{cl}}$ in $d = 1$ give

$$L_{\gamma,1}^{(1)} = 2 \left(\frac{\gamma - 1/2}{\gamma + 1/2} \right)^{\gamma - 1/2} L_{\gamma,1}^{\mathrm{cl}} \tag{5.29}$$

with the convention that $\left(\frac{\gamma-1/2}{\gamma+1/2} \right)^{\gamma-1/2} = 1$ for $\gamma = 1/2$. Therefore, the ratio $L_{\gamma,1}^{(1)}/L_{\gamma,1}^{\mathrm{cl}}$ is strictly decreasing as a function of γ and equal to one at $\gamma = 3/2$. Thus,

$$L_{\gamma,1}^{(1)} > L_{\gamma,1}^{\mathrm{cl}} \text{ if } \gamma < 3/2, \quad L_{\gamma,1}^{(1)} = L_{\gamma,1}^{\mathrm{cl}} \text{ if } \gamma = 3/2, \quad L_{\gamma,1}^{(1)} < L_{\gamma,1}^{\mathrm{cl}} \text{ if } \gamma > 3/2,$$

as claimed in Proposition 5.12.

No closed analytic expression for $L_{\gamma,d}^{(1)}$ is known if $d \geq 2$ and $\gamma > 0$. Nevertheless, we can prove that the quotient $L_{\gamma,d}^{(1)}/L_{\gamma,d}^{\mathrm{cl}}$ is strictly decreasing as a function of γ.

Lemma 5.13 *For any $d \geq 1$, the ratio $L_{\gamma,d}^{(1)}/L_{\gamma,d}^{\mathrm{cl}}$ is strictly decreasing in γ.*

Proof The first part of the proof is similar to that of Lemma 5.2. Let $\sigma > \gamma$. We shall show that, for any $V \in L^{\sigma+d/2}(\mathbb{R}^d)$ such that $-\Delta - V$ has a negative eigenvalue $-E_1$, we have

$$E_1^\sigma \le \frac{L^{(1)}_{\gamma,d}}{L^{\mathrm{cl}}_{\gamma,d}} L^{\mathrm{cl}}_{\sigma,d} \int_{\mathbb{R}^d} V(x)_+^{\sigma+d/2} \, dx. \tag{5.30}$$

Clearly, this implies that the sharp constant $L^{(1)}_{\sigma,d}$ satisfies $L^{(1)}_{\sigma,d} \le \frac{L^{(1)}_{\gamma,d}}{L^{\mathrm{cl}}_{\gamma,d}} L^{\mathrm{cl}}_{\sigma,d}$. This gives the non-strict monotonicity claimed in the lemma. At the end of the proof, we will upgrade this to strict monotonicity.

Identity (5.10) with $E = E_1$ reads

$$E_1^\sigma = \frac{1}{B(\gamma+1, \sigma-\gamma)} \int_0^\infty (E_1 - t)_+^\gamma t^{\sigma-\gamma-1} dt.$$

Note that $-(E_1 - t)$ is the lowest eigenvalue of $-\Delta - V + t$ and, by the variational principle, this is not smaller than $\inf \sigma(-\Delta - (V-t)_+)$. Therefore, by definition of $L^{(1)}_{\gamma,d}$ and by (5.8), we have

$$(E_1 - t)_+^\gamma \le L^{(1)}_{\gamma,d} \int_{\mathbb{R}^d} (V(x) - t)_+^{\gamma+d/2} \, dx$$

$$= \frac{L^{(1)}_{\gamma,d}}{L^{\mathrm{cl}}_{\gamma,d}} \iint_{\mathbb{R}^d \times \mathbb{R}^d} \left(|\xi|^2 - V(x) + t\right)_-^\gamma \frac{dx \, d\xi}{(2\pi)^d}. \tag{5.31}$$

Thus,

$$E_1^\sigma \le \frac{L^{(1)}_{\gamma,d}/L^{\mathrm{cl}}_{\gamma,d}}{B(\gamma+1, \sigma-\gamma)} \int_0^\infty \iint_{\mathbb{R}^d \times \mathbb{R}^d} \left(|\xi|^2 - V(x) + t\right)_-^\gamma \frac{dx \, d\xi}{(2\pi)^d} t^{\sigma-\gamma-1} \, dt.$$

Combining this with (5.12) gives (5.30).

We now proceed to prove the strict inequality $L^{(1)}_{\sigma,d} < \frac{L^{(1)}_{\gamma,d}}{L^{\mathrm{cl}}_{\gamma,d}} L^{\mathrm{cl}}_{\sigma,d}$. For the proof we use the fact that there is an optimal potential $0 \not\equiv V_* \in L^{\sigma+d/2}(\mathbb{R}^d)$ which achieves equality (5.14) with σ instead of γ; see Remark 5.6. We need to show that inequality (5.30) for V_* is strict. To prove this, we argue by contradiction and assume that equality holds in (5.30) for V_*. Repeating the above argument, we see that equality must hold in (5.31) for almost every $t > 0$, and therefore, for almost every $t > 0$, the function $(V_* - t)_+$ needs to achieve equality in (5.14) with the given γ. From the discussion after (5.20) we know that any optimizer for $L^{(1)}_{\gamma,d}$ is positive almost everywhere on $\mathbb{R}^d$. However,

$$(\gamma + d/2) \int_0^\infty |\{V_* > t\}| t^{\gamma+d/2-1} \, dt = \int V_*^{\gamma+d/2} \, dx < \infty,$$

so for any $t > 0$, $(V_* - t)_+$ is non-zero only on a set of finite measure. This is a contradiction. □

We remark that the *strict* monotonicity, which we have just proved for the quotient $L_{\gamma,d}^{(1)}/L_{\gamma,d}^{\mathrm{cl}}$, cannot be expected in general for the ratio $L_{\gamma,d}/L_{\gamma,d}^{\mathrm{cl}}$ in Lemma 5.2. We will see, as a consequence of the optimal Lieb–Thirring inequality, that the latter quotient is constant for $\gamma \geq 3/2$.

The following inequality for $L_{\gamma,d}^{(1)}$ is in the spirit of the method of lifting with respect to the dimension, which is the subject of the next chapter.

Lemma 5.14 *Let $d \geq 2$ and $\gamma \geq 1/2$. Then*

$$L_{\gamma,d}^{(1)} < L_{\gamma,1}^{(1)}\, L_{\gamma+\frac{1}{2},d-1}^{(1)}\,.$$

Proof We prove the lemma only with $\leq$ instead of $<$. This is enough for the application of the lemma in the proof of Proposition 5.12 and we refer to Frank et al. (2021b) for the complete proof.

According to Remark 5.6, there is a $V \geq 0$, $V \not\equiv 0$, achieving equality in (5.14). We write the variables as $x = (x_1, x') \in \mathbb{R} \times \mathbb{R}^{d-1}$ and, for $x_1 \in \mathbb{R}$, we define $W(x_1) := -\inf \sigma(-\Delta_{\mathbb{R}^{d-1}} - V(x_1, \cdot))$. Then for all $u \in H^1(\mathbb{R}^d)$,

$$\int_{\mathbb{R}^d} \left(|\nabla u|^2 - V|u|^2 \right) dx \geq \int_{\mathbb{R}^{d-1}} \int_{\mathbb{R}} \left(\left|\tfrac{du}{dx_1}\right|^2 - W|u|^2 \right) dx_1\, dx'\,,$$

and, consequently,

$$\inf \sigma(-\Delta - V) \geq \inf \sigma\left(-\tfrac{d^2}{dx_1^2} - W \right).$$

Using the optimality of V and by (5.14), we deduce

$$L_{\gamma,d}^{(1)} \int_{\mathbb{R}^d} V(x)_-^{\gamma+d/2}\, dx = (\inf \sigma(-\Delta - V))_-^{\gamma} \leq \left(\inf \sigma\left(-\tfrac{d^2}{dx_1^2} - W \right) \right)_-^{\gamma}$$

$$\leq L_{\gamma,1}^{(1)} \int_{\mathbb{R}} W(x_1)^{\gamma+1/2}\, dx_1\,.$$

On the other hand,

$$W(x_1)^{\gamma+1/2} \leq L_{\gamma+1/2,d-1}^{(1)} \int_{\mathbb{R}^{d-1}} V(x_1, x')^{\gamma+d/2}\, dx'\,.$$

Combining the last two inequalities gives the $\leq$-version of the lemma. □

We now prove the main result of this subsection.

Proof of Proposition 5.12 Step 1. We begin by showing that

$$L_{0,d}^{(1)} > L_{0,d}^{\mathrm{cl}} \text{ if } 3 \leq d \leq 7,$$

$$L_{0,d}^{(1)} < L_{0,d}^{\mathrm{cl}} \text{ if } d \geq 8.$$

Indeed, formulas (5.7) and (5.25) imply that

$$\frac{L_{0,d}^{(1)}}{L_{0,d}^{\text{cl}}} = \left(\frac{1}{d(d-2)}\right)^{\frac{d}{2}} 2^{2d-1}\pi^{-\frac{1}{2}}\Gamma(\tfrac{d+1}{2})\Gamma(1+\tfrac{d}{2}) = \left(\frac{1}{d(d-2)}\right)^{\frac{d}{2}} 2^{d-1}d!,$$
(5.32)

where we used the duplication formula for the gamma function. As a consequence,

$$\frac{L_{0,d}^{(1)}}{L_{0,d}^{\text{cl}}} < \frac{L_{0,d-1}^{(1)}}{L_{0,d-1}^{\text{cl}}} \frac{L_{\frac{d-1}{2},1}^{(1)}}{L_{\frac{d-1}{2},1}^{\text{cl}}} \qquad \text{for all } d \geq 3.$$
(5.33)

Indeed, by (5.32) and (5.29), this is equivalent to $(d-1)(d-3) < (d-2)^2$. As remarked after (5.29), $L_{\gamma,1}^{(1)}/L_{\gamma,1}^{\text{cl}} \leq 1$ if $\gamma \geq 3/2$, so (5.33) implies that $L_{0,d}^{(1)}/L_{0,d}^{\text{cl}}$ is decreasing for $d \geq 3$. Since this quotient is larger than 1 for $d = 7$ and less than 1 for $d = 8$, we obtain the claim.

Step 2. We now show that for $d = 2$ and $\gamma = 1$ we have $L_{1,2}^{(1)} > L_{1,2}^{\text{cl}}$.

Indeed, by (5.7), $L_{1,2}^{\text{cl}} = (8\pi)^{-1}$ and, by Remark 2.51 and relation (5.13),

$$L_{1,2}^{(1)} = 4^{-1}S_{4,2}^{-2} \geq (2\pi)^{-1}\frac{3+2\sqrt{3}}{12+7\sqrt{3}} > (8\pi)^{-1}.$$

Step 3. We show that for $d \geq 2$ and $\gamma \geq 3/2$ we have $L_{\gamma,d}^{(1)} < L_{\gamma,d}^{\text{cl}}$.

To see this, we combine Lemma 5.14 (in the weaker, non-strict form) with the multiplicativity property (3.72) of the semiclassical constant to get

$$\frac{L_{\gamma,d}^{(1)}}{L_{\gamma,d}^{\text{cl}}} \leq \frac{L_{\gamma,1}^{(1)}}{L_{\gamma,1}^{\text{cl}}} \frac{L_{\gamma+\frac{1}{2},d-1}^{(1)}}{L_{\gamma+\frac{1}{2},d-1}^{\text{cl}}}.$$

Iterating this, gives

$$\frac{L_{\gamma,d}^{(1)}}{L_{\gamma,d}^{\text{cl}}} = \prod_{k=0}^{d-1} \frac{L_{\gamma+\frac{k}{2},1}^{(1)}}{L_{\gamma+\frac{k}{2},1}^{\text{cl}}}.$$

If $\gamma \geq 3/2$, by the explicit form of the one-dimensional expressions the factor corresponding to $k = 0$ is at most 1 and the remaining factors are strictly smaller than 1. This proves the claim.

Step 4. We complete the proof of the proposition. The inequality $L_{0,d}^{(1)} < L_{0,d}^{\text{cl}}$ for $d \geq 8$ from Step 1, together with the monotonicity of the quotient $L_{\gamma,d}^{(1)}/L_{\gamma,d}^{\text{cl}}$ from Lemma 5.13, implies that $L_{\gamma,d}^{(1)} < L_{\gamma,d}^{\text{cl}}$ for all $\gamma \geq 0$ in dimensions $d \geq 8$. By the discussion around (5.29) it remains to consider the case $2 \leq d \leq 7$.

From Step 3 we know that $L_{\gamma,d}^{(1)} < L_{\gamma,d}^{cl}$ for all $\gamma \geq 3/2$. On the other hand, Steps 1 and 2 imply that $L_{\gamma,d}^{(1)} > L_{\gamma,d}^{cl}$ for $\gamma = 1$ in $d = 2$ and for $\gamma = 0$ in $3 \leq d \leq 7$. Thus, the strict monotonicity from Lemma 5.13 implies the existence of a $\gamma_{c,d}$ (with $1 < \gamma_{c,2} < 3/2$ in $d = 2$ and $0 < \gamma_{c,d} < 3/2$ in $3 \leq d \leq 7$) such that

$$L_{\gamma,d}^{(1)} > L_{\gamma,d}^{cl} \quad \text{if } \gamma < \gamma_{c,d}, \qquad L_{\gamma,d}^{(1)} < L_{\gamma,d}^{cl} \quad \text{if } \gamma > \gamma_{c,d}.$$

From (5.7) it is clear that $L_{\gamma,d}^{cl}$ is continuous in γ. By Remark 5.10, $L_{\gamma,d}^{(1)}$ is also continous in γ. This proves that $L_{\gamma,d}^{(1)} = L_{\gamma,d}^{cl}$ if $\gamma = \gamma_{c,d}$. $\qquad\square$

5.2 Lieb–Thirring inequalities for special classes of potentials

5.2.1 Lieb–Thirring inequalities for the Pöschl–Teller Hamiltonian

As already noticed by Lieb and Thirring (1976) in the study of the one-dimensional case of their inequality, an important role is played by the family of potentials $V(x) = v(v + 1)\cosh^{-2} x$. We have computed the spectrum of the associated Schrödinger operator explicitly in §4.2.2. In this subsection we compare both sides in the Lieb–Thirring inequality for this particular family of potentials.

Let us start by considering the lowest eigenvalue $-E_1$ of the Schrödinger operators with these potentials. Combining Sz. Nagy's sharp constant in the one-dimensional Gagliardo–Nirenberg inequality and the duality principle in Proposition 5.3 we obtained the explicit value of the constant $L_{\gamma,1}^{(1)}$. Going through the proofs of these two results one can infer that the optimal potential for $L_{\gamma,1}^{(1)}$ is of the form $v(v + 1)\cosh^{-2} x$ for a certain v depending on γ. It is instructive, however, to verify this by a direct computation.

Lemma 5.15 *Let $d = 1$ and $V_v(x) = v(v + 1)\cosh^{-2} x$. Then for all $\gamma \geq 1/2$,*

$$\sup_{v>0} \frac{E_1^\gamma}{\int_\mathbb{R} V_v^{\gamma+1/2}\, dx} = L_{\gamma,1}^{(1)}$$

and the supremum is attained at $v = \gamma - 1/2$ for $\gamma > 1/2$ and in the limit $v \to 0$ for $\gamma = 1/2$.

Proof From Proposition 4.9 we know that the lowest eigenvalue of $-\frac{d^2}{dx^2} - V_v$

is given by $-v^2$. On the other hand, a straightforward computation gives

$$\int_{\mathbb{R}} \frac{dx}{\cosh^{2\gamma+1} x} = 2^{2\gamma+1} \int_{\mathbb{R}} \frac{dx}{(e^x + e^{-x})^{2\gamma+1}} = 2^{2\gamma+1} \int_0^\infty \frac{dt}{t(t + t^{-1})^{2\gamma+1}}$$

$$= 2^{2\gamma+1} \int_0^\infty \frac{dt\, t^{2\gamma}}{(t^2 + 1)^{2\gamma+1}} = 2^{2\gamma} \int_0^\infty \frac{du\, u^{\gamma-1/2}}{(u + 1)^{2\gamma+1}}$$

$$= 2^{2\gamma} \frac{\Gamma(\gamma + 1/2)^2}{\Gamma(2\gamma + 1)} = \sqrt{\pi}\, \frac{\Gamma(\gamma + 1/2)}{\Gamma(\gamma + 1)},$$

where we used formula (3.16) for the beta function, together with a change of variables $s = (u + 1)^{-1}$, as well as the duplication formula for the gamma function, $\Gamma(z)\Gamma(z + 1/2) = 2^{1-2z} \sqrt{\pi}\, \Gamma(2z)$. Therefore

$$\frac{E_1^\gamma}{\int_{\mathbb{R}} V_v^{\gamma+1/2}\, dx} = \frac{1}{\sqrt{\pi}} \frac{\Gamma(\gamma + 1)}{\Gamma(\gamma + 1/2)} \frac{v^{\gamma-1/2}}{(v + 1)^{\gamma+1/2}} \qquad \text{for all } v > 0.$$

We optimize the left side with respect to $v > 0$ by choosing $v = \gamma - 1/2$ for $\gamma > 1/2$. For $\gamma = 1/2$ we let $v \to 0$. This shows that

$$\sup_{v>0} \frac{E_1^\gamma}{\int_{\mathbb{R}} V_v^{\gamma+1/2}\, dx} = \frac{1}{\sqrt{\pi}} \frac{\Gamma(\gamma + 1)}{\Gamma(\gamma + 1/2)} \frac{(\gamma - 1/2)^{\gamma-1/2}}{(\gamma + 1/2)^{\gamma+1/2}},$$

and we recognize the right side as $L_{\gamma,1}^{(1)}$ from (5.18).   □

It is remarkable that for $\gamma > 3/2$ the optimal potential for $L_{\gamma,1}^{(1)}$ corresponds to a Schrödinger operator with at least two negative eigenvalues.

The following lemma and remark concern moments of all negative eigenvalues of $-\frac{d^2}{dx^2} - V_v$. We begin with the case $\gamma = 3/2$, and use the notation $v = [v] + \{v\}$ with $[v] \in \mathbb{N}_0$ and $\{v\} \in [0, 1)$.

Lemma 5.16   *Let* $V_v(x) = v(v + 1)\cosh^{-2} x$ *for* $v > 0$. *Then*

$$\frac{\operatorname{Tr}\left(-\frac{d^2}{dx^2} - V_v\right)_-^{3/2}}{\int_{\mathbb{R}} V_v^2\, dx} = \frac{3}{16} - \frac{3}{16} \frac{\{v\}^2(1 - \{v\})^2}{v^2(v + 1)^2}.$$

In particular,

$$\frac{\operatorname{Tr}\left(-\frac{d^2}{dx^2} - V_v\right)_-^{3/2}}{\int_{\mathbb{R}} V_v^2\, dx} \le \frac{3}{16} = L_{3/2,1}^{cl}$$

with equality if and only if $v \in \mathbb{N}$.

Proof By the computation from the previous proof, we have

$$\int_{\mathbb{R}} V_v^2 dx = \frac{4}{3} v^2 (v+1)^2 .$$

On the other hand, from Proposition 4.9 we know that $-\frac{d^2}{dx^2} - V_v$ has simple negative eigenvalues given by

$$-E_j = -(v - j + 1)^2, \quad j = 1, \ldots, \lceil v \rceil .$$

Setting $v = \lceil v \rceil - \theta$, expanding the cube and using well-known formulas for $\sum_{k=1}^{K} k^\ell$, $\ell = 0, 1, 2, 3$, we find

$$\sum_{j=1}^{\lceil v \rceil} E_j^{\frac{3}{2}} = \sum_{j=1}^{\lceil v \rceil} (v - j + 1)^3 = \sum_{k=1}^{\lceil v \rceil} (k - \theta)^3 = \frac{1}{4} v^2 (v+1)^2 - \frac{1}{4} \theta^2 (1 - \theta)^2 .$$

Since $\theta(1 - \theta) = (1 - \{v\})\{v\}$, this proves the claimed formula. $\square$

Remark 5.17 Similarly as in the above proof, one can show that

$$\frac{\text{Tr} \left(-\frac{d^2}{dx^2} - V_v \right)_-^{1/2}}{\int_{\mathbb{R}} V_v \, dx} = \frac{1}{4} + \frac{1}{4} \frac{\{v\}(1 - \{v\})}{v(v+1)} .$$

This implies, in particular, that

$$L_{1/2,1}^{\text{cl}} = \frac{1}{4} \leq \frac{\text{Tr} \left(\frac{d^2}{dx^2} \quad V_v \right)_-^{1/2}}{\int_{\mathbb{R}} V_v \, dx} < \frac{1}{2} = 2 L_{1/2,1}^{\text{cl}} .$$

The first inequality is an equality if and only if $v \in \mathbb{N}$, and the second inequality becomes an equality in the limit $v \to 0$.

5.2.2 Lieb–Thirring inequalities for the Coulomb Hamiltonian

In this subsection we discuss bounds for Lieb–Thirring sums

$$\text{Tr}(-\Delta - \kappa |x|^{-1} + \Lambda)_-^\gamma$$

for the Coulomb Hamiltonian in dimension $d = 3$. The two parameters Λ and κ are assumed to be positive.

Proposition 5.18 *Let $d = 3$ and $1 \leq \gamma < 3/2$. Then for all $\kappa > 0$ and $\Lambda > 0$,*

$$\text{Tr}(-\Delta - \kappa |x|^{-1} + \Lambda)_-^\gamma \leq L_{\gamma,3}^{\text{cl}} \int_{\mathbb{R}^3} \left(\kappa |x|^{-1} - \Lambda \right)_+^{\gamma + 3/2} dx .$$

The upper bound $3/2$ on γ is necessary to have the right side in the inequality finite.

Proof We begin by considering the situation in general dimension $d \geq 2$. According to Proposition 4.10, the negative spectrum of $-\Delta - \kappa|x|^{-1}$ consists precisely of the eigenvalues $-\kappa^2(2k + d - 1)^{-2}$, $k \in \mathbb{N}_0$, and the corresponding multiplicity is

$$\mu_k := \frac{(d - 2 + k)! \, (d - 1 + 2k)}{(d - 1)! \, k!} .$$

Thus,

$$\mathrm{Tr}\left(-\Delta - \kappa|x|^{-1} + \Lambda\right)_-^{\gamma} = \sum_{k=0}^{\infty} \mu_k \left(\frac{\kappa^2}{(2k + d - 1)^2} - \Lambda\right)_+^{\gamma} .$$

Meanwhile, for $\gamma < \frac{d}{2}$, the semiclassical phase space integral is finite and equals

$$L_{\gamma,d}^{\mathrm{cl}} \int_{\mathbb{R}^d} \left(\kappa|x|^{-1} - \Lambda\right)_+^{\gamma + \frac{d}{2}} dx = 2^{1-d} \kappa^d \Lambda^{\gamma - \frac{d}{2}} \frac{\Gamma(\gamma + 1)\Gamma(\frac{d}{2} - \gamma)}{\Gamma(d + 1)\Gamma(\frac{d}{2})} .$$

Let us now focus on the special case $d = 3$ and $\gamma = 1$. Then,

$$\mathrm{Tr}\left(-\Delta - \kappa|x|^{-1} + \Lambda\right)_- = \sum_{k=0}^{\infty} \left(\frac{\kappa^2}{4} - \Lambda(k + 1)^2\right)_+ .$$

Since $\left(\frac{\kappa^2}{4} - \Lambda(k + 1)^2\right)_+ \leq \left(\frac{\kappa^2}{4} - \Lambda t^2\right)_+$ for all $0 \leq t \leq k + 1$, we obtain

$$\left(\frac{\kappa^2}{4} - \Lambda(k + 1)^2\right)_+ \leq \int_k^{k+1} \left(\frac{\kappa^2}{4} - \Lambda t^2\right)_+ dt \qquad \text{for all } k \geq 0,$$

and therefore

$$\mathrm{Tr}\left(-\Delta - \kappa|x|^{-1} + \Lambda\right)_- \leq \int_0^{\infty} \left(\frac{\kappa^2}{4} - \Lambda t^2\right)_+ dt = \frac{\kappa^3}{12\sqrt{\Lambda}}$$

$$= L_{1,3}^{\mathrm{cl}} \int_{\mathbb{R}^3} \left(\kappa|x|^{-1} - \Lambda\right)_+^{5/2} dx .$$

This proves the inequality for $\gamma = 1$. The case $1 < \gamma < 3/2$ follows by the Aizenman–Lieb argument in the proof of Lemma 5.2. $\qquad\square$

Numerical evidence suggests that the inequality in Proposition 5.18 holds also for γ smaller than, but close to, 1; see Schmetzer (2022). The case of dimensions $d \neq 3$ remains open.

5.2.3 The harmonic oscillator in dimension $d = 1$

In this and the following two subsections we discuss the Lieb–Thirring sums $\mathrm{Tr}(-\Delta - V)_-^{\gamma}$ in the special case of the harmonic oscillator. We will see that the inequality holds with semiclassical constant $L^{cl}_{\gamma,d}$ for all $\gamma \geq 1$ but does not hold with this constant if $\gamma < 1$.

In this subsection we deal with the one-dimensional situation. We begin with the case $\gamma \geq 1$ and prove a sharp Lieb–Thirring inequality for the harmonic oscillator with semiclassical constant.

Proposition 5.19 *Let $d = 1$ and $\gamma \geq 1$. Then for all $\omega, \Lambda > 0$,*

$$\mathrm{Tr}\left(-\frac{d^2}{dx^2} + \omega^2 x^2 - \Lambda\right)^{\gamma}_{-} \leq L^{cl}_{\gamma,1} \int_{\mathbb{R}} (\Lambda - \omega^2 x^2)^{\gamma+1/2}_{+} \, dx. \tag{5.34}$$

Moreover, if $\gamma = 1$, then equality holds if and only if $\frac{\Lambda}{2\omega} \in \mathbb{N}$. For $\gamma > 1$, the inequality is strict for all $\Lambda, \omega > 0$.

Proof We begin with the case $\gamma = 1$. Using the explicit formula for the eigenvalues of the harmonic oscillator from Proposition 4.6, we find

$$\mathrm{Tr}\left(-\frac{d^2}{dx^2} + \omega^2 x^2 - \Lambda\right)_{-} = \sum_{k \geq 1} (\omega(2k - 1) - \Lambda)_{-} = \omega\left(\Lambda^2(2\omega)^{-2} - t^2\right)$$

with $t = 1 + \left[\frac{\Lambda}{2\omega} - \frac{1}{2}\right] - \frac{\Lambda}{2\omega}$. Since

$$L^{cl}_{1,1} \int_{\mathbb{R}} (\Lambda - \omega^2 x^2)^{3/2}_{+} \, dx = \omega\Lambda^2(2\omega)^{-2},$$

this proves the inequality in (5.34). Moreover, we find that equality holds if and only if $t - 0$, which is equivalent to $\frac{\Lambda}{2\omega} \in \mathbb{N}$, as claimed.

Next, we turn to the case $\gamma > 1$. By the Aizenman–Lieb argument of Lemma 5.2 (more precisely, by (5.11) with (σ, γ) replaced by $(\gamma, 1)$) one has

$$\mathrm{Tr}\left(-\frac{d^2}{dx^2} + \omega^2 x^2 - \Lambda\right)^{\gamma}_{-} = \gamma(\gamma-1) \int_0^{\infty} \mathrm{Tr}\left(-\frac{d^2}{dx^2} + \omega^2 x^2 - \Lambda + \tau\right)_{-} \tau^{\gamma-2} \, d\tau.$$

Applying the inequality for $\gamma = 1$, which we have just shown, to the integrand and using

$$L^{cl}_{1,1}\gamma(\gamma - 1) \int_0^{\infty} \left(\Lambda - \tau - \omega x^2\right)^{3/2}_{+} \tau^{\gamma-2} \, d\tau = L^{cl}_{\gamma,1}(\Lambda - \omega^2 x^2)^{\gamma+1/2}_{+},$$

we obtain the claimed inequality for $\gamma > 1$. Since the inequality for $\gamma = 1$ is strict away from a discrete set of points τ, we see that the inequality for $\gamma > 1$ is always strict. $\qquad\square$

We emphasize that, despite the inequality being strict for $\gamma > 1$, the constant $L^{cl}_{\gamma,1}$ is best possible and is achieved in the limit $\Lambda/\omega \to \infty$. This follows either from Weyl asymptotics or by an explicit computation.

Next, we turn to the case $\gamma < 1$. We will again prove a sharp inequality, but in this case, the constant is not the semiclassical one.

Proposition 5.20   *Let $d = 1$ and $0 \le \gamma < 1$. Then*

$$\mathrm{Tr}\left(-\frac{d^2}{dx^2} + \omega^2 x^2 - \Lambda\right)_-^\gamma \le 2\left(\frac{\gamma}{\gamma+1}\right)^\gamma L_{\gamma,1}^{\mathrm{cl}} \int_{\mathbb{R}} (\Lambda - \omega^2 x^2)_+^{\gamma+1/2}\, dx$$

with $(\gamma/(\gamma+1))^\gamma = 1$ for $\gamma = 0$. Moreover, equality holds at $\Lambda = (1+\gamma)\omega$ if $\gamma > 0$ and as $\Lambda \searrow \omega$ if $\gamma = 0$.

Proof   In view of Proposition 4.6, the assertions for $\gamma = 0$ are immediate and in the remainder we focus on the case $0 < \gamma < 1$. Elementary analysis yields

$$1 \le \gamma^\gamma (1-\gamma)^{1-\gamma} s^{\gamma-1}(1+s), \quad \text{for all } 0 < \gamma < 1, \quad s > 0.$$

Substituting $s = \tau/E$ and multiplying by E^γ gives

$$E^\gamma \le \gamma^\gamma (1-\gamma)^{1-\gamma} \tau^{\gamma-1}(E+\tau), \quad \text{for all } 0 < \gamma < 1, \quad E, \tau > 0.$$

Applying this inequality for $E = \Lambda - (2j-1)\omega > 0$ and summing, we arrive at

$$\mathrm{Tr}\left(-\frac{d^2}{dx^2} + \omega^2 x^2 - \Lambda\right)_-^\gamma \le \gamma^\gamma (1-\gamma)^{1-\gamma} \tau^{\gamma-1} \mathrm{Tr}\left(-\frac{d^2}{dx^2} + \omega^2 x^2 - \Lambda - \tau\right)_-.$$

By Proposition 5.19 with γ replaced by 1 this implies

$$\mathrm{Tr}\left(-\frac{d^2}{dx^2} + \omega^2 x^2 - \Lambda\right)_-^\gamma \le \gamma^\gamma (1-\gamma)^{1-\gamma} \tau^{\gamma-1} \frac{(\Lambda+\tau)^2}{4\omega}.$$

Choosing $\tau = (1-\gamma)(1+\gamma)^{-1}\Lambda$ will minimize the right side with respect to $\tau > 0$ and so we arrive at

$$\mathrm{Tr}\left(-\frac{d^2}{dx^2} + \omega^2 x^2 - \Lambda\right)_-^\gamma \le 2\left(\frac{\gamma}{1+\gamma}\right)^\gamma \frac{\Lambda^{1+\gamma}}{2(1+\gamma)\omega}$$

$$= 2\left(\frac{\gamma}{1+\gamma}\right)^\gamma L_{\gamma,1}^{\mathrm{cl}} \int_{\mathbb{R}} (\Lambda - \omega^2 x^2)_+^{\gamma+1/2}\, dx.$$

To prove that the constant on the right side cannot be improved, we put $\Lambda = (1+\gamma)\omega$ and compute

$$\mathrm{Tr}\left(-\frac{d^2}{dx^2} + \omega^2 x^2 - (1+\gamma)\omega\right)_-^\gamma = ((1+\gamma)\omega - \omega)^\gamma = \gamma^\gamma \omega^\gamma,$$

which coincides with the value of the right side. This completes the proof.   □

5.2.4 The harmonic oscillator in higher dimensions $d \ge 2$

Similar computations in the d-dimensional case are more involved. In this subsection we deal with the case $\gamma \ge 1$ and in the next one with the case $0 \le \gamma < 1$. The following result is due to de la Brèteche (1999) and Laptev (1999).

Proposition 5.21 *Let $d \geq 2$ and $\gamma \geq 1$. Then*

$$\mathrm{Tr}\left(-\Delta + \sum_{k=1}^{d} \omega_k^2 x_k^2 - \Lambda\right)_{-}^{\gamma} \leq L_{\gamma,d}^{\mathrm{cl}} \int_{\mathbb{R}^d} \left(\Lambda - \sum_{k=1}^{d} \omega_k^2 x_k^2\right)_{+}^{\gamma+d/2} dx .$$

The proof that we give has some similarities with that of Theorem 3.36. Like that, it is based on separation of variables, which is a very special case of the lifting technique, which we will see again in Chapter 6.

Proof We abbreviate $\omega = (\omega_1, \ldots, \omega_d)$ and

$$V_\omega(x) := - \sum_{k=1}^{d} \omega_k^2 x_k^2 .$$

The idea of the proof is to separate variables, writing $x' = (x_1, \ldots, x_{d-1})$, $\omega' = (\omega_1, \ldots, \omega_{d-1})$ and

$$V_{\omega'}(x') := - \sum_{k=1}^{d-1} \omega_k^2 x_k^2 ,$$

so that $V_\omega(x) = V_{\omega'}(x') - \omega_d^2 x_d^2$. Similarly, we define $\Delta' = \sum_{k=1}^{d-1} \frac{\partial^2}{\partial x_k^2}$.

According to Proposition 4.6, the eigenvalues of $-\Delta + \sum_{k=1}^{d} \omega_k^2 x_k^2 = -\Delta - V_\omega$ are given by

$$E_\tau(\omega) := \sum_{k=1}^{d} \omega_k(2\tau_k - 1), \quad \text{for } \tau \in \mathbb{N}^d .$$

Similarly, those of $-\Delta' - V_{\omega'}$ are given by

$$E_{\tau'}(\omega') := \sum_{k=1}^{d-1} \omega_k(2\tau_k - 1), \quad \text{for } \tau' \in \mathbb{N}^{d-1} .$$

Thus

$$E_\tau(\omega) = E_{\tau'}(\omega') + \omega_d(2\tau_d - 1)$$

and

$$\mathrm{Tr}\left(-\Delta + \sum_{k=1}^{d} \omega_k^2 x_k^2 - \Lambda\right)_{-}^{\gamma} = \sum_{\tau \in \mathbb{N}^d} (E_\tau(\omega) - \Lambda)_{-}^{\gamma}$$

$$= \sum_{\tau' \in \mathbb{N}^{d-1}} \sum_{\tau_d \in \mathbb{N}} (\omega_d(2\tau_d - 1) + E_{\tau'}(\omega') - \Lambda)_{-}^{\gamma} .$$

According to the one-dimensional result (5.34), we have, for any fixed $\tau' \in \mathbb{N}^{d-1}$,

$$\sum_{\tau_d \in \mathbb{N}} \left(\omega_d(2\tau_d - 1) + E_{\tau'}(\omega') - \Lambda \right)_-^\gamma \le L_{\gamma,1}^{cl} \int_{\mathbb{R}} \left(\Lambda - E_{\tau'}(\omega') - \omega_d^2 x_d^2 \right)_+^{\gamma+1/2} dx .$$

Just like in the proof of Lemma 5.2, it will be convenient to use (5.8) to write the right side as

$$\iint_{\mathbb{R} \times \mathbb{R}} \left(\xi_d^2 + \omega_d^2 x_d^2 + E_{\tau'}(\omega') - \Lambda \right)_-^\gamma \frac{dx_d \, d\xi_d}{2\pi} .$$

Therefore, we have shown that

$$\mathrm{Tr}\left(-\Delta - V_\omega - \Lambda\right)_-^\gamma \le \sum_{\tau' \in \mathbb{N}^{d-1}} \iint_{\mathbb{R} \times \mathbb{R}} \left(\xi_d^2 + \omega_d^2 x_d^2 + E_{\tau'}(\omega') - \Lambda \right)_-^\gamma \frac{dx_d \, d\xi_d}{2\pi}$$

$$= \iint_{\mathbb{R} \times \mathbb{R}} \mathrm{Tr}\left(-\Delta' - V_{\omega'} + \xi_d^2 + \omega_d^2 x_d^2 - \Lambda \right)_-^\gamma \frac{dx_d \, d\xi_d}{2\pi} .$$

Repeating the same procedure another $d - 1$ times, we finally obtain

$$\mathrm{Tr}\left(-\Delta - V_\omega - \Lambda\right)_-^\gamma \le \iint_{\mathbb{R}^d \times \mathbb{R}^d} \left(|\xi|^2 + \sum_{k=1}^d \omega_k^2 x_k^2 - \Lambda \right)_-^\gamma \frac{dx \, d\xi}{(2\pi)^d} .$$

According to (5.8), this proves the claimed inequality.   □

5.2.5 The counterexample of Helffer and Robert

We now discuss the Lieb–Thirring sums for the harmonic oscillator in dimensions $d \ge 2$ for exponents $0 \le \gamma < 1$. Recall that for such a γ in dimension $d = 1$, we computed the sharp constant in the Lieb–Thirring inequality for the harmonic oscillator and showed that this constant is strictly larger than $L_{\gamma,d}^{cl}$. In higher dimensions the sharp constant cannot be computed explicitly, but one can still show that the constant is strictly larger than $L_{\gamma,d}^{cl}$. This result is due to Helffer and Robert (1990b).

Theorem 5.22   *Let $d \ge 2$ and $0 \le \gamma < 1$. Then*

$$\limsup_{\Lambda \to \infty} \frac{\mathrm{Tr}\left(-\Delta + |x|^2 - \Lambda\right)_-^\gamma - L_{\gamma,d}^{cl} \int_{\mathbb{R}^d} (|x|^2 - \Lambda)_+^{\gamma+d/2} \, dx}{\Lambda^{d-1}} > 0 .$$

As an immediate consequence of this theorem we infer that the optimal constant $L_{\gamma,d}$ in the Lieb–Thirring inequality is strictly larger than the semiclassical constant $L_{\gamma,d}^{cl}$ provided that $\gamma < 1$. In particular, this provides a counterexample to the Lieb–Thirring conjecture, discussed in §5.1.3, in any dimension $d \ge 3$ for γ in a certain range.

Corollary 5.23   *Let $d \geq 2$ and $0 \leq \gamma < 1$. Then*

$$L_{\gamma,d} > L_{\gamma,d}^{cl} \, .$$

To prove Theorem 5.22 we compute the asymptotics of $\mathrm{Tr}(-\Delta - |x|^2 + \Lambda)_-^{\gamma}$ as $\Lambda \to \infty$. By Weyl asymptotics (Theorems 4.28 and 4.29), the leading term will be $L_{\gamma,d}^{cl} \int_{\mathbb{R}^d} (|x|^2 - \Lambda)_+^{\gamma+d/2} \, dx$, which is equal to a constant times $\Lambda^{d+\gamma}$. In the following proof we will isolate the leading correction term of order Λ^{d-1}. This term is oscillating and attains both positive and negative values.

The computation of the asymptotics in the higher-dimensional case is based on a precise computation in the one-dimensional case. Let $\gamma \geq 0$ and let us abbreviate

$$r_\gamma(\Lambda) := \mathrm{Tr}\left(-\frac{d^2}{dx^2} + x^2 - \Lambda\right)_-^{\gamma} = \sum_{j=0}^{\infty} (2j + 1 - \Lambda)_-^{\gamma} \, ;$$

see Proposition 4.6. A key role will be played by the 2-periodic function

$$\rho_\gamma(s) := \Gamma(\gamma + 1) \sum_{j \geq 1} \frac{\cos(j\pi(s + 1) - \frac{\pi}{2}(\gamma + 1))}{(\pi j)^{1+\gamma}} \, .$$

The following lemma is due to Helffer and Sjöstrand (1990).

Lemma 5.24   *Let $d = 1$ and $\gamma \geq 0$. Then, as $\Lambda \to \infty$,*

$$r_\gamma(\Lambda) = \frac{\Lambda^{1+\gamma}}{2(\gamma + 1)} + \rho_\gamma(\Lambda) + O(\Lambda^{-1+\gamma}).$$

Proof   If $\gamma = 0$, by inspection, the function

$$\Lambda \mapsto r_0(\Lambda) - \frac{\Lambda}{2}$$

is 2-periodic and

$$r_0(\Lambda) - \frac{\Lambda}{2} = \begin{cases} -\frac{\Lambda}{2} & \text{if } 0 \leq \Lambda \leq 1, \\ 1 - \frac{\Lambda}{2} & \text{if } 1 < \Lambda \leq 2 \, . \end{cases}$$

By a straightforward computation of the Fourier coefficients, one sees that this function is equal to $\rho_0(\Lambda)$ and so the lemma holds as an identity without the $O(\Lambda^{-1})$-term.

Now let $\gamma > 0$. We will in fact prove that $\Lambda^{-1-\gamma}(r_\gamma(\Lambda) - \rho_\gamma(\Lambda))$ has a complete asymptotic expansion as $\Lambda \to \infty$ in powers of Λ^{-2}. Our starting point is the formula

$$\sigma_+^{\gamma} = \frac{\Gamma(\gamma + 1)}{2\pi i} \int_{c-i\infty}^{c+i\infty} e^{t\sigma} \frac{dt}{t^{\gamma+1}} \qquad \text{for all } \sigma \in \mathbb{R}, \tag{5.35}$$

valid for any $c > 0$. In fact, (5.35) for $\sigma \leq 0$ follows by analyticity with respect to t of the integrand in an open neighborhood of the half-plane $\{\text{Re}\, t \geq c\}$. For $\sigma > 0$ and $\gamma \in \mathbb{N}_0$ the formula follows by the residue theorem using meromorphicity in an open neighborhood of the half-plane $\{\text{Re}\, t \leq c\}$. If $\sigma > 0$ and $\gamma \notin \mathbb{N}_0$, we deform the integration to the contour C_ε introduced below and apply Hankel's integral representation (Abramowitz and Stegun, 1964, Equation (6.1.4)) to obtain (5.35).

We now apply (5.35) to $\sigma = \Lambda - 2j - 1$ and sum over j. In this way we obtain

$$r_\gamma(\Lambda) = \frac{\Gamma(\gamma + 1)}{4\pi i} \int_{c-i\infty}^{c+i\infty} \frac{e^{t\Lambda}}{\sinh t} \frac{dt}{t^{\gamma+1}} .$$

Let $0 < \varepsilon < \min\{\pi, c\}$. Consider the contour C_ε consisting of $(-\infty - i0, -\varepsilon - i0)$, of the circle of radius ε around the origin traversed counterclockwise and of $(-\varepsilon + i0, -\infty + i0)$. We apply the residue theorem and, since $1/\sinh$ has poles at $i\pi k$, $k \in \mathbb{Z}$, with residues $(-1)^k$, we obtain

$$r_\gamma(\Lambda) = \rho_\gamma(\Lambda) + I_\gamma(\Lambda)$$

with

$$I_\gamma(\Lambda) := \frac{\Gamma(\gamma + 1)}{4\pi i} \int_{C_\varepsilon} \frac{e^{t\Lambda}}{\sinh t} \frac{dt}{t^{\gamma+1}} .$$

It remains to compute the asymptotics of $I_\gamma(\Lambda)$ as $\Lambda \to \infty$. A straightforward bound gives

$$I_\gamma(\Lambda) = \frac{\Gamma(\gamma + 1)}{4\pi i} \int_{|t|=\varepsilon} \frac{e^{t\Lambda}}{\sinh t} \frac{dt}{t^{\gamma+1}} + O(\varepsilon^{-1-\gamma} e^{-\varepsilon\Lambda}) .$$

We now use the fact that there is a sequence $(\alpha_m)_{m\in\mathbb{N}_0}$ with $\alpha_0 = 1$ such that for any $M \in \mathbb{N}$,

$$\frac{t}{\sinh t} = \sum_{m=0}^{M-1} \alpha_m t^{2m} + O(t^{2M}) \qquad \text{as } t \to 0 .$$

This gives

$$I_\gamma(\Lambda) = \frac{\Gamma(\gamma + 1)}{4\pi i} \sum_{m=0}^{M-1} \alpha_m \int_{|t|=\varepsilon} e^{t\Lambda} t^{2m} \frac{dt}{t^{\gamma+2}}$$
$$+ O(\varepsilon^{2M-1-\gamma} e^{\varepsilon\Lambda}) + O(\varepsilon^{-1-\gamma} e^{-\varepsilon\Lambda}) .$$

We now replace the integral over the circle by the full contour C_ε and find

$$I_\gamma(\Lambda) = \frac{\Gamma(\gamma+1)}{4\pi i} \sum_{m=0}^{M-1} \alpha_m \left(\int_{C_\varepsilon} e^{t\Lambda} t^{2m} \frac{dt}{t^{\gamma+2}} + O(\varepsilon^{2m-1-\gamma} e^{-\varepsilon\Lambda}) \right)$$
$$+ O(\varepsilon^{2M-1-\gamma} e^{\varepsilon\Lambda}) + O(\varepsilon^{-1-\gamma} e^{-\varepsilon\Lambda})$$
$$= \frac{\Gamma(\gamma+1)}{4\pi i} \sum_{m=0}^{M-1} \alpha_m \int_{C_\varepsilon} e^{t\Lambda} t^{2m} \frac{dt}{t^{\gamma+2}} + O(\varepsilon^{2M-1-\gamma} e^{\varepsilon\Lambda}) + O(\varepsilon^{-1-\gamma} e^{-\varepsilon\Lambda}).$$

Finally, using again Hankel's integral representation,

$$\frac{1}{2\pi i} \int_{C_\varepsilon} e^{t\Lambda} t^{2m} \frac{dt}{t^{\gamma+2}} = \frac{1}{\Gamma(\gamma+2-2m)} \Lambda^{\gamma+1-2m}.$$

Thus we have shown that

$$I_\gamma(\Lambda) = \frac{\Gamma(\gamma+1)}{2} \sum_{m=0}^{M-1} \alpha_m \frac{\Lambda^{\gamma+1-2m}}{\Gamma(\gamma+2-2m)} + O(\varepsilon^{2M-1-\gamma} e^{\varepsilon\Lambda}) + O(\varepsilon^{-1-\gamma} e^{-\varepsilon\Lambda}).$$

Finally, we choose $\varepsilon = (M \ln \Lambda)/\Lambda$ (which satisfies $0 < \varepsilon < \min\{\pi, c\}$ for Λ large enough) and obtain

$$I_\gamma(\Lambda) = \frac{\Gamma(\gamma+1)}{2} \sum_{m=0}^{M-1} \alpha_m \frac{\Lambda^{\gamma+1-2m}}{\Gamma(\gamma+2-2m)} + O((\ln\Lambda)^{2M-1-\gamma} \Lambda^{-M+1+\gamma}).$$

This proves that $\Lambda^{-1-\gamma} I_\gamma(\Lambda)$ has a complete asymptotic expansion in powers of Λ^{-2} as $\Lambda \to \infty$. Its limit is $\Gamma(\gamma+1)/(2\Gamma(\gamma+2)) = 1/(2(\gamma+1))$, as claimed. In particular, choosing $M > 2$, we obtain the assertion of the lemma.   □

We next discuss the multi-dimensional case.

Lemma 5.25 Let $d \geq 2$ and $\gamma \geq 0$. Then, as $\Lambda \to \infty$,

$$\mathrm{Tr}\left(-\Delta + |x|^2 - \Lambda\right)_-^\gamma = \frac{\Lambda^{d+\gamma}}{2^d(\gamma+d)(\gamma+d-1)\cdots(\gamma+1)}$$
$$+ \frac{1}{2^{d-1}(d-1)!} \rho_\gamma(\Lambda - d + 1) \Lambda^{d-1} + O(\Lambda^{d+\gamma-2}).$$

Proof *Step 1.* Recall that $r_\gamma(\Lambda)$ was defined before Lemma 5.24 and define, for $k = 0, \ldots, d-1$,

$$s_{k,\gamma}(\mu) := \sum_{m=0}^{k} \binom{k}{m} (-1)^m \mu^{k-m} r_{\gamma+m}(\mu).$$

We claim that there are numbers $\alpha_k^{(d)}$, $k = 0, \ldots, d - 1$, such that

$$\mathrm{Tr}\left(-\Delta + |r|^2 - \Lambda\right)_-^\gamma = \sum_{k=0}^{d-1} \alpha_k^{(d)} s_{k,\gamma}(\Lambda - d + 1). \tag{5.36}$$

We also claim that

$$\alpha_{d-1}^{(d)} = \frac{1}{2^{d-1}(d-1)!} \quad \text{and} \quad \alpha_{d-2}^{(d)} = \frac{d-1}{2^{d-1}(d-2)!}. \tag{5.37}$$

To prove (5.36) and (5.37), we write, using Proposition 4.6,

$$\mathrm{Tr}\left(-\Delta + |x|^2 - \Lambda\right)_-^\gamma = \sum_{j_1=0}^\infty \cdots \sum_{j_d=0}^\infty (2(j_1 + \cdots + j_d) + d - \Lambda)_-^\gamma$$

$$= \sum_{\ell=0}^\infty \binom{\ell + d - 1}{d - 1}(2\ell + d - \Lambda)_-^\gamma,$$

where we used the fact that for each $\ell \in \mathbb{N}_0$,

$$\#\left\{(j_1, \ldots, j_d) \in \mathbb{N}_0^d : j_1 + \cdots + j_d = \ell\right\} = \binom{\ell + d - 1}{d - 1}.$$

We assume now that $\Lambda \geq d - 1$ and introduce the new variable

$$\mu := \Lambda - d + 1,$$

so that

$$\mathrm{Tr}\left(-\Delta + |x|^2 - \Lambda\right)_-^\gamma = \sum_{\ell=0}^\infty \binom{\ell + d - 1}{d - 1}(2\ell + 1 - \mu)_-^\gamma.$$

Now

$$\binom{\ell + d - 1}{d - 1} = \frac{(\ell + d - 1)(\ell + d - 2) \cdots (\ell + 1)}{(d - 1)!}$$

is a polynomial in ℓ of degree $d - 1$ whose leading coefficient is $1/(d-1)!$ and whose next term has coefficient $d/(2(d-2)!)$. Hence there are numbers $\alpha_k^{(d)}$, $k = 0, \ldots, d - 1$, such that

$$\binom{\ell + d - 1}{d - 1} = \sum_{k=0}^{d-1} \alpha_k^{(d)}(2\ell + 1)^k$$

and such that (5.37) holds. In terms of these numbers we can write

$$\mathrm{Tr}\left(-\Delta + |x|^2 - \Lambda\right)_-^\gamma = \sum_{\ell=0}^\infty \sum_{k=0}^{d-1} \alpha_k^{(d)}(2\ell + 1)^k(2\ell + 1 - \mu)_-^\gamma.$$

We now expand

$$(2\ell + 1)^k = \sum_{m=0}^{k} \binom{k}{m} (2\ell + 1 - \mu)^m \mu^{k-m},$$

so that

$$\mathrm{Tr}\left(-\Delta + |x|^2 - \Lambda\right)_{-}^{\gamma} = \sum_{\ell=0}^{\infty} \sum_{k=0}^{d-1} \sum_{m=0}^{k} \alpha_k^{(d)} \binom{k}{m} (-1)^m \mu^{k-m} (2\ell + 1 - \mu)_{-}^{\gamma+m}$$

$$= \sum_{k=0}^{d-1} \sum_{m=0}^{k} \alpha_k^{(d)} \binom{k}{m} (-1)^m \mu^{k-m} r_{\gamma+m}(\mu)$$

$$= \sum_{k=0}^{d-1} \alpha_k^{(d)} s_{k,\gamma}(\mu).$$

This proves the claimed identity (5.36).

Step 2. We now aim at finding the behavior of the quantities $s_{k,\gamma}(\mu)$ as $\mu \to \infty$. We use Lemma 5.24 for $r_{\gamma+m}$ to obtain

$$s_{d-1,\gamma}(\mu) = \sum_{m=0}^{d-1} \binom{d-1}{m} (-1)^m \mu^{d-1-m} \left(\frac{1}{2(\gamma + m + 1)} \mu^{1+m+\gamma} \right.$$

$$\left. + \rho_{\gamma+m}(\mu) + O(\mu^{-1+m+\gamma}) \right)$$

$$= S_{d-1,\gamma} \mu^{d+\gamma} + \rho_{\gamma}(\mu) \mu^{d-1} + O(\mu^{d-2+\gamma}),$$

$$s_{d-2,\gamma}(\mu) = \sum_{m=0}^{d-2} \binom{d-2}{m} (-1)^m \mu^{d-2-m} \left(\frac{1}{2(\gamma + m + 1)} \mu^{1+m+\gamma} \right.$$

$$\left. + \rho_{\gamma+m}(\mu) + O(\mu^{-1+m+\gamma}) \right)$$

$$= S_{d-2,\gamma} \mu^{d-1+\gamma} + O(\mu^{d-2+\gamma})$$

and

$$s_{k,\gamma}(\mu) = O(\mu^{d-2+\gamma}) \qquad \text{if } k \le d - 3.$$

Here

$$S_{d-1,\gamma} := \sum_{m=0}^{d-1} \binom{d-1}{m} \frac{(-1)^m}{2(\gamma + m + 1)}, \qquad S_{d-2,\gamma} := \sum_{m=0}^{d-2} \binom{d-2}{m} \frac{(-1)^m}{2(\gamma + m + 1)}.$$

In the following, we need an alternative expression for these constants; namely,

$$S_{d-1,\gamma} = \frac{(d-1)!}{?(\gamma+d)(\gamma+d-1)\cdots(\gamma+1)} \tag{5.38}$$

and

$$S_{d-2,\gamma} = \frac{(d-2)!}{2(\gamma+d-1)(\gamma+d-2)\cdots(\gamma+1)}. \tag{5.39}$$

Indeed, we have

$$\int_0^1 (1-t)^\gamma t^{d-1}\, dt = (-1)^{d-1}\int_0^1 (1-t)^\gamma (1-t-1)^{d-1}\, dt$$

$$= \sum_{m=0}^{d-1}\binom{d-1}{m}\int_0^1 (1-t)^{\gamma+m}(-1)^m\, dt$$

$$= \sum_{m=0}^{d-1}\binom{d-1}{m}\frac{(-1)^m}{\gamma+m+1} = 2S_{d-1,\gamma}.$$

The left side of this identity is equal to the beta function,

$$B(\gamma+1,d) = \frac{\Gamma(\gamma+1)\Gamma(d)}{\Gamma(\gamma+d+1)} = \frac{(d-1)!}{(\gamma+d)(\gamma+d-1)\cdots(\gamma+1)}.$$

This proves (5.38). The proof of (5.39) is similar.

Step 3. We now conclude the proof of the lemma. We rewrite the result of Step 2 in the original variable $\Lambda = \mu + d - 1$ as

$$s_{d-1,\gamma}(\mu) = S_{d-1,\gamma}\Lambda^{d+\gamma} - (d-1)(d+\gamma)S_{d-1,\gamma}\Lambda^{d-1+\gamma}$$
$$+ \rho_\gamma(\Lambda - d + 1)\Lambda^{d-1} + O(\Lambda^{d-2+\gamma}),$$

$$s_{d-2,\gamma}(\mu) = S_{d-2,\gamma}\Lambda^{d-1+\gamma} + O(\Lambda^{d-2+\gamma}),$$

$$s_{k,\gamma}(\mu) = O(\Lambda^{d-2+\gamma}) \qquad \text{if } k \le d-3.$$

We insert this expression into (5.36). The key observation is that

$$-\alpha_{d-1}^{(d)}(d-1)(d+\gamma)S_{d-1,\gamma} + \alpha_{d-2}^{(d)}S_{d-2,\gamma} = 0,$$

which follows from (5.37), (5.38) and (5.39). Using (5.37) and (5.38) once again for the main term, we obtain the claimed asymptotics. □

Remark 5.26 The fact that in the expansion in Lemma 5.25 the term of order $\Lambda^{d-1+\gamma}$ cancels is no coincidence. This is a general result about semiclassical asymptotics for Schrödinger operators (and, in fact, for general differential operators with vanishing subprincipal symbol). In their original proof, Helffer and Robert refer at this point to an earlier result of theirs about this fact (Helffer and Robert, 1983).

Proof of Theorem 5.22 We compute

$$L_{\gamma,d}^{\mathrm{cl}} \int_{\mathbb{R}^d} \left(\Lambda - |x|^2 \right)_+^{\gamma+d/2} dx = \frac{\Lambda^{d+\gamma}}{2^d (\gamma + d) \cdots (\gamma + 1)}.$$

Since, for any $j \geq 1$, $\cos(j\pi(s+1) - \pi(\gamma+1)/2)$ has mean zero with respect to s over any interval of length two, the same is true for ρ_γ. Therefore, since ρ_γ is not identically zero, there is an $s_\gamma \in [0, 2]$ such that $\rho_\gamma(s_\gamma) > 0$. By Lemma 5.25, for $\Lambda_k = s_\gamma + d - 1 + 2k$ with $k \in \mathbb{N}$,

$$\mathrm{Tr} \left(-\Delta + |x|^2 - \Lambda_k \right)_-^\gamma - L_{\gamma,d}^{\mathrm{cl}} \int_{\mathbb{R}^d} \left(\Lambda_k - |x|^2 \right)_+^{\gamma+d/2} dx$$

$$= \frac{1}{2^{d-1}(d-1)!} \rho_\gamma(s_\gamma) \Lambda_k^{d-1} \left(1 + O\left(\Lambda_k^{-1+\gamma} \right) \right).$$

Since $\gamma < 1$, this implies the assertion of the theorem. $\qquad\square$

5.3 The sharp bound for $\gamma = \frac{1}{2}$ in one dimension

In this section we prove the following theorem due to Hundertmark, Lieb and Thomas (1998)

Theorem 5.27 *Let $d = 1$ and $\gamma = \frac{1}{2}$. Then, for all $V \in L_{\mathrm{loc}}^1(\mathbb{R})$ with $V_+ \in L^1(\mathbb{R})$,*

$$\mathrm{Tr} \left(-\frac{d^2}{dx^2} - V \right)_-^{1/2} \leq \frac{1}{2} \int_{\mathbb{R}} V(x)_+ \, dx.$$

The constant $\frac{1}{2}$ is optimal.

Optimality of the constant $\frac{1}{2}$ follows from Corollary 5.4, which says that

$$\frac{1}{2} = L_{1/2,1}^{(1)};$$

see also Lemma 5.15 for an explicit family of potentials for which this constant is asymptotically attained. Note too that $L_{1/2,1}^{\mathrm{cl}} = 1/4$.

For the original proof we refer to Hundertmark et al. (1998). Here we follow a partially alternative argument due to Hundertmark, Laptev and Weidl (2000), which extends to a more general situation treated in §8.2.

5.3.1 A modified Birman–Schwinger operator

Let $W \in L^2(\mathbb{R})$ be a non-negative function. For $\varepsilon > 0$ we consider the operator

$$\mathcal{L}_\varepsilon := 2\varepsilon \, W \left(-\frac{d^2}{dx^2} + \varepsilon^2 \right)^1 W \qquad \text{on } L^2(\mathbb{R}). \qquad (5.40)$$

More precisely, this modified Birman–Schwinger operator, $\mathcal{L}_\varepsilon$, is defined via the quadratic form $2\varepsilon \|Q_\varepsilon^* f\|^2$, where Q_ε is the bounded operator given by

$$Q_\varepsilon := W \left(-\frac{d^2}{dx^2} + \varepsilon^2 \right)^{-1/2}.$$

The boundedness of Q_ε follows from Theorem 2.46 with $d = 1$ and $q = \infty$. Indeed, by (2.18) with $\ell = \varepsilon^{-1}$,

$$\|Q_\varepsilon f\|^2 \le \|W\|^2 \left\| \left(-\frac{d^2}{dx^2} + \varepsilon^2 \right)^{-1/2} f \right\|_\infty^2 \le \tilde{S}_{\infty,1}^{-1} \varepsilon^{-1} \|W\|^2 \|f\|^2.$$

Note that, by the fact that $(Q_\varepsilon^*)^* = Q_\varepsilon$, we have $\mathcal{L}_\varepsilon = 2\varepsilon \, Q_\varepsilon Q_\varepsilon^*$. In particular, $\mathcal{L}_\varepsilon$ is non-negative. Next, we compute its trace.

Lemma 5.28 *For any $\varepsilon > 0$,*

$$\operatorname{Tr} \mathcal{L}_\varepsilon = \int_{\mathbb{R}} W(x)^2 \, dx.$$

Proof The operator Q_ε is an integral operator in $L^2(\mathbb{R})$ with integral kernel

$$Q_\varepsilon(x, x') = W(x) \int_{\mathbb{R}} \frac{e^{i\xi(x-x')}}{(\xi^2 + \varepsilon^2)^{1/2}} \frac{d\xi}{2\pi}$$

and its adjoint Q_ε^* is an integral operator with integral kernel

$$Q_\varepsilon^*(x, x') = \overline{Q_\varepsilon(x', x)} = \int_{\mathbb{R}} \frac{e^{i\xi(x'-x)}}{(\xi^2 + \varepsilon^2)^{1/2}} \frac{d\xi}{2\pi} W(x').$$

Therefore, according to Lemma 1.38 with $K = Q_\varepsilon^*$ and Plancherel's theorem,

$$\begin{aligned}
\operatorname{Tr} Q_\varepsilon Q_\varepsilon^* &= \iint_{\mathbb{R}\times\mathbb{R}} |Q_\varepsilon^*(x, x')|^2 \, dx \, dx' \\
&= \iint_{\mathbb{R}\times\mathbb{R}} \left| \int_{\mathbb{R}} \frac{e^{i\xi(x'-x)}}{(\xi^2 + \varepsilon^2)^{1/2}} \frac{d\xi}{2\pi} \right|^2 W(x')^2 \, dx \, dx' \\
&= \iint_{\mathbb{R}\times\mathbb{R}} \frac{1}{\xi^2 + \varepsilon^2} W(x')^2 \frac{d\xi}{2\pi} \, dx' = (2\varepsilon)^{-1} \int_{\mathbb{R}} W(x')^2 \, dx'.
\end{aligned}$$

The last identity is a computation; see also (5.43) below. This proves the lemma. $\qquad\square$

The next lemma relates the operators $\mathcal{L}_\varepsilon$ for different values of ε. We show that for $\varepsilon > \varepsilon'$, $\mathcal{L}_\varepsilon$ can be written as an average of unitary equivalent copies of $\mathcal{L}_{\varepsilon'}$. To state this precisely, we denote by $U(\xi)$ the unitary operator in $L^2(\mathbb{R})$ of multiplication by the function $x \mapsto e^{-i\xi x}$ and for $\varepsilon > 0$ and $\xi \in \mathbb{R}$ we set

$$g_\varepsilon(\xi) := \frac{\varepsilon}{\pi(\xi^2 + \varepsilon^2)} \,. \tag{5.41}$$

As noted in the proof of Lemma 5.28, this function is a probability density.

Lemma 5.29 *Let $0 < \varepsilon' \le \varepsilon$. Then*

$$\mathcal{L}_\varepsilon = \int_{\mathbb{R}} U(\xi)^* \, \mathcal{L}_{\varepsilon'} \, U(\xi) \, g_{\varepsilon-\varepsilon'}(\xi) \, d\xi \,.$$

Proof We prove that for all $u \in L^2 \cap L^\infty(\mathbb{R})$,

$$(\mathcal{L}_\varepsilon u, u) = \int_{\mathbb{R}} (\mathcal{L}_{\varepsilon'} \, U(\xi) u, U(\xi) u) \, g_{\varepsilon-\varepsilon'}(\xi) \, d\xi \,. \tag{5.42}$$

Since $L^2 \cap L^\infty(\mathbb{R})$ is dense in $L^2(\mathbb{R})$ and since the quadratic forms on both sides are bounded, this implies the claimed identity in the quadratic form sense.

For $u \in L^2 \cap L^\infty(\mathbb{R})$ we have $f = Wu \in L^2(\mathbb{R})$ and $Q_\varepsilon^* u = \left(-\frac{d^2}{dx^2} + \varepsilon^2\right)^{-1/2} f$, so

$$(\mathcal{L}_\varepsilon u, u) = 2\varepsilon \|Q_\varepsilon^* u\|^2 = 2\varepsilon \left\| \left(-\frac{d^2}{dx^2} + \varepsilon^2\right)^{-1/2} f \right\|^2 = 2\pi \int_{\mathbb{R}} g_\varepsilon(\xi')|\widehat{f}(\xi')|^2 \, d\xi' \,.$$

Similarly, since $WU(\xi)u = U(\xi)f$ and $\overline{U(\xi)f}(\xi') = \widehat{f}(\xi' + \xi)$,

$$(\mathcal{L}_{\varepsilon'} \, U(\xi)u, U(\xi)u) = 2\pi \int_{\mathbb{R}} g_{\varepsilon'}(\xi')|\widehat{U(\xi)f}(\xi')|^2 \, d\xi'$$

$$= 2\pi \int_{\mathbb{R}} g_{\varepsilon'}(\xi')|\widehat{f}(\xi' + \xi)|^2 \, d\xi' = 2\pi \int_{\mathbb{R}} g_{\varepsilon'}(\xi' - \xi)|\mathcal{F} f(\xi')|^2 \, d\xi' \,.$$

Thus, the claimed identity (5.42) follows by Fubini's theorem using the fact that

$$g_\varepsilon(\xi') = \int_{\mathbb{R}} g_{\varepsilon'}(\xi' - \xi) g_{\varepsilon-\varepsilon'}(\xi) \, d\xi$$

for all $\xi' \in \mathbb{R}$ and $0 < \varepsilon' < \varepsilon$. This latter formula can be verified, for instance, by Fourier transformation using the fact that, for all $\varepsilon > 0$ and $x \in \mathbb{R}$,

$$\int_{\mathbb{R}} g_\varepsilon(\xi) e^{i\xi x} \, dx = e^{-\varepsilon|x|} \,, \tag{5.43}$$

which can be shown by means of the residue theorem. $\square$

Let us denote by $\mu_n(\mathcal{L}_\varepsilon)$ the eigenvalues of $\mathcal{L}_\varepsilon$, arranged in non-increasing order and repeated according to multiplicities. It follows from Lemma 5.28 that this sequence is either finite or accumulates at 0. In the first case we extend the sequence by zeros. The identity from Lemma 5.29, together with Corollary 1.34 applied with $A = -\mathcal{L}_\varepsilon$, has the following important consequence.

Corollary 5.30   *Let $0 < \varepsilon' \leq \varepsilon$. Then for any $N \in \mathbb{N}$,*

$$\sum_{n=1}^{N} \mu_n(\mathcal{L}_\varepsilon) \leq \sum_{n=1}^{N} \mu_n(\mathcal{L}_{\varepsilon'}).$$

To appreciate the inequality in the corollary, we emphasize that, in contrast to the ordinary Birman–Schwinger operators, the modified Birman–Schwinger operators are *not* monotone in ε. Remarkably, partial sums of their eigenvalues are.

5.3.2 Proof of Theorem 5.27

By the variational principle it suffices to prove the theorem for $V \geq 0$. In that case, we set $W := \sqrt{V} \in L^2(\mathbb{R})$. Consider the Birman–Schwinger operator

$$\frac{1}{2\varepsilon} \mathcal{L}_\varepsilon = W \left(-\frac{d^2}{dx^2} + \varepsilon^2 \right)^{-1} W, \quad \varepsilon > 0,$$

where $\mathcal{L}_\varepsilon$ is defined in (5.40). We recall that $\mu_n(\mathcal{L}_\varepsilon)$ denote the eigenvalues of $\mathcal{L}_\varepsilon$ in non-increasing order and repeated according to multiplicities. Moreover, denote by $-E_n$ the negative eigenvalues of the Schrödinger operator $-\frac{d^2}{dx^2} - V$ in non-decreasing order and repeated according to multiplicities. (In fact, using ODE methods, we will see in Proposition 5.34 that all eigenvalues are simple. Note that the assumptions of that proposition are satisfied according to Proposition 4.3.) The Birman–Schwinger principle in the form of Corollary 4.26 implies that

$$1 = \frac{1}{2\sqrt{E_n}} \mu_n(\mathcal{L}_{\sqrt{E_n}}) \quad \text{for all } n. \tag{5.44}$$

Based on this identity and the inequality in Corollary 5.30, we will show that

$$2 \sum_{n=1}^{N} \sqrt{E_n} \leq \sum_{n=1}^{N} \mu_n(\mathcal{L}_{\sqrt{E_N}}) \quad \text{for all } N \in \mathbb{N}. \tag{5.45}$$

Before proving this, however, let us use it to deduce the theorem. Indeed, by (5.45) and Lemma 5.28 we have, for all $N \in \mathbb{N}$,

$$2 \sum_{n=1}^{N} \sqrt{E_n} \leq \sum_{n=1}^{\infty} \mu_n(\mathcal{L}_{\sqrt{E_N}}) = \operatorname{Tr} \mathcal{L}_{\sqrt{E_N}} = \int_{\mathbb{R}} W(x)^2 \, dx = \int_{\mathbb{R}} V(x) \, dx.$$

This clearly implies the theorem.

It remains to prove (5.45), which we do by induction. For $N = 1$, this is simply (5.44). Now let $N \geq 2$ and assume the inequality has been proved for all smaller values of N. Using first the induction assumption (5.45) with N replaced by $N - 1$ and then Corollary 5.30, we find

$$2 \sum_{n=1}^{N-1} \sqrt{E_n} \leq \sum_{n=1}^{N-1} \mu_n(\mathcal{L}_{\sqrt{E_{N-1}}}) \leq \sum_{n=1}^{N-1} \mu_n(\mathcal{L}_{\sqrt{E_N}}) .$$

Multiplying (5.44) with $n = N$ by $2\sqrt{E_N}$ and adding it to the previous inequality, we obtain (5.45), as claimed. This completes the proof. □

5.4 The sharp bound for $\gamma = \frac{3}{2}$ in one dimension

5.4.1 Statement and discussion of the result

Our goal in this section is to prove the following sharp Lieb–Thirring inequality, which is due to Gardner et al. (1974, Equation 3.27); see also the original paper (Lieb and Thirring, 1976).

Theorem 5.31 *Let $d = 1$ and $\gamma = \frac{3}{2}$. Then, for all $V \in L^1_{\mathrm{loc}}(\mathbb{R})$ with $V_+ \in L^2(\mathbb{R})$,*

$$\mathrm{Tr}\left(-\frac{d^2}{dx^2} - V\right)_-^{3/2} \leq \frac{3}{16} \int_{\mathbb{R}} V(x)_+^2 \, dx . \tag{5.46}$$

The constant $\frac{3}{16}$ is optimal.

Optimality of the constant $\frac{3}{16}$ follows, for instance, from Lemma 5.16, which implies that, for any $N \in \mathbb{N}$, $V(x) = N(N + 1) \cosh^{-2} x$ attains equality in (5.46). In fact, we have

$$\frac{3}{16} = L^{\mathrm{cl}}_{3/2,1} = L^{(1)}_{3/2,1} .$$

The fact that the constant in Theorem 5.31 is the semiclassical one implies, via the Aizenman–Lieb argument (Lemma 5.2), the following sharp inequality.

Corollary 5.32 *Let $d = 1$ and $\gamma \geq \frac{3}{2}$. Then, for all $V \in L^1_{\mathrm{loc}}(\mathbb{R})$ with $V_+ \in L^{\gamma+1/2}(\mathbb{R})$,*

$$\mathrm{Tr}\left(-\frac{d^2}{dx^2} - V\right)_-^{\gamma} \leq L^{\mathrm{cl}}_{\gamma,1} \int_{\mathbb{R}} V(x)_+^{\gamma+1/2} \, dx .$$

The constant $L^{\mathrm{cl}}_{\gamma,1}$ is optimal.

This corollary in the special case $\gamma = 3/2 + k$, $k \in \mathbb{N}$, is due to Lieb and Thirring (1976). The full result first appeared in Aizenman and Lieb (1978).

We will present the original proof of Theorem 5.31 in §5.5. In this section we present an alternative proof due to Benguria and Loss (2000). We also provide a proof of the following reversed Lieb–Thirring inequality.

Theorem 5.33 *Let $d = 1$ and $\gamma = \frac{1}{2}$. Then for any $V \in L^1(\mathbb{R})$,*

$$\mathrm{Tr}\left(-\frac{d^2}{dx^2} - V\right)_-^{1/2} \geq \frac{1}{4} \int_{\mathbb{R}} V(x)\, dx.$$

The constant $\frac{1}{4}$ is optimal.

The optimality of the constant $\frac{1}{4}$ follows from Remark 5.17.

This theorem is independently due to Glaser et al. (1978, Equation (41)) and Schmincke (1978). We give the proof of Glaser et al. later in §5.5 and present here Schmincke's proof, which predates and is similar to the Benguria–Loss proof of Theorem 5.31.

5.4.2 Proof of Theorems 5.31 and 5.33

We shall prove both Theorems 5.31 and 5.33 under the additional assumption that V has compact support and, in the case of Theorem 5.31, that $V_- \in L^2(\mathbb{R})$. At the end of the proof, we explain how to remove these extra assumptions.

We note that, by Proposition 4.3 and Remark 4.4, under the assumptions of Theorems 5.31 and 5.33 the Schrödinger operator

$$H = -\frac{d^2}{dx^2} - V$$

is defined as a self-adjoint, lower semibounded operator in $L^2(\mathbb{R})$ via a lower semibounded and closed quadratic form with form domain $H^1(\mathbb{R})$. The assumption that V has compact support implies that H has finitely many eigenvalue $-E_1 \leq -E_2 \leq \cdots \leq -E_N < 0$; see Proposition 4.18. We assume that H has at least one negative eigenvalue. For if not, then Theorem 5.31 holds trivially and Theorem 5.33 is a consequence of Proposition 4.43.

We will use the following two standard facts which, for the convenience of the reader, we state and prove in Propositions 5.34 and 5.35 below. The first fact is that each eigenvalue $-E_n$ is simple and the second is that an eigenfunction ψ corresponding to the lowest eigenvalue $-E_1$ can be chosen positive.

Let us introduce the logarithmic derivative of ψ,

$$F(x) := \psi'(x)/\psi(x).$$

Since $\psi' \in L^2(\mathbb{R})$ and since the continuous (Lemma 2.6), positive function ψ is bounded away from zero in any compact interval, we have $F \in L^2_{\mathrm{loc}}(\mathbb{R})$.

The equation for ψ reads

$$\left(-\frac{d^2}{dx^2} - V(x)\right)\psi = -E_1\psi .\tag{5.47}$$

As we shall argue in the paragraph that precedes Lemma 5.36, this equation, which we know initially to hold in the weak sense (that is, when integrated against an $H^1(\mathbb{R})$ function), implies that ψ' is weakly differentiable. Additionally, (5.47) holds as an equality of L^1_{loc} functions. Since linear combinations of $e^{-\sqrt{E_1}x}$ and $e^{+\sqrt{E_1}x}$ are the only solutions of (5.47) outside the support of V, it follows that, to the left and to the right of the support, ψ is such a combination. Since $\psi \in L^2(\mathbb{R})$, we see that only $e^{-\sqrt{E_1}|x|}$ can occur in this linear combination and, consequently,

$$F(x) = \begin{cases} -\sqrt{E_1} & \text{if } x \geq \sup \operatorname{supp} V, \\ \sqrt{E_1} & \text{if } x \leq \inf \operatorname{supp} V. \end{cases}\tag{5.48}$$

Moreover, F is weakly differentiable and satisfies the Riccati equation

$$F'(x) + F(x)^2 = -V(x) + E_1 .\tag{5.49}$$

Indeed, let $\varphi \in H^1(\mathbb{R})$ with compact support. On the compact set $\operatorname{supp} \varphi$, the continuous function ψ is bounded away from zero by a positive constant, and therefore $1/\psi$ is weakly differentiable with $(1/\psi)' = -\psi'/\psi^2$ (Lemma 2.10). Hence, by the product rule for Sobolev functions (Lemma 2.14), φ/ψ is weakly differentiable with $(\varphi/\psi)' = \varphi'/\psi - \varphi\psi'/\psi^2$. By the eigenvalue equation for ψ in its weak form,

$$\int_{\mathbb{R}} \varphi' F \, dx = \int_{\mathbb{R}} \left(\left(\frac{\varphi}{\psi}\right)'\psi' + \frac{\varphi\psi'}{\psi^2}\psi'\right) dx = \int_{\mathbb{R}} \left((V - E_1)\frac{\varphi}{\psi}\psi + \frac{\varphi\psi'}{\psi^2}\psi'\right) dx$$
$$= \int_{\mathbb{R}} \varphi\left(V - E_1 + F^2\right) dx .$$

This proves that F is weakly differentiable with $F' = -V + E_1 - F^2$, as claimed.

The Riccati equation (5.49), the asymptotic behavior (5.48), and the fact that $F \in L^2_{\mathrm{loc}}(\mathbb{R})$ imply that $F' \in L^1(\mathbb{R})$. From Lemma 2.6 and (5.48) we infer that $F \in L^\infty(\mathbb{R})$. In particular, in the setting of Theorem 5.31, where we assume $V \in L^2(\mathbb{R})$, we deduce from (5.49) that $F' \in L^2(\mathbb{R})$.

Let us define the operator

$$Q := \frac{d}{dx} - F$$

with domain $H^1(\mathbb{R})$. Clearly, $\frac{d}{dx}$ with domain $H^1(\mathbb{R})$ is closed in $L^2(\mathbb{R})$ and, since F is bounded, the operator Q is closed in $L^2(\mathbb{R})$ as well. A computation as in the proof of Proposition 4.9 shows that, for all $u \in H^1(\mathbb{R})$,

$$\int_{\mathbb{R}} |Qu|^2 \, dx = \int_{\mathbb{R}} \left(|u'|^2 - V|u|^2 + E_1|u|^2 \right) dx \,.$$

This implies that

$$Q^*Q = H + E_1 \,. \tag{5.50}$$

The adjoint of Q is given by

$$Q^* = -\frac{d}{dx} - F$$

with domain $H^1(\mathbb{R})$. Moreover, since $V + 2F'$ is in $L^1(\mathbb{R})$ with compact support, just like V, a similar computation to before shows that, for all $u \in H^1(\mathbb{R})$,

$$\int_{\mathbb{R}} |Q^*u|^2 \, dx = \int_{\mathbb{R}} \left(|u'|^2 - (V + 2F')|u|^2 + E_1|u|^2 \right) dx \,.$$

The quadratic form on the right side is lower semibounded and closed with form domain $H^1(\mathbb{R})$ and defines a new Schrödinger operator

$$H^{(1)} = -\frac{d^2}{dx^2} - V^{(1)} \quad \text{with} \quad V^{(1)} := V + 2F' \,.$$

Using the closedness of Q it follows that

$$QQ^* = H^{(1)} + E_1 \,. \tag{5.51}$$

Let us show that the kernel of the operator QQ^* is trivial. Indeed, assume that $QQ^*\varphi = 0$ for some $\varphi \in \operatorname{dom} QQ^*$. Then

$$\|Q^*\varphi\|^2 = \langle QQ^*\varphi, \varphi \rangle = 0,$$

so $Q^*\varphi = 0$, that is, $\varphi' = -F\varphi$. (A remark similar to the one in the proof of Proposition 4.6, concerning the passage from weak to classical derivative, applies here.) However, since F is equal to a positive constant to the right of the support of V and equal to a negative constant to the left of the support of V, the function φ, if non-trivial, would be exponentially growing and hence cannot be an element of $L^2(\mathbb{R})$. Therefore we have $\varphi \equiv 0$ and so $\ker QQ^* = \{0\}$.

The operators Q and Q^* are closed and, by Proposition 1.23, the non-zero spectra of Q^*Q and QQ^* coincide. The kernel of QQ^* is trivial and, as we have mentioned before, the kernel of the operator $Q^*Q = H + E_1$ is one-dimensional. Therefore, the negative spectrum of the operator $H^{(1)}$ consists precisely of the eigenvalues $-E_2 < -E_3 < \cdots < -E_N < 0$. Thus, by modifying the potential, we have removed the lowest eigenvalue from the negative spectrum of a Schrödinger operator.

The potential $V^{(1)}$ of the 'new' Schrödinger operator $H^{(1)}$ satisfies the same properties as the initial one: namely, it belongs to L^1 or L^2 in the two respective cases and, since F' vanishes outside of the convex hull of the support of V, the support of the new potential $V^{(1)}$ is again bounded. Its new ground state $\psi^{(1)}$ is positive and corresponds to the eigenvalue $-E_2$ of the original operator H. Introducing its logarithmic derivative,

$$F^{(1)} := (\psi^{(1)})'/\psi^{(1)},$$

we can remove the eigenvalue $-E_2$ obtaining another potential,

$$V^{(2)} := V^{(1)} + 2(F^{(1)})',$$

with the negative eigenvalues $-E_3 < \cdots < -E_N < 0$. Repeating this procedure, we can eliminate all the eigenvalues and arrive in the Nth step at a potential,

$$V^{(N)} = V^{(N-1)} + 2(F^{(N-1)})',$$

with empty negative spectrum.

We now show how to prove Theorem 5.33 in the case of compactly supported V. For $n = 1, \ldots, N$, using the analogue of (5.48) for $F^{(n-1)}$, we have

$$\int_{\mathbb{R}} V^{(n)} \, dx = \int_{\mathbb{R}} V^{(n-1)} \, dx + 2 \, F^{(n-1)} \Big|_{-\infty}^{+\infty} = \int_{\mathbb{R}} V^{(n-1)} \, dx - 4\sqrt{E_n} \, .$$

Here we have written $V =: V^{(0)}$ and $F =: F^{(0)}$. Summing the previous identities, we obtain

$$\int_{\mathbb{R}} V^{(N)} \, dx = \int_{\mathbb{R}} V \, dx - 4 \sum_{n=1}^{N} \sqrt{E_n} \, . \tag{5.52}$$

The operator $H^{(N)} = -\frac{d^2}{dx^2} - V^{(N)}$ has no negative eigenvalues. According to Proposition 4.43, this implies that

$$\int_{\mathbb{R}} V^{(N)} \, dx \le 0 \, .$$

This inequality, together with (5.52), proves Theorem 5.33 in the case of compactly supported V.

We now proceed to the proof of Theorem 5.31 for compactly supported $V \in L^2(\mathbb{R})$. As argued above, $F' \in L^2(\mathbb{R})$. This implies that $V^{(n)} \in L^2(\mathbb{R})$ and

$F^{(n)} \in L^2(\mathbb{R})$ for all $n = 1, \ldots, N$. Using the analogue of (5.49) for $F^{(n-1)}$, we have, again for $n = 1, \ldots, N$

$$\int_{\mathbb{R}} (V^{(n)})^2 \, dx = \int_{\mathbb{R}} (V^{(n-1)} + 2(F^{(n-1)})')^2 \, dx$$

$$= \int_{\mathbb{R}} (V^{(n-1)})^2 dx + 4 \int_{\mathbb{R}} (V^{(n-1)} + (F^{(n-1)})')(F^{(n-1)})' \, dx$$

$$= \int_{\mathbb{R}} (V^{(n-1)})^2 dx + 4 \int_{\mathbb{R}} (E_n - (F^{(n-1)})^2)(F^{(n-1)})' \, dx .$$

Using the analogue of (5.48) for $F^{(n-1)}$, we find

$$4 \int_{\mathbb{R}} (E_n - (F^{(n-1)})^2)(F^{(n-1)})' \, dx = 4E_n F^{(n-1)} \Big|_{-\infty}^{+\infty} - \frac{4}{3} (F^{(n-1)})^3 \Big|_{-\infty}^{+\infty}$$

$$= -\frac{16}{3} E_n^{3/2} .$$

This gives, for $n = 1, \ldots, N$,

$$\int_{\mathbb{R}} (V^{(n)})^2 \, dx = \int_{\mathbb{R}} (V^{(n-1)})^2 \, dx - \frac{16}{3} E_n^{3/2} ,$$

and therefore, by summing these identities,

$$0 \le \int_{\mathbb{R}} (V^{(N)})^2 \, dx = \int_{\mathbb{R}} V^2 \, dx - \frac{16}{3} \sum_{n=1}^{N} E_n^{3/2} .$$

This proves Theorem 5.31 in the case of compactly supported $V \in L^2(\mathbb{R})$.

It remains to remove the extra assumptions. In the case of Theorem 5.31 we can do this immediately by Theorem 4.45. Therefore, let us discuss the situation for Theorem 5.33. We recall from Proposition 1.40 that for any $\rho > 1$ there is a $C_\rho < \infty$ such that, for any self-adjoint, lower semibounded operators A and B, we have

$$\mathrm{Tr}(A + B)_-^{1/2} \le \rho \, \mathrm{Tr} \, A_-^{1/2} + C_\rho \, \mathrm{Tr} \, B_-^{1/2} .$$

(In fact, by Rotfel'd's inequality in Proposition 1.43 one can take $\rho = C_\rho = 1$, but for our purposes, the simpler Proposition 1.40 is sufficient.)

Let $V \in L^1(\mathbb{R})$ and $0 < \varepsilon < 1$. For $W \in L^1(\mathbb{R})$ with compact support we write

$$-\frac{d^2}{dx^2} - (1 - \varepsilon)W = (1 - \varepsilon)\left(-\frac{d^2}{dx^2} - V\right) + \left(-\varepsilon \frac{d^2}{dx^2} - (1 - \varepsilon)(W - V)\right)$$

and bound, for $\rho > 1$,

$$\mathrm{Tr}\left(-\frac{d^2}{dx^2} - (1-\varepsilon)W\right)_-^{1/2} \leq \rho(1-\varepsilon)^{1/2} \mathrm{Tr}\left(-\frac{d^2}{dx^2} - V\right)_-^{1/2}$$

$$+ C_\rho \mathrm{Tr}\left(-\varepsilon\frac{d^2}{dx^2} - (1-\varepsilon)(W-V)\right)_-^{1/2}.$$

According to what we have already shown,

$$\mathrm{Tr}\left(-\frac{d^2}{dx^2} - (1-\varepsilon)W\right)_-^{1/2} \geq \frac{1-\varepsilon}{4}\int_{\mathbb{R}} W\,dx.$$

Meanwhile, by the Lieb–Thirring inequality for $\gamma = 1/2$ (Theorem 4.38),

$$\mathrm{Tr}\left(-\varepsilon\frac{d^2}{dx^2} - (1-\varepsilon)(W-V)\right)_-^{1/2} \leq L_{\gamma,1/2}\frac{1-\varepsilon}{\sqrt{\varepsilon}}\int_{\mathbb{R}} (W-V)_+\,dx.$$

Thus, we infer that

$$\mathrm{Tr}\left(-\frac{d^2}{dx^2} - V\right)_-^{1/2} \geq \frac{\sqrt{1-\varepsilon}}{4\rho}\int_{\mathbb{R}} W\,dx - L_{\gamma,1/2}\frac{C_\rho}{\rho}\frac{\sqrt{1-\varepsilon}}{\sqrt{\varepsilon}}\int_{\mathbb{R}} (W-V)_+\,dx.$$

Choosing a sequence (W_n) of compactly supported functions that converges to V in $L^1(\mathbb{R})$, we obtain

$$\mathrm{Tr}\left(-\frac{d^2}{dx^2} - V\right)_-^{1/2} \geq \frac{\sqrt{1-\varepsilon}}{4\rho}\int_{\mathbb{R}} V\,dx,$$

and, since $\varepsilon > 0$ and $\rho > 1$ are arbitrary, we obtain Theorem 5.33 for $V \in L^1(\mathbb{R})$.

5.4.3 Some results about one-dimensional Schrödinger operators

In this subsection we provide proofs of the following well-known results about one-dimensional Schrödinger operators that were used in the previous subsection. We assume in this subsection that $V \in L^1_{\mathrm{loc}}(\mathbb{R})$ satisfies the conditions of Proposition 4.1. Then the operator $-\frac{d^2}{dx^2} - V(x)$ in $L^2(\mathbb{R})$ is defined in Corollary 4.2.

Proposition 5.34 *All eigenvalues of the operator $-\frac{d^2}{dx^2} - V(x)$ are simple.*

Proposition 5.35 *Assume that $\inf \sigma\left(-\frac{d^2}{dx^2} - V(x)\right)$ is an eigenvalue of the operator $-\frac{d^2}{dx^2} - V(x)$. Then there is a corresponding eigenfunction that is everywhere positive.*

We emphasize that eigenfunctions belong to the operator domain and, therefore, to the form domain, which is contained in $H^1(\mathbb{R})$. Hence, by Lemma 2.6, they are continuous. In particular, as a consequence of Proposition 5.35, an eigenfunction belonging to $\inf \sigma\left(-\frac{d^2}{dx^2} - V(x)\right)$ is bounded away from zero on any compact interval.

For the proof of Propositions 5.34 and 5.35 we use two auxiliary lemmas. The first one states uniqueness of the Cauchy problem under rather weak assumptions on the coefficients and the solution. To formulate it, we denote by $W^{2,1}(0,T)$ the set of all weakly differentiable functions u on $(0,T)$ such that u' is weakly differentiable and such that $u, u', u'' \in L^1(0,T)$. By Remark 2.7, the boundary values $u(0)$ and $u'(0)$ are well defined for $u \in W^{2,1}(0,T)$.

Lemma 5.36 *Let $T > 0$ and $q \in L^1(0,T)$. Let $\psi \in W^{2,1}(0,T)$ be such that $\psi'' = q\psi$ in $(0,T)$ and $\psi(0) = \psi'(0) = 0$. Then $\psi = 0$ in $(0,T)$.*

Proof Let $\varphi := \psi'$ and put $x_0 := \sup\{x \le T : (\psi, \varphi) \equiv (0,0) \text{ on } [0,x]\}$. If $x_0 = T$, we are done. Thus, assume that $x_0 < T$. Since φ and ψ are continuous by Lemma 2.6, we have $\psi(x_0) = \varphi(x_0) = 0$. Then, according to Lemma 2.6, for any $x_0 \le x < T$,

$$\psi(x) = \int_{x_0}^x \varphi(t)\, dt \quad \text{and} \quad \varphi(x) = \int_{x_0}^x q(t)\psi(t)\, dt \,.$$

Thus

$$\psi(x) = \int_{x_0}^x \int_{x_0}^y q(t)\psi(t)\, dt\, dy$$

and

$$|\psi(x)| \le \int_{x_0}^x \int_{x_0}^y |q(t)|\, dt\, dy \max_{x_0 \le t \le x} |\psi(t)| = \int_{x_0}^x (x-t)|q(t)|\, dt \max_{x_0 \le t \le x} |\psi(t)| \,.$$

Since $q \in L^1(0,T)$, there is an $x_0 < x_1 \le T$ such that

$$\int_{x_0}^x (x-t)|q(t)|\, dt \le \frac{1}{2} \quad \text{for all } x_0 \le x \le x_1 \,.$$

This implies

$$|\psi(x)| \le \frac{1}{2} \max_{x_0 \le t \le x} |\psi(t)| \quad \text{for all } x_0 \le x \le x_1 \,,$$

which implies $\max_{x_0 \le x \le x_1} |\psi(x)| = 0$. This contradicts the maximality of x_0 and concludes the proof. $\qquad\square$

Lemma 5.37 *Let $q \in L^1_{loc}(\mathbb{R})$ be real and let $\psi_1, \psi_2 \in H^1(\mathbb{R})$ be such that, for any $\varphi \in H^1(\mathbb{R})$ with compact support,*

$$\int_{\mathbb{R}} \left(\psi'_j \varphi' + q\psi_j \varphi \right) dx = 0 \qquad \text{for } j = 1,2 . \tag{5.53}$$

Then ψ_1 and ψ_2 are linearly dependent.

Note that $\int_{\mathbb{R}} q\psi_j \varphi \, dx$ is well defined since the H^1 functions ψ_j and φ are bounded and $q \chi_{\text{supp}\varphi} \in L^1(\mathbb{R})$.

Proof The assumption $\psi_j \in H^1(\mathbb{R})$ implies, by Lemma 2.6, that ψ_j is continuous. Equation (5.53) implies that ψ'_j is weakly differentiable with $\psi''_j = q\psi_j$. Therefore, again by Lemma 2.6, ψ'_j is also continuous.

We consider the Wronskian $W := \psi'_1\psi_2 - \psi_1\psi'_2 \in L^1(\mathbb{R})$. By the product rule (Lemma 2.14) and equation (5.53), we find that W is weakly differentiable with

$$W' = \psi''_1\psi_2 - \psi_1\psi''_2 = q\psi_1\psi_2 - \psi_1 q\psi_2 = 0 .$$

According to Lemma 2.5, W is constant and, since $W \in L^1(\mathbb{R})$, we have

$$\psi'_1\psi_2 - \psi_1\psi'_2 = W = 0 .$$

If $\psi_1 \equiv 0$, there is nothing to prove, so we may assume that there is an $x_0 \in \mathbb{R}$ with $\psi_1(x_0) \neq 0$. Then the function

$$\psi_2 - \frac{\psi_2(x_0)}{\psi_1(x_0)} \psi_1$$

solves (5.53), vanishes at x_0 and, because of the identity $W(x_0) = 0$, its derivative also vanishes at x_0. Therefore, by Lemma 5.36, $\psi_2 - \frac{\psi_2(x_0)}{\psi_1(x_0)}\psi_1 \equiv 0$, as claimed. $\qquad\square$

Proof of Proposition 5.34 Assume that λ is an eigenvalue of the operator $-\frac{d^2}{dx^2} - V(x)$. The corresponding eigenfunctions belong to the operator domain, and therefore also to the form domain, which is contained in $H^1(\mathbb{R})$. The conclusion follows now from Lemma 5.37 with $q = -\lambda - V$. $\qquad\square$

Proof of Proposition 5.35 As an $H^1(\mathbb{R})$ function the eigenfunction φ corresponding to the eigenvalue $\lambda := \inf \sigma\left(-\frac{d^2}{dx^2} - V(x)\right)$ is continuous. Hence, it suffices to show that it does not vanish at any point. We argue by contradiction and assume that $\varphi(x_0) = 0$ for some $x_0 \in \mathbb{R}$. Put $\tilde{\varphi}(x) := \varphi(x)$ for $x \leq x_0$ and $\tilde{\varphi}(x) := 0$ for $x \geq x_0$. This function belongs to $H^1(\mathbb{R}) \cap L^2(\mathbb{R}, V_- \, dx)$, which is the form domain of the operator. Therefore, by the weak form of the eigenvalue equation for φ,

$$\int_{\mathbb{R}} (|\tilde{\varphi}'|^2 - V(x)|\tilde{\varphi}|^2) \, dx = \int_{\mathbb{R}} (\tilde{\varphi}'\overline{\varphi}' - V(x)\tilde{\varphi}\overline{\varphi}) \, dx = \lambda \int_{\mathbb{R}} \tilde{\varphi}\overline{\varphi} \, dx = \lambda \int_{\mathbb{R}} |\tilde{\varphi}|^2 \, dx .$$

Thus, if not identically zero, $\tilde{\varphi}$ is a minimizer of the quadratic form, and so, by Lemma 1.19, an eigenfunction of the operator. As an eigenfunction, $\tilde{\varphi}$ is a solution of $\tilde{\varphi}'' = q\tilde{\varphi}$ on $\mathbb{R}$ for $q = -V - \lambda$. As in the proof of Lemma 5.37, this equation, which initially holds when tested against functions in $H^1(\mathbb{R}) \cap L^2(\mathbb{R}, V_- \, dx)$, implies that $\tilde{\varphi}'$ is weakly differentiable and that it holds as an identity of $L^1_{\mathrm{loc}}(\mathbb{R})$ functions. Moreover, $\tilde{\varphi}$ vanishes on $[x_0, \infty)$. Therefore, by Lemma 5.36, it is identically equal to zero on $\mathbb{R}$ and, in particular, $\varphi(x) = 0$ for all $x \leq x_0$. Again by Lemma 5.36, this implies that φ is identically equal to zero on all of $\mathbb{R}$, which is a contradiction. $\qquad\square$

5.5 Trace formulas for one-dimensional Schrödinger operators

Our goal in this section is to give an alternative proof of the sharp Lieb–Thirring inequality for $\gamma = 3/2$ in $d = 1$ (Theorem 5.31). It is in fact the original proof due to Gardner et al. (1974, Equation 3.27); see also Lieb and Thirring (1976). It uses an identity, called the trace formula, which relates spectral characteristics coming from the discrete and continuous spectrum of the Schrödinger operator $-\frac{d^2}{dx^2} - V$ to a quantity expressed in terms of the potential V. The relevant trace formula is originally due to Zaharov and Faddeev (1971) and we refer to §5.6 for further references.

5.5.1 Statement and discussion of the trace formulas

We first state a result on the existence of *Jost solutions* of the Schrödinger equation. We denote by $\mathbb{C}_+ = \{k \in \mathbb{C} : \mathrm{Im}\, k > 0\}$ the upper half-plane.

Lemma 5.38 *Let $V \in L^1(\mathbb{R})$ and $k \in \overline{\mathbb{C}_+} \setminus \{0\}$. Then there is a unique solution ψ_k of*

$$-\psi_k'' - V\psi_k = k^2\psi_k \qquad in\ \mathbb{R} \qquad (5.54)$$

satisfying, as $x \to \infty$,

$$\psi_k(x) = e^{ikx}\,(1 + o(1)) \quad and \quad \psi_k'(x) = ike^{ikx}\,(1 + o(1)). \qquad (5.55)$$

For any fixed $x \in \mathbb{R}$, $k \mapsto \psi_k(x)$ is analytic in $\mathbb{C}_+$ and continuous in $\overline{\mathbb{C}_+} \setminus \{0\}$. Moreover, for all $k \in \overline{\mathbb{C}_+} \setminus \{0\}$, $\varphi_k(x) := e^{-ikx}\psi_k(x)$ satisfies

$$\|\varphi_k\|_\infty \leq \exp(\|V\|_1/|k|) \quad and \quad \|\varphi_k - 1\|_\infty \leq |k|^{-1}\|V\|_1 \exp(\|V\|_1/|k|). \qquad (5.56)$$

The proof of this lemma is based on the integral equation

$$\varphi_k(x) = 1 - (2ik)^{-1} \int_x^\infty \left(e^{2ik(y-x)} - 1 \right) V(y)\varphi_k(y)\,dy. \tag{5.57}$$

We defer the details of the argument until after the statement of the main results of this section. Writing (5.57) in the form

$$\psi_k(x) =$$

$$e^{ikx}\left(1 + (2ik)^{-1} \int_x^\infty V(y)\varphi_k(y)\,dy \right) - e^{-ikx}(2ik)^{-1} \int_x^\infty e^{2iky} V(y)\varphi_k(y)\,dy$$

and using the bounds on φ_k in (5.56), we see that for $k \in \mathbb{R} \setminus \{0\}$ we have, as $x \to -\infty$,

$$\begin{aligned} \psi_k(x) &= a(k)e^{ikx} + b(k)e^{-ikx} + o(1) &\text{and} \\ \psi_k'(x) &= a(k)ike^{ikx} - b(k)ike^{-ikx} + o(1). \end{aligned} \tag{5.58}$$

Here

$$\begin{aligned} a(k) &:= 1 + (2ik)^{-1} \int_{-\infty}^\infty V(y)\varphi_k(y)\,dy &\text{and} \\ b(k) &:= -(2ik)^{-1} \int_{-\infty}^\infty e^{2iky} V(y)\varphi_k(y)\,dy. \end{aligned} \tag{5.59}$$

The following property will be crucial for our application.

Lemma 5.39 *For all $k \in \mathbb{R} \setminus \{0\}$ we have*

$$|a(k)|^2 - |b(k)|^2 = 1.$$

Proof For real k, both ψ_k and ψ_{-k} are solutions of the same equation and therefore their Wronskian $W = \psi_k'\psi_{-k} - \psi_k\psi_{-k}'$ is constant. Using asymptotics (5.55) at $+\infty$, we find $W = 2ik$, while asymptotics (5.58) at $-\infty$ give

$$W = 2ik(|a(k)|^2 - |b(k)|^2).$$

This implies the claimed identity. $\qquad\square$

The following is the main result of this section.

Theorem 5.40 *Let $V \in L^1(\mathbb{R})$ and let $-E_n$ be the negative eigenvalues of $-\frac{d^2}{dx^2} - V$ in $L^2(\mathbb{R})$. Then*

$$2 \int_0^\infty \ln|a(k)|\,dk - 2\pi \sum_n E_n^{1/2} = -\frac{\pi}{2} \int_{\mathbb{R}} V(x)\,dx. \tag{5.60}$$

Moreover if, in addition, $V \in L^2(\mathbb{R})$, then

$$2 \int_0^\infty k^2 \ln|a(k)|\,dk + \frac{2\pi}{3} \sum_n E_n^{3/2} = \frac{\pi}{8} \int_{\mathbb{R}} V(x)^2\,dx. \tag{5.61}$$

Note that by Lemma 5.39, $|a| \geq 1$, and therefore the integrands in the k-integrals in (5.60) and (5.61) are pointwise non-negative. This leads to the following bounds.

Corollary 5.41   *Let $V \in L^1(\mathbb{R})$ and let $-E_n$ be the negative eigenvalues of $-\frac{d^2}{dx^2} - V$ in $L^2(\mathbb{R})$. Then*

$$\sum_n E_n^{1/2} \geq \frac{1}{4} \int_{\mathbb{R}} V(x)\,dx\,. \tag{5.62}$$

Moreover if, in addition, $V \in L^2(\mathbb{R})$, then

$$\sum_n E_n^{3/2} \leq \frac{3}{16} \int_{\mathbb{R}} V(x)^2\,dx\,. \tag{5.63}$$

As we mentioned before, the proof of the bound (5.63) via trace formulas is from Gardner et al. (1974, Equation (3.27)); see also Lieb and Thirring (1976). The proof of the bound (5.62) via trace formulas is due to Glaser et al. (1978, Equation (41)).

Remark 5.42   In fact, the method used to prove Theorem 5.40 provides a countable family of trace formulas of the form

$$2 \int_0^\infty k^{2m} \ln|a(k)|\,dk + \frac{(-1)^{m+1}\pi}{m+1/2} \sum_n E_n^{m+1/2} = \int_{\mathbb{R}} F_m(V)(x)\,dx\,,$$

where $m \in \mathbb{N}$ and where, for each $x \in \mathbb{R}$, $F_m(V)(x)$ is a polynomial in $V(x)$ and finitely many derivatives $V^{(\ell)}(x)$. For instance for $m = 2$, one can show that

$$2 \int_0^\infty k^4 \ln|a(k)|\,dk - \frac{2\pi}{5} \sum_n E_n^{5/2} = -\frac{\pi}{16} \int_{\mathbb{R}} V(x)^3\,dx + \frac{\pi}{32} \int_{\mathbb{R}} V'(x)^2\,dx\,. \tag{5.64}$$

Remark 5.43   As an application of the trace formula (5.64) and the sharp Lieb–Thirring inequality for $\gamma = 1/2$ (Theorem 5.27) one can derive the bound

$$0 \leq \frac{3}{16} \int_{\mathbb{R}} V(x)^2\,dx - \sum_n E_n^{3/2} \leq \frac{3}{16} \sqrt{\int_{\mathbb{R}} V(x)\,dx} \sqrt{\int_{\mathbb{R}} V'(x)^2\,dx} \tag{5.65}$$

for $0 \leq V \leq H^1 \cap L^1(\mathbb{R})$; see Laptev and Weidl (2000a, §5.3). In the limit of a large coupling constant, when V is replaced by αV with $\alpha \to \infty$, both terms in the middle of the chain of inequalities grow individually like α^2, whereas the right side only grows like $\alpha^{3/2}$, thus giving an explicit bound on the remainder in the Weyl asymptotics.

The left inequality in (5.65) follows from the sharp Lieb–Thirring inequality

(5.63). Let us prove the right inequality. By the trace formula (5.61) and the Cauchy–Schwarz inequality,

$$\frac{\pi}{8}\int_{\mathbb{R}} V(x)^2\, dx - \frac{2\pi}{3}\sum_n E_n^{3/2} = 2\int_0^\infty k^2 \ln|a(k)|\, dk$$

$$\leq \sqrt{2\int_0^\infty \ln|a(k)|\, dk}\ \sqrt{2\int_0^\infty k^4 \ln|a(k)|\, dk}\,,$$

and it remains to bound the two square roots on the right side. On the one hand, by the trace formula (5.60) and the sharp Lieb–Thirring inequality for $\gamma = 1/2$ (Theorem 5.27), we have

$$2\int_0^\infty \ln|a(k)|\, dk = 2\pi\sum_n E_n^{1/2} - \frac{\pi}{2}\int_{\mathbb{R}} V(x)\, dx \leq \frac{\pi}{2}\int_{\mathbb{R}} V(x)\, dx\,.$$

On the other hand, by the trace formula (5.64) and the sharp Lieb–Thirring inequality for $\gamma = 5/2$ (Corollary 5.32), we have (Lieb and Thirring, 1976, Equation (4.31))

$$2\int_0^\infty k^4 \ln|a(k)|\, dk = \frac{2\pi}{5}\sum_n E_n^{5/2} - \frac{\pi}{16}\int_{\mathbb{R}} V(x)^3\, dx + \frac{\pi}{32}\int_{\mathbb{R}} V'(x)^2\, dx$$

$$\leq \frac{\pi}{32}\int_{\mathbb{R}} V'(x)^2\, dx\,.$$

This proves the right inequality in (5.65).

5.5.2 Proof of the trace formulas

We turn now to the proof of Theorem 5.40. We assume throughout this subsection that $V \in L^1(\mathbb{R})$. We begin by proving Lemma 5.38 concerning the existence of and bounds on Jost solutions.

Proof of Lemma 5.38   Ignoring for the moment the k dependence, we denote by K the integral operator, acting on functions on $\mathbb{R}$, with integral kernel

$$K(x, y) = -(2ik)^{-1}\left(e^{2ik(y-x)} - 1\right) V(y)\chi_{\{y\geq x\}}\,.$$

In terms of this operator, equation (5.57) can be written as $\varphi_k = 1 + K\varphi_k$, and we shall show that $1 - K$ is boundedly invertible in $L^\infty(\mathbb{R})$.

Using $|e^{2ik(y-x)} - 1| \leq 2$ for $y \geq x$ and $k \in \overline{\mathbb{C}_+}$, we obtain by induction

$$|(K^n f)(x)| \leq \frac{\|f\|_\infty}{n!\,|k|^n}\left(\int_x^\infty |V(y)|\, dy\right)^n\,.$$

This implies, in particular, that, as an operator from $L^\infty(\mathbb{R})$ to itself,

$$\|K^n\| \le \frac{1}{n!|k|^n} \|V\|_1^n \,.$$

Thus the operator $1 - K$ is invertible on $L^\infty(\mathbb{R})$ with $(1 - K)^{-1} = \sum_{n=0}^\infty K^n$ satisfying

$$\left\|(1 - K)^{-1}\right\| \le \sum_{n=0}^\infty \|K^n\| \le \exp(|k|^{-1}\|V\|_1) \,. \tag{5.66}$$

The function $\varphi_k := (1 - K)^{-1}1$ solves (5.57) and the first bound in (5.56). The second bound there follows from

$$\|\varphi_k - 1\|_\infty = \left\|\sum_{n=1}^\infty K^n 1\right\|_\infty \le \sum_{n=1}^\infty \|K^n\| \le |k|^{-1}\|V\|_1 \exp(|k|^{-1}\|V\|_1) \,.$$

Since, for each fixed $x \in \mathbb{R}$, $(K^n 1)(x)$ depends analytically on k in $\mathbb{C}_+$ and continuously in $\overline{\mathbb{C}_+} \setminus \{0\}$, the uniform limit $\varphi_k(x)$ of their sum has the same properties.

It follows from equation (5.57) that $\varphi_k \in W_{\mathrm{loc}}^{2,1}(\mathbb{R})$ with

$$\varphi_k'' + 2ik\varphi_k' = -V\varphi_k$$

and that $\varphi_k(x) = 1 + o(1)$ and $\varphi_k'(x) = o(1)$ as $x \to \infty$. Thus, $\psi_k(x) := e^{ikx}\varphi_k(x)$ is the desired solution of (5.54) and (5.55).

This solution is unique, since for any two solutions of (5.54) their Wronskian is constant and, by (5.55), this Wronskian vanishes. Thus the two functions are linearly dependent and so, again by (5.55), they coincide. $\quad\square$

We will later need an improvement of the first bound in (5.56) for small $|k|$.

Lemma 5.44 *The inequality* $\limsup_{|k|\to 0} |k| \ln \|\varphi_k\|_\infty \le 0$ *holds.*

Proof We follow Hryniv and Mykytyuk (2021). With a parameter $M > 0$ we split $V = V_< + V_0 + V_>$ with $V_< := V\chi_{(-\infty,-M)}$ and $V_> := V\chi_{(M,\infty)}$. We denote by $K_<$, K_0 and $K_>$ the corresponding operators as in the proof of Lemma 5.38 and observe that $K_>K_0 = K_>K_< = K_0K_< = 0$. Thus, $1-K = (1-K_>)(1-K_0)(1-K_<)$ and

$$\left\|(1 - K)^{-1}\right\| \le \left\|(1 - K_<)^{-1}\right\| \left\|(1 - K_0)^{-1}\right\| \left\|(1 - K_>)^{-1}\right\| \,.$$

For the first and third terms on the right side, we use the bound (5.66) (with V replaced by $V_\gtrless$). For the second term we argue differently. Namely, we claim that

$$\left\|(1 - K_0)^{-1}\right\| \le \exp\left(\sqrt{(|k|^{-1} + 2M)2\|V_0\|_1}\right) \,. \tag{5.67}$$

Accepting this for the moment and using $\varphi_k = (1 - K)^{-1} 1$ by (5.57), we obtain

$$\ln \|\varphi_k\|_\infty \leq \ln \left\|(1 - K)^{-1}\right\| \leq |k|^{-1} (\|V_<\|_1 + \|V_>\|_1) + \sqrt{(|k|^{-1} + 2M)2\|V_0\|_1} .$$

Therefore, uniformly in the argument of $k \in \overline{\mathbb{C}_+} \setminus \{0\}$,

$$\limsup_{|k| \to 0} |k| \ln \|\varphi_k\|_\infty \leq \|V_<\|_1 + \|V_>\|_1 .$$

By dominated convergence, the right side tends to zero as $M \to \infty$, yielding the lemma.

It remains to prove (5.67). We use $|(e^{2ik(y-x)} - 1)/(2ik)| \leq y - x$ for $y \geq x$ and $k \in \overline{\mathbb{C}_+}$ and obtain, inductively for all $n \in \mathbb{N}$,

$$|(K_0^n f)(x)| \leq \frac{\|f\|_\infty}{|k|} \int \cdots \int_{\{x \leq x_1 \leq \cdots \leq x_n\}} (x_2 - x_1) \cdots (x_n - x_{n-1})$$
$$\times |V_0(x_1)| \cdots |V_0(x_n)| \, dx_1 \cdots dx_n .$$

Note that one of the iterates is bounded as in the proof of Lemma 5.38, which removes a factor $x_1 - x$ at the expense of the factor of $|k|^{-1}$. For an upper bound we can drop the constraint $x_1 \geq x$; that is, we can replace the domain of integration by

$$\Pi_n := \{(x_1, \ldots, x_n) \in \mathbb{R}^n : -M \leq x_1 \leq \cdots \leq x_n \leq M\}.$$

By the arithmetic–geometric mean inequality,

$$\sup_{(x_1,\ldots,x_n) \in \Pi_n} |k|^{-1}(x_2 - x_1) \cdots (x_n - x_{n-1}) \leq \sup_{(x_1,\ldots,x_n) \in \Pi_n} \left(\frac{|k|^{-1} + x_n - x_1}{n} \right)^n$$
$$\leq \left(\frac{|k|^{-1} + 2M}{n} \right)^n .$$

Thus

$$|(K_<^n f)(x)| \leq \left(\frac{|k|^{-1} + 2M}{n} \right)^n \frac{\|f\|_\infty \|V_0\|_1^n}{n!} .$$

This proves that

$$\left\|(1 - K_0)^{-1}\right\| \leq \sum_{n=0}^\infty \|K_0^n\| \leq \sum_{n=0}^\infty \left(\frac{|k|^{-1} + 2M}{n} \right)^n \frac{\|V_0\|_1^n}{n!} .$$

Using $(2n)^n n! \geq (2n)!$ we obtain

$$\left\|(1 - K_0)^{-1}\right\| \leq \sum_{n=0}^\infty \frac{\left(\sqrt{2\|V_0\|_1(|k|^{-1} + 2M)} \right)^{2n}}{(2n)!} \leq \exp \left(\sqrt{2\|V_0\|_1(|k|^{-1} + 2M)} \right),$$

which is the claimed bound. $\qquad\square$

The coefficients $a(k)$ and $b(k)$ were introduced in (5.59) for $k \in \mathbb{R} \setminus \{0\}$. Because of the bound on φ_k in (5.56) we see that the definition of a makes sense for $k \in \overline{\mathbb{C}_+} \setminus \{0\}$ and that the resulting function is analytic in $\mathbb{C}_+$ and continuous in $\overline{\mathbb{C}_+} \setminus \{0\}$.

The function a is our main object of interest. We record a weak a priori bound on its behavior at infinity, which will later be improved, but is sufficient initially. Indeed, it follows by (5.56) and (5.59) that, for all $k \in \overline{\mathbb{C}_+} \setminus \{0\}$,

$$|a(k) - 1| \le (2|k|)^{-1} \|V\|_1 \exp(\|V\|_1/|k|). \tag{5.68}$$

We next turn our attention to the zeros of a.

Lemma 5.45 *All zeros of the function a in $\overline{\mathbb{C}_+} \setminus \{0\}$ lie on the positive imaginary axis and are simple. Moreover, $-i\kappa$ is a zero of a if and only if $-\kappa^2$ is an eigenvalue of the operator $-\frac{d^2}{dx^2} - V$ in $L^2(\mathbb{R})$. Finally, the eigenvalues of $-\frac{d^2}{dx^2} - V$ are simple.*

Proof Step 1. We begin by noting that for any $k \in \overline{\mathbb{C}_+} \setminus \{0\}$ there is a unique solution $\tilde{\psi}_k$ of (5.54) satisfying, as $x \to -\infty$,

$$\tilde{\psi}_k(x) = e^{-ikx}(1 + o(1)) \quad \text{and} \quad \tilde{\psi}'_k(x) = -ike^{-ikx}(1 + o(1)). \tag{5.69}$$

This follows by Lemma 5.38, applied with $V(-x)$ instead of $V(x)$. We claim that

$$a(k) = (2ik)^{-1}W[\psi_k, \tilde{\psi}_k] \quad \text{for all } k \in \overline{\mathbb{C}_+} \setminus \{0\}, \tag{5.70}$$

where $W[f, g](x) := f'(x)g(x) - f(x)g'(x)$ is the Wronskian.

We begin by proving (5.70) for $k \in \mathbb{R} \setminus \{0\}$. In this case, as in the proof of Lemma 5.39, one finds

$$W[\tilde{\psi}_k, \tilde{\psi}_{-k}] = -2ik. \tag{5.71}$$

Consequently, $\tilde{\psi}_k$ and $\tilde{\psi}_{-k}$ are linearly independent and we can write the function ψ_k, which satisfies the same equation, as

$$\psi_k(x) = \tilde{a}(k)\tilde{\psi}_{-k}(x) + \tilde{b}(k)\tilde{\psi}_k(x) \tag{5.72}$$

with some coefficients $\tilde{a}(k)$ and $\tilde{b}(k)$. Recalling (5.71), we see that (5.72) implies

$$W[\tilde{\psi}_k, \psi_k] = -2ik\tilde{a}(k). \tag{5.73}$$

Meanwhile, by (5.69), (5.72) implies, as $x \to -\infty$,

$$\psi_k(x) = \tilde{a}(k)e^{ikx}(1 + o(1)) + \tilde{b}(k)e^{-ikx}(1 + o(1)).$$

Comparing this with (5.58), we conclude that $\tilde{a}(k) = a(k)$ and $\tilde{b}(k) = b(k)$. By (5.73), this completes the proof of (5.70) for $k \in \mathbb{R} \setminus \{0\}$.

As remarked before, a is analytic in $\mathbb{C}_+$ and continuous in $\overline{\mathbb{C}_+} \setminus \{0\}$. At the same time, by Lemma 5.38 and its proof, $\psi_k(x)$, $\psi_k'(x)$, $\tilde{\psi}_k(x)$, $\tilde{\psi}_k'(x)$ are also analytic in $\mathbb{C}_+$ and continuous in $\overline{\mathbb{C}_+} \setminus \{0\}$, so $W[\tilde{\psi}_k, \tilde{\psi}_{-k}]$ has the same properties. By a uniqueness result about analytic functions (see, for instance, Yafaev, 1992, Theorem 1.2.3), this implies that (5.70) holds for all $k \in \overline{\mathbb{C}_+} \setminus \{0\}$, as claimed.

Step 2. Note that, by Lemma 5.39, $|a(t)| \geq 1$ for $t \in \mathbb{R} \setminus \{0\}$ and so a has no real zeros in $\overline{\mathbb{C}_+} \setminus \{0\}$. Next, assume that $a(k) = 0$ for some $k \in \mathbb{C}_+$. Then, by (5.70), ψ_k and $\tilde{\psi}_k$ are linearly dependent and, by (5.55) and (5.69), $\psi_k \in H^1(\mathbb{R})$. Therefore, k^2 is an eigenvalue of $-\frac{d^2}{dx^2} - V(x)$ and, since the latter operator is self-adjoint in $L^2(\mathbb{R})$, we infer that necessarily $k \in i(0, \infty)$.

Step 3. In order to show that, if $-\kappa^2$ is an eigenvalue of $-\frac{d^2}{dx^2} - V$, then a has a simple zero at $i\kappa$, we recall that the resolvent of a self-adjoint operator H near an isolated eigenvalue $\lambda \in \mathbb{R}$ has the form

$$(H - z)^{-1} = (\lambda - z)^{-1} P + G(z) \tag{5.74}$$

with an orthogonal projection P and an operator-valued function $z \mapsto G(z)$ that is analytic near $z = \lambda$. This follows from the spectral theorem. Moreover, $\dim \operatorname{ran} P$ equals the multiplicity of the eigenvalue λ of H. Meanwhile, the resolvent of $H = -\frac{d^2}{dx^2} - V$ is an integral operator with integral kernel given by

$$(H - k^2)^{-1}(x, y) = \frac{1}{W[\tilde{\psi}_k, \psi_k]} \begin{cases} \psi_k(x)\tilde{\psi}_k(y) & \text{if } x \geq y, \\ \tilde{\psi}_k(x)\psi_k(y) & \text{if } x \leq y. \end{cases} \tag{5.75}$$

This follows either by Sturm–Liouville theory or, more directly, by verifying that, for each $f \in L^2(\mathbb{R})$, the function

$$u(x) := \frac{1}{W[\tilde{\psi}_k, \psi_k]} \psi_k(x) \int_{-\infty}^x \tilde{\psi}_k(y) f(y) \, dy + \frac{1}{W[\tilde{\psi}_k, \psi_k]} \tilde{\psi}_k(x) \int_x^\infty \psi_k(y) f(y) \, dy$$

belongs to $H^1(\mathbb{R})$ and satisfies

$$\int_{\mathbb{R}} (u'v' - Vuv) \, dx = k^2 \int_{\mathbb{R}} uv \, dx + \int_{\mathbb{R}} fv \, dx \qquad \text{for all } v \in H^1(\mathbb{R}).$$

Comparing (5.74) and (5.75), we see, if $-\kappa^2$ is an eigenvalue, then $k \mapsto W[\tilde{\psi}_k, \psi_k]$ has a simple zero at $i\kappa$. By (5.70), the same holds for a. Finally, simplicity of eigenvalues of H was proved, under less restrictive assumptions on V, in Proposition 5.34. $\qquad \square$

In the proof of the trace formula in Theorem 5.40 it will often be convenient to "remove the zeros from a"; that is, to divide a by the Blaschke product of its zeros. To be specific, let us denote by $\kappa_1 > \kappa_2 > \kappa_3 > \cdots$ the imaginary parts of the zeros of a in $\mathbb{C}_+$, so by Lemma 5.45, $-\kappa_1^2 < -\kappa_2^2 < -\kappa_3^2 < \cdots$ are the negative eigenvalues of the operator $-\frac{d^2}{dx^2} - V$ in $L^2(\mathbb{R})$. We now use the fact that $V \in L^1(\mathbb{R})$ implies that $\sum_n E_n^{1/2} < \infty$, which is a consequence of, for instance, the Lieb–Thirring inequality in Theorem 4.38. The summability of the sequence (κ_n), together with the inequality

$$\left| \frac{k - i\kappa_n}{k + i\kappa_n} - 1 \right| \le \frac{2\kappa_n}{|k|} \qquad \text{for all } k \in \overline{\mathbb{C}_+} \setminus \{0\}, \tag{5.76}$$

implies that the Blaschke product

$$B(k) := \prod_n \frac{k - i\kappa_n}{k + i\kappa_n}$$

converges uniformly on sets of the form $\overline{\mathbb{C}_+} \setminus B_\delta(0)$ with $\delta > 0$; see, for instance, Rudin (1987, Theorem 15.4). Therefore, B is analytic in $\mathbb{C}_+$ and continuous in $\overline{\mathbb{C}_+} \setminus \{0\}$. It vanishes precisely at the $i\kappa_n$. Moreover, the inequality $|k - i\kappa_n| \le |k + i\kappa_n|$ for $k \in \overline{\mathbb{C}_+}$, with equality for $k \in \mathbb{R}$, implies that $|B(k)| \le 1$ for $k \in \overline{\mathbb{C}_+} \setminus \{0\}$, with equality for $k \in \mathbb{R} \setminus \{0\}$. Finally, we note that from (5.76) one easily deduces that

$$B(k) = 1 + O(|k|^{-1}) \qquad \text{as } |k| \to \infty, \, k \in \mathbb{C}_+. \tag{5.77}$$

This will be made more precise later, but suffices initially.

The following inequality is fundamental in our argument.

Lemma 5.46 *For all $k \in \overline{\mathbb{C}_+} \setminus \{0\}$, $|a(k)| \ge |B(k)|$.*

Proof Step 1. Since the function B has zeros precisely at the zeros of a, the quotient B/a defines an analytic function in $\mathbb{C}_+$ that does not vanish. By (5.68) and (5.77), we have $B(k)/a(k) \to 1$ as $|k| \to \infty$, and, by Lemma 5.39 and the discussion preceding Lemma 5.46, we have $|B(k)/a(k)| \le 1$ on $\mathbb{R} \setminus \{0\}$. If we knew that B/a extended continuously to zero, then we could deduce from the above facts, by the maximum modulus principle for analytic functions, that $|a/B| \le 1$ in $\mathbb{C}_+$ as claimed.

If we assume, in addition to our standing assumption $V \in L^1(\mathbb{R})$, that the support of V is bounded, then we can complete the proof in this way. Indeed, in this case it follows by an easy extension of the proof of Lemma 5.38 that for each $x \in \mathbb{R}$, the map $k \mapsto \varphi_k(x)$ is entire. Using this fact and the corresponding bound in the definition (5.59) of a, one finds that $k \mapsto ka(k)$ is an entire function. Moreover, since the operator $-\frac{d^2}{dx^2} - V$ in $L^2(\mathbb{R})$ has only finitely many negative eigenvalues (see Proposition 4.18), Lemma 5.45 implies that B is a finite product and analytic in an open neighborhood of $\overline{\mathbb{C}_+}$. Therefore, B/a is also analytic in an open neighborhood of $\overline{\mathbb{C}_+}$ and we can apply the maximum modulus principle to obtain the assertion.

Step 2. For general $V \in L^1(\mathbb{R})$ we define, for $M > 0$, $V^{(M)} := \chi_{[-M,M]}V$ and claim that, as $M \to \infty$, the corresponding functions $a^{(M)}$ and $B^{(M)}$ converge pointwise in $\mathbb{C}_+$ to a and B. For the proof of these facts, we follow Hryniv and Mykytyuk (2021). As a result, the claimed inequality for V follows from those for the $V^{(M)}$.

Step 3. Let us show that $B^{(M)} \to B$ pointwise in $\mathbb{C}_+$ as $M \to \infty$. We proceed in two substeps. In the first we show that, if $(\kappa_n^{(M)})_n$ denotes the sequence of zeros of $a^{(M)}$, then $\kappa^{(M)} \to \kappa$ in $\ell^1(\mathbb{N})$ as $M \to \infty$. Here, we extend the finite sequence $(\kappa_n^{(M)})_n$ by zero to an infinite sequence. In the second substep we show that, if $\tilde{B}$ is the Blaschke product corresponding to a sequence $(\tilde{\kappa}_n)$ of decreasing non-negative numbers, then $|\tilde{B}(k) - B(k)| \leq 2\|\tilde{\kappa} - \kappa\|_1/|k|$ for all $k \in \overline{\mathbb{C}_+} \setminus \{0\}$. Combining these two steps proves the assertion.

For the first substep, one can argue as in §4.7; see, in particular, the proof of Theorem 4.46 based on (4.88) and also the approximation argument in Theorem 5.33. More specifically, we compare $\sum \kappa_n$ and $\sum \kappa_n^{(M)}$ with $\sum \tilde{\kappa}_n^{(M)}$, where the $\tilde{\kappa}_n^{(M)}$ correspond to $V_+ - V_-\chi_{[-M,M]}$. We omit the details. For an alternative, more elementary, argument that is not based on the Lieb–Thirring inequality, we refer to Hryniv and Mykytyuk (2021, Lemma 4.1).

Now let us carry out the second substep mentioned above. We define sequences $\kappa^{(j)}$, $j \in \mathbb{N}$, by

$$\kappa_n^{(j)} := \begin{cases} \kappa_n & \text{if } n < j, \\ \tilde{\kappa}_n & \text{if } n \geq j, \end{cases}$$

and denote by $B^{(j)}$ the corresponding Blaschke products. Then for all $k \in \overline{\mathbb{C}_+} \setminus \{0\}$,

$$\left|\tilde{B}(k) - B(k)\right| \leq \sum_{j=1}^{\infty} \left|B^{(j)}(k) - B^{(j+1)}(k)\right|.$$

For each $j \in \mathbb{N}$ and $k \in \overline{\mathbb{C}_+} \setminus \{0\}$,

$$\left|B^{(j)}(k) - B^{(j+1)}(k)\right| \leq \left|\frac{k - i\tilde{\kappa}_j}{k + i\tilde{\kappa}_j} - \frac{k - i\kappa_j}{k + i\kappa_j}\right| = \frac{2|k||\tilde{\kappa}_j - \kappa_j|}{|k + i\tilde{\kappa}||k + i\kappa_j|} \leq \frac{2|\tilde{\kappa}_j - \kappa_j|}{|k|}.$$

This proves the claimed inequality.

Step 4. Finally, let us show that $a^{(M)} \to a$ pointwise in $\mathbb{C}_+$. More generally, assume that $\tilde{\varphi}_k$ is as in Lemma 5.38 corresponding to a $\tilde{V} \in L^1(\mathbb{R})$. Then the difference $\tilde{\varphi}_k - \varphi_k$ satisfies the equation

$$(\tilde{\varphi}_k - \varphi_k)(x) = f(x) - (2ik)^{-1} \int_x^\infty \left(e^{2ik(y-x)} - 1 \right) \tilde{V}(y)(\tilde{\varphi}_k - \varphi_k)(y)\, dy$$

with, dropping the k dependence in the notation,

$$f(x) := -(2ik)^{-1} \int_x^\infty \left(e^{2ik(y-x)} - 1 \right) (\tilde{V}(y) - V(y))\varphi_k(y)\, dy .$$

Defining the operator $\tilde{K}$ as in the proof of Lemma 5.38 with V replaced by $\tilde{V}$, we obtain $\tilde{\varphi}_k - \varphi_k = (1 - \tilde{K})^{-1} f$, and therefore, by (5.66),

$$\|\tilde{\varphi}_k - \varphi_k\| \le \left\| (1 - \tilde{K})^{-1} \right\| \|f\|_\infty \le \exp(|k|^{-1}\|\tilde{V}\|_1)\|f\|_\infty .$$

Recalling (5.56), we have the bound

$$\|f\|_\infty \le |k|^{-1}\|\tilde{V} - V\|_1 \|\varphi_k\|_\infty \le |k|^{-1}\|\tilde{V} - V\|_1 \exp(\|V\|_1/|k|) .$$

Thus, by (5.59), for any $k \in \overline{\mathbb{C}_+} \setminus \{0\}$,

$$\begin{aligned}
&|\tilde{a}(k) - a(k)| \\
&= \left| (2ik)^{-1} \int_\mathbb{R} \left(\tilde{V}(y)\tilde{\varphi}_k(y) - V(y)\varphi_k(y) \right) dy \right| \\
&\le (2|k|)^{-1} \left(\|\tilde{V}\|_1 \|\tilde{\varphi}_k - \varphi_k\|_\infty + \|\tilde{V} - V\|_1 \|\varphi_k\|_\infty \right) \\
&\le (2|k|)^{-1} \|\tilde{V} - V\|_1 \left(|k|^{-1}\|\tilde{V}\|_1 \exp(\|\tilde{V}\|_1/|k|) + 1 \right) \exp(\|V\|_1/|k|) .
\end{aligned}$$

Applying this with $\tilde{V} = V^{(M)}$, we see that $a^{(M)}(k) \to a(k)$ as $M \to \infty$ for each $k \in \overline{\mathbb{C}_+} \setminus \{0\}$, as claimed. $\qquad \square$

We now prove a representation formula for $\ln|a|$ in $\mathbb{C}_+$ through its boundary values and its zeros. The trace formulas in Theorem 5.40 will be rather straightforward consequences of this formula.

Proposition 5.47 *It is the case that* $(1 + t^2)^{-1} \ln|a| \in L^1(\mathbb{R})$ *and, for all* $k \in \mathbb{C}_+$ *with* $k \ne \kappa_n$ *for all* n, *that*

$$\ln|a(k)| = \frac{\operatorname{Im} k}{\pi} \int_\mathbb{R} \frac{\ln|a(t)|}{|t - k|^2}\, dt + \sum_n \ln \left| \frac{k - i\kappa_n}{k + i\kappa_n} \right| .$$

Proof Step 1. By Lemma 5.46 and the same reasoning as at the beginning of its proof one finds that a/B is analytic and non-vanishing in $\mathbb{C}_+$ and has modulus bounded from below by 1. Consequently, $\ln|a/B|$ is a non-negative harmonic function in $\mathbb{C}_+$. By the representation theorem for such functions there is a constant $c \geq 0$ and a non-negative Borel measure μ on $\mathbb{R}$ satisfying $\int_{\mathbb{R}} (1 + t^2)^{-1} d\mu(t) < \infty$ such that

$$\ln \left| \frac{a(k)}{B(k)} \right| = c \operatorname{Im} k + \frac{\operatorname{Im} k}{\pi} \int_{\mathbb{R}} \frac{d\mu(t)}{|k - t|^2} \qquad \text{for all } k \in \mathbb{C}_+ ; \qquad (5.78)$$

see, e.g., Garnett (2007, Theorem I.3.5). This can also be deduced via a fractional linear map from the corresponding result for the unit disk in Rudin (1987, Theorem 11.30 (c)).

We take $k = i\kappa$ in (5.78), divide by κ and let $\kappa \to \infty$. As noted in the proof of Lemma 5.46 we have $a(i\kappa)/B(i\kappa) \to 1$ as $\kappa \to \infty$. Using dominated convergence for the integral in (5.78), we deduce that $c = 0$.

Step 2. We claim that there is a $\gamma \geq 0$ such that

$$d\mu(t) = \ln |a(t)| \, dt + \gamma \, d\delta_0(t) .$$

Indeed, for $g \in C_c(\mathbb{R})$ it is well known that

$$\frac{\varepsilon}{\pi} \int_{\mathbb{R}} \frac{g(s)}{(t - s)^2 + \varepsilon^2} \, ds \to g(t)$$

uniformly in $t \in \mathbb{R}$; see, e.g., Garnett (2007, Theorem I.3.1(d)). On the other hand, if $T := \sup\{|t| : t \in \operatorname{supp} g\}$, then for $|t| \geq 2T$ and $|s| \leq T$ one has $|t - s| \geq (1/2)|t|$ and, consequently,

$$\left| \frac{\varepsilon}{\pi} \int_{\mathbb{R}} \frac{g(s)}{(t - s)^2 + \varepsilon^2} \, ds \right| \leq \frac{\varepsilon}{\pi} \frac{\|g\|_1}{t^2 + \varepsilon^2} \qquad \text{for } |t| \geq 2T .$$

Thus, using Fubini's theorem and dominated convergence,

$$\int_{\mathbb{R}} g(s) \left(\frac{\varepsilon}{\pi} \int_{\mathbb{R}} \frac{d\mu(t)}{(t - s)^2 + \varepsilon^2} \right) ds = \int_{\mathbb{R}} \left(\frac{\varepsilon}{\pi} \int_{\mathbb{R}} \frac{g(s) \, ds}{(t - s)^2 + \varepsilon^2} \right) d\mu(t)$$
$$\to \int_{\mathbb{R}} g(t) \, d\mu(t) .$$

Meanwhile, if $0 \notin \operatorname{supp} g$, then the continuity of $a(k)/B(k)$ in $\overline{\mathbb{C}_+} \setminus \{0\}$ and the fact that $|B| = 1$ on $\mathbb{R} \setminus \{0\}$ implies that

$$\int_{\mathbb{R}} g(s) \ln \left| \frac{a(s + i\varepsilon)}{B(s + i\varepsilon)} \right| ds \to \int_{\mathbb{R}} g(s) \ln |a(s)| \, ds .$$

Thus we deduce from (5.78) that, for any $g \in C_c(\mathbb{R} \setminus \{0\})$,

$$\int_{\mathbb{R}} g(s) \, d\mu(s) = \int_{\mathbb{R}} g(s) \ln |a(s)| \, ds .$$

This implies that the measure μ is absolutely continuous on $\mathbb{R} \setminus \{0\}$ with density $d\mu = \ln|a|\, dt$ and, in particular,

$$\int_{\mathbb{R}} \frac{\ln|a(t)|}{1+t^2}\, dt \leq \int_{\mathbb{R}} \frac{d\mu(t)}{1+t^2} < \infty. \tag{5.79}$$

Moreover, $d\mu - \ln|a|\, dt$ defines a non-negative measure that is supported in $\{0\}$ and, therefore, a non-negative multiple of the Dirac measure at the origin. This proves the claim made at the beginning of this step.

Step 3. Before proceeding with the main argument, let us show, following Hryniv and Mykytyuk (2021), that

$$\lim_{\varepsilon \to 0_+} \varepsilon \int_0^\pi \ln|B(\varepsilon e^{i\theta})|\, d\theta = 0. \tag{5.80}$$

Indeed, it is easy to see that

$$\beta := \sup_{t>0} \int_0^\pi \ln\left|\frac{e^{i\theta}+it}{e^{i\theta}-it}\right|\, d\theta < \infty.$$

Consequently, for all $\kappa > 0$,

$$\int_0^\pi \ln\left|\frac{\varepsilon e^{i\theta}+i\kappa}{\varepsilon e^{i\theta}-i\kappa}\right|\, d\theta \leq \beta.$$

On the other hand, if $\kappa \leq \varepsilon/2$, then

$$\left|\frac{\varepsilon e^{i\theta}+i\kappa}{\varepsilon e^{i\theta}-i\kappa}\right| \leq 1 + \left|\frac{2i\kappa}{\varepsilon e^{i\theta}-i\kappa}\right| \leq 1 + \frac{2\kappa}{|\varepsilon|-\kappa} \leq 1 + \frac{4\kappa}{\varepsilon},$$

so

$$\int_0^\pi \ln\left|\frac{\varepsilon e^{i\theta}+i\kappa}{\varepsilon e^{i\theta}-i\kappa}\right|\, d\theta \leq \frac{4\kappa}{\varepsilon} \int_0^\pi d\theta = \frac{4\pi\kappa}{\varepsilon}.$$

Thus,

$$\varepsilon \int_0^\pi \ln\frac{1}{|B(\varepsilon e^{i\theta})|}\, d\theta = \sum_n \varepsilon \int_0^\pi \ln\left|\frac{\varepsilon e^{i\theta}+i\kappa_n}{\varepsilon e^{i\theta}-i\kappa_n}\right| \leq \sum_{\kappa_n>\varepsilon/2} \beta\varepsilon + \sum_{\kappa_n\leq\varepsilon/2} 4\pi\kappa_n$$

$$\leq \max\{2\beta, 4\pi\} \sum_n \min\{\varepsilon/2, \kappa_n\}.$$

By dominated convergence this tends to zero as $\varepsilon \to 0_+$. Since $|B| \leq 1$ as discussed before Lemma 5.46, this implies the claim.

Step 4. We claim that $\gamma = 0$. To prove this, we insert the information from Step 2 into (5.78) to obtain

$$\ln\left|\frac{a(k)}{B(k)}\right| = \frac{\mathrm{Im}\, k}{\pi} \int_{\mathbb{R}} \frac{\ln|a(t)|}{|k-t|^2}\, dt + \frac{\gamma\, \mathrm{Im}\, k}{\pi\, |k|^2} \qquad \text{for all } k \in \mathbb{C}_+,$$

and therefore, using $|a| \geq 1$ on $\mathbb{R}$ by Lemma 5.39,

$$\ln|a(k)| \geq \ln|B(k)| + \frac{\gamma \operatorname{Im} k}{\pi|k|^2} \qquad \text{for all } k \in \mathbb{C}_+ .$$

Integrating this inequality over a semicircle of radius $\varepsilon > 0$ around the origin gives

$$\int_0^\pi \ln|a(\varepsilon e^{i\theta})| \, d\theta \geq \int_0^\pi \ln|B(\varepsilon e^{i\theta})| \, d\theta + \frac{\gamma}{\pi\varepsilon} \int_0^\pi \sin\theta \, d\theta$$
$$= \int_0^\pi \ln|B(\varepsilon e^{i\theta})| \, d\theta + \frac{2\gamma}{\pi\varepsilon} .$$

In view of (5.80) we infer that

$$\liminf_{\varepsilon \to 0_+} \varepsilon \int_0^\pi \ln|a(\varepsilon e^{i\theta})| \, d\theta \geq \frac{2\gamma}{\pi} .$$

On the other hand, inserting the bound from Lemma 5.44 in the definition (5.59) of a, we obtain $\ln|a(\varepsilon e^{i\theta})| \leq o(\varepsilon^{-1})$ uniformly in $\theta \in [0, \pi]$ as $\varepsilon \to 0_+$. Thus, the lim inf in the previous formula is non-positive and consequently $0 \geq 2\gamma/\pi$. Since $\gamma \geq 0$, this shows that $\gamma = 0$, as claimed. □

Remark 5.48   Proposition 5.47 implies the following more general formula. Namely, for all $k \in \mathbb{C}_+ \setminus (i(0, \kappa_1])$,

$$\log a(k) = \frac{1}{\pi i} \int_{\mathbb{R}} \frac{\ln|a(t)|}{t - k} \, dt + \sum_n \log \frac{k - i\kappa_n}{k + i\kappa_n}, \qquad (5.81)$$

where we use the logarithm of a which is defined in the simply connected domain $\mathbb{C}_+ \setminus (i(0, \kappa_1])$ with the branch fixed by $\log a(k) \to 1$ as $|k| \to \infty$; see (5.68). The logarithm of $(k - i\kappa_n)/(k + i\kappa_n)$ is defined similarly.

To deduce (5.81) from Proposition 5.47 we define

$$f(k) := \exp\left(\frac{1}{\pi i} \int_{\mathbb{R}} \frac{\ln|a(t)|}{t - k} \, dt \right) \qquad \text{for all } k \in \mathbb{C}_+ .$$

It follows from the local integrability in (5.79) and the asymptotic $O(|t|^{-1})$-bound in (5.68) that the integral converges absolutely and defines an analytic function in $\mathbb{C}_+$. By Proposition 5.47, $f(k)B(k)/a(k)$ is an analytic function in $\mathbb{C}_+$ of constant absolute value and, by the maximum modulus principle, a constant. One easily finds that $f(is) \to 1$ as $s \to \infty$. Since also $B(is)/a(is) \to 1$ as $s \to \infty$, the constant in question is 1; that is, $fB/a \equiv 1$, as claimed.

The remaining ingredient in the proof of the trace formula in Theorem 5.40 is the asymptotic behavior of $a(k)$ as $|k| \to \infty$.

Lemma 5.49 *As $|k| \to \infty$, $k \in \overline{\mathbb{C}_+} \setminus \{0\}$,*

$$\log a(k) = (2ik)^{-1} \int_{-\infty}^{\infty} V(x)\, dx + O(|k|^{-2}).$$

If, in addition, $V \in L^2(\mathbb{R})$ and $k = is \in i\mathbb{R}_+$ then, as $s \to \infty$,

$$\log a(is) = -(2s)^{-1} \int_{-\infty}^{\infty} V(x)\, dx - (2s)^{-3} \int_{-\infty}^{\infty} V(x)^2\, dx + o(s^{-3}). \quad (5.82)$$

Proof We write the definition (5.59) of a in the form

$$a(k) = 1 + (2ik)^{-1} \int_{-\infty}^{\infty} V(x)\, dx + J_2$$

with

$$J_2 := (2ik)^{-1} \int_{-\infty}^{\infty} V(x)(\varphi_k(x) - 1)\, dx.$$

It follows from (5.56) that $J_2 = O(|k|^{-2})$ as $|k| \to \infty$. This proves the first bound in the lemma. To prove the refined asymptotics we insert (5.57) into J_2. Tis gives

$$J_2 = I_2 + J_3$$

with

$$I_2 := -(2ik)^{-2} \int_{-\infty}^{\infty} V(x) \int_{x}^{\infty} \left(e^{2ik(y-x)} - 1 \right) V(y)\, dy\, dx$$

and

$$J_3 := -(2ik)^{-2} \int_{-\infty}^{\infty} V(x) \int_{x}^{\infty} \left(e^{2ik(y-x)} - 1 \right) V(y)(\varphi_k(y) - 1)\, dy\, dx,$$

and, repeating the argument once again,

$$J_3 = I_3 + J_4$$

with

$$I_3 :=$$
$$\frac{1}{(2ik)^3} \int_{-\infty}^{\infty} V(x) \int_{x}^{\infty} \left(e^{2ik(y-x)} - 1 \right) V(y) \int_{y}^{\infty} \left(e^{2ik(z-y)} - 1 \right) V(z)\, dz\, dy\, dx$$

and, by (5.56), $J_4 = O(|k|^{-4})$. Thus, writing $v := \int_{-\infty}^{\infty} V(x)\, dx$, we have

$$a(k) = 1 + (2ik)^{-1}v + I_2 + I_3 + O(|k|^{-4}).$$

Let us study the terms I_2 and I_3 in more detail. Symmetrizing the contributions without exponentials in the above expressions, we can write

$$I_2 = \frac{1}{2}(2ik)^{-2}v^2 + \tilde{I}_2, \qquad \tilde{I}_2 := -(2ik)^{-2}\int_{-\infty}^{\infty} V(x) \int_{x}^{\infty} e^{2ik(y-x)}V(y)\,dy\,dx$$

and

$$I_3 = \frac{1}{6}(2ik)^{-3}v^3 + \tilde{I}_3$$

with

$$\tilde{I}_3 :=$$

$$\frac{1}{(2ik)^3}\int_{-\infty}^{\infty} V(x) \int_{x}^{\infty} V(y) \int_{y}^{\infty} V(z) \left(e^{2ik(z-x)} - e^{2ik(y-x)} - e^{2ik(z-y)} \right) dz\,dy\,dx .$$

We deduce, using simple a priori bounds based on $V \in L^1(\mathbb{R})$,

$\log a(k)$

$$= (2ik)^{-1}v + I_2 + I_3 - \frac{1}{2}(2ik)^{-2}v^2 - (2ik)^{-1}v\,I_2 + \frac{1}{3}(2ik)^{-3}v^3 + O(|k|^{-4})$$

$$= (2ik)^{-1}v + \tilde{I}_2 + \tilde{I}_3 - (2ik)^{-1}v\,\tilde{I}_2 + O(|k|^{-4}) . \tag{5.83}$$

We will now show that

$$\tilde{I}_3 = o(|k|^{-3}) \quad \text{and} \quad (2ik)^{-1}v\,\tilde{I}_2 = o(|k|^{-3}) . \tag{5.84}$$

This argument is in the spirit of the Riemann–Lebesgue lemma. If $W \in C_c^1(\mathbb{R})$, then

$$\int_{x}^{\infty} e^{2ik(y-x)}W(y)\,dy = -(2ik)^{-1}W(x) - (2ik)^{-1}\int_{x}^{\infty} e^{2ik(y-x)}W'(y)\,dy$$

and, consequently,

$$\tilde{I}_2 = -(2ik)^{-2}\int_{-\infty}^{\infty} V(x) \int_{x}^{\infty} e^{2ik(y-x)}(V(y) - W(y))\,dy\,dx$$

$$+ (2ik)^{-3}\int_{-\infty}^{\infty} V(x)W(x)\,dx$$

$$+ (2ik)^{-3}\int_{-\infty}^{\infty} V(x) \int_{x}^{\infty} e^{2ik(y-x)}W'(y)\,dy\,dx .$$

Thus,

$$|\tilde{I}_2| \le (2|k|)^{-2}\|V\|_1\|V - W\|_1 + (2|k|)^{-3}\|V\|_1\|W\|_\infty + (2|k|)^{-3}\|V\|_1\|W'\|_1$$

and, since $C_c^1(\mathbb{R})$ is dense in $L^1(\mathbb{R})$, we obtain $\tilde{I}_2 = o(|k|^{-2})$.

To prove the corresponding bound for $\tilde{I}_3$, we can argue in the exact same way for the exponentials $e^{2ik(z-x)}$ and $e^{2ik(x-y)}$. For the remaining exponential, we use

$$\int_x^\infty W(y)e^{2ik(y-x)}\int_y^\infty V(z)\,dz\,dy$$

$$= -(2ik)^{-1}W(x)\int_x^\infty V(z)\,dz$$

$$- (2ik)^{-1}\int_x^\infty e^{2ik(y-x)}\left(W'(y)\int_y^\infty V(z)\,dz - W(y)V(y)\right)dy.$$

Then one can argue similarly using the density. This proves (5.84).

Inserting (5.84) into (5.83), we obtain

$$\log a(k) = (2ik)^{-1}v + \tilde{I}_2 + o(|k|^{-3}).$$

We symmetrize $\tilde{I}_2$ and obtain

$$\tilde{I}_2 = -(2ik)^{-2}\frac{1}{2}\iint_{\mathbb{R}\times\mathbb{R}} V(x)e^{2ik|y-x|}V(y)\,dx\,dy.$$

In particular, for $k = is \in i\mathbb{R}_+$, using

$$e^{-2s|y-x|} = \frac{2s}{\pi}\int_{\mathbb{R}} \frac{e^{-i\xi(y-x)}}{\xi^2 + (2s)^2}\,d\xi,$$

we find

$$\tilde{I}_2 = -(2s)^{-1}\int_{\mathbb{R}} \frac{|\widehat{V}(\xi)|^2}{\xi^2 + (2s)^2}\,d\xi.$$

If we now assume, in addition, that $V \in L^2(\mathbb{R})$, we obtain, by dominated convergence and Plancherel's theorem,

$$\tilde{I}_2 = -(2s)^{-3}\int_{\mathbb{R}} V(x)^2\,dx + o(s^{-3}).$$

This proves (5.82) and concludes the proof of the lemma.   □

We are finally in position to prove the main result of this section.

Proof of Theorem 5.40   We begin with the proof of the trace formula (5.60), where we assume, as throughout this section, that $V \in L^1(\mathbb{R})$. We set $k = is \in i\mathbb{R}_+$ in the formula in Proposition 5.47 and let $s \to \infty$. By Lemma 5.49,

$$\ln|a(is)| = \text{Re}\log a(is) = -(2s)^{-1}\int_{-\infty}^\infty V(x)\,dx + O(s^{-2}).$$

Moreover, using the fact that for $|z| > 1$ we have

$$\log\frac{1-z}{1+z} = \log(1-z) - \log(1+z) = -\sum_{m=0}^\infty \frac{z^{2m+1}}{m+1/2}, \tag{5.85}$$

we find for $|k| > \kappa_1$,

$$\sum_n \log \frac{k - i\kappa_n}{k + i\kappa_n} = -i \sum_n \sum_{m=0}^{\infty} \frac{(-1)^m}{m + 1/2} \frac{\kappa_n^{2m+1}}{k^{2m+1}}.$$

In particular, by dominated convergence, since $\kappa \in \ell^1(\mathbb{N})$,

$$\sum_n \ln \left| \frac{is - i\kappa_n}{is + i\kappa_n} \right| = \sum_n \mathrm{Re} \log \frac{is - i\kappa_n}{is + i\kappa_n} = -2s^{-1} \sum_n \kappa_n + o(s^{-1}).$$

Finally, we note that by Lemma 5.49, $\ln |a(t)| = \mathrm{Re} \log a(t) = O(t^{-2})$ as $|t| \to \infty$. Combining this with (5.79) we find $\ln |a| \in L^1(\mathbb{R})$, and therefore, by dominated convergence,

$$\frac{s}{\pi} \int_{\mathbb{R}} \frac{\ln |a(t)|}{t^2 + s^2} \, dt = (\pi s)^{-1} \int_{\mathbb{R}} \ln |a(t)| \, dt + o(s^{-1}).$$

For $k \in \mathbb{R} \setminus \{0\}$, $\overline{\psi_{-k}}$ satisfies the same equation and the same asymptotics as ψ_k, and therefore, by the uniqueness assertion in Lemma 5.38, $\overline{\psi_{-k}} = \psi_k$. By (5.59), this implies $\overline{a(-k)} = a(k)$, and therefore, $\int_{\mathbb{R}} \ln |a(t)| \, dt = 2 \int_0^{\infty} \ln |a(t)| \, dt$. Combining the asymptotics of the three terms in the formula in Proposition 5.47, we obtain (5.60).

For the proof of the trace formula (5.61) under the assumption $V \in L^1 \cap L^2(\mathbb{R})$ we argue similarly, except that now we expand the formula in Proposition 5.47 to higher order. According to Lemma 5.49,

$$\ln |a(is)| = \mathrm{Re} \log a(is) = -(2s)^{-1} \int_{-\infty}^{\infty} V(x) \, dx - (2s)^{-3} \int_{-\infty}^{\infty} V(x)^2 \, dx + o(s^{-3}).$$

Moreover, using (5.85), we easily find that, as $s \to \infty$,

$$\sum_n \ln \left| \frac{is - i\kappa_n}{is + i\kappa_n} \right| = -2s^{-1} \sum_n \kappa_n - \frac{2}{3} s^{-3} \sum_n \kappa_n^3 + o(s^{-3}).$$

Inserting this into the formula in Proposition 5.47, we infer that

$$\frac{s}{\pi} \int_{\mathbb{R}} \frac{\ln |a(t)|}{t^2 + s^2} \, dt = -(2s)^{-1} \int_{-\infty}^{\infty} V(x) \, dx + 2s^{-1} \sum_n \kappa_n$$
$$- (2s)^{-3} \int_{-\infty}^{\infty} V(x)^2 \, dx + \frac{2}{3} s^{-3} \sum_n \kappa_n^3 + o(s^{-3}).$$

Using (5.60), we can rewrite this as

$$\frac{s^2}{\pi} \int_{\mathbb{R}} \frac{t^2 \ln |a(t)|}{t^2 + s^2} \, dt = -\frac{s^4}{\pi} \int_{\mathbb{R}} \frac{\ln |a(t)|}{t^2 + s^2} \, dt + \frac{s^2}{\pi} \int_{\mathbb{R}} \ln |a(t)| \, dt$$
$$= 2^{-3} \int_{-\infty}^{\infty} V(x)^2 \, dx - \frac{2}{3} \sum_n \kappa_n^3 + o(1).$$

By monotone convergence (recall $|a| \geq 1$ by Lemma 5.39) the left side converges to $\pi^{-1} \int_{\mathbb{R}} t^2 \ln |a(t)| \, dt$, yielding the claimed trace formula (5.61). $\square$

5.6 Comments

Section 5.1: Basic facts about Lieb–Thirring constants

The material presented in this section is mainly from the original Lieb and Thirring (1976) paper. The argument in Lemma 5.2 is from Aizenman and Lieb (1978).

Keller (1961) raised the question of minimizing the lowest eigenvalue of a Schrödinger operator under a constraint on the L^p norm of the potential. Assuming the existence of an optimal potential, he determined its form through a solution of the Euler–Lagrange equation. Independently of Keller, Lieb and Thirring (1976) posed the same question, argued how one can prove existence of an optimal potential and related this problem to that of finding the optimal constant in a Gagliardo–Nirenberg inequality. Uniqueness (up to symmetries) of the optimal potential, which in dimension $d \geq 2$ is not known in closed form, follows via Proposition 5.5 from the uniqueness (up to symmetries) of the optimizer in the Gagliardo–Nirenberg inequality; see the references in §2.9. For a stability result for the one-particle Lieb–Thirring inequality (5.14), see Carlen et al. (2014).

An interesting variation on the theme of Keller's problem concerns bounds on the location of eigenvalues of Schrödinger operators with *complex* potentials belonging to some L^p space with $p < \infty$. After an initial result by Abramov, Aslanyan and Davies (2001), further results were obtained in, for instance, Davies and Nath (2002), Laptev and Safronov (2009), Frank (2011, 2018a), Cuenin et al. (2014), Frank and Simon (2017), Cuenin (2020), and Bögli and Cuenin (2021). A new difficulty that appears in the non-self-adjoint context is to control the behavior of eigenvalues close to the positive real axis. Tools from harmonic analysis, like Fourier restriction estimates and uniform Sobolev inequalities (Kenig et al., 1987), are sometimes helpful for this purpose.

The smallness condition in Proposition 5.7 can be replaced by other conditions which, except for the value of the constants involved, allow for a larger class of potentials. For instance, using the optimal form of the Lorentz space embedding of $\dot{H}^1(\mathbb{R}^d)$ (Alvino, 1977), one can show that, if a function $V \geq 0$ on $\mathbb{R}^d$, $d \geq 3$, satisfies

$$\left| \{x \in \mathbb{R}^d : V(x) > \tau\} \right| \leq d^{-1} |\mathbb{S}^{d-1}| \left(\frac{d-2}{2} \right)^d \tau^{-d/2} \qquad \text{for all } \tau > 0,$$

then $-\Delta - V$ in $L^2(\mathbb{R}^d)$ has no negative eigenvalue. A simple proof of this fact can be based on Hardy's inequality (Theorem 2.74) and rearrangement inequalities, as in the proof of Theorem 2.49. Other conditions for the absence of eigenvalues can be found, for instance, in Glaser et al. (1976).

The conjecture that the optimal constant in the Lieb–Thirring inequality is the maximum between the semiclassical and the one-particle constants appears in Lieb and Thirring (1976). This conjecture is known to hold in some cases and to fail in others. This will be discussed in the remaining chapters and for a summary we refer to §8.4.

The fact that the constants $L_{\gamma,d}^{cl}$ and $L_{\gamma,d}^{(1)}$ intersect at a unique point in $d \leq 7$ and do not intersect in $d \geq 8$ (Proposition 5.12) is conjectured in Lieb and Thirring (1976) and proved in Frank et al. (2021b). An analogue of Lemma 5.14 holds when more than one dimension is split off (Frank et al., 2021b); namely, if $1 \leq k < d$ and

$$
\begin{aligned}
\gamma \geq 1/2 \quad &\text{if } k = 1, \\
\gamma > 0 \quad &\text{if } k = 2, \\
\gamma \geq 0 \quad &\text{if } k \geq 3,
\end{aligned}
$$

then

$$
L_{\gamma,d}^{(1)} \leq L_{\gamma,k}^{(1)} \, L_{\gamma+k/2,d-k}^{(1)} \, .
$$

Moreover, dividing by $L_{\gamma,d}^{cl} = L_{\gamma,k}^{cl} \, L_{\gamma+k/2,d-k}^{cl}$ we obtain

$$
\frac{L_{\gamma,d}^{(1)}}{L_{\gamma,d}^{cl}} \leq \frac{L_{\gamma,k}^{(1)}}{L_{\gamma,k}^{cl}} \, \frac{L_{\gamma+k/2,d-k}^{(1)}}{L_{\gamma+k/2,d-k}^{cl}} \, . \tag{5.86}
$$

This general result covers, in particular, (5.33). Setting $k = d - 1$ and using the fact that $L_{\gamma+(d-1)/2,1}^{(1)} < L_{\gamma+(d-1)/2,1}^{cl}$ if $\gamma + (d - 1)/2 > 3/2$, we obtain that, for all $d \geq 2$,

$$
\frac{L_{\gamma,d}^{(1)}}{L_{\gamma,d}^{cl}} < \frac{L_{\gamma,d-1}^{(1)}}{L_{\gamma,d-1}^{cl}} \qquad \text{if } \gamma + d/2 > 2, \, \gamma \geq 0 \, .
$$

As a consequence of this (plus the facts that $\gamma_{c,1} = 3/2$ and $\gamma_{c,2} > 1 > 1/2$), the intersection point $\gamma_{c,d}$ is decreasing with respect to d; see Frank et al. (2021b).

In §5.1.3 we proved the inequality $L_{\gamma,1} > L_{\gamma,1}^{cl}$ for $\gamma < 3/2$, following Lieb and Thirring (1976), by computing $L_{\gamma,1}^{(1)}$ and showing that a potential with a single negative eigenvalue beats the semiclassical constant. In addition, for the same range of γ, Helffer and Robert (1990a) showed that the semiclassical constant can also be beaten for a class of potentials in the large coupling regime, one which involves many eigenvalues.

Just like the one-particle constant, $L_{\gamma,d}^{(1)}$, one can define a two-particle constant $L_{\gamma,d}^{(2)}$; see, for instance, §7.3 – especially (7.13). In particular, for $\gamma = 1$ in $d = 1$, Benguria and Loss (2004) have an interesting geometric reformulation of the problem of determining $L_{1,1}^{(2)}$. For partial results, see Burchard and Thomas (2005), Linde (2006), Denzler (2015) and references therein.

The analogue of the Lieb–Thirring conjecture for the one-dimensional fourth-order operator fails in the critical case $\gamma = \frac{3}{4}$, as shown by Förster and Östensson (2008); there seems to be no conjecture for the optimal constant in that case.

Section 5.2: Lieb–Thirring inequalities for special classes of potentials
References for the various results in this section are mentioned in the main text. Proposition 5.18 seems to appear here for the first time in print. The precise range of γ and d for which this bound holds seems to be unknown.

Section 5.3: The sharp bound for $\gamma = \frac{1}{2}$ in one dimension
The first proof of the Lieb–Thirring inequality for $\gamma = \frac{1}{2}$ in $d = 1$ is due to Weidl (1996). The result with optimal constant in Theorem 5.27 is due to Hundertmark et al. (1998). The proof we presented is from Hundertmark et al. (2000). For an extension to the case of Jacobi matrices, see Hundertmark and Simon (2002) and also Laptev et al. (2021).

We note that, via quadratic forms, in one dimension we can define Schrödinger operators with V replaced by a finite measure. If this measure is a positive multiple of a Dirac measure, then the resulting Schrödinger operator has a unique negative eigenvalue and the bound in Theorem 5.27 is saturated.

Section 5.4: The sharp bound for $\gamma = \frac{3}{2}$ in one dimension
The sharp Lieb–Thirring inequality in $d = 1$ for $\gamma = 3/2$ appears in Gardner et al. (1974, Equation 3.27), where it is deduced from the trace formula of Zaharov and Faddeev. This trace formula and the deduction of the sharp Lieb–Thirring inequality is the topic of §5.5. Lieb and Thirring (1976) cite Gardner et al. (1974) and thank Lax for private communications in this regard. They also prove Corollary 5.32 for $\gamma = \frac{3}{2} + k$, $k \in \mathbb{N}$. This corollary for the full range of values of γ is due to Aizenman and Lieb (1978).

The proof of Theorem 5.31 that we presented in §5.4 is due to Benguria and Loss (2000). Just like the proof of Theorem 5.33 in Schmincke (1978), it is based on a commutation method, which goes back at least to Jacobi (1837) and Darboux (1882). For more recent references concerning this method and its extensions, see, for instance, Crum (1955), Deift (1978), Matveev and Salle (1991), Gesztesy (1993), Gesztesy and Teschl (1996), Gesztesy et al. (1996), and Schmincke (2003). We also mention that the representation of a Schrödinger operator $H = Q^*Q - E_1$ with $Q = \frac{d}{dx} - \varphi_1'/\varphi_1$, which forms the basis of the commutation method, is the same as the ground state substitution method, where one typically writes $Q = \varphi_1 \frac{d}{dx}(1/\varphi_1)$; see, for instance, Birman (1961, §4.5).

Similar computations leading to estimates for the eigenvalue sums $\sum_{k=1}^{n} E_k^{5/2}$ and $\sum_{k=1}^{n} E_k^{7/2}$ can be found in Pavlović (2003). For an extension of the commutation method to the case of fourth-order operators, see Hoppe et al. (2006); to the case of the semi-axis with a Robin boundary condition, see Exner et al. (2014); and to the case of Jacobi matrices, see Schimmer (2015).

Section 5.5: Trace formulas for one-dimensional Schrödinger operators
The trace formulas in §5.5 are due to Zaharov and Faddeev (1971). These trace formulas have their roots in earlier related results for the case of half-line Schrödinger operators (Levinson, 1949; Agranovič and Marčenko, 1960; Agranovich and Marchenko, 1963; Faddeev, 1957; Buslaev and Faddeev, 1960) and Schrödinger operators on finite intervals (Gel'fand and Levitan, 1953; Gel'fand, 1956). The trace formula (5.60) under the minimal assumption $V \in L^1(\mathbb{R})$ appears in Hryniv and Mykytyuk (2021). Earlier proofs required V to be sufficiently rapidly decaying, for instance, $(1 + |x|)V \in L^1(\mathbb{R})$. The main difference between the case of such potentials and that of L^1 potentials is the more erratic behavior of $a(k)$ near $k = 0$ in the latter case; see Yafaev (1982) for a detailed study of this behavior in the half-line case. Our proof uses several ideas from Hryniv and Mykytyuk (2021), but, in contrast to them, we do not use the trace formula for sufficiently rapdily decaying potentials as an a priori ingredient. A novel input here is the use of the Poisson kernel representation of $\ln |a/B|$ in Proposition 5.47.

References for trace formulas in higher dimensions are, among others, Buslaev (1962, 1966), and Colin de Verdière (1981). For trace formulas in the discrete case, we refer to Case (1974, 1975), Killip and Simon (2003), Gamboa et al. (2016), and Breuer et al. (2018).

Properties of Jost functions on the real line and, in particular, the important identity $|a(k)|^2 - |b(k)|^2 = 1$ were established in Faddeev (1958, 1964). For detailed presentations of this material we refer to Marchenko (2011), Deift and Trubowitz (1979), and Yafaev (2010). The coefficients $a(k)$ and $b(k)$ for

$k \in \mathbb{R} \setminus \{0\}$ appear in the scattering matrix $S(k)$; specifically,

$$S(k) = \begin{pmatrix} \frac{1}{a(k)} & -\frac{\overline{b(k)}}{a(k)} \\ \frac{b(k)}{a(k)} & \frac{1}{a(k)} \end{pmatrix};$$

see, e.g., Yafaev (2010, §5.1.2). The relation $|a(k)|^2 - |b(k)|^2 = 1$ is equivalent to the unitarity of the scattering matrix. It seems unclear how to use trace formulas in the half-line case or in the higher-dimensional case to prove sharp spectral inequalities. One reason is that there does not seem to be an analogue of the inequality $|a(k)| \geq 1$.

Reflectionless potentials (that is, potentials with $|a(k)| = 1$ for all $k \in \mathbb{R} \setminus \{0\}$) were characterized in Kay and Moses (1956) and Hryniv et al. (2021).

Jost and Pais (1951) showed that the scattering coefficient coincides with the perturbation determinant. That is,

$$a(k) = \det \left(1 + (\operatorname{sgn} V)\sqrt{|V|}(-\tfrac{d^2}{dx^2} - k^2)^{-1}\sqrt{|V|} \right);$$

see also Simon (2005, Proposition 5.7), and Yafaev (2010, Proposition 5.3.3). Using a modified determinant instead of the regular one provides a "renormalized" definition of the scattering coefficient even in situations where $V \notin L^1(\mathbb{R})$; see, for instance, Killip and Simon (2003); Killip et al. (2018) and Yafaev (2010, §9.1).

The trace formulas discussed here are closely related to the Korteweg–de Vries (KdV) equation and to its integrability; see, for instance, Gardner et al. (1967, 1974), Miura et al. (1968), Lax (1968), and Zaharov and Faddeev (1971). In Lieb and Thirring (1976), the authors thank Lax for pointing out the connection to the KdV equation and mention a heuristic proof of the sharp inequality for $\gamma = \frac{3}{2}$ based on the long-time behavior of solutions of this equation.

Trace formulas have also found important applications in the study of the continuous spectrum of Schrödinger operators; see Deift and Killip (1999), Molchanov et al. (2001), Killip and Simon (2003, 2009), Laptev et al. (2003), Laptev et al. (2005), Safronov (2005), Killip (2007), and references therein.

6

Sharp Lieb–Thirring Inequalities in Higher Dimensions

Our main goal in this chapter is to prove the sharp Lieb–Thirring inequality

$$\mathrm{Tr}(-\Delta - V)_-^\gamma \leq L_{\gamma,d}^{\mathrm{cl}} \int_{\mathbb{R}^d} V(x)_+^{\gamma+\frac{d}{2}}\, dx \qquad \text{for all } \gamma \geq 3/2,\, d \geq 2, \quad (6.1)$$

which involves the semiclassical constant $L_{\gamma,d}^{\mathrm{cl}}$. This is the extension to higher dimensions of the one-dimensional case of this inequality, which was proved in the previous chapter in Theorem 5.31.

The technique that we use in its proof is called 'lifting in dimensions' and originates in Laptev and Weidl (2000b). As motivation, it is useful to write the right side of (6.1) as a phase space integral:

$$L_{\gamma,d}^{\mathrm{cl}} \int_{\mathbb{R}^d} V(x)_+^{\gamma+\frac{d}{2}}\, dx = \iint_{\mathbb{R}^d \times \mathbb{R}^d} \left(|\xi|^2 - V(x)\right)_-^\gamma \frac{dx\, d\xi}{(2\pi)^d}.$$

Thus the Lieb–Thirring inequality is, in some sense, a uniform version of the semiclassical approximation

$$\mathrm{Tr}\left(-\Delta - V\right)_-^\gamma \approx \iint_{\mathbb{R}^d \times \mathbb{R}^d} \left(|\xi|^2 - V(x)\right)_-^\gamma \frac{dx\, d\xi}{(2\pi)^d}.$$

Recall that we have justified this approximation in the limit of a large coupling constant in Theorem 4.29. Let us write the variables $x = (x_1, x') \in \mathbb{R} \times \mathbb{R}^{d-1}$.

Then the phase space integral becomes

$$\iint_{\mathbb{R}^d \times \mathbb{R}^d} (|\xi|^2 - V(x))_-^\gamma \, \frac{dx \, d\xi}{(2\pi)^d}$$

$$= \iint_{\mathbb{R} \times \mathbb{R}} \iint_{\mathbb{R}^{d-1} \times \mathbb{R}^{d-1}} (\xi_1^2 + |\xi'|^2 - V(x_1, x'))_-^\gamma \, \frac{dx' \, d\xi'}{(2\pi)^{d-1}} \frac{dx_1 \, d\xi_1}{2\pi}$$

$$- L_{\gamma,1}^{\mathrm{cl}} \int_{\mathbb{R}} \iint_{\mathbb{R}^{d-1} \times \mathbb{R}^{d-1}} (|\xi'|^2 - V(x_1, x'))_-^{\gamma+\frac{1}{2}} \, \frac{dx' \, d\xi'}{(2\pi)^{d-1}} \, dx_1$$

$$\approx L_{\gamma,1}^{\mathrm{cl}} \int_{\mathbb{R}} \mathrm{Tr}(-\Delta' - V(x_1, \cdot))_-^{\gamma+\frac{1}{2}} \, dx_1 \,,$$

where we used again the semiclassical approximation, this time in dimension $d - 1$. Thus, in the spirit of the Lieb–Thirring inequality it is natural to ask whether there is an inequality of the form

$$\mathrm{Tr}(-\Delta - V)_-^\gamma \le \mathrm{const} \times \int_{\mathbb{R}} \mathrm{Tr}(-\Delta' - V(x_1, \cdot))_-^{\gamma+\frac{1}{2}} \, dx_1 \,, \tag{6.2}$$

and what the sharp constant in this inequality is.

An important observation is that this argument can be iterated and, after $d - 1$ steps, leads to a Lieb–Thirring inequality of the form

$$\mathrm{Tr}(-\Delta - V)_-^\gamma \le \mathrm{const} \times \int_{\mathbb{R}^d} V(x)_+^{\gamma+\frac{d}{2}} \, dx \,, \tag{6.3}$$

where the constant in (6.3) is proportional to a product of the constants in (6.2) for $\gamma, \gamma + \frac{1}{2}, \ldots, \gamma + \frac{d-1}{2}$.

The reason why this argument is relevant for the proof of *sharp* Lieb–Thirring inequalities with semiclassical constants is that, if (6.2) holds with constant $= L_{\gamma,1}^{\mathrm{cl}}$ for the values of γ in question, then the resulting inequality (6.3) holds with the optimal constant constant $= L_{\gamma,d}^{\mathrm{cl}}$. This follows essentially by the phase space heuristics above.

In this chapter, we will carry out this approach and, essentially, prove (6.2) with constant $= L_{\gamma,1}^{\mathrm{cl}}$ for all $\gamma \ge 3/2$ in Theorem 6.11.

Note that 'lifting in dimension' can be thought of as an extension of the technique of separation of variables. We have seen it before, for instance, when it was applied to prove sharp semiclassical spectral inequalities for the eigenvalues of the Laplacian on product domains (Theorem 3.36) and for multi-dimensional harmonic oscillators (Proposition 5.21). In these cases, however, an exact separation of variables was possible. In the case of general multi-dimensional Schrödinger operators this cannot be expected to be possible. The idea of the method is to formally write

$$-\Delta - V = -\frac{d^2}{dx_1^2} \otimes \mathbb{I}_{L^2(\mathbb{R}^{d-1})} - Q \,,$$

where Q acts as a multiplication operator in $L^2(\mathbb{R}, L^2(\mathbb{R}^{d-1}))$ with values in operators in $L^2(\mathbb{R}^{d-1})$, namely,

$$Q(x_1) = \Delta' + V(x_1, \cdot).$$

Inequality (6.2) then reads

$$\mathrm{Tr}(-\tfrac{d^2}{dx_1^2} \otimes \mathbb{I}_{\mathcal{G}} - Q)_-^\gamma \le \mathrm{const} \times \int_{\mathbb{R}^d} \mathrm{Tr}_{\mathcal{G}} \, Q(x_1)_+^{\gamma+\frac{1}{2}} \, dx_1 \qquad (6.4)$$

with $\mathcal{G} = L^2(\mathbb{R}^{d-1})$. This is a Lieb–Thirring inequality for a one-dimensional Schrödinger operator with a potential Q that takes values in the self-adjoint operators in $\mathcal{G}$. The question whether this inequality holds with constant $= L^{\mathrm{cl}}_{\gamma,1}$ is an extension of the same question in the scalar case $\mathcal{G} = \mathbb{C}$.

By an approximation argument (see the proof of Lemma 6.13), if we want to prove (6.4) for $\mathcal{G} = L^2(\mathbb{R}^{d-1})$, it is enough to prove it for finite-dimensional spaces $\mathcal{G} = \mathbb{C}^M$ with constants independent of the dimension M. Thus, in the remainder of this chapter, we will concentrate on Schrödinger operators with matrix-valued potentials.

Let us describe the outline of this chapter. In §6.1 we introduce Schrödinger operators with matrix-valued potentials and briefly discuss their basic properties, which are similar to those in the scalar case. The main result of this chapter, namely the sharp inequality (6.1), is contained in §6.2. There, we explain the idea of lifting in dimension rigorously and in more detail. As mentioned before, the key ingredient is inequality (6.4) with constant $= L^{\mathrm{cl}}_{\gamma,1}$ for all $\gamma \ge 3/2$. This inequality originally comes from Laptev and Weidl (2000b), who proved it using trace formulas. The proof we present in §6.3 is due to Benguria and Loss (2000). It follows the same basic strategy as in the scalar case, but several aspects of its technical realization are more involved. The trace formula approach will be briefly sketched in §6.4.

The technique of lifting in dimension can be used not only to prove Lieb–Thirring inequalities with sharp constants, but also to obtain good, even though not optimal, constants. This will be explained in Chapter 8.

Convention. In this and the remaining chapters of the book we always assume that V is either real-valued or takes values in Hermitian matrices, even if this is not explicitly mentioned.

6.1 Schrödinger operators with matrix-valued potentials

In this section we define Schrödinger operators with matrix-valued potentials acting on vector-valued functions on $\mathbb{R}^d$, and show that many of the results about their scalar counterparts carry over to this setting.

Let us introduce some notation. For $M \in \mathbb{N}$, we denote by $\langle \cdot, \cdot \rangle$ and $|\cdot|$ the inner product and norm in $\mathbb{C}^M$. Measurability of functions from $\mathbb{R}^d$ to $\mathbb{C}^M$ is defined componentwise. We denote by $L^2(\mathbb{R}^d, \mathbb{C}^M) = L^2(\mathbb{R}^d) \otimes \mathbb{C}^M$ the Hilbert space of (a.e. equivalence classes of) measurable square-integrable functions on $\mathbb{R}^d$ with values in $\mathbb{C}^M$. Weak differentiability of a function u in $L^1_{\mathrm{loc}}(\mathbb{R}^d, \mathbb{C}^M)$ is defined componentwise and in this case $\partial_j u \in L^1_{\mathrm{loc}}(\mathbb{R}^d, \mathbb{C}^M)$. We use the notation $|\nabla u| = (\sum_{j=1}^d |\partial_j u|^2)^{1/2}$ for the norm of the gradient of u. We denote by $H^1(\mathbb{R}^d, \mathbb{C}^M)$ the space of (a.e. equivalence classes of) weakly differentiable functions $u \in L^2(\mathbb{R}^d, \mathbb{C}^M)$ with $\partial_j u \in L^2(\mathbb{R}^d, \mathbb{C}^M)$ for all $j = 1, \ldots, d$. Similarly, existence of a weak Laplacian is defined componentwise.

We denote by $\mathcal{H}_M$ the real space of Hermitian $M \times M$ matrices and by $\|\cdot\|$ the operator norm in $\mathbb{C}^M$. Measurability of functions from $\mathbb{R}^d$ to $\mathcal{H}_M$ is defined entrywise.

For $1 \le p \le \infty$, we denote by $L^p(\mathbb{R}^d, \mathcal{H}_M)$ the real space of (a.e. equivalence classes of) measurable functions V on $\mathbb{R}^d$ taking values in $\mathcal{H}_M$ and such that $x \mapsto \|V(x)\|$ belongs to $L^p(\mathbb{R}^d)$. The spaces $L^p_{\mathrm{loc}}(\mathbb{R}^d, \mathcal{H}_M)$ are defined similarly. Note that $x \mapsto \|V(x)\|$ is measurable if $x \mapsto V(x)$ is measurable.

The functions $V_\pm$ are defined by taking pointwise positive and negative parts in the spectral sense of Hermitian matrices, that is, $V_\pm(x) := (V(x))_\pm$ for all $x \in \mathbb{R}$. These two functions are measurable provided V is measurable. For a function $W \in L^1_{\mathrm{loc}}(\mathbb{R}^d, \mathcal{H}_M)$ taking values in the non-negative $M \times M$ Hermitian matrices, we set

$$H^1(\mathbb{R}^d, \mathbb{C}^M) \cap L^2(\mathbb{R}^d, \mathbb{C}^M, W\, dx)$$

$$:= \left\{ u \in H^1(\mathbb{R}^d, \mathbb{C}^M) : \int_{\mathbb{R}^d} \langle W(x) u(x), u(x) \rangle \, dx < \infty \right\}.$$

Proposition 6.1   *Let* $V \in L^1_{\mathrm{loc}}(\mathbb{R}^d, \mathcal{H}_M)$ *and assume that there are constants* $\theta < 1$ *and* $C < \infty$ *such that*

$$\int_{\mathbb{R}^d} \langle V_+ u, u \rangle \, dx \le \theta \int_{\mathbb{R}^d} \left(|\nabla u|^2 + \langle V_- u, u \rangle \right) dx + C \int_{\mathbb{R}^d} |u|^2 \, dx$$

$$\text{for all } u \in H^1(\mathbb{R}^d, \mathbb{C}^M) \cap L^2(\mathbb{R}^d, \mathbb{C}^M, V_-\, dx). \qquad (6.5)$$

Then the quadratic form

$$\int_{\mathbb{R}^d} \left(|\nabla u|^2 - \langle V u, u \rangle \right) dx$$

with form domain $H^1(\mathbb{R}^d, \mathbb{C}^M) \cap L^2(\mathbb{R}^d, \mathbb{C}^M, V_- \, dx)$ is lower semibounded and closed in $L^2(\mathbb{R}^d, \mathbb{C}^M)$, and $C_0^\infty(\mathbb{R}^d, \mathbb{C}^M)$ is a form core.

This proposition is proved similarly to its scalar analogue, Proposition 4.1. An important consequence is the following.

Corollary 6.2 *Let V be as in Proposition 6.1. Then there is a unique, lower semibounded and self-adjoint operator H in $L^2(\mathbb{R}^d, \mathbb{C}^M)$ that satisfies*

$$\mathrm{dom}\, H \subset H^1(\mathbb{R}^d, \mathbb{C}^M) \cap L^2(\mathbb{R}^d, \mathbb{C}^M, V_- \, dx)$$

and

$$\int_{\mathbb{R}^d} (\langle \nabla u, \nabla v \rangle - \langle Vu, v \rangle)\, dx = (Hu, v)$$

$$\text{for all } u \in \mathrm{dom}\, H, \; v \in H^1(\mathbb{R}^d, \mathbb{C}^M) \cap L^2(\mathbb{R}^d, \mathbb{C}^M, V_- \, dx).$$

Its domain is given by

$$\mathrm{dom}\, H = \big\{ u \in H^1(\mathbb{R}^d, \mathbb{C}^M) \cap L^2(\mathbb{R}^d, \mathbb{C}^M, V_- \, dx) \colon u \text{ has a weak Laplacian}$$

$$\text{and } -\Delta u - Vu \in L^2(\mathbb{R}^d, \mathbb{C}^M) \big\}$$

and, for $u \in \mathrm{dom}\, H$, we have $Hu = -\Delta u - Vu$.

Because of the last part we write $H = -\Delta \otimes \mathbb{I}_{\mathbb{C}^M} - V$ in what follows.

Lemma 6.3 *Let $V \in L^1_{\mathrm{loc}}(\mathbb{R}^d, \mathcal{H}_M)$ and assume that there are constants $\theta < 1$ and $C < \infty$ such that*

$$\int_{\mathbb{R}^d} \|V_+\| |u|^2 \, dx \leq \theta \int_{\mathbb{R}^d} |\nabla u|^2 \, dx + C \int_{\mathbb{R}^d} |u|^2 \, dx \qquad \text{for all } u \in H^1(\mathbb{R}^d).$$

$$(6.6)$$

Then (6.5) holds and, in the sense of operators,

$$-\Delta \otimes \mathbb{I}_{\mathbb{C}^M} - V \geq (-\Delta - \|V_+\|) \otimes \mathbb{I}_{\mathbb{C}^M}. \tag{6.7}$$

Proof According to Corollary 4.2, the operator $-\Delta - \|V_+\|$ is well-defined with quadratic form domain $H^1(\mathbb{R}^d)$ and, hence, the operator $(-\Delta - \|V_+\|) \otimes \mathbb{I}_{\mathbb{C}^M}$ is well-defined with form domain $H^1(\mathbb{R}^d, \mathbb{C}^M)$. Using the pointwise inequalities

$$\langle V_+(x)u(x), u(x) \rangle \leq \|V_+(x)\| |u(x)|^2, \quad \text{and} \quad \langle V_-(x)u(x), u(x) \rangle \geq 0, \tag{6.8}$$

we conclude from assumption (6.6), applied to each component of $u \in H^1(\mathbb{R}^d, \mathbb{C}^M)$ that inequality (6.5) holds. Consequently, by Corollary 6.2, the operator $-\Delta \otimes \mathbb{I}_{\mathbb{C}^M} - V$ is well-defined with form domain

$$H^1(\mathbb{R}^d, \mathbb{C}^M) \cap L^2(\mathbb{R}^d, \mathbb{C}^M, V_- \, dx).$$

This form domain is contained in that of the operator $(-\Delta - \|V_+\|) \otimes \mathbb{I}_{\mathbb{C}^M}$ and, in view of (6.8), we have the required inequality between the quadratic forms of the two operators.   □

The following proposition gives sufficient conditions for the inequality in Proposition 6.1.

Proposition 6.4   *Let*

$$V_- \in L^1_{\mathrm{loc}}(\mathbb{R}^d, \mathcal{H}_M) \quad \text{and} \quad V_+ \in L^\infty(\mathbb{R}^d, \mathcal{H}_M) + L^p(\mathbb{R}^d, \mathcal{H}_M),$$

where

$$\begin{cases} p = 1 & \text{if } d = 1, \\ p > 1 & \text{if } d = 2, \\ p = d/2 & \text{if } d \geq 3. \end{cases}$$

Then, for any $\theta > 0$, there is a $C > 0$ such that

$$\int_{\mathbb{R}^d} \langle V_+ u, u \rangle \, dx \leq \theta \int_{\mathbb{R}^d} |\nabla u|^2 \, dx + C \int_{\mathbb{R}^d} |u|^2 \, dx \qquad \text{for all } u \in H^1(\mathbb{R}^d, \mathbb{C}^M).$$

Consequently, Proposition 6.1 and Corollary 6.2 define a self-adjoint, lower semibounded operator $H = -\Delta \otimes \mathbb{I}_{\mathbb{C}^M} - V$ in $L^2(\mathbb{R}^d, \mathbb{C}^M)$ with form domain $H^1(\mathbb{R}^d, \mathbb{C}^M) \cap L^2(\mathbb{R}^d, \mathbb{C}^M, V_- \, dx)$ and form core $C_0^\infty(\mathbb{R}^d, \mathbb{C}^M)$.

Proof   Inequality (6.6) follows from Proposition 4.3, and therefore the assertion follows from Lemma 6.3.   □

Remark 6.5   Assume that, in addition to the assumptions in Proposition 6.4, we have $V_- \in L^\infty(\mathbb{R}^d, \mathcal{H}_M) + L^p(\mathbb{R}^d, \mathcal{H}_M)$ with p as in the proposition. Then the form domain of the operator in the proposition is $H^1(\mathbb{R}^d, \mathbb{C}^M)$. This follows as in the scalar case; see Remark 4.4.

Sufficient conditions for the discreteness and finiteness of the negative spectrum of $-\Delta \otimes \mathbb{I}_{\mathbb{C}^M} - V$ follow immediately via Lemma 6.3 from corresponding results in the scalar case. Indeed, as a consequence of the operator inequality (6.7), if the negative spectrum of $-\Delta - \|V_+\|$ in $L^2(\mathbb{R}^d)$ is discrete or finite, then the same holds for $-\Delta \otimes \mathbb{I}_{\mathbb{C}^M} - V$ in $L^2(\mathbb{R}^d, \mathbb{C}^M)$. Conditions for the discreteness and finiteness of the negative spectrum of $-\Delta - \|V_+\|$ are given, for example, in §§4.3.1 and 4.3.2.

As another application of Lemma 6.3, we now prove bounds on the lowest eigenvalue of $-\Delta \otimes \mathbb{I}_{\mathbb{C}^M} - V$ in terms of the L^p norm of V. We emphasize that the constants are independent of M.

Proposition 6.6   *Let*

$$\gamma \geq 1/2 \quad \text{if } d = 1,$$
$$\gamma > 0 \quad \text{if } d \geq 2.$$

If $V_- \in L^1_{\text{loc}}(\mathbb{R}^d, \mathcal{H}_M)$ *and* $V_+ \in L^{\gamma+d/2}(\mathbb{R}^d, \mathcal{H}_M)$, *then*

$$\inf \sigma(-\Delta \otimes \mathbb{I}_{\mathbb{C}^M} - V) \geq -\left(L^{(1)}_{\gamma,d} \int_{\mathbb{R}^d} \|V(x)_+\|^{\gamma+d/2} \, dx \right)^{1/\gamma}$$

with the same constant $L^{(1)}_{\gamma,d}$ *as in Proposition 5.3.*

Proposition 6.7   *Let* $d \geq 3$. *If* $V \in L^1_{\text{loc}}(\mathbb{R}^d, \mathcal{H}_M)$ *and*

$$L^{(1)}_{0,d} \int_{\mathbb{R}^d} \|V(x)_+\|^{d/2} \, dx \leq 1$$

with the same constant $L^{(1)}_{0,d}$ *as in Proposition 5.7, then the Schrödinger operator* $-\Delta \otimes \mathbb{I}_{\mathbb{C}^M} - V$ *in* $L^2(\mathbb{R}^d, \mathbb{C}^M)$ *is non-negative.*

Propositions 6.6 and 6.7 follow immediately, via (6.7), from their scalar analogues, Propositions 5.3 and 5.7.

Proposition 6.8   *Let* $d = 1, 2$ *and* $V \in L^1(\mathbb{R}^d, \mathcal{H}_M)$ *and assume that the matrix* $\int_{\mathbb{R}^d} V \, dx$ *has* $K \geq 1$ *positive eigenvalues. (If* $d = 2$ *assume, in addition, that* V *is as in Proposition 6.1.) Then*

$$N(0, -\Delta \otimes \mathbb{I}_{\mathbb{C}^M} - V) \geq K.$$

This means, in particular, that the analogue of Proposition 6.7 fails in one and two dimensions.

Proof   Let $e \in \mathbb{C}^M$ be from the positive spectral subspace of the matrix $W := \int_{\mathbb{R}^d} V \, dx$. Then, using the same functions β_R as in the proof of Proposition 4.43, we see that there is an R_0, independent of e, such that for all $R \geq R_0$,

$$\int_{\mathbb{R}^d} \left(|\nabla \beta_R \otimes e|^2 - \langle V \beta_R \otimes e, \beta_R \otimes e \rangle \right) dx < 0.$$

Since the space of $\beta_{R_0} \otimes e$, where e ranges over the positive spectral subspace of W, has dimension K, the claim follows by the variational principle (Theorem 1.25). $\square$

Using again Lemma 6.3, together with Theorem 4.31 and 4.38, we obtain the following versions of the Cwikel–Lieb–Rozenblum and Lieb–Thirring inequalities. Assume that

$$\begin{aligned}
\gamma &\geq 1/2 \quad \text{if } d = 1, \\
\gamma &> 0 \quad \text{if } d = 2, \\
\gamma &\geq 0 \quad \text{if } d \geq 3.
\end{aligned}$$

Then

$$\operatorname{Tr}(-\Delta \otimes \mathbb{I}_{\mathbb{C}^M} - V)_-^{\gamma} \leq L_{\gamma,d}\, M \int_{\mathbb{R}^d} \|V(x)_+\|^{\gamma+\frac{d}{2}}\, dx, \tag{6.9}$$

where $L_{\gamma,d}$ is the constant in the corresponding inequality in the scalar case.

The constant $L_{\gamma,d}\, M$ here on the right side is best possible. Indeed, for $V \in L^{\gamma+\frac{d}{2}}(\mathbb{R}^d, \mathbb{C}^M)$ of the form $V(x) = v(x)\mathbb{I}_{\mathbb{C}^M}$ with $v \in L^{\gamma+\frac{d}{2}}(\mathbb{R}^d, \mathbb{R})$, the inequality with this constant is equivalent to the corresponding scalar inequality (5.1).

To motivate the following, let us consider a matrix-valued potential of the form

$$V(x) = \begin{pmatrix} v_1(x) & 0 & 0 & \cdots \\ 0 & v_2(x) & 0 & \cdots \\ \cdots & \cdots & \cdots & \cdots \\ \cdots & \cdots & \cdots & v_M(x) \end{pmatrix} \tag{6.10}$$

with $v_1, \ldots, v_M \in L^{\gamma+\frac{d}{2}}(\mathbb{R}^d, \mathbb{R})$. Then

$$-\Delta \otimes \mathbb{I}_{\mathbb{C}^M} - V = \bigoplus_{m=1}^{M} (-\Delta - v_m), \tag{6.11}$$

and therefore,

$$\operatorname{Tr}(-\Delta \otimes \mathbb{I}_{\mathbb{C}^M} - V)_-^{\gamma} = \sum_{m=1}^{M} \operatorname{Tr}(-\Delta - v_m)_-^{\gamma}.$$

Applying the scalar Lieb–Thirring inequality (5.1) to each term, we get

$$\operatorname{Tr}\left(-\Delta \otimes \mathbb{I}_{\mathbb{C}^M} - V\right)_-^{\gamma} \leq L_{\gamma,d} \sum_{m=1}^{M} \int_{\mathbb{R}^d} v_m(x)_+^{\gamma+\frac{d}{2}}\, dx$$

$$= L_{\gamma,d} \int_{\mathbb{R}^d} \operatorname{Tr}_{\mathbb{C}^M}\left(V(x)_+^{\gamma+\frac{d}{2}}\right) dx.$$

This inequality differs from (6.9) in two respects. First, the quantity $\|V(x)_+\|^{\gamma+\frac{d}{2}}$ on the right side is replaced by $\operatorname{Tr}_{\mathbb{C}^M}\left(V(x)_+^{\gamma+\frac{d}{2}}\right)$ and, second, the resulting constant is independent of M. These two aspects are not independent of each other, since

$$\operatorname{Tr}_{\mathbb{C}^M}\left(V(x)_+^{\gamma+\frac{d}{2}}\right) \le M\,\|V(x)_+\|^{\gamma+\frac{d}{2}}$$

with equality if $V(x)_+$ is a multiple of the identity matrix in $\mathbb{C}^M$.

This discussion motivates the question of whether an inequality

$$\operatorname{Tr}\left(-\Delta \otimes \mathbb{I}_{\mathbb{C}^M} - V\right)_-^{\gamma} \le \tilde{L}_{\gamma,d}\int_{\mathbb{R}^d} \operatorname{Tr}_{\mathbb{C}^M}\left(V(x)_+^{\gamma+\frac{d}{2}}\right)dx \qquad (6.12)$$

holds for *arbitrary* $V \in L^{\gamma+\frac{d}{2}}(\mathbb{R}^d, \mathcal{H}_M)$, not necessarily of the form (6.10). The constant $\tilde{L}_{\gamma,d}$ may be different from $L_{\gamma,d}$, but should be independent of M.

Further evidence for why such an inequality would be natural comes from Weyl asymptotics, which we discuss next. Returning again to the diagonal case (6.10), we conclude from (6.11) and the scalar Weyl asymptotics in Theorems 4.28 and 4.29 that

$$\lim_{\alpha\to\infty} \alpha^{-\gamma-\frac{d}{2}} \operatorname{Tr}\left(-\Delta \otimes \mathbb{I}_{\mathbb{C}^M} - \alpha V\right)_-^{\gamma} = L_{\gamma,d}^{\mathrm{cl}} \sum_{m=1}^{M} \int_{\mathbb{R}^d} v_m(x)_+^{\gamma+\frac{d}{2}}\,dx$$

$$= L_{\gamma,d}^{\mathrm{cl}} \int_{\mathbb{R}^d} \operatorname{Tr}_{\mathbb{C}^M}\left(V(x)_+^{\gamma+\frac{d}{2}}\right)dx.$$

This proves Weyl asymptotics for diagonal potentials. Note that the right side here coincides with that in (6.12), at least up to the value of the constant.

Next we note that, if there is a unitary $M \times M$ matrix U such that $U^*V(x)U =: D(x)$ is diagonal for all x, then

$$-\Delta \otimes \mathbb{I}_{\mathbb{C}^M} - V = \left(\mathbb{I}_{L^2(\mathbb{R}^d)} \otimes U\right)^*\left(-\Delta \otimes \mathbb{I}_{\mathbb{C}^M} - D(x)\right)\left(\mathbb{I}_{L^2(\mathbb{R}^d)} \otimes U\right),$$

and therefore,

$$\operatorname{Tr}\left(-\Delta \otimes \mathbb{I}_{\mathbb{C}^M} - \alpha V\right)_-^{\gamma} = \operatorname{Tr}\left(-\Delta \otimes \mathbb{I}_{\mathbb{C}^M} - \alpha D\right)_-^{\gamma}$$

and

$$\operatorname{Tr}_{\mathbb{C}^M}\left(D(x)_+^{\gamma+\frac{d}{2}}\right) = \operatorname{Tr}_{\mathbb{C}^M}\left(V(x)_+^{\gamma+\frac{d}{2}}\right).$$

This proves Weyl asymptotics for potentials that are diagonal in some x-independent basis.

Of course, for every $x \in \mathbb{R}^d$ there is a unitary $M \times M$ matrix $U(x)$ such that $U(x)^*V(x)U(x)$ is diagonal; but, if U depends on x, the Schrödinger operator is, in general, not unitarily equivalent to a direct sum of scalar Schrödinger operators, so the above argument breaks down. Nevertheless, we now show that Weyl's asymptotics formula holds for general continuous V, not just those of the form (6.10). The basic idea is that, locally, one can choose an x-independent unitary matrix U that approximately diagonalizes V.

Theorem 6.9 *Let $V : \mathbb{R}^d \to \mathcal{H}_M$ be continuous with compact support. Then, for all $\gamma \geq 0$,*

$$\lim_{\alpha \to \infty} \alpha^{-\gamma - \frac{d}{2}} \operatorname{Tr} \left(- \Delta \otimes \mathbb{I}_{\mathbb{C}^M} - \alpha V \right)_-^\gamma = L_{\gamma,d}^{cl} \int_{\mathbb{R}^d} \operatorname{Tr}_{\mathbb{C}^M} \left(V(x)_+^{\gamma + \frac{d}{2}} \right) dx .$$

Proof We only give the details for $\gamma = 0$. A straightforward modification of the argument in the proof of Theorem 4.29 then yields the assertion for $\gamma > 0$.

For $\gamma = 0$ we follow the same overall strategy as in the proof of Theorem 4.28 and we use the same notation as there. One difference is the choice of the values of V in each cube $Q_L(a)$. In the proof of Theorem 4.28 we chose either maximal or minimal values in each cube. Now, in the matrix case, we fix an arbitrary value of the matrix V. Uniform continuity guarantees that for any $\varepsilon > 0$ there is a $\delta > 0$ such that, if $|x - x'| \leq \delta$, then $\|V(x) - V(x')\| \leq \varepsilon$. Thus, for any $L \leq \delta/\sqrt{d}$ and any choice of points x_a such that $x_a \in Q_L(a)$, we have, for almost every $x \in \mathbb{R}^d$ in the sense of operators on $\mathbb{C}^M$,

$$\chi_{\operatorname{supp} \|V\|}(x) \sum_{a \in L\mathbb{Z}^d} (V(x_a) - \varepsilon \mathbb{I}_{\mathbb{C}^M}) \chi_{Q_L(a)}(x) \leq V(x)$$

$$\leq \chi_{\operatorname{supp} \|V\|}(x) \sum_{a \in L\mathbb{Z}^d} (V(x_a) + \varepsilon \mathbb{I}_{\mathbb{C}^M}) \chi_{Q_L(a)}(x) .$$

Now we argue as before using Dirichlet–Neumann bracketing to reduce the problem to the study of $-\Delta_Q^D \otimes \mathbb{I}_{\mathbb{C}^M} - \alpha W$ and $-\Delta_Q^N \otimes \mathbb{I}_{\mathbb{C}^M} - \alpha W$ in $L^2(Q, \mathbb{C}^M)$ with a cube Q and a constant matrix $W \in \mathcal{H}_M$. This matrix is equal to $V(x_a) \pm \varepsilon \mathbb{I}_{\mathbb{C}^M}$ with the $-$ sign in the Dirichlet and the $+$ sign in the Neumann case. There is a unitary $M \times M$ matrix U such that

$$U^* W U = \begin{pmatrix} w_1 & 0 & 0 & \cdots \\ 0 & w_2 & 0 & \cdots \\ \cdots & \cdots & \cdots & \cdots \\ \cdots & \cdots & \cdots & w_M \end{pmatrix} =: D,$$

where $w_1, \ldots, w_M$ are the eigenvalues of W. Then

$$-\Delta_Q^{D/N} \otimes \mathbb{I}_{\mathbb{C}^M} - \alpha W = \left(\mathbb{I}_{L^2(Q)} \otimes U \right)^* \left(-\Delta_Q^{D/N} \otimes \mathbb{I}_{\mathbb{C}^M} - \alpha D \right) \left(\mathbb{I}_{L^2(Q)} \otimes U \right),$$

and therefore, by the result in the scalar case,

$$N\left(0, -\Delta_Q^{D/N} \otimes \mathbb{I}_{\mathbb{C}^M} - \alpha W\right) = N\left(0, -\Delta_Q^{D/N} \otimes \mathbb{I}_{\mathbb{C}^M} - \alpha D\right)$$

$$= \sum_{m=1}^{M} N\left(0, -\Delta_Q^{D/N} - \alpha w_m\right)$$

$$= L_{0,d}^{cl} |Q| \sum_{m=1}^{d} (w_m)_+^{d/2} \alpha^{d/2} \left(1 + o(1)\right)$$

$$= L_{0,d}^{cl} |Q| \operatorname{Tr} \left(W_+^{d/2}\right) \alpha^{d/2} \left(1 + o(1)\right).$$

As in the proof of Theorem 4.28, we sum these formulas over cubes $Q_L(a)$, with $a \in L\mathbb{Z}^d$, that intersect the support of the potential. We then first let $\alpha \to \infty$, then $L \to 0$, and finally $\varepsilon \to 0$ to obtain the claimed asymptotics for $\gamma = 0$. $\qquad \square$

Remark 6.10   Using the inequalities (6.9) and the technique from the proof of Theorem 4.46, one can show that the Weyl-type asymptotics hold for all $V \in L^{\gamma + d/2}(\mathbb{R}^d, \mathcal{H}_M)$ if $\gamma \geq 1/2$ in $d = 1$, $\gamma > 0$ in $d = 2$, and $\gamma \geq 0$ in $d \geq 3$.

6.2 The Lieb–Thirring inequality with the semiclassical constant

The main result of this chapter is the following theorem of Laptev and Weidl (2000b).

Theorem 6.11   *Let $d \geq 1$ and $\gamma \geq \frac{3}{2}$. Then, for all $V \in L_{\mathrm{loc}}^1(\mathbb{R}^d)$ with $V_+ \in L^{\gamma + \frac{d}{2}}(\mathbb{R}^d)$,*

$$\operatorname{Tr}(-\Delta - V)_-^\gamma \leq L_{\gamma,d}^{cl} \int_{\mathbb{R}^d} V(x)_+^{\gamma + \frac{d}{2}} \, dx \, .$$

The constant $L_{\gamma,d}^{cl}$ is optimal.

Optimality of the constant $L_{\gamma,d}^{cl}$ follows from Weyl asymptotics (Theorem 4.29).

As we will explain in this section, the proof of Theorem 6.11 proceeds by induction in the dimension d. Recall that the result for $d = 1$ was established in Corollary 5.32. The main ingredient in the induction step is the following sharp Lieb–Thirring inequality for one-dimensional Schrödinger operators with matrix-valued potentials, which is the matrix generalization of Theorem 5.31 and which is also due to Laptev and Weidl (2000b).

Theorem 6.12 *Let $d = 1$ and $\gamma = \frac{3}{2}$. Then, for all $M \in \mathbb{N}$ and all $V \in L^1_{\mathrm{loc}}(\mathbb{R}, \mathcal{H}_M)$ with $V_+ \in L^2(\mathbb{R}, \mathcal{H}_M)$,*

$$\mathrm{Tr}\left(-\frac{d^2}{dx^2} \otimes \mathbb{I}_{\mathbb{C}^M} - V\right)_-^{3/2} \leq L^{\mathrm{cl}}_{3/2,1} \int_{\mathbb{R}} \mathrm{Tr}_{\mathbb{C}^M}\left(V(x)_+^2\right) dx \,.$$

The constant $L^{\mathrm{cl}}_{3/2,1} = \frac{3}{16}$ is optimal.

Optimality of the constant $L^{\mathrm{cl}}_{3/2,1}$ for any $M \in \mathbb{N}$ follows from Weyl asymptotics (Theorem 6.9).

We present the proof of Theorem 6.12 in the following section. Here, we explain how to deduce Theorem 6.11 from Theorem 6.12. The proof of Theorem 6.11 is based on an inductive argument where one variable is split off at each step, generalizing a separation of variables procedure. The splitting-off of a single dimension, together with an approximation argument, is the content of the following lemma.

Lemma 6.13 *Let $\gamma \geq 1/2$. Assume that there is a constant L such that for all $M \in \mathbb{N}$ one has*

$$\mathrm{Tr}_{L^2(\mathbb{R}) \otimes \mathbb{C}^M}\left(-\frac{d^2}{dx^2} \otimes \mathbb{I}_{\mathbb{C}^M} - Q\right)_-^{\gamma} \leq L \int_{\mathbb{R}} \mathrm{Tr}_{\mathbb{C}^M}\left(Q(x)_+^{\gamma+\frac{1}{2}}\right) dx$$

$$\textit{for all } Q \in L^{\gamma+\frac{1}{2}}(\mathbb{R}, \mathcal{H}_M). \qquad (6.13)$$

Then, for all $d \geq 2$,

$$\mathrm{Tr}_{L^2(\mathbb{R}^d)}\left(-\Delta - V\right)_-^{\gamma} \leq L \int_{\mathbb{R}} \mathrm{Tr}_{L^2(\mathbb{R}^{d-1})}\left(-\Delta' - V(x_1, \cdot)\right)_-^{\gamma+\frac{1}{2}} dx_1$$

$$\textit{for all } V \in L^{\gamma+\frac{d}{2}}(\mathbb{R}^d).$$

The proof of this lemma is not difficult, but somewhat lengthy and we defer it to the end of this section. We now use it to prove Theorem 6.11.

Proof of Theorem 6.11 We prove Theorem 6.11 by induction on d. For $d = 1$, the theorem for any $\gamma \geq 3/2$ holds according to Corollary 5.32. Now let $d \geq 2$ and assume that the theorem holds for all $\gamma \geq 3/2$ with d replaced by $d - 1$. First let $\gamma = 3/2$. Then the assumption (6.13) holds according to Theorem 6.12 with $L = L^{\mathrm{cl}}_{3/2,1}$ and therefore, by Lemma 6.13, we have

$$\mathrm{Tr}_{L^2(\mathbb{R}^d)}\left(-\Delta - V\right)_-^{3/2} \leq L^{\mathrm{cl}}_{3/2,1} \int_{\mathbb{R}} \mathrm{Tr}_{L^2(\mathbb{R}^{d-1})}\left(-\Delta' - V(x_1, \cdot)\right)_-^{2} dx_1$$

for all $V \in L^{(d+3)/2}(\mathbb{R}^d)$. By induction hypothesis (with $\gamma = 2$), we have

$$\mathrm{Tr}_{L^2(\mathbb{R}^{d-1})}\left(-\Delta' - V(x_1, \cdot)\right)_-^{2} \leq L^{\mathrm{cl}}_{2,d-1} \int_{\mathbb{R}^{d-1}} V\left(x_1, x'\right)_+^{2+(d-1)/2} dx' \,.$$

Combining the last two inequalities gives

$$\mathrm{Tr}_{L^2(\mathbb{R}^d)}(-\Delta - V)_-^{3/2} \le L_{3/2,1}^{\mathrm{cl}} L_{2,d-1}^{\mathrm{cl}} \int_{\mathbb{R}^d} V(x)_+^{3/2+d/2} \, dx \,.$$

We now recall from (3.72) that

$$L_{3/2,1}^{\mathrm{cl}} L_{2,d-1}^{\mathrm{cl}} = L_{3/2,d}^{\mathrm{cl}} \,.$$

This proves the claimed inequality for $\gamma = 3/2$ in dimension d for $0 \le V \in L^{(d+3)/2}(\mathbb{R}^d)$. The validity under the assumptions of the theorem follows by the variational principle. The claimed inequality for $\gamma > 3/2$ is a consequence of that for $\gamma = 3/2$ by the Aizenman–Lieb argument (Lemma 5.2). This proves the theorem. □

It remains to give the proof of Lemma 6.13.

Proof of Lemma 6.13 For any u from the form domain of the operator $-\Delta - V$ in $L^2(\mathbb{R}^d)$ we have

$$\int_{\mathbb{R}^d} \left(\sum_{j=1}^d \left| \frac{\partial u}{\partial x_j} \right|^2 - V|u|^2 \right) dx \ge \int_{\mathbb{R}} \left(\left\| \frac{\partial u}{\partial x_1} \right\|^2_{L^2(\mathbb{R}^{d-1})} - (W(x_1)u, u)_{L^2(\mathbb{R}^{d-1})} \right) dx_1 \,,$$

where

$$W(x_1) := (-\Delta' - V(x_1, \cdot))_- $$

is the negative part of a Schrödinger operator in $L^2(\mathbb{R}^{d-1})$. Thus, from the variational principle, for any $\kappa > 0$,

$$\mathrm{Tr}_{L^2(\mathbb{R}^d)}(-\Delta - V + \kappa)_-^\gamma \le \mathrm{Tr}_{L^2(\mathbb{R}) \otimes L^2(\mathbb{R}^{d-1})} \left(-\frac{d^2}{dx_1^2} \otimes \mathbb{I}_{L^2(\mathbb{R}^{d-1})} - W + \kappa \right)_-^\gamma \,.$$

Next, we recall from Proposition 1.40 that for every $\varepsilon > 0$ there is a constant $C_{\varepsilon,\gamma}$ such that for any self-adjoint, lower semibounded operators A and B,

$$\mathrm{Tr}(A + B)_-^\gamma \le (1 + \varepsilon) \mathrm{Tr}\, A_-^\gamma + C_{\varepsilon,\gamma} \mathrm{Tr}\, B_-^\gamma \,. \tag{6.14}$$

Let P_n be a sequence of orthogonal projections in $L^2(\mathbb{R}^{d-1})$ with finite-dimensional range such that $P_n \to \mathbb{I}_{L^2(\mathbb{R}^{d-1})}$ strongly as $n \to \infty$. (Such a sequence exists by separability of $L^2(\mathbb{R}^{d-1})$.) We fix a parameter $0 < \delta < 1$ and apply (6.14) with

$$A = -(1-\delta)\frac{d^2}{dx_1^2} \otimes \mathbb{I}_{L^2(\mathbb{R}^{d-1})} - P_n W P_n \,, \quad B = -\delta \frac{d^2}{dx_1^2} \otimes \mathbb{I}_{L^2(\mathbb{R}^{d-1})} - W + P_n W P_n + \kappa \,.$$

The operator A is equal to the direct sum of $-(1-\delta)\frac{d^2}{dx_1^2} \otimes \mathbb{I}_{\mathrm{ran}\, P_n} - P_n W P_n$ and $-(1-\delta)\frac{d^2}{dx_1^2} \otimes \mathbb{I}_{(\mathrm{ran}\, P_n)^\perp}$. The latter is non-negative and the former is a Schrödinger operator with a matrix-valued potential. Therefore, by the assumed inequality (6.13), we have

$$\mathrm{Tr}\, A_-^\gamma \le (1-\delta)^{-\gamma+\frac{1}{2}} L \int_{\mathbb{R}} \mathrm{Tr}_{L^2(\mathbb{R}^{d-1})}(P_n W(x_1) P_n)^{\gamma+\frac{1}{2}}\, dx_1\,.$$

By Lemma 1.44, with $A = -W(x_1)$ and $T = P_n$,

$$\mathrm{Tr}_{L^2(\mathbb{R}^{d-1})}(P_n W(x_1) P_n)^{\gamma+\frac{1}{2}} \le \mathrm{Tr}_{L^2(\mathbb{R}^{d-1})} W(x_1)^{\gamma+\frac{1}{2}}\,,$$

and so

$$\mathrm{Tr}\, A_-^\gamma \le (1-\delta)^{-\gamma+\frac{1}{2}} L \int_{\mathbb{R}} \mathrm{Tr}_{L^2(\mathbb{R}^{d-1})} \big(-\Delta' - V(x_1,\cdot)\big)_-^{\gamma+\frac{1}{2}}\, dx_1\,. \qquad (6.15)$$

We now turn our attention to the operator B and show that

$$\mathrm{Tr}\, B_-^\gamma = 0 \qquad \text{for all sufficiently large } n \text{ (depending on } \delta \text{ and } \kappa\text{)}. \qquad (6.16)$$

For any $u \in L^2(\mathbb{R}^d)$ that has a weak derivative with respect to x_1 in $L^2(\mathbb{R}^d)$ we have, with $\|W(x_1) - P_n W(x_1) P_n\|$ denoting the operator norm of the operator $W(x_1) - P_n W(x_1) P_n$ in $L^2(\mathbb{R}^{d-1})$,

$$\int_{\mathbb{R}} \left(\delta \left\| \frac{\partial u}{\partial x_1} \right\|^2_{L^2(\mathbb{R}^{d-1})} - \Big(\big(W(x_1) - P_n W(x_1) P_n\big) u(x_1), u(x_1) \Big)_{L^2(\mathbb{R}^{d-1})} \right) dx_1$$

$$\ge \int_{\mathbb{R}} \left(\delta \left\| \frac{\partial u}{\partial x_1} \right\|^2_{L^2(\mathbb{R}^{d-1})} - \|W(x_1) - P_n W(x_1) P_n\| \|u(x_1)\|^2_{L^2(\mathbb{R}^{d-1})} \right) dx_1$$

$$= \int_{\mathbb{R}^{d-1}} \int_{\mathbb{R}} \left(\delta \left| \frac{\partial u}{\partial x_1}(x_1, x') \right|^2 - \|W(x_1) - P_n W(x_1) P_n\| |u(x_1, x')|^2 \right) dx_1\, dx'\,.$$

By Proposition 5.3, we have for fixed $x' \in \mathbb{R}^{d-1}$,

$$\int_{\mathbb{R}} \left(\delta \left| \frac{\partial u}{\partial x_1}(x_1, x') \right|^2 - \|W(x_1) - P_n W(x_1) P_n\| |u(x_1, x')|^2 \right) dx_1$$

$$\ge -\delta^{-\frac{1}{2\gamma}} \left(L_{\gamma,1}^{(1)} \int_{\mathbb{R}} \|W(x_1) - P_n W(x_1) P_n\|^{\gamma+\frac{1}{2}}\, dx_1 \right)^{\frac{1}{\gamma}} \int_{\mathbb{R}} |u(x_1, x')|^2\, dx_1\,.$$

Therefore, if

$$L_{\gamma,1}^{(1)} \int_{\mathbb{R}} \|W(x_1) - P_n W(x_1) P_n\|^{\gamma+\frac{1}{2}}\, dx_1 \le \kappa^\gamma \delta^{\frac{1}{2}}\,, \qquad (6.17)$$

then

$$\int_{\mathbb{R}} \left(\delta \left\| \frac{\partial u}{\partial x_1}(x_1) \right\|^2_{L^2(\mathbb{R}^{d-1})} - \left((W(x_1) - P_n W(x_1) P_n) u(x_1), u(x_1) \right)_{L^2(\mathbb{R}^{d-1})} \right) dx_1$$

$$\geq -\kappa \int_{\mathbb{R}} \|u(x_1)\|^2_{L^2(\mathbb{R}^{d-1})} \, dx_1 \,,$$

which means that $B \geq 0$. Thus, in order to prove (6.16), we need to show that (6.17) holds for all sufficiently large n (depending on δ and κ). In fact, we shall show that

$$\lim_{n \to \infty} \int_{\mathbb{R}} \|W(x_1) - P_n W(x_1) P_n\|^{\gamma + \frac{1}{2}} \, dx_1 = 0 \,. \tag{6.18}$$

In order to deduce this by dominated convergence, we first show that the integrand is dominated by an integrable function independent of n. We have $-P_n W(x_1) P_n \leq W(x_1) - P_n W(x_1) P_n \leq W(x_1)$. Clearly, $W(x_1) \leq \|W(x_1)\|$ and $-P_n W(x_1) P_n \geq -\|P_n W(x_1) P_n\| \geq -\|W(x_1)\|$. Therefore

$$\|W(x_1) - P_n W(x_1) P_n\|^{\gamma + \frac{1}{2}} \leq \|W(x_1)\|^{\gamma + \frac{1}{2}} \,.$$

Moreover, $\|W(x_1)\| = |\inf \sigma(-\Delta' - V(x_1, \cdot))|$ and therefore, by Proposition 5.3,

$$\|W(x_1)\|^{\gamma + \frac{1}{2}} \leq L^{(1)}_{\gamma + \frac{1}{2}, d-1} \int_{\mathbb{R}^{d-1}} V(x_1, x')_+^{\gamma + \frac{d}{2}} \, dx' \,.$$

Since, by assumption $V \in L^{\gamma + \frac{d}{2}}(\mathbb{R}^d)$, the right side is integrable with respect to x_1.

Thus, to deduce the claim (6.18) from dominated convergence, we need to show that $\|W(x_1) - P_n W(x_1) P_n\| \to 0$ as $n \to \infty$ for almost every $x_1 \in \mathbb{R}$. This follows from the fact, shown below, that $W(x_1)$ is compact for almost every $x_1 \in \mathbb{R}$, together with Lemma 1.4.

To justify the claimed almost-everywhere compactness of $W(x_1)$, we can argue in two ways. Either, we use the fact that, by Fubini's theorem, for almost every $x_1 \in \mathbb{R}^{d-1}$ we have $V(x_1, \cdot) \in L^{\gamma + \frac{d}{2}}(\mathbb{R}^{d-1})$, and therefore, by Proposition 4.14, the negative spectrum of $-\Delta' - V(x_1, \cdot)$ is discrete; that is, $W(x_1)$ is compact. Alternatively, we notice that the claim of the lemma is trivially correct if $\int_{\mathbb{R}} \mathrm{Tr}_{L^2(\mathbb{R}^{d-1})} W(x_1)^{\gamma + \frac{1}{2}} \, dx_1 = \infty$, and therefore we may assume that this integral is finite, so, for almost every $x_1 \in \mathbb{R}$, we have $\mathrm{Tr}_{L^2(\mathbb{R}^{d-1})} W(x_1)^{\gamma + \frac{1}{2}} < \infty$ and, in particular, $W(x_1)$ is compact.

To summarize, dominated convergence proves (6.18) and therefore (6.16).

We now conclude the proof of Lemma 6.13. By combining (6.14) with the above choices of A and B with (6.15) and (6.16), we obtain that

$$\operatorname{Tr}_{L^2(\mathbb{R}^d)} \left(-\Delta - V + \kappa \right)_-^\gamma$$
$$\leq (1 + \varepsilon)(1 - \delta)^{-\gamma + \frac{1}{2}} L \int_{\mathbb{R}} \operatorname{Tr}_{L^2(\mathbb{R}^{d-1})} \left(-\Delta' - V(x_1, \cdot) \right)_-^{\gamma + \frac{1}{2}} dx_1 .$$

We first let ε and δ tend to zero and then, using monotone convergence, we let κ tend to zero to obtain the statement of the lemma. $\square$

6.3 The sharp bound in the matrix-valued case for $\gamma = \frac{3}{2}$

Our goal in this section is to prove Theorem 6.12, namely that, for all $M \in \mathbb{N}$ and all $V \in L^1_{\text{loc}}(\mathbb{R}, \mathcal{H}_M)$ with $V_+ \in L^2(\mathbb{R}, \mathcal{H}_M)$,

$$\operatorname{Tr} \left(-\frac{d^2}{dx^2} \otimes \mathbb{I}_{\mathbb{C}^M} - V \right)_-^{3/2} \leq \frac{3}{16} \int_{\mathbb{R}} \operatorname{Tr}_{\mathbb{C}^M} \left(V(x)_+^2 \right) dx . \tag{6.19}$$

This is the matrix generalization of Theorem 5.31, due to Laptev and Weidl (2000b), and represents, as we explained in the previous section, the main step in the proof of Theorem 6.11.

The proof of inequality (6.19) that we present in this section follows an argument of Benguria and Loss (2000). The original proof from Laptev and Weidl (2000b) is sketched in §6.4 below. We have encountered the analogues of both proofs before in the scalar case in §§5.4 and 5.5.

6.3.1 The matrix Riccati equation

Let V be a continuous function on $\mathbb{R}$ with values in $\mathcal{H}_M$ and support in $[-a, a]$ for some $a > 0$. We assume that the negative spectrum of the Schrödinger operator $-\frac{d^2}{dx^2} \otimes \mathbb{I}_{\mathbb{C}^M} - V$ in $L^2(\mathbb{R}) \otimes \mathbb{C}^M$ is non-empty and denote

$$-E_1 := \inf \sigma \left(-\frac{d^2}{dx^2} \otimes \mathbb{I}_{\mathbb{C}^M} - V \right) .$$

Moreover, let

$$K := \dim \ker \left(-\frac{d^2}{dx^2} \otimes \mathbb{I}_{\mathbb{C}^M} - V + E_1 \right) .$$

In the scalar case, $M = 1$, we have $K = 1$ but this need not be so in general; for instance, take (6.10) with $v_1 = \cdots = v_M$. The fact that K can exceed 1 is a source of difficulties in the matrix case.

Fix an invertible $M \times M$ matrix Ψ and consider the $M \times M$ matrix-valued solution Φ of the differential equation

$$\left(-\frac{d^2}{dx^2} \otimes \mathbb{I}_{\mathbb{C}^M} - V(x)\right)\Phi(x) = -E_1\Phi(x) \quad \text{for all } x \in \mathbb{R},$$

$$\text{satisfying } \Phi(x) = e^{\sqrt{E_1}x}\Psi \quad \text{for all } x \le -a \,.$$

(6.20)

Because of the assumed continuity of V, the existence of a C^2 solution Φ follows from standard ODE results.

Lemma 6.14 *For every $x \in \mathbb{R}$, $\Phi(x)$ is invertible.*

Note that the proof has some similarities with that of Proposition 5.35, which states that an eigenfunction corresponding to the lowest eigenvalue of a scalar Schrödinger operator can be chosen positive.

Proof We argue by contradiction and assume that there are $x_0 \in \mathbb{R}$ and $0 \neq v \in \mathbb{C}^M$ such that $\Phi(x_0)v = 0$. Let $\varphi(x) := \Phi(x)v$, which solves the differential equation

$$\left(-\frac{d^2}{dx^2} \otimes \mathbb{I}_{\mathbb{C}^M} - V\right)\varphi = -E_1\varphi \quad \text{satisfying} \quad \varphi(x) = e^{\sqrt{E_1}x}\psi \text{ for } x \le -a,$$

(6.21)

with $\psi := \Psi v$. Note that $\psi \neq 0$ since Ψ is invertible and $v \neq 0$. Put

$$\tilde{\varphi}(x) := \begin{cases} \varphi(x) & \text{if } x \le x_0, \\ 0 & \text{if } x > x_0, \end{cases}$$

where 0 is the vector in $\mathbb{C}^M$ with all components equal to zero. By Lemmas 2.27 and 2.28, it follows from $\varphi(x_0) = 0$ that $\tilde{\varphi}$ belongs to $H^1(\mathbb{R}) \otimes \mathbb{C}^M$, which is the form domain of the operator $-\frac{d^2}{dx^2} \otimes \mathbb{I}_{\mathbb{C}^M} - V$, and satisfies

$$\int_{\mathbb{R}} \left(|\tilde{\varphi}'|^2 - \langle V(x)\tilde{\varphi}, \tilde{\varphi}\rangle\right) dx = \int_{-\infty}^{x_0} \left(|\varphi'|^2 - \langle V(x)\varphi, \varphi\rangle\right) dx$$

$$= \int_{-\infty}^{x_0} \langle -\varphi'' - V(x)\varphi, \varphi\rangle \, dx = -E_1 \int_{-\infty}^{x_0} |\varphi|^2 \, dx = -E_1 \|\tilde{\varphi}\|^2 \,.$$

Hence $\tilde{\varphi}$ minimizes the quadratic form corresponding to $-\frac{d^2}{dx^2} \otimes \mathbb{I}_{\mathbb{C}^M} - V$. Thus, by Lemma 1.19, it belongs to the operator domain and is an eigenfunction of the corresponding operator. As an eigenfunction, $\tilde{\varphi}$ is a solution of $\tilde{\varphi}'' = q\tilde{\varphi}$ on $\mathbb{R}$ for $q := -V + E_1 \mathbb{I}_{\mathbb{C}^M}$ that vanishes on $[x_0, \infty)$. Therefore, by a generalization of Lemma 5.36 to the matrix-valued case, it is identically equal to zero on $\mathbb{R}$ and, in particular, $\varphi(x) = 0$ for all $x \le x_0$. Again by Lemma 5.36, this implies that φ is identically equal to zero on all of $\mathbb{R}$, which contradicts $\psi \neq 0$ and completes the proof. $\qquad\qquad\square$

In view of the previous lemma, the logarithmic derivative

$$F(x) := \Phi'(x)\Phi(x)^{-1},$$

is well defined for all $x \in \mathbb{R}$. We shall see below in Lemma 6.17 that F is independent of the choice of Ψ and that, for every $x \in \mathbb{R}$, the matrix $F(x)$ is Hermitian.

Similarly as in the scalar case, F satisfies a Riccati equation.

Lemma 6.15 *We have*

$$F'(x) + (F(x))^2 = -V(x) + E_1\mathbb{I}_{\mathbb{C}^M} \qquad \text{for all } x \in \mathbb{R}.$$

Proof Under our continuity assumption on V, Φ is twice continuously differentiable. Moreover, recalling Lemma 6.14, we see that Φ^{-1} is continuously differentiable and we easily find

$$\frac{d}{dx}\Phi(x)^{-1} = -\Phi(x)^{-1}\Phi'(x)\Phi(x)^{-1}.$$

Thus, in view of (6.20),

$$F' + F^2 = (\Phi'\Phi^{-1})' + (\Phi'\Phi^{-1})^2 = \Phi''\Phi^{-1} - \Phi'\Phi^{-1}\Phi'\Phi^{-1} + \Phi'\Phi^{-1}\Phi'\Phi^{-1}$$
$$= (E_1\mathbb{I}_{\mathbb{C}^M} - V)\Phi\Phi^{-1} = E_1\mathbb{I}_{\mathbb{C}^M} - V,$$

which proves the claimed equation. $\qquad\qquad\square$

6.3.2 The asymptotic behavior of F

We next discuss the behavior of $F(x)$ as $x \to \pm\infty$. Because of $\Phi(x) = e^{\sqrt{E_1}x}\Psi$ for $x \le -a$, we have

$$F(x) = \sqrt{E_1}\mathbb{I}_{\mathbb{C}^M} \quad \text{for} \quad x \le -a. \tag{6.22}$$

The key difference between the scalar and the matrix case is the behavior of $F(x)$ as $x \to +\infty$. In the scalar case, $F(x)$ is constant to the right of the support of V. In the matrix case, this is, in general, not so and the behavior of F depends on the multiplicity K of the eigenvalue $-E_1$. The remainder of this subsection is devoted to the proof of the following result.

Proposition 6.16 *There is an orthogonal projection P on $\mathbb{C}^M$ with* $\dim \operatorname{ran} P = K$ *such that*

$$\lim_{x \to +\infty} F(x) = -\sqrt{E_1}P + \sqrt{E_1}(\mathbb{I}_{\mathbb{C}^M} - P).$$

Moreover, $F'(x)$ decays exponentially as $x \to +\infty$.

Once we have proved this proposition, apart from an additional approximation argument, the proof of (6.19) will follow by steps similar to those in the
scalar case.

Lemma 6.17 *The function F is independent of the choice of Ψ, and for any
$x \in \mathbb{R}$ the matrix $F(x)$ is Hermitian.*

Proof Let Ψ_1 and Ψ_2 be invertible $N \times N$ matrices and let Φ_1 and Φ_2 be
solutions of the equation

$$-\Phi_j'' - V\Phi_j = -E_1\Phi_j \quad \text{satisfying} \quad \Phi_j(x) = e^{\sqrt{E_1}x}\Psi_j \text{ for } x \leq -a. \quad (6.23)$$

Note that, since $V(x)^* = V(x)$, we also have

$$-(\Phi^*)_j'' - \Phi_j^*V = -E_1\Phi_j^*. \quad (6.24)$$

As in the scalar case, the Wronskian

$$W(x) := \Phi_1^*(x)\Phi_2'(x) - (\Phi_1^*)'(x)\Phi_2(x)$$

is independent of x. Indeed, by (6.23) for $j = 2$ and (6.24) for $j = 1$, its
derivative is

$$W' = (\Phi_1^*)'\Phi_2' + \Phi_1^*\Phi_2'' - (\Phi_1^*)''\Phi_2 - (\Phi_1^*)'\Phi_2' = \Phi_1^*\Phi_2'' - (\Phi_1^*)''\Phi_2$$
$$= \Phi_1^*(E_1\Phi_2 - V\Phi_2) - (\Phi_1^*E_1 - \Phi_1^*V)\Phi_2 = 0.$$

Since $\Phi_j(x) = e^{\sqrt{E_1}x}\Psi_j$ for $x \leq -a$, we find that $W(x) = 0$ for $x \leq -a$, and
therefore $W(x) = 0$ for all $x \in \mathbb{R}$. For $F_j(x) := \Phi_j'(x)\Phi_j(x)^{-1}$ this gives

$$F_2 = \Phi_2'\Phi_2^{-1} = (\Phi_1^*)^{-1}(\Phi_1^*)' = F_1^*. \quad (6.25)$$

In particular, choosing $\Psi_1 = \Psi_2 = \Psi$, we find that

$$F = F_2 = \Phi_2'(\Phi_2)^{-1} = (\Phi_1^*)^{-1}(\Phi_1^*)' = F_1^* = F^*;$$

that is, F is Hermitian for any Ψ. In view of that, (6.25) turns into $F_2 = F_1^* = F_1$
for arbitrary Ψ_1 and Ψ_2; that is, F is independent of the choice of Ψ.   □

We next derive an explicit expression of the matrix solution Φ for $x \geq a$
in terms of its value and its derivative at $x = a$. Since Φ solves, in (a, ∞),
a Schrödinger equation without potential, its entries are linear combinations
of the appropriately scaled exponential functions and the coefficients can be
obtained by matching the function and its derivate at the point $x = a$. After an

elementary computation, we obtain

$$\Phi(x) = \cosh(\sqrt{E_1}(x-a))\Phi(a) + \frac{1}{\sqrt{E_1}}\sinh(\sqrt{E_1}(x-a))\Phi'(a)$$

$$= \left(\cosh(\sqrt{E_1}(x-a))\mathbb{I}_{\mathbb{C}^M} + \frac{1}{\sqrt{E_1}}\sinh(\sqrt{E_1}(x-a))F(a)\right)\Phi(a). \quad (6.26)$$

In the following, the matrix $F(a)$ will play an important role. The next lemma says that the multiplicity K of the eigenvalue $-E_1$ of the Schrödinger operator $-\frac{d^2}{dx^2}\otimes\mathbb{I}_{\mathbb{C}^M} - V$ is encoded in the spectrum of $F(a)$.

Lemma 6.18   *We have* $K = \dim\ker(F(a)+\sqrt{E_1}\mathbb{I}_{\mathbb{C}^M})$.

Proof   Since Φ is a fundamental system of solutions of the Cauchy problem (6.21) and since $\Phi(a)$ is invertible by Lemma 6.14, any particular solution φ of (6.21) can be uniquely represented as $\varphi(x) = \Phi(x)(\Phi(a))^{-1}v$ with some $v \in \mathbb{C}^M$. By (6.26), in this representation we have, for $x \geq a$,

$$\varphi(x) = \cosh(\sqrt{E_1}(x-a))v + \frac{1}{\sqrt{E_1}}\sinh(\sqrt{E_1}(x-a))F(a)v$$

$$= \frac{1}{2\sqrt{E_1}}e^{\sqrt{E_1}(x-a)}\left(\sqrt{E_1}v + F(a)v\right) + \frac{1}{2\sqrt{E_1}}e^{-\sqrt{E_1}(x-a)}\left(\sqrt{E_1}v - F(a)v\right).$$

Since, by construction, this function decays exponentially at $-\infty$, it is an eigenfunction of $-\frac{d^2}{dx^2}\otimes\mathbb{I}_{\mathbb{C}^M} - V$ in $L^2(\mathbb{R},\mathbb{C}^M)$ if and only if $F(a)v = -\sqrt{E_1}v$. Therefore the dimension of the space of eigenfunctions corresponding to $-E_1$ coincides with the dimension of the space of $v \in \mathbb{C}^M$ with $F(a)v = -\sqrt{E_1}v$, as claimed. $\qquad\square$

Proof of Proposition 6.16   In view of (6.26), we have

$$\Phi'(x) = \left(\sqrt{E_1}\sinh(\sqrt{E_1}(x-a))\mathbb{I}_{\mathbb{C}^M} + \cosh(\sqrt{E_1}(x-a))F(a)\right)\Phi(a),$$

$$\Phi(x)^{-1} = \Phi(a)^{-1}\left(\cosh(\sqrt{E_1}(x-a))\mathbb{I}_{\mathbb{C}^M} + \frac{1}{\sqrt{E_1}}\sinh(\sqrt{E_1}(x-a))F(a)\right)^{-1}.$$

Thus,

$$F(x) = \left(\sqrt{E_1}\sinh(\sqrt{E_1}(x-a))\mathbb{I}_{\mathbb{C}^M} + \cosh(\sqrt{E_1}(x-a))F(a)\right)$$

$$\times \left(\cosh(\sqrt{E_1}(x-a))\mathbb{I}_{\mathbb{C}^M} + \frac{1}{\sqrt{E_1}}\sinh(\sqrt{E_1}(x-a))F(a)\right)^{-1}$$

$$= f(x, F(a)), \quad (6.27)$$

where we used the spectral theorem for the Hermitian $N \times N$ matrix $F(a)$ and set

$$f(x,\mu) := \sqrt{E_1} \frac{\sqrt{E_1}\tanh(\sqrt{E_1}(x-a)) + \mu}{\sqrt{E_1} + \mu\tanh(\sqrt{E_1}(x-a))}.$$

Let P be the spectral projection of $F(a)$ corresponding to the point $-\sqrt{E_1}$. Since $f(x, -\sqrt{E_1}) = -\sqrt{E_1}$, we have

$$F(x) = -\sqrt{E_1}P + f(x, F(a))(\mathbb{I}_{\mathbb{C}^M} - P).$$

Since $f(x,\mu) \to +\sqrt{E_1}$ as $x \to \infty$ for $\mu \neq -\sqrt{E_1}$, we obtain

$$\lim_{x\to\infty} F(x) = -\sqrt{E_1}P + \sqrt{E_1}(\mathbb{I}_{\mathbb{C}^M} - P),$$

as claimed. The formula for dim ran P follows from Lemma 6.18.

An explicit computation shows that $\frac{\partial}{\partial x}f(x,\mu)$ decays exponentially fast to zero as $x \to +\infty$, uniformly in bounded subsets of μ. Because of (6.27), this implies that $F'(x)$ tends exponentially fast to zero. This completes the proof. □

Remark 6.19 Combining the asymptotic behavior of F near $+\infty$ from Proposition 6.16 with its behavior near $-\infty$ given in (6.22) and its continuous differentiability, we obtain global information on F. Namely, we have $F \in L^\infty(\mathbb{R}, \mathcal{H}_M)$ and $F' \in L^1(\mathbb{R}, \mathcal{H}_M) \cap L^\infty(\mathbb{R}, \mathcal{H}_M)$. These two properties will be used in what follows.

6.3.3 The removal of the ground state

Let us introduce the operator

$$Q := \frac{d}{dx} \otimes \mathbb{I}_{\mathbb{C}^M} - F \quad \text{in } L^2(\mathbb{R}, \mathbb{C}^M)$$

with domain $H^1(\mathbb{R}, \mathbb{C}^M)$. Since F is bounded (Remark 6.19), this operator is closed and its adjoint is given by

$$Q^* = -\frac{d}{dx} \otimes \mathbb{I}_{\mathbb{C}^M} - F$$

with domain $H^1(\mathbb{R}, \mathbb{C}^M)$. Integrating by parts in the corresponding quadratic forms as in the proofs of (5.50) and (5.51) in the scalar case, we obtain

$$Q^*Q = -\frac{d^2}{dx^2} \otimes \mathbb{I}_{\mathbb{C}^M} + F' + F^2 = -\frac{d^2}{dx^2} \otimes \mathbb{I}_{\mathbb{C}^M} - V + E_1\mathbb{I}_{\mathbb{C}^M},$$

$$QQ^* = -\frac{d^2}{dx^2} \otimes \mathbb{I}_{\mathbb{C}^M} - F' + F^2 = -\frac{d^2}{dx^2} \otimes \mathbb{I}_{\mathbb{C}^M} - V - 2F' + E_1\mathbb{I}_{\mathbb{C}^M}.$$

In the last identity we used Riccati's equation from Lemma 6.15. Note that $QQ^* - E_1 \mathbb{I}_{\mathbb{C}^M}$ is again a Schrödinger operator with potential

$$V^{(1)} := V + 2F'.$$

This potential takes again values in the Hermitian matrices (Lemma 6.17), is continuous and belongs to $L^1(\mathbb{R}, \mathcal{H}_M) \cap L^\infty(\mathbb{R}, \mathcal{I}_M)$ (Remark 6.19).

Lemma 6.20   *We have* $\ker(QQ^*) = \{0\}$.

Proof   It suffices to show that $\ker Q^* = \{0\}$. Let $\varphi \in H^1(\mathbb{R}, \mathbb{C}^M)$ with $Q^*\varphi = 0$. Then $\varphi' = -F\varphi$ on $\mathbb{R}$. Since $F(x) = \sqrt{E_1}\mathbb{I}_{\mathbb{C}^M}$ for $x \le -a$, this means $\varphi(x) = e^{-\sqrt{E_1}(x+a)}\varphi(-a)$ for $x \le -a$. This function is not in L^2 unless $\varphi(a) = 0$, and thus, $\varphi(x) = 0$ for all $x \le -a$. By uniqueness of the Cauchy problem (see, e.g., a matrix version of Lemma 5.36), we conclude that $\varphi(x) = 0$ for all $x \in \mathbb{R}$.   □

We now write $-E_n(V)$, respectively $-E_n(V^{(1)})$, for the negative eigenvalues of $-\frac{d^2}{dx^2} \otimes \mathbb{I}_{\mathbb{C}^M} - V$, respectively $-\frac{d^2}{dx^2} \otimes \mathbb{I}_{\mathbb{C}^M} - V^{(1)}$, where these eigenvalues are arranged in non-decreasing order and repeated according to multiplicities. We denote by $N(V)$ and $N(V^{(1)})$ the respective numbers of negative eigenvalues. While it is not really needed in the proof that follows, we mention that $N(V)$ is finite. This follows from inequality (6.7) and from Proposition 4.18.

It follows from Proposition 1.23 that the non-zero spectra of Q^*Q and QQ^* coincide. Hence, the negative spectra of $-\frac{d^2}{dx^2} \otimes \mathbb{I}_{\mathbb{C}^M} - V$ and $-\frac{d^2}{dx^2} \otimes \mathbb{I}_{\mathbb{C}^M} - V^{(1)}$ coincide except for $-E_1 = -E_1(V)$. This number belongs to the spectrum of the former but, according to Lemma 6.20, not to the latter operator. We emphasize that, in the case of a K-fold degenerate eigenvalue $-E_1(V) = \cdots = -E_K(V)$, this method removes *all* these eigenvalues $-E_1(V), \ldots, -E_K(V)$ from the spectrum of $-\frac{d^2}{dx^2} \otimes \mathbb{I}_{\mathbb{C}^M} - V$. That is,

$$E_n(V^{(1)}) = E_{n+K}(V) \qquad \text{for all } n \le N(V_1) = N(V) - K$$

and, in particular, for all $K \le N \le N(V)$,

$$\sum_{n=1}^{N-K} E_n(V^{(1)})^{3/2} = \sum_{n=1}^{N} E_n(V)^{3/2} - \sum_{n=1}^{K} E_n(V)^{3/2} = \sum_{n=1}^{N} E_n(V)^{3/2} - KE_1(V)^{3/2}.$$

$$(6.28)$$

We now compare the integrals of $\mathrm{Tr}_{\mathbb{C}^M}(V)^2$ and $\mathrm{Tr}_{\mathbb{C}^M}(V^{(1)})^2$ with each other.

Note that both integrals exist (see Remark 6.19). We have

$$\int_{\mathbb{R}} \mathrm{Tr}_{\mathbb{C}^M} (V^{(1)})^2 \, dx = \int_{\mathbb{R}} \mathrm{Tr}_{\mathbb{C}^M} (V + 2F')^2 \, dx$$

$$= \int_{\mathbb{R}} \left(\mathrm{Tr}_{\mathbb{C}^M} V^2 + 4 \, \mathrm{Tr}_{\mathbb{C}^M} F'(V + F') \right) dx$$

$$= \int_{\mathbb{R}} \left(\mathrm{Tr}_{\mathbb{C}^M} V^2 + 4 \, \mathrm{Tr}_{\mathbb{C}^M} F'(E_1 - F^2) \right) dx$$

$$= \int_{\mathbb{R}} \mathrm{Tr}_{\mathbb{C}^M} V^2 \, dx + 4 \int_{\mathbb{R}} \frac{d}{dx} \mathrm{Tr}_{\mathbb{C}^M} \left(E_1 F - \tfrac{1}{3} F^3 \right) dx.$$

Here we used Riccati's equation, the cyclicity of the trace, and the identity

$$\frac{d}{dx} \mathrm{Tr}_{\mathbb{C}^M} F^3 = \mathrm{Tr}_{\mathbb{C}^M} \frac{d}{dx} F^3 = \mathrm{Tr}_{\mathbb{C}^M} (F'F^2 + FF'F + F^2F') = 3 \, \mathrm{Tr}_{\mathbb{C}^M} F'F^2.$$

To evaluate the integral of the derivative of $\mathrm{Tr} \left(E_1 F - \tfrac{1}{3} F^3 \right)$, we note that, by (6.22), we have, if $x \leq -a$,

$$\mathrm{Tr}_{\mathbb{C}^M} \left(E_1 F - \frac{1}{3} F^3 \right) = M \left(E_1^{\frac{3}{2}} - \tfrac{1}{3} E_1^{\frac{3}{2}} \right)$$

and, by Proposition 6.16,

$$\lim_{x \to \infty} \mathrm{Tr}_{\mathbb{C}^M} \left(E_1 F - \tfrac{1}{3} F^3 \right) = K \left(-E_1^{\frac{3}{2}} + \tfrac{1}{3} E_1^{\frac{3}{2}} \right) + (M - K) \left(E_1^{\frac{3}{2}} - \tfrac{1}{3} E_1^{\frac{3}{2}} \right).$$

This implies that

$$\int_{\mathbb{R}} \mathrm{Tr}_{\mathbb{C}^M} (V^{(1)})^2 \, dx = \int_{\mathbb{R}} \mathrm{Tr}_{\mathbb{C}^M} V^2 \, dx - \frac{16}{3} K E_1(V)^{\frac{3}{2}}.$$

Combining this identity with (6.28), we obtain for any $K \leq N \leq N(V)$,

$$\frac{3}{16} \int_{\mathbb{R}} \mathrm{Tr}_{\mathbb{C}^M} (V^{(1)})^2 dx - \sum_{n=1}^{N-K} E_n(V^{(1)})^{3/2} = \frac{3}{16} \int_{\mathbb{R}} \mathrm{Tr}_{\mathbb{C}^M} V^2 dx - \sum_{n=1}^{N} E_n(V)^{3/2}.$$
$$(6.29)$$

This identity is the main result of one step of the commutation procedure.

6.3.4 The proof of Theorem 6.12

We are now in position to prove the main result of this section.

Proof We fix $M \in \mathbb{N}$ and show that for all $N \in \mathbb{N}$ and all $V \in L^2(\mathbb{R}, \mathcal{H}_M)$,

$$\sum_{n=1}^{\min\{N, N(V)\}} E_n(V)^{3/2} \leq \frac{3}{16} \int_{\mathbb{R}} \mathrm{Tr}_{\mathbb{C}^M} \left(V(x)_+^2 \right) dx. \qquad (6.30)$$

Since N is arbitrary, once we have proved this, we obtain the assertion of the theorem. The assumption on V_- can be relaxed to $L^1_{\text{loc}}(\mathbb{R}, \mathcal{H}_M)$ by the variational principle.

We shall prove (6.30) by induction on N. For $N = 0$ the statement is trivial. Now assume that $N \geq 1$ and that (6.30) holds for all $V \in L^2(\mathbb{R}, \mathcal{H}_M)$ if N is replaced by an integer smaller than N. We aim at proving (6.30) for the given N. By an approximation argument given below in Lemma 6.21, this would follow if we could prove (6.30) for a dense set of potentials in $L^2(\mathbb{R}, \mathcal{H}_M)$, and therefore we may assume that V is a continuous function on $\mathbb{R}$ with values in $\mathcal{H}_M$ and has compact support. Moreover, by the variational principle, we may assume that $V(x) \geq 0$ for all $x \in \mathbb{R}$. Finally, we may assume that $N \leq N(V)$, for otherwise the inequality follows already from the induction hypothesis.

As before, let K be the multiplicity of the lowest eigenvalue of $-\frac{d^2}{dx^2} \otimes \mathbb{I}_{CM} - V$. We distinguish two cases according to whether $N \leq K$ or $N > K$. In the first case, we have, by (6.29) with $N = K$,

$$
\begin{aligned}
\sum_{n=1}^{N} E_n(V)^{3/2} &\leq \sum_{n=1}^{K} E_n(V)^{3/2} \\
&= \frac{3}{16} \int_{\mathbb{R}} \text{Tr}_{CM}\, V(x)^2\, dx - \frac{3}{16} \int_{\mathbb{R}} \text{Tr}_{CM}\, (V^{(1)}(x))^2\, dx \\
&\leq \frac{3}{16} \int_{\mathbb{R}} \text{Tr}_{CM}\, V(x)^2\, dx,
\end{aligned}
$$

where the last inequality comes from the fact that $(V^{(1)}(x))^2 \geq 0$ for all $x \in \mathbb{R}$. This proves the claim in the case $N \leq K$.

Now assume that $N > K$. Then, again by (6.29),

$$
\begin{aligned}
\sum_{n=1}^{N} E_n(V)^{3/2} &= \frac{3}{16} \int_{\mathbb{R}} \text{Tr}_{CM}\, V(x)^2\, dx + \sum_{n=1}^{N-K} E_n(V^{(1)})^{3/2} \\
&\quad - \frac{3}{16} \int_{\mathbb{R}} \text{Tr}_{CM}\, (V^{(1)}(x))^2\, dx.
\end{aligned}
$$

By the induction hypothesis we have

$$
\sum_{n=1}^{N-K} E_n(V^{(1)})^{3/2} \leq \frac{3}{16} \int_{\mathbb{R}} \text{Tr}_{CM}\, V^{(1)}(x)_+^2\, dx \leq \frac{3}{16} \int_{\mathbb{R}} \text{Tr}_{CM}\, (V^{(1)}(x))^2\, dx
$$

and, inserting this into the previous inequality, we obtain

$$
\sum_{n=1}^{N} E_n(V)^{3/2} \leq \frac{3}{16} \int_{\mathbb{R}} \text{Tr}_{CM}\, V(x)^2\, dx,
$$

which proves the claim in the case $N > K$. This completes the proof of
(6.30). $\square$

The proof in the matrix case is somewhat more complicated than in the scalar
case. One reason is that the compact support property of the potential is, in
general, not preserved under the commutation procedure. Therefore we have to
apply the approximation argument in each step of the induction procedure, in
contrast to the scalar case where the approximation argument is applied only
once at the end of the proof.

6.3.5 An approximation argument

In this subsection, we provide the details of an approximation argument that
was used in the proof of Theorem 6.12. Since this is a general argument, we
consider also the higher-dimensional case. We denote by $-E_m(-\Delta \otimes \mathbb{I}_{\mathbb{C}M} - V)$
the negative eigenvalues of $-\Delta \otimes \mathbb{I}_{\mathbb{C}M} - V$ arranged in non-decreasing order
and repeated according to multiplicities.

Lemma 6.21   *Let* $\gamma \geq 1/2$ *if* $d = 1$ *and* $\gamma > 0$ *if* $d \geq 2$. *Assume that there is
an* $N \in \mathbb{N}$, *a constant* $C < \infty$, *and a dense subspace* $\mathcal{D}$ *of* $L^{\gamma + \frac{d}{2}}(\mathbb{R}^d, \mathcal{H}_M)$ *such
that, for all* $V \in \mathcal{D}$,

$$\sum_{m=1}^{\min\{N, N(0, -\Delta\otimes\mathbb{I}_{\mathbb{C}M} - V)\}} E_m(-\Delta\otimes\mathbb{I}_{\mathbb{C}M} - V)^{\gamma} \leq C \int_{\mathbb{R}^d} \mathrm{Tr}_{\mathbb{C}M}\left(V(x)_+^{\gamma + \frac{d}{2}}\right) dx. \quad (6.31)$$

Then (6.31) *holds for any* $V \in L^{\gamma + \frac{d}{2}}(\mathbb{R}^d, \mathcal{H}_M)$.

Proof   Let $V \in L^{\gamma + \frac{d}{2}}(\mathbb{R}^d, \mathcal{H}_M)$. For any $0 < \delta < 1$ and $W \in \mathcal{D}$ we write the
quadratic form of $-\Delta\otimes\mathbb{I}_{\mathbb{C}M} - V$ as in (4.91) with $\alpha = 1$. If we apply Proposition
6.6 with the potential $\delta^{-1}(V - W)$, we estimate the second term as follows,

$$\delta\left(\|\nabla u\|^2 - \delta^{-1}\int_{\mathbb{R}^d}\langle(V - W)u, u\rangle\, dx\right) \geq -\delta^{-d/2\gamma}\left(L_{\gamma,d}^{(1)}\right)^{1/\gamma} I(V - W)^{1/\gamma}\|u\|^2$$

with a certain constant $L_{\gamma,d}^{(1)}$ and

$$I(V - W) := \int_{\mathbb{R}^d} \|(V(x) - W(x))_+\|^{\gamma + d/2}\, dx.$$

The variational principle in the form of Proposition 1.29 implies that

$$E_m(-\Delta \otimes \mathbb{I}_{\mathbb{C}M} - V) \leq (1 - \delta)\, E_m(-\Delta \otimes \mathbb{I}_{\mathbb{C}M} - (1 - \delta)^{-1}W)$$
$$+ \delta^{-d/2\gamma}\left(L_{\gamma,d}^{(1)}\right)^{1/\gamma} I(V - W)^{1/\gamma}. \quad (6.32)$$

Here we set $E_m(-\Delta \otimes \mathbb{I}_{\mathbb{C}M} - (1-\delta)^{-1}W) := 0$ if $m > N(0, -\Delta \otimes \mathbb{I}_{\mathbb{C}M} - (1-\delta)^{-1}W)$.

In order to proceed, we recall that for any $\gamma > 0$ and $\rho > 1$ there is a C_ρ such that

$$(a+b)^\gamma \leq \rho a^\gamma + C_\rho b^\gamma \qquad \text{for all } a, b \geq 0.$$

Indeed, if $\gamma \leq 1$ one can even choose $\rho = C_\rho = 1$. If $\gamma > 1$, then one can choose $C_\rho = \rho(\rho^{1/(\gamma-1)} - 1)^{1-\gamma}$.

If we raise (6.32) to the power γ, use the above inequality and sum in $m = 1, \ldots, \min\{N, N(0, -\Delta \otimes \mathbb{I}_{\mathbb{C}M} - V)\}$, we arrive at

$$\sum_{m=1}^{\min\{N, N(0, -\Delta \otimes \mathbb{I}_{\mathbb{C}M} - V)\}} E_m(-\Delta \otimes \mathbb{I}_{\mathbb{C}M} - V)^\gamma$$

$$\leq \rho(1-\delta)^\gamma \sum_{m=1}^{N} E_m(-\Delta \otimes \mathbb{I}_{\mathbb{C}M} - (1-\delta)^{-1}W)^\gamma$$

$$+ C_\rho N \delta^{-d/2} L_{\gamma,d}^{(1)} I(V - W).$$

We can now apply assumption (6.31) to $(1-\delta)^{-1}W \in \mathcal{D}$ to get

$$\sum_{m=1}^{\min\{N, N(0, -\Delta \otimes \mathbb{I}_{\mathbb{C}M} - V)\}} E_m(-\Delta \otimes \mathbb{I}_{\mathbb{C}M} - V)^\gamma$$

$$\leq \rho(1-\delta)^{-\frac{d}{2}} C \int_{\mathbb{R}^d} \mathrm{Tr}_{\mathbb{C}M}\left(W(x)_+^{\gamma+\frac{d}{2}}\right) dx + C_\rho N \delta^{-\frac{d}{2}} L_{\gamma,d}^{(1)} I(V - W).$$

We now choose a sequence $(W_n) \subset \mathcal{D}$ with $W_n \to V$ in $L^{\gamma+\frac{d}{2}}(\mathbb{R}^d, \mathcal{H}_M)$ and note that

$$\lim_{n \to \infty} \int_{\mathbb{R}^d} \mathrm{Tr}_{\mathbb{C}M}\left(W_n(x)_+^{\gamma+\frac{d}{2}}\right) dx = \int_{\mathbb{R}^d} \mathrm{Tr}_{\mathbb{C}M}\left(V(x)_+^{\gamma+\frac{d}{2}}\right) dx. \qquad (6.33)$$

Accepting this for the moment, we deduce that

$$\sum_{m=1}^{\min\{N, N(0, -\Delta \otimes \mathbb{I}_{\mathbb{C}M} - V)\}} E_m(-\Delta \otimes \mathbb{I}_{\mathbb{C}M} - V)^\gamma \leq \rho(1-\delta)^{-\frac{d}{2}} C \int_{\mathbb{R}^d} \mathrm{Tr}_{\mathbb{C}M}\left(V(x)_+^{\gamma+\frac{d}{2}}\right) dx.$$

Since $0 < \delta < 1$ and $\rho > 1$ are arbitrary, this proves the claimed inequality.

Let us prove (6.33). We use inequality (4.88), which implies that for any $\sigma > 1$ there is a C'_σ such that, for all $x \in \mathbb{R}^d$ and all $n \in \mathbb{N}$,

$$\mathrm{Tr}_{\mathbb{C}M}\left(W_n(x)_+^{\gamma+\frac{d}{2}}\right) \leq \sigma \, \mathrm{Tr}_{\mathbb{C}M}\left(V(x)_+^{\gamma+\frac{d}{2}}\right) + C'_\sigma \, \mathrm{Tr}_{\mathbb{C}M}\left((W_n(x) - V(x))_+^{\gamma+\frac{d}{2}}\right)$$

and

$$\mathrm{Tr}_{\mathbb{C}M}\left(V(x)_+^{\gamma+\frac{d}{2}}\right) \leq \sigma \, \mathrm{Tr}_{\mathbb{C}M}\left(W_n(x)_+^{\gamma+\frac{d}{2}}\right) + C'_\sigma \, \mathrm{Tr}_{\mathbb{C}M}\left((V(x) - W_n(x))_+^{\gamma+\frac{d}{2}}\right).$$

The last term on the right side of both inequalities does not exceed $C'_\sigma M \|W_n(x) - V(x)\|^{\gamma + \frac{d}{2}}$, and therefore we obtain, as $n \to \infty$,

$$\limsup_{n \to \infty} \int_{\mathbb{R}^d} \mathrm{Tr}_{\mathbb{C}^M} \left(W_n(x)_+^{\gamma + \frac{d}{2}} \right) dx \leq \sigma \int_{\mathbb{R}^d} \mathrm{Tr}_{\mathbb{C}^M} \left(V(x)_+^{\gamma + \frac{d}{2}} \right) dx$$

and

$$\liminf_{n \to \infty} \int_{\mathbb{R}^d} \mathrm{Tr}_{\mathbb{C}^M} \left(W_n(x)_+^{\gamma + \frac{d}{2}} \right) dx \geq \sigma^{-1} \int_{\mathbb{R}^d} \mathrm{Tr}_{\mathbb{C}^M} \left(V(x)_+^{\gamma + \frac{d}{2}} \right) dx .$$

Since $\sigma > 1$ is arbitrary we obtain (6.33). This concludes the proof. $\qquad \square$

6.4 Trace formulas in the matrix case

We recall that the proof of sharp Lieb–Thirring inequalities for $\gamma \geq 3/2$ in higher dimensions relies on a sharp Lieb–Thirring inequality for a one-dimensional Schrödinger operator with matrix-valued potential. In §6.3 we have presented a proof of the latter inequality by Benguria and Loss (2000) based on a commutation method. Here we briefly sketch the original argument of Laptev and Weidl (2000b) based on trace formulas. This is the matrix analogue of the argument in §5.5.

Let $V \in L^1(\mathbb{R}, \mathcal{H}_M)$ for some $M \in \mathbb{N}$. For any $k \in \overline{\mathbb{C}_+} \setminus \{0\}$ there is a unique $M \times M$ matrix-valued solution Ψ_k of

$$-\Psi_k'' - V\Psi_k = k^2 \Psi_k \qquad \text{in } \mathbb{R}$$

satisfying, as $x \to \infty$,

$$\Psi_k(x) = e^{ikx} \mathbb{I}_{\mathbb{C}^M} (1 + o(1)) \quad \text{and} \quad \Psi_k'(x) = ike^{ikx} \mathbb{I}_{\mathbb{C}^M} (1 + o(1)). \quad (6.34)$$

This can be shown by solving a matrix-valued Volterra equation as in Lemma 5.38. This argument also shows that for $k \in \mathbb{R} \setminus \{0\}$ there are $M \times M$ matrices $A(k)$ and $B(k)$ such that, as $x \to -\infty$,

$$\begin{aligned} \Psi_k(x) &= A(k)e^{ikx} + B(k)e^{-ikx} + o(1) \quad \text{and} \\ \Psi_k'(x) &= A(k)ike^{ikx} - B(k)ike^{-ikx} + o(1). \end{aligned} \qquad (6.35)$$

The following property will be crucial for our application.

Lemma 6.22 *For all $k \in \mathbb{R} \setminus \{0\}$,*

$$A(k)^* A(k) - B(k)^* B(k) = \mathbb{I}_{\mathbb{C}^M}$$

and, in particular, $|\det A(k)| \geq 1$.

Proof Note that for real k, using the fact that $V(x)$ is Hermitian for each $x \in \mathbb{R}$,

$$-(\Psi_k^*)'' - \Psi_k^* V = k^2 \Psi_k^* \qquad \text{in } \mathbb{R} .$$

Thus, setting

$$W_k := \Psi_k^* \Psi_k' - (\Psi_k^*)' \Psi_k ,$$

we find $W_k' = 0$; that is, W_k is constant. By (6.34),

$$W_k(x) = 2ik \, \mathbb{I}_{\mathbb{C}^M} + o(1) \qquad \text{as } x \to \infty$$

and, by (6.35),

$$W_k(x) = 2ik \, (A(k)^* A(k) - B(k)^* B(k)) + o(1) \qquad \text{as } x \to -\infty .$$

This, together with the constancy of W_k, implies the claimed identity.

Moreover,

$$|\det A(k)|^2 = \det A(k)^* A(k) = \det(\mathbb{I}_{\mathbb{C}^M} + B(k)^* B(k)) \geq \det \mathbb{I}_{\mathbb{C}^M} = 1 ,$$

as claimed. $\square$

The following theorem extends the trace formulas in Theorem 5.40 to the case of matrix-valued potentials.

Theorem 6.23 *Let $V \in L^1(\mathbb{R}, \mathcal{H}_M)$ and let $-E_n$ be the negative eigenvalues of $-\frac{d^2}{dx^2} \otimes \mathbb{I}_{\mathbb{C}^M} - V$ in $L^2(\mathbb{R}, \mathbb{C}^M)$, repeated according to multiplicities. Then*

$$2 \int_0^\infty \ln |\det A(k)| \, dk - 2\pi \sum_n E_n^{1/2} = -\frac{\pi}{2} \int_{\mathbb{R}} \mathrm{Tr}_{\mathbb{C}^M} \left(V(x) \right) dx .$$

Moroever if, in addition, $V \in L^2(\mathbb{R}, \mathcal{H}_M)$, then

$$2 \int_0^\infty k^2 \ln |\det A(k)| \, dk + \frac{2\pi}{3} \sum_n E_n^{3/2} = \frac{\pi}{8} \int_{\mathbb{R}} \mathrm{Tr}_{\mathbb{C}^M} \left(V(x)^2 \right) dx .$$

Note that by Lemma 6.22, $|\det A| \geq 1$, and therefore the integrands in the k-integrals are pointwise non-negative. This leads to the following bounds.

Corollary 6.24 *Let $V \in L^1(\mathbb{R}, \mathcal{H}_M)$ and let $-E_n$ be the negative eigenvalues of $-\frac{d^2}{dx^2} \otimes \mathbb{I}_{\mathbb{C}^M} - V$ in $L^2(\mathbb{R}, \mathbb{C}^M)$, repeated according to multiplicities. Then*

$$\sum_n E_n^{1/2} \geq \frac{1}{4} \int_{\mathbb{R}} \mathrm{Tr}_{\mathbb{C}^M} \left(V(x) \right) dx .$$

Moreover if, in addition, $V \in L^2(\mathbb{R}, \mathcal{H}_M)$, then

$$\sum_n E_n^{3/2} \leq \frac{3}{16} \int_{\mathbb{R}} \mathrm{Tr}_{\mathbb{C}^M} \left(V(x)^2 \right) dx .$$

Remark 6.25   Just like in the scalar case, the method used to prove Theorem 6.23 provides a countable family of trace formulas of the form

$$2 \int_0^\infty k^{2m} \ln |\det A(k)| \, dk + \frac{(-1)^{m+1}\pi}{m+1/2} \sum_n E_n^{m+1/2} = \int_{\mathbb{R}} \mathrm{Tr}_{\mathbb{C}M} \left(F_m(V)(x) \right) dx \, ,$$

where $m \in \mathbb{N}$ and where, for each $x \in \mathbb{R}$, $F_m(V)(x)$ is a polynomial in $V(x)$ and finitely many derivatives $V^{(\ell)}(x)$. For instance for $m = 2$, one can show that

$$2 \int_0^\infty k^4 \ln |\det A(k)| \, dk - \frac{2\pi}{5} \sum_n E_n^{5/2}$$

$$= -\frac{\pi}{16} \int_{\mathbb{R}} \mathrm{Tr}_{\mathbb{C}M} \left(V(x)^3 \right) dx + \frac{\pi}{32} \int_{\mathbb{R}} \mathrm{Tr}_{\mathbb{C}M} \left(V'(x)^2 \right) dx \, .$$

Theorem 6.23 is proved for $V \in C_0^\infty(\mathbb{R}, \mathcal{H}_M)$ in Laptev and Weidl (2000b). The result as stated follows by combining the methods in Laptev and Weidl (2000b) with the arguments in §5.5 in the scalar case. There are some complications, however, due to the matrix nature of the problem. One of the issues is that eigenvalues of $-\frac{d^2}{dx^2} \otimes \mathbb{I}_{\mathbb{C}M} - V$ need not be simple, unlike in the scalar case. Furthermore, the function $\det A$, which extends to an analytic function in $\mathbb{C}_+$ (in fact, A does), may have zeros that are not simple. The crucial observation, however, is that the multiplicity of the eigenvalues can be characterized by the order of the zeros. Thus, the following lemma replaces Lemma 5.45 in the scalar case.

Lemma 6.26   *All zeros of the function* $\det A$ *in* $\overline{\mathbb{C}_+} \setminus \{0\}$ *lie on the positive imaginary axis. Moreover,* $-i\kappa$ *is a zero of* $\det A$ *of order* K *if and only if* $-\kappa^2$ *is an eigenvalue of the operator* $-\frac{d^2}{dx^2} \otimes \mathbb{I}_{\mathbb{C}M} - V$ *in* $L^2(\mathbb{R}, \mathbb{C}^M)$ *of multiplicity* K.

The proof of this lemma uses properties of the Ψ_k and Sturm–Liouville theory for differential operators with matrix-valued coefficients, where the noncommutativity needs to be taken into account; see Laptev and Weidl (2000b, Eq. (1.31)) for $V \in C_0^\infty(\mathbb{R}, \mathcal{H}_M)$.

Let us denote by $\kappa_1 \geq \kappa_2 \geq \kappa_3 \geq \cdots$ the imaginary parts of the zeros of $\det A$ in $\mathbb{C}_+$, repeated according to multiplicities. It follows from the rough bound (6.9) that the sequence (κ_n) is summable, and therefore we can form the corresponding Blaschke product B. By analyzing the analytic function $(\det A)/B$ in $\mathbb{C}_+$, we can prove the analogue of Proposition 5.47. Namely, $(1 + t^2)^{-1} \ln |\det A| \in L^1(\mathbb{R})$ and, for all $k \in \mathbb{C}_+$ with $k \neq \kappa_n$ for all n,

$$\ln |\det A(k)| = \frac{\mathrm{Im}\, k}{\pi} \int_{\mathbb{R}} \frac{\ln |\det A(t)|}{|t - k|^2} \, dt + \sum_n \ln \left| \frac{k - i\kappa_n}{k + i\kappa_n} \right| ; \qquad (6.36)$$

see Laptev and Weidl (2000b, Eq. (1.57)) for $V \in C_0^\infty(\mathbb{R}, \mathcal{H}_M)$.

Next, we show that, as $|k| \to \infty$, $\ln |\det A(k)|$ has asymptotic behavior similar to that in Lemma 5.49. The argument here is slightly more complicated than in the scalar case because of non-commutativity. Indeed, by iterating the Volterra equation for $e^{-ikx}\Psi_k$, we can compute the asymptotic behavior of $A(k)$ as in the proof of Lemma 5.49. However, we cannot symmetrize to simplify the expressions for I_j, $j = 2, 3$, as in the scalar case, and therefore the expression for $\log A(k)$ is more complicated. The crucial observation is that, because of the trace, for $\mathrm{Tr}_{\mathbb{C}^M} \log A(k) = \log \det A(k)$ we find the same simplification as in the scalar case. The remaining part of the proof of Lemma 5.49 then goes through as before. We note that the proof of the asymptotic behavior of $\ln |\det A(k)|$ in Laptev and Weidl (2000b) (see, in particular, Eqs. (1.49) and (1.54) therein) for $V \in C_0^\infty(\mathbb{R}, \mathcal{H}_M)$ is different and makes use of the regularity of V.

Combining the asymptotic behavior of $\ln |\det A(k)|$ as $|k| \to \infty$ with formula (6.36), Theorem 6.23 follows by arguments similar to those in the scalar case. This concludes our sketch of the proof of Theorem 6.23.

6.5 Comments

Section 6.1: Schrödinger operators with matrix-valued potentials

Schrödinger operators with operator-valued potentials appear in different contexts. In particular, they arise as technical tools in connection with Weyl (or non-Weyl) asymptotics in Vulis and Solomjak (1974), Laptev (1981, 1993), and Simon (1983), where they play a role in an 'asymptotic separation of variables'.

The material in §6.1 is a straightforward extension of the scalar case. Instead of matrix-valued potentials one could also consider the case of operator-valued potentials, acting on possibly infinite-dimensional Hilbert spaces. We omit this to avoid technicalities.

Proposition 6.8 concerning the existence of negative eigenvalues in dimensions $d = 1, 2$ in the matrix case is from Weidl (1999b), where a sharper result also appears.

Concerning the Lieb–Thirring inequality (6.12), we state the following result.

Theorem 6.27   *Let*

$$\gamma \geq 1/2 \quad \text{if } d = 1,$$
$$\gamma > 0 \quad \text{if } d = 2,$$
$$\gamma \geq 0 \quad \text{if } d \geq 3.$$

Then there is a constant $L_{\gamma,d}^{(\mathrm{mat})}$ such that, for all $M \in \mathbb{N}$, $V \in L_{\mathrm{loc}}^1(\mathbb{R}, \mathcal{H}_M)$ with

$V_+ \in L^{\gamma+d/2}(\mathbb{R}, \mathcal{H}_M)$,

$$\mathrm{Tr}\,(-\Delta \otimes \mathbb{I}_{\mathbb{C}^M} - V)_-^\gamma \le L_{\gamma,d}^{(\mathrm{mat})} \int_{\mathbb{R}^d} \mathrm{Tr}_{\mathbb{C}^M}\left(V(x)_+^{\gamma+d/2}\right) dx\,.$$

In particular, we see that the Lieb–Thirring inequality (6.12) with matrix-valued potentials is valid whenever the Lieb–Thirring inequality is valid with scalar potentials.

In this book, we will prove Theorem 6.27 only in the restricted parameter range $\gamma \ge 1/2$ for all d; see Theorem 8.8 and Lemma 8.4. In the range $\gamma > 0$ for $d \ge 2$ the original proof strategy of Lieb and Thirring (1976) is applicable. The first proof for $\gamma = 0$ and $d \ge 3$ is due to Hundertmark (2002).

Having settled the question about the validity of the matrix Lieb–Thirring inequality (6.12), it is natural to ask about its sharp constants. This will be further discussed in Chapter 8.

Section 6.2: The Lieb–Thirring inequality with the semiclassical constant
Theorem 6.11 and its proof via Theorem 6.12 are due to Laptev and Weidl (2000b); see also Laptev and Weidl (2000a).

As mentioned before, Schrödinger operators with operator-valued potentials appear, in particular, in connection with Weyl (or non-Weyl) asymptotics. Simon (1983) showed that, for the heat kernel, the asymptotics are accompanied by a universal, non-asymptotic inequality coming from splitting off dimensions. An inequality that follows easily from Simon (1983) and the techniques used therein is that, for $W : \mathbb{R}^d \to \mathcal{H}_M$,

$$\mathrm{Tr}_{L^2(\mathbb{R}^d,\mathbb{C}^M)}\exp(-t(-\Delta + W)) \le \iint_{\mathbb{R}^d \times \mathbb{R}^d} \mathrm{Tr}_{\mathbb{C}^M}\, e^{-t(|\xi|^2 + W(x))}\,\frac{dx\,d\xi}{(2\pi)^d}\,;$$

this follows by taking the trace in Simon (1983, Theorem 2.4) and applying the standard Golden–Thompson inequality. Simon refers to inequalities of this type as "sliced bread inequalities". For comparison, we note that Theorem 6.12 can be written as

$$\mathrm{Tr}_{L^2(\mathbb{R},\mathbb{C}^M)}(-\tfrac{d^2}{dx^2} - V)_-^{3/2} \le \iint_{\mathbb{R} \times \mathbb{R}} \mathrm{Tr}_{\mathbb{C}^M}\left(|\xi|^2 - V(x)\right)_-^{3/2}\frac{dx\,d\xi}{2\pi}\,.$$

Despite the similarities, the proofs are rather different and, in particular, Theorem 6.12 depends much more on special features of one-dimensional Schrödinger operators.

Apart from the one-dimensional inequality in Theorem 6.12, an important idea in Laptev and Weidl (2000b) is the strategy of lifting in dimension. Note that instead of splitting off one dimension, as in Lemma 6.13, we can also split off several at once. This is used in Hundertmark (2002), for instance.

The techniques in this section give, in fact, a stronger result than stated in Theorem 6.11, namely that the Lieb–Thirring inequality for $\gamma \geq 3/2$ holds with semiclassical constant even for matrix-valued potentials. This appears later in Theorem 8.1.

The lifting method has also been used for discrete Schrödinger operators in Hundertmark and Simon (2002).

In Chapter 8 we will use the method of lifting in dimension again to obtain the current best values in Lieb–Thirring inequalities in situations where optimal constants are not known. Let us mention here some other applications of this method, which in many cases yields optimal constants:

- Lieb–Thirring inequalities for magnetic Schrödinger operators (Laptev and Weidl, 2000b; Barseghyan et al., 2016);
- Remainder terms in Berezin–Li–Yau inequalities for eigenvalues of the Dirichlet Laplacian on a domain (Weidl, 2008; Geisinger et al., 2011; Larson, 2017);
- Eigenvalue bounds for Dirichlet Laplacians in domains of infinite volume (Geisinger and Weidl, 2011);
- Bounds on eigenvalue moments for Schrödinger operators in an asymptotically straight waveguide (Exner et al., 2004);
- Lieb–Thirring bounds for potentials supported on hyperplanes (Frank and Laptev, 2008).

Section 6.3: The sharp bound in the matrix-valued case for $\gamma = \frac{3}{2}$

The original proof of Theorem 6.12 appears in Laptev and Weidl (2000b) and is presented in §6.4. Soon afterwards, Benguria and Loss (2000) found the alternative proof presented in §6.3. For references on the commutation methods see the comments on §5.4. Benguria and Loss seem to be the first who applied this method in the matrix-valued case.

The reversed Lieb–Thirring inequality in Corollary 6.24 for $\gamma = \frac{1}{2}$ can also be proved using the commutation method in §6.3; see §5.4 for the scalar case.

Section 6.4: Trace formulas in the matrix case

For references on trace formulas in the scalar case see the comments on §5.5. Laptev and Weidl seem to be the first who applied trace formulas in the matrix-valued case.

7

More on Sharp Lieb–Thirring Inequalities

In this chapter, we collect several additional results concerning the optimal constants in Lieb–Thirring inequalities. Some of these results are of a positive nature, in that they give alternative proofs of sharp inequalities or sharp inequalities for restricted classes of potentials. Other results are of a negative nature, in that they provide counterexamples to earlier conjectures about the sharp constants. The different sections of this chapter are independent of each other.

In §7.1 we present a theorem of Stubbe (2010) that shows that the function $\alpha \mapsto \alpha^{-2-d/2} \operatorname{Tr}(-\Delta - \alpha V)^2_-$ is non-decreasing. Since, by Weyl asymptotics (Theorem 4.29), its limit as $\alpha \to \infty$ is $L^{cl}_{2,d} \int_{\mathbb{R}^d} V(x)^{2+d/2}_+ dx$, we obtain another proof of the sharp Lieb–Thirring inequality

$$\operatorname{Tr}(-\Delta - V)^\gamma_- \leq L^{cl}_{\gamma,d} \int_{\mathbb{R}^d} V(x)^{\gamma+d/2}_+ dx$$

for $\gamma = 2$. The inequality for $\gamma > 2$ then follows by the Aizenman–Lieb argument in Lemma 5.2. While the range $\gamma \geq 2$ is smaller than the range $\gamma \geq 3/2$ in Theorem 6.11, the monotonicity with respect to the coupling constant is interesting in its own right.

In §7.2 we study Cwikel–Lieb–Rozenblum inequalities for radial V following Glaser, Grosse and Martin (1978). In particular, as an application of the sharp Lieb–Thirring inequality in $d = 1$ with $\gamma = 3/2$, one obtains the optimal constant in the CLR inequality for radial V in dimension $d = 4$; see Theorem 7.6. Another result from their paper is that the original conjecture of Lieb and Thirring (1976) does not hold for $\gamma = 0$ in dimensions $d \geq 7$; see Proposition 7.10.

Section 7.3 is also concerned with counterexamples to the Lieb–Thirring conjecture. Namely, we show that $L_{\gamma,d} > L^{(1)}_{\gamma,d}$ if $\gamma > \max\{0, 2 - \frac{d}{2}\}$. We only sketch the overall strategy of the proof, which relies on an exponentially small attraction between two distant parts of a potential, and refer for details to Frank et al. (2021b).

In §7.4 we present an equivalent formulation of the Lieb–Thirring inequality for $\gamma = 1$. This so-called kinetic (or dual) form was, in fact, the one that was used in the proof of stability of matter by Lieb and Thirring (1975). The kinetic form is of fundamental importance in the analysis of fermionic quantum many-body systems and questions related to the validity of density functional theories. Additionally, it has proved useful in connection with the dimension of attractors for the Navier–Stokes and other non-linear evolution equations and has served as the blueprint for extensions of other inequalities in harmonic analysis to orthonormal systems; see the references in §7.5. Remarkably, the recent advances on the optimal constants $L_{1,d}$, including the proof of the current best bound, which we present in the next chapter, are based on this dual formulation.

7.1 Monotonicity with respect to the semiclassical parameter

Let $\hbar > 0$, let V be as in Proposition 4.1 satisfying (4.5) for some $\theta < \hbar^2$ and some $C < \infty$, and consider the Schrödinger operator

$$H(\hbar) = -\hbar^2 \Delta - V \quad \text{in} \quad L^2(\mathbb{R}^d).$$

We denote its negative eigenvalues, in non-decreasing order and repeated with multiplicities, by $-E_j(\hbar)$, $j \geq 1$. By Weyl asymptotics (Theorem 4.46), if $V \in L^{\gamma + \frac{d}{2}}(\mathbb{R}^d)$ the function

$$(0, +\infty) \ni \hbar \mapsto \hbar^d \operatorname{Tr} H(\hbar)_-^\gamma = \hbar^d \sum_j E_j(\hbar)^\gamma$$

satisfies

$$\lim_{\hbar \to 0+} \hbar^d \operatorname{Tr} H(\hbar)_-^\gamma = L_{\gamma,d}^{\mathrm{cl}} \int_{\mathbb{R}^d} V_+^{\gamma + \frac{d}{2}} \, dx . \tag{7.1}$$

The following theorem of Stubbe (2010) says that, for $\gamma \geq 2$, this function is non-increasing in $\hbar$.

Theorem 7.1 *Let $d \geq 1$, $\gamma \geq 2$ and $V \in L_{\mathrm{loc}}^1(\mathbb{R}^d)$ such that, for any $\theta > 0$, there is $C < \infty$ such that (4.5) holds. Assume that the negative spectrum of the operators $-\hbar^2\Delta - V$ is discrete for all $\hbar > 0$. Then*

$$(0, +\infty) \ni \hbar \mapsto \hbar^d \operatorname{Tr} H(\hbar)_-^\gamma = \hbar^d \sum_j E_j(\hbar)^\gamma$$

is non-increasing.

The assumption $\gamma \geq 2$ in Theorem 7.1 is optimal. Indeed, let $V(x) = -|x|^2 + 1$, so that by Proposition 4.6,

$$\hbar^d \operatorname{Tr} H(\hbar)_-^\gamma = \hbar^d \left((1 - d\hbar)^\gamma + d(1 - (d + 2)\hbar)^\gamma\right) \quad \text{if } (d+4)^{-1} \leq \hbar \leq (d+2)^{-1}.$$

An explicit computation shows that the left-sided derivative at $\hbar = (d + 2)^{-1}$ is positive for $\gamma < 2$. For $\gamma = 2$ this derivative vanishes at $\hbar = (d + 2)^{-1}$.

Theorem 7.1, combined with the asymptotics (7.1), gives an alternative proof of the sharp Lieb–Thirring inequality in Theorem 6.11 in the range $\gamma \geq 2$. Note that assumptions of Theorem 7.1 are satisfied if $V_+ \in L^{\gamma + \frac{d}{2}}(\mathbb{R}^d)$ according to Propostions 4.3 and 4.14.

In order to prove Theorem 7.1, we need the following auxiliary result from abstract operator theory concerning commutators.

Lemma 7.2 *Let H and G be self-adjoint operators in a Hilbert space $\mathcal{H}$ with domains* dom H *and* dom G *and assume that G dom $H \subset$ dom $H \subset$ dom G. Moreover, assume that the negative spectral subspace of the operator H is finite-dimensional and let (φ_j) be an orthonormal basis of this subspace consisting of eigenfunctions. Let $(-E_j)$ be the corresponding eigenvalues. Then*

$$\sum_j E_j^2 \left([G, [H, G]]\varphi_j, \varphi_j\right) \leq 2 \sum_j E_j \left\|[H, G]\varphi_j\right\|^2.$$

The expression on the left side of the inequality in the lemma is slightly informal. Before explaining its precise meaning, let us explain that of the right side. The operator $[H, G]$ is defined to be $HG - GH$ with domain

$$\{\varphi \in \operatorname{dom} G \cap \operatorname{dom} H : G\varphi \in \operatorname{dom} H \text{ and } H\varphi \in \operatorname{dom} G\}.$$

According to the assumptions of the lemma, an eigenvector φ_j of H belongs to this domain and we have

$$[H, G]\varphi_j = (H + E_j)G\varphi_j. \tag{7.2}$$

In general, we do *not* understand the expression $[G, [H, G]]$ on the left side as an operator, but rather as the corresponding quadratic form. Namely, we define, for any vector φ in the domain of $[H, G]$,

$$([G, [H, G]]\varphi, \varphi) = ([H, G]\varphi, G\varphi) + (G\varphi, [H, G]\varphi).$$

In particular, for an eigenvector φ_j of H we have, by (7.2),

$$([G, [H, G]]\varphi_j, \varphi_j) = 2\left((H + E_j)G\varphi_j, G\varphi_j\right). \tag{7.3}$$

Proof We denote by $-\lambda_n$ the negative eigenvalues of H *without* repetitions. Let P_λ be the spectral measure of H and, for any n, write

$$d\mu_n(\lambda) := \sum_{j:\, E_j = \lambda_n} d(P_\lambda G\varphi_j, G\varphi_j).$$

Then we obtain from (7.3) that, for each n,

$$\sum_{j:\, E_j = \lambda_n} ([G, [H, G]]\varphi_j, \varphi_j) = 2 \int_{\mathbb{R}} (\lambda + \lambda_n) \, d\mu_n(\lambda)$$

$$= 2 \sum_m (\lambda_n - \lambda_m) w_{n,m} + 2 \int_{[0,\infty)} (\lambda + \lambda_n) \, d\mu_n(\lambda),$$

with $w_{n,m} := \mu_n(\{-\lambda_m\})$. Thus

$$\sum_j E_j^2 \left([G, [H, G]]\varphi_j, \varphi_j\right)$$

$$= 2 \sum_{n,m} \lambda_n^2 (\lambda_n - \lambda_m) w_{n,m} + 2 \sum_n \lambda_n^2 \int_{[0,\infty)} (\lambda + \lambda_n) \, d\mu_n(\lambda).$$

Note that

$$w_{n,m} = \mu_n(\{-\lambda_m\})$$

$$= \sum_{j,k:\, E_j = \lambda_n,\, E_k = \lambda_m} |(G\varphi_j, \varphi_k)|^2 = \sum_{j,k:\, E_j = \lambda_n,\, E_k = \lambda_m} |(\varphi_j, G\varphi_k)|^2$$

$$= \mu_m(\{-\lambda_n\}) = w_{m,n}.$$

Using this symmetry twice,

$$2 \sum_{n,m} \lambda_n^2 (\lambda_n - \lambda_m) w_{n,m} = \sum_{n,m} (\lambda_n^2 - \lambda_m^2)(\lambda_n - \lambda_m) w_{n,m}$$

$$= \sum_{n,m} (\lambda_n + \lambda_m)(\lambda_n - \lambda_m)^2 w_{n,m}$$

$$= 2 \sum_{n,m} \lambda_n (\lambda_n - \lambda_m)^2 w_{n,m}.$$

Thus, we have shown that

$$\sum_j E_j^2 \left([G, [H, G]]\varphi_j, \varphi_j\right)$$

$$= 2 \sum_{n,m} \lambda_n (\lambda_n - \lambda_m)^2 w_{n,m} + 2 \sum_n \lambda_n^2 \int_{[0,\infty)} (\lambda + \lambda_n) \, d\mu_n(\lambda).$$

Meanwhile, by (7.2), for all n,

$$\sum_{j: E_j = \lambda_n} \left\| [H, G] \varphi_j \right\|^2 = \int_{\mathbb{R}} (\lambda + \lambda_n)^2 \, d\mu_n(\lambda)$$

$$= \sum_m (\lambda_n - \lambda_m)^2 w_{n,m} + \int_{[0,\infty)} (\lambda + \lambda_n)^2 \, d\mu_n(\lambda).$$

Thus

$$\sum_j E_j \left\| [H, G] \varphi_j \right\|^2 = \sum_{n,m} \lambda_n (\lambda_n - \lambda_m)^2 w_{n,m} + \sum_n \lambda_n \int_{[0,\infty)} (\lambda + \lambda_n)^2 \, d\mu_n(\lambda).$$

To summarize, we have

$$\sum_j E_j^2 \left([G, [H, G]] \varphi_j, \varphi_j \right) - 2 \sum_j E_j \left\| [H, G] \varphi_j \right\|^2$$

$$= 2 \sum_n \lambda_n^2 \int_{[0,\infty)} (\lambda + \lambda_n) \, d\mu_n(\lambda) - 2 \sum_n \lambda_n \int_{[0,\infty)} (\lambda + \lambda_n)^2 \, d\mu_n(\lambda)$$

$$= -2 \sum_n \lambda_n \int_{[0,\infty)} \lambda (\lambda + \lambda_n) \, d\mu_n(\lambda) \le 0,$$

where the inequality at the end comes from $\lambda \ge 0$ for λ in the domain of integration and $\lambda_n \ge 0$. This proves the claimed bound. □

Lemma 7.3 *Let* $\hbar > 0$, *let* $V \in L^1_{\mathrm{loc}}(\mathbb{R}^d)$ *be such that* (4.5) *holds for some* $\theta < \hbar^2$ *and* $C < \infty$, *and assume that the negative spectrum of the operator* $-\hbar^2 \Delta - V$ *in* $L^2(\mathbb{R}^d)$ *is discrete. Let* $(-E_j(\hbar))$ *be its negative eigenvalues, repeated according to multiplicities, and* (φ_j) *the corresponding orthonormal eigenfunctions. Then, for all* $\tau > 0$,

$$\sum_j \left(E_j(\hbar) - \tau \right)_+^2 \le \frac{4}{d} \hbar^2 \sum_j \left(E_j(\hbar) - \tau \right)_+ \int_{\mathbb{R}^d} \left| \nabla \varphi_j \right|^2 \, dx.$$

Heuristically, this lemma follows by using Lemma 7.2, with $H = -\hbar^2 \Delta - V + \tau$ and G as multiplication by x_k, and then summing with respect to $k = 1, \ldots, d$.

Proof Fix $k \in \{1, \ldots, d\}$. For $\varepsilon > 0$, let G_ε be the operator of multiplication by $g_\varepsilon(x) := x_k / (1 + \varepsilon^2 x_k^2)^{1/2}$. For later purposes we record the derivatives:

$$\frac{\partial g_\varepsilon}{\partial x_k} = \frac{1}{(1 + \varepsilon^2 x_k^2)^{3/2}} \quad \text{and} \quad \frac{\partial^2 g_\varepsilon}{\partial x_k^2} = \frac{-3\varepsilon^2 x_k}{(1 + \varepsilon^2 x_k^2)^{5/2}} . \tag{7.4}$$

Since g_ε is a bounded function, we have dom $H(\hbar) \subset$ dom $G_\varepsilon = L^2(\mathbb{R}^d)$. Let us prove that $G_\varepsilon \operatorname{dom} H(\hbar) \subset \operatorname{dom} H(\hbar)$. We recall that, by definition of $H(\hbar)$ via its quadratic form,

$$\operatorname{dom} H(\hbar)$$

$$= \left\{ \psi \in H^1(\mathbb{R}^d) \cap L^2(\mathbb{R}^d, V_- dx) : \text{there is an } f \in L^2(\mathbb{R}^d) \text{ such that, for all} \right.$$

$$\left. \varphi \in H^1(\mathbb{R}^d) \cap L^2(\mathbb{R}^d, V_- dx), \int_{\mathbb{R}^d} \left(\hbar^2 \nabla \psi \cdot \nabla \overline{\varphi} + V \psi \overline{\varphi} \right) dx = \int_{\mathbb{R}^d} f \overline{\varphi} \, dx \right\}.$$

For $\psi \in \operatorname{dom} H(\hbar)$, the f on the right side is unique and satisfies $f = H(\hbar)\psi$. We have to prove that $G_\varepsilon \psi \in \operatorname{dom} H(\hbar)$ for $\psi \in \operatorname{dom} H(\hbar)$. Since g_ε and $|\nabla g_\varepsilon|$ are bounded, we see that $g_\varepsilon \psi \in H^1(\mathbb{R}^d) \cap L^2(\mathbb{R}^d, V_- dx)$. Set $f := H(\hbar)\psi \in L^2(\mathbb{R}^d)$ and write

$$u_\varepsilon := -\hbar^2 \psi \Delta g_\varepsilon - 2\hbar^2 \nabla \psi \cdot \nabla g_\varepsilon + f g_\varepsilon \,.$$

Since g_ε, $|\nabla g_\varepsilon|$ and Δg_ε are bounded, and ψ and $|\nabla \psi|$ are in $L^2(\mathbb{R}^d)$, we conclude that $u_\varepsilon \in L^2(\mathbb{R}^d)$. Moreover, for all $\varphi \in H^1(\mathbb{R}^d) \cap L^2(\mathbb{R}^d, V_- dx)$,

$$\int_{\mathbb{R}^d} \left(\hbar^2 \nabla (g_\varepsilon \psi) \cdot \nabla \overline{\varphi} + V g_\varepsilon \psi \overline{\varphi} \right) dx$$

$$= \int_{\mathbb{R}^d} \left(\hbar^2 g_\varepsilon \nabla \psi \cdot \nabla \overline{\varphi} + V g_\varepsilon \psi \overline{\varphi} \right) dx + \int_{\mathbb{R}^d} \hbar^2 \psi \nabla g_\varepsilon \cdot \nabla \overline{\varphi} \, dx$$

$$= \int_{\mathbb{R}^d} \left(\hbar^2 \nabla \psi \cdot \nabla \overline{(g_\varepsilon \varphi)} + V \psi \overline{g_\varepsilon \varphi} \right) dx - \hbar^2 \int_{\mathbb{R}^d} \left(\overline{\varphi} \nabla \psi \cdot \nabla g_\varepsilon - \psi \nabla g_\varepsilon \cdot \nabla \overline{\varphi} \right) dx$$

$$= \int_{\mathbb{R}^d} f g_\varepsilon \overline{\varphi} \, dx - \hbar^2 \int_{\mathbb{R}^d} \left(2 \overline{\varphi} \nabla \psi \cdot \nabla g_\varepsilon + \psi \Delta g_\varepsilon \overline{\varphi} \right) dx$$

$$= \int_{\mathbb{R}^d} u_\varepsilon \overline{\varphi} \, dx \,.$$

Thus we have shown that $g_\varepsilon \psi \in \operatorname{dom} H(\hbar)$.

Let $\tau > 0$ and $H_\tau := H(\hbar) + \tau$. By assumption, the negative spectral subspace of H_τ is finite-dimensional. Moreover, $\operatorname{dom} H_\tau = \operatorname{dom} H(\hbar)$. Thus we can apply Lemma 7.2 to the operators H_τ and G_ε and obtain, using $[H_\tau, G_\varepsilon] = [H(\hbar), G_\varepsilon]$,

$$\sum_j (E_j(\hbar) - \tau)_+^2 \left([G_\varepsilon, [H(\hbar), G_\varepsilon]] \varphi_j, \varphi_j \right) \le 2 \sum_j (E_j(\hbar) - \tau)_+ \left\| [H(\hbar), G_\varepsilon] \varphi_j \right\|^2 .$$

$$(7.5)$$

Let us compute the expressions appearing in this formula. We claim that

$\operatorname{dom}[H(\hbar), G_\varepsilon] = \operatorname{dom} H(\hbar)$ and that, for all φ from this domain,

$$[H(\hbar), G_\varepsilon]\varphi = -2\hbar^2 \frac{1}{(1 + \varepsilon^2 x_k^2)^{3/2}} \frac{\partial}{\partial x_k}\varphi + 3\hbar^2 \frac{\varepsilon^2 x_k}{(1 + \varepsilon^2 x_k^2)^{5/2}}\varphi, \quad (7.6)$$

$$[G_\varepsilon, [H(\hbar), G_\varepsilon]]\varphi = \frac{2\hbar^2}{(1 + \varepsilon^2 x_k^2)^3}\varphi. \quad (7.7)$$

Indeed, since G_ε is bounded and $G_\varepsilon \operatorname{dom} H(\hbar) \subset \operatorname{dom} H(\hbar)$, we have $\operatorname{dom}[H(\hbar), G_\varepsilon] = \operatorname{dom} H(\hbar)$. We note that, as a consequence of this, we can consider $[G_\varepsilon, [H(\hbar), G_\varepsilon]]$ as an operator (and not only in the sense explained before the proof of Lemma 7.2). Its domain, which by definition equals

$$\{\varphi \in \operatorname{dom} G_\varepsilon \cap \operatorname{dom}[H(\hbar), G_\varepsilon] : [H(\hbar), G_\varepsilon]\varphi \in \operatorname{dom} G_\varepsilon,$$
$$G_\varepsilon \varphi \in \operatorname{dom}[H(\hbar), G_\varepsilon]\},$$

coincides with $\operatorname{dom}[H(\hbar), G_\varepsilon] = \operatorname{dom} H(\hbar)$.

For any $\varphi, f \in \operatorname{dom} H(\hbar)$, we have

$$([H(\hbar), G_\varepsilon]\varphi, f) = (H(\hbar)G_\varepsilon\varphi, f) - (H(\hbar)\varphi, G_\varepsilon f)$$
$$= \hbar^2 \int_{\mathbb{R}^d} \left(\nabla(g_\varepsilon\varphi) \cdot \nabla\overline{f} - \nabla\varphi \cdot \nabla(g_\varepsilon\overline{f})\right) dx$$
$$= \hbar^2 \int_{\mathbb{R}^d} \nabla g_\varepsilon \cdot \left(\varphi\nabla\overline{f} - \overline{f}\nabla\varphi\right) dx$$
$$= -\hbar^2 \int_{\mathbb{R}^d} (\nabla \cdot (\varphi\nabla g_\varepsilon) + \nabla g_\varepsilon \cdot \nabla\varphi)\overline{f}\, dx$$
$$= -\hbar^2 \int_{\mathbb{R}^d} (\varphi\Delta g_\varepsilon + 2\nabla g_\varepsilon \cdot \nabla\varphi)\overline{f}\, dx$$
$$= -\hbar^2 \int_{\mathbb{R}^d} \left(\varphi\frac{\partial^2 g_\varepsilon}{\partial x_k^2} + 2\frac{\partial g_\varepsilon}{\partial x_k}\frac{\partial\varphi}{\partial x_k}\right)\overline{f}\, dx.$$

Here we used the product rule and the fact that φ and f belong to the form domain of $H(\hbar)$, which is $H^1(\mathbb{R}^d) \cap L^2(\mathbb{R}^d, V_-dx)$. Since the domain of $H(\hbar)$ is dense, this shows that for all $\varphi \in \operatorname{dom}[H(\hbar), G_\varepsilon]$,

$$[H(\hbar), G_\varepsilon]\varphi = -\hbar^2 \frac{\partial^2 g_\varepsilon}{\partial x_k^2}\varphi - 2\hbar^2 \frac{\partial g_\varepsilon}{\partial x_k}\frac{\partial\varphi}{\partial x_k}.$$

This, together with (7.4), implies the claimed formula (7.6).

For the proof of (7.7) we use the above-mentioned fact that, if $\varphi \in \operatorname{dom} H(\hbar)$, then $g_\varepsilon \varphi \in \operatorname{dom} H(\hbar)$. Thus

$$[G_\varepsilon, [H(\hbar), G_\varepsilon]]\, \varphi$$
$$= g_\varepsilon [H(\hbar), G_\varepsilon] \varphi - [H(\hbar), G_\varepsilon](g_\varepsilon \varphi)$$
$$= -\hbar^2 \left(g_\varepsilon \left(\frac{\partial^2 g_\varepsilon}{\partial x_k^2} \varphi + 2 \frac{\partial g_\varepsilon}{\partial x_k} \frac{\partial \varphi}{\partial x_k} \right) - \left(\frac{\partial^2 g_\varepsilon}{\partial x_k^2} g_\varepsilon \varphi + 2 \frac{\partial g_\varepsilon}{\partial x_k} \frac{\partial (g_\varepsilon \varphi)}{\partial x_k} \right) \right)$$
$$= 2\hbar^2 \left(\frac{\partial g_\varepsilon}{\partial x_k} \right)^2 \varphi,$$

which, together with (7.4), implies the claimed formula (7.7).

Inserting (7.6) and (7.7) into (7.5), we obtain

$$\sum_j \left(E_j(\hbar) - \tau \right)_+^2 \int_{\mathbb{R}^d} |\varphi_j|^2 \frac{dx}{(1 + \varepsilon^2 x_k^2)^3}$$
$$\leq 4\hbar^2 \sum_j \left(E_j(\hbar) - \tau \right)_+ \int_{\mathbb{R}^d} \left| \frac{\partial}{\partial x_k} \varphi_j - \frac{3}{2} \frac{\varepsilon^2 x_k}{1 + \varepsilon^2 x_k^2} \varphi_j \right|^2 \frac{dx}{(1 + \varepsilon^2 x_k^2)^3}.$$

By dominated convergence and recalling that both sums are finite, we find

$$\sum_j \left(E_j(\hbar) - \tau \right)_+^2 \int_{\mathbb{R}^d} |\varphi_j|^2 \, dx \leq 4\hbar^2 \sum_j \left(E_j(\hbar) - \tau \right)_+ \int_{\mathbb{R}^d} \left| \frac{\partial}{\partial x_k} \varphi_j \right|^2 dx.$$

Recalling that φ_j is normalized and summing with respect to $k \in \{1, \ldots, d\}$, we obtain the claimed inequality. $\square$

Proof of Theorem 7.1 We will prove that, for any $\tau > 0$, the function

$$\hbar \mapsto \hbar^d \operatorname{Tr}(H(\hbar) + \tau)_-^2$$

is non-increasing; that is, for any $\hbar \geq \hbar' > 0$, we have

$$\hbar^d \operatorname{Tr}(H(\hbar) + \tau)_-^2 \leq (\hbar')^d \operatorname{Tr}(H(\hbar') + \tau)_-^2.$$

Letting $\tau \to 0+$ in this inequality, we obtain the monotonicity in the theorem for $\gamma = 2$. Integrating this inequality against $\tau^{\gamma-3}\, d\tau$ over $(0, \infty)$ and using the fact that

$$\lambda_-^\gamma = c_\gamma \int_0^\infty (\lambda + \tau)_-^2 \tau^{\gamma-3}\, d\tau$$

for some $c_\gamma > 0$ if $\gamma > 2$, we obtain the claimed inequality for $\gamma > 2$.

We begin by showing that $\operatorname{Tr}(H(\hbar) + \tau)_-^2$ is a convex function of $\hbar^2$ and that for all $\hbar, \hbar_0 > 0$,

$$\mathrm{Tr}(H(\hbar) + \tau)_-^2 \geq \mathrm{Tr}(H(\hbar_0) + \tau)_-^2 + 2(\hbar_0^2 - \hbar^2) \sum_j (E_j(\hbar_0) - \tau)_+ \int_{\mathbb{R}^d} |\nabla \varphi_j|^2 \, dx,$$

(7.8)

where the (φ_j) are orthonormal eigenfunctions corresponding to the negative eigenvalues of $H(\hbar_0)$. Indeed, for any $\kappa > 0$, we have, according to Corollary 1.35,

$$- \mathrm{Tr}\,(H(\hbar) + \kappa)_-$$

$$= \inf \left\{ \sum_j \int_{\mathbb{R}^d} \left(\hbar^2 |\nabla \psi_j|^2 - V |\psi_j|^2 + \kappa |\psi_j|^2 \right) \, dx : \right.$$

$$\left. (\psi_j) \subset H^1(\mathbb{R}^d) \cap L^2(\mathbb{R}^d, V_- dx), \text{ orthonormal in } L^2(\mathbb{R}^d) \right\}.$$

Since, for any given (ψ_j), the sum $\sum_j \int_{\mathbb{R}^d} \left(\hbar^2 |\nabla \psi_j|^2 - V |\psi_j|^2 + \kappa |\psi_j|^2 \right) dx$ is linear in $\hbar^2$ and since the infimum of linear functions is concave, we infer that $- \mathrm{Tr}\,(H(\hbar) + \kappa)_-$ is concave in $\hbar^2$. Moreover, by choosing the ψ_j to be those eigenfunctions φ_j of $H(\hbar_0)$ corresponding to eigenvalues less than $-\kappa$, we see that

$$- \mathrm{Tr}\,(H(\hbar) + \kappa)_- \leq - \mathrm{Tr}\,(H(\hbar_0) + \kappa)_- + (\hbar^2 - \hbar_0^2) \sum_{j:\, E_j(\hbar_0^2) > \kappa} \int_{\mathbb{R}^d} |\nabla \varphi_j|^2 \, dx.$$

In view of the formula

$$\mathrm{Tr}\,(H(\hbar) + \tau)_-^2 = 2 \int_\tau^\infty \mathrm{Tr}\,(H(\hbar) + \kappa)_- \, d\kappa$$

we see that $\mathrm{Tr}\,(H(\hbar) + \tau)_-^2$ is convex in $\hbar^2$ as an integral of convex functions and, in addition, that

$$-\frac{1}{2} \mathrm{Tr}\,(H(\hbar) + \tau)_-^2 \leq -\frac{1}{2} \mathrm{Tr}\,\left(H(\hbar_0) + \tau \right)_-^2$$

$$+ (\hbar^2 - \hbar_0^2) \sum_j \left(E_j(\hbar_0) - \tau \right)_+ \int_{\mathbb{R}^d} |\nabla \varphi_j|^2 \, dx.$$

This proves the intermediate claim.

We now insert the inequality from Lemma 7.3 for the parameter $\hbar_0$ into (7.8) and obtain that for all $\hbar \leq \hbar_0$,

$$\mathrm{Tr}(H(\hbar) + \tau)_-^2 \geq \left(1 + \frac{d}{2} \left(1 - \frac{\hbar^2}{\hbar_0^2} \right) \right) \mathrm{Tr}(H(\hbar_0) + \tau)_-^2.$$

Equivalently,

$$\frac{\hbar^d \operatorname{Tr}(H(\hbar) + \tau)_-^2 - \hbar_0^d \operatorname{Tr}(H(\hbar_0) + \tau)_-^2}{\hbar^2 - \hbar_0^2}$$

$$\leq \frac{\left(\hbar^d - \hbar_0^d\right) + \frac{d}{2}\hbar^d \left(1 - \frac{\hbar^2}{\hbar_0^2}\right)}{\hbar^2 - \hbar_0^2} \operatorname{Tr}(H(\hbar_0) + \tau)_-^2,$$

and therefore, by L'Hôpital's rule,

$$\limsup_{\hbar \to \hbar_0-} \frac{\hbar^d \operatorname{Tr}(H(\hbar) + \tau)_-^2 - \hbar_0^d \operatorname{Tr}(H(\hbar_0) + \tau)_-^2}{\hbar^2 - \hbar_0^2}$$

$$\leq \limsup_{\hbar \to \hbar_0-} \frac{\left(\hbar^d - \hbar_0^d\right) + \frac{d}{2}\hbar^d \left(1 - \frac{\hbar^2}{\hbar_0^2}\right)}{\hbar^2 - \hbar_0^2} \operatorname{Tr}(H(\hbar_0) + \tau)_-^2 = 0.$$

The convexity of $\operatorname{Tr}(H(\hbar) + \tau)_-^2$ in $\hbar^2$ implies that this function is locally absolutely continuous, and therefore $\hbar \mapsto \hbar^d \operatorname{Tr}(H(\hbar) + \tau)_-^2$ is locally absolutely continuous as well and, in particular, almost everywhere differentiable. The above inequality shows that the left-sided derivative is almost everywhere nonpositive. Since an absolutely continuous function can be written as the integral over its derivative, the non-positivity of the derivative implies the monotonicity, as claimed. □

7.2 Bounds for radial potentials

Our goal in this section is to derive upper and lower bounds on the number of eigenvalues of Schrödinger operators on $\mathbb{R}^d$ with radial potentials. All the results in this section are due to Glaser, Grosse and Martin (1978) and we give a detailed exposition of their ideas.

Our starting point is the fact that a Schrödinger operator with radial potential is unitarily equivalent to an orthogonal sum of half-line Schrödinger operators. Via a logarithmic change of variables, one can transform these half-line operators into whole-line operators. The fundamental observation now is that under this transformation the number of negative eigenvalues is preserved. This is remarkable, given that the transformation is not unitary and therefore the individual negative eigenvalues need not be preserved. Once we have arrived at whole-line Schrödinger operators we can use known results about them to obtain information about the original d-dimensional operator with radial potential.

Throughout this section we assume that $d \geq 2$.

7.2.1 A formula for the number of negative eigenvalues for radial potentials

The following result, which is key for everything else in this section, expresses the number of negative eigenvalues of Schrödinger operators in $L^2(\mathbb{R}^d)$ with a radial potential in terms of the distribution of negative eigenvalues of one-dimensional Schrödinger operators. The statement involves the numbers v_ℓ, $\ell \in \mathbb{N}_0$, from Theorem 3.49: in other words, the dimension of the space of spherical harmonics of degree ℓ.

Proposition 7.4 *Let $V \in L^1_{\mathrm{loc}}(\mathbb{R}^d)$ be radial and as in Proposition 4.1, and let*

$$F(t) := e^{2t} V(e^t), \qquad t \in \mathbb{R}.$$

Then

$$N(0, -\Delta - V) = \sum_{\ell \in \mathbb{N}_0} v_\ell \, N\left(0, -\frac{d^2}{dt^2} + \left(\ell + \frac{d-2}{2}\right)^2 - F\right).$$

Our proof of the proposition relies on the change of variables $t = \ln r$.

Let $u \in L^1_{\mathrm{loc}}(\mathbb{R}_+)$ and let $w \in L^1_{\mathrm{loc}}(\mathbb{R})$ be such that $u(r) = r^{-(d-2)/2}w(\ln r)$ and let W and G be non-negative, measurable functions on $\mathbb{R}_+$ and $\mathbb{R}$, respectively, such that $W(r) = r^{-2}G(\ln r)$. Then

$$\int_0^\infty W(r)|u(r)|^2 r^{d-1}\, dr - \int_{\mathbb{R}} G(t)|w(t)|^2\, dt.$$

We recall that the behavior of derivatives under the change of variables $t = \ln r$ was discussed in Lemma 2.34.

We now turn to the proof of Proposition 7.4. We proceed by separation of variables as in §§3.8.4 and 4.2.3. In particular, if for a given u, the $u_{\ell,m}$ are defined as in §3.8.3, then, by (3.73) and the assumption that V is radial,

$$\int_{\mathbb{R}^d} V|u|^2\, dx = \sum_{\ell,m} \int_0^\infty V(r)|u_{\ell,m}|^2 r^{d-1}\, dr.$$

Following the previous arguments, we see that the operator $-\Delta - V$ in $L^2(\mathbb{R}^d)$ is unitarily equivalent to the orthogonal sum

$$\bigoplus_{\ell \in \mathbb{N}_0, m \in \mathcal{M}_\ell} (h_\ell - V(r)), \tag{7.9}$$

where, for each $\ell \in \mathbb{N}_0$, $\mathcal{M}_\ell$ is an index set of cardinality v_ℓ, and where, for brevity, we have written

$$h_\ell = -\frac{d^2}{dr^2} - \frac{d-1}{r}\frac{d}{dr} + \frac{\ell(\ell+d-2)}{r^2}.$$

The operators $h_\ell - V(r)$ in (7.9) act in $L_d^2 := L^2(\mathbb{R}_+, r^{d-1} dr)$ and are defined through the closure of the quadratic form

$$\int_0^\infty \left(|u'|^2 + \frac{\ell(\ell + d - 2)}{r^2} |u|^2 - V|u|^2 \right) r^{d-1} dr$$

with form domain consisting of all weakly differentiable functions $u \in L_d^2$ for which

$$\int_0^\infty \left(|u'|^2 + \frac{\ell(\ell + d - 2)}{r^2} |u|^2 + V_-|u|^2 \right) r^{d-1} dr < \infty .$$

The fact that V_+ is $(h_\ell + V_-)$-form bounded with form bound < 1 follows as in §4.2.3 from the corresponding assumption on the function V_+ on $\mathbb{R}^d$ in Proposition 4.1. The operators in (7.9) are then defined via Lemma 1.47, similar to the situation in the proofs of Proposition 4.1 and Corollary 4.2.

Proof of Proposition 7.4 Since $-\Delta - V$ is unitarily equivalent to the orthogonal sum (7.9), we have

$$N(0, -\Delta - V) = \sum_{\ell, m} N(0, h_\ell - V(r)) = \sum_{\ell \in \mathbb{N}_0} \nu_l \, N(0, h_\ell - V(r)) .$$

We denote by $\mathcal{F}$ the form domain of the operator $h_\ell - V_-$ in L_d^2 if $d \geq 3$ and the intersection of this form domain with $L^2(\mathbb{R}_+, r^{-1} dr)$ if $d = 2$. Moreover, let $\mathcal{G}$ be the intersection of the form domain of the operator $-\frac{d^2}{dt^2} + (\ell + (d-2)/2)^2 + F_-$ in $L^2(\mathbb{R})$ with $L^2(\mathbb{R}, e^{2t} dt)$. Then, as recalled after the statement of Proposition 7.4, there is a bijection between functions $u \in \mathcal{F}$ and functions $w \in \mathcal{G}$, and we have

$$\int_0^\infty \left(|u'(r)|^2 + \frac{\ell(\ell + d - 2)}{r^2} |u(r)|^2 - V(r)|u(r)|^2 \right) r^{d-1} dr$$
$$= \int_{\mathbb{R}} \left(|w'(t)|^2 + \left(\ell + \frac{d - 2}{2} \right)^2 |w(t)|^2 - F(t)|w(t)|^2 \right) dt .$$

We claim that $\mathcal{F}$ is a form core of the operator $h_\ell + V$ in L_d^2. This is clear for $d \geq 3$ and, for $d = 2$, follows from Lemma 2.33. (For $V_- \neq 0$ one needs to argue as in Remark 2.24.) Also, by Proposition 4.1, $\mathcal{G}$ is a form core of the operator $-\frac{d^2}{dt^2} + \frac{(d-2)^2}{4} + F_-$ in $L^2(\mathbb{R})$. Thus, by Glazman's lemma (Theorem 1.25),

$$N(0, h_\ell - V(r)) = N\left(0, -\frac{d^2}{dt^2} + \left(\ell + \frac{d - 2}{2} \right)^2 - F \right) .$$

This proves the proposition. □

We emphasize that the operators $h_\ell - V$ in L_d^2 and $-\frac{d^2}{dt^2} + \left(\ell + \frac{d-2}{2} \right)^2 - F$

in $L^2(\mathbb{R})$ are *not* unitarily equivalent. In general, their negative eigenvalues do not coincide, and therefore the identity in Proposition 7.4 does not generalize to Riesz means of positive order.

7.2.2 Upper and lower bounds for radial potentials

Let us use Proposition 7.4 to establish upper and lower bounds on the number of eigenvalues of $N(0, -\Delta - V)$ with radial V in dimensions two and four.

Theorem 7.5 *Let $V \in L^1(\mathbb{R}^2)$ be radial and as in Proposition 4.1. Then*

$$N(0, -\Delta - V) \geq \frac{1}{8\pi} \int_{\mathbb{R}^2} V(x)\, dx\,.$$

The constant $1/(8\pi)$ in this bound is best possible and, as we will see in Lemma 7.8 below, it is attained for $V(x) = 8/(1 + |x|^2)^2$. Note that this constant is half of the semiclassical one, which is $L_{0,2}^{\mathrm{cl}} = 1/(4\pi)$.

Proof Let F be defined as in Proposition 7.4 and let us denote the negative eigenvalues of $-\frac{d^2}{dt^2} - F$ in $L^2(\mathbb{R})$ by $-E_n$. Then, by Proposition 7.4,

$$N(0, -\Delta - V) = \sum_{\ell \in \mathbb{N}_0} \nu_\ell\, N\left(0, -\frac{d^2}{dt^2} + \ell^2 - F\right) = \sum_{\ell \in \mathbb{N}_0} \nu_\ell \sum_{n:\, E_n > \ell^2} 1$$

$$= \sum_n \sum_{\ell \in \mathbb{N}_0:\, \ell < E_n^{1/2}} \nu_\ell\,.$$

Recall that $\nu_0 = 1$ and $\nu_\ell = 2$ if $\ell \geq 1$. We have, for any $\kappa > 0$,

$$\sum_{\ell \in \mathbb{N}_0:\, \ell < \kappa} \nu_\ell = \sum_{k \in \mathbb{Z}:\, |k| < \kappa} 1 \geq \kappa\,.$$

Thus

$$N(0, -\Delta - V) \geq \sum_n E_n^{1/2}\,,$$

and the bound from Theorem 5.33 (or Corollary 5.41) on the sum of the square roots of eigenvalues of one-dimensional Schrödinger operators yields

$$\sum_n E_n^{1/2} \geq \frac{1}{4} \int_{\mathbb{R}} F(t)\, dt = \frac{1}{8\pi} \int_{\mathbb{R}^2} V(x)\, dx\,.$$

This proves the claimed bound. $\qquad\square$

Next, we prove an upper bound in dimension four.

Theorem 7.6 *Let $V \in L^1_{loc}(\mathbb{R}^4)$ be radial with $V_+ \in L^2(\mathbb{R}^4)$. Then*

$$N(0, -\Delta - V) \le \frac{3}{32\,\pi^2} \int_{\mathbb{R}^4} V(x)_+^2 \, dx \,.$$

The constant coincides with $L_{0,4}^{(1)}$ (see (5.25)) and is, therefore, best possible. Equality is attained asymptotically for $V^{(L)}(x) = 4(L+1)(L+2)(1 + |x|^2)^{-2}$ as $L \to 0$; see Lemma 7.8 below.

Proof We present the first part of the proof in general dimension d. Again we define F as in Proposition 7.4, we denote the negative eigenvalues of $-\frac{d^2}{dt^2} - F$ in $L^2(\mathbb{R})$ by $-E_n$, and we use Proposition 7.4 to get

$$N(0, -\Delta - V) = \sum_{\ell \in \mathbb{N}_0} \nu_\ell \, N\left(0, -\frac{d^2}{dt^2} + \left(\ell + \frac{d-2}{2}\right)^2 - F\right)$$

$$= \sum_{\ell \in \mathbb{N}_0} \nu_\ell \sum_{n:\, E_n > (\ell + (d-2)/2)^2} 1$$

$$= \sum_n \sum_{\ell \in \mathbb{N}_0:\, \ell + (d-2)/2 < E_n^{1/2}} \nu_\ell$$

$$= \sum_n P([(E_n^{1/2} - (d-2)/2)_+]) \,,$$

where

$$P(L) := \frac{(d - 2 + L)!\,(d - 1 + 2L)}{(d - 1)!\,L!}$$

and where we used the second equality in (4.35). We therefore get

$$N(0, -\Delta - V) \le C_d \sum_{n \ge 1} E_n^{(d-1)/2}$$

with

$$C_d := \sup_{\kappa \ge 0} \frac{P([\kappa])}{(\kappa + (d-2)/2)^{d-1}} \,.$$

Note that, since x^{d-1} is increasing, we have

$$C_d = \sup_{L \in \mathbb{N}_0} \frac{P(L)}{(L + (d-2)/2)^{d-1}} \,.$$

Applying the one-dimensional Lieb–Thirring inequality (Theorem 4.38), we finally get

$$N(0, -\Delta - V) \le C_d \, L_{(d-1)/2,1} \int_{\mathbb{R}} F(t)_+^{d/2} \, dt$$

$$= C_d \, |\mathbb{S}^{d-1}|^{-1} L_{(d-1)/2,1} \int_{\mathbb{R}^d} V(x)_+^{d/2} \, dx . \qquad (7.10)$$

We now specialize to the case $d = 4$. An easy computation shows that

$$C_4 = 1$$

and the supremum is attained at $L = 0$. Using the sharp bound on $L_{3/2,1}$ from Theorem 5.31, we obtain the bound in the theorem. $\qquad \square$

Remark 7.7 Similarly, for $d = 3$, the bound (7.10) together with the simple fact that $C_3 = 4$ yields, for radial potentials V,

$$N(0, -\Delta - V) \le \pi^{-1} L_{1,1} \int_{\mathbb{R}^3} V(x)_+^{3/2} \, dx .$$

Note that

$$\pi^{-1} L_{1,1}^{(1)} = \pi^{-1} \left(\frac{4}{\pi \, 3^{3/2}} \right) = \frac{4}{\pi^2 3^{3/2}} = L_{0,3}^{(1)}$$

with $L_{0,3}^{(1)}$ from (5.25). Therefore if the conjectured equality $L_{1,1} = L_{1,1}^{(1)}$ is true, then the above bound yields the best possible constant $L_{0,3}^{(1)}$ in the CLR inequality for radial functions.

7.2.3 The number of negative eigenvalues for a certain family of potentials

In this subsection, we will use Proposition 7.4 to explicitly compute the number of negative eigenvalues of Schrödinger operators in $L^2(\mathbb{R}^d)$ with potential given by a constant times $(1 + |x|^2)^{-2}$. Interestingly, for large dimensions this will lead to counterexamples to a bound that Lieb and Thirring conjectured.

Lemma 7.8 *For $L > 0$, let*

$$V^{(L)}(x) := \left(L + \frac{d-2}{2} \right) \left(L + \frac{d}{2} \right) \left(\frac{2}{1 + |x|^2} \right)^2 , \qquad x \in \mathbb{R}^d .$$

Then

$$N(0, -\Delta - V^{(L)}) = \frac{2}{d!} \frac{(\lceil L - 1 \rceil + d - 1)! \, (\lceil L - 1 \rceil + \frac{d}{2})}{\lceil L - 1 \rceil!}$$

and

$$\int_{\mathbb{R}^d} \left(V^{(L)}(x)\right)^{\frac{d}{2}} dx = \left((L + \tfrac{d-2}{2})(L + \tfrac{d}{2})\right)^{\frac{d}{2}} \frac{2\pi^{\frac{d+1}{2}}}{\Gamma(\frac{d+1}{2})}.$$

Here we use the notation $\lceil v \rceil = N$ is $v = N - \theta$ with $0 \le \theta < 1$ and $N \in \mathbb{Z}$.

Proof Let $F^{(L)}$ correspond to $V^{(L)}$ as in Proposition 7.4; namely,

$$F^{(L)}(t) := \left(L + \frac{d-2}{2}\right)\left(L + \frac{d}{2}\right)\frac{1}{\cosh^2 t}.$$

According to Proposition 4.9 with $v = L + \frac{d-2}{2}$, the negative spectrum of $-\frac{d^2}{dt^2} - F^{(L)}$ in $L^2(\mathbb{R})$ consists precisely of the simple eigenvalues $-(L + \frac{d}{2} - n)^2$, where $n = 1, \ldots, \lceil L + \frac{d-2}{2} \rceil$. Thus, for any $\kappa \ge 0$,

$$N\left(0, -\frac{d^2}{dt^2} + \kappa^2 - F^{(L)}\right) = \#\left\{n \in \mathbb{N} : L + \frac{d}{2} - n > \kappa\right\} = \left\lceil L + \frac{d-2}{2} - \kappa \right\rceil_+.$$

Hence, by Proposition 7.4,

$$N(0, -\Delta - V^{(L)}) = \sum_{\ell \in \mathbb{N}_0} v_\ell \lceil L - \ell \rceil_+ = \sum_{\ell=0}^{\lceil L-1 \rceil} v_\ell \left(\lceil L \rceil - \ell\right) = \sum_{\ell=0}^{\lceil L-1 \rceil} \sum_{k=\ell}^{\lceil L-1 \rceil} v_\ell$$

$$= \sum_{k=0}^{\lceil L-1 \rceil} \sum_{\ell=0}^{k} v_\ell = \sum_{k=0}^{\lceil L-1 \rceil} \frac{(d-2+k)!\,(d-1+2k)}{(d-1)!\,k!},$$

where we used (4.35). Note that

$$\frac{(d-2+k)!\,(d-1+2k)}{(d-1)!\,k!} = \tilde{v}_k,$$

where $\tilde{v}_k$ is the same as v_k, but with d replaced by $d+1$. Therefore, applying (4.35) in dimension $d+1$, we obtain

$$\sum_{k=0}^{\lceil L-1 \rceil} \frac{(d-2+k)!\,(d-1+2k)}{(d-1)!\,k!} = \frac{(d-1+\lceil L-1 \rceil)!\,(d+2\lceil L-1 \rceil)}{d!\,\lceil L-1 \rceil!},$$

as claimed.

Finally,

$$\int_{\mathbb{R}^d} \left(\frac{2}{1+|x|^2}\right)^d dx = |\mathbb{S}^{d-1}| \int_0^\infty \left(\frac{2r}{1+r^2}\right)^d \frac{dr}{r} = |\mathbb{S}^{d-1}| \int_{\mathbb{R}} \cosh^{-d} t \, dt$$

$$= \sqrt{\pi}\, \frac{\Gamma(\frac{d}{2})}{\Gamma(\frac{d+1}{2})} |\mathbb{S}^{d-1}|,$$

where we used the same computation as in the proof of Lemma 5.15 with

$\gamma = (d-1)/2$. Using $|\mathbb{S}^{d-1}| = 2\pi^{d/2}/\Gamma(d/2)$, this leads to the claimed formula for the integral of $(V^{(L)})^{d/2}$ and concludes the proof. $\qquad\square$

We now use this lemma to derive a lower bound on the sharp constant $L_{0,d}$ in the Cwikel–Lieb–Rozenblum inequality.

Corollary 7.9 *For all $d \geq 3$,*

$$L_{0,d} \geq L_{0,d}^{\mathrm{cl}} \sup_{L \in \mathbb{N}_0} \frac{(L+d-1)!\,(L+\frac{d}{2})}{L!\left((L+\frac{d-2}{2})(L+\frac{d}{2})\right)^{d/2}}.$$

Proof Lemma 7.8 implies that

$$L_{0,d} \geq \sup_{L \geq 0} \frac{N(0, -\Delta - V^{(L)})}{\int_{\mathbb{R}^d}(V^{(L)})^{d/2}\,dx} = L_{0,d}^{\mathrm{cl}} \sup_{L \in \mathbb{N}_0} \frac{(L+d-1)!\,(L+\frac{d}{2})}{L!\left((L+\frac{d-2}{2})(L+\frac{d}{2})\right)^{d/2}}.$$

Here we used $L_{0,d}^{\mathrm{cl}} = (4\pi)^{-d/2}\Gamma(1+d/2)^{-1}$ and $\Gamma(\frac{d}{2})\Gamma(\frac{d+1}{2}) = \frac{\sqrt{\pi}}{2^{d-1}}\Gamma(d)$. $\qquad\square$

As we discussed in §5.1, we always have $L_{0,d} \geq \max\left\{L_{0,d}^{(1)}, L_{0,d}^{\mathrm{cl}}\right\}$: Lieb and Thirring conjectured that equality holds in this inequality. The following proposition shows that this is not the case in dimensions $d \geq 7$.

Proposition 7.10 *If $d \geq 7$, then $L_{0,d} > \max\left\{L_{0,d}^{(1)}, L_{0,d}^{\mathrm{cl}}\right\}$.*

We recall from Proposition 5.12 that $L_{0,d}^{\mathrm{cl}} < L_{0,d}^{(1)}$ for $d \leq 7$ and $L_{0,d}^{\mathrm{cl}} > L_{0,d}^{(1)}$ if $d \geq 8$.

Proof In view of Corollary 7.9 it suffices to analyze the sequence

$$a_L := \frac{(L+d-1)!\,(L+\frac{d}{2})}{L!\left((L+\frac{d-2}{2})(L+\frac{d}{2})\right)^{d/2}}, \qquad L \in \mathbb{N}_0.$$

Note that $a_L \to 1$ as $L \to \infty$. Using $\ln(1+x) = x + O(x^2)$ as $x \to 0$, one sees that

$$\ln a_L = \tfrac{d}{2} L^{-1} + O(L^{-2}) \qquad \text{as } L \to \infty,$$

so $a_L > 1$ for all sufficiently large L. In view of Corollary 7.9, this implies that $L_{0,d} > L_{0,d}^{\mathrm{cl}}$ for all $d \geq 3$. Note that this is a special case of Corollary 5.23.

On the other hand, we claim that $a_1 > a_0$ for all $d \geq 7$. (For $d = 6$ we have equality and for $d = 3, 4, 5$ the inequality reverses.) Indeed, the inequality $a_1 > a_0$ is equivalent to the inequality $2(x-1)^x(x+1)^{1-x} > 1$ for $x = d/2$. Since $2(x-1)^x(x+1)^{1-x} = 1$ for $x = 3$, it suffices to prove that $x \mapsto 2(x-1)^x(x+1)^{1-x}$

is increasing for $x \geq 3$. To see this, we note that the derivative of its logarithm is

$$-\ln((x+1)/(x-1)) + x/(x-1) - (x-1)/(x+1).$$

Bounding $\ln((x+1)/(x-1)) < 2/(x-1)$, we see that the derivative is positive for $x > 3$.

Note that, by the explicit forms (5.25) and (5.7) of the one-particle constant and the semiclassical constant, we have

$$L_{0,d}^{(1)} = L_{0,d}^{cl} \, a_0 \, .$$

In view of Corollary 7.9, the inequality $\sup_{L \in \mathbb{N}_0} a_L \geq a_1 > a_0$ implies that $L_{0,d} > L_{0,d}^{(1)}$ for $d \geq 7$. This completes the proof. $\square$

Glaser et al. (1978) conjectured that equality holds in Corollary 7.9: that is, the optimal constant in the CLR inequality is obtained by optimizing with respect to $L \in \mathbb{N}_0$ over the family of potentials in Lemma 7.8; see Conjecture 8.23.

7.3 More on the one-particle constants

We recall that the one-particle constant $L_{\gamma,d}^{(1)}$ was defined in §5.1.2 and that for the optimal Lieb–Thirring constant $L_{\gamma,d}$ one always has the inequality $L_{\gamma,d} \geq L_{\gamma,d}^{(1)}$. Moreover, in the previous section we saw that, if $\gamma = 0$, then $L_{0,d} > L_{0,d}^{(1)}$ for $d \geq 7$. In this section, we extend the regime of γ for which the inequality $L_{\gamma,d} > L_{\gamma,d}^{(1)}$ holds, following Frank et al. (2021b).

Theorem 7.11   *If $\gamma > \max\{2 - \frac{d}{2}, 0\}$, then $L_{\gamma,d} > L_{\gamma,d}^{(1)}$.*

Here we only sketch the basic idea of the proof and refer to Frank et al. (2021b) for the details. The strategy is to construct a trial potential and to bound its lowest two eigenvalues from above and the integral involving the potential from below. The potential will be made out of two copies of the optimal one-particle potential placed very far from each other.

More precisely, as discussed in Remark 5.6, for all γ as in Proposition 5.5 there is a $0 \leq V \in L^{\gamma + \frac{d}{2}}(\mathbb{R}^d)$ such that the Schrödinger operator $-\Delta - V$ in $L^2(\mathbb{R}^d)$ has a lowest eigenvalue $-E_1 < 0$ that satisfies

$$E_1^\gamma = L_{\gamma,d}^{(1)} \int_{\mathbb{R}^d} V^{\gamma + \frac{d}{2}} \, dx \, . \tag{7.11}$$

The function V is radially symmetric, although this is not really necessary for the proof. We fix $v \in \mathbb{S}^{d-1}$ and define for any $R > 0$, which will eventually go to infinity,

$$V_R(x) := \left(V(x + Rv)^{\gamma + \frac{d}{2} - 1} + V(x - Rv)^{\gamma + \frac{d}{2} - 1} \right)^{\frac{1}{\gamma + \frac{d}{2} - 1}} .$$

Let us define

$$A_R := \frac{1}{2} \int_{\mathbb{R}^d} V_R(x)^{\gamma + \frac{d}{2}} \, dx - \int_{\mathbb{R}^d} V(x)^{\gamma + \frac{d}{2}} \, dx$$

$$= \frac{1}{2} \int_{\mathbb{R}^d} \left(\left(V(x + Rv)^{\gamma + \frac{d}{2} - 1} + V(x - Rv)^{\gamma + \frac{d}{2} - 1} \right)^{\frac{\gamma + \frac{d}{2}}{\gamma + \frac{d}{2} - 1}} \right.$$

$$\left. - V(x + Rv)^{\gamma + d/2} - V(x - Rv)^{\gamma + d/2} \right) dx .$$

From this, it is easy to see that $A_R > 0$ and

$$A_R \to 0 \quad \text{as} \quad R \to \infty .$$

Let us denote by $-E_1(R)$ and $-E_2(R)$ the two lowest eigenvalues of the operator $-\Delta - V_R$. In fact, it is not hard to see that, for all sufficiently large R, this operator has at least two negative eigenvalues and that the lowest two eigenvalues converge to $-E_1$, the lowest eigenvalue of $-\Delta - V$, as R tends to infinity. The main work in the proof of Theorem 7.11 consists in quantifying this convergence. Specifically, one shows that

$$E_1(R)^\gamma + E_2(R)^\gamma \geq 2 \left(1 + \frac{\gamma + \frac{d}{2}}{\int_{\mathbb{R}^d} V^{\gamma + \frac{d}{2}} \, dx} A_R + o(A_R) \right) E_1^\gamma \qquad \text{as } R \to \infty . \quad (7.12)$$

The proof of this bound uses the assumption $\gamma > \max\{2 - d/2, 0\}$. It proceeds by the variational principle using orthonormalized copies of the ground states of $-\Delta - V(x + Rv)$ and of $-\Delta - V(x - Rv)$ as trial functions for the eigenvalues of $-\Delta - V_R$. For the details we refer to Frank et al. (2021b).

Accepting (7.12), it is easy to complete the proof of Theorem 7.11. Indeed, we clearly have

$$\int_{\mathbb{R}^d} V_R^{\gamma + \frac{d}{2}} \, dx = 2 \int_{\mathbb{R}^d} V^{\gamma + \frac{d}{2}} \, dx + 2 A_R ,$$

so, with the notation $I := \int_{\mathbb{R}^d} V^{\gamma + \frac{d}{2}} \, dx$ and recalling (7.11),

$$\frac{E_1(R)^\gamma + E_2(R)^\gamma}{\int_{\mathbb{R}^d} V_R^{\gamma + \frac{d}{2}} \, dx} \geq \frac{E_1^\gamma}{\int_{\mathbb{R}^d} V^{\gamma + \frac{d}{2}} \, dx} \frac{1 + (\gamma + \frac{d}{2}) I^{-1} A_R + o(A_R)}{1 + I^{-1} A_R}$$

$$= L_{\gamma,d}^{(1)} \left(1 + (\gamma + \frac{d}{2} - 1) I^{-1} A_R + o(A_R) \right) .$$

For large R, the right side is strictly larger than $L_{\gamma,d}^{(1)}$, which completes our sketch of the proof of Theorem 7.11.

In fact, one can show a stronger conclusion than that in Theorem 7.11. Namely, let us define $L_{\gamma,d}^{(N)}$ to be the best constant in the inequality

$$\sum_{n=1}^{N} E_n^{\gamma} \leq L_{\gamma,d}^{(N)} \int_{\mathbb{R}^d} V(x)_+^{\gamma + \frac{d}{2}} \, dx, \tag{7.13}$$

where $-E_n$ are the negative eigenvalues of $-\Delta - V$ in non-decreasing order, counting multiplicities, and with the convention that $E_n = 0$ if $-\Delta - V$ has less than n negative eigenvalues. Then, clearly, $L_{\gamma,d}^{(N)} \leq L_{\gamma,d}^{(N+1)}$ and $L_{\gamma,d} = \lim_{N \to \infty} L_{\gamma,d}^{(N)}$. Moreover, in terms of these constants, the sketched proof of Theorem 7.11 actually shows that

$$L_{\gamma,d}^{(2)} > L_{\gamma,d}^{(1)} \qquad \text{if } \gamma > \max\left\{2 - \tfrac{d}{2}, 0\right\}.$$

The stronger conclusion is that, for any $N \geq 1$, one has

$$L_{\gamma,d}^{(2N)} > L_{\gamma,d}^{(N)} \qquad \text{if } \gamma > \max\left\{2 - \tfrac{d}{2}, 0\right\}. \tag{7.14}$$

Therefore, in particular, one has

$$L_{\gamma,d} > L_{\gamma,d}^{(N)} \qquad \text{for all } N \geq 1, \text{ if } \gamma > \max\left\{2 - \tfrac{d}{2}, 0\right\}.$$

This shows that, if there is a potential attaining the best Lieb–Thirring constant, then this potential has necessarily infinitely many eigenvalues.

The proof of (7.14) consists of two steps. First, one shows that there is an optimal potential for $L_{\gamma,d}^{(N)}$. (This part of the argument works for $\gamma > 1/2$ in $d = 1$ and for $\gamma > 0$ in $d \geq 2$.) Second, one uses an argument similar to that in the proof of Theorem 7.11 to construct a potential consisting of two widely separated bumps made out of the optimizers of the $L_{\gamma,d}^{(N)}$ problem to deduce (7.14). For the details we refer to Frank et al. (2021c).

7.4 The dual Lieb–Thirring inequality

7.4.1 A duality principle for $\gamma = 1$

In order to motivate the following discussion, let us recall the special case of Proposition 5.3 corresponding to $\gamma = 1$. It says that the fact that the inequality

$$\lambda \geq -L \int_{\mathbb{R}^d} V^{1 + d/2} \, dx$$

holds for any $0 \le V \in L^{1+d/2}(\mathbb{R}^d)$ and for any eigenvalue λ of the Schrödinger operator $-\Delta - V$ in $L^2(\mathbb{R}^d)$, is equivalent to the fact that the Gagliardo–Nirenberg inequality

$$\int_{\mathbb{R}^d} |\nabla \psi|^2 \, dx \ge K \int_{\mathbb{R}^d} |\psi|^{2+4/d} \, dx \qquad (7.15)$$

holds for any $\psi \in H^1(\mathbb{R}^d)$ with $\|\psi\|_{L^2(\mathbb{R}^d)} = 1$; furthermore, it says that the optimal constants K and L are related by the identity

$$\left(\left(1 + \tfrac{d}{2}\right)L\right)^{1+\frac{2}{d}} \left(\left(1 + \tfrac{2}{d}\right)K\right)^{1+\frac{d}{2}} = 1 .$$

The following theorem generalizes this equivalence to sums of eigenvalues.

Theorem 7.12 *Let $d \ge 1$. Then the inequality*

$$\operatorname{Tr}(-\Delta - V)_- \le L \int_{\mathbb{R}^d} V^{1+d/2} \, dx \qquad \text{for all } 0 \le V \in L^{1+d/2}(\mathbb{R}^d) \qquad (7.16)$$

is equivalent to

$$\sum_{n=1}^{N} \int_{\mathbb{R}^d} |\nabla \psi_n|^2 \, dx \ge K \int_{\mathbb{R}^d} \left(\sum_{n=1}^{N} |\psi_n|^2 \right)^{1+2/d} \, dx$$

$$\text{for all } N \in \mathbb{N} \text{ and all } L^2\text{-orthonormal } \psi_n \in H^1(\mathbb{R}^d), \qquad (7.17)$$

in the sense that the optimal constants K and L are related by the identity

$$\left(\left(1 + \tfrac{d}{2}\right)L\right)^{1+\frac{2}{d}} \left(\left(1 + \tfrac{2}{d}\right)K\right)^{1+\frac{d}{2}} = 1 . \qquad (7.18)$$

Proof First, assume that (7.16) holds. Let $\psi_1, \dots, \psi_N \in H^1(\mathbb{R}^d)$ be L^2-orthonormal functions. Then, by the variational principle for sums of eigenvalues (Corollary 1.35), for all $0 \le V \in L^{1+d/2}(\mathbb{R}^d)$,

$$\sum_{n=1}^{N} \int_{\mathbb{R}^d} \left(|\nabla \psi_n|^2 - V|\psi_n|^2 \right) dx \ge -\operatorname{Tr}(-\Delta - V)_- \ge -L \int_{\mathbb{R}^d} V^{1+d/2} \, dx ;$$

that is, writing $\rho := \sum_{n=1}^{N} |\psi_n|^2$,

$$\sum_{n=1}^{N} \int_{\mathbb{R}^d} |\nabla \psi_n|^2 \, dx \ge \int_{\mathbb{R}^d} \left(V\rho - LV^{1+d/2} \right) dx .$$

We maximize the right side by choosing $V = (\rho/(L(1 + d/2)))^{2/d}$ and obtain

$$\sum_{n=1}^{N} \int_{\mathbb{R}^d} |\nabla \psi_n|^2 \, dx \ge \frac{d/2}{(1 + d/2)^{1+2/d}} L^{-2/d} \int_{\mathbb{R}^d} \rho^{1+2/d} \, dx .$$

This proves (7.17) with constant (7.18).

Conversely, assume now that (7.17) holds. Let $0 \leq V \in L^{1+d/2}(\mathbb{R}^d)$ and let ψ_n be the normalized eigenfunctions corresponding to the negative eigenvalues $-E_n$ of the Schrödinger operator $-\Delta - V$ in $L^2(\mathbb{R}^d)$. Let us set $\rho = \sum_n |\psi_n|^2$. Then, by (7.17) (which, by monotone convergence, extends to infinitely many functions), we have

$$-\sum_n E_n = \sum_n \int_{\mathbb{R}^d} \left(|\nabla \psi_n|^2 - V |\psi_n|^2 \right) dx \geq \int_{\mathbb{R}^d} \left(K \rho^{1+2/d} - V \rho \right) dx \,.$$

By Hölder's inequality we have

$$\int_{\mathbb{R}^d} \left(K \rho^{1+2/d} - V \rho \right) dx \geq X - K^{-d/(2+d)} \left(\int_{\mathbb{R}^d} V^{1+d/2} dx \right)^{2/(2+d)} X^{d/(2+d)}$$

with $X = K \int_{\mathbb{R}^d} \rho(x)^{1+2/d} dx$. Since

$$X - a X^{d/(2+d)} \geq -\frac{(d/2)^{d/2}}{(1+d/2)^{1+d/2}} a^{1+d/2} \qquad \text{for all } a \geq 0,$$

we obtain

$$-\sum_n E_n \geq -\frac{(d/2)^{d/2}}{(1+d/2)^{1+d/2}} K^{-d/2} \int_{\mathbb{R}^d} V^{1+d/2} dx \,.$$

This proves (7.16) with constant (7.18).   □

In the same way one proves the corresponding result in the matrix case, except that one uses Hölder's inequality for Hermitian, non-negative matrices (Lemma 8.12). When V is a measurable function on $\mathbb{R}^d$ taking values in the $M \times M$ Hermitian matrices, we shall write $V \geq 0$ to mean that $V(x)$ is a non-negative matrix for a.e. $x \in \mathbb{R}^d$.

Theorem 7.13 *Let $d \geq 1$ and $M \geq 1$. Then the inequality*

$$\mathrm{Tr}\,(-\Delta - V)_- \leq L \int_{\mathbb{R}^d} \mathrm{Tr}_{\mathbb{C}^M} V^{1+d/2} dx \qquad \text{for all } 0 \leq V \in L^{1+d/2}(\mathbb{R}^d, \mathcal{H}_M)$$

is equivalent to

$$\sum_{n=1}^{N} \int_{\mathbb{R}^d} |\nabla \psi_n|^2 dx \geq K \int_{\mathbb{R}^d} \left(\sum_{n=1}^{N} |\psi_n|^2 \right)^{1+2/d} dx$$

for all $N \in \mathbb{N}$ and all L^2-orthonormal $\psi_n \in H^1(\mathbb{R}^d, \mathbb{C}^M)$,

in the sense that the optimal constants K and L are related by (7.18).

Remark 7.14   It is clear from the proofs of Theorems 7.12 and 7.13 that the validity of the results rests on the choice neither of the Laplacian as unperturbed operator nor of $\mathbb{R}^d$ as underlying space. More generally, consider the following setting. Let X be a separable measure space with integration with respect to the underlying measure denoted by dx. Let t be a non-negative, closed quadratic form with form domain $d[t]$ in $L^2(X)$ and corresponding operator T. Then for any $r, r' > 1$ with $1/r + 1/r' = 1$ the following are equivalent:

1. There is a constant $L > 0$ such that for any function $0 \leq V \in L^r(X)$ and any $\theta > 0$ there is a $C < \infty$ such that for all $u \in d[t]$, $\int_X V|u|^2 \, dx \leq \theta t[u] + C\|u\|^2$, and the negative eigenvalues of the operator $T - V$ in $L^2(X)$ satisfy

$$\operatorname{Tr}(T - V)_- \leq L \int_X V^r \, dx \, .$$

2. There is a constant $K > 0$ such that for all $N \in \mathbb{N}$ and all $L^2(X)$-orthonormal $\psi_n \in d[t]$ one has

$$\sum_{n=1}^{N} t[\psi_n] \geq K \int_X \left(\sum_{n=1}^{N} |\psi_n|^2 \right)^{r'} dx \, .$$

Moreover, the optimal constants are related by

$$(rL)^{r'} (r'K)^r = 1 \, .$$

The following relaxation of inequality (7.17) will become important in the next subsection.

Lemma 7.15   *Assume that (7.17) holds with some constant $K > 0$. Then, with the same constant K, one has*

$$\sum_n v_n \int_{\mathbb{R}^d} |\nabla \psi_n|^2 \, dx \, \|v\|_\infty^{2/d} \geq K \int_{\mathbb{R}^d} \left(\sum_n v_n |\psi_n|^2 \right)^{1+2/d} dx$$

for all $0 \leq v \in \ell^\infty(\mathbb{N})$ and all L^2-orthonormal $\psi_n \in H^1(\mathbb{R}^d)$.   (7.19)

In other words, if one can prove (7.19) with v_n taking values in $\{0, 1\}$, then the inequality holds automatically, with the same constant, for general v_n.

Proof   We write $M := \|v\|_\infty$ for short and observe that

$$\sum_n v_n \|\nabla \psi_n\|^2 = \int_0^M \sum_{v_n > \tau} \|\nabla \psi_n\|^2 \, d\tau \, .$$

By assumption, we have for each fixed $\tau > 0$,

$$\sum_{\nu_n > \tau} \|\nabla \psi_n\|^2 \geq K \int_{\mathbb{R}^d} \left(\sum_{\nu_n > \tau} |\psi_n|^2 \right)^{1+2/d} dx,$$

and consequently, by Fubini's theorem,

$$\sum_n \nu_n \|\nabla \psi_n\|^2 \geq K \int_{\mathbb{R}^d} \int_0^M \left(\sum_{\nu_n > \tau} |\psi_n|^2 \right)^{1+2/d} d\tau \, dx.$$

The claimed inequality now follows from the pointwise inequality

$$\int_0^M \left(\sum_{\nu_n > \tau} |\psi_n|^2 \right)^{1+2/d} d\tau \geq M^{-2/d} \left(\int_0^M \sum_{\nu_n > \tau} |\psi_n|^2 \, d\tau \right)^{1+2/d}$$

$$= M^{-2/d} \left(\sum_n \nu_n |\psi_n|^2 \right)^{1+2/d},$$

which, in turn, follows from Hölder's inequality.   □

7.4.2 The Lieb–Thirring inequality for antisymmetric functions

We now describe the form in which the Lieb–Thirring inequality was originally used in the proof of stability of matter by Lieb and Thirring (1975).

Let $N \geq 2$. We denote points in $\mathbb{R}^{dN} = (\mathbb{R}^d)^N$ by $X = (X_1, \ldots, X_N)$ with $X_n \in \mathbb{R}^d$. For a function $\psi \in L^2(\mathbb{R}^{dN})$ we define a non-negative function ρ_ψ on $\mathbb{R}^d$ by

$$\rho_\psi(x) := \sum_{n=1}^N \int_{\mathbb{R}^{d(N-1)}} |\psi(\ldots, X_{n-1}, x, X_{n+1}, \ldots)|^2 \, d\hat{X}_n, \qquad x \in \mathbb{R}^d,$$

where $d\hat{X}_n$ denotes $dX_1 \cdots dX_{n-1} dX_{n+1} \cdots dX_N$. It follows by Fubini's theorem that ρ_ψ is well-defined almost everywhere and that

$$\int_{\mathbb{R}^d} \rho_\psi(x) \, dx = N \|\psi\|^2.$$

Moreover, let S_N be the symmetric group of $\{1, \ldots, N\}$. A function $\psi : \mathbb{R}^{dN} \to \mathbb{C}$ is called *antisymmetric* if, for all $\sigma \in S_N$,

$$\psi(X_{\sigma(1)}, \ldots, X_{\sigma(N)}) = (\operatorname{sgn} \sigma) \, \psi(X_1, \ldots, X_N) \quad \text{for all } (X_1, \ldots, X_N) \in \mathbb{R}^{dN}.$$

The following theorem is due to Lieb and Thirring (1975, 1976).

Theorem 7.16 *Let $\psi \in H^1(\mathbb{R}^{dN})$ be antisymmetric. Then*

$$\int_{\mathbb{R}^{dN}} |\nabla\psi|^2 \, dX \, \|\psi\|^{4/d} \geq K_d \int_{\mathbb{R}^d} \rho_\psi(x)^{1+2/d} \, dx \qquad (7.20)$$

where K_d is related to the optimal constant $L_{1,d}$ by

$$\left((1 + \tfrac{d}{2}) L_{1,d} \right)^{1+\frac{2}{d}} \left((1 + \tfrac{2}{d}) K_d \right)^{1+\frac{d}{2}} = 1. \qquad (7.21)$$

The main point of this theorem is that the constant K_d can be chosen independently of N. This is a consequence of the assumed antisymmetry of ψ.

It is instructive to discuss briefly the inequality without this assumption. For general $\psi \in H^1(\mathbb{R}^{dN})$ we have

$$\int_{\mathbb{R}^{dN}} |\nabla\psi|^2 \, dX \, \|\psi\|^{4/d} \geq K_d^{(1)} N^{-2/d} \int_{\mathbb{R}^d} \rho_\psi^{1+2/d} \, dx, \qquad (7.22)$$

where $K_d^{(1)}$ is the best constant in (7.15). This follows from (7.15) by an inequality due to Hoffmann-Ostenhof and Hoffmann-Ostenhof (1977) which states that, for any $\psi \in H^1(\mathbb{R}^{dN})$, we have

$$\int_{\mathbb{R}^{dN}} |\nabla\psi|^2 \, dX \geq \int_{\mathbb{R}^d} |\nabla\sqrt{\rho_\psi}|^2 \, dx. \qquad (7.23)$$

Note that (7.23) is an equality for any product function $\psi(X) = \varphi(X_1) \cdots \varphi(X_N)$ with $\varphi \in H^1(\mathbb{R}^d)$. Taking φ to be an optimizer for inequality (7.15) (which exists), we see that for each N the constant $K_d^{(1)} N^{-2/d}$ in (7.22) is optimal. In particular, this constant tends to zero as $N \to \infty$, in contrast to the constant in Theorem 7.16.

For the proof of Theorem 7.16 we need to introduce an auxiliary object. For a function $\psi \in L^2(\mathbb{R}^{dN})$ we define an integral operator γ_ψ in $L^2(\mathbb{R}^d)$ through its integral kernel

$$\gamma_\psi(x,y) := \sum_{n=1}^{N} \int_{\mathbb{R}^{d(N-1)}} \psi(\ldots, X_{n-1}, x, X_{n+1}, \ldots) \overline{\psi(\ldots, X_{n-1}, y, X_{n+1}, \ldots)} \, d\hat{X}_n,$$

with $x, y \in \mathbb{R}^d$.

Lemma 7.17 *Let $\psi \in L^2(\mathbb{R}^{dN})$. Then the operator γ_ψ in $L^2(\mathbb{R}^d)$ is nonnegative and compact with $\operatorname{Tr} \gamma_\psi = N \|\psi\|^2$. Moreover, if*

$$\gamma_\psi = \sum_j \nu_j (\cdot, \varphi_j) \varphi_j \qquad (7.24)$$

with orthonormal φ_j and with $\nu_j > 0$, then

$$\rho_\psi = \sum_j \nu_j |\varphi_j|^2.$$

Finally, if $\psi \in H^1(\mathbb{R}^{dN})$, then $\varphi_j \in H^1(\mathbb{R}^d)$ for all j and

$$\int_{\mathbb{R}^{dN}} |\nabla \psi|^2 \, dX = \sum_j v_j \|\nabla \varphi_j\|^2 \, .$$

The so-called *Schmidt expansion* (7.24) of γ_ψ exists according to Lemma 1.15.

Proof We define an operator $K : L^2(\mathbb{R}^d) \to L^2(\mathbb{R}^{d(N-1)}, \mathbb{C}^N)$ by

$$(K\varphi)_n(\ldots, X_{n-1}, X_{n+1}, \ldots) := \int_{\mathbb{R}^d} \overline{\psi(\ldots, X_{n-1}, y, X_{n+1}, \ldots)} \, \varphi(y) \, dy \, .$$

Then

$$(\gamma_\psi \varphi, \varphi) = \sum_{n=1}^N \int_{\mathbb{R}^{d(N-1)}} |(K\varphi)_n(\ldots, X_{n-1}, X_{n+1}, \ldots)|^2 \, d\hat{X}_n = \|K\varphi\|^2 \, ,$$

so $\gamma_\psi = K^* K$. This proves that γ_ψ is non-negative. Moreover, Lemma 1.38 shows that γ_ψ is compact with

$$\operatorname{Tr} \gamma_\psi = \operatorname{Tr} K^* K = \sum_{n=1}^N \iint_{\mathbb{R}^{d(N-1)} \times \mathbb{R}^d} \left| \overline{\psi(\ldots, X_{n-1}, y, X_{n+1}, \ldots)} \right|^2 \, d\hat{X}_n \, dy$$

$$= N\|\psi\|^2 \, ,$$

as claimed.

As preparation for the proof of the second part of the theorem, we note that for any bounded operator T on $L^2(\mathbb{R}^d)$,

$$\sum_{n=1}^N \int_{\mathbb{R}^{dN}} |T_n \psi|^2 \, dX = \operatorname{Tr} T \gamma_\psi T^* \, ,$$

where $T_n := I \otimes \cdots \otimes I \otimes T \otimes I \otimes \cdots \otimes I$ in $L^2(\mathbb{R}^{dN}) = \bigotimes_{n=1}^N L^2(\mathbb{R}^d)$ with T at the nth place. Indeed, this follows by Lemma 1.38, applied to the operator KT^*.

If we use the Schmidt expansion of γ_ψ as in the lemma then, by Lemma 1.36 with an arbitrary complete orthonormal system (u_k),

$$\operatorname{Tr} T \gamma_\psi T^* = \sum_k (T \gamma_\psi T^* u_k, u_k) = \sum_k \sum_j v_j |(T^* u_k, \varphi_j)|^2 = \sum_j v_j \|T \varphi_j\|^2 \, .$$

Here we interchanged the order of summation using positivity and then applied Parseval's identity. Thus we obtain the identity

$$\sum_{n=1}^N \int_{\mathbb{R}^{dN}} |T_n \psi|^2 \, dX = \sum_j v_j \|T \varphi_j\|^2 \qquad (7.25)$$

for any bounded operator T on $L^2(\mathbb{R}^d)$.

Taking T to be multiplication by $w \in L^\infty(\mathbb{R}^d)$, we obtain

$$\int_{\mathbb{R}^d} |w(x)|^2 \rho_\psi(x)\, dx = \sum_j v_j \int_{\mathbb{R}^d} |w(x)|^2 |\varphi_j(x)|^2\, dx \,.$$

This implies the stated identity for ρ_ψ.

Now assume that $\psi \in H^1(\mathbb{R}^{dN})$. The identity in the lemma involving gradients would follow immediately if we could choose $T = \sqrt{-\Delta}$ in (7.25) (or $T = \partial_\ell$ and then sum over $\ell = 1, \ldots, d$). Since these operators are not bounded, we use an approximation argument, applying the above identity with

$$T = \left(t^{-1}(1 - e^{t\Delta})\right)^{1/2}$$

and a parameter $t > 0$. This gives

$$\sum_j v_j \int_{\mathbb{R}^d} \frac{1 - e^{-t|\xi|^2}}{t} |\widehat{\varphi}_j(\xi)|^2\, d\xi = \sum_{n=1}^{N} \int_{\mathbb{R}^{dN}} \frac{1 - e^{-t|\Xi_n|^2}}{t} |\widehat{\psi}(\Xi)|^2\, d\Xi \,.$$

Since $a \mapsto (1 - e^{-a})/a$ is monotone decreasing and tends to 1 as $a \to 0$ and, since all quantities involved are non-negative, we can pass to the limit $t \to 0$ and obtain

$$\sum_j v_j \int_{\mathbb{R}^d} |\xi|^2 |\widehat{\varphi}_j(\xi)|^2\, d\xi = \sum_{n=1}^{N} \int_{\mathbb{R}^{dN}} |\Xi_n|^2 |\widehat{\psi}(\Xi)|^2\, d\Xi \,.$$

We recall the Fourier characterization of H^1 (Lemma 2.31). Since the right side is finite by assumption, so is the left side and we obtain the claimed identity. □

The identity $\mathrm{Tr}\, \gamma_\psi = N\|\psi\|^2$ implies the bound $\|\gamma_\psi\| \leq N\|\psi\|^2$. This bound is saturated in the case $\psi(X) = \varphi(X_1) \cdots \varphi(X_N)$. We now show that under an antisymmetry assumption, the bound can be improved significantly.

Lemma 7.18 *Let $\psi \in L^2(\mathbb{R}^{dN})$ be antisymmetric. Then $\|\gamma_\psi\| \leq \|\psi\|^2$.*

Proof We need to show that

$$(\gamma_\psi f, f) \leq \|\psi\|^2 \|f\|^2 \qquad \text{for all } f \in L^2(\mathbb{R}^d). \tag{7.26}$$

For the proof of (7.26) we may assume that $\|f\| = 1$. Then we may choose an orthonormal basis $(e_\alpha)_{\alpha \in \mathbb{N}}$ of $L^2(\mathbb{R}^d)$ with $e_1 = f$. For $j \in \mathbb{N}^N$ let

$$\psi_j^\sharp := \int_{\mathbb{R}^{dN}} \psi(X)\, \overline{e_{j_1}(X_1) \cdots e_{j_N}(x_N)}\, dX \,,$$

so that

$$\psi = \sum_{j \in \mathbb{N}^N} \psi_j^\sharp\, e_{j_1} \otimes \cdots \otimes e_{j_N} \,,$$

and therefore, by orthonormality,

$$(\gamma_\psi f, f) = \sum_{n=1}^{N} \sum_{j,j' \in \mathbb{N}^N} \psi_j^\sharp \overline{\psi_{j'}^\sharp} \delta_{j_1,j_1'} \cdots \delta_{j_n,1} \delta_{j_n',1} \cdots \delta_{j_N,j_N'}$$

$$= \sum_{n=1}^{N} \sum_{j \in \mathbb{N}^N} |\psi_j^\sharp|^2 \delta_{j_n,1} = \sum_{j \in \mathbb{N}^N} \#\{n \colon j_n = 1\} |\psi_j^\sharp|^2 .$$

The antisymmetry of ψ implies that

$$\psi_{(\sigma(j_1),...,\sigma(j_N))}^\sharp = (\mathrm{sgn}\,\sigma)\,\psi_{(j_1,...,j_N)}^\sharp \qquad \text{for all } \sigma \in S_N,\, j \in \mathbb{N}^N .$$

In particular, if $\psi_j^\sharp \neq 0$ then $j_n = 1$ for at most one n and thus

$$\sum_{j \in \mathbb{N}^N} \#\{n \colon j_n = 1\} |\psi_j^\sharp|^2 \leq \sum_{j \in \mathbb{N}^N} |\psi_j^\sharp|^2 = \|\psi\|^2 .$$

This proves (7.26) and completes the proof. □

Finally, we are in position to prove the main result of this subsection.

Proof of Theorem 7.16   We shall prove that for any $\psi \in H^1(\mathbb{R}^{dN})$, not necessarily antisymmetric, we have

$$\int_{\mathbb{R}^{dN}} |\nabla \psi|^2 \, dX \, \|\gamma_\psi\|^{2/d} \geq K_d \int_{\mathbb{R}^d} \rho_\psi(x)^{1+2/d} \, dx .$$

By Lemma 7.18, this implies the statement of the theorem.

Using the Schmidt expansion of γ_ψ as in Lemma 7.17, we see that the claimed inequality is equivalent to the inequality

$$\sum_j v_j \|\nabla \varphi_j\|^2 \, \|v\|_\infty^{2/d} \geq K_d \int_{\mathbb{R}^d} \left(\sum_j v_j |\varphi_j|^2 \right)^{1+2/d} dx .$$

The latter inequality holds according to Theorems 4.38 and 7.12, and Lemma 7.15. This completes the proof. □

Remark 7.19   Conversely, if (7.20) holds with some constant K_d, then (7.17) holds with $K = K_d$. Indeed, for L^2-orthonormal $\psi_1, \ldots, \psi_N \in H^1(\mathbb{R}^d)$, we can apply (7.20) to

$$\psi(X_1, \ldots, X_N) := \det \left(\psi_n(X_m) \right)_{n,m} ,$$

which is an antisymmetric function in $H^1(\mathbb{R}^{dN})$ with $\|\psi\| = 1$. Moreover, $\rho_\psi = \sum_{n=1}^{N} |\psi_n|^2$ and

$$\int_{\mathbb{R}^{dN}} |\nabla \psi|^2 \, dX = \sum_{n=1}^{N} \int_{\mathbb{R}^d} |\nabla \psi_n|^2 \, dx .$$

Thus, by Theorem 7.12, (7.20) with constant K_d implies (7.16) with $L = L_{1,d}$ defined by (7.21).

Remark 7.20   The proof of Theorem 7.16 shows, in fact, the following inequality, which is often useful in applications. For any non-negative compact operator γ in $L^2(\mathbb{R}^d)$ with ran $\gamma \subset H^1(\mathbb{R}^d)$ one has

$$(\mathrm{Tr}(-\Delta)\gamma)\, \|\gamma\|^{2/d} \geq K_d \int_{\mathbb{R}^d} \rho_\gamma(x)^{1+2/d}\, dx. \qquad (7.27)$$

Here we *define* $\mathrm{Tr}(-\Delta)\gamma := \sum_j v_j \|\nabla \varphi_j\|^2$ and $\rho_\gamma := \sum_j v_j |\varphi_j|^2$ in terms of the Schmidt expansion $\gamma = \sum_j v_j (\cdot, \varphi_j)\varphi_j$. One can easily check that these quantities are well defined, in the sense that they are independent of the choice of the φ_j in the case of a degenerate eigenvalue v_j. Inequality (7.27) generalizes Theorem 7.16 since, if $\psi \in H^1(\mathbb{R}^{dN})$ is antisymmetric, then $\|\nabla \psi\|^2 = \mathrm{Tr}(-\Delta)\gamma_\psi$ and $\rho_\psi = \rho_{\gamma_\psi}$.

7.4.3 One-particle and semiclassical constants in the dual formulation

In §5.1 we discussed the Lieb–Thirring conjecture, which says, in particular, that the optimal constant $L_{1,d}$ for $\gamma = 1$ is given by $\max\{L_{1,d}^{(1)}, L_{1,d}^{cl}\}$. We now discuss the corresponding conjecture in the dual formulation. In this subsection we denote by K_d the optimal constant in (7.17) and recall that, by Theorem 7.12, this constant is related to the optimal constant $L_{1,d}$ in (7.16) by

$$\left((1 + \tfrac{d}{2})L_{1,d}\right)^{1+\frac{2}{d}} \left((1 + \tfrac{2}{d})K_d\right)^{1+\frac{d}{2}} = 1. \qquad (7.28)$$

The discussion of duality for the one-particle constant appears in Proposition 5.3 and was briefly mentioned at the beginning of §7.4.1. Here we discuss it in more detail. We consider the optimal constant $K_d^{(1)}$ in the inequality

$$\int_{\mathbb{R}^d} |\nabla \psi|^2\, dx \geq K_d^{(1)} \int_{\mathbb{R}^d} |\psi|^{2+4/d}\, dx$$

for all $\psi \in H^1(\mathbb{R}^d)$ with $\|\psi\|_{L^2(\mathbb{R}^d)} = 1$. This inequality is the case $N = 1$ of (7.17), and consequently,

$$K_d^{(1)} \geq K_d.$$

Moreover, if $L_{1,d}^{(1)}$ continues to denote the optimal constant in the inequality

$$E_1 \leq L_{1,d}^{(1)} \int_{\mathbb{R}^d} V_+^{1+d/2}\, dx,$$

where $-E_1 = \min\{\inf \sigma(-\Delta - V), 0\}$, then, by Proposition 5.3,

$$\left(\left(1 + \tfrac{d}{2}\right)L_{1,d}^{(1)}\right)^{1+\frac{2}{d}}\left(\left(1 + \tfrac{2}{d}\right)K_d^{(1)}\right)^{1+\frac{d}{2}} = 1.$$

This should be compared with (7.28).

In the remainder of this subsection we consider semiclassical constants. Recall that

$$L_{1,d}^{\mathrm{cl}} = \frac{1}{(4\pi)^{d/2}\,\Gamma(2 + d/2)} = \frac{2}{d+2}\,\frac{\omega_d}{(2\pi)^d},$$

where ω_d is the volume of the unit ball in $\mathbb{R}^d$. We define the constant

$$K_d^{\mathrm{cl}} := \frac{d}{d+2}\,\frac{(2\pi)^2}{\omega_d^{2/d}}$$

and note that

$$\left(\left(1 + \tfrac{d}{2}\right)L_{1,d}^{\mathrm{cl}}\right)^{1+\frac{2}{d}}\left(\left(1 + \tfrac{2}{d}\right)K_d^{\mathrm{cl}}\right)^{1+\frac{d}{2}} = 1.$$

In the remainder of this subsection we prove the following inequality.

Lemma 7.21 $K_d^{\mathrm{cl}} \geq K_d$

Of course, by the duality relations for K_d^{cl} and K_d this lemma follows from the inequality $L_{1,d} \geq L_{1,d}^{\mathrm{cl}}$. We proved the latter inequality in §5.1.1 using Weyl asymptotics, but our point here is that the inequality in the lemma can be proved directly in the dual formulation. This is closely related to the Thomas–Fermi approximation in mathematical physics.

Proof   Let $Q_L := (-L/2, L/2)^d$ and $0 \leq \zeta \in C_c^1(\mathbb{R}^d)$ with $\int_{\mathbb{R}^d} \zeta\, dx = 1$. For $k \in (2\pi/L)\mathbb{Z}^d$, we consider

$$\psi_k(x) := L^{-d/2}\sqrt{\chi_{Q_L} * \zeta}\; e^{ik\cdot x}.$$

These functions are orthonormal in $L^2(\mathbb{R}^d)$ since

$$(\psi_k, \psi_{k'}) = L^{-d}\int_{\mathbb{R}^d} \chi_{Q_L} * \zeta\; e^{i(k-k')\cdot x}\, dx = (2\pi)^{d/2}L^{-d}\widehat{\chi_{Q_L} * \zeta}(k' - k)$$

$$= (2\pi)^d L^{-d}\widehat{\chi_{Q_L}}(k' - k)\widehat{\zeta}(k' - k) = \delta_{k,k'}.$$

The last identity follows from the fact that $\widehat{\zeta}(0) = (2\pi)^{-d/2}$ and

$$\widehat{\chi_{Q_L}}(\xi) = (2\pi)^{-d/2}L^d\prod_{\ell=1}^{d}\operatorname{sinc}(L\xi_\ell/2)$$

where $\operatorname{sinc}(\tau) := \tau^{-1}\sin\tau$ if $\tau \neq 0$ and $\operatorname{sinc}(0) := 1$.

We will fix a parameter $\mu > 0$ and consider the functions ψ_k with $|k|^2 < \mu$ as trial functions in the Lieb–Thirring inequality. Let

$$N := \#\{k \in (2\pi/L)\mathbb{Z}^d : |k|^2 < \mu\}.$$

In the limit $L \to \infty$ (with μ fixed) we have

$$N = L^d \int_{|\xi|^2 < \mu} \frac{d\xi}{(2\pi)^d}(1 + o(1)) = \frac{\omega_d}{(2\pi)^d}(\mu L^2)^{d/2}(1 + o(1)).$$

Moreover,

$$\sum_{|k|^2 < \mu} |\psi_k|^2 = L^{-d}N\,\chi_{Q_L} * \zeta.$$

If

$$\ell := \inf\{\rho > 0 : \operatorname{supp}\zeta \subset \overline{B_\rho(0)}\},$$

then $\chi_{Q_{L-\ell}} \leq \chi_{Q_L} * \zeta \leq \chi_{Q_{L+\ell}}$, and therefore as $L \to \infty$ (with ℓ fixed)

$$\int_{\mathbb{R}^d}\left(\sum_{|k|^2 < \mu}|\psi_k|^2\right)^{1+2/d} dx = L^{-2}N^{1+2/d}\left(1 + O(L^{-1})\right)$$

$$= \frac{\omega_d^{1+2/d}}{(2\pi)^{d+2}}\mu^{1+d/2}L^d(1 + o(1)).$$

Meanwhile, a simple computation shows that

$$\|\nabla\psi_k\|^2 = k^2 + L^{-d}\left\|\nabla\sqrt{\chi_{Q_k} * \zeta}\right\|^2.$$

Using the fact that $\nabla\sqrt{\chi_{Q_k} * \zeta}$ is supported in $\overline{Q_{L+\ell}} \setminus Q_{L-\ell}$, one easily sees that $L^{-d}\|\nabla\sqrt{\chi_{Q_k} * \zeta}\|^2 = O(L^{-1})$. Consequently,

$$\sum_{|k|^2 < \mu}\|\nabla\psi_k\|^2 = \sum_{|k|^2 < \mu}k^2 + O(NL^{-1})$$

$$= L^d\int_{|\xi|^2 < \mu}|\xi|^2\frac{d\xi}{(2\pi)^d}(1 + o(1)) + O(L^{d-1})$$

$$= \frac{d}{d+2}\frac{\omega_d}{(2\pi)^d}\mu^{1+d/2}L^d(1 + o(1)).$$

Inserting these bounds into the Lieb–Thirring inequality and letting $L \to \infty$, we see that the optimal constant K_d satisfies

$$\frac{d}{d+2}\frac{\omega_d}{(2\pi)^d}\mu^{1+d/2} \geq K_d\frac{\omega_d^{1+2/d}}{(2\pi)^{d+2}}\mu^{1+d/2},$$

which is the same as $K_d \leq \frac{d}{d+2}(2\pi)^2\omega_d^{-2/d} = K_d^{\mathrm{cl}}$, as claimed. $\qquad\square$

7.5 Comments

Section 7.1: Monotonicity with respect to the semiclassical parameter

Theorem 7.1 for $\gamma = 2$ is due to Stubbe (2010). The proof that we presented essentially follows Stubbe's, but is somewhat more elementary. The idea of multiplying by g_ε instead of by the unbounded coordinate functions is from Demirel (2012). Moreover, here we use a convexity argument to avoid differentiating eigenvalues. The result of Theorem 7.1, with the same proof, extends to the case of matrix-valued potentials.

A monotonicity argument based on trace identities as in Lemma 7.2 appeared already in Harrell and Stubbe (1997) in the context of the Dirichlet Laplacian on a domain. Further references for trace identities and monotonicity arguments are Harrell and Stubbe (2010, 2011).

It is interesting to note that, whereas the proof of sharp Lieb–Thirring inequalities in higher dimensions in Theorem 6.11 proceeds via the one-dimensional result in Theorem 6.12, we can prove Theorem 7.1 directly in any dimension.

Section 7.2: Bounds for radial potentials

The material in §7.2 is based on Glaser et al. (1978).

The proof of Proposition 7.4 in Glaser et al. (1978) uses a characterization of the number of negative eigenvalues by means of Sturm–Liouville theory. We presented an alternative proof from Frank (2021), based on its variational characterization. A not unrelated use of Glazman's lemma appears in Birman (1961, Theorem 2.7).

Theorem 7.5 remains valid without the assumption of V being radially symmetric; see Theorem 4.53 and the comments in §4.9.

In §7.2.3 we fill in some details in the arguments in Glaser et al. (1978) following partly Frank (2021, §4.6).

Section 7.3: More on the one-particle constants

The material in §7.3 is based on Frank et al. (2021b,c). The details that we omitted in the proof of Theorem 7.11 are mostly concerned with the exponential behavior at infinity of the optimal one-particle potential.

Frank et al. (2021b) raises the question of whether the optimal Lieb–Thirring potentials V_N for the inequality (7.13) with given N, suitably normalized, converge to a non-trivial function V that does not belong to $L^{\gamma + d/2}(\mathbb{R}^d)$, for instance, a periodic function. Both analytical and numerical investigations in this direction can be found in Frank et al. (2021a).

Section 7.4: The dual Lieb–Thirring inequality

While it is the kinetic energy version of the Lieb–Thirring inequality (Theorem 7.16) that was the main ingredient in the proof of stability of matter by Lieb and Thirring (1975), it was the spectral version that they proved and then transferred via the duality argument explained in this section. The first proof of the Lieb–Thirring inequality that worked directly in the kinetic version was given by Eden and Foias (1991) in the one-dimensional setting. A proof in arbitrary dimension was found only much later by Rumin (2011).

Recently, progress on good values of the constants in the Lieb–Thirring inequalities was made mostly in the dual formulation, combined with the idea of lifting of dimension discussed in the previous chapter. This will be further discussed in the next chapter.

Nam (2018) proved an interesting form of the dual Lieb–Thirring inequality with a constant that is arbitrarily close to the semiclassical constant, at the expense of a gradient correction. More precisely, for any $d \geq 1$ there is a constant $C_d < \infty$ such that for all $\psi_n \in H^1(\mathbb{R}^d)$ that are orthogonal in $L^2(\mathbb{R}^d)$ and all $\varepsilon > 0$,

$$\sum_n \int_{\mathbb{R}^d} |\nabla \psi_n|^2 \, dx \geq (1 - \varepsilon) K_d^{\mathrm{cl}} \int_{\mathbb{R}^d} \left(\sum_n |\psi_n|^2 \right)^{1+2/d} dx$$
$$- \varepsilon^{-3-4/d} C_d \int_{\mathbb{R}^3} \left| \nabla \sqrt{\sum_n |\psi_n|^2} \right|^2 dx \,.$$

Theorem 7.12 gives an equivalent form of the Lieb–Thirring inequality for $\gamma = 1$. A dual form in the same spirit for all $\gamma \geq 1$ appears in the appendix to Lions and Paul (1993). It is also discussed from the broader perspective of Legendre transforms in Dolbeault et al. (2006). There, in particular, a duality is established between the Golden–Thompson inequality and a log-Sobolev inequality for orthonormal functions. There is also a dual version of the spectral Lieb–Thirring inequality (7.13) with a fixed number of eigenvalues; see Frank et al. (2021b).

The relaxation in Lemma 7.15 to "occupation numbers" ν_n that are not necessarily zero or one is closely related to a relaxation of the variational principle for the sum of negative eigenvalues (Proposition 1.33 and Corollary 1.35) mentioned in the comments to §1.2.

The proof of Lemma 7.21 is from Gontier et al. (2021). A more traditional proof considers the Lieb–Thirring inequality in the version of Remark 7.20 and the trial density matrix γ in $L^2(\mathbb{R}^d)$ with integral kernel

$$\gamma(x, y) = \eta(x/L) \int_{|\xi|^2 < \mu} e^{i\xi \cdot (x-y)} \frac{d\xi}{(2\pi)^d} \, \eta(y/L),$$

where $0 \leq \eta \in C_c^1(\mathbb{R}^d)$. Then the validity of the dual Lieb–Thirring inequality implies the inequality

$$\frac{d}{d+2} \int_{\mathbb{R}^d} \eta^2 \, dx + \mu^{-1} L^{-2} \int_{\mathbb{R}^d} |\nabla \eta|^2 \, dx \geq K_d \left(\frac{\omega_d}{(2\pi)^d} \right)^{2/d} \int_{\mathbb{R}^d} \eta^{1+2/d} \, dx .$$

First taking the limit $\mu^{-1} L^{-2} \to 0$ and then letting η approach a characteristic function, one obtains the inequality in Lemma 7.21.

The Lieb–Thirring inequality in the dual formulation is related to the Thomas–Fermi approximation in atomic physics. For the discussion of this and other density functional theories we refer, for instance, to Lieb and Simon (1977), Lieb (1981, 1983a), Lewin et al. (2018, 2020a,b), and references therein.

An extension of the Lieb–Thirring inequality to the case of a homogeneous background density appears in Frank et al. (2013). The motivation, statement, and proof of this inequality are easier in the kinetic version. For a partial translation into the spectral version, see Frank and Pushnitski (2015).

The kinetic version of the Lieb–Thirring inequality can be generalized to the case where interactions between the particles are present, a situation that does not have an analogue in the spectral formulation of Lieb–Thirring inequalities. This extension is motivated, for instance, by the consideration of non-fermionic (quasi)particles like anyons. These developments started with Lundholm and Solovej (2013) and were further pursued, for instance, in Frank and Seiringer (2012), Lundholm and Solovej (2014), Lundholm et al. (2015, 2016), Lundholm and Seiringer (2018), and Larson et al. (2021).

While the main applications of the Lieb–Thirring inequalities concern the analysis of quantum many-body systems, they have also proved useful in connection with the dimension of attractors for the Navier–Stokes and other nonlinear evolution equations; see for instance Lieb (1984), Constantin et al. (1985), Temam (1997), Chepyzhov and Ilyin (2004), Il'in (2005)[1], and references therein.

In the spirit of the interpretation of the Lieb–Thirring inequality as a Gagliardo–Nirenberg inequality for orthonormal functions, Lieb proved a version of the Hardy–Littlewood–Sobolev inequality (Lieb, 1983b). Recently, partially motivated by questions from many-body quantum dynamics (Lewin and Sabin, 2015, 2014) and independently by the spectral theory of Schrödinger operators with complex potentials, there have been extensions of various inequalities from harmonic analysis to the case of orthonormal systems; see, for instance, Frank et al. (2014), Frank and Sabin (2017a,c,b), Bez et al. (2019, 2020, 2021), and Nakamura (2020).

---

[1] Here we use the transliteration Il'in from the original paper, while elsewhere we use the current convention for the spelling of his name.

8

More on the Lieb–Thirring Constants

Apart from those treated in Chapters 5 and 6, there are no other cases where the sharp constants in the Lieb–Thirring inequalities are known. In this chapter, we discuss the currently best known values for these constants in cases where the optimal constants are not known. Such good, even if non-optimal values are important in some applications. All results that we prove in this chapter are obtained using the technique of lifting in dimension described in Chapter 6.

In §8.1, we formulate a version of this method suitable for these applications. The input for the method are Lieb–Thirring inequalities with matrix-valued potentials in low dimensions and the output are Lieb–Thirring inequalities in higher dimensions with matrix-valued potentials. The crucial observation is that the quality of the constants relative to the semiclassical constants does not deteriorate as one passes from low to high dimensions. While this is not a necessity, the input inequalities that we will use in this book will all be in one dimension.

In §8.2, we discuss a Lieb–Thirring inequality for the sum of the square roots of the eigenvalues in one dimension with matrix-valued potential. We will see that the inequality holds with the same constant as in the scalar case, namely with the one-particle constant. The proof follows rather closely that in the scalar case.

In §8.3, we discuss a Lieb–Thirring inequality for the sum of the eigenvalues in one dimension with matrix-valued potential. The constant in this inequality is probably not optimal, but, via the method of lifting in dimension, leads to the currently best known values for the Lieb–Thirring constants $L_{1,d}$ in the important special case $\gamma = 1$ for any dimension d. The inequality in this section is proved in the dual formulation discussed in §7.4.3.

In §8.4, we summarize the sharp and non-sharp results about the constants in the Lieb–Thirring inequality that we have discussed in this book.

8.1 More on Lieb–Thirring inequalities in the matrix-valued case

In this section, we revisit the method of lifting in dimension from §6.2. We will prove three theorems. The first one is a generalization of Theorem 6.11 and states that the Lieb–Thirring inequality with semiclassical constant holds for $\gamma \geq 3/2$ in any dimension even for matrix-valued potentials.

Theorem 8.1 *Let $d \in \mathbb{N}$ and $\gamma \geq \frac{3}{2}$. Then, for any $M \in \mathbb{N}$ and any $V \in L^{\gamma+\frac{d}{2}}(\mathbb{R}^d, \mathcal{H}_M)$,*

$$\operatorname{Tr}(-\Delta \otimes \mathbb{I}_{\mathbb{C}^M} - V)^{\gamma}_{-} \leq L^{\mathrm{cl}}_{\gamma,d} \int_{\mathbb{R}^d} \operatorname{Tr}_{\mathbb{C}^M} V(x)^{\gamma+\frac{d}{2}}_{+} \, dx \,.$$

The following two theorems say that, in the matrix-valued setting, one gets higher-dimensional Lieb–Thirring inequalities from one-dimensional ones without deterioration of the constant relative to the semiclassical one.

Theorem 8.2 *Let $1 \leq \gamma < 3/2$ and assume that there is a constant $R_{\gamma,1}$ such that, for all $N \in \mathbb{N}$, one has*

$$\operatorname{Tr}_{L^2(\mathbb{R}) \otimes \mathbb{C}^N} \left(-\frac{d^2}{dx^2} \otimes \mathbb{I}_{\mathbb{C}^N} - Q \right)^{\gamma}_{-} \leq R_{\gamma,1} L^{\mathrm{cl}}_{\gamma,1} \int_{\mathbb{R}} \operatorname{Tr}_{\mathbb{C}^N} Q(x)^{\gamma+\frac{1}{2}}_{+} \, dx$$

$$\textit{for all } Q \in L^{\gamma+\frac{1}{2}}(\mathbb{R}, \mathcal{H}_N) \,. \tag{8.1}$$

Then, for all $d \in \mathbb{N}$ and all $M \in \mathbb{N}$,

$$\operatorname{Tr}_{L^2(\mathbb{R}^d) \otimes \mathbb{C}^M} (-\Delta \otimes \mathbb{I}_{\mathbb{C}^M} - V)^{\gamma}_{-} \leq R_{\gamma,1} L^{\mathrm{cl}}_{\gamma,d} \int_{\mathbb{R}^d} \operatorname{Tr}_{\mathbb{C}^M} V(x)^{\gamma+\frac{d}{2}}_{+} \, dx$$

$$\textit{for all } V \in L^{\gamma+\frac{d}{2}}(\mathbb{R}^d, \mathcal{H}_M) \,.$$

Theorem 8.3 *Let $1/2 \leq \gamma < 1$ and assume that there are constants $R_{\gamma,1}, R_{\gamma+\frac{1}{2},1}$ such that, for each $v = 0,1$ and all $N \in \mathbb{N}$, one has*

$$\operatorname{Tr}_{L^2(\mathbb{R}) \otimes \mathbb{C}^N} \left(-\frac{d^2}{dx^2} \otimes \mathbb{I}_{\mathbb{C}^N} - Q \right)^{\gamma+\frac{v}{2}}_{-} \leq R_{\gamma+\frac{v}{2},1} L^{\mathrm{cl}}_{\gamma+\frac{v}{2},1} \int_{\mathbb{R}} \operatorname{Tr}_{\mathbb{C}^N} Q(x)^{\gamma+\frac{v+1}{2}}_{+} \, dx$$

$$\textit{for all } Q \in L^{\gamma+\frac{v+1}{2}}(\mathbb{R}, \mathcal{H}_N) \,. \tag{8.2}$$

Then, for all $d \in \mathbb{N}$ and all $M \in \mathbb{N}$,

$$\operatorname{Tr}_{L^2(\mathbb{R}^d) \otimes \mathbb{C}^M} (-\Delta \otimes \mathbb{I}_{\mathbb{C}^M} - V)^{\gamma}_{-} \leq R_{\gamma,1} R_{\gamma+\frac{1}{2},1} L^{\mathrm{cl}}_{\gamma,d} \int_{\mathbb{R}^d} \operatorname{Tr}_{\mathbb{C}^M} V(x)^{\gamma+\frac{d}{2}}_{+} \, dx$$

$$\textit{for all } V \in L^{\gamma+\frac{d}{2}}(\mathbb{R}^d, \mathcal{H}_M) \,.$$

Theorems 8.2 and 8.3 will be used in the following two sections to derive "good" constants in Lieb–Thirring inequalities in higher dimensions.

The proofs of all three theorems are based on the same method. Let us prepare the ingredients. We begin with the generalization of the Aizenman–Lieb argument to the case of matrix-valued potentials.

Lemma 8.4 *Let $d \in \mathbb{N}$, $M \in \mathbb{N}$, and $\gamma \geq 0$ and assume that for some $R_{\gamma,d}$ one has*

$$\operatorname{Tr}_{L^2(\mathbb{R}^d) \otimes \mathbb{C}^M} (-\Delta \otimes \mathbb{I}_{\mathbb{C}^M} - Q)^\gamma_- \leq R_{\gamma,d} L^{\mathrm{cl}}_{\gamma,d} \int_{\mathbb{R}^d} \operatorname{Tr}_{\mathbb{C}^M} Q(x)^{\gamma+\frac{d}{2}}_+ \, dx$$

$$\text{for all } Q \in L^{\gamma+\frac{d}{2}}(\mathbb{R}^d, \mathcal{H}_M).$$

Then, for all $\sigma \geq \gamma$,

$$\operatorname{Tr}_{L^2(\mathbb{R}^d) \otimes \mathbb{C}^M} (-\Delta \otimes \mathbb{I}_{\mathbb{C}^M} - V)^\sigma_- \leq R_{\gamma,d} L^{\mathrm{cl}}_{\sigma,d} \int_{\mathbb{R}^d} \operatorname{Tr}_{\mathbb{C}^M} V(x)^{\sigma+\frac{d}{2}}_+ \, dx$$

$$\text{for all } V \in L^{\sigma+\frac{d}{2}}(\mathbb{R}^d, \mathcal{H}_M).$$

The proof of this lemma is similar to that in the scalar case, except that one replaces the formula (5.8) by its matrix generalization

$$L^{\mathrm{cl}}_{\gamma,d} \int_{\mathbb{R}^d} \operatorname{Tr}_{\mathbb{C}^M} Q(x)^{\gamma+d/2}_+ \, dx = \iint_{\mathbb{R}^d \times \mathbb{R}^d} \operatorname{Tr}_{\mathbb{C}^M} \left(|\xi|^2 \otimes \mathbb{I}_{\mathbb{C}^M} - Q(x) \right)^\gamma_- \frac{dx \, d\xi}{(2\pi)^d}.$$

This formula holds, in fact, pointwise in x and follows, by diagonalizing $Q(x)$, from the corresponding scalar formula (5.8), which also holds pointwise in x.

Next we turn to the method of lifting in dimension. We split the variables $x = (x', x'') \in \mathbb{R}^k \times \mathbb{R}^{d-k}$ and denote by $-\Delta''$ the Laplacian with respect to x''.

Lemma 8.5 *Let $k \in \mathbb{N}$ and*

$$\begin{aligned} \gamma &\geq 1/2 &&\text{if } k = 1, \\ \gamma &> 0 &&\text{if } k = 2, \\ \gamma &\geq 0 &&\text{if } k \geq 3. \end{aligned}$$

Assume that there is a constant L such that, for all $N \in \mathbb{N}$, one has

$$\operatorname{Tr}_{L^2(\mathbb{R}^k) \otimes \mathbb{C}^N} (-\Delta \otimes \mathbb{I}_{\mathbb{C}^N} - Q)^\gamma_- \leq L \int_{\mathbb{R}^k} \operatorname{Tr}_{\mathbb{C}^N} Q(x)^{\gamma+\frac{k}{2}}_+ \, dx$$

$$\text{for all } Q \in L^{\gamma+\frac{k}{2}}(\mathbb{R}^k, \mathcal{H}_N). \qquad (8.3)$$

Then, for all $d \geq k + 1$ and all $M \in \mathbb{N}$,

$$\operatorname{Tr}_{L^2(\mathbb{R}^d) \otimes \mathbb{C}^M} (-\Delta \otimes \mathbb{I}_{\mathbb{C}^M} - V)^\gamma_-$$

$$\leq L \int_{\mathbb{R}^k} \operatorname{Tr}_{L^2(\mathbb{R}^{d-k}) \otimes \mathbb{C}^M} (-\Delta'' \otimes \mathbb{I}_{\mathbb{C}^M} - V(x', \cdot))^{\gamma+\frac{k}{2}}_- \, dx'$$

$$\text{for all } V \in L^{\gamma+\frac{d}{2}}(\mathbb{R}^d, \mathcal{H}_M).$$

This lemma generalizes Lemma 6.13 in two independent respects. First, the assumed inequality (8.3) holds on $\mathbb{R}^k$ rather than just on $\mathbb{R}$ and second, the conclusion holds for matrix-valued potentials. Nevertheless, the proof of Lemma 6.13 carries over to this more general assertion without any change and is therefore not repeated.

Proof of Theorem 8.1 We prove Theorem 8.1 by induction on d. For $d = 1$, the assertion holds for any $\gamma \geq 3/2$ by combining the sharp Lieb–Thirring inequality for $\gamma = 3/2$ from Theorem 6.12 with the Aizenman–Lieb argument in Lemma 8.4.

Now let $d \geq 2$ and assume that the assertion holds for all $\gamma \geq 3/2$ with d replaced by $d - 1$. Let us fix a $\gamma \geq 3/2$. Then, as we have just shown, (8.3) holds for all $N \in \mathbb{N}$ with $k = 1$ and $L = L_{\gamma,1}^{\mathrm{cl}}$, and therefore, by Lemma 8.5, we have, for all $M \in \mathbb{N}$ and $V \in L^{\gamma+d/2}(\mathbb{R}^d, \mathcal{H}_M)$,

$$\mathrm{Tr}_{L^2(\mathbb{R}^d)\otimes\mathbb{C}^M}(-\Delta \otimes \mathbb{I}_{\mathbb{C}^M} - V)_-^\gamma$$
$$\leq L_{\gamma,1}^{\mathrm{cl}} \int_{\mathbb{R}} \mathrm{Tr}_{L^2(\mathbb{R}^{d-1})\otimes\mathbb{C}^M}(-\Delta' \otimes \mathbb{I}_{\mathbb{C}^M} - V(x_1,\cdot))_-^{\gamma+1/2}\, dx_1 \, .$$

By the induction hypothesis (with γ replaced by $\gamma + 1/2$), we have

$$\mathrm{Tr}_{L^2(\mathbb{R}^{d-1})\otimes\mathbb{C}^M}(-\Delta' \otimes \mathbb{I}_{\mathbb{C}^M} - V(x_1,\cdot))_-^{\gamma+1/2}$$
$$\leq L_{\gamma+1/2,d-1}^{\mathrm{cl}} \int_{\mathbb{R}^{d-1}} \mathrm{Tr}_{\mathbb{C}^M} V(x_1,x')_+^{\gamma+d/2}\, dx' \, .$$

Combining the last two inequalities gives

$$\mathrm{Tr}_{L^2(\mathbb{R}^d)\otimes\mathbb{C}^M}(-\Delta \otimes \mathbb{I}_{\mathbb{C}^M} - V)_-^\gamma \leq L_{\gamma,1}^{\mathrm{cl}} L_{\gamma+1/2,d-1}^{\mathrm{cl}} \int_{\mathbb{R}^d} \mathrm{Tr}_{\mathbb{C}^M} V(x)_+^{\gamma+d/2}\, dx \, .$$

We now recall from (3.72) that

$$L_{\gamma,1}^{\mathrm{cl}} L_{\gamma+1/2,d-1}^{\mathrm{cl}} = L_{\gamma,d}^{\mathrm{cl}} \, . \tag{8.4}$$

This proves the claim in dimension d and completes the proof. $\qquad\square$

Proof of Theorem 8.2 By assumption (8.1), assertion (8.3) holds for all $N \in \mathbb{N}$ with $k = 1$ and $L = R_{\gamma,1} L_{\gamma,1}^{\mathrm{cl}}$ and therefore by Lemma 8.5 we have, for all $M \in \mathbb{N}$ and $V \in L^{\gamma+d/2}(\mathbb{R}^d, \mathcal{H}_M)$,

$$\mathrm{Tr}_{L^2(\mathbb{R}^d)\otimes\mathbb{C}^M}(-\Delta \otimes \mathbb{I}_{\mathbb{C}^M} - V)_-^\gamma$$
$$\leq R_{\gamma,1} L_{\gamma,1}^{\mathrm{cl}} \int_{\mathbb{R}} \mathrm{Tr}_{L^2(\mathbb{R}^{d-1})\otimes\mathbb{C}^M}(-\Delta' \otimes \mathbb{I}_{\mathbb{C}^M} - V(x_1,\cdot))_-^{\gamma+1/2}\, dx_1 \, .$$

Since $\gamma + 1/2 \geq 3/2$ under the assumptions of the theorem, we can apply

Theorem 8.1 and obtain

$$\mathrm{Tr}_{L^2(\mathbb{R}^{d-1})\otimes \mathbb{C}^M}\left(-\Delta' \otimes \mathbb{I}_{\mathbb{C}^M} - V(x_1,\cdot)\right)_-^{\gamma+1/2}$$
$$\leq L_{\gamma+1/2,d-1}^{\mathrm{cl}} \int_{\mathbb{R}^{d-1}} \mathrm{Tr}_{\mathbb{C}^M} V(x_1,x')_+^{\gamma+d/2}\, dx'.$$

Combining the last two inequalities gives

$$\mathrm{Tr}_{L^2(\mathbb{R}^d)\otimes \mathbb{C}^M}\left(-\Delta \otimes \mathbb{I}_{\mathbb{C}^M} - V\right)_-^{\gamma} \leq R_{\gamma,1} L_{\gamma,1}^{\mathrm{cl}} L_{\gamma+1/2,d-1}^{\mathrm{cl}} \int_{\mathbb{R}^d} \mathrm{Tr}_{\mathbb{C}^M} V(x)_+^{\gamma+d/2}\, dx,$$

which, in view of (8.4), proves the claimed inequality. $\qquad\square$

Proof of Theorem 8.3 The proof is similar to that of Theorem 8.2, except that instead of Theorem 8.1 we apply Theorem 8.2. We omit the details. $\qquad\square$

This completes the proof of the three main theorems in this section. We end with a remark on the dependence of the matrix-valued Lieb–Thirring constants on the spatial dimension d. To state the result more succinctly, we denote by $L_{\gamma,d}^{(\mathrm{mat})}$ the optimal constant such that, for all $M \in \mathbb{N}$ and all $Q \in L^{\gamma+\frac{d}{2}}(\mathbb{R}^d,\mathcal{H}_M)$, we have

$$\mathrm{Tr}_{L^2(\mathbb{R}^d)\otimes \mathbb{C}^M}\left(-\Delta \otimes \mathbb{I}_{\mathbb{C}^M} - Q\right)_-^{\gamma} \leq L_{\gamma,d}^{(\mathrm{mat})} \int_{\mathbb{R}^d} \mathrm{Tr}_{\mathbb{C}^M} Q(x)_+^{\gamma+\frac{d}{2}}\, dx.$$

Although not needed in what follows, we refer to Theorem 6.27 for the fact that $L_{\gamma,d}^{(\mathrm{mat})} < \infty$ if $\gamma \geq 1/2$ in $d = 1$, $\gamma > 0$ if $d = 2$ and $\gamma \geq 0$ if $d \geq 3$.

Lemma 8.6 *Let $1 \leq k < d$ and assume that*

$$\gamma \geq 1/2 \quad \text{if } k = 1,$$
$$\gamma > 0 \quad \text{if } k = 2,$$
$$\gamma \geq 0 \quad \text{if } k \geq 3.$$

Then

$$\frac{L_{\gamma,d}^{(\mathrm{mat})}}{L_{\gamma,d}^{\mathrm{cl}}} \leq \frac{L_{\gamma,k}^{(\mathrm{mat})}}{L_{\gamma,k}^{\mathrm{cl}}} \frac{L_{\gamma+\frac{k}{2},d-k}^{(\mathrm{mat})}}{L_{\gamma+\frac{k}{2},d-k}^{\mathrm{cl}}}.$$

Proof For the proof of the lemma we need to show that, if there are constants $R_{\gamma,k}$ and $R_{\gamma+k/2,d-k}$ such that, for all $N \in \mathbb{N}$, we have

$$\mathrm{Tr}_{L^2(\mathbb{R}^k)\otimes \mathbb{C}^N}\left(-\Delta \otimes \mathbb{I}_{\mathbb{C}^N} - Q\right)_-^{\gamma} \leq R_{\gamma,k} L_{\gamma,k}^{\mathrm{cl}} \int_{\mathbb{R}^k} \mathrm{Tr}_{\mathbb{C}^N} Q(x)_+^{\gamma+\frac{k}{2}}\, dx$$

$$\text{for all } Q \in L^{\gamma+\frac{k}{2}}(\mathbb{R}^k,\mathcal{H}_N)$$

and

$$\mathrm{Tr}_{L^2(\mathbb{R}^{d-k})\otimes\mathbb{C}^N}\,(-\Delta\otimes\mathbb{I}_{\mathbb{C}^N}-Q)_-^{\gamma+\frac{k}{2}}$$

$$\leq R_{\gamma+\frac{k}{2},d-k}L_{\gamma+\frac{k}{2},d-k}^{\mathrm{cl}}\int_{\mathbb{R}^{d-k}}\mathrm{Tr}_{\mathbb{C}^N}\,Q(x)_+^{\gamma+\frac{d}{2}}\,dx\ \text{ for all }\ Q\in L^{\gamma+\frac{d}{2}}(\mathbb{R}^{d\ k},\mathcal{H}_N),$$

then, for all $M\in\mathbb{N}$, we have

$$\mathrm{Tr}_{L^2(\mathbb{R}^d)\otimes\mathbb{C}^M}(-\Delta\otimes\mathbb{I}_{\mathbb{C}^M}-V)_-^\gamma\leq R_{\gamma,k}\,R_{\gamma+\frac{k}{2},d-k}\,L_{\gamma,d}^{\mathrm{cl}}\int_{\mathbb{R}^d}\mathrm{Tr}_{\mathbb{C}^M}\,V(x)_-^{\gamma\,|\,\frac{d}{2}}\,dx$$

$$\text{for all }\ V\in L^{\gamma+\frac{d}{2}}(\mathbb{R}^d,\mathcal{H}_M).$$

The proof of this uses Lemma 8.5 and the techniques in the proof of the theorems in this section, in particular, (3.72) with $d_1=k$ and $d_2=d-k$. We omit the details. $\square$

In the proofs of Theorems 8.1, 8.2, and 8.3 we implicitly used this lemma with $k=1$. Using it with $k=d-1$ and recalling from Theorem 8.1 that $L_{\gamma+\frac{d-1}{2},1}^{(\mathrm{mat})}/L_{\gamma+\frac{d-1}{2},1}^{\mathrm{cl}}=1$ if $\gamma+\frac{d}{2}\geq 2$, gives the following monotonicity result.

Corollary 8.7   *The quotient $L_{\gamma,d}^{(\mathrm{mat})}/L_{\gamma,d}^{\mathrm{cl}}$ is monotone decreasing in d for $d\geq 3$ if $0\leq\gamma<1/2$, for $d\geq 2$ if $1/2\leq\gamma<1$ and for $d\geq 1$ if $\gamma\geq 1$.*

8.2 The sum of the square roots of the eigenvalues

The following theorem is the matrix generalization of Theorem 5.27 and is due to Hundertmark, Laptev and Weidl (2000).

Theorem 8.8   *Let $d=1$ and $\gamma=\frac{1}{2}$. Then, for all $M\in\mathbb{N}$ and all $V\in L^1(\mathbb{R},\mathcal{H}_M)$,*

$$\mathrm{Tr}\left(-\frac{d^2}{dx^2}\otimes\mathbb{I}_{\mathbb{C}^M}-V\right)_-^{1/2}\leq\frac{1}{2}\int_{\mathbb{R}}\mathrm{Tr}_{\mathbb{C}^M}\,V(x)_+\,dx\,.$$

The constant $1/2$ in the theorem is best possible since it is so already in the scalar case $M=1$, as discussed after Theorem 5.27. It is *twice* the semiclassical constant.

The proof of Theorem 8.8 is very similar to that in the scalar case and we only focus on the newly arising issues in the matrix case.

Proof   By the variational principle we may assume that, for almost every

$x \in \mathbb{R}$, the matrix $V(x)$ is non-negative on $\mathbb{C}^M$. Let $W(x)$ be its non-negative square root. As in the scalar case, we put

$$\mathcal{L}_\varepsilon := 2\varepsilon \, W \left(-\frac{d^2}{dx^2} \otimes \mathbb{1}_{\mathbb{C}^M} + \varepsilon^2 \right)^{-1} W \qquad \text{in } L^2(\mathbb{R}, \mathbb{C}^M).$$

This operator is non-negative and, as in Lemma 5.28, we get, from Lemma 1.38,

$$\operatorname{Tr} \mathcal{L}_\varepsilon = \int \operatorname{Tr}_{\mathbb{C}^M} (W(x))^2 \, dx.$$

As in Lemma 5.29, with the probability density g_ε from (5.41), we find, for $0 < \varepsilon' < \varepsilon$,

$$\mathcal{L}_\varepsilon = \int_{\mathbb{R}} U(\xi)^* \, \mathcal{L}_{\varepsilon'} \, U(\xi) \, g_{\varepsilon - \varepsilon'}(\xi) \, d\xi, \tag{8.5}$$

where $U(\xi)$ is the unitary operator in $L^2(\mathbb{R}, \mathbb{C}^M)$ of multiplication by $e^{-i\xi \cdot} \otimes \mathbb{1}_{\mathbb{C}^M}$.

Let $\mu_n(\mathcal{L}_\varepsilon)$ be the eigenvalues of $\mathcal{L}_\varepsilon$ in non-increasing order and repeated according to multiplicities. Then, by Corollary 1.34, representation (8.5) implies

$$\sum_{n=1}^N \mu_n(\mathcal{L}_\varepsilon) \le \sum_{n=1}^N \mu_n(\mathcal{L}_{\varepsilon'}) \qquad \text{for } 0 < \varepsilon' \le \varepsilon \text{ and all } N \in \mathbb{N}.$$

As in the scalar case, by the Birman–Schwinger principle (Corollary 1.57), we have

$$1 = \frac{1}{2\sqrt{E_n}} \mu_n(\mathcal{L}_{\sqrt{E_n}}) \qquad \text{for all } n \in \mathbb{N}.$$

Therefore, we are in the same setting as the scalar case and the proof is completed by the same arguments. $\qquad\square$

8.3 The sum of the eigenvalues

The main result of this section is the following one-dimensional Lieb–Thirring inequality for $\gamma = 1$ with matrix-valued potentials (Frank et al., 2021d).

Theorem 8.9 *Let $d = 1$ and $\gamma = 1$. Then, for all $M \in \mathbb{N}$ and all $V \in L^{3/2}(\mathbb{R}, \mathcal{H}_M)$,*

$$\operatorname{Tr} \left(-\frac{d^2}{dx^2} \otimes \mathbb{1}_{\mathbb{C}^M} - V \right)_- \le R_{1,1} \, L_{1,1}^{\mathrm{cl}} \int_{\mathbb{R}} \operatorname{Tr}_{\mathbb{C}^M} V(x)_+^{3/2} \, dx.$$

with

$$R_{1,1} - \frac{3^{5/2}}{8} \inf \left\{ \left(\int_0^\infty w(s)^2 \, ds \right)^{1/2} \int_0^\infty \frac{(1 - g(t))^2}{t^{3/2}} \, dt : f, w \geq 0, \right.$$

$$\left. \int_0^\infty f(s)^2 \, ds = 1, \; g(t) = \int_0^\infty w(s) f(st) \, ds \right\}.$$

Moreover,

$$R_{1,1} \leq 1.456 \,.$$

We shall deduce this theorem from a Sobolev-type inequality for orthonormal functions. We formulate this inequality for functions defined on $\mathbb{R}^d$ for general $d \geq 1$, although in the application to Theorem 8.9 we will choose $d = 1$.

Let $\psi_1, \ldots, \psi_N \in H^1(\mathbb{R}^d, \mathbb{C}^M)$ and, for $x \in \mathbb{R}^d$ (recalling that $\langle \cdot, \cdot \rangle$ denotes the inner product in $\mathbb{C}^M$), write

$$R(x) := \sum_{n=1}^N \langle \cdot, \psi_n(x) \rangle \psi_n(x) \,; \tag{8.6}$$

that is, $R(x)$ is the $M \times M$ Hermitian matrix $R(x) = \left(R_{m,m'}(x) \right)_{m,m'=1}^M$ with entries

$$R_{m,m'}(x) = \sum_{n=1}^N \psi_n(x, m) \overline{\psi_n(x, m')},$$

where $\psi_n(x) = \left(\psi_n(x, m) \right)_{m=1}^M$. This definition appears naturally in our context since for an $\mathcal{H}_M$-valued V on $\mathbb{R}^d$ we have, for all $x \in \mathbb{R}^d$,

$$\sum_{n=1}^N \langle V(x) \psi_n(x), \psi_n(x) \rangle = \mathrm{Tr}_{\mathbb{C}^M} \left(V(x) R(x) \right). \tag{8.7}$$

Theorem 8.10 *Let $d \geq 1$, $M \geq 1$ and $N \in \mathbb{N}$. Let $\psi_1, \ldots, \psi_N \in H^1(\mathbb{R}^d, \mathbb{C}^M)$ be orthonormal in $L^2(\mathbb{R}^d, \mathbb{C}^M)$ and $R(x) = \sum_{n=1}^N \langle \cdot, \psi_n(x) \rangle \psi_n(x)$ for $x \in \mathbb{R}^d$. Then*

$$\sum_{n=1}^N \int_{\mathbb{R}^d} |\nabla \psi_n|^2 \, dx \geq K_d \int_{\mathbb{R}^d} \mathrm{Tr}_{\mathbb{C}^M} R(x)^{1+2/d} \, dx,$$

where

$$K_d := \frac{2^{6/d} d^2 (2\pi)^2}{(d+2)^{2+4/d} |\mathbb{S}^{d-1}|^{2/d}} I_d^{-2/d}$$

and

$$\mathcal{I}_d := \inf \left\{ \left(\int_0^\infty w(s)^2 \, ds \right)^{d/2} \int_0^\infty \frac{(1 - g(t))^2}{t^{1+d/2}} \, dt : f, w \geq 0, \right.$$

$$\left. \int_0^\infty f(s)^2 \, ds = 1, \ g(t) = \int_0^\infty w(s) f(st) \, ds \right\}.$$

Proof Step 1. Let f be a non-negative function on $(0, \infty)$ with $\int_0^\infty f(s)^2 \, ds = 1$. For any $E > 0$ we define $\mathbb{C}^M$-valued functions $\psi_1^E, \ldots, \psi_N^E$ on $\mathbb{R}^d$ by

$$\widehat{\psi_n^E}(\xi) := f(E/|\xi|^2) \widehat{\psi_n}(\xi) \qquad \text{for all } \xi \in \mathbb{R}^d \, .$$

Then, since

$$|\xi|^2 = \int_0^\infty f(E/|\xi|^2)^2 \, dE \qquad \text{for all } \xi \in \mathbb{R}^d \, ,$$

we have

$$\sum_{n=1}^N \int_{\mathbb{R}^d} |\nabla \psi_n(x)|^2 \, dx = \sum_{n=1}^N \int_{\mathbb{R}^d} |\xi|^2 |\widehat{\psi_n}(\xi)|^2 \, d\xi = \sum_{n=1}^N \int_{\mathbb{R}^d} \int_0^\infty |\widehat{\psi_n^E}(\xi)|^2 \, dE \, d\xi$$

$$= \int_{\mathbb{R}^d} \sum_{n=1}^N \int_0^\infty |\psi_n^E(x)|^2 \, dE \, dx \, . \tag{8.8}$$

Our goal will be to bound $\sum_{n=1}^N \int_0^\infty |\psi_n^E(x)|^2 \, dE$ from below pointwise in x.

Step 2. Let $R^E(x)$ be the Hermitian non-negative $M \times M$ matrix given by

$$R^E(x) := \sum_{n=1}^N \langle \cdot, \psi_n^E(x) \rangle \psi_n^E(x) \, .$$

Moreover, let w be a non-negative, square-integrable function on $(0, \infty)$ and let

$$g(t) := \int_0^\infty w(s) f(st) \, ds \, .$$

The crucial step in the proof will be to show that for any $x \in \mathbb{R}^d$, $\varepsilon > 0$, and $\mu > 0$, one has, in the sense of matrices,

$$R(x) \leq (1 + \varepsilon) \mu^{-1} \|w\|^2 \int_0^\infty R^E(x) \, dE + (1 + \varepsilon^{-1}) A \mu^{d/2}, \tag{8.9}$$

where

$$A := 2^{-1} (2\pi)^{-d} |\mathbb{S}^{d-1}| \int_0^\infty (1 - g(t))^2 t^{-1-d/2} \, dt \, .$$

In order to prove (8.9) let $\varphi_1^E, \ldots, \varphi_N^E \in L^2(\mathbb{R}^d, \mathbb{C}^M)$, for any $E > 0$, be defined by

$$\widehat{\varphi_n^F}(\xi) := g(E/|\xi|^2)\widehat{\psi_n}(\xi) \qquad \text{for all } \xi \in \mathbb{R}^d .$$

For $e \in \mathbb{C}^M$, $n \in \{1, \ldots, N\}$, and $\mu > 0$ we bound

$$|\langle \psi_n(x), e \rangle|^2 = |\langle \varphi_n^\mu(x), e \rangle|^2 + 2 \operatorname{Re} \overline{\langle \varphi_n^\mu(x), e \rangle} \, \langle \psi_n(x) - \varphi_n^\mu(x), e \rangle$$

$$+ |\langle \psi_n(x) - \varphi_n^\mu(x), e \rangle|^2$$

$$\leq (1 + \varepsilon)|\langle \varphi_n^\mu(x), e \rangle|^2 + (1 + \varepsilon^{-1})|\langle \psi_n(x) - \varphi_n^\mu(x), e \rangle|^2 . \quad (8.10)$$

For the first term on the right side we have, by the Cauchy–Schwarz inequality,

$$|\langle \varphi_n^\mu(x), e \rangle|^2 = (2\pi)^{-d} \left| \int_{\mathbb{R}^d} \int_0^\infty e^{i\xi \cdot x} w(s) f(s\mu/|\xi|^2)\langle \widehat{\psi_n}(\xi), e \rangle \, ds \, d\xi \right|^2$$

$$\leq \|w\|^2 (2\pi)^{-d} \int_0^\infty \left| \int_{\mathbb{R}^d} e^{i\xi \cdot x} f(s\mu/|\xi|^2)\langle \widehat{\psi_n}(\xi), e \rangle \, d\xi \right|^2 ds$$

$$= \|w\|^2 \int_0^\infty |\langle \psi_n^{s\mu}(x), e \rangle|^2 \, ds$$

$$= \mu^{-1} \|w\|^2 \int_0^\infty |\langle \psi_n^E(x), e \rangle|^2 \, dE .$$

Inserting this into (8.10) and summing over n, we obtain

$$\langle R(x)e, e \rangle = \sum_n |\langle \psi_n(x), e \rangle|^2$$

$$\leq (1 + \varepsilon)\mu^{-1} \|w\|^2 \int_0^\infty \sum_{n=1}^N |\langle \psi_n^E(x), e \rangle|^2 \, dE$$

$$+ (1 + \varepsilon^{-1}) \sum_{n=1}^N |\langle \psi_n(x) - \varphi_n^\mu(x), e \rangle|^2$$

$$= (1 + \varepsilon)\mu^{-1} \|w\|^2 \int_0^\infty \langle R^E(x)e, e \rangle \, dE$$

$$+ (1 + \varepsilon^{-1}) \sum_{n=1}^N |\langle \psi_n(x) - \varphi_n^\mu(x), e \rangle|^2 .$$

To bound the second term on the right side, we write

$$\langle \psi_n(x) - \varphi_n^\mu(x), e \rangle = (2\pi)^{-d/2} \int_{\mathbb{R}^d} e^{i\xi \cdot x} (1 - g(\mu/|\xi|^2))\langle \widehat{\psi_n}(\xi), e \rangle \, d\xi$$

$$= \left(\widehat{\psi_n}, \chi_{x,\mu} e \right),$$

where the last inner product is in $L^2(\mathbb{R}^d, \mathbb{C}^M)$ and where

$$\chi_{x,\mu}(\xi) := (2\pi)^{-d/2} e^{-i\xi \cdot x} (1 - g(\mu/|\xi|^2)) \qquad \text{for all } \xi \in \mathbb{R}^d.$$

Since the $\widehat{\psi_n}$ are orthonormal, we obtain, by Bessel's inequality,

$$\sum_{n=1}^{N} \left| \langle \psi_n(x) - \varphi_n^\mu(x), e \rangle \right|^2 \leq \|\chi_{x,\mu} e\|^2 = A\mu^{d/2} |e|^2,$$

where, in the last equality, we used

$$\mu^{-d/2} \|\chi_{x,\mu}\|^2 = (2\pi)^{-d} \int_{\mathbb{R}^d} \left(1 - g(1/|\eta|^2)\right)^2 d\eta = A.$$

To summarize, we have shown that

$$\langle R(x)e, e \rangle \leq (1 + \varepsilon)\mu^{-1} \|w\|^2 \int_0^\infty \langle R^E(x)e, e \rangle \, dE + (1 + \varepsilon^{-1}) A\mu^{d/2} |e|^2,$$

which is the same as (8.9).

Step 3. We denote by $\lambda_m(H)$, $m = 1, \ldots, M$, the eigenvalues of a Hermitian $M \times M$ matrix H, arranged in non-increasing order and counted according to multiplicities. Then, by the variational principle, the matrix inequality (8.9) implies that, for $m = 1, \ldots, M$,

$$\lambda_m(R(x)) \leq (1 + \varepsilon)\mu^{-1} \|w\|^2 \lambda_m \left(\int_0^\infty R^E(x) \, dE \right) + (1 + \varepsilon^{-1}) A\mu^{d/2}.$$

Optimizing with respect to $\varepsilon > 0$ and $\mu > 0$ for each j, we obtain

$$\lambda_m(R(x)) \leq \left(\frac{2}{d}\right)^{\frac{2d}{d+2}} \left(1 + \frac{d}{2}\right)^2 \|w\|^{\frac{2d}{d+2}} A^{\frac{2}{d+2}} \left(\lambda_m \left(\int_0^\infty R^E(x) \, dE\right)\right)^{\frac{d}{d+2}},$$

which is the same as

$$\lambda_m \left(\int_0^\infty R^E(x) \, dE\right) \geq \left(\frac{d}{2}\right)^2 \left(1 + \frac{d}{2}\right)^{-\frac{2(d+2)}{d}} \|w\|^{-2} A^{-\frac{2}{d}} \left(\lambda_m(R(x))\right)^{1+\frac{2}{d}}.$$

Thus

$$\int_0^\infty \sum_{n=1}^{N} |\psi_n^E(x)|^2 \, dE = \operatorname{Tr}_{\mathbb{C}^M} \int_0^\infty R^E(x) \, dE = \sum_{m=1}^{M} \lambda_m \left(\int_0^\infty R^E(x) \, dE\right)$$

$$\geq \left(\frac{d}{2}\right)^2 \left(1 + \frac{d}{2}\right)^{-\frac{2(d+2)}{d}} \|w\|^{-2} A^{-\frac{2}{d}} \sum_{m=1}^{M} \left(\lambda_m(R(x))\right)^{1+\frac{2}{d}}$$

$$= \left(\frac{d}{2}\right)^2 \left(1 + \frac{d}{2}\right)^{-\frac{2(d+2)}{d}} \|w\|^{-2} A^{-\frac{2}{d}} \operatorname{Tr}_{\mathbb{C}^M} R(x)^{1+\frac{2}{d}}.$$

Inserting this bound into (8.8), we obtain the claimed bound. $\qquad \square$

Lemma 8.11 *If $d = 1$, then*

$$\frac{2}{3} \leq \mathcal{J}_1 \leq 0.747112 \,.$$

The upper bound on $\mathcal{J}_1$ leads to an explicit constant in the Lieb–Thirring inequality. The lower bound shows the limitation of the method.

Proof Let $f, w \geq 0$ with $\int_0^\infty f(s)^2 \, ds = 1$ and denote $a := \int_0^\infty w(s)^2 \, ds$. Then

$$g(t) = \int_0^\infty w(s) f(st) \, ds \leq \left(\int_0^\infty w(s)^2 \, ds \right)^{1/2} \left(\int_0^\infty f(st)^2 \, ds \right)^{1/2} = a^{1/2} t^{-1/2} \,,$$

and therefore

$$\int_0^\infty \frac{(1 - g(t))^2}{t^{3/2}} \, dt \geq \int_0^\infty \frac{(1 - a^{1/2} t^{-1/2})_+^2}{t^{3/2}} \, dt = \frac{2}{3} a^{-1/2} \,.$$

Thus $\mathcal{J}_1 \geq \frac{2}{3}$, as claimed.

In order to prove the upper bound on $\mathcal{J}_1$, we choose

$$f(s) = (1 + \mu_0 s^{4.5})^{-0.25} \,, \qquad w(s) = c_0 \frac{(1 - s^{0.36})^{2.1}}{1 + s} \chi_{[0,1]}(s) \,,$$

where μ_0 and c_0 are determined by $\int_0^\infty f(s)^2 \, ds = \int_0^\infty w(s) \, ds = 1$. A numerical computation leads to the claimed bound on $\mathcal{J}_1$. □

Proof of Theorem 8.9 Theorem 8.9 follows from Theorem 8.10 by Theorem 7.13. Since we omitted the proof of the latter theorem, let us give the details of this argument. Let $-E_n$ be the negative eigenvalues of $-\frac{d^2}{dx^2} \otimes \mathbb{I}_{\mathbb{C}^M} - V$ in non-decreasing order and counting multiplicities, and let ψ_n be a corresponding system of orthonormal eigenfunctions. Then, for any $N \in \mathbb{N}$, defining $R(x)$ by (8.6) and using (8.7),

$$\sum_{n=1}^N \int_{\mathbb{R}} |\psi_n|^2 \, dx = \int_{\mathbb{R}} \operatorname{Tr}_{\mathbb{C}^M} (V(x) R(x)) \, dx - \sum_{n=1}^N E_n \,.$$

Meanwhile, by Theorem 8.10,

$$\sum_{n=1}^N \int_{\mathbb{R}} |\psi_n'|^2 \, dx \geq K_1 \int_{\mathbb{R}} \operatorname{Tr}_{\mathbb{C}^M} R(x)^3 \, dx \,.$$

Combining these two relations gives

$$\sum_{n=1}^N E_n \leq \int_{\mathbb{R}} \operatorname{Tr}_{\mathbb{C}^M} \left(V(x) R(x) - K_1 R(x)^3 \right) dx \,.$$

For fixed $x \in \mathbb{R}$ we have, by Hölder's inequality for $M \times M$ Hermitian matrices (see Lemma 8.12 below),

$$\operatorname{Tr}_{\mathbb{C}^M}\left(V(x)R(x) - K_1 R(x)^3\right) \leq \operatorname{Tr}_{\mathbb{C}^M}\left(V(x)_+ R(x) - K_1 R(x)^3\right)$$

$$\leq \left(\operatorname{Tr}_{\mathbb{C}^M} V(x)_+^{3/2}\right)^{2/3} r(x) - K_1 r(x)^3,$$

where $r(x) := \left(\operatorname{Tr}_{\mathbb{C}^M} R(x)^3\right)^{1/3}$. Here we also used $R(x) \geq 0$. Since

$$Ar - Kr^3 \leq \frac{2}{3\sqrt{3}} K^{-1/2} A^{3/2} \qquad \text{for all } A, r, K \geq 0,$$

we obtain

$$\operatorname{Tr}_{\mathbb{C}^M}\left(V(x)R(x) - K_1 R(x)^3\right) \leq \frac{2}{3\sqrt{3}} K_1^{-1/2} \operatorname{Tr}_{\mathbb{C}^M} V(x)_+^{3/2}.$$

Note that, so far, we have not used the value of the constant K_1 in Theorem 8.10. Inserting this expression and recalling that $L_{1,1}^{\text{cl}} = 2/(3\pi)$, we arrive at the claimed bound for $R_{1,1}$. Moreover, inserting the bound from Lemma 8.11 for $\mathcal{I}_1$, we obtain the claimed numerical value. □

Lemma 8.12 *Let A, B be Hermitian, non-negative matrices and let $1 < p < \infty$ and $p' = p/(p-1)$. Then*

$$0 \leq \operatorname{Tr} AB \leq \left(\operatorname{Tr} A^p\right)^{1/p}\left(\operatorname{Tr} B^{p'}\right)^{1/p'}.$$

Proof Write $A = \sum_j a_j(\cdot, f_j)f_j$ and $B = \sum_k b_k(\cdot, g_k)g_k$ with orthonormal bases (f_j) and (g_k) and eigenvalues $a_j \geq 0$ and $b_k \geq 0$. Then, by Hölder's inequality for sequences and the completeness of the eigenvectors,

$$\operatorname{Tr} AB = \sum_{j,k} a_j b_k |(g_k, f_j)|^2 \leq \left(\sum_{j,k} a_j^p |(g_k, f_j)|^2\right)^{1/p}\left(\sum_{j,k} b_k^{p'} |(g_k, f_j)|^2\right)^{1/p'}$$

$$= \left(\sum_j a_j^p\right)^{1/p}\left(\sum_k b_k^{p'}\right)^{1/p'} = \left(\operatorname{Tr} A^p\right)^{1/p}\left(\operatorname{Tr} B^{p'}\right)^{1/p'}.$$

This completes the proof. □

8.4 Summary on constants in Lieb–Thirring inequalities

In this final section, we summarize what is currently known about the values of the constants in Lieb–Thirring inequalities. Throughout this section, $L_{\gamma,d}$ denotes the optimal value of the constant in the Lieb–Thirring inequality

$$\operatorname{Tr}(-\Delta - V)_-^\gamma \leq L_{\gamma,d} \int_{\mathbb{R}^d} V(x)_+^{\gamma + \frac{d}{2}} dx.$$

We know that $L_{\gamma,d} < \infty$ if $\gamma \geq 1/2$ in $d = 1$, if $\gamma > 0$ in $d = 2$ and if $\gamma \geq 0$ in $d \geq 3$. Moreover, from Weyl asymptotics we know that (see Lemma 5.1)

$$L_{\gamma,d} \geq L_{\gamma,d}^{\mathrm{cl}}$$

with the semiclassical constant

$$L_{\gamma,d}^{\mathrm{cl}} = \frac{\Gamma(\gamma + 1)}{(4\pi)^{\frac{d}{2}} \Gamma(\gamma + 1 + \frac{d}{2})}. \tag{8.11}$$

At the same time, we know from Sobolev inequalities (see Lemma 5.11)

$$L_{\gamma,d} \geq L_{\gamma,d}^{(1)},$$

where $L_{\gamma,d}^{(1)}$ is the one-particle constant, characterized as the optimal constant in the inequality

$$\left(\inf \sigma(-\Delta - V)\right)_{-}^{\gamma} \leq L_{\gamma,d}^{(1)} \int_{\mathbb{R}^d} V(x)_{+}^{\gamma + \frac{d}{2}} \, dx.$$

It is in one-to-one correspondence with an optimal constant in a Sobolev interpolation inequality.

We know from Proposition 5.12 that for any $1 \leq d \leq 7$ there is a $\gamma_{c,d} > 0$ such that

$$\begin{cases} L_{\gamma,d}^{(1)} > L_{\gamma,d}^{\mathrm{cl}} & \text{if } \gamma < \gamma_{c,d}, \\ L_{\gamma,d}^{(1)} = L_{\gamma,d}^{\mathrm{cl}} & \text{if } \gamma = \gamma_{c,d}, \\ L_{\gamma,d}^{(1)} < L_{\gamma,d}^{\mathrm{cl}} & \text{if } \gamma > \gamma_{c,d}. \end{cases}$$

In dimension $d = 1$, the explicit formulas for $L_{\gamma,1}^{(1)}$ and $L_{\gamma,1}^{\mathrm{cl}}$ in (5.18) and (8.11) imply that $\gamma_{c,1} = 3/2$. Moreover, if $d \geq 8$, we have

$$L_{\gamma,d}^{(1)} < L_{\gamma,d}^{\mathrm{cl}} \qquad \text{for all } \gamma \geq 0.$$

Originally, Lieb and Thirring (1976) conjectured that the optimal constant $L_{\gamma,d}$ is given by the larger of the two constants $L_{\gamma,d}^{\mathrm{cl}}$ and $L_{\gamma,d}^{(1)}$. This conjecture has motivated the study of sharp constants in Lieb–Thirring inequalities. It has been verified in a variety of cases, but in other cases it has been disproved. In the following subsections we summarize what is known with regard to this conjecture.

8.4.1 Optimal results for the Lieb–Thirring constants

Let us restate Theorems 6.11 and 5.27, which are the two cases where the sharp constants in the Lieb–Thirring inequality are known.

Theorem 8.13 *Let $d \geq 1$ and $\gamma \geq 3/2$. Then*

$$L_{\gamma,d} = L_{\gamma,d}^{cl} = \frac{\Gamma(\gamma + 1)}{(4\pi)^{\frac{d}{2}} \Gamma(\gamma + 1 + \frac{d}{2})}.$$

In Theorem 8.13, the sharp constant is the semiclassical one. In the following theorem the sharp constant is the one-particle constant.

Theorem 8.14 *Let $d = 1$ and $\gamma = 1/2$. Then*

$$L_{1/2,1} = L_{1/2,1}^{(1)} = 2 L_{1/2,1}^{cl} = \frac{1}{2}.$$

The case $d = 1$ and $\gamma = 3/2$ is peculiar since in this case

$$L_{3/2,1} = L_{3/2,1}^{cl} = L_{3/2,1}^{(1)} = \frac{3}{16}.$$

8.4.2 Current best bounds on the Lieb–Thirring constants

The sharp values of the constants $L_{\gamma,d}$ are not known in the cases not covered by Theorems 8.13 and 8.14. In this subsection, we give the current best known bounds on the sharp values.

We begin with the case $\gamma \geq 1/2$ where one can obtain bounds using the lifting method, starting from a one-dimensional, operator-valued inequality.

Theorem 8.15 *Let $d \geq 1$. Then*

$$L_{\gamma,d} \leq \begin{cases} 2 L_{\gamma,d}^{cl} & \text{for all } \gamma \geq 1/2 \text{ and } d = 1, \\ 2 R_{1,1} L_{\gamma,d}^{cl} & \text{for all } \gamma \geq 1/2 \text{ and } d \geq 2, \\ R_{1,1} L_{\gamma,d}^{cl} & \text{for all } \gamma \geq 1, \end{cases}$$

with $R_{1,1}$ from Theorem 8.9 and, in particular, $R_{1,1} \leq 1.456$.

Proof The inequality for $\gamma = 1$ follows from Theorem 8.2 since assumption (8.1) is satisfied by Theorem 8.9. The inequality for $\gamma = 1/2$ in $d = 1$ follows from Theorem 5.27 and that in $d \geq 2$ follows from Theorem 8.3 since assumption (8.2) for $\nu = 0, 1$ is satisfied by Theorems 8.8 and 8.9, respectively. The inequalities for intermediate values of γ follow by the Aizenman–Lieb argument in Lemma 5.2. $\qquad \square$

With the exception of the case $\gamma = 1/2$ in $d = 1$, the constants given in Theorem 8.15 for $1/2 \leq \gamma < 3/2$ are probably not optimal. Some of the bounds, in particular for parameters close to those where sharp bounds are known, can be slightly improved using interpolation methods; see §8.5.

We now turn to the remaining cases, $\gamma < 1/2$. Our proofs of the Cwikel–Lieb–Rozenblum and Lieb–Thirring inequalities in Theorems 4.31 and 4.38 give in principle some computable values of the constants $L_{\gamma,d}$. They are, however, quite far from optimal. (Note that the proofs of Theorems 4.31 and 4.38 use Proposition 2.62 only in the special case where Ω is a cube, in which case the non-computable constant in (2.38) can be replaced by the explicit one in Lemma 2.57.)

We next state without proof a result, due to Lieb (1976, 1980), which gives, in particular in dimensions $d \le 4$ for $\gamma < 1/2$, the currently best-known values of the constants.

Theorem 8.16 *Let*

$$\begin{aligned}
\gamma > 1/2 \quad &\textit{if } d = 1, \\
\gamma > 0 \quad &\textit{if } d = 2, \\
\gamma \ge 0 \quad &\textit{if } d \ge 3.
\end{aligned}$$

Then

$$L_{\gamma,d} \le \Gamma(\gamma + \tfrac{d}{2} - 1) \left(\sup_{a>0} a^{\gamma + \frac{d}{2}} \int_a^\infty t^{-2} e^{-t} \, dt \right)^{-1} L_{\gamma,d}^{\mathrm{cl}}.$$

Because of its special importance, we state separately what is currently known about the best constants in the CLR inequality.

Theorem 8.17 *Let $d \ge 3$ and $\gamma = 0$. Then*

$$L_{0,d} \le R_{0,d} \, L_{0,d}^{\mathrm{cl}}$$

with

$$\begin{aligned}
R_{0,3} &= 6.847, \\
R_{0,4} &= 6.034, \\
R_{0,5} &= 5.955, \\
R_{0,6} &= 5.771.
\end{aligned}$$

Moreover, $R_{0,d} \le 5.771$ for $d \ge 7$.

The constants for $d = 3, 4$ are due to Lieb (1976, 1980) and can be obtained from his bound in Theorem 8.16. The constants for $d = 5, 6$, as well as the bound for $d \ge 7$ are due to Hundertmark et al. (2018).

8.4.3 Counterexamples

Theorem 8.13 says that $L_{\gamma,d} = L_{\gamma,d}^{\mathrm{cl}}$ for $\gamma \ge 3/2$ and all $d \ge 1$. It is an open problem to find *all* γ and d for which this equality holds. By Proposition 5.12, in dimensions $1 \leq d \leq 7$, this equality is only possible if $\gamma \geq \gamma_{c,d}$. We know that $\gamma_{c,d} > 1$ for $d = 1, 2$ and $\gamma_{c,d} < 1$ for $3 \leq d \leq 7$.

We now restate Proposition 5.20 and Corollary 5.23. They show that the equality $L_{\gamma,d} = L^{\mathrm{cl}}_{\gamma,d}$ does not hold for $\gamma_{c,d} \leq \gamma < 1$ in dimensions $3 \leq d \leq 7$ and for $0 \leq \gamma < 1$ in dimensions $d \geq 8$, in contrast to the original conjecture of Lieb and Thirring.

Theorem 8.18   *Let $d \geq 1$ and $\gamma < 1$. Then*

$$L_{\gamma,d} > L^{\mathrm{cl}}_{\gamma,d} \, .$$

We recall that Theorems 8.13 and 8.14 say that $L_{\gamma,1} = L^{(1)}_{\gamma,1}$ for $\gamma = 1/2$ and $\gamma = 3/2$ in $d = 1$. We now discuss the problem of finding *all* γ and d for which $L_{\gamma,d} = L^{(1)}_{\gamma,d}$. By Proposition 5.12, this can only happen in dimensions $1 \leq d \leq 7$ and only for $\gamma \leq \gamma_{c,d}$. The following two theorems, which are restatements of Proposition 7.10 and Theorem 7.11, show that the equality $L_{\gamma,d} = L^{(1)}_{\gamma,d}$ does *not* hold for

$$
\begin{aligned}
1 < \gamma \leq \gamma_{c,2} \qquad & \text{if } d = 2, \\
1/2 < \gamma \leq \gamma_{c,3} \qquad & \text{if } d = 3, \\
0 < \gamma \leq \gamma_{c,d} \qquad & \text{if } d = 4, 5, 6, \\
0 \leq \gamma \leq \gamma_{c,7} \qquad & \text{if } d = 7,
\end{aligned}
$$

in contrast to the original conjecture of Lieb and Thirring.

Theorem 8.19   *Let $d \geq 7$ and $\gamma = 0$. Then*

$$L_{0,d} > L^{(1)}_{0,d} \, .$$

Theorem 8.20   *Let $d \geq 1$ and $\gamma > \max\{2 - d/2, 0\}$. Then*

$$L_{\gamma,d} > L^{(1)}_{\gamma,d} \, .$$

While the counterexamples in Theorems 8.18, 8.19 and 8.20 disprove part of the original Lieb–Thirring conjecture, they only give a limited amount of intuition about what could, in fact, be the sharp constant.

8.4.4 The Lieb–Thirring conjecture

We single out two important open instances of the Lieb–Thirring conjecture that are generally believed to be true. The first one concerns the semiclassical regime and, in particular, the physically most relevant case of $\gamma = 1$ in dimension $d = 3$.

Conjecture 8.21 (Lieb–Thirring)   *Let $d \geq 3$ and $1 \leq \gamma < 3/2$. Then*

$$L_{\gamma,d} = L_{\gamma,d}^{\mathrm{cl}} \, .$$

The second one concerns the one-dimensional case, where we recall that $\gamma_{c,1} = 3/2$.

Conjecture 8.22 (Lieb–Thirring)   *Let $d = 1$ and $1/2 < \gamma < 3/2$. Then*

$$L_{\gamma,1} = L_{\gamma,1}^{(1)} \, .$$

We wish the reader good luck in proving this!

8.5 Comments

Section 8.1: More on Lieb–Thirring inequalities in the matrix-valued case

The matrix-valued Lieb–Thirring inequality in arbitrary dimension with semi-classical constant for $\gamma \geq 3/2$ in Theorem 8.1 appeared first in Laptev and Weidl (2000b). The two ingredients in that paper are a matrix-valued Lieb–Thirring inequality in one-dimension and the method of lifting in dimension. In Hundertmark et al. (2000), it is observed that these two ingredients can be used to obtain good, although not necessarily sharp constants in Lieb–Thirring inequalities in higher dimensions. This led, in particular, to the constant $4L_{1,3}^{\mathrm{cl}}$ in the Lieb–Thirring inequality for the sum of negative eigenvalues in three dimensions, the then best known result.

We mention that, by an approximation argument similar to that in the proof of Lemma 6.13, Theorem 8.1 can be show to hold also for potentials taking values in operators on a separable Hilbert space (instead of in matrices acting on $\mathbb{C}^M$).

It is an open question whether the optimal constant $L_{\gamma,d}^{(\mathrm{mat})}$ in the matrix case coincides with the optimal constant $L_{\gamma,d}$ in the scalar case. In those cases where $L_{\gamma,d}^{(\mathrm{mat})}$ is known, namely for $\gamma = 1/2$ in $d = 1$ and for $\gamma \geq 3/2$ in $d \geq 1$, it coincides with $L_{\gamma,d}$. Note that the analogue of the one-particle constant in the matrix case coincides with its scalar analogue, as shown in Propositions 6.6 and 6.7.

It is also an open question whether the analogues of Lemma 8.6 and Corollary 8.7 for the scalar Lieb–Thirring constants are valid. Note that a straightforward generalization of the argument in §6.2 gives, under the assumptions of Lemma 8.6, on k and γ,

$$\frac{L_{\gamma,d}}{L_{\gamma,d}^{\mathrm{cl}}} \leq \frac{L_{\gamma,k}^{(\mathrm{mat})}}{L_{\gamma,k}^{\mathrm{cl}}} \, \frac{L_{\gamma+\frac{k}{2},d-k}}{L_{\gamma+\frac{k}{2},d-k}^{\mathrm{cl}}} \, .$$

Note the similarity between Corollary 8.7 and (5.86).

In relation to the monotonicity result in Corollary 8.7 we mention the inequality

$$\frac{L_{\delta,d}}{L^{\mathrm{cl}}_{\delta,d}} \le \frac{L_{\gamma,d+1}}{L^{\mathrm{cl}}_{\gamma,d+1}} \qquad \text{for all } \delta \ge \gamma + \frac{1}{2}$$

in Martin (1990), which is proved by constructing a trial potential in dimension $d + 1$ from a potential in dimension d by adding a homogeneous function of the additional variable with a large coupling constant.

Section 8.2: The sum of the square roots of the eigenvalues
Theorem 8.8 is due to Hundertmark et al. (2000). As mentioned before, this paper was the first to use the method of lifting in dimension to obtain good (but not optimal) constants in Lieb–Thirring inequalities in higher dimensions.

Section 8.3: The sum of the eigenvalues
Theorem 8.9 is due to Frank et al. (2021d). This theorem slightly improves on a previous result in Dolbeault et al. (2008), which was obtained by a generalization to the matrix-valued setting of the method in Eden and Foias (1991); for a partially alternative proof, see also Ilyin et al. (2016). All these proofs proceed via proving Sobolev inequalities for orthonormal functions (the kinetic form of the Lieb–Thirring inequality).

Section 8.4: Summary on constants in Lieb–Thirring inequalities
Theorem 8.16 is due to Lieb (1976, 1980); see also Lieb (1984) for $\gamma > 0$. The same formula appears in Daubechies (1983, Eq. (2.14)), with a typographical error, and in Blanchard and Stubbe (1996, Corollary 5.3). A small improvement appears in Blanchard and Stubbe (1996, §5.4).

For the use of (real) interpolation methods to further improve constants, see Weidl (1996) and Schmetzer (2022).

For certain classes of potentials, one can prove the Lieb–Thirring inequality with semiclassical constant. One example is the Berezin–Li–Yau inequality (Theorem 3.25), which can be understood as a Lieb–Thirring inequality with $\gamma \ge 1$ for a potential that is constant on an open set of finite measure and $-\infty$ off that set. Another class consists of harmonic oscillator potentials for $\gamma \ge 1$; see Propositions 5.19 and 5.21. Yet another class, again for $\gamma \ge 1$, consists of potentials of the form $V(x) = V_1(x_1) + V_2(x_2)$ with $x = (x_1, x_2) \in \mathbb{R}^{d_1} \times \mathbb{R}^{d_2}$, where V_1 is such that, for the given $\gamma \ge 1$, the Lieb–Thirring inequality with semiclassical constant holds for $V_1 + t$ in dimension d_1 for any $t \in \mathbb{R}$; see Laptev and Weidl (2000a). Note the similarity between this result and Theorem

3.36. The assumption on V_1 is satisfied, for instance, for harmonic oscillator potentials by Propositions 5.19 and 5.21.

One can show that, for instance in dimensions $d = 3$ and 4, the CLR inequality holds with a constant that is better than the value given in Theorem 8.17 if one considers only V that are multiples of characteristic functions (Laptev, 1995). Namely, one has, for all $d \geq 3$, all $\lambda > 0$, and all measurable $\Omega \subset \mathbb{R}^d$ of finite measure,

$$N(0, -\Delta - \lambda\chi_\Omega) \leq \frac{\omega_{d-1}}{\omega_d} \frac{d^{d/2}}{(d-2)^{d/2}} L_{0,d}^{\mathrm{cl}} \lambda^{d/2} |\Omega|,$$

where ω_d denotes the volume of the unit ball in $\mathbb{R}^d$.

The proof of the CLR inequality in Hundertmark et al. (2018) extends to the case of matrix-valued potentials and, moreover, the constants given in Theorem 8.17 for $d \geq 5$ remain valid in the matrix-valued setting. The constants obtained in this way in dimensions $d = 3, 4$ in the matrix-valued setting are slightly worse than those given in Theorem 8.17, but currently are the best known in the matrix case. So far, an adaptation of Lieb's proof (Lieb, 1976, 1980) to the matrix-valued setting has resulted in a deterioration of the constant (Frank et al., 2007). The first proof of the CLR inequality in the matrix-valued setting was due to Hundertmark (2002). For an alternative approach, see Frank (2014).

As discussed in §7.2, Glaser, Grosse and Martin (1978) disproved the Lieb–Thirring conjecture for $\gamma = 0$ in dimensions $d \geq 7$ and made the following alternative conjecture; see also Frank (2021).

Conjecture 8.23   *Let $d \geq 3$ and $\gamma = 0$. Then*

$$L_{0,d} = L_{0,d}^{\mathrm{cl}} \sup_{L \in \mathbb{N}_0} \frac{(L + d - 1)! \, (L + \frac{d}{2})}{L! \left((L + \frac{d-2}{2})(L + \frac{d}{2}) \right)^{d/2}} .$$

In particular, they conjectured that $L_{0,d} = L_{0,d}^{(1)}$ if $d \leq 6$, as in the original Lieb–Thirring conjecture. The results in Morpurgo (1999) can be viewed as evidence for the validity of Conjecture 8.23 in general dimension.

References

Abramov, A. A., Aslanyan, A., and Davies, E. B. 2001. Bounds on complex eigenvalues and resonances. *J. Phys. A*, **34**(1), 57–72.

Abramowitz, M., and Stegun, I. A. 1964. *Handbook of Mathematical Functions with Formulas, Graphs, and Mathematical Tables*. National Bureau of Standards Applied Mathematics Series, vol. 55.

Adams, D. R., and Hedberg, L. I. 1996. *Function Spaces and Potential Theory*. Grundlehren der Mathematischen Wissenschaften, vol. 314. Springer-Verlag, Berlin.

Adams, R. A., and Fournier, J. J. F. 2003. *Sobolev Spaces*. Second edn. Pure and Applied Mathematics (Amsterdam), vol. 140. Elsevier/Academic Press, Amsterdam.

Agmon, S. 1965. On kernels, eigenvalues, and eigenfunctions of operators related to elliptic problems. *Comm. Pure Appl. Math.*, **18**, 627–663.

Agmon, S. 1967/68. Asymptotic formulas with remainder estimates for eigenvalues of elliptic operators. *Arch. Rational Mech. Anal.*, **28**, 165–183.

Agmon, S., and Kannai, Y. 1967. On the asymptotic behavoir of spectral functions and resolvant kernels of elliptic operators. *Israel J. Math.*, **5**, 1–30.

Agranovič, Z. S., and Marčenko, V. A. 1960. Reconstruction of the potential energy from the dispersion matrix. *Amer. Math. Soc. Transl. (2)*, **16**, 355–357.

Agranovich, Z. S., and Marchenko, V. A. 1963. *The Inverse Problem of Scattering Theory*. Translated from the Russian by B. D. Seckler. Gordon and Breach Science Publishers, New York and London.

Aizenman, M., and Lieb, E. H. 1978. On semiclassical bounds for eigenvalues of Schrödinger operators. *Phys. Lett. A*, **66**(6), 427–429.

Aizenman, M., and Simon, B. 1982. Brownian motion and Harnack inequality for Schrödinger operators. *Comm. Pure Appl. Math.*, **35**(2), 209–273.

Aizenman, M., Elgart, A., Naboko, S., Schenker, J. H., and Stolz, G. 2006. Moment analysis for localization in random Schrödinger operators. *Invent. Math.*, **163**(2), 343–413.

Akhiezer, N. I., and Glazman, I. M. 1963. *Theory of Linear Operators in Hilbert Space*, in two volumes. English translation, with a preface by M. Nestell, 1993. Two volumes bound as one. Dover Publications, Inc., New York.

Alama, S., Deift, P. A., and Hempel, R. 1989. Eigenvalue branches of the Schrödinger operator $H - \lambda W$ in a gap of $\sigma(H)$. *Comm. Math. Phys.*, **121**(2), 291–321.

Allen, M., Kriventsov, D., and Neumayer, R. 2021. Sharp quantitative Faber–Krahn inequalities and the Alt–Caffarelli–Friedman monotonicity formula. Preprint, available at ArXiv:2107.03505

Alonso, A., and Simon, B. 1980, The Birman–Kreĭn–Vishik theory of selfadjoint extensions of semibounded operators. *J. Operator Theory*, **4**(2), 251–270,

Alvino, A. 1977. Sulla diseguaglianza di Sobolev in spazi di Lorentz. *Boll. Un. Mat. Ital. A (5)*, **14**(1), 148–156.

Andrews, B., and Clutterbuck, J. 2011. Proof of the fundamental gap conjecture. *J. Amer. Math. Soc.*, **24**(3), 899–916.

Araki, H. 1990. On an inequality of Lieb and Thirring. *Lett. Math. Phys.*, **19**(2), 167–170.

Arazy, J., and Zelenko, L. 2006. Virtual eigenvalues of the high order Schrödinger operator. I. *Integral Equations Operator Theory*, **55**(2), 189–231.

Ashbaugh, M. S., and Benguria, R. D. 1992. A sharp bound for the ratio of the first two eigenvalues of Dirichlet Laplacians and extensions. *Ann. of Math. (2)*, **135**(3), 601–628.

Ashbaugh, M. S., and Benguria, R. D. 2007. Isoperimetric inequalities for eigenvalues of the Laplacian. Pages 105–139 of: *Spectral Theory and Mathematical Physics: a Festschrift in Honor of Barry Simon's 60th Birthday.* F. Gesztesy, P. Deift, C. Galvez, P. Perry, and W. Schlag (editors). Proc. Sympos. Pure Math., vol. 76. Amer. Math. Soc., Providence, RI.

Ashbaugh, M. S., Gesztesy, F., Mitrea, M., Shterenberg, R., and Teschl, G. 2010a. The Krein–von Neumann extension and its connection to an abstract buckling problem. *Math. Nachr.*, **283**(2), 165–179.

Ashbaugh, M. S., Gesztesy, F., Mitrea, M., and Teschl, G. 2010b. Spectral theory for perturbed Krein Laplacians in nonsmooth domains. *Adv. Math.*, **223**(4), 1372–1467.

Ashu, A. M. 2013. *Some properties of Bessel functions with applications to Neumann eigenvalues in the unit disc.* Student Paper, Lund University. https://lup.lub.lu.se/student-papers/search/publication/7370411.

Aubin, T.. 1976. Problèmes isopérimétriques et espaces de Sobolev. *J. Differential Geometry*, **11**(4), 573–598.

Aubin, T.. 1998. *Some Nonlinear Problems in Riemannian Geometry.* Springer Monographs in Mathematics. Springer-Verlag, Berlin.

Avakumović, Vojislav G. 1956. Über die Eigenfunktionen auf geschlossenen Riemannschen Mannigfaltigkeiten. *Math. Z.*, **65**, 327–344.

Aviles, P.. 1986. Symmetry theorems related to Pompeiu's problem. *Amer. J. Math.*, **108**(5), 1023–1036.

Avron, J., Herbst, I., and Simon, B. 1978. Schrödinger operators with magnetic fields. I. General interactions. *Duke Math. J.*, **45**(4), 847–883.

Balinsky, A. A., Evans, W. D., and Lewis, R. T. 2015. *The Analysis and Geometry of Hardy's Inequality.* Universitext. Springer, Cham.

Bandle, C.. 1980. *Isoperimetric Inequalities and Applications.* Monographs and Studies in Mathematics, vol. 7. Pitman (Advanced Publishing Program), Boston, MA and London.

Bargmann, V. 1952. On the number of bound states in a central field of force. *Proc. Nat. Acad. Sci. U.S.A.*, **38**, 961–966.

Barsegyan, D. S. 2007. On inequalities of Lieb–Thirring type. *Mat. Zametki*, **82**(4), 504–514. English translation in *Math. Notes* **82** (2007), no. 3–4, 451–460.

Barseghyan, D., Exner, P., Kovařík, H., and Weidl, T. 2016. Semiclassical bounds in magnetic bottles. *Rev. Math. Phys.*, **28**(1), 1650002, 29.

Beals, R. 1967. Classes of compact operators and eigenvalue distributions for elliptic operators. *Amer. J. Math.*, **89**, 1056–1072.

Bebendorf, M. 2003. A note on the Poincaré inequality for convex domains. *Z. Anal. Anwendungen*, **22**(4), 751–756.

Beckner, W. 2004. Estimates on Moser embedding. *Potential Anal.*, **20**(4), 345–359.

Benguria, R. D. 2011. Isoperimetric inequalities for eigenvalues of the Laplacian. Pages 21–60 of: *Entropy and the Quantum II*. Contemp. Math., vol. 552. Amer. Math. Soc., Providence, RI.

Benguria, R., and Loss, M. 2000. A simple proof of a theorem of Laptev and Weidl. *Math. Res. Lett.*, **7**(2-3), 195–203.

Benguria, R. D., and Loss, M. 2004. Connection between the Lieb–Thirring conjecture for Schrödinger operators and an isoperimetric problem for ovals on the plane. Pages 53–61 of: *Partial Differential Equations and Inverse Problems*. C. Conca, R. Manásevich, G. Uhlmann, and M. S. Vogelius (editors). Contemp. Math., vol. 362. Amer. Math. Soc., Providence, RI.

Benguria, R. D., Linde, H., and Loewe, B. 2012. Isoperimetric inequalities for eigenvalues of the Laplacian and the Schrödinger operator. *Bull. Math. Sci.*, **2**(1), 1–56.

Bérard, P. H. 1977. On the wave equation on a compact Riemannian manifold without conjugate points. *Math. Z.*, **155**(3), 249–276.

Bérard, P. H. 1980. Spectres et groupes cristallographiques. I. Domaines euclidiens. *Invent. Math.*, **58**(2), 179–199.

Bérard, P. H. 1986. *Spectral Geometry: Direct and Inverse Problems*. With appendices by G. Besson, and by P. Bérard and M. Berger. Lecture Notes in Mathematics, vol. 1207. Springer-Verlag, Berlin.

Berezin, F. A. 1972a. Convex functions of operators. *Mat. Sb. (N.S.)*, **88(130)**, 268–276. English translation in *Math. USSR-Sb.*, **17**(2), (1972), 269–277.

Berezin, F. A. 1972b. Covariant and contravariant symbols of operators. *Izv. Akad. Nauk SSSR Ser. Mat.*, **36**, 1134–1167. English translation in *Math. USSR-Izv.* **6**, (1972), 1117–1151.

Berry, M. V. 1980. Some geometric aspects of wave motion: wavefront dislocations, diffraction catastrophes, diffractals. Pages 13–28 of: *Geometry of the Laplace Operator (Proc. Sympos. Pure Math., Univ. Hawaii, Honolulu, Hawaii, 1979)*. R. Osserman and A. Weinstein (editors). Proc. Sympos. Pure Math., XXXVI. Amer. Math. Soc., Providence, RI.

Bez, N., Hong, Y., Lee, S., Nakamura, S., and Sawano, Y. 2019. On the Strichartz estimates for orthonormal systems of initial data with regularity. *Adv. Math.*, **354**, 106736, 37.

Bez, N., Lee, S., and Nakamura, S. 2020. Maximal estimates for the Schrödinger equation with orthonormal initial data. *Selecta Math. (N.S.)*, **26**(4), Paper No. 52, 24.

Bez, N., Lee, S., and Nakamura, S. 2021. Strichartz estimates for orthonormal families of initial data and weighted oscillatory integral estimates. *Forum Math. Sigma*, **9**, Paper No. e1, 52.

Bianchi, G., and Egnell, H. 1991. A note on the Sobolev inequality. *J. Funct. Anal.*, **100**(1), 18–24.

Birman, M. Š. 1959. Perturbations of quadratic forms and the spectrum of singular boundary value problems. *Dokl. Akad. Nauk SSSR*, **125**, 471–474.

Birman, M. Š. 1961. On the spectrum of singular boundary-value problems. *Mat. Sb. (N.S.)*, **55 (97)**, 125–174. English translation: Pages 23–80 of: *Eleven Papers on Analysis*, Amer. Math. Soc. Transl. 53, (1966). Amer. Math. Soc., Providence, RI.

Birman, M. Š., and Borzov, V. V. 1971. The asymptotic behavior of the discrete spectrum of certain singular differential operators. Pages 24–28 of: *Spectral Theory*. M. Sh. Birman (editor). Problems of Mathematical Physics, No. 5. Izdat. Leningrad. Univ., Leningrad. English translation in *Spectral Theory*, Topics in Mathematical Physics **5**, 1–18 (1972). Consultants Bureau, New York and London.

Birman, M. Sh., and Laptev, A. 1996. The negative discrete spectrum of a two-dimensional Schrödinger operator. *Comm. Pure Appl. Math.*, **49**(9), 967–997.

Birman, M. Š., and Solomjak, M. Z. 1966. Approximation of functions of the W_p^α-classes by piece-wise-polynomial functions. *Dokl. Akad. Nauk SSSR*, **171**, 1015–1018. English translation in *Sov. Math. Dokl.*, **7**, 1573-1577, 1966. Corrections *ibid.* **8**(1), vi, (1967).

Birman, M. Š., and Solomjak, M. Z. 1967. Piecewise polynomial approximations of functions of classes W_p^α. *Mat. Sb. (N.S.)*, **73 (115)**, 331–355. English translation in *Math USSR Sb.*, **2** (1967), 295–317.

Birman, M. Š., and Solomjak, M. Z. 1970. The principal term of the spectral asymptotics for "non-smooth" elliptic problems. *Funkcional. Anal. i Priložen.*, **4**(4), 1–13. English translation in *Functional Anal. Appl.*, **4** (1970), 265–275.

Birman, M. Š., and Solomjak, M. Z. 1971. The asymptotics of the spectrum of "non-smooth" elliptic equations. *Funkcional. Anal. i Priložen.*, **5**(1), 69–70. English translation in *Functional Anal. Appl.*, **5**, (1971), 56–57.

Birman, M. Š., and Solomjak, M. Z. 1972. Spectral asymptotics of nonsmooth elliptic operators. I, II. *Trudy Moskov. Mat. Obšč.*, **27**, 3–52; ibid. **28** (1973), 3–34. English translation in *Trans. Moscow Math. Soc.*, **27**, (1972), 1–52 (1975); *ibid.*, **28**, (1973), 1–32 (1975).

Birman, M. Š., and Solomjak, M. Z. 1977a. Asymptotic properties of the spectrum of differential equations. Pages 5–58, i. (loose errata) of: *Mathematical Analysis, Vol. 14*.

Birman, M. Š., and Solomjak, M. Z. 1977b. Estimates for the singular numbers of integral operators. *Uspehi Mat. Nauk*, **32**(1(193)), 17–84, 271. English translation in *Russian Math. Surveys* **32**, (1977), 15–89.

Birman, M. Š., and Solomjak, M. Z. 1980. *Quantitative Analysis in Sobolev Imbedding Theorems and Applications to Spectral Theory*. Translated from the Russian by F. A. Cezus. Amer. Math. Soc. Transl., Series 2, vol. 114. Amer. Math. Soc., Providence, RI.

Birman, M. Sh., and Solomjak, M. Z. 1987. *Spectral Theory of Selfadjoint Operators in Hilbert Space*. Translated from the 1980 Russian original by S. Khrushchëv and

V. Peller. Mathematics and its Applications (Soviet Series). Dordrecht: D. Reidel Publishing Co.

Birman, M. Sh., and Solomyak, M. Z. 1991. Estimates for the number of negative eigenvalues of the Schrödinger operator and its generalizations. Pages 1–55 of: *Estimates and Asymptotics for Discrete Spectra of Integral and Differential Equations (Leningrad, 1989–90)*. M. Sh. Birman (editor). Adv. Soviet Math., vol. 7. Amer. Math. Soc., Providence, RI.

Birman, M. Sh., and Solomyak, M. Z. 1992. Schrödinger operator. Estimates for number of bound states as function-theoretical problem. Pages 1–54 of: *Spectral Theory of Operators (Novgorod, 1989)*. S. G. Gindikin (editor). Amer. Math. Soc. Transl. Ser. 2, vol. 150. Amer. Math. Soc., Providence, RI.

Birman, M. Š., Koplienko, L. S., and Solomjak, M. Z. 1975. Estimates of the spectrum of a difference of fractional powers of selfadjoint operators. *Izv. Vysš. Učebn. Zaved. Matematika*, 3–10. English translation in *Soviet Math. (Iz. VUZ)*, **19**(3), (1975), 1–6.

Birman, M. Sh., Karadzhov, G. E., and Solomyak, M. Z. 1991. Boundedness conditions and spectrum estimates for the operators $b(X)a(D)$ and their analogs. Pages 85–106 of: *Estimates and Asymptotics for Discrete Spectra of Integral and Differential Equations (Leningrad, 1989–90)*. M. Sh. Birman (editor). Adv. Soviet Math., vol. 7. Amer. Math. Soc., Providence, RI.

Birman, M. Sh., Laptev, A., and Solomyak, M. 1997. The negative discrete spectrum of the operator $(-\Delta)^l - \alpha V$ in $L_2(\mathbf{R}^d)$ for d even and $2l \geq d$. *Ark. Mat.*, **35**(1), 87–126.

Blanchard, Ph., and Stubbe, J. 1996. Bound states for Schrödinger Hamiltonians: phase space methods and applications. *Rev. Math. Phys.*, **8**(4), 503–547.

Blanchard, Ph., Stubbe, J., and Rezende, J. 1987. New estimates on the number of bound states of Schrödinger operators. *Lett. Math. Phys.*, **14**(3), 215–225.

Blankenbecler, R., Goldberger, M.L., and Simon, B. 1977. The bound states of weakly coupled long-range one-dimensional quantum Hamiltonians. *Annals of Physics*, **108**(1), 69–78.

Bliss, G. A. 1930. An integral inequality. *J. Lond. Math. Soc.*, **5**(1), 40–46.

Bögli, S. 2017. Schrödinger operator with non-zero accumulation points of complex eigenvalues. *Comm. Math. Phys.*, **352**(2), 629–639.

Bögli, S., and Cuenin, J-C. 2021. Counterexample to the Laptev–Safronov conjecture. Preprint, available at ArXiv:2109.06135.

Bögli, S,, and Štampach, F. 2020. On Lieb–Thirring inequalities for one-dimensional non-self-adjoint Jacobi and Schrödinger operators. *J. Spec. Theory*, to appear. Preprint, available at ArXiv:2004.09794.

Bohr, N. 1913. The spectra of helium and hydrogen. *Nature*, **92**, 231–232.

Borichev, A., Golinskii, L., and Kupin, S. 2009. A Blaschke-type condition and its application to complex Jacobi matrices. *Bull. Lond. Math. Soc.*, **41**(1), 117–123.

Borichev, A., Frank, R., and Volberg, A. 2022. Counting eigenvalues of Schrödinger operator with complex fast decreasing potential. *Adv. Math.*, **397**, 108115.

Born, M., and Jordan, P. 1925. Zur Quantenmechanik. *Z. Phys.*, **34**, 858–888.

Brasco, L., De Philippis, G., and Velichkov, B. 2015. Faber–Krahn inequalities in sharp quantitative form. *Duke Math. J.*, **164**(9), 1777–1831.

Breuer, J., Simon, B., and Zeitouni, O. 2018. Large deviations and sum rules for spectral theory: a pedagogical approach. *J. Spectr. Theory*, **8**(4), 1551–1581.

Brezis, H. 2011. *Functional Analysis, Sobolev Spaces and Partial Differential equations.* Universitext. Springer, New York.

Brezis, H., and Lieb, E. H. 1985. Sobolev inequalities with remainder terms, *J. Funct. Anal.*, **62**(1), 73–86.

Brossard, J., and Carmona, R. 1986. Can one hear the dimension of a fractal? *Comm. Math. Phys.*, **104**(1), 103–122.

Brownell, F. H., and Clark, C. W. 1961. Asymptotic distribution of the eigenvalues of the lower part of the Schrödinger operator spectrum. *J. Math. Mech.*, **10**(31–70; addendum), 525–527.

Bucur, D., and Henrot, A. 2019. Maximization of the second non-trivial Neumann eigenvalue. *Acta Math.*, **222**(2), 337–361.

Bugliaro, L., Fefferman, C., Fröhlich, J., Graf, G. M., and Stubbe, J. 1997. A Lieb–Thirring bound for a magnetic Pauli Hamiltonian. *Comm. Math. Phys.*, **187**(3), 567–582.

Burchard, A., and Thomas, L. E. 2005. On an isoperimetric inequality for a Schrödinger operator depending on the curvature of a loop. *J. Geom. Anal.*, **15**(4), 543–563.

Buslaev, V. S. 1962. Trace formulas for the Schrödinger operator in a three-dimensional space. *Dokl. Akad. Nauk SSSR*, **143**, 1067–1070.

Buslaev, V. S. 1966. The trace formulae and certain asymptotic estimates of the kernel of the resolvent for the Schrödinger operator in three-dimensional space. Pages 82–101 of: *Spectral Theory and Wave Processes*. Problems of Mathematical Physics, No. 1. Izdat. Leningrad. Univ., Leningrad. English translation in *Spectral Theory and Wave Processes*. M. Sh. Birman (editor). Topics in Mathematical Physics, vol 1. Springer, Boston, MA.

Buslaev, V. S., and Faddeev, L. D. 1960. Formulas for traces for a singular Sturm–Liouville differential operator. *Dokl. Akad. Nauk SSSR*, **132**, 13–16. English translation in *Soviet Math. Dokl.*, **1**, (1960), 451–454.

Buttazzo, G., Guarino Lo Bianco, S., and Marini, M. 2018. Sharp estimates for the anisotropic torsional rigidity and the principal frequency. *J. Math. Anal. Appl.*, **457**(2), 1153–1172.

Calderón, A.-P. 1961. Lebesgue spaces of differentiable functions and distributions. Pages 33–49 of: *Proc. Sympos. Pure Math., Vol. IV*. Amer. Math. Soc., Providence, RI.

Calogero, F. 1965. Upper and lower limits for the number of bound states in a given central potential. *Comm. Math. Phys.*, **1**, 80–88.

Carleman, T. 1935. Propriétés asymptotiques des fonctions fondamentales des membranes vibrantes. Pages 34–44 of: *8. Skand. Mat.-Kongr.*.

Carleman, T. 1936. Über die asymptotische Verteilung der Eigenwerte partieller Differentialgleichungen. *Ber. Sächs. Akad. Wiss. Leipzig, Math.-Phys. Kl.*, **88**, 119–132.

Carlen, E. A. 2010. Trace inequalities and quantum entropy: an introductory course. Pages 73–140 of: *Entropy and the Quantum*. Contemp. Math., vol. 529. Amer. Math. Soc., Providence, RI.

Carlen, E. A., Carrillo, J. A., and Loss, M. 2010. Hardy–Littlewood–Sobolev inequalities via fast diffusion flows. *Proc. Natl. Acad. Sci. USA*, **107**(46), 19696–19701.

Carlen, E. A., Frank, R. L., and Lieb, E. H. 2014. Stability estimates for the lowest eigenvalue of a Schrödinger operator. *Geom. Funct. Anal.*, **24**(1), 63–84.

Case, K. M. 1974. Orthogonal polynomials from the viewpoint of scattering theory. *J. Mathematical Phys.*, **15**, 2166–2174.

Case, K. M. 1975. Orthogonal polynomials. II. *J. Mathematical Phys.*, **16**, 1435–1440.

Chadan, K. M. 1968. The asymptotic behaviour of the number of bound states of a given potential in the limit of large coupling. *Nuovo Cimento*, **58 A**(1), 191 ff.

Chadan, K. M., Khuri, N. N., Martin, A., and Wu, T. T. 2003. Bound states in one and two spatial dimensions. *J. Math. Phys.*, **44**(2), 406–422.

Chang, S.-Y. A., Wilson, J. M., and Wolff, T. H. 1985. Some weighted norm inequalities concerning the Schrödinger operators. *Comment. Math. Helv.*, **60**(2), 217–246.

Chemin, J.-Y., and Xu, C.-J. 1997. Inclusions de Sobolev en calcul de Weyl–Hörmander et champs de vecteurs sous-elliptiques. *Ann. Sci. École Norm. Sup. (4)*, **30**(6), 719–751.

Chepyzhov, V. V., and Ilyin, A. A. 2004. On the fractal dimension of invariant sets: applications to Navier–Stokes equations. *Discrete and Continuous Dynamical Systems. Series A*, **10**(1–2), 117–135. .

Ciesielski, Z. 1970. On the spectrum of the Laplace operator. *Comment. Math. Prace Mat.*, **14**, 41–50.

Clark, C. 1967. The asymptotic distribution of eigenvalues and eigenfunctions for elliptic boundary value problems. *SIAM Rev.*, **9**, 627–646.

Coffman, C. V. 1972. Uniqueness of the ground state solution for $\Delta u - u + u^3 = 0$ and a variational characterization of other solutions. *Arch. Rational Mech. Anal.*, **46**, 81–95.

Colin de Verdière, Y. 1977. Nombre de points entiers dans une famille homothétique de domains de $\mathbf{R}^n$. *Ann. Sci. École Norm. Sup. (4)*, **10**(4), 559–575.

Colin de Verdière, Y. 1981. Une formule de traces pour l'opérateur de Schrödinger dans $\mathbf{R}^3$. *Ann. Sci. École Norm. Sup. (4)*, **14**(1), 27–39.

Colin de Verdière, Y. 1987. Construction de Laplaciens dont une partie finie du spectre est donnée. *Ann. Sci. École Norm. Sup. (4)*, **20**(4), 599–615.

Colin de Verdière, Y. 2010–2011. On the remainder in the Weyl formula for the Euclidean disk. *Séminaire de Théorie Spectrale et Géométrie*, **29**, 1–13.

Conlon, J. G. 1985. A new proof of the Cwikel–Lieb–Rosenbljum bound. *Rocky Mountain J. Math.*, **15**(1), 117–122.

Constantin, P., Foias, C., and Temam, R. 1985. Attractors representing turbulent flows. *Mem. Amer. Math. Soc.*, **53**(314).

Cordero-Erausquin, D., Nazaret, B., and Villani, C. 2004. A mass-transportation approach to sharp Sobolev and Gagliardo–Nirenberg inequalities. *Adv. Math.*, **182**(2), 307–332.

Cornfeld, I. P., Fomin, S. V., and Sinaĭ, Ya. G. 1982. *Ergodic Theory*. Translated from the Russian by A. B. Sosinskiĭ. Grundlehren der Mathematischen Wissenschaften, vol. 245. Springer-Verlag, New York.

Courant, R. 1920. Über die Eigenwerte bei den Differentialgleichungen der mathematischen Physik. *Math. Z.*, **7**(1-4), 1–57.

Courant, R. 1922. Über die Lösungen der Differentialgleichungen der Physik. *Math. Ann.*, **85**(1), 280–325.

Courant, R. 1925. Über direkte Methoden bei Variations- und Randwertproblemen. *Jahresber. Dtsch. Math.-Ver.*, **34**, 90–117.

Courant, R., and Hilbert, D. 1953. *Methods of Mathematical Physics. Vol. I*. Interscience Publishers, Inc., New York.

Courant, R., and Hilbert, D. 1962. *Methods of Mathematical Physics, Vol. II*. Partial Differential Equations. Reprinted 1989 in paperback in Wiley Classics Library. John Wiley & Sons, Inc., New York.

Crum, M. M. 1955. Associated Sturm–Liouville systems. *Q. J. Math., Oxf. II. Ser.*, **6**, 121–125.

Cuenin, J.-C. 2020. Improved eigenvalue bounds for Schrödinger operators with slowly decaying potentials. *Comm. Math. Phys.*, **376**(3), 2147–2160.

Cuenin, J.-C., Laptev, A., and Tretter, C. 2014. Eigenvalue estimates for non-selfadjoint Dirac operators on the real line. *Ann. Henri Poincaré*, **15**(4), 707–736.

Cwikel, M. 1977. Weak type estimates for singular values and the number of bound states of Schrödinger operators. *Ann. Math. (2)*, **106**(1), 93–100.

Cycon, H. L., Froese, R. G., Kirsch, W., and Simon, B. 1987. *Schrödinger Operators with Application to Quantum Mechanics and Global Geometry*. Texts and Monographs in Physics. Springer-Verlag, Berlin.

Damanik, D., and Remling, C. 2007. Schrödinger operators with many bound states. *Duke Math. J.*, **136**(1), 51–80.

Darboux, G. 1882. Sur une proposition relative aux équations linéaires. *C. R. Acad. Sci., Paris*, **94**, 1456–1459.

Daubechies, I. 1983. An uncertainty principle for fermions with generalized kinetic energy. *Comm. Math. Phys.*, **90**(4), 511–520.

Davies, E. B. 1995. *Spectral Theory and Differential Operators*. Cambridge Studies in Advanced Mathematics, vol. 42. Cambridge University Press, Cambridge.

Davies, E. B., and Nath, J. 2002. Schrödinger operators with slowly decaying potentials. *J. Computat. Appl. Math.*. **148**(1), 1–28.

Davies, E. B., and Simon, B. 1992. Spectral properties of Neumann Laplacian of horns. *Geom. Funct. Anal.*, **2**(1), 105–117.

de Guzmán, M. 1975. *Differentiation of Integrals in R^n*. Lecture Notes in Mathematics, Vol. 481. With appendices by Antonio Córdoba, and Robert Fefferman, and two by Roberto Moriyón. Springer-Verlag, Berlin and New York.

de la Bretèche, R. 1999. Preuve de la conjecture de Lieb-Thirring dans le cas des potentiels quadratiques strictement convexes. *Ann. Inst. H. Poincaré Phys. Théor.*, **70**(4), 369–380.

de Wet, J. S., and Mandl, F. 1950. On the asymptotic distribution of eigenvalues. *Proc. Roy. Soc. London Ser. A*, **200**, 572–580.

Deift, P., and Killip, R. 1999. On the absolutely continuous spectrum of one-dimensional Schrödinger operators with square summable potentials. *Comm. Math. Phys.*, **203**(2), 341–347.

Deift, P., and Trubowitz, E. 1979. Inverse scattering on the line. *Comm. Pure Appl. Math.*, **32**(2), 121–251.

Deift, P. A. 1978. Applications of a commutation formula. *Duke Math. J.*, **45**(2), 267–310.

Demirel, S. 2012. *Spectral Theory of Quantum Graphs*. Dissertation. University of Stuttgart.

Demuth, M., Hansmann, M., and Katriel, G. 2013. Eigenvalues of non-selfadjoint operators: a comparison of two approaches. Pages 107–163 of: *Mathematical Physics, Spectral Theory and Stochastic Analysis*. M. Demuth and W. Kirsch (editors). Oper. Theory Adv. Appl., vol. 232. Birkhäuser/Springer Basel AG, Basel.

Deny, J., and Lions, J. L. 1954. Les espaces du type de Beppo Levi. *Ann. Inst. Fourier (Grenoble)*, **5**, 305–370.

Denzler, J. 2015. Existence and regularity for a curvature dependent variational problem. *Trans. Amer. Math. Soc.*, **367**(6), 3829–3845.

Dimassi, M., and Sjöstrand, J. 1999. *Spectral Asymptotics in the Semi-Classical Limit*. London Mathematical Society Lecture Note Series, vol. 268. Cambridge University Press, Cambridge.

Dolbeault, J., Felmer, P., Loss, M., and Paturel, E. 2006. Lieb–Thirring type inequalities and Gagliardo–Nirenberg inequalities for systems. *J. Funct. Anal.*, **238**(1), 193–220.

Dolbeault, J., Laptev, A., and Loss, M. 2008. Lieb–Thirring inequalities with improved constants. *J. Eur. Math. Soc. (JEMS)*, **10**(4), 1121–1126.

Duistermaat, J. J., and Guillemin, V. W. 1975. The spectrum of positive elliptic operators and periodic bicharacteristics. *Invent. Math.*, **29**(1), 39–79.

Dyson, F. J., and Lenard, A. 1967. Stability of matter. I. *J. Math. Phys.*, **8**(3), 423–434.

Eckart, C. 1930. The penetration of a potential barrier by electrons. *Phys. Rev., II. Ser.*, **35**, 1303–1309.

Eden, A., and Foias, C. 1991. A simple proof of the generalized Lieb–Thirring inequalities in one-space dimension. *J. Math. Anal. Appl.*, **162**(1), 250–254.

Edmunds, D. E., and Evans, W. D. 2018. *Elliptic Differential Operators and Spectral Analysis*. Springer Monographs in Mathematics. Springer, Cham.

Egorov, Yu. V., and Kondrat'ev, V. A. 1992. Estimates of the negative spectrum of an elliptic operator. Pages 111–140 of: *Spectral Theory of Operators (Novgorod, 1989)*. S. G. Gindikin (editor). Amer. Math. Soc. Transl. Ser. 2, vol. 150. Amer. Math. Soc., Providence, RI.

Egorov, Yu. V., and Kondratiev, V. A. 1995. On moments of negative eigenvalues of an elliptic operator. *Math. Nachr.*, **174**, 73–79.

Egorov, Yu., and Kondratiev, V. 1996. *On Spectral Theory of Elliptic Operators*. Oper. Theory Adv. Appl., vol. 89. Birkhäuser Verlag, Basel.

Ekholm, T., and Frank, R. L. 2006. On Lieb–Thirring inequalities for Schrödinger operators with virtual level. *Comm. Math. Phys.*, **264**(3), 725–740.

Ekholm, T., and Frank, R. L. 2008. Lieb–Thirring inequalities on the half-line with critical exponent. *J. Eur. Math. Soc. (JEMS)*, **10**(3), 739–755.

El Soufi, A., and Ilias, S. 1986. Immersions minimales, première valeur propre du Laplacien et volume conforme. *Math. Ann.*, **275**(2), 257–267.

Epstein, P. S. 1930. Reflection of waves in an inhomogeneous absorbing medium. *Proc. Natl. Acad. Sci. USA*, **16**, 627–637.

Erdős, L. 1995. Magnetic Lieb–Thirring inequalities. *Comm. Math. Phys.*, **170**(3), 629–668.

Erdős, L., and Solovej, J. P. 1997. Semiclassical eigenvalue estimates for the Pauli operator with strong non-homogeneous magnetic fields. II. Leading order asymptotic estimates. *Comm. Math. Phys.*, **188**(3), 599–656.

Erdős, L., and Solovej, J.P. 1999. Semiclassical eigenvalue estimates for the Pauli operator with strong nonhomogeneous magnetic fields. I. Nonasymptotic Lieb–Thirring-type estimate. *Duke Math. J.*, **96**(1), 127–173.

Erdős, L., Loss, M., and Vougalter, V. 2000. Diamagnetic behavior of sums of Dirichlet eigenvalues. *Ann. Inst. Fourier (Grenoble)*, **50**(3), 891–907.

Evans, L. C. 2010. *Partial Differential Equations*. Second edn. Graduate Studies in Mathematics, vol. 19. Amer. Math. Soc., Providence, RI.

Exner, P., and Šeba, P. 1989. Bound states in curved quantum waveguides. *J. Math. Phys.*, **30**(11), 2574–2580.

Exner, P., Linde, H., and Weidl, T. 2004. Lieb-Thirring inequalities for geometrically induced bound states. *Lett. Math. Phys.*, **70**(1), 83–95.

Exner, P., Laptev, A., and Usman, M. 2014. On some sharp spectral inequalities for Schrödinger operators on semiaxis. *Comm. Math. Phys.*, **326**(2), 531–541.

Faber, G. 1923. Beweis, daß unter allen homogenen Membranen von gleicher Fläche und gleicher Spannung die kreisförmige den tiefsten Grundton gibt. *Münch. Ber.*, 169–172.

Faddeev, L. D. 1957. An expression for the trace of the difference between two singular differential operators of the Sturm–Liouville type. *Dokl. Akad. Nauk SSSR (N.S.)*, **115**, 878–881.

Faddeev, L. D. 1958. On the relation between S-matrix and potential for the one-dimensional Schrödinger operator. *Dokl. Akad. Nauk SSSR (N.S.)*, **121**, 63–66.

Faddeev, L. D. 1964. Properties of the S-matrix of the one-dimensional Schrödinger equation. *Trudy Mat. Inst. Steklov.*, **73**, 314–336. English translation in: Amer. Math. Soc. Transl. (Ser. 2), **65**, (1967), 139–166.

Fan, K. 1949. On a theorem of Weyl concerning eigenvalues of linear transformations. I. *Proc. Nat. Acad. Sci. USA*, **35**, 652–655.

Farmer, J. D., Ott, E., and Yorke, J. A. 1983. The dimension of chaotic attractors. *Phys. D*, **7**(1-3), 153–180.

Federer, H., and Fleming, W. H. 1960. Normal and integral currents. *Ann. of Math. (2)*, **72**, 458–520.

Fefferman, C. L. 1983. The uncertainty principle. *Bull. Amer. Math. Soc. (N.S.)*, **9**(2), 129–206.

Filonov, N. 2004. On an inequality for the eigenvalues of the Dirichlet and Neumann problems for the Laplace operator. *Algebra i Analiz*, **16**(2), 172–176. English translation in *St. Petersburg Math. J.*, **16**(2), 413–416, (2005).

Filonov, N. 2022. On the Pólya conjecture for circular sectors and for balls. Preprint, available at ArXiv:2208.03463.

Fischer, E. 1905. Über quadratische Formen mit reellen Koeffizienten. *Monatsh. Math. Phys.*, **16**(1), 234–249.

Fleckinger-Pellé, J., and Vassiliev, D. G. 1993. An example of a two-term asymptotics for the "counting function" of a fractal drum. *Trans. Amer. Math. Soc.*, **337**(1), 99–116.

Flügge, S. 1999. *Practical Quantum Mechanics*. Springer-Verlag, Berlin and Heidelberg.

Folland, G. B. 1999. *Real Analysis. Modern Techniques and their Applications*. Second edn. John Wiley & Sons, Inc., New York.

Förster, C., and Weidl, T. 2011. Trapped modes in an elastic plate with a hole. *Algebra i Analiz*, **23**(1), 255–288. English translation in *St. Petersburg Math. J.*, **3**(1), 179–202, (2012).

Förster, C. 2008. Trapped modes for an elastic plate with a perturbation of Young's modulus. *Comm. Partial Differential Equations*, **33**(7-9), 1339–1367.

Förster, C., and Östensson, J. 2008. Lieb–Thirring inequalities for higher order differential operators. *Math. Nachr.*, **281**(2), 199–213.

Frank, R. L. 2009a. Remarks on eigenvalue estimates and semigroup domination. Pages 63–86 of: *Spectral and Scattering Theory for Quantum Magnetic Systems*. Philippe Briet, François Germinet, and Georgi Raikov (editors). Contemp. Math., vol. 500. Amer. Math. Soc., Providence, RI.

Frank, R. L. 2009b. A simple proof of Hardy–Lieb–Thirring inequalities. *Comm. Math. Phys.*, **290**(2), 789–800.

Frank, R. L. 2011. Eigenvalue bounds for Schrödinger operators with complex potentials. *Bull. Lond. Math. Soc.*, **43**(4), 745–750.

Frank, R. L. 2014. Cwikel's theorem and the CLR inequality. *J. Spectr. Theory*, **4**(1), 1–21.

Frank, R. L. 2018a. Eigenvalue bounds for Schrödinger operators with complex potentials. III. *Trans. Amer. Math. Soc.*, **370**(1), 219–240.

Frank, R. L. 2018b. Eigenvalue bounds for the fractional Laplacian: a review. Pages 210–235 of: *Recent Developments in Nonlocal Theory*. G. Palatucci and T. Kuusi (editors). De Gruyter, Berlin.

Frank, Rupert L. 2021. The Lieb–Thirring inequalities: Recent results and open problems. Pages 45–86 of: *Nine Mathematical Challenges – An Elucidation*. Proc. Sympos. Pure Math., vol. 104. A. Kechris, N. Makarov, D. Ramakrishnan, and X. Zhu (editors). Amer. Math. Soc., Providence, RI.

Frank, R. L. 2022. Weyl's law under minimal assumptions. Preprint, available at ArXiv:2202.00323.

Frank, R. L., and Geisinger, L. 2011. Two-term spectral asymptotics for the Dirichlet Laplacian on a bounded domain. Pages 138–147 of: *Mathematical Results in Quantum Physics*. P. Exner (editor). World Sci. Publ., Hackensack, NJ.

Frank, R. L., and Geisinger, L. 2016. Refined semiclassical asymptotics for fractional powers of the Laplace operator. *J. Reine Angew. Math.*, **712**, 1–37.

Frank, R. L., and Laptev, A. 2008. Spectral inequalities for Schrödinger operators with surface potentials. Pages 91–102 of: *Spectral Theory of Differential Operators*. T. Suslina and D. Yafaev (editors). Amer. Math. Soc. Transl. Ser. 2, vol. 225. Amer. Math. Soc., Providence, RI.

Frank, R. L., and Laptev, A. 2010. Inequalities between Dirichlet and Neumann eigenvalues on the Heisenberg group. *Int. Math. Res. Not.*, **2010**(15), 2889–2902.

Frank, R. L., and Larson, S. 2020. Two-term spectral asymptotics for the Dirichlet Laplacian in a Lipschitz domain. *J. Reine Angew. Math.*, **766**, 195–228.

Frank, R. L., and Lieb, E. H. 2010. Inversion positivity and the sharp Hardy-Littlewood-Sobolev inequality. *Calc. Var. Partial Differential Equations*, **39**(1-2), 85–99.

Frank, R. L., and Lieb, E. H. 2012. A new, rearrangement-free proof of the sharp Hardy–Littlewood–Sobolev inequality. Pages 55–67 of: *Spectral Theory, Function Spaces and Inequalities*. B. M. Brown, J. Lang, and I. G. Wood (editors). Oper. Theory Adv. Appl., vol. 219. Birkhäuser/Springer Basel AG, Basel.

Frank, R. L., and Pushnitski, A. 2015. Trace class conditions for functions of Schrödinger operators. *Comm. Math. Phys.*, **335**(1), 477–496.

Frank, R. L., and Sabin, J. 2017a. Restriction theorems for orthonormal functions, Strichartz inequalities, and uniform Sobolev estimates. *Amer. J. Math.*, **139**(6), 1649–1691.

Frank, R. L., and Sabin, J. 2017b. Spectral cluster bounds for orthonormal systems and oscillatory integral operators in Schatten spaces. *Adv. Math.*, **317**, 157–192.

Frank, R. L., and Sabin, J. 2017c. The Stein–Tomas inequality in trace ideals. Pages Exp. No. XV, 12 of: *Séminaire Laurent Schwartz – Équations aux Dérivées Partielles et Applications. Année 2015–2016.* Ed. Éc. Polytech., Palaiseau.

Frank, R. L., and Sabin, J. 2022. Sharp Weyl laws with singular potentials. *Pure and Appl. Anal.*, to appear. Preprint, available at ArXiv:2007.04284.

Frank, R. L., and Seiringer, R. 2008. Non-linear ground state representations and sharp Hardy inequalities. *J. Funct. Anal.*, **255**(12), 3407–3430.

Frank, R. L., and Seiringer, R. 2012. Lieb–Thirring inequality for a model of particles with point interactions. *J. Math. Phys.*, **53**(9), 095201, 11.

Frank, R. L., and Simon, B. 2011. Critical Lieb–Thirring bounds in gaps and the generalized Nevai conjecture for finite gap Jacobi matrices. *Duke Math. J.*, **157**(3), 461–493.

Frank, R. L., and Simon, B. 2017. Eigenvalue bounds for Schrödinger operators with complex potentials. II. *J. Spectr. Theory*, **7**(3), 633–658.

Frank, R. L., Laptev, A., Lieb, E. H., and Seiringer, R. 2006. Lieb–Thirring inequalities for Schrödinger operators with complex-valued potentials. *Lett. Math. Phys.*, **77**(3), 309–316.

Frank, R. L., Lieb, E. H., and Seiringer, R. 2007. Number of bound states of Schrödinger operators with matrix-valued potentials. *Lett. Math. Phys.*, **82**(2–3), 107–116.

Frank, R. L., Simon, B., and Weidl, T. 2008a. Eigenvalue bounds for perturbations of Schrödinger operators and Jacobi matrices with regular ground states. *Comm. Math. Phys.*, **282**(1), 199–208.

Frank, R. L., Lieb, E. H., and Seiringer, R. 2008b. Hardy–Lieb–Thirring inequalities for fractional Schrödinger operators. *J. Amer. Math. Soc.*, **21**(4), 925–950.

Frank, R. L., Loss, M., and Weidl, T. 2009. Pólya's conjecture in the presence of a constant magnetic field. *J. Eur. Math. Soc. (JEMS)*, **11**(6), 1365–1383.

Frank, R. L., Lieb, E. H., and Seiringer, R. 2010. Equivalence of Sobolev inequalities and Lieb–Thirring inequalities. Pages 523–535 of: *XVIth International Congress on Mathematical Physics*. P. Exner (editor). World Sci. Publ., Hackensack, NJ.

Frank, R. L., Lewin, M., Lieb, E. H., and Seiringer, R. 2013. A positive density analogue of the Lieb–Thirring inequality. *Duke Math. J.*, **162**(3), 435–495.

Frank, R. L., Lewin, M., Lieb, E. H., and Seiringer, R. 2014. Strichartz inequality for orthonormal functions. *J. Eur. Math. Soc. (JEMS)*, **16**(7), 1507–1526.

Frank, R. L., Laptev, A., and Safronov, O. 2016. On the number of eigenvalues of Schrödinger operators with complex potentials. *J. Lond. Math. Soc. (2)*, **94**(2), 377–390.

Frank, R. L., Gontier, D., and Lewin, M. 2021a. The periodic Lieb–Thirring inequality. Pages 135–154 of: *Partial Differential Equations, Spectral Theory, and Mathematical Physics: The Ari Laptev Anniversary Volume.* P. Exner, R. L. Frank, F.

Gesztesy, H. Holden, T. Weidl (editors). European Mathematical Society Publishing House, Zürich.

Frank, R. L., Gontier, D., and Lewin, M. 2021b. The nonlinear Schrödinger equation for orthonormal functions II: Application to Lieb–Thirring Inequalities. *Comm. Math. Phys.*, **384**(3), 1783–1828.

Frank, R. L., Gontier, D., and Lewin, M. 2021c. Optimizers for the finite-rank Lieb–Thirring inequality. Preprint, available at ArXiv:2109.05984.

Frank, R. L., Hundertmark, D., Jex, M., and Nam, P.T. 2021d. The Lieb–Thirring inequality revisited. *J. Eur. Math. Soc. (JEMS)*, **23**(8), 2583–2600.

Frank, R. L., Laptev, A., and Weidl, T. 2022. An improved one-dimensional Hardy inequality. *J. Math. Sciences*, to appear. Preprint, available at ArXiv:2204.00877.

Friedlander, L. 1991. Some inequalities between Dirichlet and Neumann eigenvalues. *Arch. Rational Mech. Anal.*, **116**(2), 153–160.

Friedman, A. 1970. *Foundations of Modern Analysis*. Reprinted 1982 in paperback by Dover Publications, Inc., New York.

Friedrichs, K. 1928. Die Randwert-und Eigenwertprobleme aus der Theorie der elastischen Platten. (Anwendung der direkten Methoden der Variationsrechnung). *Math. Ann.*, **98**(1), 205–247.

Friedrichs, K. 1934a. Spektraltheorie halbbeschränkter Operatoren und Anwendung auf die Spektralzerlegung von Differentialoperatoren. *Math. Ann.*, **109**(1), 465–487.

Friedrichs, K. 1934b. Spektraltheorie halbbeschränkter Operatoren und Anwendung auf die Spektralzerlegung von Differentialoperatoren. *Math. Ann.*, **109**(1), 685–713.

Gagliardo, E. 1958. Proprietà di alcune classi di funzioni in più variabili. *Ric. Mat.*, **7**, 102–137.

Gagliardo, E. 1959. Ulteriori proprieta di alcune classi di funzioni in piu variabili. *Ric. Mat.*, **8**, 24–51.

Gamboa, F., Nagel, J., and Rouault, A. 2016. Sum rules via large deviations. *J. Funct. Anal.*, **270**(2), 509–559.

Gårding, L. 1951. The asymptotic distribution of the eigenvalues and eigenfunctions of a general vibration problem. *Kungl. Fysiografiska Sällskapets i Lund Förhandlingar [Proc. Roy. Physiog. Soc. Lund]*, **21**(11), 10.

Gårding, L. 1953. Dirichlet's problem for linear elliptic partial differential equations. *Math. Scand.*, **1**, 55–72.

Gardner, C. S., Greene, J. M., Kruskal, M. D., and Miura, R. M. 1967. Method for solving the Korteweg–deVries equation. *Phys. Rev. Lett.*, **19**(Nov), 1095–1097.

Gardner, C. S., Greene, J. M., Kruskal, M. D., and Miura, R. M. 1974. Korteweg–deVries equation and generalization. VI. Methods for exact solution. *Comm. Pure Appl. Math.*, **27**, 97–133.

Garnett, J. B. 2007. *Bounded Analytic Functions*. Revised first edn. Graduate Texts in Mathematics, vol. 236. Springer, New York.

Geisinger, L., and Weidl, T. 2011. Sharp spectral estimates in domains of infinite volume. *Rev. Math. Phys.*, **23**(6), 615–641.

Geisinger, L., Laptev, A., and Weidl, T. 2011. Geometrical versions of improved Berezin–Li–Yau inequalities. *J. Spectr. Theory*, **1**(1), 87–109.

Gel'fand, I. M. 1956. On identities for eigenvalues of a differential operator of second order. *Uspehi Mat. Nauk (N.S.)*, **11**(1(67)), 191–198.

Gel'fand, I. M., and Levitan, B. M. 1953. On a simple identity for the characteristic values of a differential operator of the second order. *Doklady Akad. Nauk SSSR (N.S.)*, **88**, 593–596.

Gesztesy, F. 1993. A complete spectral characterization of the double commutation method. *J. Funct. Anal.*, **117**(2), 401–446.

Gesztesy, F., and Teschl, G. 1996. On the double commutation method. *Proc. Amer. Math. Soc.*, **124**(6), 1831–1840.

Gesztesy, F., Simon, B., and Teschl, G. 1996. Spectral deformations of one-dimensional Schrödinger operators. *J. Anal. Math.*, **70**, 267–324.

Gidas, B., Ni, W. M., and Nirenberg, L. 1981. Symmetry of positive solutions of nonlinear elliptic equations in $\mathbf{R}^n$. Pages 369–402 of: *Mathematical Analysis and Applications, Part A*. L. Nachbin (editor). Adv. in Math. Suppl. Stud., vol. 7. Academic Press, New York and London.

Gilbarg, D., and Trudinger, N. S. 1998. *Elliptic Partial Differential Equations of Second Order*. Paperback reprint 1998. Classics in Mathematics. Springer-Verlag, Berlin.

Glaser, V., Martin, A., Grosse, H., and Thirring, W. 1976. A family of optimal conditions for the absence of bound states in a potential. Pages 169–194 of: *Studies in Mathematical Physics, Essays in Honor of Valentine Bargmann*. E. H. Lieb, B. Simon, and A. S. Wightman (editors). Princeton Series in Physics vol. 58. Princeton University Press, Princeton, NJ.

Glaser, V., Grosse, H., and Martin, A. 1978. Bounds on the number of eigenvalues of the Schrödinger operator. *Comm. Math. Phys.*, **59**(2), 197–212.

Glazman, I. M. 1966. *Direct Methods of Qualitative Spectral Analysis of Singular Differential Operators*. Translated from the Russian by the IPST staff. Israel Program for Scientific Translations, Jerusalem, 1965; Daniel Davey & Co., Inc., New York.

Gordon, C., Webb, D., and Wolpert, S. 1992a. Isospectral plane domains and surfaces via Riemannian orbifolds. *Invent. Math.*, **110**(1), 1–22.

Gordon, C., Webb, D. L., and Wolpert, S. 1992b. One cannot hear the shape of a drum. *Bull. Amer. Math. Soc. (N.S.)*, **27**(1), 134–138.

Gontier, D., Lewin, M., and Nazar, F. Q. 2021. The nonlinear Schrödinger equation for orthonormal functions: existence of ground states. *Arch. Rational Mech. Anal.*, **240**(3), 1203–1254.

Grebenkov, D. S., and Nguyen, B.-T. 2013. Geometrical structure of Laplacian eigenfunctions. *SIAM Rev.*, **55**(4), 601–667.

Grigor'yan, A., and Nadirashvili, N. 2015. Negative eigenvalues of two-dimensional Schrödinger operators. *Arch. Rational Mech. Anal.*, **217**(3), 975–1028.

Grigor'yan, A., Netrusov, Y., and Yau, S.-T. 2004. Eigenvalues of elliptic operators and geometric applications. Pages 147–217 of: *Surveys in Differential Geometry. Vol. IX*. A. Grigor'yan and S.-T. Yau (editors). International Press, Somerville, MA.

Grigor'yan, A., Nadirashvili, N., and Sire, Y. 2016. A lower bound for the number of negative eigenvalues of Schrödinger operators. *J. Differential Geom.*, **102**(3), 395–408.

Grisvard, P. 1985. *Elliptic Problems in Nonsmooth Domains*. Paperback reprint, with a foreword by Susanne C. Brenner 2011. Classics in Applied Mathematics, vol. 69. SIAM, Philadelphia, PA.

Grünbaum, B., and Shephard, G. C. 1987. *Tilings and Patterns*. W. H. Freeman and Company, New York.

Hänel A., and Weidl, T. 2017. Spectral asymptotics for the Dirichlet Laplacian with a Neumann window via a Birman–Schwinger analysis of the Dirichlet-to-Neumann operator. Pages 315–352 of: *Functional Analysis and Operator Theory for Quantum Physics*. J. Dittrich, H. Kovarik. and A. Laptev (editors). EMS Ser. Congr. Rep. Eur. Math. Soc., Zürich.

Hainzl, C., Hamza, E., Seiringer, R., and Solovej, J. P. 2008. The BCS functional for general pair interactions. *Comm. Math. Phys.*, **281**(2), 349–367.

Hajłasz, P., Koskela, P., and Tuominen, H. 2008. Sobolev embeddings, extensions and measure density condition. *J. Funct. Anal.*, **254**(5), 1217–1234.

Hansmann, M. 2011. An eigenvalue estimate and its application to non-selfadjoint Jacobi and Schrödinger operators. *Lett. Math. Phys.*, **98**(1), 79–95.

Hansson, A. M., and Laptev, A. 2008. Sharp spectral inequalities for the Heisenberg Laplacian. Pages 100–115 of: *Groups and Analysis*. K. Tent (editor). London Math. Soc. Lecture Note Ser., vol. 354. Cambridge University Press, Cambridge.

Hardy, G. H. 1919. Notes on some points in the integral calculus. LI. (On Hilbert's double-series theorem, and some connected theorems concerning the convergence of infinite series and integrals.). *Messenger of Mathematics*, **48**, 107–112.

Hardy, G. H. 1925. Notes on some points in the integral calculus. LX. *Messenger of Mathematics*, **54**, 150–156.

Hardy, G. H. 1928. Note on some points in the integral calculus LXIV. *Messenger of Mathematics*, **57**, 12–16.

Hardy, G. H., Littlewood, J. E., and Pólya, G. 1942. *Inequalities*. Reprinted in paperback in Cambridge Mathematical Library (1988). Cambridge University Press, Cambridge.

Harrell, II, E. M., and Stubbe, J. 1997. On trace identities and universal eigenvalue estimates for some partial differential operators. *Trans. Amer. Math. Soc.*, **349**(5), 1797–1809.

Harrell, II, E. M., and Stubbe, J. 2010. Universal bounds and semiclassical estimates for eigenvalues of abstract Schrödinger operators. *SIAM J. Math. Anal.*, **42**(5), 2261–2274.

Harrell, II, E. M., and Stubbe, J. 2011. Trace identities for commutators, with applications to the distribution of eigenvalues. *Trans. Amer. Math. Soc.*, **363**(12), 6385–6405.

Harrell, II, E. M., and Stubbe, J. 2018. Two-term, asymptotically sharp estimates for eigenvalue means of the Laplacian. *J. Spectr. Theory*, **8**(4), 1529–1550.

Hebey, E. 1996. *Sobolev Spaces on Riemannian Manifolds*. Lecture Notes in Mathematics, vol. 1635. Springer-Verlag, Berlin.

Heisenberg, W. 1925. Über quantentheoretische Umdeutung kinematischer und mechanischer Beziehungen. *Z. Phys.*, **33**, 879–893.

Helffer, B., and Robert, D. 1983. Calcul fonctionnel par la transformation de Mellin et opérateurs admissibles. *J. Funct. Anal.*, **53**(3), 246–268.

Helffer, B., and Robert, D. 1990a. Riesz means of bound states and semiclassical limit connected with a Lieb–Thirring's conjecture. *Asymptotic Anal.*, **3**(2), 91–103.

Helffer, B., and Robert, D. 1990b. Riesz means of bounded states and semi-classical limit connected with a Lieb–Thirring conjecture. II. *Ann. Inst. H. Poincaré Phys. Théor.*, **53**(2), 139–147.

Helffer, B., and Sjöstrand, J. 1990. On diamagnetism and de Haas-van Alphen effect. *Ann. Inst. H. Poincaré Phys. Théor.*, **52**(4), 303–375.

Helffer, B.. 2013. *Spectral Theory and its Applications*. Cambridge Studies in Advanced Mathematics, vol. 139. Cambridge University Press, Cambridge.

Helffer, B , and Persson Sundqvist, M. 2016. On nodal domains in Euclidean balls. *Proc. Amer. Math. Soc.*, **144**(11), 4777–4791.

Hempel, R., Seco, L. A., and Simon, B. 1991. The essential spectrum of Neumann Laplacians on some bounded singular domains. *J. Funct. Anal.*, **102**(2), 448–483.

Henrot, A. 2006. *Extremum Problems for Eigenvalues of Elliptic Operators*. Frontiers in Mathematics. Birkhäuser Verlag, Basel.

Henrot, A. (editor). 2017. *Shape Optimization and Spectral Theory*. De Gruyter, Berlin.

Herbst, I. W., and Zhao, Z. X. 1988. Sobolev spaces, Kac-regularity, and the Feynman–Kac formula. Pages 171–191 of: *Seminar on Stochastic Processes, (Princeton, NJ, 1987)*. E. Çinlar, K. L. Chung, R. K. Getoor, and J. Glover (editors). Progr. Probab. Statist., vol. 15. Birkhäuser Boston, Boston, MA.

Hersch, J. 1960. Sur la fréquence fondamentale d'une membrane vibrante: évaluations par défaut et principe de maximum. *Z. Angew. Math. Phys.*, **11**, 387–413.

Hilbert, D. 1906. Grundzüge einer allgemeinen Theorie der linearen Integralgleichungen. Vierte Mitteilung. *Nachr. Ges. Wiss. Göttingen, Math.-Phys. Kl.*, **1906**, 157–227.

Hoang, V., Hundertmark, D., Richter, J., and Vugalter, S. 2017. *Quantitative bounds versus existence of weakly coupled bound states for Schrödinger type operators*. Preprint, available at ArXiv:1610.09891.

Hoffmann-Ostenhof, M., and Hoffmann-Ostenhof, T. 1977. "Schrödinger inequalities" and asymptotic behavior of the electron density of atoms and molecules. *Phys. Rev. A (3)*, **16**(5), 1782–1785.

Hong, I. 1954. On an inequality concerning the eigenvalue problem of membrane. *Kōdai Math. Sem. Rep.*, **6**, 113–114.

Hoppe, J., Laptev, A., and Östensson, J. 2006. Solitons and the removal of eigenvalues for fourth-order differential operators. *Int. Math. Res. Not.*, Art. ID 85050, 14.

Hörmander, L. 1968. The spectral function of an elliptic operator. *Acta Math.*, **121**, 193–218.

Hörmander, L. 1983. *The Analysis of Linear Partial Differential operators. II, Differential Operators with Constant Coefficients*. Reprinted 2005 in Classics in Mathematics. Springer-Verlag, Berlin.

Hörmander, L. 1990. *The Analysis of Linear Partial Differential operators. I, Distribution Theory and Fourier Analysis*. Second edn. Reprinted 2003 in Classics in Mathematics. Springer-Verlag, Berlin.

Hörmander, L. 1994a. *The Analysis of Linear Partial Differential operators. III, Pseudodifferential Operators*. Reprinted 2007 in Classics in Mathematics. Springer-Verlag, Berlin.

Hörmander, L. 1994b. *The Analysis of Linear Partial Differential operators. IV, Fourier Integral Operators*. Reprinted 2009 in Classics in Mathematics. Springer-Verlag, Berlin.

Hryniv, R., and Mykytyuk, Y. 2021. On the first trace formula for Schrödinger operators. *J. Spect. Theory*, **11**, 489–507.

Hryniv, R., Melnyk, B., and Mykytyuk, Y. 2021. Inverse scattering for reflectionless Schrödinger operators with integrable potentials and generalized soliton solutions for the KdV equation. *Ann. Henri Poincaré*, **22**(2), 487–527.

Hundertmark, D. 2002. On the number of bound states for Schrödinger operators with operator-valued potentials. *Ark. Mat.*, **40**(1), 73–87.

Hundertmark, D., and Simon, B. 2002. Lieb–Thirring inequalities for Jacobi matrices. *J. Approx. Theory*, **118**(1), 106–130.

Hundertmark, D., Lieb, E. H., and Thomas, L. E. 1998. A sharp bound for an eigenvalue moment of the one-dimensional Schrödinger operator. *Adv. Theor. Math. Phys.*, **2**(4), 719–731.

Hundertmark, D., Laptev, A., and Weidl, T. 2000. New bounds on the Lieb–Thirring constants. *Invent. Math.*, **140**(3), 693–704.

Hundertmark, D., Kunstmann, P., Ried, T., and Vugalter, S.. 2018. *Cwikel's bound reloaded*. Preprint, available at ArXiv:1809.05069.

Il'in, A. A. 2005. Lieb–Thirring integral inequalities and their applications to attractors of Navier–Stokes equations. *Mat. Sb.*, **196**(1), 33–66. English translation in *Sb. Math.* **196**(1–2), 29–61, (2005).

Il'in, A. A. 2009. On the spectrum of the Stokes operator. *Funktsional. Anal. i Prilozhen.*, **43**(4), 14–25. English translation in *Funct. Anal. Appl.* **43**(4), 254–263, (2009).

Ilyin, A., Laptev, A., Loss, M., and Zelik, S. 2016. One-dimensional interpolation inequalities, Carlson–Landau inequalities, and magnetic Schrödinger operators. *Int. Math. Res. Not.*, 1190–1222.

Ivriĭ, V. Ja. 1980a. The second term of the spectral asymptotics for a Laplace–Beltrami operator on manifolds with boundary. *Funktsional. Anal. i Prilozhen.*, **14**(2), 25–34. English translation in *Functional Anal. Appl.*, **14**(2), 98–106, (1980).

Ivriĭ, V. Ja. 1980b. The second term of the spectral asymptotics for the Laplace–Beltrami operator on manifolds with boundary and for elliptic operators acting in vector bundles. *Dokl. Akad. Nauk SSSR*, **250**(6), 1300–1302. English translation in *Soviet Math. Dokl.*, **20**(1), 300–302, (1980).

Ivrii, V. 1998. *Microlocal Analysis and Precise Spectral Asymptotics*. Springer Monographs in Mathematics. Springer-Verlag, Berlin.

Ivrii, V. 2019a. *Microlocal Analysis, Sharp Spectral Asymptotics and Applications. I, Semiclassical Microlocal Analysis and Local and Microlocal Semiclassical Asymptotics*. Springer, Cham.

Ivrii, V. 2019b. *Microlocal Analysis, Sharp Spectral Asymptotics and Applications. II, Functional Methods and Eigenvalue Asymptotics*. Springer, Cham.

Ivrii, V. 2019c. *Microlocal Analysis, Sharp Spectral Asymptotics and Applications. III, Magnetic Schrödinger Operator 1*. Springer, Cham.

Ivrii, V. 2019d. *Microlocal Analysis, Sharp Spectral Asymptotics and Applications. IV, Magnetic Schrödinger operator 2*. Springer, Cham.

Jacobi, C. G. J. 1837. Zur Theorie der Variationsrechnung und der Differentialgleichungen. *J. Reine Angew. Math.*, **17**, 68–82.

Jakšić, V., Molčanov, S., and Simon, B. 1992. Eigenvalue asymptotics of the Neumann Laplacian of regions and manifolds with cusps. *J. Funct. Anal.*, **106**(1), 59–79.

Jensen, A. 1980. Spectral properties of Schrödinger operators and time-decay of the wave functions results in $L^2(\mathbf{R}^m)$, $m \geq 5$. *Duke Math. J.*, **47**(1), 57–80.

Jensen, A. 1984. Spectral properties of Schrödinger operators and time-decay of the wave functions. Results in $L^2(\mathbf{R}^4)$. *J. Math. Anal. Appl.*, **101**(2), 397–422.

Jensen, A., and Kato, T. 1979. Spectral properties of Schrödinger operators and time-decay of the wave functions. *Duke Math. J.*, **46**(3), 583–611.

Jones, P. W. 1981. Quasiconformal mappings and extendability of functions in Sobolev spaces. *Acta Math.*, **147**(1-2), 71–88.

Jost, R., and Pais, A. 1951. On the scattering of a particle by a static potential. *Phys. Rev. (2)*, **82**, 840–851.

Kac, I. S., and Kreĭn, M. G. 1958. Criteria for the discreteness of the spectrum of a singular string. *Izv. Vysš. Učebn. Zaved. Matematika*, **1958**(2 (3)), 136–153.

Kac, M. 1966. Can one hear the shape of a drum? *Amer. Math. Monthly*, **73**(4, part II), 1–23.

Kalf, H., Schmincke, U.-W., Walter, J., and Wüst, R. 1975. On the spectral theory of Schrödinger and Dirac operators with strongly singular potentials. Pages 182–226 of: *Spectral Theory and Differential Equations, (Proc. Sympos., Dundee, 1974; dedicated to Konrad Jörgens)*. W. N. Everitt (editor). Lecture Notes in Math., Vol. 448. Springer-Verlag, Berlin.

Karpukhin, M., Nadirashvili, N., Penskoi, A. V., and Polterovich, I. 2021. An isoperimetric inequality for Laplace eigenvalues on the sphere. *J. Differential Geom.*, **118**(2), 313–333.

Kashin, B. S. 2006. On a class of inequalities for orthonormal systems. *Mat. Zametki*, **80**(2), 204–208. English translation in *Math. Notes*, **80**(1–2), 199–203, (2006).

Kato, T. 1951. Fundamental properties of Hamiltonian operators of Schrödinger type. *Trans. Amer. Math. Soc.*, **70**, 195–211.

Kato, T. 1972. Schrödinger operators with singular potentials. *Israel J. Math.*, **13**, 135–148 (1973).

Kato, T. 1978. Remarks on Schrödinger operators with vector potentials. *Integral Equations Operator Theory*, **1**(1), 103–113.

Kato, T. 1980. *Perturbation Theory for Linear Operators*. Reprinted 1995 in Classics in Mathematics. Springer-Verlag, Berlin.

Kay, I., and Moses, H. E. 1956. Reflectionless transmission through dielectrics and scattering potentials. *J. Appl. Phys.*, **27**(12), 1503–1508.

Keller, J. B. 1961. Lower bounds and isoperimetric inequalities for eigenvalues of the Schrödinger equation. *J. Math. Phys.*, **2**, 262–266.

Kellner, R. 1966. On a theorem of Pólya. *Amer. Math. Monthly*, **73**, 856–858.

Kenig, C. E., Ruiz, A., and Sogge, C. D. 1987. Uniform Sobolev inequalities and unique continuation for second order constant coefficient differential operators. *Duke Math. J.*, **55**(2), 329–347.

Khuri, N. N., Martin, A., and Wu, T. T. 2002. Bound states in n dimensions (especially $n = 1$ and $n = 2$). *Few-Body Systems*, **31**, 83–89.

Killip, R. 2007. Spectral theory via sum rules. Pages 907–930 of: *Spectral Theory and Mathematical Physics: a Festschrift in Honor of Barry Simon's 60th Birthday*. F. Gesztesy, P. Deift, C. Galvez, P. Perry, and W. Schlag (editors). Proc. Sympos. Pure Math., vol. 76. Amer. Math. Soc., Providence, RI.

Killip, R., and Simon, B. 2003. Sum rules for Jacobi matrices and their applications to spectral theory. *Ann. of Math. (2)*, **158**(1), 253–321.

Killip, R., and Simon, B. 2009. Sum rules and spectral measures of Schrödinger operators with L^2 potentials. *Ann. of Math. (2)*, **170**(2), 739–782.

Killip, R., Vişan, M., and Zhang, X. 2018. Low regularity conservation laws for integrable PDE. *Geom. Funct. Anal.*, **28**(4), 1062–1090.

Klaus, M. 1977. On the bound state of Schrödinger operators in one dimension. *Ann. Phys.*, **108**(2), 288–300.

Klaus, M., and Simon, B. 1980. Coupling constant thresholds in nonrelativistic quantum mechanics. I. Short-range two-body case. *Ann. Physics*, **130**(2), 251–281.

Klaus, M. 1982/83. Some applications of the Birman–Schwinger principle. *Helv. Phys. Acta*, **55**(1), 49–68.

Kondrachov, W. 1945. Sur certaines propriétés des fonctions dans l'espace. *C. R. (Dokl.) Acad. Sci. URSS, n. Ser.*, **48**, 535–538.

Korevaar, N. 1993. Upper bounds for eigenvalues of conformal metrics. *J. Diff. Geom.*, **37**(1), 73–93.

Kovařík, H., and Weidl, T. 2015. Improved Berezin–Li–Yau inequalities with magnetic field. *Proc. Roy. Soc. Edinburgh Sect. A*, **145**(1), 145–160.

Kovařík, H., Vugalter, S., and Weidl, T. 2007. Spectral estimates for two-dimensional Schrödinger operators with application to quantum layers. *Comm. Math. Phys.*, **275**(3), 827–838.

Kovařík, H., Vugalter, S., and Weidl, T. 2009. Two-dimensional Berezin–Li–Yau inequalities with a correction term. *Comm. Math. Phys.*, **287**(3), 959–981.

Krahn, E. 1925. Über eine von Rayleigh formulierte Minimaleigenschaft des Kreises. *Math. Ann.*, **94**, 97–100.

Krahn, E. 1926. Über Minimaleigenschaften der Kugel in drei und mehr Dimensionen. *Acta Univ. Dorpat A*, **9**, 1-44. English translation in *Edgar Krahn 1894–1961. A Centenary Volume*. Ü. Lumiste and J. Peetre (editors). Amsterdam: IOS Press; Tartu: The Estonian Mathematical Society, 139–175, (1994).

Kreĭn, M. G. 1951. Determination of the density of a nonhomogeneous symmetric cord by its frequency spectrum. *Doklady Akad. Nauk SSSR (N.S.)*, **76**, 345–348.

Kröger, P. 1992. Upper bounds for the Neumann eigenvalues on a bounded domain in Euclidean space. *J. Funct. Anal.*, **106**(2), 353–357.

Kufner, A., Maligranda, L., and Persson, L.-E. 2006. The prehistory of the Hardy inequality. *Amer. Math. Monthly*, **113**(8), 715–732.

Kufner, A., Maligranda, L., and Persson, L.-E. 2007. *The Hardy Inequality. About its History and Some Related Results*. Vydavatelský Servis, Plzeň.

Kuznetsov, N. V., and Fedosov, B. V. 1967. Asymptotische Formel für die Eigenwerte einer Kreismembran. *Differ. Equation*, **1**, 1326–1329.

Kwaśnicki, M., Laugesen, R. S., and Siudeja, B. A. 2019. Pólya's conjecture fails for the fractional Laplacian. *J. Spectr. Theory*, **9**(1), 127–135.

Kwong, M. K. 1989. Uniqueness of positive solutions of $\Delta u - u + u^p = 0$ in $\mathbf{R}^n$. *Arch. Rational Mech. Anal.*, **105**(3), 243–266.

Lamé, G. 1833. Mémoire sur la propagation de la chaleur dans les polyèdres. *J. Éc. Polytec. Math.*, **22**, 194–251.

Landau, L. D., and Lifshitz, E. M. 1958. *Quantum Mechanics: Non-Relativistic Theory. Course of Theoretical Physics, Vol. 3.* Translated from the Russian by J. B. Sykes and J. S. Bell. Pergamon Press Ltd., London and Paris; and Addison-Wesley Publishing Co., Inc., Reading, MA.

Langmann, E., Laptev, A., and Paufler, C. 2006. Singular factorizations, self-adjoint extensions and applications to quantum many-body physics. *J. Phys. A*, **39**(5), 1057–1071.

Lapidus, M. L., and Pomerance, C. 1993. The Riemann zeta-function and the one-dimensional Weyl–Berry conjecture for fractal drums. *Proc. Lond. Math. Soc. (3)*, **66**(1), 41–69.

Lapidus, M. L., and Pomerance, C. 1996. Counterexamples to the modified Weyl–Berry conjecture on fractal drums. *Math. Proc. Camb. Philos. Soc.*, **119**(1), 167–178.

Laptev, A. A. 1981. Spectral asymptotics of a class of Fourier integral operators. *Trudy Moskov. Mat. Obshch.*, **43**, 92–115. English translation in *Trans. Mosc. Math. Soc*, **1**, 101–127 (1983).

Laptev, A. 1993. Asymptotics of the negative discrete spectrum of a class of Schrödinger operators with large coupling constant. *Proc. Amer. Math. Soc.*, **119**(2), 481–488.

Laptev, A. 1995. On inequalities for the bound states of Schrödinger operators. Pages 221–225 of: *Partial Differential Operators and Mathematical Physics (Holzhau, 1994).* M. Demuth and B.-W. Schulze (editors). Oper. Theory Adv. Appl., vol. 78. Birkhäuser, Basel.

Laptev, A. 1997. Dirichlet and Neumann eigenvalue problems on domains in Euclidean spaces. *J. Funct. Anal.*, **151**(2), 531–545.

Laptev, A. 1999. On the Lieb–Thirring conjecture for a class of potentials. Pages 227–234 of: *The Maz'ya Anniversary Collection, Vol. 2 (Rostock, 1998).* J. Rossmann, P. Takáč, and G. Wildenhain (editors). Oper. Theory Adv. Appl., vol. 110. Birkhäuser, Basel.

Laptev, A. 2021. On factorisation of a class of Schrödinger operators. *Complex Var. Elliptic Equ.*, **66**(6-7), 1100–1107.

Laptev, A., and Safarov, Yu. 1996. A generalization of the Berezin–Lieb inequality. Pages 69–79 of: *Contemporary Mathematical Physics*. R. A. Minlos, M. A. Shubin, and A. M. Vershik (editors). Amer. Math. Soc. Transl. Ser. 2, vol. 175. Amer. Math. Soc., Providence, RI.

Laptev, A., and Safronov, O. 2009. Eigenvalue estimates for Schrödinger operators with complex potentials. *Comm. Math. Phys.*, **292**(1), 29–54.

Laptev, A., and Solomyak, M. 2012. On the negative spectrum of the two-dimensional Schrödinger operator with radial potential. *Comm. Math. Phys.*, **314**(1), 229–241.

Laptev, A., and Solomyak, M. 2013. On spectral estimates for two-dimensional Schrödinger operators. *J. Spectr. Theory*, **3**(4), 505–515.

Laptev, A., and Weidl, T. 2000a. Recent results on Lieb–Thirring inequalities. Exp. No. XX of: *Journ. Équ. Dériv. Partielles (La Chapelle sur Erdre, 2000).* Univ. Nantes, Nantes.

Laptev, A., and Weidl, T. 2000b. Sharp Lieb–Thirring inequalities in high dimensions. *Acta Math.*, **184**(1), 87–111.

Laptev, A., Naboko, S., and Safronov, O. 2003. On new relations between spectral properties of Jacobi matrices and their coefficients. *Comm. Math. Phys.*, **241**(1), 91–110.

Laptev, A., Naboko, S., and Safronov, O. 2005. Absolutely continuous spectrum of Schrödinger operators with slowly decaying and oscillating potentials. *Comm. Math. Phys.*, **253**(3), 611–631.

Laptev, A., Loss, M., and Schimmer, L. 2021. A remark on a paper by Hundertmark and Simon. Preprint, available at ArXiv:2012.13793.

Larson, S. 2017. On the remainder term of the Berezin inequality on a convex domain. *Proc. Amer. Math. Soc.*, **145**(5), 2167–2181.

Larson, S., Lundholm, D., and Nam, P. T. 2021. Lieb–Thirring inequalities for wave functions vanishing on the diagonal set. *Ann. H. Lebesgue*, **4**, 251–282.

Latushkin, Y., and Sukhtayev, A. 2010. The algebraic multiplicity of eigenvalues and the Evans function revisited. *Math. Model. Nat. Phenom.*, **5**(4), 269–292.

Lax, P. D. 1968. Integrals of nonlinear equations of evolution and solitary waves. *Comm. Pure Appl. Math.*, **21**, 467–490.

Lax, P. D. 2002. *Functional Analysis*. John Wiley & Sons, New York.

Leinfelder, H., and Simader, C. G. 1981. Schrödinger operators with singular magnetic vector potentials. *Math. Z.*, **176**(1), 1–19.

Lenard, A., and Dyson, F. J. 1968. Stability of matter. II. *J. Math. Phys.*, **9**(5), 698–711.

Leoni, G. 2017. *A First Course in Sobolev Spaces*. Second edn. Graduate Studies in Mathematics, vol. 181. Amer. Math. Soc., Providence, RI.

Leoni, G., and Morini, M. 2007. Necessary and sufficient conditions for the chain rule in $W^{1,1}_{\mathrm{loc}}(\mathbb{R}^N;\mathbb{R}^d)$ and $\mathrm{BV}_{\mathrm{loc}}(\mathbb{R}^N;\mathbb{R}^d)$. *J. Eur. Math. Soc. (JEMS)*, **9**(2), 219–252.

Levin, D., and Solomyak, M. 1997. The Rozenblum–Lieb–Cwikel inequality for Markov generators. *J. Anal. Math.*, **71**, 173–193.

Levine, H. A., and Weinberger, H. F. 1986. Inequalities between Dirichlet and Neumann eigenvalues. *Arch. Rational Mech. Anal.*, **94**(3), 193–208,

Levinson, N. 1949. On the uniqueness of the potential in a Schrödinger equation for a given asymptotic phase. *Danske Vid. Selsk. Mat.-Fys. Medd.*, **25**(9), 29.

Levitan, B. M. 1952. On the asymptotic behavior of the spectral function of a self-adjoint differential equation of the second order. *Izvestiya Akad. Nauk SSSR. Ser. Mat.*, **16**, 325–352.

Levitin, M., and Vassiliev, D. 1996. Spectral asymptotics, renewal theorem, and the Berry conjecture for a class of fractals. *Proc. Lond. Math. Soc. (3)*, **72**(1), 188–214.

Levitin, M., Polterovich, I., and Sher, D. A. 2022. Pólya's conjecture for the disk: a computer-assisted proof. Preprint, available at ArXiv.2203.07696.

Lewin, M. 2022. *Théorie Spectrale et Mécanique Quantique*. Mathématiques et Applications. Springer.

Lewin, M., and Sabin, J. 2014. The Hartree equation for infinitely many particles, II: Dispersion and scattering in 2D. *Anal. PDE*, **7**(6), 1339–1363.

Lewin, M., and Sabin, J. 2015. The Hartree equation for infinitely many particles I. Well-posedness theory. *Comm. Math. Phys.*, **334**(1), 117–170.

Lewin, M., Lieb, E. H., and Seiringer, R. 2018. Statistical mechanics of the uniform electron gas. *J. Éc. Polytech. Math.*, **5**, 79–116.

Lewin, M., Lieb, E. H., and Seiringer, R. 2020a. The local density approximation in density functional theory. *Pure Appl. Anal.*, **2**(1), 35–73.

Lewin, M., Lieb, E. H., and Seiringer, R. 2020b. Universal functionals in density functional theory. Preprint, available at ArXiv:1912.10424.

Li, P., and Yau, S. T. 1980. Estimates of eigenvalues of a compact Riemannian manifold. Pages 205–239 of: *Geometry of the Laplace Operator (Proc. Sympos. Pure Math., Univ. Hawaii, Honolulu, Hawaii, 1979)*. R. Osserman and A. Weinstein (editors). Proc. Sympos. Pure Math., XXXVI, Amer. Math. Soc., Providence, RI.

Li, P., and Yau, S. T. 1983. On the Schrödinger equation and the eigenvalue problem. *Comm. Math. Phys.*, **88**(3), 309–318.

Lieb, E. H. 1976. Bounds on the eigenvalues of the Laplace and Schrödinger operators. *Bull. Amer. Math. Soc.*, **82**(5), 751–753.

Lieb, E. H. 1973. The classical limit of quantum spin systems. *Comm. Math. Phys.*, **31**, 327–340.

Lieb, E. H. 1976/77. Existence and uniqueness of the minimizing solution of Choquard's nonlinear equation. *Studies in Appl. Math.*, **57**(2), 93–105.

Lieb, E. H. 1980. The number of bound states of one-body Schrödinger operators and the Weyl problem. Pages 241–252 of: *Geometry of the Laplace Operator (Proc. Sympos. Pure Math., Univ. Hawaii, Honolulu, Hawaii, 1979)*. R. Osserman and A. Weinstein (editors). Proc. Sympos. Pure Math., XXXVI. Amer. Math. Soc., Providence, RI.

Lieb, E. H. 1981. Thomas–Fermi and related theories of atoms and molecules. *Rev. Modern Phys.*, **53**(4), 603–641. Erratum *ibid.* **54**(1), p. 311 (1982).

Lieb, E. H. 1983a. Density functionals for Coulomb systems. *Int. J. Quantum Chem.*, **24**(3), 243–277.

Lieb, E. H. 1983b. An L^p bound for the Riesz and Bessel potentials of orthonormal functions. *J. Funct. Anal.*, **51**(2), 159–165.

Lieb, E. H. 1983c. On the lowest eigenvalue of the Laplacian for the intersection of two domains. *Invent. Math.*, **74**(3), 441–448.

Lieb, E. H. 1983d. Sharp constants in the Hardy–Littlewood–Sobolev and related inequalities. *Ann. of Math. (2)*, **118**(2), 349–374.

Lieb, E. H. 1984. On characteristic exponents in turbulence. *Comm. Math. Phys.*, **92**(4), 473–480.

Lieb, E. H., and Loss, M. 2001. *Analysis*. Second edn. Graduate Studies in Mathematics, vol. 14. Amer. Math. Soc., Providence, RI.

Lieb, E. H., and Seiringer, R. 2010. *The Stability of Matter in Quantum Mechanics*. Cambridge University Press, Cambridge.

Lieb, E. H., and Simon, B. 1977. The Thomas–Fermi theory of atoms, molecules and solids. *Adv. Math.*, **23**(1), 22–116.

Lieb, E. H., and Thirring, W. E. 1975. Bound for the kinetic energy of fermions which proves the stability of matter. *Phys. Rev. Lett.*, **35**(Sep), 687–689.

Lieb, E. H., and Thirring, W. E. 1976. Inequalities for the moments of the eigenvalues of the Schrödinger Hamiltonian and their relation to Sobolev inequalities. Pages 269–303 of: *Studies in Mathematical Physics (Essays in Honor of Valentine Bargmann)*. E. H. Lieb, B. Simon, and A. S. Wightman (editors). Princeton University Press, Princeton, NJ.

Lieb, E. H., Siedentop, H., and Solovej, J. P. 1997. Stability and instability of relativistic electrons in classical electromagnetic fields. *J. Stat. Phys.*, **89**, 37–59.

Linde, H. 2006. A lower bound for the ground state energy of a Schrödinger operator on a loop. *Proc. Amer. Math. Soc.*, **134**(12), 3629–3635.

Lions, P.-L. 1984a. The concentration–compactness principle in the calculus of variations. The locally compact case. I. *Ann. Inst. H. Poincaré Anal. Non Linéaire*, **1**(2), 109–145.

Lions, P.-L. 1984b. The concentration–compactness principle in the calculus of variations. The locally compact case. II. *Ann. Inst. H. Poincaré Anal. Non Linéaire*, **1**(4), 223–283.

Lions, P.-L., and Paul, T. 1993. Sur les mesures de Wigner. *Rev. Mat. Iberoamericana*, **9**(3), 553–618.

Lorentz, H. A. 1910. Alte und neue Fragen der Physik. *Phys. Z.*, **11**, 1234–1257.

Lundholm, D., and Seiringer, R. 2018. Fermionic behavior of ideal anyons. *Lett. Math. Phys.*, **108**(11), 2523–2541.

Lundholm, D., and Solovej, J. P. 2013. Hardy and Lieb–Thirring inequalities for anyons. *Comm. Math. Phys.*, **322**(3), 883–908.

Lundholm, D., and Solovej, J. P. 2014. Local exclusion and Lieb–Thirring inequalities for intermediate and fractional statistics. *Ann. Henri Poincaré*, **15**(6), 1061–1107.

Lundholm, D., Portmann, F., and Solovej, J. P. 2015. Lieb–Thirring bounds for interacting Bose gases. *Comm. Math. Phys.*, **335**(2), 1019–1056.

Lundholm, D., Nam, P. T., and Portmann, F. 2016. Fractional Hardy–Lieb–Thirring and related inequalities for interacting systems. *Arch. Rational Mech. Anal.*, **219**(3), 1343–1382.

Makai, E. 1965. A lower estimation of the principal frequencies of simply connected membranes. *Acta Math. Acad. Sci. Hungar.*, **16**, 319–323.

Marchenko, V. A. 2011. *Sturm–Liouville Operators and Applications*. Revised edn. AMS Chelsea Publishing, Providence, RI.

Martin, A. 1972. Bound states in the strong coupling limit. *Helv. Phys. Acta*, **45**, 140–148.

Martin, A. 1990. New results on the moments of the eigenvalues of the Schrödinger Hamiltonian and applications. *Comm. Math. Phys.*, **129**(1), 161–168.

Matveev, V. B., and Salle, M. A. 1991. *Darboux Transformations and Solitons*. Springer Series in Nonlinear Dynamics. Springer-Verlag, Berlin.

Maz'ya, V. G. 1960. Classes of domains and imbedding theorems for function spaces. *Sov. Math., Dokl.*, **1**, 882–885.

Maz'ja, V. G. 1962. The negative spectrum of the higher-dimensional Schrödinger operator. *Dokl. Akad. Nauk SSSR*, **144**, 721–722.

Maz'ja, V. G. 1964. On the theory of the higher-dimensional Schrödinger operator. *Izv. Akad. Nauk SSSR Ser. Mat.*, **28**, 1145–1172.

Maz'ya, V. G. 1969. On Neumann's problem in domains with nonregular boundaries. *Sib. Math. J.*, **9**, 990–1012.

Maz'ya, V. G. 1974. On connection of two kinds of capacity. *Vestn. Leningr. Univ., Mat. Mekh. Astron.*, **1974**(2), 33–40.

Maz'ya, V. 2003. Lectures on isoperimetric and isocapacitary inequalities in the theory of Sobolev spaces. Pages 307–340 of: *Heat Kernels and Analysis on Manifolds, Graphs, and Metric Spaces (Paris, 2002)*. P. Auscher, T. Coulhon, and A. Grigoryan (editors). Contemp. Math., vol. 338. Amer. Math. Soc., Providence, RI.

Maz'ya, V. 2007. Analytic criteria in the qualitative spectral analysis of the Schrödinger operator. Pages 257–288 of: *Spectral Theory and Mathematical Physics: a Festschrift in Honor of Barry Simon's 60th Birthday*. F. Gesztesy, P. Deift, C.

494 *References*

Galvez, P. Perry, and W. Schlag (editors). Proc. Sympos. Pure Math., vol. 76. Amer. Math. Soc., Providence, RI.

Maz'ya, V. 2011. *Sobolev Spaces with Applications to Elliptic Partial Differential Equations*. Augmented edn. Grundlehren der Mathematischen Wissenschaften, vol. 342. Springer-Verlag, Heidelberg.

Maz'ja, V. G., and Otelbaev, M. 1977. Imbedding theorems and the spectrum of a certain pseudodifferential operator. *Sibirsk. Mat. Ž.*, **18**(5), 1073–1087, 1206. English translation in *Siberian Math. J.*, **18**(5), 758–770, (1977).

Maz'ya, V., and Shubin, M. 2005a. Can one see the fundamental frequency of a drum? *Lett. Math. Phys.*, **74**(2), 135–151.

Maz'ya, V., and Shubin, M. 2005b. Discreteness of spectrum and positivity criteria for Schrödinger operators. *Ann. of Math. (2)*, **162**(2), 919–942.

Maz'ya, V. G., and Verbitsky, I. E. 2002. The Schrödinger operator on the energy space: boundedness and compactness criteria. *Acta Math.*, **188**(2), 263–302.

Maz'ya, V. G., and Verbitsky, I. E. 2005. Infinitesimal form boundedness and Trudinger's subordination for the Schrödinger operator. *Invent. Math.*, **162**(1), 81–136.

McLeod, J. B. 1961. The distribution of the eigenvalues for the hydrogen atom and similar cases. *Proc. Lond. Math. Soc. (3)*, **11**, 139–158.

McLeod, K. 1993. Uniqueness of positive radial solutions of $\Delta u + f(u) = 0$ in $\mathbf{R}^n$. II. *Trans. Amer. Math. Soc.*, **339**(2), 495–505.

McLeod, K., and Serrin, J. 1987. Uniqueness of positive radial solutions of $\Delta u + f(u) = 0$ in $\mathbf{R}^n$. *Arch. Rational Mech. Anal.*, **99**(2), 115–145.

Melas, A. D. 2003. A lower bound for sums of eigenvalues of the Laplacian. *Proc. Amer. Math. Soc.*, **131**(2), 631–636.

Melrose, R. B. 1980. Weyl's conjecture for manifolds with concave boundary. Pages 257–274 of: *Geometry of the Laplace operator (Proc. Sympos. Pure Math., Univ. Hawaii, Honolulu, Hawaii, 1979)*. R. Osserman and A. Weinstein (editors). Proc. Sympos. Pure Math., XXXVI. Amer. Math. Soc., Providence, RI.

Metivier, G. 1982. *Estimation du reste en théorie spectrale*. Journ. Équ. Dériv. Partielles, Saint-Jean-De-Monts 1982, Exp. No. 1, 5 p. http://eudml.org/doc/93075.

Meyers, N. G., and Serrin, J. 1964. H = W. *Proc. Nat. Acad. Sci. USA.*, **51**, 1055–1056.

Mikhaïlov, V. P. 1978. *Partial Differential Equations*. Translated from the Russian by P. C. Sinha. "MIR", Moscow; distributed by Imported Publications, Inc., Chicago, Ill.

Milnor, J. 1964. Eigenvalues of the Laplace operator on certain manifolds. *Proc. Nat. Acad. Sci. U.S.A.*, **51**, 542.

Minakshisundaram, S., and Pleijel, Å. 1949. Some properties of the eigenfunctions of the Laplace-operator on Riemannian manifolds. *Canad. J. Math.*, **1**, 242–256.

Mitrinović, D. S. 1970. *Analytic Inequalities*. In cooperation with P. M. Vasić. Grundlehren der Mathematischen Wissenschaften, Band 165. Springer-Verlag, New York and Berlin.

Miura, R. M., Gardner, C. S., and Kruskal, M. D. 1968. Korteweg–de Vries equation and generalizations. II. Existence of conservation laws and constants of motion. *J. Math. Phys.*, **9**, 1204–1209.

Molčanov, A. M. 1953. On conditions for discreteness of the spectrum of self-adjoint differential equations of the second order. *Trudy Moskov. Mat. Obšč.*, **2**, 169–199.

Molchanov, S., and Vainberg, B. 1997. On spectral asymptotics for domains with fractal boundaries. *Comm. Math. Phys.*, **183**(1), 85–117.

Molchanov, S., and Vainberg, B. 1998. On spectral asymptotics for domains with fractal boundaries of cabbage type. *Math. Phys. Anal. Geom.*, **1**(2), 145–170.

Molchanov, S., and Vainberg, B. 2010. On general Cwikel–Lieb–Rozenblum and Lieb–Thirring inequalities. Pages 201–246 of: *Around the Research of Vladimir Maz'ya. III. Analysis and Applications*. A. Laptev (editor). Springer, Dordrecht; Tamara Rozhkovskaya Publisher, Novosibirsk.

Molchanov, S., and Vainberg, B. 2012. Bargmann type estimates of the counting function for general Schrödinger operators. *J. Math. Sciences*, **184**(4), 457–508.

Molchanov, S., Novitskii, M., and Vainberg, B. 2001. First KdV integrals and absolutely continuous spectrum for 1-D Schrödinger operator. *Comm. Math. Phys.*, **216**(1), 195–213.

Montiel, S,, and Ros, A. 1986. Minimal immersions of surfaces by the first eigenfunctions and conformal area. *Invent. Math.*, **83**(1), 153–166.

Morpurgo, C. 1999. Sharp trace inequalities for intertwining operators on S^n and $\mathbf{R}^n$. *Internat. Math. Res. Notices*, **1999**(20), 1101–1117.

Morpurgo, C. 2002. Sharp inequalities for functional integrals and traces of conformally invariant operators. *Duke Math. J.*, **114**(3), 477–553.

Moser, T., and Seiringer, R. 2017. Stability of a fermionic $N + 1$ particle system with point interactions. *Comm. Math. Phys.*, **356**(1), 329–355.

Muckenhoupt, B. 1972. Hardy's inequality with weights. *Studia Math.*, **44**, 31–38.

Naimark, K., and Solomyak, M. 1997. Regular and pathological eigenvalue behavior for the equation $-\lambda u'' = V u$ on the semiaxis. *J. Funct. Anal.*, **151**(2), 504–530.

Nakamura, S. 2020. The orthonormal Strichartz inequality on torus. *Trans. Amer. Math. Soc.*, **373**(2), 1455–1476.

Nam, P. T. 2018. Lieb–Thirring inequality with semiclassical constant and gradient error term. *J. Funct. Anal.*, **274**(6), 1739–1746.

Nam, P. T. 2022. A proof of the Lieb–Thirring inequality via the Besicovitch covering lemma. Preprint, available at ArXiv:2206.15368.

Naumann, J. 2002. *Remarks on the Prehistory of Sobolev Spaces*. Humboldt-Universität zu Berlin, Mathematisch-Naturwissenschaftliche Fakultät II, Institut für Mathematik.

Netrusov, Y., and Weidl, T. 1996. On Lieb–Thirring inequalities for higher order operators with critical and subcritical powers. *Comm. Math. Phys.*, **182**(2), 355–370.

Netrusov, Yu., and Safarov, Yu. 2005. Weyl asymptotic formula for the Laplacian on domains with rough boundaries. *Comm. Math. Phys.*, **253**(2), 481–509.

Nirenberg, L. 1959. On elliptic partial differential equations. *Ann. Sc. Norm. Super. Pisa, Sci. Fis. Mat., III. Ser.*, **13**, 115–162.

Opic, B., and Kufner, A. 1990. *Hardy-Type Inequalities*. Pitman Research Notes in Mathematics Series, vol. 219. Longman Scientific & Technical, Harlow.

Pauli, W. 1926. Über das Wasserstoffspektrum vom Standpunkt der neuen Quantenmechanik. *Z. Phys.*, **36**(5), 336–363.

Pavlov, B. S. 1966. On a non-selfadjoint Schrödinger operator. Pages 102–132 of: *Spectral Theory and Wave Processes*. Problems of Mathematical Physics, No. 1. Izdat. Leningrad. Univ., Leningrad.

Pavlov, B. S. 1967. On a non-selfadjoint Schrödinger operator. II. Pages 133–157 of: *Spectral Theory, Diffraction Problems*. Problems of Mathematical Physics, No. 2. Izdat. Leningrad. Univ., Leningrad.

Pavlović, N. 2003. Bounds for sums of powers of eigenvalues of Schrödinger operators via the commutation method. Pages 271–281 of: *Advances in Differential Equations and Mathematical Physics (Birmingham, AL, 2002)*. Y. Karpeshina, G. Stolz, R. Weikard, and Y. Zeng (editors). Contemp. Math., vol. 327. Amer. Math. Soc., Providence, RI.

Payne, L. E. 1955. Inequalities for eigenvalues of membranes and plates. *J. Rational Mech. Anal.*, **4**, 517–529.

Payne, L. E., and Stakgold, I. 1973. On the mean value of the fundamental mode in the fixed membrane problem. *Applicable Anal.*, **3**, 295–306.

Payne, L. E., and Weinberger, H. F. 1960. An optimal Poincaré inequality for convex domains. *Arch. Rational Mech. Anal.*, **5**, 286–292 (1960).

Petkov, V. M., and Stojanov, L. N. 1988. On the number of periodic reflecting rays in generic domains. *Ergodic Theory Dynam. Systems*, **8**(1), 81–91.

Pinsky, M. A. 1980. The eigenvalues of an equilateral triangle. *SIAM J. Math. Anal.*, **11**(5), 819–827.

Pinsky, M. A. 1985. Completeness of the eigenfunctions of the equilateral triangle. *SIAM J. Math. Anal.*, **16**(4), 848–851.

Pockels, F. C. A. 1891. *Über die Partielle Differentialgleichung $\Delta u + k^2 u = 0$ und deren Auftreten in der Mathematischen Physik*. B. G. Teubner, Leipzig.

Poincaré, H. 1890. Sur les équations aux dérivées partielles de la physique mathématique. *Amer. J. Math.*, **12**(3), 211–294.

Poincaré, H. 1894. Sur les équations de la physique mathématique. *Rend. Circ. Mat. Palermo*, **8**, 57–156.

Poincaré, H. 1897. La méthode de Neumann et le problème de Dirichlet. *Acta Math.*, **20**, 59–142.

Pólya, G. 1952. Remarks on the foregoing paper. *J. Math. Phys.*, **31**, 55–57.

Pólya, G. 1954. *Patterns of Plausible Inference. Mathematics and Plausible Reasoning, vol. II*. Princeton University Press, Princeton, NJ.

Pólya, G. 1955. On the characteristic frequencies of a symmetric membrane. *Math. Z.*, **63**, 331–337.

Pólya, G. 1961. On the eigenvalues of vibrating membranes. *Proc. Lond. Math. Soc. (3)*, **11**, 419–433.

Pöschl, G., and Teller, E. 1933. Bemerkungen zur Quantenmechanik des anharmonischen Oszillators. *Z. Phys.*, **83**, 143–151.

Protter, M. H. 1981. A lower bound for the fundamental frequency of a convex region. *Proc. Amer. Math. Soc.*, **81**(1), 65–70.

Pushnitski, A. 2009. Operator theoretic methods for the eigenvalue counting function in spectral gaps. *Ann. Henri Poincaré*, **10**(4), 793–822.

Pushnitski, A. 2011. The Birman–Schwinger principle on the essential spectrum. *J. Funct. Anal.*, **261**(7), 2053–2081.

Radin, C. 1994. The pinwheel tilings of the plane. *Ann. of Math. (2)*, **139**(3), 661–702.

Rayleigh, John William Strutt, Lord. 1877. *The Theory of Sound*. MacMillan and Co. London.

Reed, M., and Simon, B. 1972. *Methods of Modern Mathematical Physics. I. Functional Analysis.* Academic Press, New York and London.

Reed, M., and Simon, B. 1975. *Methods of Modern Mathematical Physics. II. Fourier Analysis, Self-Adjointness.* Academic Press, New York and London.

Reed, M., and Simon, B. 1978. *Methods of Modern Mathematical Physics. IV. Analysis of Operators.* Academic Press, New York and London.

Reed, M., and Simon, B. 1979. *Methods of Modern Mathematical Physics. III. Scattering Theory.* Academic Press, New York and London.

Rellich, F. 1930. Ein Satz über mittlere Konvergenz. *Nachr. Ges. Wiss. Göttingen, Math.-Phys. Kl.*, **1930**, 30–35.

Rellich, F. 1942. Störungstheorie der Spektralzerlegung. V. *Math. Ann.*, **118**, 462–484.

Ritz, W. 1908. Über eine neue Methode zur Lösung gewisser Variationsprobleme der mathematischen Physik. *J. Reine Angew. Math.*, **135**, 1–61.

Robinson, D. W. 1971. *The Thermodynamic Pressure in Quantum Statistical Mechanics.* Lecture Notes in Physics, Vol. 9. Springer-Verlag, Berlin and New York.

Rodemich, E. 1966. The Sobolev inequalities with best possible constants. *Analysis Seminar at California Institute of Technology*, 1–25.

Rogers, L. G. 2006. Degree-independent Sobolev extension on locally uniform domains. *J. Funct. Anal.*, **235**(2), 619–665.

Rosen, G. 1971. Minimum value for c in the Sobolev inequality $\|\varphi^3\| \leq c\|\nabla\varphi\|^3$. *SIAM J. Appl. Math.*, **21**, 30–32.

Rotfel'd, S. Ju. 1967. Remarks on the singular values of a sum of completely continuous operators. *Funkcional. Anal. i Priložen.*, **1**(3), 95–96. English translation in *Functional Analysis and its Applications*, **1**(3), 252–253 (1967).

Rotfel'd, S. Ju. 1968. The singular values of the sum of completely continuous operators. Pages 81–87 of: *Spectral Theory*. M. Sh. Birman (editor). Problems of Mathematical Physics, No. 3 Izdat. Leningrad. Univ., Leningrad. English translation in *Spectral Theory*. Topics in Mathematical Physics 3, 73–78 (1969). Consultants Bureau, New York and London.

Royden, H. L. 1963. *Real Analysis.* The Macmillan Co., New York; Collier–Macmillan Ltd., London.

Rozenbljum, G. V. 1971. The distribution of the eigenvalues of the first boundary value problem in unbounded domains. *Dokl. Akad. Nauk SSSR*, **200**, 1034–1036. English translation in *Soviet Math. Dokl.*, **12**, 1539–1542, (1971).

Rozenbljum, G. V. 1972a. Distribution of the discrete spectrum of singular differential operators. *Dokl. Akad. Nauk SSSR*, **202**, 1012–1015. English translation in *Soviet Math. Dokl.*, **13**, 245–249, (1972).

Rozenbljum, G. V. 1972b. The eigenvalues of the first boundary value problem in unbounded domains. *Mat. Sb. (N.S.)*, **89 (131)**, 234–247, 350. English translation in *Math USSR Sb.*, **18**(2), 235–248, (1972).

Rozenbljum, G. V. 1973. The calculation of the spectral asymptotics for the Laplace operator in domains of infinite measure. Pages 95–106, 144 of: *Problems of Mathematical Analysis, No. 4: Integral and Differential Operators. Differential Equations*, English translation in *J. Soviet Math.*, **6**, 64–71, (1976).

Rozenbljum, G. V. 1976. Distribution of the discrete spectrum of singular differential operators. *Izv. Vysš. Učebn. Zaved. Matematika*, **1**, 75–86. English translation in *Soviet Math. (Iz. VUZ)*, **20**(1), 63–71, (1976).

Rozenblum, G. 2021. Lieb–Thirring estimates for singular measures. Preprint, available at ArXiv:2108.11429.

Rozenblum, G., and Shargorodsky, E. 2021. Eigenvalue estimates and asymptotics for weighted pseudodifferential operators with singular measures in the critical case. Preprint, available at ArXiv:2011.14877.

Rozenblyum, G., and Solomyak, M. 1997. The Cwikel–Lieb–Rozenblyum estimator for generators of positive semigroups and semigroups dominated by positive semigroups. *Algebra i Analiz*, **9**(6), 214–236. English translation in *St. Petersburg Math. J.*, **9**(6), 1195–1211 (1998).

Rozenblum, G., and Tashchiyan, G. 2021. *Eigenvalues of the Birman–Schwinger operator for singular measures: the noncritical case*. Preprint, available at ArXiv:2107.04682.

Rozenblum, G. V., Shubin, M. A., and Solomyak, M. Z. 1989. *Partial Differential Equations VII: Spectral Theory of Differential Operators*. Encyclopedia of Mathematical Sciences. vol. 64. Translated from the Russian by T. Zastawniak. Springer-Verlag, Berlin.

Rudin, W. 1987. *Real and Complex Analysis*. Third edn. McGraw-Hill Book Co., New York.

Rudin, W. 1991. *Functional Analysis*. Second edn. McGraw-Hill, Inc., New York.

Rumin, M. 2011. Balanced distribution-energy inequalities and related entropy bounds. *Duke Math. J.*, **160**(3), 567–597.

Sabin, J. 2016. Littlewood–Paley decomposition of operator densities and application to a new proof of the Lieb–Thirring inequality. *Math. Phys. Anal. Geom.*, **19**(2), Art. 11, 11.

Safarov, Yu., and Vassiliev, D. 1997. *The Asymptotic Distribution of Eigenvalues of Partial Differential Operators*. Translations of Mathematical Monographs, vol. 155. Amer. Math. Soc., Providence, RI.

Safronov, O. L. 1996. The discrete spectrum in gaps of the continuous spectrum for indefinite-sign perturbations with a large coupling constant. *Algebra i Analiz*, **8**(2), 162–194. English translation in *St. Petersburg Math. J.*, **8**(2), 307–331, (1997).

Safronov, O. 2005. On the absolutely continuous spectrum of multi-dimensional Schrödinger operators with slowly decaying potentials. *Comm. Math. Phys.*, **254**(2), 361–366.

Schimmer, L. 2015. Spectral inequalities for Jacobi operators and related sharp Lieb–Thirring inequalities on the continuum. *Comm. Math. Phys.*, **334**(1), 473–505.

Schmetzer, T. 2022. New interpolation results for Lieb–Thirring inequalities. M.Sc thesis, University of Stuttgart. Available at https://codeberg.org/attachments/a1a0d333-e7fb-400f-a7d3-a0345161a8f8.

Schmincke, U.-W. 1978. On Schrödinger's factorization method for Sturm–Liouville operators. *Proc. Roy. Soc. Edinburgh Sect. A*, **80**(1-2), 67–84.

Schmincke, U.-W. 2003. On a paper by F. Gesztesy, B. Simon, and G. Teschl concerning isospectral deformations of ordinary Schrödinger operators: "Spectral deformations of one-dimensional Schrödinger operators" [J. Anal. Math. **70** (1996), 267–324; MR1444263 (98m:34171)]. *J. Math. Anal. Appl.*, **277**(1), 51–78.

Schrödinger, E. 1926a. Quantisierung als Eigenwertproblem. *Ann. der Phys.*, **384**(4), 361–376.

Schrödinger, E. 1926b. Quantisierung als Eigenwertproblem. II. *Ann. der Phys.*, **79**(4), 489–527.

Schwartz, L. 1950. *Théorie des Distributions. Tome I.* Actualités Scientifiques et Industrielles [Current Scientific and Industrial Topics], No. 1091. Hermann & Cie., Paris.

Schwartz, L. 1951. *Théorie des Distributions. Tome II.* Actualités Scientifiques et Industrielles [Current Scientific and Industrial Topics], No. 1122. Hermann & Cie., Paris.

Schwinger, J. 1961. On the bound states of a given potential. *Proc. Nat. Acad. Sci. USA.*, **47**, 122–129.

Seeley, R. 1978. A sharp asymptotic remainder estimate for the eigenvalues of the Laplacian in a domain of $\mathbf{R}^3$. *Adv. Math.*, **29**, 244–269.

Seeley, R. 1980. An estimate near the boundary for the spectral function of the Laplace operator. *Amer. J. Math.*, **102**(5), 869–902.

Seiler, E., and Simon, B. 1975. Bounds in the Yukawa$_2$ quantum field theory: upper bound on the pressure, Hamiltonian bound and linear lower bound. *Comm. Math. Phys.*, **45**(2), 99–114.

Serrin, J., and Tang, M. 2000. Uniqueness of ground states for quasilinear elliptic equations. *Indiana Univ. Math. J.*, **49**(3), 897–923.

Shargorodsky, E. 2013. An estimate for the Morse index of a Stokes wave. *Arch. Rational Mech. Anal.*, **209**(1), 41–59.

Shargorodsky, E. 2014. On negative eigenvalues of two-dimensional Schrödinger operators. *Proc. Lond. Math. Soc. (3)*, **108**(2), 441–483.

Shubin, M. A. 2001. *Pseudodifferential Operators and Spectral Theory.* Second edn. Translated from the 1978 Russian original by S. I. Andersson. Springer-Verlag, Berlin.

Sigal, I. M. 1983. Geometric parametrices and the many-body Birman–Schwinger principle. *Duke Math. J.*, **50**(2), 517–537.

Simon, B. 1976a. Analysis with weak trace ideals and the number of bound states of Schrödinger operators. *Trans. Amer. Math. Soc.*, **224**(2), 367–380.

Simon, B. 1976b. The bound state of weakly coupled Schrödinger operators in one and two dimensions. *Ann. Physics*, **97**(2), 279–288.

Simon, B. 1979a. *Functional Integration and Quantum Physics.* Pure and Applied Mathematics, vol. 86. Academic Press, New York and London.

Simon, B. 1979b. Maximal and minimal Schrödinger forms. *J. Operator Theory*, **1**(1), 37–47.

Simon, B. 1980. The classical limit of quantum partition functions. *Comm. Math. Phys.*, **71**(3), 247–276.

Simon, B. 1982. Schrödinger semigroups. *Bull. Amer. Math. Soc. (N.S.)*, **7**(3), 447–526.

Simon, B. 1983. Nonclassical eigenvalue asymptotics. *J. Funct. Anal.*, **53**(1), 84–98.

Simon, B. 1992. The Neumann Laplacian of a jelly roll. *Proc. Amer. Math. Soc.*, **114**(3), 783–785.

Simon, B. 2005. *Trace Ideals and their Applications.* Second edn. Mathematical Surveys and Monographs, vol. 120. Amer. Math. Soc., Providence, RI.

Simon, B. 2011. *Convexity: An Analytic Viewpoint.* Cambridge Tracts in Mathematics, vol. 187. Cambridge University Press, Cambridge.

Simon, B. 2015a. *A Comprehensive Course in Analysis: Part 1, Real Analysis.* Amer. Math. Soc., Providence, RI.

Simon, B. 2015b. *A Comprehensive Course in Analysis: Part 3, Harmonic Analysis.* Amer. Math. Soc., Providence, RI.

Simon, B. 2015c. *A Comprehensive Course in Analysis: Part 4, Operator Theory.* Amer. Math. Soc., Providence, RI.

Smirnov, V. I. 1964. *A Course of Higher Mathematics. Vol. V [Integration and Functional Analysis].* Translated by D. E. Brown; translation edited by I.N. Sneddon. Pergamon Press, Oxford and New York; Addison-Wesley Publishing Co., Inc., Reading, MA and London.

Sobolev, A. V. 1993. The Efimov effect. Discrete spectrum asymptotics. *Comm. Math. Phys.*, **156**(1), 101–126.

Sobolev, A. V. 1996. On the Lieb–Thirring estimates for the Pauli operator. *Duke Math. J.*, **82**(3), 607–635.

Sobolev, S. L. 1935. Le problème de Cauchy dans l'espace des fonctionelles. *Dokl. Akad. Nauk SSSR*, **3**(7), 291–294.

Sobolev, S. L. 1936. On some estimates relating to families of functions having derivatives that are square integrable. *Dokl. Akad. Nauk SSSR*, **1**, 267–270.

Soboleff, S. 1938. Sur un théorème d'analyse fonctionnelle. *Rec. Math. Moscou, n. Ser.*, **4**, 471–497.

Sobolev, S. L. 1963. *Some Applications of Functional Analysis in Mathematical Physics.* Translated from the third Russian edition by H. H. McFaden, with comments by V. P. Palamodov, 1963. Translations of Mathematical Monographs, vol. 90. Amer. Math. Soc., Providence, RI.

Solomyak, M. 1994. Piecewise-polynomial approximation of functions from $H^l((0, 1)^d)$, $2l = d$, and applications to the spectral theory of the Schrödinger operator. *Israel J. Math.*, **86**(1-3), 253–275.

Solomyak, M. 1998. On the discrete spectrum of a class of problems involving the Neumann Laplacian in unbounded domains. Pages 233–251 of: *Voronezh Winter Mathematical Schools.* P. Kuchment and V. Lin (editors). Amer. Math. Soc. Transl. Ser. 2, vol. 184. Amer. Math. Soc., Providence, RI.

Sommerfeld, A. 1910. Die Greensche Funktion der Schwingungsgleichung für ein beliebiges Gebiet. *Phys. Z.*, **11**, 1057–1066.

Stein, E. M. 1970. *Singular Integrals and Differentiability Properties of Functions.* Princeton Mathematical Series, No. 30. Princeton University Press, Princeton, NJ.

Stein, E. M., and Shakarchi, R. 2005. *Real Analysis: Measure Theory, Integration, and Hilbert Spaces.* Princeton Lectures in Analysis, vol. 3. Princeton University Press, Princeton, NJ.

Stein, E. M., and Weiss, G. 1971. *Introduction to Fourier Analysis on Euclidean Spaces.* Princeton Mathematical Series, No. 32. Princeton University Press, Princeton, N.J.

Steklov, V. A. 1896-97. On the differential equations of mathematical physics. *Mat. Sb.*, **19**(1), 469–585.

Stepin, S. A. 2001. The Birman–Schwinger principle and the Nelkin conjecture in neutron transport theory. *Dokl. Akad. Nauk*, **380**(1), 19–22. English translation in *Dokl. Math.*, **64**(2), 152–155. (2001).

Strichartz, R. S. 1996. Estimates for sums of eigenvalues for domains in homogeneous spaces. *J. Funct. Anal.*, **137**(1), 152–190.

Stubbe, J. 2010. Universal monotonicity of eigenvalue moments and sharp Lieb–Thirring inequalities. *J. Eur. Math. Soc. (JEMS)*, **12**(6), 1347–1353.

Stummel, F. 1956. Singuläre elliptische Differentialoperatoren in Hilbertschen Räumen. *Math. Ann.*, **132**, 150–176.

Sunada, T. 1985. Riemannian coverings and isospectral manifolds. *Ann. of Math. (2)*, **121**(1), 169–186.

Sz. Nagy, B. 1941. Über Integralungleichungen zwischen einer Funktion und ihrer Ableitung. *Acta Sci. Math.*, **10**, 64–74.

Szegö, G. 1954. Inequalities for certain eigenvalues of a membrane of given area. *J. Rational Mech. Anal.*, **3**, 343–356.

Talenti, G. 1969. Osservazioni sopra una classe di disuguaglianze. *Rend. Sem. Mat. Fis. Milano*, **39**, 171–185.

Talenti, G. 1976. Best constant in Sobolev inequality. *Ann. Mat. Pura Appl. (4)*, **110**, 353–372.

Tamura, H. 1974a. The asymptotic distribution of the lower part eigenvalues for elliptic operators. *Proc. Japan Acad.*, **50**, 185–187.

Tamura, H. 1974b. The asymptotic eigenvalue distribution for non-smooth elliptic operators. *Proc. Japan Acad.*, **50**, 19–22.

Taylor, M E. 1981. *Pseudodifferential Operators*. Princeton Mathematical Series, vol. 34. Princeton University Press, Princeton, NJ.

Temam, R. 1997. *Infinite-Dimensional Dynamical Systems in Mechanics and Physics*. Second edn. Applied Mathematical Sciences, vol. 68. Springer-Verlag, New York.

Teschl, G. 2014. *Mathematical Methods in Quantum Mechanics: With Applications to Schrödinger Operators*, Second edn. Graduate Studies in Mathematics, vol. 157. Amer. Math. Soc., Providence, RI.

Titchmarsh, E. C. 1958. *Eigenfunction Expansions Associated with Second-order Differential Equations*. Vol. 2. Clarendon Press, Oxford.

Tomaselli, G. 1969. A class of inequalities. *Boll. Un. Mat. Ital. (4)*, **2**, 622–631.

Urakawa, H. 1984. Lower bounds for the eigenvalues of the fixed vibrating membrane problems. *Tohoku Mathematical Journal*, **36**(2), 185–189.

van den Berg, M. 1984. On the spectrum of the Dirichlet Laplacian for horn-shaped regions in $\mathbf{R}^n$ with infinite volume. *J. Funct. Anal.*, **58**(2), 150–156.

van den Berg, M. 1992. On the spectral counting function for the Dirichlet Laplacian. *J. Funct. Anal.*, **107**(2), 352–361.

van den Berg, M., and Lianantonakis, M. 2001. Asymptotics for the spectrum of the Dirichlet Laplacian on horn-shaped regions. *Indiana Univ. Math. J.*, **50**(1), 299–333.

Vasil'ev, D. G. 1984. Two-term asymptotic behavior of the spectrum of a boundary value problem in interior reflection of general form. *Funktsional. Anal. i Prilozhen.*, **18**(4), 1–13, 96. English translation in *Functional Anal. Appl.*, **18**, 267–277, (1984).

Vasil'ev, D. G. 1986. Two-term asymptotic behavior of the spectrum of a boundary value problem in the case of a piecewise smooth boundary. *Dokl. Akad. Nauk SSSR*, **286**(5), 1043–1046. English translation in *Soviet Math. Dokl.*, **33**(1), 227–230, (1986).

von Neumann, J. 1930. Allgemeine Eigenwerttheorie Hermitescher Funktionaloperatoren. *Math. Ann.*, **102**(1), 49–131.

Vulis, I. L., and Solomjak, M. Z. 1974. Spectral asymptotic analysis for degenerate second order elliptic operators. *Izv. Akad. Nauk SSSR Ser. Mat.*, **38**, 1362–1392. English translation in *Math. USSR Izv.*, **8**(6), 1343–1371, (1974).

Watson, G. N. 1944. *A Treatise on the Theory of Bessel Functions.* Second edn. Reprinted 1995 in Cambridge Mathematical Library. Cambridge University Press, Cambridge.

Weidl, T. 1996. On the Lieb–Thirring constants $L_{\gamma,1}$ for $\gamma \geq 1/2$. *Comm. Math. Phys.*, **178**(1), 135–146.

Weidl, T. 1999a. Another look at Cwikel's inequality. Pages 247–254 of: *Differential Operators and Spectral Theory.* Amer. Math. Soc. Transl. Ser. 2, vol. 189. Providence, RI: Amer. Math. Soc.

Weidl, T. 1999b. Remarks on virtual bound states for semi-bounded operators. *Comm. Partial Diff Equations*, **24**(1-2), 25–60.

Weidl, T. 2008. Improved Berezin–Li–Yau inequalities with a remainder term. Pages 253–263 of: *Spectral Theory of Differential Operators.* Amer. Math. Soc. Transl. Ser. 2, vol. 225. Amer. Math. Soc., Providence, RI.

Weinberger, H. F. 1956. An isoperimetric inequality for the N-dimensional free membrane problem. *J. Rational Mech. Anal.*, **5**, 633–636.

Weinstein, M. I. 1983. Nonlinear Schrödinger equations and sharp interpolation estimates. *Comm. Math. Phys.*, **87**(4), 567–576.

Weyl, H. 1909. Über beschränkte quadratische Formen, deren Differenz vollstetig ist. *Palermo Rend.*, **27**(1), 373–392.

Weyl, H. 1911. Über die asymptotische Verteilung der Eigenwerte. *Nachr. Ges. Wiss. Göttingen, Math.-Phys. Kl.*, **1911**, 110–117.

Weyl, H. 1912a. Das asymptotische Verteilungsgesetz der Eigenwerte linearer partieller Differentialgleichungen (mit einer Anwendung auf die Theorie der Hohlraumstrahlung). *Math. Ann.*, **71**(4), 441–479.

Weyl, H. 1912b. Über die Abhängigkeit der Eigenschwingungen einer Membran von deren Begrenzung. *J. Reine Angew. Math.*, **141**, 1–11.

Weyl, H. 1912c. Über das Spektrum der Hohlraumstrahlung. *J. Reine Angew. Math.*, **141**, 163–181.

Weyl, H. 1913. Über die Randwertaufgabe der Strahlungstheorie und asymptotische Spektralgesetze. *J. Reine Angew. Math.*, **143**, 177–202.

Yafaev, D. R. 1982. The low energy scattering for slowly decreasing potentials. *Comm. Math. Phys.*, **85**(2), 177–196.

Yafaev, D. R. 1992. *Mathematical Scattering Theory: General Theory.* Translated from the Russian by J. R. Schulenberger. Translations of Mathematical Monographs, vol. 105. Amer. Math. Soc., Providence, RI.

Yafaev, D. R. 2010. *Mathematical Scattering Theory: Analytic Theory.* Mathematical Surveys and Monographs, vol. 158. Amer. Math. Soc., Providence, RI.

Zaharov, V. E., and Faddeev, L. D. 1971. The Korteweg–de Vries equation is a fully integrable Hamiltonian system. *Funkcional. Anal. i Priložen.*, **5**(4), 18–27. English translation in *Funct. Anal. Appl.*, **5**, 280–287, (1972).

Zelditch, S. 2004. Inverse spectral problem for analytic domains. I. Balian–Bloch trace formula. *Comm. Math. Phys.*, **248**(2), 357–407.

Zelditch, S. 2009. Inverse spectral problem for analytic domains. II. $\mathbb{Z}_2$-symmetric domains. *Ann. of Math. (2)*, **170**(1), 205–269.

Zworski, M. 2012. *Semiclassical Analysis*. Graduate Studies in Mathematics, vol. 138. Amer. Math. Soc., Providence, RI.

Index